Introduction to
Geophysical Fluid Dynamics

This is Volume 101 in the
INTERNATIONAL GEOPHYSICS SERIES
A series of monographs and textbooks
Edited by RENATA DMOWSKA, DENNIS HARTMANN and H. THOMAS ROSSBY

Introduction to Geophysical Fluid Dynamics

Physical and Numerical Aspects

Second Edition

Benoit Cushman-Roisin
Thayer School of Engineering
Dartmouth College
Hanover, New Hampshire 03755
USA

Jean-Marie Beckers
Département d'Astrophysique,
Géophysique et Océanographie
Université de Liège
B-4000 Liège
Belgium

AMSTERDAM • BOSTON • HEIDELBERG • LONDON
NEW YORK • OXFORD • PARIS • SAN DIEGO
SAN FRANCISCO • SINGAPORE • SYDNEY • TOKYO

Academic Press is an imprint of Elsevier

Academic Press is an imprint of Elsevier
225 Wyman Street, Waltham, MA 02451, USA
The Boulevard, Langford Lane, Kidlington, Oxford OX5 1GB, UK
Radarweg 29, PO Box 211, 1000 AE Amsterdam, The Netherlands

Notice
No responsibility is assumed by the publisher for any injury and/or damage to persons
or property as a matter of products liability, negligence or otherwise, or from any use or
operation of any methods, products, instructions or ideas contained in the material
herein

Library of Congress Cataloging-in-Publication Data
A catalog record for this book is available from the Library of Congress

British Library Cataloguing-in-Publication Data
A catalogue record for this book is available from the British Library.

ISBN: 978-0-12-088759-0

For information on all Academic Press publications
visit our Web site at *www.books.elsevier.com*

**Working together to grow
libraries in developing countries**

www.elsevier.com | www.bookaid.org | www.sabre.org

ELSEVIER   BOOK AID
             International   Sabre Foundation

Contents

Part I
Fundamentals

1. Introduction

2. The Coriolis Force

Part II
Rotation Effects

Part III
Stratification Effects

11. Stratification

12. Layered Models

13. Internal Waves

14. Turbulence in Stratified Fluids

Part IV
Combined Rotation and Stratification Effects

15. Dynamics of Stratified Rotating Flows

16. Quasi-Geostrophic Dynamics

17. Instabilities of Rotating Stratified Flows

18. Fronts, Jets and Vortices

Part V
Special Topics

19. Atmospheric General Circulation

20. Oceanic General Circulation

21. Equatorial Dynamics

22. Data Assimilation

Part VI
Web site Information

Appendix A   Elements of Fluid Mechanics

Appendix B   Wave Kinematics

Appendix C   Recapitulation of Numerical Schemes

On reading this wonderful text on Geophysical Fluid Dynamics (GFD) by Benoit Cushman-Roisin and Jean-Marie Beckers, Antoine de Saint-Exupéry's memorable quote regarding artful simplification seems very appropriate:

In anything at all, perfection is finally attained not when there is no longer anything to add, but when there is no longer anything to take away.

Any scientific endeavor, particularly one that addresses a system as important and complex as the fluid earth, demands a hierarchy of approaches. One must not only strip away extraneous detail to expose what lies beneath, but also study the emergent behavior that results from the interaction of myriad components. Today, sophisticated computer models simulate virtual earths so comprehensively that even the effect of a cloud's shadow cast on the ocean can be represented. Such models are used to synthesize observations, make projections about the vagaries of the weather or the likely future evolution of earth's atmosphere and ocean under anthropogenic forcing.

But, as Jorge Luis Borges' one-paragraph parable on "Exactitude in Science" warns us, we should be wary of the danger of plunging headlong into complexity:

In that Empire the art of Cartography attained such Perfection that the map of a single Province occupied the entirety of a city, and the map of the Empire the entirety of a Province...In time it was realized that the vast Map was Useless.

Like the Empire's perfect cartography, our virtual earths, although far from useless!, are often not the most appropriate tools to figure out where we are, or build understanding and intuition about what matters and what does not. In short, complex models are rather poor pedagogical tools, yet that pedagogy is vital if we are to make wise inferences from them.

In their updated GFD text, the art of intelligent simplification and clear exposition is used by Cushman-Roisin and Beckers. Carefully chosen models are presented and tailored to the phenomenon at hand so that the reader learns by being taken up and down the modeling hierarchy. Moreover, the parallel development of physical and numerical aspects of GFD, both reinforcing and echoing the other, succeeds in breaking down artificial barriers between analytical and numerical approaches.

John Marshall
Massachusetts Institute of Technology
August 2009

The intent of *Introduction to Geophysical Fluid Dynamics—Physical and Numerical Aspects* is to introduce its readers to the principles governing air and water flows on large terrestrial scales and to the methods by which these flows can be simulated on the computer. First and foremost, the book is directed to students and scientists in dynamical meteorology and physical oceanography. In addition, the environmental concerns raised by the possible impact of industrial activities on climate and the accompanying variability of the atmosphere and oceans create a strong desire on the part of atmospheric chemists, biologists, engineers, and many others to understand the basic concepts of atmospheric and oceanic dynamics. It is hoped that those will find here a readable reference text that will provide them with the necessary fundamentals.

The present volume is a significantly enlarged and updated revision of *Introduction to Geophysical Fluid Dynamics* published by Prentice-Hall in 1994, but the objective has not changed, namely to provide an introductory textbook and an approachable reference book. Simplicity and clarity have therefore remained the guiding principles in writing the text. Whenever possible, the physical principles are illustrated with the aid of the simplest existing models, and the computer methods are shown in juxtaposition with the equations to which they apply. The terminology and notation have also been selected to alleviate to a maximum the intellectual effort necessary to extract the meaning from the text. For example, the expressions planetary wave and stratification frequency are preferred to Rossby wave and Brunt-Väisälä frequency, respectively.

The book is divided in five parts. Following a presentation of the fundamentals in Part I, the effects of rotation and of stratification are explored in Parts II and III, respectively. Then, Part IV investigates the combined effects of rotation and stratification, which are at the core of geophysical fluid dynamics. The book closes with Part V, which gathers a group of more applied topics of contemporary interest. Each part is divided into relatively well-contained chapters to provide flexibility of coverage to the professor and ease of access to the researcher. Physical principles and numerical topics are interspersed in order to show the relation of the latter to the former, but a clear division in sections and subsections makes it possible to separate the two if necessary.

Used as a textbook, the present volume should meet the needs of two courses, which are almost always taught sequentially in oceanography and meteorology curricula, namely Geophysical Fluid Dynamics and Numerical Modeling of Geophysical Flows. The integration of both subjects here under a single cover makes it possible to teach both courses with a unified notation and clearer connection of one part to the other than the traditional use of two textbooks, one for each subject. To facilitate the use as a textbook, a number of exercises are offered at the end of every chapter, some more theoretical to reinforce the understanding of the physical principles and others requiring access to a computer to apply the numerical methods. An accompanying Web site (http://booksite.academicpress.com/9780120887590/) contains an assortment of data sets and MATLAB™ codes that permit instructors to ask students to perform realistic and challenging exercises. At the end of every chapter, the reader will also find short biographies, which together form a history of the intellectual developments of the subject matter and should inspire students to achieve similar levels of distinction.

A general remark about notation is appropriate. Because mathematical physics in general and this discipline in particular involve an array of symbols to represent a multitude of variables and constants, with and without dimensions, some conventions are desirable in order to maximize clarity and minimize ambiguity. To this end, a systematic effort has been made to reserve classes of symbols for certain types of variables: Dimensional variables are denoted by lowercase Roman letters (such as u, v, and w for the three velocity components), dimensional constants and parameters use uppercase Roman letters (such as H for domain height and L for length scale), and dimensionless quantities are assigned lowercase Greek letters (such as α for an angle and ϵ for a small dimensionless ratio). In keeping with a well established convention in fluid mechanics, dimensionless numbers credited to particular scientists are denoted by the first two letters of their name (e.g., Ro for the Rossby number and Ek for the Ekman number). Numerical notation is borrowed from Patrick J. Roache, and numerical variables are represented by tildas (˜). Of course, rules breed exceptions (e.g., g for the gravitational acceleration, ω for frequency, and ψ for streamfunction).

We the authors wish to acknowledge the assistance from numerous colleagues across the globe, too many to permit an exhaustive list here. There is one person, however, who deserves a very special note of recognition. Prof. Eric Deleersnijder of the Université catholique de Louvain, Belgium, suggested that the numerical aspects be intertwined with the physics of Geophysical Fluid Dynamics. He also provided significant assistance during the writing of these numerical topics. An additional debt of gratitude goes to our students, who provided us not only with a testing ground for the teaching of this material but also with numerous and valuable comments. The following people are acknowledged for their pertinent remarks and suggestions made on earlier versions of the text, all of which have improved the clarity and accuracy of the presentation: Aida Alvera-Azcárate, Alexander Barth, Emmanuel Boss, Pierre Brasseur, Hans Burchard, Pierre Lermusiaux, Evan Mason, Anders Omstedt, Tamay Özgökmen, Thomas Rossby, Charles Troupin, and Lars Umlauf. We also would like to thank our wives Mary and Françoise for their patience and support.

Benoit Cushman-Roisin
Jean-Marie Beckers
January 2011

Preface of the First Edition

The intent of *Introduction to Geophysical Fluid Dynamics* is to introduce readers to this developing field. In the late 1950s, this discipline emerged as a few scientists, building on a miscellaneous heritage of fluid mechanics, meteorology, and oceanography, began to model complex atmospheric and oceanic flows by relatively simple mathematical analysis, thereby unifying atmospheric and oceanic physics. Turning from art to science, the discipline then matured during the 1970s. Appropriately, a first treatise titled *Geophysical Fluid Dynamics* by Joseph Pedlosky (Springer-Verlag) was published in 1979. Since then, several other authoritative textbooks have become available, all aimed at graduate students and researchers dedicated to the physics of the atmosphere and oceans. It is my opinion that the teaching of geophysical fluid dynamics is now making its way into science graduate curricula outside of meteorology and oceanography (e.g., physics and engineering). Simultaneously and in view of today's concerns regarding global change, acid precipitations, sea-level rise, and so forth, there is also a growing desire on the part of biologists, atmospheric chemists, and engineers to understand the rudiments of climate and ocean dynamics. In this perspective, I believe that the time has come for an introductory text aimed at upper-level undergraduate students, graduate students, and researchers in environmental fluid dynamics.

In the hope of fulfilling this need, simplicity and clarity have been the guiding principles in preparing this book. Whenever possible, the physical principles are illustrated with the aid of the simplest existing models, and the terminology and notation have been selected to maximize the physical interpretation of the concepts and equations. For example, the expression *planetary wave* is preferred to *Rossby wave*, and subscripts are avoided whenever not strictly indispensable.

The book is divided in five parts. After the fundamentals have been established in Part I, the effects of rotation and stratification are explored separately in the following two parts. Then, Part IV analyzes the combined effects of rotation and stratification, and the book closes with Part V, on miscellaneous topics of contemporary interest. Each part is divided into short, relatively well-contained chapters to provide flexibility in the choice of materials to be discussed, according to the needs of the curriculum or the reader's interests. Each chapter corresponds to one or two lectures, occasionally three, and the length is deemed suitable for a one-semester course (45 lectures). Although it is also an inevitable reflection of my personal choices, the selection of materials has been guided by the desire to emphasize the physical principles at work behind observed phenomena. Such emphasis is also much in keeping with the traditional teaching of geophysical fluid dynamics. The scientist interested in the description of atmospheric and oceanic phenomena will find available an abundance of introductory texts in meteorology and oceanography.

Unlike existing texts in geophysical fluid dynamics, this book offers a number of exercises at the end of every chapter. There, the reader/teacher will also find short biographies and suggestions for laboratory demonstrations. Finally, the text ends with an appendix on wave kinematics, for it is my experience that not all students are familiar with the concepts of wavenumber, dispersion relation, and group velocity, whereas these are central to the understanding of geophysical wave phenomena.

A general remark on the notation is appropriate. Because mathematical physics in general and this discipline in particular involve symbols representing variables and constants, with and without dimensions, I believe that clarity is brought to the mathematical description of the subject when certain classes of symbols are reserved for certain types of terms. In that spirit, a systematic effort has been placed to assign the notation according to the following rules: Dimensional variables are denoted by lowercase Roman letters (such as u, v, and w for the velocity components), dimensional constants and parameters use uppercase Roman letters (such as H for the domain depth, L for length scale), and dimensionless quantities are assigned lowercase Greek letters (such as θ for an angle). In keeping with a well-established convention of fluid mechanics, dimensionless numbers credited to particular scientists are denoted by the first two letters of those scientists' names (e.g., Ro for the Rossby number). Of course, conventions breed exceptions (e.g., g for the constant gravitational acceleration, ω for frequency, and ψ for streamfunction).

In closing, I wish to acknowledge inspiration from numerous colleagues from across the globe, too many to permit an exhaustive list here. I am also particularly indebted to my students at Dartmouth College; their thirst for knowledge prompted the present text. Don L. Boyer, Arizona State University, Pijush K. Kundu, Nova University, Peter D. Killworth, Robert Hooke Institute, Fred Lutgens, Central Illinois College, Joseph Pedlosky, Woods Hole Oceanographic Institution, and George Veronis, Yale University, made many detailed and invaluable suggestions, which have improved both the clarity and accuracy of the presentation. Finally, deep gratitude goes to Lori Terino for her expertise and patience in typing the text.

Benoit Cushman-Roisin
1993

Part I

Fundamentals

Chapter 1

Introduction

ABSTRACT

This opening chapter defines the discipline known as geophysical fluid dynamics, stresses its importance, and highlights its most distinctive attributes. A brief history of numerical simulations in meteorology and oceanography is also presented. Scale analysis and its relationship with finite differences are introduced to show how discrete numerical grids depend on the scales under investigation and how finite differences permit the approximation of derivatives at those scales. The problem of unresolved scales is introduced as an aliasing problem in discretization.

1.1 OBJECTIVE

The object of geophysical fluid dynamics is the study of naturally occurring, large-scale flows on Earth and elsewhere, but mostly on Earth. Although the discipline encompasses the motions of both fluid phases – liquids (waters in the ocean, molten rock in the outer core) and gases (air in our atmosphere, atmospheres of other planets, ionized gases in stars) – a restriction is placed on the *scale* of these motions. Only the large-scale motions fall within the scope of geophysical fluid dynamics. For example, problems related to river flow, micro-turbulence in the upper ocean, and convection in clouds are traditionally viewed as topics specific to hydrology, oceanography, and meteorology, respectively. Geophysical fluid dynamics deals exclusively with those motions observed in various systems and under different guises but nonetheless governed by similar dynamics. For example, large anticyclones of our weather are dynamically germane to vortices spun off by the Gulf Stream and to Jupiter's Great Red Spot. Most of these problems, it turns out, are at the large-scale end, where either the ambient rotation (of Earth, planet, or star) or density differences (warm and cold air masses, fresh and saline waters), or both assume some importance. In this respect, geophysical fluid dynamics comprises rotating-stratified fluid dynamics.

Typical problems in geophysical fluid dynamics concern the variability of the atmosphere (weather and climate dynamics), ocean (waves, vortices, and currents), and, to a lesser extent, the motions in the earth's interior responsible for the dynamo effect, vortices on other planets (such as Jupiter's Great

Red Spot and Neptune's Great Dark Spot), and convection in stars (the sun, in particular).

1.2 IMPORTANCE OF GEOPHYSICAL FLUID DYNAMICS

Without its atmosphere and oceans, it is certain that our planet would not sustain life. The natural fluid motions occurring in these systems are therefore of vital importance to us, and their understanding extends beyond intellectual curiosity—it is a necessity. Historically, weather vagaries have baffled scientists and laypersons alike since times immemorial. Likewise, conditions at sea have long influenced a wide range of human activities, from exploration to commerce, tourism, fisheries, and even wars.

Thanks in large part to advances in geophysical fluid dynamics, the ability to predict with some confidence the paths of hurricanes (Figs. 1.1 and 1.2) has led to the establishment of a warning system that, no doubt, has saved numerous lives at sea and in coastal areas (Abbott, 2004). However, warning systems are only useful if sufficiently dense observing systems are implemented, fast prediction capabilities are available, and efficient flow of information is ensured. A dreadful example of a situation in which a warning system was not yet adequate to save lives was the earthquake off Indonesia's Sumatra Island on

FIGURE 1.1 Hurricane Frances during her passage over Florida on 5 September 2004. The diameter of the storm was about 830 km, and its top wind speed approached 200 km per hour. (*Courtesy of NOAA, Department of Commerce, Washington, D.C.*)

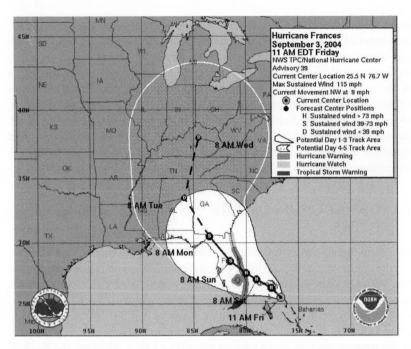

FIGURE 1.2   Computer prediction of the path of Hurricane Frances. The calculations were performed on Friday, 3 September 2004, to predict the hurricane path and characteristics over the next 5 days (until Wednesday, 8 September). The outline surrounding the trajectory indicates the level of uncertainty. Compare the position predicted for Sunday, 5 September, with the actual position shown on Fig. 1.1. (*Courtesy of NOAA, Department of Commerce, Washington, D.C.*)

26 December 2004. The tsunami generated by the earthquake was not detected, its consequences not assessed, and authorities not alerted within the 2 h needed for the wave to reach beaches in the region. On a larger scale, the passage every 3–5 years of an anomalously warm water mass along the tropical Pacific Ocean and the western coast of South America, known as the El-Niño event, has long been blamed for serious ecological damage and disastrous economical consequences in some countries (Glantz, 2001; O'Brien, 1978). Now, thanks to increased understanding of long oceanic waves, atmospheric convection, and natural oscillations in air–sea interactions (D'Aleo, 2002; Philander, 1990), scientists have successfully removed the veil of mystery on this complex event, and numerical models (e.g., Chen, Cane, Kaplan, Zebiak & Huang, 2004) offer reliable predictions with at least one year of *lead time*, that is, there is a year between the moment the prediction is made and the time to which it applies.

Having acknowledged that our industrial society is placing a tremendous burden on the planetary atmosphere and consequently on all of us, scientists, engineers, and the public are becoming increasingly concerned about the fate of pollutants and greenhouse gases dispersed in the environment and especially about their cumulative effect. Will the accumulation of greenhouse gases in the atmosphere lead to global climatic changes that, in turn, will affect our lives and societies? What are the various roles played by the oceans in maintaining our present climate? Is it possible to reverse the trend toward depletion of the ozone in the upper atmosphere? Is it safe to deposit hazardous wastes on the ocean floor? Such pressing questions cannot find answers without, first, an in-depth understanding of atmospheric and oceanic dynamics and, second, the development of predictive models. In this twin endeavor, geophysical fluid dynamics assumes an essential role, and the numerical aspects should not be underestimated in view of the required predictive tools.

1.3   DISTINGUISHING ATTRIBUTES OF GEOPHYSICAL FLOWS

Two main ingredients distinguish the discipline from traditional fluid mechanics: the effects of rotation and those of stratification. The controlling influence of one, the other, or both leads to peculiarities exhibited only by geophysical flows. In a nutshell, this book can be viewed as an account of these peculiarities.

The presence of an ambient rotation, such as that due to the earth's spin about its axis, introduces in the equations of motion two acceleration terms that, in the rotating framework, can be interpreted as forces. They are the Coriolis force and the centrifugal force. Although the latter is the more palpable of the two, it plays no role in geophysical flows; however, surprising this may be.[1] The former and less intuitive of the two turns out to be a crucial factor in geophysical motions. For a detailed explanation of the Coriolis force, the reader is referred to the following chapter in this book or to the book by Stommel and Moore (1989). A more intuitive explanation and laboratory illustrations can be found in Chapter 6 of Marshall and Plumb (2008).

In anticipation of the following chapters, it can be mentioned here (without explanation) that a major effect of the Coriolis force is to impart a certain vertical rigidity to the fluid. In rapidly rotating, homogeneous fluids, this effect can be so strong that the flow displays strict columnar motions; that is, all particles along the same vertical evolve in concert, thus retaining their vertical alignment over long periods of time. The discovery of this property is attributed to Geoffrey I. Taylor, a British physicist famous for his varied contributions to fluid dynamics. (See the short biography at the end of Chapter 7.) It is said that Taylor first arrived at the rigidity property with mathematical arguments alone. Not believing that this could be correct, he then performed laboratory experiments that revealed, much to his amazement, that the theoretical prediction was indeed correct. Drops of dye released in such rapidly rotating, homogeneous

---

[1] Here we speak about the centrifugal force associated with the earth's planetary rotation, not to be confused with the centrifugal force associated with the strong rotation of eddies or hurricanes.

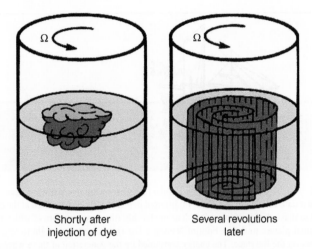

Shortly after
injection of dye

Several revolutions
later

FIGURE 1.3 Experimental evidence of the rigidity of a rapidly rotating, homogeneous fluid. In a spinning vessel filled with clear water, an initially amorphous cloud of aqueous dye is transformed in the course of several rotations into perfectly vertical sheets, known as *Taylor curtains*.

fluids form vertical streaks, which, within a few rotations, shear laterally to form spiral sheets of dyed fluid (Fig. 1.3). The vertical coherence of these sheets is truly fascinating!

In large-scale atmospheric and oceanic flows, such state of perfect vertical rigidity is not realized chiefly because the rotation rate is not sufficiently fast and the density is not sufficiently uniform to mask other, ongoing processes. Nonetheless, motions in the atmosphere, in the oceans, and on other planets manifest a tendency toward columnar behavior. For example, currents in the western North Atlantic have been observed to extend vertically over 4000 m without significant change in amplitude and direction (Schmitz, 1980).

Stratification, the other distinguishing attribute of geophysical fluid dynamics, arises because naturally occurring flows typically involve fluids of different densities (e.g., warm and cold air masses, fresh and saline waters). Here, the gravitational force is of great importance, for it tends to lower the heaviest fluid and to raise the lightest. Under equilibrium conditions, the fluid is stably stratified, consisting of vertically stacked horizontal layers. However, fluid motions disturb this equilibrium, in which gravity systematically strives to restore. Small perturbations generate internal waves, the three-dimensional analogue of surface waves, with which we are all familiar. Large perturbations, especially those maintained over time, may cause mixing and convection. For example, the prevailing winds in our atmosphere are manifestations of the planetary convection driven by the pole-to-equator temperature difference.

It is worth mentioning the perplexing situation in which a boat may experience strong resistance to forward motion while sailing under apparently calm conditions. This phenomenon, called *dead waters* by mariners, was first

FIGURE 1.4   A laboratory experiment by Ekman (1904) showing internal waves generated by a model ship in a tank filled with two fluids of different densities. The heavier fluid at the bottom has been colored to make the interface visible. The model ship (the superstructure of which was drawn onto the original picture to depict Fridtjof Nansen's *Fram*) is towed from right to left, causing a wake of waves on the interface. The energy consumed by the generation of these waves produces a drag that, for a real ship, would translate into a resistance to forward motion. The absence of any significant surface wave has prompted sailors to call such situations *dead waters*. (*From Ekman, 1904, and adapted by Gill, 1982*)

documented by the Norwegian oceanographer Fridtjof Nansen, famous for his epic expedition on the *Fram* through the Arctic Ocean, begun in 1893. Nansen reported the problem to his Swedish colleague Vagn Walfrid Ekman who, after performing laboratory simulations (Ekman, 1904), affirmed that internal waves were to blame. The scenario is as follows: During times of dead waters, Nansen must have been sailing in a layer of relatively fresh water capping the more saline oceanic waters and of thickness, coincidently, comparable to the ship draft; the ship created a wake of internal waves along the interface (Fig. 1.4), unseen at the surface but radiating considerable energy and causing the noted resistance to the forward motion of the ship.

1.4   SCALES OF MOTIONS

To discern whether a physical process is dynamically important in any particular situation, geophysical fluid dynamicists introduce *scales of motion*. These are dimensional quantities expressing the overall magnitude of the variables under consideration. They are estimates rather than precisely defined quantities and are understood solely as *orders of magnitude* of physical variables. In most situations, the key scales are those for time, length, and velocity. For example, in the dead-water situation investigated by V.W. Ekman (Fig. 1.4), fluid motions comprise a series of waves whose dominant wavelength is about the length of the submerged ship hull; this length is the natural choice for the length scale L of the problem; likewise, the ship speed provides a reference velocity that can be taken as the velocity scale U; finally, the time taken for the ship to travel the distance L at its speed U is the natural choice of time scale: $T = L/U$.

As a second example, consider Hurricane Frances during her course over the southeastern United States in early September 2004 (Fig. 1.1). The satellite picture reveals a nearly circular feature spanning approximately 7.5° of latitude (830 km). Sustained surface wind speeds of a category-4 hurricane such as Frances range from 59 to 69 m/s. In general, hurricane tracks display appreciable change in direction and speed of propagation over 2-day intervals. Altogether, these elements suggest the following choice of scales for a hurricane: $L = 800$ km, $U = 60$ m/s, and $T = 2 \times 10^5$ s ($= 55.6$ h).

As a third example, consider the famous Great Red Spot in Jupiter's atmosphere (Fig. 1.5), which is known to have existed at least several hundred years. The structure is an elliptical vortex centered at 22°S and spanning approximately 12° in latitude and 25° in longitude; its highest wind speeds exceed 110 m/s, and the entire feature slowly drifts zonally at a speed of 3 m/s (Dowling & Ingersoll, 1988; Ingersoll et al., 1979). Knowing that the planet's equatorial radius is 71,400 km, we determine the vortex semi-major and semi-minor axes (14,400 km and 7,500 km, respectively) and deem $L = 10,000$ km to be an appropriate length scale. A natural velocity scale for the fluid is $U = 100$ m/s. The selection of a timescale is somewhat problematic in view of the nearly

FIGURE 1.5   Southern hemisphere of Jupiter as seen by the spacecraft *Cassini* in 2000. The Jupiter moon Io, of size comparable to our moon, projects its shadow onto the zonal jets between which the Great Red Spot of Jupiter is located (on the left). For further images, visit http://photojournal.jpl.nasa.gov/target/Jupiter. (*Image courtesy of NASA/JPL/University of Arizona*)

steady state of the vortex; one choice is the time taken by a fluid particle to cover the distance L at the speed U ($T = L/U = 10^5$ s), whereas another is the time taken by the vortex to drift zonally over a distance equal to its longitudinal extent ($T = 10^7$ s). Additional information on the physics of the problem is clearly needed before selecting a timescale. Such ambiguity is not uncommon because many natural phenomena vary on different temporal scales (e.g., the terrestrial atmosphere exhibits daily weather variation as well as decadal climatic variations, among others). The selection of a timescale then reflects the particular choice of physical processes being investigated in the system.

There are three additional scales that play important roles in analyzing geophysical fluid problems. As we mentioned earlier, geophysical fluids generally exhibit a certain degree of density heterogeneity, called stratification. The important parameters are then the average density ρ_0, the range of density variations $\Delta\rho$, and the height H over which such density variations occur. In the ocean, the weak compressibility of water under changes of pressure, temperature, and salinity translates into values of $\Delta\rho$ always much less than ρ_0, whereas the compressibility of air renders the selection of $\Delta\rho$ in atmospheric flows somewhat delicate. Because geophysical flows are generally bounded in the vertical direction, the total depth of the fluid may be substituted for the height scale H. Usually, the smaller of the two height scales is selected.

As an example, the density and height scales in the dead-water problem (Fig. 1.4) can be chosen as follows: $\rho_0 = 1025$ kg/m^3, the density of either fluid layer (almost the same); $\Delta\rho = 1$ kg/m^3, the density difference between lower and upper layers (much smaller than ρ_0), and $H = 5$ m, the depth of the upper layer.

As the person new to geophysical fluid dynamics has already realized, the selection of scales for any given problem is more an art than a science. Choices are rather subjective. The trick is to choose quantities that are relevant to the problem, yet simple to establish. There is freedom. Fortunately, small inaccuracies are inconsequential because the scales are meant only to guide in the clarification of the problem, whereas grossly inappropriate scales will usually lead to flagrant contradictions. Practice, which forms intuition, is necessary to build confidence.

1.5 IMPORTANCE OF ROTATION

Naturally, we may wonder at which scales the ambient rotation becomes an important factor in controlling the fluid motions. To answer this question, we must first know the ambient rotation rate, which we denote by Ω and define as

$$\Omega = \frac{2\pi \text{ radians}}{\text{time of one revolution}}. \tag{1.1}$$

Since our planet Earth actually rotates in two ways simultaneously, once per day about itself and once a year around the sun, the terrestrial value of Ω consists of two terms, $2\pi/24$ hours $+ 2\pi/365.24$ days $= 2\pi/1$ sidereal day $= 7.2921 \times 10^{-5}$ s^{-1}. The *sidereal day*, equal to 23 h 56 min and 4.1 s, is the period of time spanning the moment when a fixed (distant) star is seen one day and the moment on the next day when it is seen at the same angle from the same point on Earth. It is slightly shorter than the 24-hour solar day, the time elapsed between the sun reaching its highest point in the sky two consecutive times, because the earth's orbital motion about the sun makes the earth rotate slightly more than one full turn with respect to distant stars before reaching the same Earth–Sun orientation.

If fluid motions evolve on a timescale comparable to or longer than the time of one rotation, we anticipate that the fluid does feel the effect of the ambient rotation. We thus define the dimensionless quantity

$$\omega = \frac{\text{time of one revolution}}{\text{motion timescale}} = \frac{2\pi/\Omega}{T} = \frac{2\pi}{\Omega T}, \tag{1.2}$$

where T is used to denote the timescale of the flow. Our criterion is as follows: If ω is on the order of or less than unity ($\omega \lesssim 1$), rotation effects should be considered. On Earth, this occurs when T exceeds 24 h.

Yet, motions with shorter timescales ($\omega \gtrsim 1$) but sufficiently large spatial extent could also be influenced by rotation. A second and usually more useful criterion results from considering the velocity and length scales of the motion. Let us denote these by U and L, respectively. Naturally, if a particle traveling at the speed U covers the distance L in a time longer than or comparable to a rotation period, we expect the trajectory to be influenced by the ambient rotation, so we write

$$\epsilon = \frac{\text{time of one revolution}}{\text{time taken by particle to cover distance } L \text{ at speed } U}$$
$$= \frac{2\pi/\Omega}{L/U} = \frac{2\pi U}{\Omega L}. \tag{1.3}$$

If ϵ is on the order of or less than unity ($\epsilon \lesssim 1$), we conclude that rotation is important.

Let us now consider a variety of possible length scales, using the value Ω for Earth. The corresponding velocity criteria are listed in Table 1.1.

Obviously, in most engineering applications, such as the flow of water at a speed of 5 m/s in a turbine 1 m in diameter ($\epsilon \sim 4 \times 10^5$) or the air flow past a 5-m wing on an airplane flying at 100 m/s ($\epsilon \sim 2 \times 10^6$), the inequality is not met, and the effects of rotation can be ignored. Likewise, the common task of emptying a bathtub (horizontal scale of 1 m, draining speed in the order of 0.01 m/s and a lapse of about 1000 s, giving $\omega \sim 90$ and $\epsilon \sim 900$) does not fall under the scope of geophysical fluid dynamics. On the contrary, geophysical flows (such as an ocean current flowing at 10 cm/s and meandering over a distance of 10 km or a wind blowing at 10 m/s in a 1000-km-wide anticyclonic formation) do meet the inequality. This demonstrates that rotation is usually important in geophysical flows.

TABLE 1.1 Length and Velocity Scales of Motions in Which Rotation Effects are Important

$L = 1$ m	$U \leq 0.012$ mm/s
$L = 10$ m	$U \leq 0.12$ mm/s
$L = 100$ m	$U \leq 1.2$ mm/s
$L = 1$ km	$U \leq 1.2$ cm/s
$L = 10$ km	$U \leq 12$ cm/s
$L = 100$ km	$U \leq 1.2$ m/s
$L = 1000$ km	$U \leq 12$ m/s
$L =$ Earth radius $= 6371$ km	$U \leq 74$ m/s

1.6 IMPORTANCE OF STRATIFICATION

The next question concerns the condition under which stratification effects are expected to play an important dynamical role. Geophysical fluids typically consist of fluid masses of different densities, which under gravitational action tend to arrange themselves in vertical stacks (Fig. 1.6), corresponding to a state of minimal potential energy. But, motions continuously disturb this equilibrium, tending to raise dense fluid and lower light fluid. The corresponding increase of potential energy is at the expense of kinetic energy, thereby slowing the flow. On occasions, the opposite happens: Previously disturbed stratification returns toward equilibrium, potential energy converts into kinetic energy, and the flow gains momentum. In sum, the dynamical importance of stratification can be evaluated by comparing potential and kinetic energies.

If $\Delta\rho$ is the scale of density variations in the fluid and H is its height scale, a prototypical perturbation to the stratification consists in raising a fluid element of density $\rho_0 + \Delta\rho$ over the height H and, in order to conserve volume, lowering a lighter fluid element of density ρ_0 over the same height. The corresponding change in potential energy, per unit volume, is $(\rho_0 + \Delta\rho)\, gH - \rho_0 gH = \Delta\rho gH$. With a typical fluid velocity U, the kinetic energy available per unit volume is $\frac{1}{2}\rho_0 U^2$. Accordingly, we construct the comparative energy ratio

$$\sigma = \frac{\frac{1}{2}\rho_0 U^2}{\Delta\rho gH}, \tag{1.4}$$

to which we can give the following interpretation. If σ is on the order of unity ($\sigma \sim 1$), a typical potential-energy increase necessary to perturb the stratification consumes a sizable portion of the available kinetic energy, thereby modifying the flow field substantially. Stratification is then important. If σ is much less

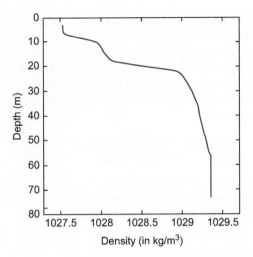

FIGURE 1.6 Vertical profile of density in the northern Adriatic Sea (43°32′N, 14°03′E) on 27 May 2003. Density increases downward by leaps and bounds, revealing the presence of different water masses stacked on top of one another in such a way that lighter waters float above denser waters. A region where the density increases significantly faster than above and below, marking the transition from one water mass to the next, is called a *pycnocline*. (*Data courtesy of Drs. Hartmut Peters and Mirko Orlić*)

than unity ($\sigma \ll 1$), there is insufficient kinetic energy to perturb significantly the stratification, and the latter greatly constrains the flow. Finally, if σ is much greater than unity ($\sigma \gg 1$), potential-energy modifications occur at very little cost to the kinetic energy, and stratification hardly affects the flow. In conclusion, stratification effects cannot be ignored in the first two cases—that is, when the dimensionless ratio defined in Eq. (1.4) is on the order of or much less than unity ($\sigma \lesssim 1$). In other words, σ is to stratification what the number ϵ, defined in Eq. (1.3), is to rotation.

A most interesting situation arises in geophysical fluids when rotation and stratification effects are simultaneously important, yet neither dominates over the other. Mathematically, this occurs when $\epsilon \sim 1$ and $\sigma \sim 1$ and yields the following relations among the various scales:

$$L \sim \frac{U}{\Omega} \quad \text{and} \quad U \sim \sqrt{\frac{\Delta\rho}{\rho_0}gH}. \tag{1.5}$$

(The factors 2π and $\frac{1}{2}$ have been omitted because they are secondary in a scale analysis.) Elimination of the velocity U yields a fundamental length scale:

$$L \sim \frac{1}{\Omega}\sqrt{\frac{\Delta\rho}{\rho_0}gH}. \tag{1.6}$$

In a given fluid, of mean density ρ_0 and density variation $\Delta\rho$, occupying a height H on a planet rotating at rate Ω and exerting a gravitational acceleration g, the scale L arises as a preferential length over which motions take place. On Earth ($\Omega = 7.29 \times 10^{-5}$ s^{-1} and $g = 9.81$ m/s^2), typical conditions in the atmosphere ($\rho_0 = 1.2$ kg/m^3, $\Delta\rho = 0.03$ kg/m^3, $H = 5000$ m) and in the ocean ($\rho_0 = 1028$ kg/m^3, $\Delta\rho = 2$ kg/m^3, $H = 1000$ m) yield the following natural length and velocity scales:

$$L_{\text{atmosphere}} \sim 500 \text{ km} \quad U_{\text{atmosphere}} \sim 30 \text{ m/s}$$
$$L_{\text{ocean}} \quad \sim \quad 60 \text{ km} \quad U_{\text{ocean}} \quad \sim \ 4 \text{ m/s}.$$

Although these estimates are relatively crude, we can easily recognize here the typical size and wind speed of weather patterns in the lower atmosphere and the typical width and speed of major currents in the upper ocean.

1.7 DISTINCTION BETWEEN THE ATMOSPHERE AND OCEANS

Generally, motions of the air in our atmosphere and of seawater in the oceans that fall under the scope of geophysical fluid dynamics occur on scales of several kilometers up to the size of the earth. Atmospheric phenomena comprise the coastal sea breeze, local to regional processes associated with topography, the cyclones, anticyclones, and fronts that form our daily weather, the general atmospheric circulation, and the climatic variations. Oceanic phenomena of interest include estuarine flow, coastal upwelling and other processes associated with the presence of a coast, large eddies and fronts, major ocean currents such as the Gulf Stream, and the large-scale circulation. Table 1.2 lists the typical velocity, length and time scales of these motions, whereas Fig. 1.7 ranks a sample of atmospheric and oceanic processes according to their spatial and temporal scales. As we can readily see, the general rule is that oceanic motions are slower and slightly more confined than their atmospheric counterparts. Also, the ocean tends to evolve more slowly than the atmosphere.

Besides notable scale disparities, the atmosphere and oceans also have their own peculiarities. For example, a number of oceanic processes are caused by the presence of lateral boundaries (continents, islands), a constraint practically nonexistent in the atmosphere, except in stratified flows where mountain ridges can sometimes play such a role, exactly as do mid-ocean ridges for stratified ocean currents. On the other hand, atmospheric motions are sometimes strongly dependent on the moisture content of the air (clouds, precipitation), a characteristic without oceanic counterpart.

Flow patterns in the atmosphere and oceans are generated by vastly different mechanisms. By and large, the atmosphere is thermodynamically driven, that is, its primary source of energy is the solar radiation. Briefly, this shortwave solar

TABLE 1.2 Length, Velocity and Time Scales in the Earth's Atmosphere and Oceans

Phenomenon	Length Scale L	Velocity Scale U	Timescale T
Atmosphere			
Microturbulence	10–100 cm	5–50 cm/s	few seconds
Thunderstorms	few km	1–10 m/s	few hours
Sea breeze	5–50 km	1–10 m/s	6 h
Tornado	10–500 m	30–100 m/s	10–60 min
Hurricane	300–500 km	30–60 m/s	Days to weeks
Mountain waves	10–100 km	1–20 m/s	Days
Weather patterns	100–5000 km	1–50 m/s	Days to weeks
Prevailing winds	Global	5–50 m/s	Seasons to years
Climatic variations	Global	1–50 m/s	Decades and beyond
Ocean			
Microturbulence	1–100 cm	1–10 cm/s	10–100 s
Internal waves	1–20 km	0.05–0.5 m/s	Minutes to hours
Tides	Basin scale	1–100 m/s	Hours
Coastal upwelling	1–10 km	0.1–1 m/s	Several days
Fronts	1–20 km	0.5–5 m/s	Few days
Eddies	5–100 km	0.1–1 m/s	Days to weeks
Major currents	50–500 km	0.5–2 m/s	Weeks to seasons
Large-scale gyres	Basin scale	0.01–0.1 m/s	Decades and beyond

radiation traverses the air layer to be partially absorbed by the continents and oceans, which in turn re-emit a radiation at longer wavelengths. This second-hand radiation effectively heats the atmosphere from below, and the resulting convection drives the winds.

In contrast, the oceans are forced by a variety of mechanisms. In addition to the periodic gravitational forces of the moon and sun that generate the tides, the ocean surface is subjected to a wind stress that drives most ocean currents. Finally, local differences between air and sea temperatures generate heat fluxes,

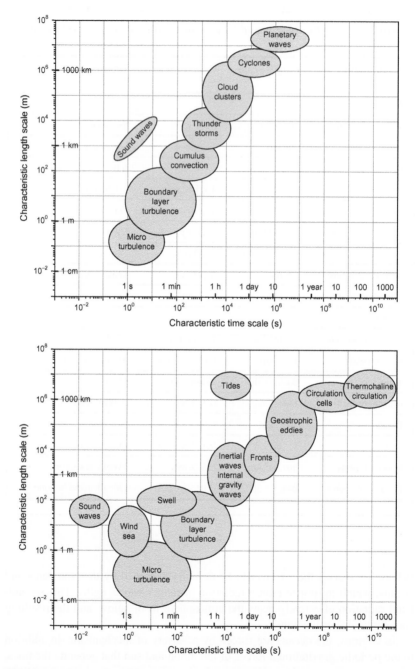

FIGURE 1.7 Various types of processes and structures in the atmosphere (top panel) and oceans (bottom panel), ranked according to their respective length and time scales. (*Diagram courtesy of Hans von Storch*)

evaporation, and precipitation, which in turn act as thermodynamical forcings capable of modifying the wind-driven currents or producing additional currents.

In passing, while we are contrasting the atmosphere with the oceans, it is appropriate to mention an enduring difference in terminology. Because meteorologists and laypeople alike are generally interested in knowing from where the winds are blowing, it is common in meteorology to refer to air velocities by their direction of origin, such as easterly (from the east—that is, toward the west). On the contrary, sailors and navigators are interested in knowing where ocean currents may take them. Hence, oceanographers designate currents by their downstream direction, such as westward (from the east or to the west). However, meteorologists and oceanographers agree on the terminology for vertical motions: upward or downward.

1.8   DATA ACQUISITION

Because geophysical fluid dynamics deals exclusively with naturally occurring flows and, moreover, those of rather sizable proportions, full-scale experimentation must be ruled out. Indeed, how could one conceive of changing the weather, even locally, for the sake of a scientific inquiry? Also, the Gulf Stream determines its own fancy path, irrespective of what oceanographers wish to study about it. In that respect, the situation is somewhat analogous to that of the economist who may not ask the government to prompt a disastrous recession for the sake of determining some parameters of the national economy. The inability to control the system under study is greatly alleviated by simulations. In geophysical fluid dynamics, these investigations are conducted via laboratory experiments and numerical models.

As well as being reduced to noting the whims of nature, observers of geophysical flows also face length and timescales that can be impractically large. A typical challenge is the survey of an oceanic feature several hundred kilometers wide. With a single ship (which is already quite expensive, especially if the feature is far away from the home shore), a typical survey can take several weeks, a time interval during which the feature might translate, distort, or otherwise evolve substantially. A faster survey might not reveal details with a sufficiently fine horizontal representation. Advances in satellite imagery and other methods of remote sensing (Conway & the Maryland Space Grant Consortium, 1997; Marzano & Visconti, 2002) do provide synoptic (i.e., quasi-instantaneous) fields, but those are usually restricted to specific levels in the vertical (e.g., cloud tops and ocean surface) or provide vertically integrated quantities. Also, some quantities simply defy measurement, such as the heat flux and vorticity. Those quantities can only be derived by an analysis on sets of proxy observations.

Finally, there are processes for which the timescale is well beyond the span of human life if not the age of civilization. For example, climate studies require a certain understanding of glaciation cycles. Our only recourse here is to be clever and to identify today some traces of past glaciation events, such as geological records. Such an indirect approach usually requires a number of assumptions, some of which may never be adequately tested. Finally, exploration of other planets and of the sun is even more arduous.

At this point one may ask: What can we actually measure in the atmosphere and oceans with a reasonable degree of confidence? First and foremost, a number of scalar properties can be measured directly and with conventional instruments. For both the atmosphere and ocean, it is generally not difficult to measure the pressure and temperature. In fact, in the ocean, the pressure can be measured so much more accurately than depth that, typically, depth is calculated from measured pressure on instruments that are gradually lowered into the sea. In the atmosphere, one can also accurately measure the water vapor, rainfall, and some radiative heat fluxes (Marzano & Visconti, 2002; Rao, Holmes, Anderson, Winston & Lehr, 1990). Similarly, the salinity of seawater can be either determined directly or inferred from electrical conductivity (Pickard & Emery, 1990). Also, the sea level can be monitored at shore stations. The typical problem, however, is that the measured quantities are not necessarily those preferred from a physical perspective. For example, one would prefer direct measurements of the vorticity field, Bernoulli function, diffusion coefficients, and turbulent correlation quantities.

Vectorial quantities are usually more difficult to obtain than scalars. Horizontal winds and currents can now be determined routinely by anemometers and current meters of various types, including some without rotating components (Lutgens & Tarbuck, 1986; Pickard & Emery, 1990) although usually not with the desired degree of spatial resolution. Fixed instruments, such as anemometers atop buildings and oceanic current meters at specific depths along a mooring line, offer fine temporal coverage, but adequate spatial coverage typically requires a prohibitive number of such instruments. To remedy the situation, instruments on drifting platforms (e.g., balloons in the atmosphere and drifters or floats in the ocean) are routinely deployed. However, these instruments provide information that is mixed in time and space and thus is not ideally suited to most purposes. A persistent problem is the measurements of the vertical velocity. Although vertical speeds can be measured with acoustic Doppler current profilers, the meaningful signal is often buried below the level of ambient turbulence and instrumental error (position and sensitivity). Measuring the vector vorticity, so dear to theoreticians, is out of the question as is the three-dimensional heat flux.

Also, some uncertainty resides in the interpretation of the measured quantities. For example, can the wind measured in the vicinity of a building be taken as representative of the prevailing wind over the city and so be used in weather forecasting, or is it more representative of a small-scale flow pattern resulting from the obstruction of the wind by the building?

Finally, sampling frequencies might not always permit the unambiguous identification of a process. Measuring quantities at a given location every week might well lead to a data set that includes also residual information on faster processes than weekly variations or a slower signal that we would like to capture with our measurements. For example, if we measure temperature on Monday at 3 o'clock in the afternoon one week and Monday at 7 o'clock in the morning the next week, the measurement will include a diurnal heating component super-imposed on the weekly variation. The measurements are thus not necessarily representative of the process of interest.

1.9   THE EMERGENCE OF NUMERICAL SIMULATIONS

Given the complexity of weather patterns and ocean currents, one can easily anticipate that the equations governing geophysical fluid motions, which we are going to establish in this book, are formidable and not amenable to analytical solution except in rare instances and after much simplification. Thus, one faces the tall challenge of having to solve the apparently unsolvable. The advent of the electronic computer has come to the rescue, but at a definite cost. Indeed, com-puters cannot solve differential equations but can only perform the most basic arithmetic operations. The partial differential equations (PDEs) of geophysical fluid dynamics (GFD) need therefore to be transformed into a sequence of arith-metic operations. This process requires careful transformations and attention to details.

The purpose of numerical simulations of GFD flows is not limited to weather prediction, operational ocean forecasting, and climate studies. There are situa-tions when one desires to gain insight and understanding of a specific process, such as a particular form of instability or the role of friction under particular con-ditions. Computer simulations are our only way to experiment with the planet. Also, there is the occasional need to experiment with a novel numerical technique in order to assess its speed and accuracy. Simulations increasingly go hand in hand with observations in the sense that the latter can point to places in which the model needs refinements while model results can suggest optimal placing of observational platforms or help to define sampling strategies. Finally, simulations can be a retracing of the past (*hindcasting*) or a smart interpolation of scattered data (*nowcasting*), as well as the prediction of future states (*forecasting*).

Models of GFD flows in meteorology, oceanography, and climate studies come in all types and sizes, depending on the geographical domain of interest (local, regional, continental, basinwide, or global) and the desired level of phys-ical realism. Regional models are far too numerous to list here, and we only mention the existence of Atmospheric General Circulations Models (AGCMs), Oceanic General Circulation Models (OGCMs) and coupled General Circula-tion Models (GCMs). A truly comprehensive model does not exist because the coupling of air, sea, ice, and land physics over the entire planet is always open to the inclusion of yet one more process heretofore excluded from the model. In developing a numerical model of some GFD system, the question immediately arises as to what actually needs to be simulated. The answer largely dictates the level of details necessary and, therefore also, the level of physical approximation and the degree of numerical resolution.

Geophysical flows are governed by a set of coupled, nonlinear equations in four-dimensional space-time and exhibit a high sensitivity to details. In mathematical terms, it is said that the system possesses chaotic properties, and the consequence is that geophysical flows are inherently unpredictable as Lorenz demonstrated for the atmosphere several decades ago (Lorenz, 1963). The physical reality is that geophysical fluid systems are replete with instabilities, which amplify in a finite time minor details into significant structures (the butterfly-causing-a-tempest syndrome). The cyclones and anticyclones of midlatitude weather and the meandering of the coastal currents are but a couple of examples among many. Needless to say, the simulation of atmospheric and oceanographic fluid motions is a highly challenging task.

The initial impetus for geophysical fluid simulations was, not surprisingly, weather prediction, an aspiration as old as mankind. More recently, climate studies have become another leading force in model development because of their need for extremely large and complex models.

The first decisive step in the quest for weather prediction was made by Vilhelm Bjerknes (1904) in a paper titled "The problem of weather prediction considered from the point of view of mechanics and physics." He was the first to pose the problem as a set of time-dependent equations derived from physics and to be solved from a given, and hopefully complete, set of initial conditions. Bjerknes immediately faced the daunting task of integrating complicated partial differential equations, and, because this was well before electronic computers, resorted to graphical methods of solution. Unfortunately, these had little practical value and never surpassed the art of subjective forecasting by a trained person pouring over weather charts.

Taking a different tack, Lewis Fry Richardson (1922; see biography at the end of Chapter 14) decided that it would be better to reduce the differential equations to a set of arithmetic operations (additions, subtractions, multiplications, and divisions exclusively) so that a step-by-step method of solution may be followed and performed by people not necessarily trained in meteorology. Such reduction could be accomplished, he reasoned, by seeking the solution at only selected points in the domain and by approximating spatial derivatives of the unknown variables by finite differences across those points. Likewise, time could be divided into finite intervals, and temporal derivatives be approximated as differences across those time intervals, and thus was born numerical analysis. Richardson's work culminated in his 1922 book entitled *Weather Prediction by Numerical Process*. His first grid, to forecast weather over western Europe, is reproduced here as Fig. 1.8. After the equations of motion had been dissected into a sequence of individual arithmetic operations, the first algorithm before the word existed, computations were performed by a large group of people,

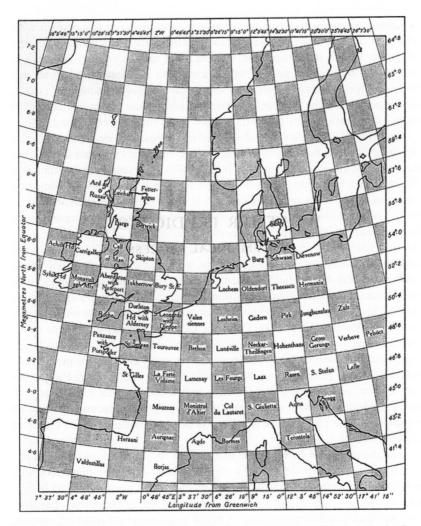

FIGURE 1.8   Model grid used by Lewis Fry Richardson as reported in his 1922 book *Weather Prediction by Numerical Process*. The grid was designed to optimize the fit between cells and existing meteorological stations, with observed surface pressures being used at the center of every shaded cell and winds at the center of every white cell.

called *computers*, sitting around an auditorium equipped with slide rules and passing their results to their neighbors. Synchronization was accomplished by a leader in the pit of the auditorium as a conductor leads an orchestra. Needless to say, the work was tedious and slow, requiring an impractically large number of people to conduct the calculations quickly enough so that a 24-h forecast could be obtained in less than 24 h.

Despite an enormous effort on Richardson's part, the enterprise was a failure, with predicted pressure variations rapidly drifting away from meteorologically acceptable values. In retrospective, we now know that Richardson's model was improperly initiated for lack of upper level data and that its 6-h time step was exceeding the limit required by numerical stability, of which, of course, he was not aware. The concept of numerical stability was not known until 1928 when it was elucidated by Richard Courant, Karl Friedrichs and Hans Lewy.

The work of Richardson was abandoned and relegated to the status of a curiosity or, as he put it himself, "a dream," only to be picked up again seriously at the advent of electronic computers. In the 1940s, the mathematician John von Neumann (see biography at end of Chapter 5) became interested in hydrodynamics and was seeking mathematical aids to solve nonlinear differential equations. Contact with Alan Turing, the inventor of the electronic computer, gave him the idea to build an automated electronic machine that could perform sequential calculations at a speed greatly surpassing that of humans. He collaborated with Howard Aiken at Harvard University, who built the first electronic calculator, named the Automatic Sequence Controlled Calculator (ASCC). In 1943, von Neumann helped build the Electronic Numerical Integrator and Computer (ENIAC) at the University of Pennsylvania and, in 1945, the Electronic Discrete Variable Calculator (EDVAC) at Princeton University. Primarily because of the wartime need for improved weather forecasts and also out of personal challenge, von Neumann paired with Jule Charney (see biography at end of Chapter 16) and selected weather forecasting as the scientific challenge. But, unlike Richardson before them, von Neumann and Charney started humbly with a much reduced set of dynamics, a single equation to predict the pressure at mid-level in the troposphere. The results (Charney, Fjörtoft & von Neumann, 1950) exceeded expectations.

Success with a much reduced set of dynamics only encouraged further development. Phillips (1956) developed a two-layer quasi-geostrophic[2] model over a hemispheric domain. The results did not predict actual weather but did behave like weather, with realistic cyclones generated at the wrong places and times. This was nonetheless quite encouraging. A major limitation of the quasi-geostrophic simplification is that it fails near the equator, and the only remedy was a return to the full equations (called *primitive equations*), back to where Richardson started. The main problem, it was found by then, is that primitive equations retain fast-moving gravity waves, and although these hold only a small amount of energy, their resolution demands both a much shorter time step of integration and a far better set of initial conditions than were available at the time.

From then on, the major intellectual challenges were overcome, and steady progress (Fig. 1.9) has been achieved, thanks to ever-faster and larger computers

---

[2] Quasi-geostrophic dynamics are described in Chapter 16. It suffices here to say that the formalism eliminates the velocity components under the assumption that rotational effects are very strong. The result is a drastic reduction in the number of equations.

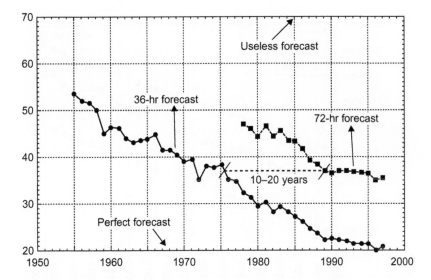

FIGURE 1.9   Historical improvement of weather forecasting skill over North America. The S1 score shown here is a measure of the relative error in the pressure gradient predictions at mid-height in the troposphere. (*From Kalnay, Lord & McPherson, 1998, reproduction with the kind permission of the American Meteorological Society*)

(Fig. 1.10) and to the gathering of an ever-denser array of data around the globe. The reader interested in the historical developments of weather forecasting will find an excellent book-length account in Nebeker (1995).

1.10   SCALES ANALYSIS AND FINITE DIFFERENCES

In the preceding section, we saw that computers are used to solve numerically equations otherwise difficult to apprehend. Yet, even with the latest supercomputers and unchanged physical laws, scientists are requesting more computer power than ever, and we may rightfully ask what is the root cause of this unquenchable demand. To answer, we introduce a simple numerical technique (*finite differences*) that shows the strong relationship between scale analysis and numerical requirement. It is a prototypical example foreshowing a characteristic of more elaborate numerical methods that will be introduced in later chapters for more realistic problems.

When performing a timescale analysis, we assume that a physical variable u changes significantly over a timescale T by a typical value U (Fig. 1.11). With this definition of scales, the time derivative is on the order of

$$\frac{\mathrm{d}u}{\mathrm{d}t} \sim \frac{U}{T}. \tag{1.7}$$

If we then assume that the timescale over which the function u changes is also the one over which its derivative changes (in other words, we assume the

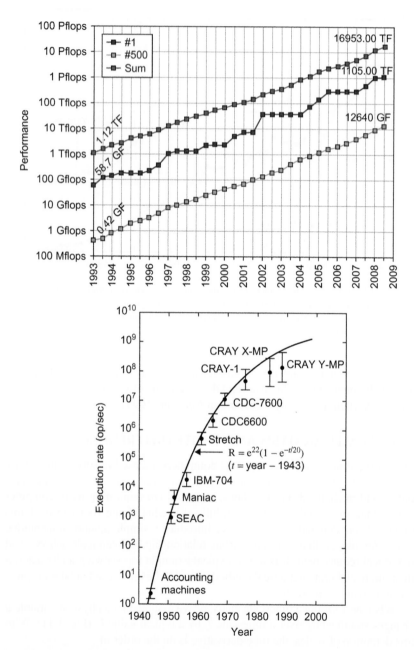

FIGURE 1.10 Historical increase of computational speed, as measured by the number of operations performed per second. (*Adapted and supplemented from Hack (1992), who gives credit to a 1987 personal communication with Worlton (left panel) and with recent data from http://www.top500.org, upper panel*)

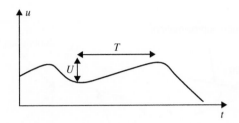

FIGURE 1.11   Timescale analysis of a variable u. The timescale T is the time interval over which the variable u exhibits variations comparable to its standard deviation U.

timescale T to be representative of all types of variabilities, including differentiated fields), we can also estimate the order of magnitude of variations of the second derivative

$$\frac{d^2u}{dt^2} = \frac{d}{dt}\left(\frac{du}{dt}\right) \sim \frac{U/T}{T} = \frac{U}{T^2}, \tag{1.8}$$

and so on for higher-order derivatives. This approach is the basis for estimating the relative importance of different terms in time-marching equations, an exercise we will repeat several times in the following chapters.

We now turn our attention to the question of estimating derivatives with more accuracy than by a mere order of magnitude. Typically, this problem arises on discretizing equations, a process by which all derivatives are replaced by algebraic approximations based on a few discrete values of the function u (Fig. 1.12). Such *discretization* is necessary because computers possess a finite memory and are incapable of manipulating derivatives. We then face the following problem: Having stored only a few values of the function, how can we retrieve the value of the function's derivatives that appear in the equations?

First, it is necessary to discretize the independent variable time t, since the first dynamical equations that we shall solve numerically are time-evolving equations. For simplicity, we shall suppose that the discrete time moments t^n, at which the function values are to be known, are uniformly distributed with a constant *time step* Δt

$$t^n = t^0 + n\Delta t, \quad n = 1, 2, \ldots. \tag{1.9}$$

where the superscript index (not an exponent) n identifies the discrete time. Then, we note by u^n the value of u at time t^n, that is, $u^n = u(t^n)$. We now would like to determine the value of the derivative du/dt at time t^n knowing only the discrete values u^n. From the definition of a derivative

$$\frac{du}{dt} = \lim_{\Delta t \to 0} \frac{u(t + \Delta t) - u(t)}{\Delta t}, \tag{1.10}$$

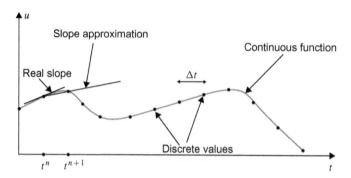

FIGURE 1.12 Representation of a function by a finite number of sampled values and approximation of a first derivative by a finite difference over Δt.

we could directly deduce an approximation by allowing Δt to remain the finite time step

$$\frac{du}{dt} \simeq \frac{u(t+\Delta t) - u(t)}{\Delta t} \rightarrow \left.\frac{du}{dt}\right|_{t^n} \simeq \frac{u^{n+1} - u^n}{\Delta t}. \tag{1.11}$$

The accuracy of this approximation can be determined with the help of a Taylor series:

$$u(t+\Delta t) = u(t) + \Delta t \left.\frac{du}{dt}\right|_t + \underbrace{\frac{\Delta t^2}{2} \left.\frac{d^2 u}{dt^2}\right|_t}_{\Delta t^2 \frac{U}{T^2}} + \underbrace{\frac{\Delta t^3}{6} \left.\frac{d^3 u}{dt^3}\right|_t}_{\Delta t^3 \frac{U}{T^3}} + \underbrace{\mathcal{O}(\Delta t^4)}_{\Delta t^4 \frac{U}{T^4}}. \tag{1.12}$$

To the leading order for small Δt, we obtain the following estimate

$$\frac{du}{dt} = \frac{u(t+\Delta t) - u(t)}{\Delta t} + \mathcal{O}\left(\frac{\Delta t}{T} \frac{U}{T}\right). \tag{1.13}$$

The *relative* error on the derivative (the difference between the finite-difference approximation and the actual derivative, divided by the scale U/T) is therefore of the order $\Delta t/T$. For the approximation to be acceptable, this relative error should be much smaller than 1, which demands that the time step Δt be sufficiently short compared to the timescale at hand:

$$\Delta t \ll T. \tag{1.14}$$

This condition can be visualized graphically by considering the effect of various values of Δt on the resulting estimation of the time derivative (Fig. 1.13). In the following, we write the formal approximation as

$$\left.\frac{du}{dt}\right|_{t^n} = \frac{u^{n+1} - u^n}{\Delta t} + \mathcal{O}(\Delta t), \tag{1.15}$$

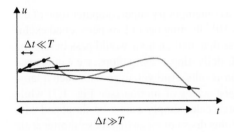

FIGURE 1.13 Finite differencing with various Δt values. Only when the time step is sufficiently short compared to the time scale, $\Delta t \ll T$, is the finite-difference slope close to the derivative, that is, the true slope.

where it is understood that the measure of whether or not Δt is "small enough" must be based on the timescale T of the variability of the variable u. Since in the simple finite difference (1.15), the error, called *truncation error*, is proportional to Δt, the approximation is said to be of first order. For an error proportional to Δt^2, the approximation is said of second order and so on.

For spatial derivatives, the preceding analysis is easily applicable, and we obtain a condition on the horizontal grid size Δx relatively to the horizontal length scale L, while the vertical grid space Δz is constrained by the vertical length scale H of the variable under investigation:

$$\Delta x \ll L, \quad \Delta z \ll H. \tag{1.16}$$

With these constraints on time steps and grid sizes, we can begin to understand the need for significant computer resources in GFD simulations: The number of grid points M in a 3D domain of surface S and height H is

$$M = \frac{H}{\Delta z} \frac{S}{\Delta x^2}, \tag{1.17}$$

while the total number of time steps N needed to cover a time period P is

$$N = \frac{P}{\Delta t}. \tag{1.18}$$

For a model covering the Atlantic Ocean ($S \sim 10^{14} \, \mathrm{m}^2$), resolving geostrophic eddies (see Fig. 1.7: $\Delta x \sim \Delta y \le 10^4$ m) and stratified water masses ($H/\Delta z \sim 50$), the number of grid points is about $M \sim 5 \times 10^7$. Then, at each of these points, several variables need to be stored and calculated (three-dimensional velocity, pressure, temperature, etc.). Since each variable takes 4 or 8 bytes of memory depending on the desired number of significant digits, 2 Gigabytes of RAM is required. The number of floating point operations to be executed to simulate a single year can be estimated by taking a time step resolving the rotational period of Earth $\Delta t \sim 10^3$ s, leading to $N \sim 30000$ time steps. The total number of operations to simulate a full year can then be estimated by observing that for every grid point and time step, a series of calculations must be performed (typically several hundreds) so that the total number of calculations amounts to 10^{14}–10^{15}. Therefore, on a contemporary supercomputer (one of the top 500 machines) with 1 Teraflops $= 10^{12}$ floating operations per second exclusively dedicated to the simulation, less than half an hour would pass before the response is available, whereas on a PC delivering 1-2 Gigaflops, we would need to wait several days before getting our results. And yet, even with such a large model, we can only resolve the largest scales of motion (see Fig. 1.7) while motions on shorter spatial and temporal scales simply cannot be simulated with this level of grid resolution. However, this does not mean that those shorter scale motions may altogether be neglected and, as we will see (e.g., Chapter 14), one of the problems of large-scale oceanic and atmospheric models is the need for appropriate *parameterization* of shorter scale motions so that they may properly bear their effects onto the larger scale motions.

Should we dream to avoid such a parameterization by explicitly calculating all scales, we would need about $M \sim 10^{24}$ grid points demanding 5×10^{16} Gigabytes of computer memory and $N \sim 3 \times 10^7$ time steps, for a total number of operations on the order of 10^{34}. Willing to wait only for 10^6 s before obtaining the results, we would need a computer delivering 10^{28} flops. This is a factor $10^{16} = 2^{53}$ higher than the present capabilities, both for speed and memory requirements. Using Moore's Law, the celebrated rule that forecasts a factor 2 in gain of computing power every 18 months, we would have to wait 53 times 18 months, that is, for about 80 years before computers could handle such a task.

Increasing resolution will therefore continue to call for the most powerful computers available, and models will need to include parameterization of turbulence or other unresolved motions for quite some time. Grid spacing will thus remain a crucial aspect of all GFD models, simply because of the large domain sizes and broad range of scales.

1.11 HIGHER-ORDER METHODS

Rather than to increase resolution to better represent structures, we may wonder whether using other approximations for derivatives than our simple finite difference (1.11) would allow larger time steps or higher quality approximations and improved model results. Based on a Taylor series

$$u^{n+1} = u^n + \Delta t \left. \frac{du}{dt} \right|_{t^n} + \frac{\Delta t^2}{2} \left. \frac{d^2 u}{dt^2} \right|_{t^n} + \frac{\Delta t^3}{6} \left. \frac{d^3 u}{dt^3} \right|_{t^n} + \mathcal{O}(\Delta t^4) \qquad (1.19)$$

$$u^{n-1} = u^n - \Delta t \left. \frac{du}{dt} \right|_{t^n} + \frac{\Delta t^2}{2} \left. \frac{d^2 u}{dt^2} \right|_{t^n} - \frac{\Delta t^3}{6} \left. \frac{d^3 u}{dt^3} \right|_{t^n} + \mathcal{O}(\Delta t^4), \qquad (1.20)$$

we can imagine that instead of using a *forward-difference* approximation of the time derivative (1.11), we try a backward Taylor series (1.20) to design a *backward-difference* approximation. This approximation is obviously still of first order because of its truncation error:

$$\frac{du}{dt}\bigg|_{t^n} = \frac{u^n - u^{n-1}}{\Delta t} + \mathcal{O}(\Delta t). \tag{1.21}$$

Comparing Eq. (1.19) with Eq. (1.20), we observe that the truncation errors of the first-order forward and backward finite differences are the same but have opposite signs so that by averaging both, we obtain a second-order truncation error (you can verify this statement by taking the difference between Eqs. (1.19) and (1.20)):

$$\frac{du}{dt}\bigg|_{t^n} = \frac{u^{n+1} - u^{n-1}}{2\Delta t} + \mathcal{O}(\Delta t^2). \tag{1.22}$$

Before considering higher-order approximations, let us first check whether the increase in order of approximation actually leads to improved approximations of the derivatives. To do so, consider the sinusoidal function of period T (and associated frequency ω)

$$u = U\sin\left(2\pi\frac{t}{T}\right) = U\sin(\omega t), \quad \omega = \frac{2\pi}{T}. \tag{1.23}$$

Knowing that the exact derivative is $\omega U\cos(\omega t)$, we can calculate the errors made by the various finite-difference approximations (Fig. 1.14). Both the forward and backward finite differences converge toward the exact value for $\omega\Delta t \to 0$, with errors decreasing proportionally to Δt. As expected, the second-order approximation (1.22) exhibits a second-order convergence (the slope is 2 in a log–log graph).

The convergence rate obeys our theoretical estimate for $\omega\Delta t \ll 1$. However, when the time step is relatively large (Fig. 1.15), the error associated with the finite-difference approximations can be as large as the derivative itself. For coarse resolution, $\omega\Delta t \sim \mathcal{O}(1)$, the relative error is of order 1 so that we expect a 100% error on the finite-difference approximation. Obviously, even with a second-order finite difference, we need at least $\omega\Delta t \le 0.8$ to keep the relative error below 10%. In terms of the period of the signal $T = (2\pi)/\omega$, we need a time step not larger than $\Delta t \lesssim T/8$, which implies that 8 points are needed along one period to resolve its derivatives within a 10% error level. Even a fourth-order method (to be shown shortly) cannot reconstruct derivatives correctly from a function sampled with fewer than several points per period.

The design of the second-order difference was accomplished simply by inspection of a Taylor series, a technique that cannot be extended to obtain higher-order approximations. An alternate method exists to obtain in a systematic way finite-difference approximations to any desired order, and it can be

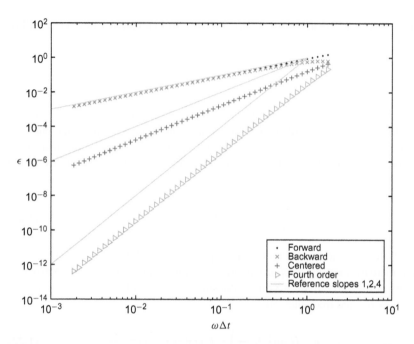

FIGURE 1.14 Relative error ϵ of various finite-difference approximations of the first derivative of the sinusoidal function as function of $\omega \Delta t$ when $\omega t = 1$. Scales are logarithmic and continuous lines of slope 1, 2, and 4 are added. First-order methods have a slope of 1, and the second-order method a slope of 2. The error behaves as expected for decreasing Δt.

illustrated with the design of a fourth-order centered finite-difference approximation of the first derivative. Expecting that higher-order approximations need more information about a function in order to estimate its derivative at time t^n, we will combine values over a longer time interval, including $t^{n-2}, t^{n-1}, t^n, t^{n+1}$, and t^{n+2}:

$$\left.\frac{du}{dt}\right|_{t^n} \simeq a_{-2} u^{n-2} + a_{-1} u^{n-1} + a_0 u^n + a_1 u^{n+1} + a_2 u^{n+2}. \qquad (1.24)$$

Expanding u^{n+2} and the other values around t^n by Taylor series, we can write

$$\left.\frac{du}{dt}\right|_{t^n} = (a_{-2} + a_{-1} + a_0 + a_1 + a_2)\, u^n$$

$$+ (-2a_{-2} - a_{-1} + a_1 + 2a_2)\, \Delta t \left.\frac{du}{dt}\right|_{t^n}$$

$$+ (4a_{-2} + a_{-1} + a_1 + 4a_2)\, \frac{\Delta t^2}{2} \left.\frac{d^2 u}{dt^2}\right|_{t^n}$$

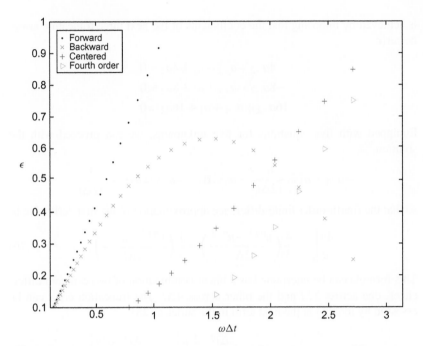

FIGURE 1.15   Relative error ϵ of various finite-difference approximations of the first derivative of the sinusoidal function as function of $\omega\Delta t$ when $\omega t = 1$. For coarse resolution $\omega\Delta t \sim \mathcal{O}(1)$, the relative error is of order 1 so that we expect a 100% error on the finite-difference approximation.

$$+ (-8a_{-2} - a_{-1} + a_1 + 8a_2) \frac{\Delta t^3}{6} \left. \frac{d^3 u}{dt^3} \right|_{t^n}$$

$$+ (16a_{-2} + a_{-1} + a_1 + 16a_2) \frac{\Delta t^4}{24} \left. \frac{d^4 u}{dt^4} \right|_{t^n}$$

$$+ (-32a_{-2} - a_{-1} + a_1 + 32a_2) \frac{\Delta t^5}{120} \left. \frac{d^5 u}{dt^5} \right|_{t^n}$$

$$+ \mathcal{O}(\Delta t^6). \tag{1.25}$$

There are five coefficients, a_{-2} to a_2, to be determined. Two conditions must be satisfied to obtain an approximation that tends to be the first derivative as $\Delta t \to 0$:

$$a_{-2} + a_{-1} + a_0 + a_1 + a_2 = 0,$$
$$(-2a_{-2} - a_{-1} + a_1 + 2a_2)\,\Delta t = 1.$$

After satisfying these two necessary conditions, we have three parameters that can be freely chosen so as to obtain the highest possible level of accuracy. This is achieved by imposing that the coefficients of the next three truncation errors be zero:

$$4a_{-2}+a_{-1}+a_1+4a_2=0$$
$$-8a_{-2}-a_{-1}+a_1+8a_2=0$$
$$16a_{-2}+a_{-1}+a_1+16a_2=0.$$

Equipped with five equations for five unknowns, we can proceed with the solution:

$$-a_{-1}=a_1=\frac{8}{12\Delta t}, \quad a_0=0, \quad -a_{-2}=a_2=-\frac{1}{12\Delta t},$$

so that the fourth-order finite-difference approximation of the first derivative is

$$\frac{du}{dt}\bigg|_{t^n}\simeq\frac{4}{3}\left(\frac{u^{n+1}-u^{n-1}}{2\Delta t}\right)-\frac{1}{3}\left(\frac{u^{n+2}-u^{n-2}}{4\Delta t}\right). \tag{1.26}$$

This formula can be interpreted as a linear combination of two centered differences, one across $2\Delta t$ and the other across $4\Delta t$. The truncation error can be assessed by looking at the next term in the series (1.25)

$$(-32a_{-2}-a_{-1}+a_1+32a_2)\frac{\Delta t^5}{120}\frac{d^5u}{dt^5}\bigg|_{t^n}=-\frac{\Delta t^4}{30}\frac{d^5u}{dt^5}\bigg|_{t^n}, \tag{1.27}$$

which shows that the approximation is indeed of fourth order.

The method can be generalized to approximate a derivative of any order p at time t^n using the current value u^n, m points in the past (before t^n) and m points in the future (after t^n):

$$\frac{d^p u}{dt^p}\bigg|_{t_n}=a_{-m}u^{n-m}+\cdots+a_{-1}u^{n-1}+a_0u^n+a_1u^{n+1}+\cdots+a_mu^{n+m}. \tag{1.28}$$

The discrete points $n-m$ to $n+m$ involved in the approximation define the so-called *numerical stencil* of the operator. Using a Taylor expansion for each term

$$u^{n+q}=u^n+q\Delta t\,\frac{du}{dt}\bigg|_{t^n}+q^2\frac{\Delta t^2}{2}\frac{d^2u}{dt^2}\bigg|_{t^n}+\cdots+q^p\frac{\Delta t^p}{p!}\frac{d^p u}{dt^p}\bigg|_{t^n}+\mathcal{O}(\Delta t^{p+1})$$

$$(1.29)$$

and injecting Eq. (1.29) for $q=-m,...,m$ into the approximation (1.28), we have on the left-hand side the derivative we want to approximate and on the right a sum of derivatives. We impose that the sum of coefficients multiplying a derivative lower than order p be zero, whereas the sum of the coefficients multiplying the pth derivative be 1. This forms a set of $p+1$ equations for the $2m+1$ unknown coefficients a_q ($q=-m,\ldots,m$). All constraints can be satisfied simultaneously only if we use a number $2m+1$ of points equal to or greater than

$p+1$, that is, $2m \geq p$. When there are more points than necessary, we can take advantage of the remaining degrees of freedom to cancel the next few terms in the truncation errors. With $2m+1$ points, we can then obtain a finite difference of order $2m-p+1$. For example, with $m=1$ and $p=1$, we obtained Eq. (1.22), a second-order approximation of the first derivative, and with $m=2$ and $p=1$, Eq. (1.26), a fourth-order approximation.

Let us now turn to the second derivative, a very common occurrence, at least when considering spatial derivatives. With $p=2$, m must be at least 1, that is, three values of the function are required as a minimum: one old, one current, and one future values. Applying the preceding method, we immediately obtain

$$\frac{d^2 u}{dt^2}\bigg|_{t_n} \simeq \left(\frac{u^{n-1} - 2u^n + u^{n+1}}{\Delta t^2} \right), \tag{1.30}$$

a result we could also have obtained by direct inspection of Eqs. (1.19) and (1.20).

Appendix C recapitulates a variety of discretization schemes for different orders of derivatives and various levels of accuracy. It also includes skewed schemes, which are not symmetric between past and future values but can be constructed in a way similar to the fourth-order finite-difference approximation of the first derivative.

1.12 ALIASING

We learned that the accuracy of a finite-difference approximation of the first derivative degrades rapidly when the time step Δt is not kept much shorter than the timescale T of the variable, and we might wonder what would happen if Δt should by some misfortune be larger than T. To answer this question, we return to a physical signal u of period T,

$$u = U \sin(\omega t + \phi), \quad \omega = \frac{2\pi}{T}, \tag{1.31}$$

sampled on a uniform grid of time step Δt,

$$u^n = U \sin(n\,\omega \Delta t + \phi), \tag{1.32}$$

and assume that there exists another signal v of higher frequency $\tilde{\omega}$ given by

$$v = U \sin(\tilde{\omega} t + \phi), \quad \tilde{\omega} = \omega + \frac{2\pi}{\Delta t}. \tag{1.33}$$

The sampling of this other function at the same time intervals yields a discrete set of values

$$v^n = U \sin(n\,\tilde{\omega} \Delta t + \phi) = U \sin(n\,\omega \Delta t + 2n\pi + \phi) = u^n, \tag{1.34}$$

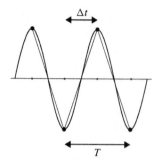

FIGURE 1.16 Shortest wave (at cutoff frequency $\pi/\Delta t$ or period $2\Delta t$) resolved by uniform grid in time.

which cannot be distinguished from the discrete values u^n of the first signal although the two signals are clearly not equal to each other. Thus, frequencies ω and $\omega + 2\pi/\Delta t$ cannot be separated in a sampling with time interval Δt because the higher-frequency signal masquerades as the lower-frequency signal. This unavoidable consequence of sampling is called *aliasing*.

Since signals of frequency $\omega + 2\pi/\Delta t$ and ω cannot be distinguished from each other, it appears that only frequencies within the following range

$$-\frac{\pi}{\Delta t} \leq \omega \leq \frac{\pi}{\Delta t} \tag{1.35}$$

can be recognized with a sampling interval Δt, and all other frequencies should preferably be absent, lest they contaminate the sampling process.

Since a negative frequency corresponds to a $180°$ phase shift, because $\sin(-\omega t + \phi) = \sin(\omega t - \phi + \pi)$, the useful range is actually $0 \leq \omega \leq \pi/\Delta t$, and to sample a wave of frequency ω, the time step Δt may not exceed $\Delta t_{max} = \pi/\omega = T/2$, which implies that at least two samples of the signal must be taken per period. This minimum required sampling frequency is called the *Nyquist frequency*. Looking at the problem in a different way, with a given sampling interval Δt (rather than a given frequency), we recognize that the highest resolved frequency is $\pi/\Delta t$, called the *cutoff frequency* (Fig. 1.16).

Should higher frequencies be present and sampled, aliasing inevitably occurs, as illustrated by a sinusoidal function sampled with increasingly fewer points per period (Fig. 1.17). The reader is invited to experiment with MATLAB™ script aliasanim.m. Up to $\Delta t = T/2$, the signal is recognizable, but, beyond that, lines connecting consecutive sampled values appear to tunnel through crests and troughs, giving the impression of a signal with longer period.

Aliasing is a major concern, and the danger it poses is often underestimated. This is because we do not know whether the signal being represented by the discretization scheme contains frequencies higher than the cutoff frequency, precisely because variability at those frequencies is not retained and computed. In geophysical situations, the time step and grid spacing are most often set not by the physics of the problem but by computer-hardware limits. This forces the modeler to discard variability at unresolved frequencies and wavelengths and

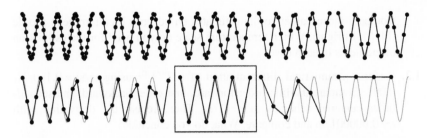

FIGURE 1.17 Aliasing illustrated by sampling a given signal (gray sinusoidal curve) with an increasing time interval. A high sampling rate (top row of images) resolves the signal properly. The boxed image on the bottom row corresponds to the cutoff frequency, and the sampled signal appears as a seesaw. The last two images correspond to excessively long time intervals that alias the signal, making it appear as if it had a longer period than it actually has.

creates aliasing. Methods to overcome the undesired effects of aliasing will be presented in subsequent chapters.

ANALYTICAL PROBLEMS

1.1. Name three naturally occurring flows in the atmosphere.

1.2. How did geophysical flows contribute to Christopher Columbus' discovery of the New World and to the subsequent exploration of the eastern shore of North America? (Think of both large-scale winds and major ocean currents.)

1.3. The sea breeze is a light wind blowing from the sea as the result of a temperature difference between land and sea. As this temperature difference reverses from day to night, the daytime sea breeze turns into a nighttime land breeze. If you were to construct a numerical model of the sea–land breeze, should you include the effects of the planetary rotation?

1.4. The Great Red Spot of Jupiter, centered at 22°S and spanning 12° in latitude and 25° in longitude, exhibits wind speeds of about 100 m/s. The planet's equatorial radius and rotation rates are, respectively, 71,400 km and 1.763×10^{-4} s^{-1}. Is the Great Red Spot influenced by planetary rotation?

1.5. Can you think of a technique for measuring wind speeds and ocean velocities with an instrument that has no rotating component? (*Hint*: Think of measurable quantities whose values are affected by translation.)

NUMERICAL EXERCISES

1.1. Using the temperature measurements of Nansen (Fig. 1.18 left), estimate typical vertical temperature gradient values and typical temperature values.

Compare those values to the estimates based on a Mediterranean profile (Fig. 1.18 right). Which temperature scale do you need for the estimation of gradients in each case?

1.2. Perform a numerical differentiation of $e^{-\omega(t+|t|)}$ using $\omega\Delta t = 0.1, 0.01, 0.001$. Compare the first-order forward, first-order backward, and second-order centered schemes at $t = 1/\omega$. Then, repeat the derivation with $t = 0.000001/\omega$ and compare. What do you conclude?

1.3. Apply forward, backward, second-order, and fourth-order discretizations to $\sinh(kx)$ at $x = 1/k$ for values of $k\Delta x$ covering the range $[10^{-4}, 1]$. Plot errors on a logarithmic scale and verify the convergence rates. Repeat the exercise for $x = 0$. What strange effect do you observe and why?

1.4. Establish a purely forward, finite-difference approximation of a first derivative that is of second or higher-order in accuracy. How many sampling points are required as a function of the order of accuracy?

1.5. Suppose you need to evaluate $\partial u/\partial x$ not at grid node i, but at mid-distance between nodes $x_i = i\Delta x$ and $x_{i+1} = (i+1)\Delta x$. Establish second-order and fourth-order finite-difference approximations to do so and compare the truncation errors to the corresponding discretizations centered on the nodal point i. What does this analysis suggest?

1.6. Assume that a spatial two-dimensional domain is covered by a uniform grid with spacing Δx in the x-direction and Δy in the y-direction. How can you discretize $\partial^2 u/\partial x\partial y$ to second order? Does the approximation satisfy a similar property as its mathematical counterpart $\partial^2 u/\partial x\partial y = \partial^2 u/\partial y\partial x$?

1.7. Determine how a wave of wavelength $\frac{4}{3}\Delta x$ and period $\frac{5}{3}\Delta t$ is interpreted in a uniform grid of mesh Δx and time step Δt. How does the computed propagation speed resulting from the discrete sampling compare to the true speed?

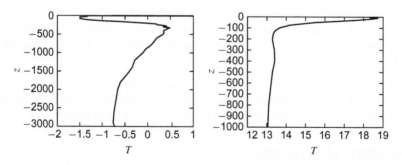

FIGURE 1.18 Temperature values measured by Nansen during his 1894 North Pole expedition (left) and a typical temperature profile in the Mediterranean Sea (from Medar© data base, right).

1.8. Suppose you use numerical finite differencing to estimate derivatives of a function that was sampled with some noise. Assuming the noise is uncorrelated (i.e., purely randomly distributed, independently of the sampling interval), what do you expect would happen during the finite differencing of first-order derivatives, second derivatives etc.? Devise a numerical program that verifies your assertion, by adding a random noise of intensity $10^{-5}A$ to the function $A\sin(\omega t)$, where the frequency ω is well resolved by the numerical sampling ($\omega\Delta t = 0.05$). Did you correctly guess what would happen? Now plot the convergence as a function of $\omega\Delta t$ for noise levels of $10^{-5}A$ and $10^{-4}A$.

Walsh Cottage, Woods Hole, Massachusetts
1962–present

Every summer since 1962, this unassuming building of the Woods Hole
Oceanographic Institution (Falmouth, Massachusetts, USA) has been home
to the Geophysical Fluid Dynamics Summer Program, which has gathered
oceanographers, meteorologists, physicists, and mathematicians from around
the world. This program (begun in 1959) has single-handedly been responsible
for many of the developments of geophysical fluid dynamics, from its humble
beginnings to its present status as a recognized discipline in physical sciences.
(*Drawing by Ryuji Kimura, reproduced with permission*)

UK Meteorological Office, Exeter, England
1854–present

The Meteorological Office in the United Kingdom was established in 1854 to provide meteorological information and sea currents by telegraph to people at sea and, within a few years, began also to issue storm warnings to seaports and weather forecasts to the press. In 1920, separate meteorological military services established during World War I were merged with the civilian office under the Air Ministry. World War II led to another large increase in both personnel and resources, including the use of balloons. The "Met Office" has an outstanding record of capitalizing on new technologies, beginning broadcasts on radio (in 1922) and then on television (in 1936 by means of simple captions and live broadcasts in 1954). The first electronic computer was installed in 1962, and satellite imagery was incorporated in 1964.

As weather forecasts began to depend less on trained meteorologists drawing weather maps and more on computational models, the need for the latest and best performing computer platform became a driving force, leading to the acquisition of a Cyber supercomputer in 1981 and a series of ever faster Cray supercomputers in the 1990s.

The impact of the Met Office can hardly be underestimated: Its numerical activities have contributed enormously not only to the field of meteorology but also to the development of computational fluid dynamics and physical oceanography, while the scope of its data analyses and forecasts has spread well beyond tomorrow's weather to other areas such as the impact of the weather on the environment and human health. (*For additional information, see http://www.metoffice.gov.uk/about-us/who/our-history*)

Chapter 2

The Coriolis Force

ABSTRACT

The objective of this chapter is to examine the Coriolis force, a fictitious force arising from the choice of a rotating framework of reference. Some physical considerations are offered to provide insight on this nonintuitive but essential element of geophysical flows. The numerical section of this chapter treats time stepping introduced in the particular case of inertial oscillations and generalized afterwards.

2.1 ROTATING FRAMEWORK OF REFERENCE

From a theoretical point of view, all equations governing geophysical fluid processes could be stated with respect to an inertial framework of reference, fixed with respect to distant stars. But, we people on Earth observe fluid motions with respect to this rotating system. Also, mountains and ocean boundaries are stationary with respect to Earth. Common sense therefore dictates that we write the governing equations in a reference framework rotating with our planet. (The same can be said for other planets and stars.) The trouble arising from the additional terms in the equations of motion is less than that which would arise from having to reckon with moving boundaries and the need to subtract systematically the ambient rotation from the resulting flow.

To facilitate the mathematical developments, let us first investigate the two-dimensional case (Fig. 2.1). Let the X– and Y–axes form the inertial framework of reference and the x– and y–axes be those of a framework with the same origin but rotating at the angular rate Ω (defined as positive in the trigonometric sense). The corresponding unit vectors are denoted $(\mathbf{I}, \mathbf{J})$ and $(\mathbf{i}, \mathbf{j})$. At any time t, the rotating x–axis makes an angle Ωt with the fixed X–axis. It follows that

$$\mathbf{i} = + \mathbf{I} \cos \Omega t + \mathbf{J} \sin \Omega t \quad (2.1a)$$

$$\mathbf{j} = - \mathbf{I} \sin \Omega t + \mathbf{J} \cos \Omega t \quad (2.1b)$$

$$\mathbf{I} = +\mathbf{i} \cos \Omega t - \mathbf{j} \sin \Omega t \quad (2.2a)$$

$$\mathbf{J} = +\mathbf{i} \sin \Omega t + \mathbf{j} \cos \Omega t, \quad (2.2b)$$

Introduction to Geophysical Fluid Dynamics

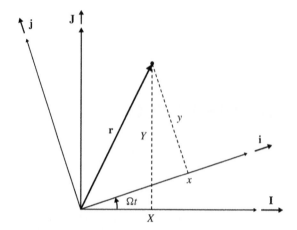

FIGURE 2.1 Fixed (X, Y) and rotating (x, y) frameworks of reference.

and that the coordinates of the position vector $\mathbf{r} = X\mathbf{I} + Y\mathbf{J} = x\mathbf{i} + y\mathbf{j}$ of any point in the plane are related by

$$x = +X\cos\Omega t + Y\sin\Omega t \tag{2.3a}$$

$$y = -X\sin\Omega t + Y\cos\Omega t. \tag{2.3b}$$

The first time derivative of the preceding expressions yields

$$\frac{dx}{dt} = +\frac{dX}{dt}\cos\Omega t + \frac{dY}{dt}\sin\Omega t \overbrace{-\Omega X\sin\Omega t + \Omega Y\cos\Omega t}^{+\Omega y} \tag{2.4a}$$

$$\frac{dy}{dt} = -\frac{dX}{dt}\sin\Omega t + \frac{dY}{dt}\cos\Omega t \underbrace{-\Omega X\cos\Omega t - \Omega Y\sin\Omega t}_{-\Omega x}. \tag{2.4b}$$

The quantities dx/dt and dy/dt give the rates of change of the coordinates relative to the moving frame as time evolves. They are thus the components of the relative velocity:

$$\mathbf{u} = \frac{dx}{dt}\mathbf{i} + \frac{dy}{dt}\mathbf{j} = u\mathbf{i} + v\mathbf{j}. \tag{2.5}$$

Similarly, dX/dt and dY/dt give the rates of change of the absolute coordinates and form the absolute velocity:

$$\mathbf{U} = \frac{dX}{dt}\mathbf{I} + \frac{dY}{dt}\mathbf{J}.$$

Writing the absolute velocity in terms of the rotating unit vectors, we obtain [using Eq. (2.2)]

$$\mathbf{U} = \left(\frac{\mathrm{d}X}{\mathrm{d}t} \cos \Omega t + \frac{\mathrm{d}Y}{\mathrm{d}t} \sin \Omega t \right) \mathbf{i} + \left(-\frac{\mathrm{d}X}{\mathrm{d}t} \sin \Omega t + \frac{\mathrm{d}Y}{\mathrm{d}t} \cos \Omega t \right) \mathbf{j}$$

$$= U\mathbf{i} + V\mathbf{j}. \tag{2.6}$$

Thus, $\mathrm{d}X/\mathrm{d}t$ and $\mathrm{d}Y/\mathrm{d}t$ are the components of the absolute velocity $\mathbf{U}$ in the inertial frame, whereas U and V are the components of the same vector in the rotating frame. Use of Eqs. (2.4) and (2.3) in the preceding expression yields the following relations between absolute and relative velocities:

$$U = u - \Omega y, \quad V = v + \Omega x. \tag{2.7}$$

These equalities simply state that the absolute velocity is the relative velocity plus the entraining velocity due to the rotation of the reference framework.

A second derivative with respect to time provides in a similar manner:

$$\frac{\mathrm{d}^2 x}{\mathrm{d}t^2} = \left(\frac{\mathrm{d}^2 X}{\mathrm{d}t^2} \cos \Omega t + \frac{\mathrm{d}^2 Y}{\mathrm{d}t^2} \sin \Omega t \right) + 2\Omega \underbrace{\left(-\frac{\mathrm{d}X}{\mathrm{d}t} \sin \Omega t + \frac{\mathrm{d}Y}{\mathrm{d}t} \cos \Omega t \right)}_{V}$$

$$- \Omega^2 \underbrace{(X \cos \Omega t + Y \sin \Omega t)}_{x} \tag{2.8a}$$

$$\frac{\mathrm{d}^2 y}{\mathrm{d}t^2} = \left(-\frac{\mathrm{d}^2 X}{\mathrm{d}t^2} \sin \Omega t + \frac{\mathrm{d}^2 Y}{\mathrm{d}t^2} \cos \Omega t \right) - 2\Omega \underbrace{\left(\frac{\mathrm{d}X}{\mathrm{d}t} \cos \Omega t + \frac{\mathrm{d}Y}{\mathrm{d}t} \sin \Omega t \right)}_{U}$$

$$- \Omega^2 \underbrace{(-X \sin \Omega t + Y \cos \Omega t)}_{y}. \tag{2.8b}$$

Expressed in terms of the relative and absolute accelerations

$$\mathbf{a} = \frac{\mathrm{d}^2 x}{\mathrm{d}t^2}\mathbf{i} + \frac{\mathrm{d}^2 y}{\mathrm{d}t^2}\mathbf{j} = \frac{\mathrm{d}u}{\mathrm{d}t}\mathbf{i} + \frac{\mathrm{d}v}{\mathrm{d}t}\mathbf{j} = a\mathbf{i} + b\mathbf{j}$$

$$\mathbf{A} = \frac{\mathrm{d}^2 X}{\mathrm{d}t^2}\mathbf{I} + \frac{\mathrm{d}^2 Y}{\mathrm{d}t^2}\mathbf{J}$$

$$= \left(\frac{\mathrm{d}^2 X}{\mathrm{d}t^2} \cos \Omega t + \frac{\mathrm{d}^2 Y}{\mathrm{d}t^2} \sin \Omega t \right) \mathbf{i} + \left(\frac{\mathrm{d}^2 Y}{\mathrm{d}t^2} \cos \Omega t - \frac{\mathrm{d}^2 X}{\mathrm{d}t^2} \sin \Omega t \right) \mathbf{j} = A\mathbf{i} + B\mathbf{j},$$

expressions (2.8) condense to

$$a = A + 2\Omega V - \Omega^2 x, \quad b = B - 2\Omega U - \Omega^2 y.$$

In analogy with the absolute velocity vector, d^2X/dt^2 and d^2Y/dt^2 are the components of the absolute acceleration $\mathbf{A}$ in the inertial frame, whereas A and B are the components of the same vector in the rotating frame. The absolute acceleration components, necessary later to formulate Newton's law, are obtained by solving for A and B and using Eq. (2.7):

$$A = a - 2\Omega v - \Omega^2 x, \quad B = b + 2\Omega u - \Omega^2 y. \tag{2.9}$$

We now see that the difference between absolute and relative acceleration consists of two contributions. The first, proportional to Ω and to the relative velocity, is called the Coriolis acceleration; the other, proportional to Ω^2 and to the coordinates, is called the centrifugal acceleration. When placed on the other side of the equality in Newton's law, these terms can be assimilated to forces (per unit mass). The centrifugal force acts as an outward pull, whereas the Coriolis force depends on the direction and magnitude of the relative velocity.

Formally, the preceding results could have been derived in a vector form. Defining the vector rotation

$$\boldsymbol{\Omega} = \Omega\mathbf{k},$$

where $\mathbf{k}$ is the unit vector in the third dimension (which is common to both systems of reference), we can write Eqs. (2.7) and (2.9) as

$$\mathbf{U} = \mathbf{u} + \boldsymbol{\Omega} \times \mathbf{r}$$
$$\mathbf{A} = \mathbf{a} + 2\boldsymbol{\Omega} \times \mathbf{u} + \boldsymbol{\Omega} \times (\boldsymbol{\Omega} \times \mathbf{r}), \tag{2.10}$$

where the symbol $\times$ indicates the vectorial product. This implies that taking a time derivative of a vector with respect to the inertial framework is equivalent to applying the operator

$$\frac{d}{dt} + \boldsymbol{\Omega}\times$$

in the rotating framework of reference.

A very detailed exposition of the Coriolis and centrifugal accelerations can be found in the book by Stommel and Moore (1989). In addition, the reader will find a historical perspective in Ripa (1994) and laboratory illustrations in Marshall and Plumb (2008).

2.2   UNIMPORTANCE OF THE CENTRIFUGAL FORCE

Unlike the Coriolis force, which is proportional to the velocity, the centrifugal force depends solely on the rotation rate and the distance of the particle to the rotation axis. Even at rest with respect to the rotating planet, particles experience an outward pull. Yet, on the earth as on other planetary bodies, objects don't fly out to space. How is that possible? Obviously, gravity keeps everything together.

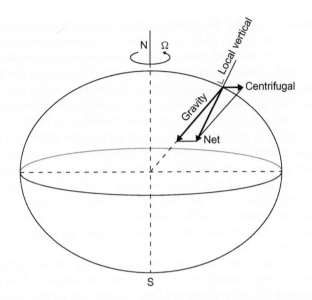

FIGURE 2.2   How the flattening of the rotating earth (grossly exaggerated in this drawing) causes the gravitational and centrifugal forces to combine into a net force aligned with the local vertical, so that equilibrium is reached.

In the absence of rotation, gravitational forces keep the matter together to form a spherical body (with the denser materials at the center and the lighter ones on the periphery). The outward pull caused by the centrifugal force distorts this spherical equilibrium, and the planet assumes a slightly flattened shape. The degree of flattening is precisely that necessary to keep the planet in equilibrium for its rotation rate.

The situation is depicted in Fig. 2.2. By its nature, the centrifugal force is directed outward, perpendicular to the axis of rotation, whereas the gravitational force points toward the planet's center. The resulting force assumes an intermediate direction, and this direction is precisely the direction of the local vertical. Indeed, under this condition, a loose particle would have no tendency of its own to fly away from the planet. In other words, every particle at rest on the surface will remain at rest unless it is subjected to additional forces.

The flattening of the earth, as well as that of other celestial bodies in rotation, is important to neutralize the centrifugal force. But, this is not to say that it greatly distorts the geometry. On the earth, for example, the distortion is very slight because gravity by far exceeds the centrifugal force; the terrestrial equatorial radius is 6378 km, slightly greater than its polar radius of 6357 km. The shape of the rotating oblate earth is treated in detail by Stommel and Moore (1989) and by Ripa (1994).

For the sake of simplicity in all that follows, we will call the gravitational force the resultant force, aligned with the vertical and equal to the sum of the true gravitational force and the centrifugal force. Due to inhomogeneous distributions of rocks and magma on Earth, the true gravitational force is not directed toward the center of the earth. For the same reason as the centrifugal force has rendered the earth surface oblate, this inhomogeneous true gravity has deformed the earth surface until the total (apparent) gravitational force is perpendicular to it. The surface so obtained is called a *geoid* and can be interpreted as the surface of an ocean at rest (with a continuous extension on land). This virtual continuous surface is perpendicular at every point to the direction of gravity (including the centrifugal force) and forms an *equipotential* surface, meaning that a particle moving on that surface undergoes no change in potential energy. The value of this potential energy per unit mass is called the geopotential, and the geoid is thus a surface of constant geopotential. This surface will be the reference surface from which land elevations, (dynamic) sea surface elevations, and ocean depth will be defined. For more on the geoid, the reader is referred to Robinson (2004), Chapter 11.

In a rotating laboratory tank, the situation is similar but not identical. The rotation causes a displacement of the fluid toward the periphery. This proceeds until the resulting inward pressure gradient prevents any further displacement. Equilibrium then requires that at any point on the surface, the downward gravitational force and the outward centrifugal force combine into a resultant force normal to the surface (Fig. 2.3), so that the surface becomes an equipotential surface. Although the surface curvature is crucial in neutralizing the centrifugal force, the vertical displacements are rather small. In a tank rotating at the rate of one revolution every 2 s (30 rpm) and 40 cm in diameter, the difference in fluid height between the rim and the center is a modest 2 cm.

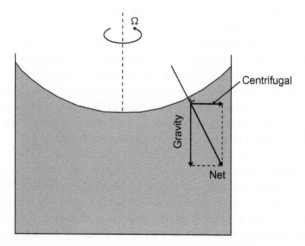

FIGURE 2.3 Equilibrium surface of a rotating fluid in an open container. The surface slope is such that gravitational and centrifugal forces combine into a net force everywhere aligned with the local normal to the surface.

2.3   FREE MOTION ON A ROTATING PLANE

The preceding argument allows us to combine the centrifugal force with the gravitational force, but the Coriolis force remains. To have an idea of what this force can cause, let us examine the motion of a free particle, that is, a particle not subject to any external force other than apparent gravity (true gravity combined with centrifugal force) on the horizontal plane extending from the North Pole.

If the particle is free of any force in this plane, its acceleration in the inertial frame is nil, by Newton's law. According to Eq. (2.9), with the centrifugal-acceleration terms no longer present, the equations governing the velocity components of the particle are

$$\frac{du}{dt} - 2\Omega v = 0, \quad \frac{dv}{dt} + 2\Omega u = 0. \tag{2.11}$$

The general solution to this system of linear equations is

$$u = V\sin(ft+\phi), \quad v = V\cos(ft+\phi), \tag{2.12}$$

where $f = 2\Omega$, called the Coriolis parameter, has been introduced for convenience, and V and ϕ are two arbitrary constants of integration. Without loss of generality, V can always be chosen as nonnegative. (Do not confuse this constant V with the y–component of the absolute velocity introduced in Section 2.1.) A first result is that the particle speed $(u^2+v^2)^{1/2}$ remains unchanged in time. It is equal to V, a constant determined by the initial conditions.

Although the speed remains unchanged, the components u and v do depend on time, implying a change in direction. To document this curving effect, it is most instructive to derive the trajectory of the particle. The coordinates of the particle position change, by definition of the vector velocity, according to $dx/dt = u$ and $dy/dt = v$, and a second time integration provides

$$x = x_0 - \frac{V}{f}\cos(ft + \phi) \tag{2.13a}$$

$$y = y_0 + \frac{V}{f}\sin(ft + \phi), \tag{2.13b}$$

where x_0 and y_0 are additional constants of integration to be determined from the initial coordinates of the particle. From the last relations, it follows directly that

$$(x-x_0)^2 + (y-y_0)^2 = \left(\frac{V}{f}\right)^2. \tag{2.14}$$

This implies that the trajectory is a circle centered at (x_0, y_0) and of radius $V/|f|$. The situation is depicted in Fig. 2.4.

In the absence of rotation $(f = 0)$, this radius is infinite, and the particle follows a straight path, as we could have anticipated. But, in the presence of

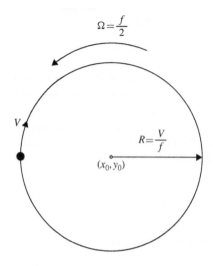

$$\Omega = \frac{f}{2}$$

$$R = \frac{V}{f}$$

(x_0, y_0)

FIGURE 2.4 Inertial oscillation of a free particle on a rotating plane. The orbital period is exactly half of the ambient revolution period. This figure has been drawn with a positive Coriolis parameter, f, representative of the northern hemisphere. If f were negative (as in the southern hemisphere), the particle would veer to the left.

rotation ($f \neq 0$), the particle turns constantly. A quick examination of Eq. (2.13) reveals that the particle turns to the right (clockwise) if f is positive or to the left (counterclockwise) if f is negative. In sum, the rule is that the particle turns in the sense opposite to that of the ambient rotation.

At this point, we may wonder whether this particle rotation is none other than the negative of the ambient rotation, in such a way as to keep the particle at rest in the absolute frame of reference. But, there are at least two reasons why this is not so. The first is that the coordinates of the center of the particle's circular path are arbitrary and are therefore not required to coincide with those of the axis of rotation. The second and most compelling reason is that the two frequencies of rotation are simply not the same: the ambient rotating plane completes one revolution in a time equal to $T_a = 2\pi / \Omega$, whereas the particle covers a full circle in a time equal to $T_p = 2\pi / f = \pi / \Omega$, called *inertial period*. Thus, the particle goes around its orbit twice as the plane accomplishes a single revolution.

The spontaneous circling of a free particle endowed with an initial velocity in a rotating environment bears the name of *inertial oscillation*. Note that, since the particle speed can vary, so can the inertial radius, $V/|f|$, whereas the frequency, $|f| = 2|\Omega|$, is a property of the rotating environment and is independent of the initial conditions.

The preceding exercise may appear rather mathematical and devoid of any physical interpretation. There exists, however, a geometric argument and a physical analogy. Let us first discuss the geometric argument. Consider a rotating table and, on it, a particle initially ($t = 0$) at a distance R from the axis of rotation, approaching the latter at a speed u (Fig. 2.5). At some later time t, the particle has approached the axis of rotation by a distance ut while it has covered

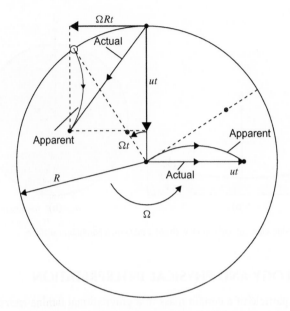

FIGURE 2.5   Geometrical interpretation of the apparent veering of a particle trajectory viewed in a rotating framework. The veering is to the right when the ambient rotation is counterclockwise, as shown here for two particular trajectories, one originating from the rim and the other from the axis of rotation.

the distance $\Omega R t$ laterally. It now lies at the position indicated by a solid dot. During the lapse t, the table has rotated by an angle Ωt, and to an observer rotating with the table, the particle seems to have originated from the point on the rim indicated by the open circle. The construction shows that, although the actual trajectory is perfectly straight, the apparent path as noted by the observer rotating with the table curves to the right. A similar conclusion holds for a particle radially pushed away from the center with a speed u. In absolute axes, the trajectory is a straight line covering a distance ut from the center. During the lapse t, the table has rotated and for an observer on the rotating platform, the particle, instead of arriving in the location of the asterisk, apparently veered to the right.

The problem with this argument is that to construct the absolute trajectory, we chose a straight path, that is, we implicitly considered the total absolute acceleration, which in the rotating framework includes the centrifugal acceleration. The latter, however, should not have been retained for consistency with the case of terrestrial rotation, but because it is a radial force, it does not account for the transverse displacement. Therefore, the apparent veering is, at least for a short interval of time, entirely due to the Coriolis effect.

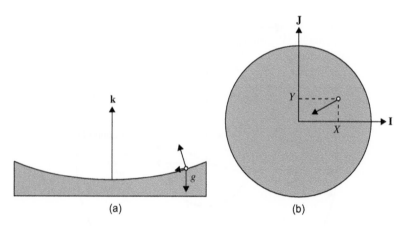

FIGURE 2.6   Side view (a) and top view (b) of a mass on a paraboloid surface.

2.4   ANALOGY AND PHYSICAL INTERPRETATION

Consider[1] a particle of a certain mass in a gravitational field g on a paraboloid surface (Fig. 2.6) of elevation Z given by

$$Z = \frac{\Omega^2}{2g}\left(X^2 + Y^2\right). \tag{2.15}$$

Provided that the paraboloid is sufficiently flat compared to its radius R ($\Omega^2 R/2g \ll 1$), the equations of motions of the mass are easily derived

$$\frac{d^2X}{dt^2} = -g\frac{\partial Z}{\partial X} = -\Omega^2 X, \quad \frac{d^2Y}{dt^2} = -g\frac{\partial Z}{\partial Y} = -\Omega^2 Y, \tag{2.16}$$

and describe a pendulum motion.

The frequency Ω measures the curvature of the surface and is the pendulum's natural frequency of oscillation. Note how the gravitational restoring force takes on the form of a negative centrifugal force. Without loss of generality, we can choose the initial position of the particle as $X = X_0$, $Y = 0$. In that location, we launch the particle with an initial velocity of $dX/dt = U_0$ and $dY/dt = V_0$ in absolute axes. The trajectory in absolute axes is easily found as the solution of Eq. (2.16)

$$X = X_0 \cos \Omega t + \frac{U_0}{\Omega} \sin \Omega t \tag{2.17a}$$

$$Y = \frac{V_0}{\Omega} \sin \Omega t. \tag{2.17b}$$

---

[1]A similar analogy was suggested to the authors by Prof. Satoshi Sakai at Kyoto University.

Two particular solutions are noteworthy. If the initial condition is a pure radial displacement ($V_0 = 0$), the particle forever oscillates along the line $Y = 0$. The oscillation, of period $2\pi / \Omega$, takes it to the center twice per period, that is, every π / Ω time interval. At the other extreme, the particle can be imparted an initial azimuthal velocity of magnitude such that the outward centrifugal force of the ensuing circling motion exactly cancels the inward gravitational pull at that radial distance:

$$U_0 = 0, \quad V_0 = \pm \Omega X_0, \tag{2.18}$$

in which case the particle remains at a fixed distance from the center ($X^2 + Y^2 = X_0^2$) and circles at a constant angular rate Ω, counterclockwise or clockwise, depending on the direction of the initial azimuthal velocity.

Outside of these two extreme behaviors, the particle describes an elliptical trajectory of size, eccentricity, and phase related to the initial condition. The orbit does not take it through the center but brings it, twice per period, to a distance of closest approach (*perigee*) and, twice per period, to a distance of largest excursion (*apogee*).

At this point, the reader may rightfully wonder: Where is the analogy with the motion of a particle subject to the Coriolis force? To show this analogy, let us now view the particle motion in a rotating frame, but, of course, not any rotating frame: Let us select the angular rotation rate Ω equal to the particle's frequency of oscillation. This choice is made so that, in the rotating frame of reference, the outward centrifugal force is everywhere and at all times exactly canceled by the inward gravitational pull experienced on the parabolic surface. Thus, the equations of motion expressed in the rotating frame include only the relative acceleration and the Coriolis force, that is, are none other than Eqs. (2.11).

Let us now consider the oscillations as seen by an observer in the rotating frame (Figs. 2.7 and 2.8). When the particle oscillates strictly back and forth, the rotating observer sees a curved trajectory. Because the particle passes by the origin twice per oscillation, the orbit seen by the rotating observer also passes by the origin twice per period. When the particle reaches its extreme displacement on one side, it reaches an apogee on its orbit as viewed in the rotating frame; then, by the time it reaches its maximum displacement on the opposite side, π / Ω later, the rotating framework has rotated exactly by half a turn, so that this second apogee of the orbit coincides with the first. Therefore, the reader can readily be convinced that the orbit in the rotating frame is drawn twice per period of oscillation. Algebraic or geometric developments reveal that the orbit in the rotating framework is circular (Fig. 2.8a).

In the other extreme situation, when the particle circles at a constant distance from the origin, two cases must be distinguished, depending on whether it circles in the direction of or opposite to the observer's rotating frame. If the direction is the same [positive sign in Eq. (2.18)], the observer simply chases the particle, which then appears stationary, and the orbit reduces to a single point (Fig. 2.8b). This case corresponds to the state of rest of a particle

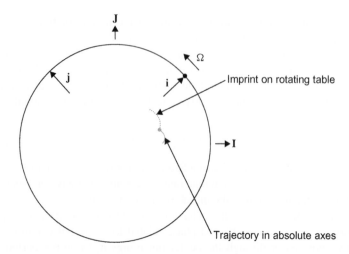

FIGURE 2.7 Oscillation of the paraboloid pendulum viewed in absolute axes. Dots represent the imprint of the mass on the paraboloid that rotates with the rotation rate Ω.

in a rotating environment [$V = 0$ in Eq. (2.12) through Eq. (2.13)]. If the sense of rotation is opposite [minus sign in Eq. (2.18)], the reference frame rotates at the rate Ω in one direction, whereas the particle circles at the same rate in the opposite direction. To the observer, the particle appears to rotate at the rate 2Ω. The orbit is obviously a circle centered at the origin and of radius equal to the particle's radial displacement; it is covered twice per revolution of the rotating frame (Fig. 2.8c). Finally, for arbitrary oscillations, the orbit in the rotating frame is a circle of finite radius that is not centered at the origin, does not pass by the origin, and may or may not include the origin (Fig. 2.8d). The reader may experiment with MATLAB™ code `parabolic.m` for further explorations of trajectories.

Looking at the system from an inertial framework, we observe that the oscillation of the particle is due to the restoring force of gravity. In particular, its projection on the parabolic surface is responsible for the tendency to move toward the center of the paraboloid. If we look from the rotating framework, this component of gravity is always cancelled by the centrifugal force associated with the rotation, and the restoring force responsible for the oscillation is now the Coriolis force.

2.5 ACCELERATION ON A THREE-DIMENSIONAL ROTATING PLANET

For all practical purposes, except as outlined earlier when the centrifugal force was discussed (Section 2.2), the earth can be taken as a perfect sphere. This sphere rotates about its North Pole–South Pole axis. At any given latitude φ,

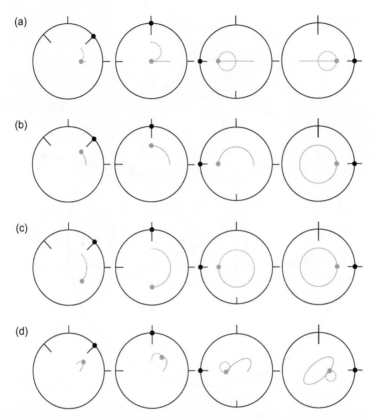

FIGURE 2.8   Orbits (full line) in absolute axes with imprint of trajectories (dots) on rotating framework (apparent trajectory). Each row shows the situation for a different initial condition and after $1/8$, $1/4$, $1/2$, and a full period $2\pi\Omega^{-1}$. Orbits differ according to the initial velocity: the first row (a) shows oscillations obtained without initial velocity, the second row (b) was created with initial velocity $U_0 = 0$, $V_0 = X_0\Omega$, the third row (c) corresponds to the opposite initial velocity $U_0 = 0$, $V_0 = -X_0\Omega$, and the last row (d) corresponds to an arbitrary initial velocity.

the north–south direction departs from the local vertical, and the Coriolis force assumes a form different from that established in the preceding section.

Figure 2.9 depicts the traditional choice for a local Cartesian framework of reference: the x-axis is oriented eastward, the y-axis, northward, and the z-axis, upward. In this framework, the earth's rotation vector is expressed as

$$\boldsymbol{\Omega} = \Omega\cos\varphi\,\mathbf{j} + \Omega\sin\varphi\,\mathbf{k}. \tag{2.19}$$

The absolute acceleration minus the centrifugal component,

$$\frac{d\mathbf{u}}{dt} + 2\boldsymbol{\Omega}\times\mathbf{u},$$

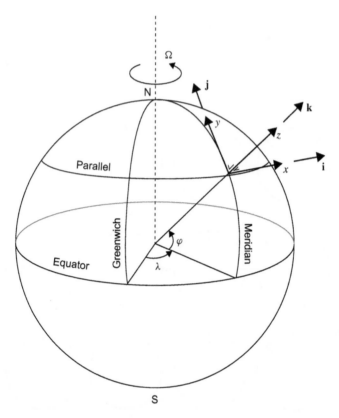

FIGURE 2.9 Definition of a local Cartesian framework of reference on a spherical Earth. The coordinate x is directed eastward, y northward, and z upward.

has the following three components:

$$x: \quad \frac{du}{dt} + 2\Omega \cos\varphi\, w - 2\Omega\,\sin\varphi\, v \qquad (2.20a)$$

$$y: \quad \frac{dv}{dt} + 2\Omega \sin\varphi\, u \qquad (2.20b)$$

$$z: \quad \frac{dw}{dt} - 2\Omega \cos\varphi\, u. \qquad (2.20c)$$

With x, y, and z everywhere aligned with the local eastward, northward, and vertical directions, the coordinate system is curvilinear, and additional terms arise in the components of the relative acceleration. These terms will be dismissed in Section 3.2 because of their relatively small size in most instances.

For convenience, we define the quantities

$$f = 2\Omega \sin\varphi \tag{2.21}$$

$$f_* = 2\Omega \cos\varphi. \tag{2.22}$$

The coefficient f is called the *Coriolis parameter*, whereas f_* has no traditional name and will be called here the *reciprocal Coriolis parameter*. In the northern hemisphere, f is positive; it is zero at the equator and negative in the southern hemisphere. In contrast, f_* is positive in both hemispheres and vanishes at the poles. An examination of the relative importance of the various terms (Section 4.3) will reveal that, generally, the f terms are important, whereas the f_* terms may be neglected.

Horizontal, unforced motions are described by

$$\frac{du}{dt} - fv = 0 \tag{2.23a}$$

$$\frac{dv}{dt} + fu = 0 \tag{2.23b}$$

and are still characterized by solution (2.12). The difference resides in the value of f, now given by Eq. (2.21). Thus, inertial oscillations on Earth have periodicities equal to $2\pi/f = \pi/\Omega\sin\varphi$, ranging from 11 h 58' at the poles to infinity along the equator. Pure inertial oscillations are, however, quite rare because of the usual presence of pressure gradients and other forces. Nonetheless, inertial oscillations are not uncommonly found to contribute to observations of oceanic currents. An example of such an occurrence, where the inertial oscillations made up almost the entire signal, was reported by Gustafson and Kullenberg (1936). Current measurements in the Baltic Sea showed periodic oscillations about a mean value. When added to one another to form a so-called progressive vector diagram (Fig. 2.10), the currents distinctly showed a mean drift, on which were superimposed quite regular clockwise oscillations. The theory of inertial oscillation predicts clockwise rotation in the northern hemisphere with period of $2\pi/f = \pi/\Omega\sin\varphi$, or 14 h at the latitude of observations, thus confirming the interpretation of the observations as inertial oscillations.

2.6 NUMERICAL APPROACH TO OSCILLATORY MOTIONS

The equations of free motion on a rotating plane (2.11) have been considered in some detail in Section 2.3, and it is now appropriate to consider their discretization, as the corresponding terms are part of all numerical models of geophysical flows. Upon introducing the time increment Δt, an approximation to the components of the velocity will be determined at the discrete instants $t^n = n\Delta t$ with $n = 1, 2, 3, \ldots$, which are denoted $\tilde{u}^n = \tilde{u}(t_n)$ and $\tilde{v}^n = \tilde{v}(t_n)$, with tildes used to distinguish the discrete solution from the exact one. The so-called *Euler method*

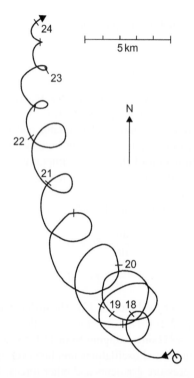

FIGURE 2.10 Evidence of inertial oscillations in the Baltic Sea, as reported by Gustafson and Kullenberg (1936). The plot is a progressive vector diagram constructed by the successive addition of velocity measurements at a fixed location. For weak or uniform velocities, such a curve approximates the trajectory that a particle starting at the point of observation would have followed during the period of observation. Numbers indicate days of the month. Note the persistent veering to the right, at a period of about 14 h, which is the value of $2\pi/f$ at that latitude (57.8°N). (*From Gustafson & Kullenberg, 1936, as adapted by Gill, 1982*)

based on first-order forward differencing yields the simplest discretization of Eqs. (2.11):

$$\frac{du}{dt} - fv = 0 \longrightarrow \frac{\tilde{u}^{n+1} - \tilde{u}^n}{\Delta t} - f\tilde{v}^n = 0$$

$$\frac{dv}{dt} + fu = 0 \longrightarrow \frac{\tilde{v}^{n+1} - \tilde{v}^n}{\Delta t} + f\tilde{u}^n = 0.$$

The latter pair can be cast into a recursive form as follows:

$$\tilde{u}^{n+1} = \tilde{u}^n + f\Delta t\,\tilde{v}^n \tag{2.24a}$$

$$\tilde{v}^{n+1} = \tilde{v}^n - f\Delta t\,\tilde{u}^n. \tag{2.24b}$$

Thus, given initial values $\tilde{u}^0$ and $\tilde{v}^0$ at t^0, the solution can be computed easily at time t^1

$$\tilde{u}^1 = \tilde{u}^0 + f\Delta t\, \tilde{v}^0 \tag{2.25}$$

$$\tilde{v}^1 = \tilde{v}^0 - f\Delta t\, \tilde{u}^0. \tag{2.26}$$

Then, by means of the same algorithm, the solution can be obtained iteratively at times t^2, t^3, and so on (do not confuse the temporal index with an exponent here and in the following). Clearly, the main advantage of the preceding scheme is its simplicity, but it is not sufficient to render it acceptable, as we shall soon learn.

To explore the numerical error generated by the Euler method, we carry out Taylor expansions of the type

$$\tilde{u}^{n+1} = \tilde{u}^n + \Delta t \left[\frac{d\tilde{u}}{dt}\right]_{t=t^n} + \frac{\Delta t^2}{2}\left[\frac{d^2\tilde{u}}{dt^2}\right]_{t=t^n} + \mathcal{O}\left(\Delta t^3\right)$$

and similarly for $\tilde{v}$ to obtain the following expressions from Eqs. (2.24)

$$\left[\frac{d\tilde{u}}{dt} - f\tilde{v}\right]_{t=t^n} = -\left[\frac{d^2\tilde{u}}{dt^2}\right]_{t=t^n}\frac{\Delta t}{2} + \mathcal{O}\left(\Delta t^2\right) \tag{2.27a}$$

$$\left[\frac{d\tilde{v}}{dt} + f\tilde{u}\right]_{t=t^n} = -\left[\frac{d^2\tilde{v}}{dt^2}\right]_{t=t^n}\frac{\Delta t}{2} + \mathcal{O}\left(\Delta t^2\right). \tag{2.27b}$$

Differentiation of Eq. (2.27a) with respect to time and use of Eq. (2.27b) to eliminate $d\tilde{v}/dt$ allow us to recast Eq. (2.27a) into a simpler form, and similarly for Eq. (2.27b):

$$\frac{d\tilde{u}^n}{dt} - f\tilde{v}^n = \frac{f^2\Delta t}{2}\tilde{u}^n + \mathcal{O}\left(\Delta t^2\right) \tag{2.28a}$$

$$\frac{d\tilde{v}^n}{dt} + f\tilde{u}^n = \frac{f^2\Delta t}{2}\tilde{v}^n + \mathcal{O}\left(\Delta t^2\right). \tag{2.28b}$$

Obviously, the numerical scheme mirrors the original equations, except that an additional term appears in each right-hand side. This additional term takes the form of antifriction (friction would have a minus sign instead) and will therefore increase the discrete velocity over time.

The *truncation error* of the Euler scheme—the right-hand side of the preceding expressions—tends to zero as Δt vanishes, which is why the scheme is said to be *consistent*. The truncation is on the order of Δt at the first power, and the scheme is therefore said to be *first-order accurate*, which is the lowest possible level of accuracy. Nonetheless, this is not the chief weakness of the present scheme, since we must expect that the introduction of antifriction will create an unphysical acceleratation. Indeed, elementary manipulations of the time-stepping algorithm (2.24) lead to $(\tilde{u}^{n+1})^2+(\tilde{v}^{n+1})^2 = (1+f^2\Delta t^2)\{(\tilde{u}^n)^2+(\tilde{v}^n)^2\}$ so that by recursion

$$||\tilde{\mathbf{u}}||^2 = (\tilde{u}^n)^2+(\tilde{v}^n)^2 = (1+f^2\Delta t^2)^n\left\{(\tilde{u}^0)^2+(\tilde{v}^0)^2\right\}. \tag{2.29}$$

Thus, although the kinetic energy (directly proportional to the squared norm $||\tilde{\mathbf{u}}||^2$) of the inertial oscillation must remain constant, as was seen in Section 2.3, the kinetic energy of the discrete solution increases without bound[2] even if the time step Δt is taken much smaller than the characteristic time $1/f$. Algorithm (2.24) is *unstable*. Because such a behavior is not acceptable, we need to formulate an alternative type of discretization.

In our first scheme, the time derivative was taken by going forward from time level t^n to t^{n+1} and the other terms at t^n, and the scheme became a recursive algorithm to calculate the next values from the current values. Such a discretization is called an *explicit scheme*. By contrast, in an *implicit scheme*, the terms other than the time derivatives are taken at the new time t^{n+1} (which is similar to taking a backward difference for the time derivative):

$$\frac{\tilde{u}^{n+1}-\tilde{u}^n}{\Delta t}-f\tilde{v}^{n+1}=0 \tag{2.30a}$$

$$\frac{\tilde{v}^{n+1}-\tilde{v}^n}{\Delta t}+f\tilde{u}^{n+1}=0. \tag{2.30b}$$

In this case, the norm of the discrete solution decreases monotonically toward zero, according to

$$(\tilde{u}^n)^2+(\tilde{v}^n)^2=(1+f^2\Delta t^2)^{-n}\left\{(\tilde{u}^0)^2+(\tilde{v}^0)^2\right\}. \tag{2.31}$$

This scheme can be regarded as *stable*, but as the kinetic energy should neither decrease or increase, it may rather be considered as *overly stable*.

Of interest is the family of algorithms based on a weighted average between explicit and implicit schemes:

$$\frac{\tilde{u}^{n+1}-\tilde{u}^n}{\Delta t}-f\left[(1-\alpha)v^n+\alpha\tilde{v}^{n+1}\right]=0 \tag{2.32a}$$

$$\frac{\tilde{v}^{n+1}-\tilde{v}^n}{\Delta t}+f\left[(1-\alpha)u^n+\alpha\tilde{u}^{n+1}\right]=0, \tag{2.32b}$$

with $0 \le \alpha \le 1$. The numerical scheme is explicit when $\alpha=0$ and implicit when $\alpha=1$. Hence, the coefficient α may be regarded as the degree of implicitness in the scheme. It has a crucial impact on the time evolution of the

---

[2] From the context, it should be clear that ^{n} in $(1+f^2\Delta t^2)^n$ is an exponent, whereas in $\tilde{u}^n$, it is the time index. In the following text, we will not point out this distinction again, leaving it to the reader to verify the context.

kinetic energy:

$$(\tilde{u}^n)^2 + (\tilde{v}^n)^2 = \left[\frac{1 + (1-\alpha)^2 f^2 \Delta t^2}{1 + \alpha^2 f^2 \Delta t^2}\right]^n \left\{(\tilde{u}^0)^2 + (\tilde{v}^0)^2\right\}. \qquad (2.33)$$

According to whether α is less than, equal to, or greater than $1/2$, the kinetic energy increases, remains constant, or decreases over time. It seems therefore appropriate to select the scheme with $\alpha = 1/2$, which is usually said to be *semi-implicit*.

It is now instructive to compare the semi-implicit approximate solution with the exact solution (2.12). For this to be relevant, the same initial conditions are prescribed, that is, $\tilde{u}^0 = V\sin\phi$ and $\tilde{v}^0 = V\cos\phi$. Then, at any time t^n, the discrete velocity may be shown (see Numerical Exercise 2.9) to be

$$\tilde{u}^n = V\sin(\tilde{f}t^n + \phi)$$
$$\tilde{v}^n = V\cos(\tilde{f}t^n + \phi),$$

with the angular frequency $\tilde{f}$ given by

$$\tilde{f} = \frac{1}{\Delta t}\arctan\left(\frac{f\Delta t}{1 - f^2\Delta t^2/4}\right). \qquad (2.34)$$

Although the amplitude of the oscillation (V) is correct, the numerical angular frequency, $\tilde{f}$, differs from the true value f. However, the smaller the dimensionless product $f\Delta t$, the smaller the error:

$$\tilde{f} \to f\left(1 - \frac{f^2\Delta t^2}{12}\right) \quad \text{as} \quad f\Delta t \to 0.$$

In other words, selecting a time increment Δt much shorter than $1/f$, the time scale of inertial oscillations, leads to a frequency that is close to the exact one.

2.7   NUMERICAL CONVERGENCE AND STABILITY

A Taylor-series expansion performed on the discrete equations of the inertial oscillations revealed that the truncation error vanishes as Δt tends to zero. However, we are not so much interested in verifying that the limit of the discretized equation for increasing resolution returns the exact equation (*consistency*) as we are in making sure that the *solution* of the discretized equation tends to the *solution* of the differential equations (i.e., the exact solution). If the difference between the exact and discrete solutions tends to zero as Δt vanishes, then the discretization is said to *converge*.

Unfortunately proving convergence is not a trivial task, especially as we generally do not know the exact solution, in which case the use of numerical discretization would be superfluous. Furthermore, the exact solution of the discrete equation can very rarely be written in a closed form because the discretization only provides a method, an *algorithm*, to construct the solution in time.

Without knowing precisely the solutions of either the continuous problem or its discrete version, direct proofs of convergence involve mathematics well beyond the scope of the present book and will be not pursued here. We will, however, rely on a famous theorem called the *Lax–Richtmyer equivalence theorem* (Lax & Richtmyer, 1956), which states that

A consistent finite-difference scheme for a linear partial differential equation for which the initial value problem is well posed is convergent if and only if it is stable.

So, while proof of convergence is a mathematical exercise for researchers well versed in functional analysis, we will restrict ourselves here and in every other instance across the book to verify consistency and stability and will then invoke the theorem to claim convergence. This is a particularly interesting approach not only because checking stability and consistency is much easier than proving convergence but also because stability analysis provides further insight in propagation properties of the numerical scheme (see Section 5.4). There remains, however, to define stability and to design efficient methods to verify the stability of numerical schemes. Our analysis of the explicit Euler scheme (2.24) for the discretization of inertial oscillations led us to conclude that it is unstable because the velocity norm, and hence the energy of the system, gradually increases with every time step.

The adjective *unstable* seems quite natural in this context but lacks precision, and an exact definition is yet to be given. Imagine, for example, the use of an implicit Euler scheme (generally taken as the archetype of a stable scheme) on a standard linear differential equation:

$$\frac{\partial u}{\partial t} = \gamma u \quad \rightarrow \quad \frac{\tilde{u}^{n+1} - \tilde{u}^n}{\Delta t} = \gamma \, \tilde{u}^{n+1}. \tag{2.35}$$

We readily see that for $0 < \gamma \, \Delta t < 1$, the norm of $\tilde{u}$ increases:

$$\tilde{u}^n = \left(\frac{1}{1 - \gamma \, \Delta t} \right)^n \tilde{u}^0. \tag{2.36}$$

We would however hardly disqualify the scheme as unstable, since the numerical solution increases its norm simply because the exact solution $u = u^0 e^{\gamma t}$ does so. In the present case, we can even show that the numerical solution actually converges to the exact solution:

$$\lim_{\Delta t \to 0} \tilde{u}^n = \tilde{u}^0 \lim_{\Delta t \to 0} \left(\frac{1}{1 - \gamma \, \Delta t} \right)^n = \tilde{u}^0 \lim_{\Delta t \to 0} \left(\frac{1}{1 - \gamma \, \Delta t} \right)^{t/\Delta t} = \tilde{u}^0 e^{\gamma t}. \tag{2.37}$$

with $t = n\Delta t$.

Stability is thus a concept that should be related not only to the behavior of the discrete solution but also to the behavior of the exact solution. Loosely speaking, we will qualify a numerical scheme as unstable if its solution grows much faster than the exact solution and, likewise, overstable if its solution decreases much faster than the exact solution.

2.7.1   Formal Stability Definition

A mathematical definition of stability, one which allows the discrete solution to grow but only to a certain extent, is as follows. If the discrete state variable is represented by an array $\mathbf{x}$ (collecting into a single vector the values of all variables at all spatial grid points), which is stepped in time by an algorithm based on the selected discretization, the corresponding numerical scheme is said to be stable over a fixed time interval T if there exists a constant C such that

$$\|\mathbf{x}^n\| \leq C\,\|\mathbf{x}^0\| \tag{2.38}$$

for all $n\Delta t \leq T$. A scheme is thus stable if regardless of Δt ($\leq T$), the numerical solution remains bounded for $t \leq T$.

This definition of stability leaves the numerical solution quite some room for growth, very often well beyond what a modeler is willing to tolerate. This definition of stability is, however, the necessary and sufficient stability used in the Lax–Richtmyer equivalence theorem and is thus the one utilized to ascertain convergence. If we permit a slower rate of growth in the numerical solution, we will not destroy convergence. In particular, we could decide to use the so-called strict stability condition.

2.7.2   Strict Stability

For a system conserving one or several integral norms (such as total energy or wave action), we may naturally impose that the corresponding norm of the numerical solution does not grow at all over time:

$$\|\mathbf{x}^n\| \leq \|\mathbf{x}^0\|\,. \tag{2.39}$$

Obviously, a scheme that is stable in the sense of Eq. (2.39) is also stable in the sense of Eq. (2.38), whereas the inverse is not necessarily true. The more stringent definition (2.39) will be called *strict stability condition* and refers to the condition that the norm of the numerical solution is not allowed to increase at all.

2.7.3   Choice of a Stability Criterion

The choice of stability criterion will depend largely on the mathematical and physical problem at hand. For a wave propagation problem, for example, strict stability will be the natural choice (assuming some norm is conserved in the physical process), whereas for physically unbounded problems, the less stringent numerical stability definition (2.38) may be used.

We can now examine two previous discretization schemes in the light of these two stability definitions. For the explicit Euler discretization (2.24) of inertial oscillation, the scheme is unstable in the sense of Eq. (2.39) (and deserves this label in view of the required energy conservation), although it is technically

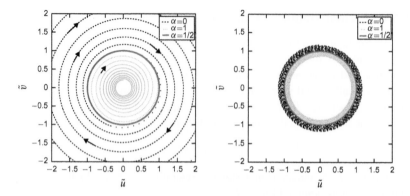

FIGURE 2.11 Representation (called a hodograph) of the numerical solution $(\tilde{u}, \tilde{v})$ (2.32a)–(2.32b) of the explicit discretization of the inertial oscillation ($\alpha = 0$), the implicit version ($\alpha = 1$), and the semi-implicit scheme ($\alpha = 1/2$). The hodograph on the left was obtained with $f\Delta t = 0.05$ and the one on the right panel with $f\Delta t = 0.005$. The inertial oscillation (Fig. 2.4) is clearly visible, but the explicit scheme induces spiralling out and the implicit scheme spiralling in. When the time step is reduced (moving from left panel to right panel), the solution approaches the exact solution. In both cases, 10 inertial periods were simulated.

stable in the sense of Eq. (2.38), as we will proceed to show. Since the norm of the velocity is, according to Eq. (2.29),

$$\|\tilde{\mathbf{u}}^n\| = \left(1 + f^2 \Delta t^2\right)^{n/2} \|\tilde{\mathbf{u}}^0\|, \tag{2.40}$$

we simply need to demonstrate[3] that the amplification is limited by a constant independent of n and Δt:

$$\left(1 + f^2 \Delta t^2\right)^{n/2} \leq \left(1 + f^2 \Delta t^2\right)^{T/(2\Delta t)} \leq e^{\frac{f^2 \Delta t T}{2}} \leq e^{\frac{f^2 T^2}{2}}. \tag{2.42}$$

The scheme is thus stable in the sense of Eq. (2.38) and even if growth of the norm can be quite important, according to the Lax–Richtmyer equivalence theorem, the solution will converge as the time step is reduced. This is indeed what is observed (Fig. 2.11) and can be proved explicitly (see Numerical Exercise 2.5). In practice, however, the time step is never allowed to be very small for obvious computer constraints. Also, the time window T over which simulations take place can be very large, and any increase of the velocity norm is unacceptable even if the solution is guaranteed to converge for smaller time steps. For

---

[3] For the demonstration, we use the inequality

$$(1 + a)^b \leq e^{ab} \text{ for } a, b \geq 0, \tag{2.41}$$

which can be easily be proved by observing that $(1 + a)^b = e^{b \ln(1+a)}$ and that $\ln(1 + a) \leq a$ when $a \geq 0$.

this reason, the strict stability condition (2.39) is preferred, and the semi-implicit Euler discretization is chosen.

In the second example, that of the implicit Euler scheme applied to the growth equation, the scheme (2.35) is stable in the sense of Eq. (2.38) (since it converges) but allows growth in the numerical solution in accordance with the exact solution.

Recapitulating the different concepts encountered in the numerical discretization, we now have a recipe to construct a convergent method: Design a discretization for which consistency (an equation-related property) can be verified by straightforward Taylor-series expansion, then check stability of the numerical scheme (some practical methods will be provided later), and finally invoke the Lax–Richtmyer equivalence theorem to prove convergence (a solution-related property). But, as the equivalence theorem is strictly valid only for linear equations, surprises may arise in nonlinear systems. We also have to mention that establishing convergence by this indirect method demands that initial and boundary conditions, too, converge to those of the continuous differential system. Finally, convergence is assured only for well-posed initial value problems. This, however, is not a concern here, since all geophysical fluid models we consider are physically well posed.

2.8   PREDICTOR-CORRECTOR METHODS

Till now, we have illustrated numerical discretizations on the linear equations describing inertial oscillations. The methods can be easily generalized to equations with a nonlinear source term Q in the equation governing the variable u, as

$$\frac{du}{dt} = Q(t, u).$$ (2.43)

For simplicity, we consider here a scalar variable u, but extension to a state vector $\mathbf{x}$, such as $\mathbf{x} = (u, v)$, is straightforward.

The previous methods can be recapitulated as follows:

- The explicit Euler method (*forward scheme*):

$$\tilde{u}^{n+1} = \tilde{u}^n + \Delta t\, Q^n$$ (2.44)

- The implicit Euler method (*backward scheme*):

$$\tilde{u}^{n+1} = \tilde{u}^n + \Delta t\, Q^{n+1}$$ (2.45)

- The semi-implicit Euler scheme (*trapezoidal scheme*):

$$\tilde{u}^{n+1} = \tilde{u}^n + \frac{\Delta t}{2}\left(Q^n + Q^{n+1}\right)$$ (2.46)

- A general two-points scheme (with $0 \leq \alpha \leq 1$):

$$\tilde{u}^{n+1} = \tilde{u}^n + \Delta t \left[(1-\alpha)Q^n + \alpha Q^{n+1} \right]. \tag{2.47}$$

Note that these schemes may be interpreted either as finite-difference approximations of the time derivative or as finite-difference approximations of the time integration of the source term. Indeed,

$$u(t^{n+1}) = u(t^n) + \int_{t^n}^{t^{n+1}} Q \, dt, \tag{2.48}$$

and the various schemes can be viewed as different ways of approximating the integral, as depicted in Fig. 2.12. All discretization schemes based on the exclusive use of Q^n and Q^{n+1} to evaluate the integral between t^n and t^{n+1}, which are called *two-point methods*, are inevitably first-order methods, except the semi-implicit (or trapezoidal) scheme, which is of second order. Second order is thus the highest order achievable with a two-point method. To achieve an order higher than two, denser sampling of the Q term must be used to approximate the time integration.

Before considering this, however, a serious handicap should be noted: The source term Q depends on the unknown variable $\tilde{u}$, and we face the problem

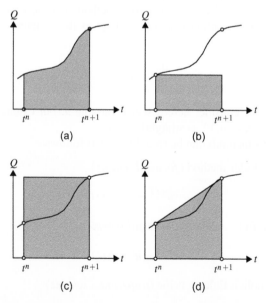

FIGURE 2.12 Time integration of the source term Q between t^n and t^{n+1}: (a) exact integration, (b) explicit scheme, (c) implicit scheme, and (d) semi-implicit, trapezoidal scheme.

of not being able to calculate Q^{n+1} *before* we know $\tilde{u}^{n+1}$, which is to be calculated from the value of Q^{n+1}. There is a vicious circle here! In the original case of inertial oscillations, the circular dependence was overcome by an algebraic manipulation of the equations prior to solution (gathering all $n+1$ terms on the left), but when the source term is nonlinear, as is often the case, such preliminary manipulation is generally not possible, and we need to circumvent the exact calculation by searching for a good approximation.

Such an approximation may proceed by using a first guess $\tilde{u}^\star$ in the Q term:

$$Q^{n+1} \simeq Q(t^{n+1}, \tilde{u}^\star), \tag{2.49}$$

as long as $\tilde{u}^\star$ is a sufficiently good estimate of $\tilde{u}^{n+1}$. The closer $\tilde{u}^\star$ is to $\tilde{u}^{n+1}$, the more faithful is the scheme to the ideal implicit value. If this estimate $\tilde{u}^\star$ is provided by a preliminary explicit (forward) step, according to:

$$\tilde{u}^\star = \tilde{u}^n + \Delta t\, Q(t^n, \tilde{u}^n) \tag{2.50a}$$

$$\tilde{u}^{n+1} = \tilde{u}^n + \frac{\Delta t}{2}\left[Q(t^n, \tilde{u}^n) + Q(t^{n+1}, \tilde{u}^\star)\right], \tag{2.50b}$$

we obtain a two-step algorithm, called the *Heun method*. It can be shown to be second-order accurate.

This second-order method is actually a particular member of a family of so-called *predictor-corrector methods*, in which a first guess $\tilde{u}^\star$ is used as a proxy for $\tilde{u}^{n+1}$ in the computation of complicated terms.

2.9   HIGHER-ORDER SCHEMES

If we want to go beyond second-order methods, we need to take into account a greater number of values of the Q term than those at t^n and t^{n+1}. We have two basic possibilities: either to include intermediate points between t^n and t^{n+1} or to use Q values at previous steps $n-1, n-2, \dots$. The first approach leads to the so-called family of *Runge–Kutta methods* (or *multistage methods*), whereas the second generates the so-called *multistep methods*.

The simplest method, using a single intermediate point, is the so-called *midpoint method*. In this case (Fig. 2.13), the integration is achieved by first calculating the value $\tilde{u}^{n+1/2}$ (playing the role of $\tilde{u}^\star$) at an intermediate stage $t^{n+1/2}$ and then integrating for the whole step based on this midpoint estimate:

$$\tilde{u}^{n+1/2} = \tilde{u}^n + \frac{\Delta t}{2} Q\left(t^n, \tilde{u}^n\right) \tag{2.51a}$$

$$\tilde{u}^{n+1} = \tilde{u}^n + \Delta t\, Q\left(t^{n+1/2}, \tilde{u}^{n+1/2}\right). \tag{2.51b}$$

This method, however, is only second-order accurate and offers no improvement over the earlier Heun method (2.50).

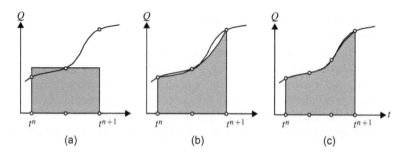

FIGURE 2.13   Runge–Kutta schemes of increasing complexity: (a) midpoint integration, (b) integration with parabolic interpolation, and (c) with cubic interpolation.

A popular fourth-order method can be constructed by using a parabolic interpolation between the values of Q with two successive estimates at the central point before proceeding with the full step:

$$\tilde{u}_a^{n+1/2} = \tilde{u}^n + \frac{\Delta t}{2} Q\left(t^n, \tilde{u}^n\right) \tag{2.52a}$$

$$\tilde{u}_b^{n+1/2} = \tilde{u}^n + \frac{\Delta t}{2} Q\left(t^{n+1/2}, \tilde{u}_a^{n+1/2}\right) \tag{2.52b}$$

$$\tilde{u}^\star = \tilde{u}^n + \Delta t\, Q\left(t^{n+1/2}, \tilde{u}_b^{n+1/2}\right) \tag{2.52c}$$

$$\tilde{u}^{n+1} = \tilde{u}^n + \Delta t \left(\frac{1}{6} Q\left(t^n, \tilde{u}^n\right) + \frac{2}{6} Q\left(t^{n+1/2}, \tilde{u}_a^{n+1/2}\right)\right.$$
$$\left. + \frac{2}{6} Q\left(t^{n+1/2}, \tilde{u}_b^{n+1/2}\right) + \frac{1}{6} Q\left(t^{n+1}, \tilde{u}^\star\right)\right). \tag{2.52d}$$

We can increase the order by using higher polynomial interpolations (Fig. 2.13).

As mentioned earlier, instead of using intermediate points to increase the order of accuracy, we can exploit already available evaluations of Q from previous steps (Fig. 2.14). The most popular method in GFD models is the *leapfrog method*, which simply reuses the value at time step $n-1$ to "jump over" the Q term at t^n in a $2\Delta t$ step:

$$\tilde{u}^{n+1} = \tilde{u}^{n-1} + 2\Delta t\, Q^n. \tag{2.53}$$

This algorithm offers second-order accuracy while being fully explicit. An alternative second-order method using the value at $n-1$ is the so-called *Adams–Bashforth method*:

$$\tilde{u}^{n+1} = \tilde{u}^n + \Delta t\frac{\left(3Q^n - Q^{n-1}\right)}{2}, \tag{2.54}$$

which can be interpreted in the light of Fig. 2.14.

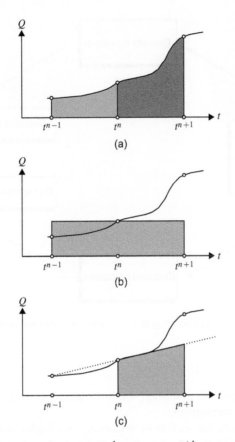

FIGURE 2.14   (a) Exact integration from t^{n-1} or t^n toward t^{n+1}, (b) leapfrog integration starts from t^{n-1} to reach t^{n+1}, whereas (c) Adams–Bashforth integration starts from t^n to reach t^{n+1}, using previous values to extrapolate Q over the integration interval t^n, t^{n+1}.

Higher-order methods can be constructed by recalling more points from the past $(n-2, n-3, \ldots)$, but we will not pursue this approach further for the following two reasons. First, using anterior points creates a problem at the start of the calculation from the initial condition. The first step must be different in order to avoid using one or several points that do not exist, and an explicit Euler scheme is usually performed. One such step is sufficient to initiate the leapfrog and Adams–Bashforth schemes, but methods that use earlier values (at $n-2, n-3, \ldots$) require more cumbersome care, which can amount to considerable effort in a GFD code. Second, the use of several points in the past demands a proportional increase in computer storage because values cannot be discarded as quickly before making room for newer values. Again, for a single equation, this is not much of a trouble, but in actual applications, size matters and only a few past values can be stored in the central memory of the machine. A similar

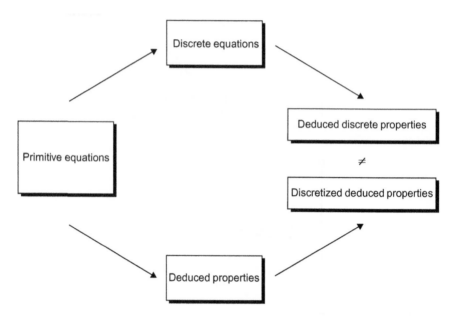

FIGURE 2.15   Schematic representation of discretization properties and mathematical properties interplay.

problem arises also with multistage methods, although these do not need any particular starting mechanism.

We can conclude the section by remarking that higher-order methods can always be designed but at the price of more frequent evaluations of the right-hand side of the equation (potentially a very complicated term) and/or greater storage of numerical values at different time steps. Since higher-order methods create more burden on the computation, we ought to ask whether they at least provide better numerical solutions than lower-order methods. We have therefore to address the question of accuracy of these methods, which will be considered in Section 4.8.

A fundamental difference between analytical solutions and numerical approximations emerges. For some equations, properties of the solution can be derived without actually solving the equations. It is easy to prove, for example, that the velocity magnitude remains constant during an inertial oscillation. The numerical solution on the other hand is generally not guaranteed to satisfy the same property as its analytical counterpart (the explicit Euler discretization did not conserve the velocity norm). Therefore, we cannot be sure that mathematical properties of the analytical solutions will also be present in the numerical solution. This might appear as a strong drawback of numerical methods but can actually be used to assess the quality of numerical schemes. Also, for numerical schemes with adjustable parameters (as the implicit factor), those parameters

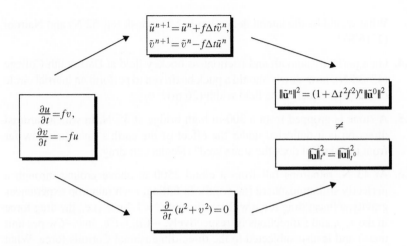

FIGURE 2.16   Schematic representation of discretization properties and mathematical properties interplay exemplified in the case of inertial oscillation.

can be chosen so that the numerical solution respects as best as possible the exact properties.

We can summarize by recognizing the fact that numerical solutions generally do not inherit the mathematical properties of the exact solution (Fig. 2.15), a handicap particularly easy to understand in the case of inertial oscillation and its discretization by an explicit scheme (Fig. 2.16). Later, we will encounter other properties (energy conservation, potential vorticity conservation, positiveness of concentrations, etc.) that can be used to guide the choice of parameter values in numerical schemes.

ANALYTICAL PROBLEMS

2.1. On Jupiter, a day lasts 9.9 Earth hours and the equatorial circumference is 448,600 km. Knowing that the measured gravitational acceleration at the equator is 26.4 m/s^2, deduce the true gravitational acceleration and the centrifugal acceleration.

2.2. The Japanese Shinkansen train (bullet train) zips from Tokyo to Ozaka (both at approximately 35°N) at a speed of 185 km/h. In the design of the train and tracks, do you think that engineers had to worry about the Earth's rotation? (*Hint*: The Coriolis effect induces an oblique force, the lateral component of which could produce a tendency of the train to lean sideways.)

2.3. Determine the lateral deflection of a cannonball that is shot in London (51°31′N) and flies for 25 s at an average horizontal speed of 120 m/s.

What would be the lateral deflection in Murmansk (68°52'N) and Nairobi (1°18'S)?

2.4. On a perfectly smooth and frictionless hockey field at Dartmouth College (43°38'N), how slowly should a puck be driven to perform an inertial circle of diameter equal to the field width (26 m)?

2.5. A stone is dropped from a 300-m high bridge at 35°N. In which cardinal direction is it deflected under the effect of the earth's rotation? How far from the vertical does the stone land? (Neglect air drag.)

2.6. At 43°N, raindrops fall from a cloud 2500 m above ground through a perfectly still atmosphere (no wind). In falling, each raindrop experiences gravity, a linear drag force with coefficient $C = 1.3$ s^{-1} (i.e., the drag force in the x, y, and z directions is, respectively, $-Cu$, $-Cv$, and $-Cw$ per unit mass) and is also subjected to the three-dimensional Coriolis force. What is the trajectory of one raindrop? How far eastward and northward has the Coriolis force deflected the raindrop by the time it hits the ground? (*Hint*: It can be shown that the terminal velocity is reached very quickly relatively to the total falling time.)

2.7. A set of two identical solid particles of mass M attached to each other by a weightless rigid rod of length L are moving on a horizontal rotating plane in the absence of external forces (Fig. 2.17). As in geophysical fluid dynamics, ignore the centrifugal force caused by the ambient rotation. Establish the equations governing the motion of the set of particles, derive the most general solution, and discuss its physical implications.

2.8. At $t = 0$, two particles of equal mass M but opposite electrical charges q are released from rest at a distance L from each other on a rotating plane (constant rotation rate $\Omega = f/2$). Assuming as in GFD that the centrifugal force

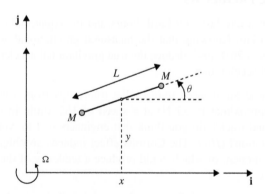

FIGURE 2.17   Two linked masses on a rotating plane (Problem 2.7).

caused by the ambient rotation is externally balanced, write the equations of motion of the two particles and the accompanying initial conditions. Then, show that the center of mass (the midpoint between the particles) is not moving, and write a differential equation governing the evolution of the distance $r(t)$ between the two particles. Is it possible that, as on a nonrotating plane, the electrical attraction between the two particles will make them collide ($r = 0$)?

2.9. Study the trajectory of a free particle of mass M released from a state of rest on a rotating, sloping, rigid plane (Fig. 2.18). The angular rotation rate is Ω, and the angle formed by the plane with the horizontal is α. Friction and the centrifugal force are negligible. What is the maximum speed acquired by the particle, and what is its maximum downhill displacement?

2.10. The curve reproduced in Fig. 2.19 is a progressive vector diagram constructed from current-meter observations at latitude $43°09'$N in the Mediterranean Sea. Under the assumption of a uniform but time-dependent

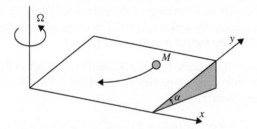

FIGURE 2.18   A free particle on a rotating, frictionless slope (Problem 2.9).

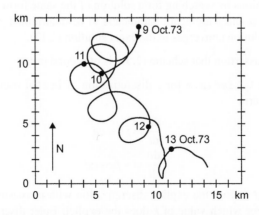

FIGURE 2.19   Progressive vector diagram constructed from current-meter observation in the Mediterranean Sea taken in October 1973 (Problem 2.10). (*Courtesy of Martin Mork, University of Bergen, Norway*)

flow field in the vicinity of the mooring, the curve can be interpreted as the trajectory of a water parcel. Using the marks counting the days along the curve, show that this set of observations reveals the presence of inertial oscillations. What is the average orbital velocity in these oscillations?

NUMERICAL EXERCISES

2.1. When using the semi-implicit scheme (2.32a)–(2.32b) with $\alpha = 1/2$, how many time steps are required per complete cycle (period of $2\pi/f$) to guarantee a relative error on f not exceeding 1%?

2.2. Develop an Euler scheme to calculate the position coordinates x and y of a particle undergoing inertial oscillations from its velocity components u and v, themselves calculated with an Euler scheme. Graph the trajectory $[\tilde{x}^n, \tilde{y}^n]$ for $n = 1, 2, 3, \ldots$ of the particle. What do you notice?

2.3. For the semi-implicit discretization of the inertial oscillation, calculate the number of complete cycles it takes before the exact solution and its numerical approximation are in phase opposition (180° phase shift). Express this number of cycles as a function of the parameter $f\Delta t$. What can you conclude for a scheme for which $f\Delta t = 0.1$ in terms of time windows that can be analyzed before the solution is out of phase?

2.4. Devise a leapfrog scheme for inertial oscillations and analyze its stability and angular frequency properties by searching for a numerical solution of the following form[4]:

$$\tilde{u}^n = V\varrho^n \sin(\tilde{f}n\Delta t + \phi), \quad \tilde{v}^n = V\varrho^n \cos(\tilde{f}n\Delta t + \phi).$$

2.5. Calculate the discrete solution of the explicit Euler scheme applied to inertial oscillations by searching for a solution of the same form as in Problem 2.4 where ϱ and $\tilde{f}$ are again parameters to be determined. Show that the discrete solution converges to the exact solution (2.12).

2.6. Prove the assertion that scheme (2.51) is of second order.

2.7. Adapt `coriolisdis.m` for a discretization of inertial oscillation with a frictional term

$$\frac{du}{dt} = fv - cu \tag{2.55}$$

$$\frac{dv}{dt} = -fu - cv, \tag{2.56}$$

where $c = f\, k$. Run the explicit discretization with increasing values of k in $[0, 1]$. For which value of k does the explicit Euler discretization give

---

[4]For ϱ^n, n is an exponent, not an index.

you a solution with constant norm? Can you interpret this result in view of Eq. (2.28)?

2.8. Try several time discretization methods on the following set of equations:

$$\frac{du}{dt} = fv \qquad (2.57)$$

$$\frac{dv}{dt} = -fu + fk(1 - u^2)v, \qquad (2.58)$$

with initial condition $u = 2$, $v = 0$ at $t = 0$. Use two values for the parameter k: first, $k = 0.1$ and then $k = 1$. Finally, try $k = 5$. What do you notice?

2.9. Prove that an inertial oscillation with modified angular frequency (2.34) is the exact solution of the semi-implicit scheme (2.32). (*Hint*: Insert the solution into the finite-difference equations and find a condition on $\tilde{f}$.)

Pierre Simon Marquis de Laplace
1749–1827

From humble roots in rural France, Pierre Simon Laplace distinguished himself early by his abilities and went on to Paris. There, at the Académie des Sciences, Jean D'Alembert recognized the talents of the young Laplace and secured for him a position in the military school. Set with this appointment, Laplace began a study of planetary motions, which led him to make advances in integral calculus and differential equations. Skillful at changing his political views during the turbulent years of the French Revolution, Laplace managed to survive and continued his research almost without interruption. In 1799, he published the first volume of a substantial memoir titled *Mécanique Céleste*, which later grew into a five-volume treatise and has since been regarded as a cornerstone of classical physics. Some have said that this work is Isaac Newton's *Principia* (of 1687) translated in the language of differential calculus with the clarification of many important points that had remained puzzling to Newton. One such aspect is the theory of ocean tides, which Laplace was the first to establish on firm mathematical grounds.

The name Laplace is attached today to a differential operator (the sum of second derivatives), which arises in countless problems of physics, including geophysical fluid dynamics (see Chapter 16). (*Portrait taken from a nineteenth century colored engraving, The Granger Collection, New York*)

Gaspard Gustave de Coriolis
1792–1843

Born in France and trained as an engineer, Gaspard Gustave de Coriolis began a career in teaching and research at age 24. Fascinated by problems related to rotating machinery, he was led to derive the equations of motion in a rotating framework of reference. The result of these studies was presented to the Académie des Sciences in the summer of 1831. In 1838, Coriolis stopped teaching to become director of studies at the Ecole Polytechnique, but his health declined quickly and he died a few short years later.

The world's largest experimental rotating table, at the Institut de Mécanique in Grenoble, France, is named after him and has been used in countless simulations of geophysical fluid phenomena. (*Photo from the archives of the Académie des Sciences, Paris*)

Chapter 3

Equations of Fluid Motion

ABSTRACT

The objective of this chapter is to establish the equations governing the movement of a stratified fluid in a rotating environment. These equations are then simplified somewhat by taking advantage of the so-called Boussinesq approximation. This chapter concludes by introducing finite-volume discretizations and showing their relation with the budget calculations used to establish the mathematical equations of motion.

3.1 MASS BUDGET

A necessary statement in fluid mechanics is that mass be conserved. That is, any imbalance between convergence and divergence in the three spatial directions must create a local compression or expansion of the fluid. Mathematically, the statement takes the following form:

$$\frac{\partial \rho}{\partial t} + \frac{\partial}{\partial x}(\rho u) + \frac{\partial}{\partial y}(\rho v) + \frac{\partial}{\partial z}(\rho w) = 0, \tag{3.1}$$

where ρ is the density of the fluid (in kg/m^3), and (u, v, w) are the three components of velocity (in m/s). All four variables generally vary in the three spatial directions, x and y in the horizontal, z in the vertical, as well as time t.

This equation, often called the *continuity equation*, is classical in traditional fluid mechanics. Sturm (2001, page 4) reports that Leonardo da Vinci (1452–1519) had derived a simplified form of the statement of mass conservation for a stream with narrowing width. However, the three-dimensional differential form provided here was most likely written much later and credit ought probably to go to Leonhard Euler (1707–1783). For a detailed derivation, the reader is referred to Batchelor (1967), Fox and McDonald (1992), or Appendix A of the present text.

Note that spherical geometry introduces additional curvature terms, which we neglect to be consistent with our previous restriction to length scales substantially shorter than the global scale.

Introduction to Geophysical Fluid Dynamics

3.2 MOMENTUM BUDGET

For a fluid, Isaac Newton's second law *"mass times acceleration equals the sum of forces"* is better stated per unit volume with density replacing mass and, in the absence of rotation ($\Omega = 0$), the resulting equations are called the Navier-Stokes equations. For geophysical flows, rotation is important, and acceleration terms must be augmented as done in (2.20):

$$x: \quad \rho\left(\frac{du}{dt} + f_* w - f v\right) = -\frac{\partial p}{\partial x} + \frac{\partial \tau^{xx}}{\partial x} + \frac{\partial \tau^{xy}}{\partial y} + \frac{\partial \tau^{xz}}{\partial z} \tag{3.2a}$$

$$y: \quad \rho\left(\frac{dv}{dt} + f u\right) = -\frac{\partial p}{\partial y} + \frac{\partial \tau^{xy}}{\partial x} + \frac{\partial \tau^{yy}}{\partial y} + \frac{\partial \tau^{yz}}{\partial z} \tag{3.2b}$$

$$z: \quad \rho\left(\frac{dw}{dt} - f_* u\right) = -\frac{\partial p}{\partial z} - \rho g + \frac{\partial \tau^{xz}}{\partial x} + \frac{\partial \tau^{yz}}{\partial y} + \frac{\partial \tau^{zz}}{\partial z}, \tag{3.2c}$$

where the x-, y-, and z-axes are directed eastward, northward, and upward, respectively, $f = 2\Omega \sin\varphi$ is the Coriolis parameter, $f_* = 2\Omega \cos\varphi$ is the reciprocal Coriolis parameter, ρ is the density, p is the pressure, g is the gravitational acceleration, and the τ terms represent the normal and shear stresses due to friction.

That the pressure force is equal and opposite to the pressure gradient, and that the viscous force involves the derivatives of a stress tensor should be familiar to the student who has had an introductory course in fluid mechanics. Appendix A retraces the formulation of those terms for the student new to fluid mechanics.

The effective gravitational force (sum of true gravitational force and the centrifugal force; see Section 2.2) is ρg per unit volume and is directed vertically downward. So, the corresponding term occurs only in the third equation for the vertical direction.

Because the acceleration in a fluid is not counted as the rate of change in velocity at a fixed location but as the change in velocity of a fluid particle as it moves along with the flow, the time derivatives in the acceleration components, du/dt, dv/dt and dw/dt, consist of both the local time rate of change and the so-called advective terms:

$$\frac{d}{dt} = \frac{\partial}{\partial t} + u\frac{\partial}{\partial x} + v\frac{\partial}{\partial y} + w\frac{\partial}{\partial z}. \tag{3.3}$$

This derivative is called the *material derivative*.

The preceding equations assume a Cartesian system of coordinates and thus hold only if the dimension of the domain under consideration is much shorter than the earth's radius. On Earth, a length scale not exceeding 1000 km is usually acceptable. The neglect of the curvature terms is in some ways analogous to the distortion introduced by mapping the curved earth's surface onto a plane.

Should the dimensions of the domain under consideration be comparable with the size of the planet, the x-, y-, and z-axes need to be replaced by spherical coordinates, and curvature terms enter all equations. See Appendix A for those equations. For simplicity in the exposition of the basic principles of geophysical fluid dynamics, we shall neglect throughout this book the extraneous curvature terms and use Cartesian coordinates exclusively.

Equations (3.2a) through (3.2c) can be viewed as three equations providing the three velocity components, u, v, and w. They implicate, however, two additional quantities, namely, the pressure p and the density ρ. An equation for ρ is provided by the conservation of mass (3.1), and one additional equation is still required.

3.3   EQUATION OF STATE

The description of the fluid system is not complete until we also provide a relation between density and pressure. This relation is called the *equation of state* and tells us about the nature of the fluid. To go further, we need to distinguish between air and water.

For an incompressible fluid such as pure water at ordinary pressures and temperatures, the statement can be as simple as $\rho = $ constant. In this case, the preceding set of equations is complete. In the ocean, however, water density is a complicated function of pressure, temperature, and salinity. Details can be found in Gill (1982, Appendix 3), but for most applications, it can be assumed that the density of seawater is independent of pressure (incompressibility) and linearly dependent upon both temperature (warmer waters are lighter) and salinity (saltier waters are denser) according to:

$$\rho = \rho_0[1 - \alpha(T - T_0) + \beta(S - S_0)], \tag{3.4}$$

where T is the temperature (in degrees Celsius or Kelvin), and S is the salinity (defined in the past as grams of salt per kilogram of seawater, i.e., in parts per thousand, denoted by ‰, and more recently by the so-called practical salinity unit "psu," derived from measurements of conductivity and having no units). The constants ρ_0, T_0, and S_0 are reference values of density, temperature, and salinity, respectively, whereas α is the coefficient of thermal expansion, and β is called, by analogy, the coefficient of saline contraction[1]. Typical seawater values are $\rho_0 = 1028$ kg/m^3, $T_0 = 10°$C $= 283$ K, $S_0 = 35$, $\alpha = 1.7 \times 10^{-4}$ K^{-1}, and $\beta = 7.6 \times 10^{-4}$.

For air, which is compressible, the situation is quite different. Dry air in the atmosphere behaves approximately as an ideal gas, and so we write:

$$\rho = \frac{p}{RT}, \tag{3.5}$$

---

[1] The latter expression is a misnomer, since salinity increases density not by contraction of the water but by the added mass of dissolved salt.

where R is a constant, equal to $287 \, \mathrm{m}^2 \mathrm{s}^{-2} \mathrm{K}^{-1}$ at ordinary temperatures and pressures. In the preceding equation, T is the absolute temperature (temperature in degrees Celsius $+ 273.15$).

Air in the atmosphere most often contains water vapor. For moist air, the preceding equation is generalized by introducing a factor that varies with the *specific humidity q*:

$$\rho = \frac{p}{RT(1+0.608q)} . \tag{3.6}$$

The specific humidity q is defined as

$$q = \frac{\text{mass of water vapor}}{\text{mass of air}} = \frac{\text{mass of water vapor}}{\text{mass of dry air} + \text{mass of water vapor}}. \tag{3.7}$$

For details, the reader is referred to Curry and Webster (1999).

Unfortunately, our set of governing equations is not yet complete. Although we have added one equation, by doing so, we have also introduced additional variables, namely temperature and, depending on the nature of the fluid, either salinity or specific humidity. Additional equations are clearly necessary.

3.4   ENERGY BUDGET

The equation governing temperature arises from conservation of energy. The principle of energy conservation, also known as the first law of thermodynamics, states that the internal energy gained by a parcel of matter is equal to the heat it receives minus the mechanical work it performs. Per unit mass and unit time, we have

$$\frac{de}{dt} = Q - W, \tag{3.8}$$

where d/dt is the material derivative introduced in (3.3), e is the internal energy, Q is the rate of heat gain, and W is the rate of work done by the pressure force onto the surrounding fluid, all per unit mass. The internal energy, a measure of the thermal agitation of the molecules inside the fluid parcel, is proportional to the temperature:

$$e = C_v T$$

where C_v is the heat capacity at constant volume, and T is the absolute temperature. For air at sea-level pressure and ambient temperatures, $C_v = 718 \, \mathrm{J \, kg}^{-1} \mathrm{K}^{-1}$, whereas for seawater, $C_v = 3990 \, \mathrm{J \, kg}^{-1} \mathrm{K}^{-1}$.

In the ocean, there is no internal heat source[2], whereas in the atmosphere release of latent heat by water-vapor condensation or, conversely,

---

[2] In most cases, the absorption of solar radiation in the first meters of the upper ocean is treated as a surface flux, though occasionally it must be taken into account as a radiative absorption.

uptake of latent heat by evaporation constitute internal sources. Leaving such complication for more advanced textbooks in dynamical and physical meteorology (Curry & Webster, 1999), the Q term in (3.8) includes only the heat gained by a parcel through its contact with its neighbors through the process of diffusion. Using the Fourier law of heat conduction, we write

$$Q = \frac{k_T}{\rho} \nabla^2 T,$$

where k_T is the thermal conductivity of the fluid, and the Laplace operator ∇^2 is defined as the sum of double derivatives:

$$\nabla^2 = \frac{\partial^2}{\partial x^2} + \frac{\partial^2}{\partial y^2} + \frac{\partial^2}{\partial z^2}.$$

The work done by the fluid is the pressure force ($=$ pressure $\times$ area) multiplied by the displacement in the direction of the force. Counting area times displacement as volume, the work is pressure multiplied by the change in volume and on a per–mass and per–time basis:

$$W = p \frac{\mathrm{d}\upsilon}{\mathrm{d}t},$$

where υ is the volume per mass, i.e., $\upsilon = 1/\rho$.

With its pieces assembled, Eq. (3.8) becomes

$$C_\upsilon \frac{\mathrm{d}T}{\mathrm{d}t} = \frac{k_T}{\rho} \nabla^2 T - p \frac{\mathrm{d}\upsilon}{\mathrm{d}t}$$

$$= \frac{k_T}{\rho} \nabla^2 T + \frac{p}{\rho^2} \frac{\mathrm{d}\rho}{\mathrm{d}t}. \tag{3.9}$$

Elimination of $\mathrm{d}\rho/\mathrm{d}t$ with the continuity Eq. (3.1) leads to:

$$\rho C_\upsilon \frac{\mathrm{d}T}{\mathrm{d}t} + p \left(\frac{\partial u}{\partial x} + \frac{\partial v}{\partial y} + \frac{\partial w}{\partial z} \right) = k_T \nabla^2 T. \tag{3.10}$$

This is the energy equation, which governs the evolution of temperature.

For water, which is nearly incompressible, the divergence term ($\partial u/\partial x + \partial v/\partial y + \partial w/\partial z$) can be neglected (to be shown later), whereas for air, one may introduce the *potential temperature* θ defined as

$$\theta = T \left(\frac{\rho_0}{\rho} \right)^{R/C_\upsilon}, \tag{3.11}$$

for which, the physical interpretation will be given later (Section 11.3). Taking its material derivative and using Eqs. (3.5) and (3.9), lead successively to

$$C_v \frac{d\theta}{dt} = \left(\frac{\rho_0}{\rho}\right)^{R/C_v} \left(C_v \frac{dT}{dt} - \frac{RT}{\rho} \frac{d\rho}{dt}\right)$$

$$C_v \frac{d\theta}{dt} = \frac{\theta}{T} \left(C_v \frac{dT}{dt} - \frac{p}{\rho^2} \frac{d\rho}{dt}\right)$$

$$\rho C_v \frac{d\theta}{dt} = k_T \frac{\theta}{T} \nabla^2 T. \tag{3.12}$$

The net effect of this transformation of variables is the elimination of the divergence term.

When k_T is zero or negligible, the right-hand side of the equation vanishes, leaving only

$$\frac{d\theta}{dt} = 0. \tag{3.13}$$

Unlike the actual temperature T, which is subject to the compressibility effect (through the divergence term), the potential temperature θ of an air parcel is conserved in the absence of heat diffusion.

3.5   SALT AND MOISTURE BUDGETS

The set of equations is not yet complete because there is a remaining variable for which a last equation is required: salinity in the ocean and specific humidity in the atmosphere.

For seawater, density varies with salinity as stated in Eq. (3.4). Its evolution is governed by the salt budget:

$$\frac{dS}{dt} = \kappa_S \nabla^2 S, \tag{3.14}$$

which simply states that a seawater parcel conserves its salt content, except for redistribution by diffusion. The coefficient κ_S is the coefficient of salt diffusion, which plays a role analogous to the heat diffusivity k_T.

For air, the remaining variable is specific humidity and, because of the possibility of evaporation and condensation, its budget is complicated. Leaving this matter for more advanced texts in meteorology, we simply write an equation similar to that of salinity:

$$\frac{dq}{dt} = \kappa_q \nabla^2 q, \tag{3.15}$$

which states that specific humidity is redistributed by contact with neighboring parcels of different moisture contents, and in which the diffusion coefficient κ_q is the analog of k_T and κ_S.

3.6   SUMMARY OF GOVERNING EQUATIONS

Our set of governing equations is now complete. For air (or any ideal gas), there are seven variables (u, v, w, p, ρ, T, and q), for which we have a continuity equation (3.1), three momentum equations (3.2a) through (3.2c), an equation of state (3.5), an energy equation (3.10), and a humidity equation (3.15). Vilhelm Bjerknes (see biography at the end of this chapter) is credited for having been the first to recognize that atmospheric physics can, in theory, be fully described by a set of equations governing the evolution of the seven aforementioned variables (Bjerknes, 1904; see also Nebeker, 1995, chapter 5).

For seawater, the situation is similar. There are again seven variables (u, v, w, p, ρ, T, and S), for which we have the same continuity, momentum and energy equations, the equation of state (3.4), and the salt equation (3.14). No particular person is credited with this set of equations.

3.7   BOUSSINESQ APPROXIMATION

Although the equations established in the previous sections already contain numerous simplifying approximations, they are still too complicated for the purpose of geophysical fluid dynamics. Additional simplifications can be obtained by the so-called *Boussinesq approximation* without appreciable loss of accuracy.

In most geophysical systems, the fluid density varies, but not greatly, around a mean value. For example, the average temperature and salinity in the ocean are $T = 4°C$ and $S = 34.7$, respectively, to which corresponds a density $\rho = 1028 \, kg/m^3$ at surface pressure. Variations in density within one ocean basin rarely exceed $3 \, kg/m^3$. Even in estuaries where fresh river waters ($S = 0$) ultimately turn into salty seawaters ($S = 34.7$), the relative density difference is less than 3%.

By contrast, the air of the atmosphere becomes gradually more rarefied with altitude, and its density varies from a maximum at ground level to nearly zero at great heights, thus covering a 100% range of variations. Most of the density changes, however, can be attributed to hydrostatic pressure effects, leaving only a moderate variability caused by other factors. Furthermore, weather patterns are confined to the lowest layer, the troposphere (approximately 10 km thick), within which the density variations responsible for the winds are usually no more than 5%.

As it appears justifiable in most instances[3] to assume that the fluid density, ρ, does not depart much from a mean reference value, ρ_0, we take the liberty to write the following:

$$\rho = \rho_0 + \rho'(x, y, z, t) \quad \text{with} \quad |\rho'| \ll \rho_0, \tag{3.16}$$

---

[3] The situation is obviously somewhat uncertain on other planets that are known to possess a fluid layer (Jupiter and Neptune, for example), and on the sun.

where the variation ρ' caused by the existing stratification and/or fluid motions is small compared with the reference value ρ_0. Armed with this assumption, we proceed to simplify the governing equations.

The continuity equation, (3.1), can be expanded as follows:

$$
\rho_0 \left(\frac{\partial u}{\partial x} + \frac{\partial v}{\partial y} + \frac{\partial w}{\partial z} \right) + \rho' \left(\frac{\partial u}{\partial x} + \frac{\partial v}{\partial y} + \frac{\partial w}{\partial z} \right)
$$
$$
+ \left(\frac{\partial \rho'}{\partial t} + u \frac{\partial \rho'}{\partial x} + v \frac{\partial \rho'}{\partial y} + w \frac{\partial \rho'}{\partial z} \right) = 0.
$$

Geophysical flows indicate that relative variations of density in time and space are not larger than—and usually much less than—the relative variations of the velocity field. This implies that the terms in the third group are on the same order as—if not much less than—those in the second. But, terms in this second group are always much less than those in the first because $|\rho'| \ll \rho_0$. Therefore, only that first group of terms needs to be retained, and we write

$$
\frac{\partial u}{\partial x} + \frac{\partial v}{\partial y} + \frac{\partial w}{\partial z} = 0. \tag{3.17}
$$

Physically, this statement means that conservation of mass has become conservation of volume. The reduction is to be expected because volume is a good proxy for mass when mass per volume (= density) is nearly constant. A hidden implication of this simplification is the elimination of sound waves, which rely on compressibility for their propagation.

The x- and y-momentum equations (3.2a) and (3.2b), being similar to each other, can be treated simultaneously. There, ρ occurs as a factor only in front of the left-hand side. So, wherever ρ' occurs, ρ_0 is there to dominate. It is thus safe to neglect ρ' next to ρ_0 in that pair of equations. Further, the assumption of a Newtonian fluid (viscous stresses proportional to velocity gradients), with the use of the reduced continuity equation, (3.17), permits us to write the components of the stress tensor as

$$
\tau^{xx} = \mu \left(\frac{\partial u}{\partial x} + \frac{\partial u}{\partial x} \right), \quad \tau^{xy} = \mu \left(\frac{\partial u}{\partial y} + \frac{\partial v}{\partial x} \right), \quad \tau^{xz} = \mu \left(\frac{\partial u}{\partial z} + \frac{\partial w}{\partial x} \right)
$$
$$
\tau^{yy} = \mu \left(\frac{\partial v}{\partial y} + \frac{\partial v}{\partial y} \right), \quad \tau^{yz} = \mu \left(\frac{\partial v}{\partial z} + \frac{\partial w}{\partial y} \right)
$$
$$
\tau^{zz} = \mu \left(\frac{\partial w}{\partial z} + \frac{\partial w}{\partial z} \right), \tag{3.18}
$$

where μ is called the coefficient of dynamic viscosity. A subsequent division by ρ_0 and the introduction of the *kinematic viscosity* $\nu = \mu/\rho_0$ yield

$$\frac{du}{dt} + f_* w - f v = -\frac{1}{\rho_0}\frac{\partial p}{\partial x} + \nu\nabla^2 u \tag{3.19}$$

$$\frac{dv}{dt} + f u = -\frac{1}{\rho_0}\frac{\partial p}{\partial y} + \nu\nabla^2 v. \tag{3.20}$$

The next candidate for simplification is the z-momentum equation, (3.2c). There, ρ appears as a factor not only in front of the left-hand side but also in a product with g on the right. On the left, it is safe to neglect ρ' in front of ρ_0 for the same reason as previously, but on the right, it is not. Indeed, the term ρg accounts for the weight of the fluid, which, as we know, causes an increase of pressure with depth (or, a decrease of pressure with height, depending on whether we think of the ocean or atmosphere). With the ρ_0 part of the density goes a hydrostatic pressure p_0, which is a function of z only:

$$p = p_0(z) + p'(x, y, z, t) \quad \text{with} \quad p_0(z) = P_0 - \rho_0 g z, \tag{3.21}$$

so that $dp_0/dz = -\rho_0 g$, and the vertical-momentum equation at this stage reduces to

$$\frac{dw}{dt} - f_* u = -\frac{1}{\rho_0}\frac{\partial p'}{\partial z} - \frac{\rho' g}{\rho_0} + \nu\nabla^2 w, \tag{3.22}$$

after a division by ρ_0 for convenience. No further simplification is possible because the remaining ρ' term no longer falls in the shadow of a neighboring term proportional to ρ_0. Actually, as we will see later, the term $\rho' g$ is the one responsible for the buoyancy forces that are such a crucial ingredient of geophysical fluid dynamics.

Note that the hydrostatic pressure $p_0(z)$ can be subtracted from p in the reduced momentum equations, (3.19) and (3.20), because it has no derivatives with respect to x and y, and is dynamically inactive.

For water, the treatment of the energy equation, (3.8), is straightforward. First, continuity of volume, (3.17), eliminates the middle term, leaving

$$\rho C_v \frac{dT}{dt} = k_T \nabla^2 T.$$

Next, the factor ρ in front of the first term can be replaced once again by ρ_0, for the same reason as it was done in the momentum equations. Defining the heat kinematic diffusivity $\kappa_T = k_T/\rho_0 C_v$, we then obtain

$$\frac{dT}{dt} = \kappa_T \nabla^2 T, \tag{3.23}$$

which is isomorphic to the salt equation, (3.14).

For seawater, the pair of Eqs. (3.14) and (3.23) for salinity and temperature, respectively, combine to determine the evolution of density. A simplification results if it may be assumed that the salt and heat diffusivities, κ_S and κ_T, can be considered equal. If diffusion is primarily governed by molecular processes, this assumption is invalid. In fact, a substantial difference between the rates of salt and heat diffusion is responsible for peculiar small-scale features, such as salt fingers, which are studied in the discipline called *double diffusion* (Turner, 1973, Chapter 8). However, molecular diffusion generally affects only small-scale processes, up to a meter or so, whereas turbulence regulates diffusion on larger scales. In turbulence, efficient diffusion is accomplished by eddies, which mix salt and heat at equal rates. The values of diffusivity coefficients in most geophysical applications may not be taken as those of molecular diffusion; instead, they should be taken much larger and equal to each other. The corresponding turbulent diffusion coefficient, also called *eddy diffusivity*, is typically expressed as the product of a turbulent eddy velocity with a mixing length (Pope, 2000; Tennekes & Lumley, 1972) and, although there exists no single value applicable to all situations, the value $\kappa_S = \kappa_T = 10^{-2}$ m^2/s is frequently adopted. Noting $\kappa = \kappa_S = \kappa_T$ and combining Eqs. (3.14) and (3.23) with the equation of state (3.4), we obtain

$$\frac{d\rho'}{dt} = \kappa \nabla^2 \rho', \tag{3.24}$$

where $\rho' = \rho - \rho_0$ is the density variation. In sum, the energy and salt conservation equations have been merged into a density equation, which is not to be confused with mass conservation (3.1).

For air, the treatment of the energy equation (3.10) is more subtle, and the reader interested in a rigorous discussion is referred to the article by Spiegel and Spiegel and Veronis (1960). Here, for the sake of simplicity, we limit ourselves to suggestive arguments. First, the change of variable in Eq. (3.11) from actual temperature to potential temperature eliminates the divergence term in Eq. (3.10) and takes care of the compressibility effect. Then, for weak departures from a reference state, the relation between actual and potential temperatures and the equation of state can both be linearized. Finally, assuming that heat and moisture are diffused by turbulent motions at the same rate, we can combine their respective budget into a single equation, (3.24).

In summary, the Boussinesq approximation, rooted in the assumption that the density does not depart much from a mean value, has allowed the replacement of the actual density ρ by its reference value ρ_0 everywhere, except in front of the gravitational acceleration and in the energy equation, which has become an equation governing density variations.

At this point, since the original variables ρ and p no longer appear in the equations, it is customary to drop the primes from ρ' and p' without risk of ambiguity. So, from here on, the variables ρ and p will be used exclusively to denote the perturbation density and perturbation pressure, respectively. This perturbation pressure is sometimes called the *dynamic pressure* because it is usually a main contributor to the flow field. The only place where total pressure comes into play is the equation of state.

3.8   FLUX FORMULATION AND CONSERVATIVE FORM

The preceding equations form a complete set of equations, and there is no need to invoke further physical laws. Nevertheless, we can manipulate the equations to write them in another form, which, though mathematically equivalent, has some practical advantages. Consider, for example, the equation for temperature (3.23), which was deduced from the energy equation using the Boussinesq approximation. If we now expand its material derivative using Eq. (3.3)

$$\frac{\partial T}{\partial t} + u\frac{\partial T}{\partial x} + v\frac{\partial T}{\partial y} + w\frac{\partial T}{\partial z} = \kappa_T \nabla^2 T, \tag{3.25}$$

and then use volume conservation (3.17), which was obtained under the same Boussinesq approximation, we obtain

$$\frac{\partial T}{\partial t} + \frac{\partial}{\partial x}(uT) + \frac{\partial}{\partial y}(vT) + \frac{\partial}{\partial z}(wT)$$
$$- \frac{\partial}{\partial x}\left(\kappa_T \frac{\partial T}{\partial x}\right) - \frac{\partial}{\partial y}\left(\kappa_T \frac{\partial T}{\partial y}\right) - \frac{\partial}{\partial z}\left(\kappa_T \frac{\partial T}{\partial z}\right) = 0. \tag{3.26}$$

The latter form is called a *conservative formulation*, the reason for which will become clear upon applying the divergence theorem. This theorem, also known as Gauss's Theorem, states that for any vector (q_x, q_y, q_z), the volume integral of its divergence is equal to the integral of the flux over the enclosing surface:

$$\int_V \left(\frac{\partial q_x}{\partial x} + \frac{\partial q_y}{\partial y} + \frac{\partial q_z}{\partial z}\right) dx\,dy\,dz = \int_S (q_x n_x + q_y n_y + q_z n_z)dS \tag{3.27}$$

where the vector (n_x, n_y, n_z) is the outward unit vector normal to the surface S delimiting the volume V (Fig. 3.1). Integrating the conservative form (3.26) over a fixed volume is then particularly simple and leads to an expression for the evolution of the heat content in the volume as a function of the fluxes entering and leaving the volume:

$$\frac{d}{dt}\int_V T\,dt + \int_S \mathbf{q}\cdot\mathbf{n}dS = 0. \tag{3.28}$$

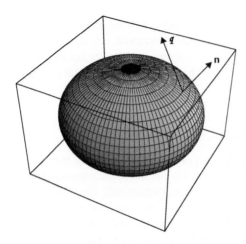

FIGURE 3.1 The divergence theorem allows the replacement of the integral over the volume $\mathcal{V}$ of the divergence $\partial q_x/\partial x + \partial q_y/\partial y + \partial q_z/\partial z$ of a flux vector $\boldsymbol{q} = (q_x, q_y, q_z)$ by the integral, over the surface $\mathcal{S}$ containing the volume, of the scalar product of the flux vector and the normal vector $\mathbf{n}$ to this surface.

The flux $\boldsymbol{q}$ of temperature is composed of an advective flux (uT, vT, wT) due to flow across the surface and a diffusive (conductive) flux $-\kappa_T(\partial T/\partial x, \partial T/\partial y, \partial T/\partial z)$. If the value of each flux is known on a closed surface, the evolution of the average temperature inside the volume can be calculated without knowing the detailed distribution of temperature. This property will be used now for the development of a particular discretization method.

3.9 FINITE-VOLUME DISCRETIZATION

The conservative form (3.26) naturally leads to a numerical method with a clear physical interpretation, the so-called *finite-volume* approach. To illustrate the concept, we consider the equation for temperature in a one-dimensional (1D) case

$$\frac{\partial T}{\partial t} + \frac{\partial q}{\partial x} = 0, \tag{3.29}$$

in which the flux q for temperature $T = T(x, t)$ includes both advection uT and diffusion $-\kappa_T \partial T/\partial x$:

$$q = uT - \kappa_T \frac{\partial T}{\partial x}. \tag{3.30}$$

We can integrate (3.29) over a given interval (labeled by index i) with boundaries noted by indices $i - 1/2$ and $i + 1/2$, so that we integrate over x in the range $x_{i-1/2} < x < x_{i+1/2}$. Though the interval of integration is of finite size

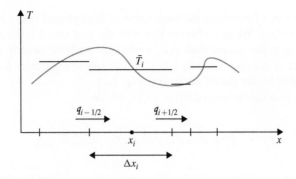

FIGURE 3.2   Replacement of the continuous function T by its cell-averaged discrete values $\bar{T}_i$. The evolution of the finite-volume averaged temperature is given by the difference of the flux q between the surrounding interfaces at $x_{i-1/2}$ and $x_{i+1/2}$.

$\Delta x_i = x_{i+1/2} - x_{i-1/2}$ (Fig. 3.2), the integration is performed exactly:

$$\frac{\mathrm{d}}{\mathrm{d}t} \int_{x_{i-1/2}}^{x_{i+1/2}} T \mathrm{d}x + q_{i+1/2} - q_{i-1/2} = 0. \tag{3.31}$$

By defining the cell-average temperature $\bar{T}_i$ for cell i as

$$\bar{T}_i = \frac{1}{\Delta x_i} \int_{x_{i-1/2}}^{x_{i+1/2}} T \mathrm{d}x, \tag{3.32}$$

we obtain the evolution equation of the discrete field $\bar{T}_i$:

$$\frac{\mathrm{d}\bar{T}_i}{\mathrm{d}t} + \frac{q_{i+1/2} - q_{i-1/2}}{\Delta x_i} = 0. \tag{3.33}$$

Although we seem to have fallen back on a *discretization* of the spatial derivative, of q in this instance, the equation we just obtained is *exact*. This seems to be paradoxical compared with our previous discussions on inevitable errors associated with discretization. At first sight, it appears that we found a discretization method without errors, but we must realize that (3.33) is still incomplete in the sense that two different variables appear in a single equation, the discretized average $\bar{T}_i$ and the discretized flux, $q_{i-1/2}$ and $q_{i+1/2}$. These two, however, are related to the *local* value of the continuous temperature field [the advective flux at $x_{i\pm1/2}$ is $uT(x_{i\pm1/2}, t)$], whereas the integrated equation is written for the *average* value of temperature. The averaging of the equation prevents us from retrieving information at the local level, and only average values (over the spatial scale Δx_i) can be determined. Therefore, we have to find an approximate way of assessing the local value of fluxes based solely on average temperature values. We also observe that with the grid size Δx_i we only retain information at scales longer than Δx_i, a property we have already mentioned in the context of aliasing (Section 1.12). The shorter spatial scales have simply been eliminated by the spatial averaging (Fig. 3.2).

A further exact time-integration of (3.33) yields

$$\bar{T}_i^{n+1} - \bar{T}_i^n + \frac{\int_{t^n}^{t^{n+1}} q_{i+1/2}\, dt - \int_{t^n}^{t^{n+1}} q_{i-1/2}\, dt}{\Delta x_i} = 0,$$

expressing that the difference in average temperature (i.e., heat content) is given by the net flux entering the finite cell during the given time interval. Again, to this stage, no approximation is needed, and the equation is *exact* and can be formulated in terms of time-averaged fluxes $\hat{q}$

$$\hat{q} = \frac{1}{\Delta t_n} \int_{t^n}^{t^{n+1}} q\, dt \qquad (3.34)$$

to yield an equation for discrete averaged quantities:

$$\frac{\bar{T}_i^{n+1} - \bar{T}_i^n}{\Delta t_n} + \frac{\hat{q}_{i+1/2} - \hat{q}_{i-1/2}}{\Delta x_i} = 0. \qquad (3.35)$$

This equation is still *exact* but to be useful needs to be supplemented with a scheme to calculate the average fluxes $\hat{q}$ as functions of average temperatures $\bar{T}$. Only at that point are discretization approximations required, and discretization errors introduced.

It is noteworthy also to realize how easy the introduction of nonuniform grid spacing and timestepping has been up to this point. Though we refer the interfaces by index $i \pm 1/2$, the position of an interface does not need to lie at mid-distance between consecutive grid nodes x_i. Only their logical, topological position must be ordered in the sense that grid nodes and interfaces must be interleaved.

Without further investigation of the way average fluxes can be computed, we interpret different discretization methods in relation to the mathematical budget formulation used to establish the governing equations (Fig. 3.3). From brute-force replacement of differential operators by finite differences to the establishment of equations for finite volumes and subsequent discretization of fluxes, all methods aim at replacing the continuous problem by a finite set of discrete equations.

One of the main advantages of the finite-volume approach presented here is its conservation property. Consider the set of integrated equations for

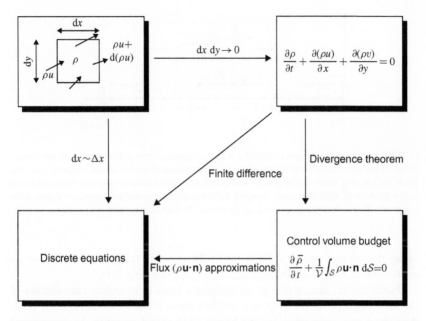

FIGURE 3.3 Schematic representation of several discretization methods. From the budget calculations (upper-left box), the limit to infinitesimal values of dx, dy leads to the continuous equations (upper-right box), whereas keeping differentials formally at finite values lead to crude finite differencing (downward path from upper left to bottom left). If the operators in the continuous equations are discretized using Taylor expansions, higher-quality finite-difference methods are obtained (diagonal path from upper right to lower left). Finally, by preliminary integration of the continuous equations over a finite volume and then discretization of fluxes (path from upper right to lower right and then to lower left), discrete equations satisfying conservation properties can be designed.

consecutive cells:

$$\Delta x_1 \bar{T}_1^{n+1} = \Delta x_1 \bar{T}_1^n + \Delta t_n \hat{q}_{1/2} - \Delta t_n \hat{q}_{1+1/2}$$

$$\dots$$

$$\Delta x_{i-1} \bar{T}_{i-1}^{n+1} = \Delta x_{i-1} \bar{T}_{i-1}^n + \Delta t_n \hat{q}_{i-1-1/2} - \Delta t_n \hat{q}_{i-1/2}$$

$$\Delta x_i \bar{T}_i^{n+1} = \Delta x_i \bar{T}_i^n + \Delta t_n \hat{q}_{i-1/2} - \Delta t_n \hat{q}_{i+1/2}$$

$$\Delta x_{i+1} \bar{T}_{i+1}^{n+1} = \Delta x_{i+1} \bar{T}_{i+1}^n + \Delta t_n \hat{q}_{i+1/2} - \Delta t_n \hat{q}_{i+1+1/2}$$

$$\dots$$

$$\Delta x_m \bar{T}_m^{n+1} = \Delta x_m \bar{T}_m^n + \Delta t_n \hat{q}_{m-1/2} - \Delta t_n \hat{q}_{m+1/2}.$$

Since every flux appears in two consecutive equations with opposite sign, the flux leaving a cell enters its neighbor, and there is no loss or gain of the quantity

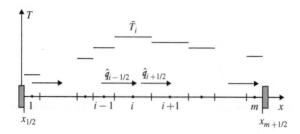

FIGURE 3.4 Within a domain covered by m finite volumes, fluxes at the interface ensure conservation of the relevant property (heat in the case of temperature) between finite volumes, since the fluxes are uniquely defined at interfaces. Fluxes therefore redistribute the property from cell to cell across the domain, without actually changing their total content, except for import and export at the end points. The finite-volume approach easily ensures both local and global conservation.

being transported across cells (heat in the case of temperature). This is an expression of *local conservation* between grid cells (Fig. 3.4).

Furthermore, summation of all equations leads to complete cancellation of the fluxes except for the very first and last ones. What we obtain is none other than the exact expression for evolution of the total quantity. In the case of temperature, this is a global heat budget:

$$\frac{d}{dt} \int_{x_{1/2}}^{x_{m+1/2}} T \, dx = q_{1/2} - q_{m+1/2} \tag{3.36}$$

which states that the total heat content of the system increases or decreases over time according to the import or export of heat at the extremities of the domain. In particular, if the domain is insulated ($q = 0$ at both boundaries), the total heat content is conserved in the numerical scheme and in the original mathematical model. Moreover, this holds irrespectively of the way by which the fluxes are evaluated from the cell-averaged temperatures, provided that they are uniquely defined at every cell interface $x_{i+1/2}$. Therefore, the finite-volume approach also ensures *global conservation*.

We will show later how advective and diffusive fluxes can be approximated using the cell-averaged discrete values $\bar{T}_i$, but will have to remember then that the conservative character of the finite-volume approach is ensured simply by using a unique flux estimate at each volume interface.

ANALYTICAL PROBLEMS

3.1. Derive the energy equation (3.10) from Eqs. (3.1) and (3.9).

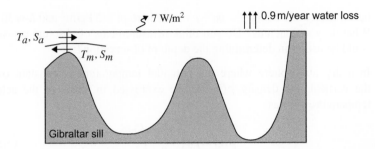

T_a, S_a

T_m, S_m

7 W/m²

0.9 m/year water loss

Gibraltar sill

FIGURE 3.5 Schematic representation of the Mediterranean basin and its exchanges across the Strait of Gibraltar in order to perform budget calculations, relating Atlantic Water characteristics T_a, S_a, and losses over the basin, to Mediterranean outflow characteristics T_m, S_m.

3.2. Derive the continuity equation (3.17) from first principles by invoking conservation of volume. (*Hint*: State that the volume in a cube of dimensions $\Delta x \Delta y \Delta z$ is unchanged as fluid is imported and exported through all six sides.)

3.3. A laboratory tank consists of a cylindrical container 30 cm in diameter, filled while at rest with 20 cm of fresh water and then spun at 30 rpm. After a state of solid-body rotation is achieved, what is the difference in water level between the rim and the center? How does this difference compares with the minimum depth at the center?

3.4. Consider the Mediterranean Sea of surface $S = 2.5 \times 10^{12}$ m² over which an average heat loss of 7 W/m² is observed. Because of an average surface water loss of 0.9 m/year (evaporation being more important than rain and river runoff combined), salinity would increase, water level would drop, and temperature would decrease, if it were not for a compensation by exchange with the Atlantic Ocean through the Strait of Gibraltar. Assuming that water, salt, and heat contents of the Mediterranean do not change over time and that exchange across Gibraltar is accomplished by a two-layer process (Fig. 3.5), establish sea-wide budgets of water, salt, and heat. Given that the Atlantic inflow is characterized by $T_a = 15.5°C$, $S_a = 36.2$, and a volume flow of 1.4 Sv (1 Sv = 10^6 m³/s), what are the outflow characteristics? Is the outflow at the surface or at the bottom?

3.5. Within the Boussinesq approximation and for negligible diffusion in (3.24), show that for an ocean at rest, density can only be a function of depth: $\rho = \rho(z)$. (*Hint*: The situation at rest is characterized by the absence of movement and temporal variations.)

3.6. Neglecting atmospheric pressure, calculate the pressure $p_0(z) = -\rho_0 g z$ at 500 m depth in the ocean. Compare it with the dynamic pressure of an ocean at rest of density $\rho = \rho_0 - \rho' e^{z/h}$, where $\rho' = 5\,\text{kg/m}^3$ and $h = 30\,\text{m}$. What do you conclude? Do you think measurements of absolute pressure could be useful in determining the depth of observation?

3.7. In a dry atmosphere where the potential temperature is constant over the vertical, the density $\rho(z)$ can be expressed in terms of the actual temperature $T(z)$ as

$$\rho(z) = \rho_0 \left(\frac{T(z)}{\theta} \right)^{C_v/R} \tag{3.37}$$

according to (3.11). This allows the equation of state (3.5) to be expressed in terms of only pressure $p(z)$ and temperature. By taking the vertical derivative of this expression and using the hydrostatic balance ($dp/dz = -\rho g$), show that the vertical temperature gradient dT/dz is constant. Of what sign is this constant?

NUMERICAL EXERCISES

3.1. Compare values of density obtained with the full equation of state for sea-water found in MATLAB™ file ies80.m with values obtained from the linearized version (3.4), for various trial values of T and S. Then, compare density differences between two different water masses, calculated again with both state equations. Finally, using numerical derivatives of the full equation of state with the help of MATLAB™ file ies80.m, can you provide a numerical estimate for the expansion coefficients α and β introduced in (3.4) for a Mediterranean water mass of $T_0 = 12.8°C$, $S_0 = 38.4$?

3.2. Generalize the finite-volume method to a two-dimensional system. In particular, what kind of fluxes do you have to define and how do you interpret them? Is local and global conservation still ensured?

3.3. Derive a conservative form of the momentum equations without friction in spherical coordinates and outline a finite-volume discretization. (*Hint*: Use volume conservation expressed in spherical coordinates and volume integrals in spherical coordinates according to $\int_V u\,dV = \int_r \int_\lambda \int_\varphi u r^2 \cos\varphi \, d\varphi \, d\lambda \, dr$.)

3.4. Using the finite-volume approach of the one-dimensional temperature evolution assuming that only advection is present, with a flow directed towards increasing $x\,(u > 0)$, discretize the average fluxes. What kind of hypothesis do you need to make to obtain an algorithm allowing you to calculate $\bar{T}_i^{n+1}$ knowing the values of $\bar{T}$ at the preceding time-step?

3.5. For flux calculations, interpolations at the cell interface are generally used. Analyze how a linear interpolation using the two neighbor points behaves compared with a cubic interpolation using four points. To do so, sample the function e^x at $x = -1.5, -0.5, 0.5, 1.5$ and interpolate at $x = 0$. Compare with the exact value. What happens if you calculated the interpolation not at the center but at $x = 3$ or $x = -3$ (extrapolation)? Redo the exercise but add an alternating error of $+0.1$ and -0.1 to the four sampled values.

Joseph Valentin Boussinesq
1842–1929

Perhaps not as well known as he deserves, Joseph Boussinesq was a French physicist who made significant contributions to the theory of hydrodynamics, vibration, light, and heat. One possible reason for this relative obscurity is the ponderous style of his writings. Among his subjects of study was hydraulics, which led to his research on turbulent flow. In 1896, the work of Osborne Reynolds (see biography at the end of the following chapter) was barely a year old when it was picked up by Boussinesq, who applied the partitioning between average and fluctuating quantities to observations of pipe and river flows. This led him to identify correctly that the cause of turbulence in those instances is friction against boundaries. This paved the way for Ludwig Prandtl's theory of boundary layers (see biography at end of Chapter 8).

It can almost be claimed that the word *turbulence* itself is owed in large part to Boussinesq. Indeed, although Osborne Reynolds spoke of "sinuous motion," Boussinesq used the more expressive phrase "écoulement tourbillonnant et tumultueux," which was reduced by one of his followers to "régime turbulent,' hence turbulence. (*Photo from Ambassade de France au Canada*)

Vilhelm Frimann Koren Bjerknes
1862–1951

Early in his career, Bjerknes became interested in applying the then-recent work of Lord Kelvin and Hermann von Helmholtz on energy and vorticity dynamics to motions in the atmosphere and ocean. He argued that the dynamics of air and water flows on geophysical scales could be framed as a problem of physics and that, given a particular state of the atmosphere, one should be able to compute its future states. In other words, weather forecasting is reducible to seek the solution of a mathematical problem. This statement, self-evident today, was quite revolutionary at the time (1904).

When in 1917, he was offered a professorship at the University of Bergen in Norway, Bjerknes founded the Bergen Geophysical Institute and began systematic efforts at developing a self-contained mathematical model for the evolution of weather based on measurable quantities. Faced by the complexity of these equations, he gradually shifted his efforts toward more qualitative aspects of weather description, and out of this work came the now familiar concepts of air masses, cyclones, and fronts.

Throughout his work, Bjerknes projected enthusiasm for his ideas and was able to attract and stimulate young scientists to follow in his footsteps, including his son Jacob Bjerknes. (*Photo courtesy of the Bergen Geophysical Institute*)

Chapter 4

Equations Governing Geophysical Flows

ABSTRACT

This chapter continues the development of the equations that form the basis of dynamical meteorology and physical oceanography. Averaging is performed over turbulent fluctuations and further simplifications are justified based on a scale analysis. In the process, some important dimensionless numbers are introduced. The need for an appropriate set of initial and boundary conditions is also explored from mathematical, physical, and numerical points of view.

4.1 REYNOLDS-AVERAGED EQUATIONS

Geophysical flows are typically in a state of turbulence, and most often we are only interested in the statistically averaged flow, leaving aside all turbulent fluctuations. To this effect and following Reynolds (1894), we decompose each variable into a mean, denoted with a set of brackets, and a fluctuation, denoted by a prime:

$$u = \langle u \rangle + u', \tag{4.1}$$

such that $\langle u' \rangle = 0$ by definition.

There are several ways to define the averaging process, some more rigorous than others, but we shall not be concerned here with those issues, preferring to think of the mean as a temporal average over rapid turbulent fluctuations, on a time interval long enough to obtain a statistically significant mean, yet short enough to retain the slower evolution of the flow under consideration. Our hypothesis is that such an intermediate time interval exists.

Quadratic expressions such as the product uv of two velocity components have the following property:

$$\langle uv \rangle = \langle \langle u \rangle \langle v \rangle \rangle + \langle \langle u \rangle v' \rangle^{=0} + \langle \langle v \rangle u' \rangle^{=0} + \langle u'v' \rangle$$
$$= \langle u \rangle \langle v \rangle + \langle u'v' \rangle \tag{4.2}$$

and similarly for $\langle uu \rangle$, $\langle uw \rangle$, $\langle up \rangle$, etc. We recognize here that the average of a product is not equal to the product of the averages. This is a double-edged sword: On one hand, it generates mathematical complications but on the other hand, it also creates interesting situations.

Our objective is to establish equations governing the mean quantities, $\langle u \rangle$, $\langle v \rangle$, $\langle w \rangle$, $\langle p \rangle$, and $\langle \rho \rangle$. Starting with the average of the x-momentum equation (3.19), we have

$$\frac{\partial \langle u \rangle}{\partial t} + \frac{\partial \langle uu \rangle}{\partial x} + \frac{\partial \langle vu \rangle}{\partial y} + \frac{\partial \langle wu \rangle}{\partial z} + f_* \langle w \rangle - f \langle v \rangle = -\frac{1}{\rho_0}\frac{\partial \langle p \rangle}{\partial x} + \nu \nabla^2 \langle u \rangle, \quad (4.3)$$

which becomes

$$\frac{\partial \langle u \rangle}{\partial t} + \frac{\partial (\langle u \rangle \langle u \rangle)}{\partial x} + \frac{\partial (\langle u \rangle \langle v \rangle)}{\partial y} + \frac{\partial (\langle u \rangle \langle w \rangle)}{\partial z} + f_* \langle w \rangle - f \langle v \rangle$$
$$= -\frac{1}{\rho_0}\frac{\partial \langle p \rangle}{\partial x} + \nu \nabla^2 \langle u \rangle - \frac{\partial \langle u'u' \rangle}{\partial x} - \frac{\partial \langle u'v' \rangle}{\partial y} - \frac{\partial \langle u'w' \rangle}{\partial z}. \quad (4.4)$$

We note that this last equation for the mean velocity looks identical to the original equation, except for the presence of three new terms at the end of the right-hand side. These terms represent the effects of the turbulent fluctuations on the mean flow. Combining these terms with corresponding frictional terms

$$\frac{\partial}{\partial x}\left(\nu\frac{\partial \langle u \rangle}{\partial x} - \langle u'u' \rangle \right), \quad \frac{\partial}{\partial y}\left(\nu\frac{\partial \langle u \rangle}{\partial y} - \langle u'v' \rangle \right), \quad \frac{\partial}{\partial z}\left(\nu\frac{\partial \langle u \rangle}{\partial z} - \langle u'w' \rangle \right)$$

indicates that the averages of velocity fluctuations add to the viscous stresses (e.g., $-\langle u'w' \rangle$ adds to $\nu \partial \langle u \rangle / \partial z$) and can therefore be considered frictional stresses caused by turbulence. To give credit to Osborne Reynolds who first decomposed the flow into mean and fluctuating components, the expressions $-\langle u'u' \rangle$, $-\langle u'v' \rangle$, and $-\langle u'w' \rangle$ are called *Reynolds stresses*. Since they do not have the same form as the viscous stresses, it can be said that the mean turbulent flow behaves as a fluid governed by a frictional law other than that of viscosity. In other words, a turbulent flow behaves as a non-Newtonian fluid.

Similar averages of the y- and z-momentum equations (3.20)–(3.22) over the turbulent fluctuations yield

$$\frac{\partial \langle v \rangle}{\partial t} + \frac{\partial (\langle u \rangle \langle v \rangle)}{\partial x} + \frac{\partial (\langle v \rangle \langle v \rangle)}{\partial y} + \frac{\partial (\langle v \rangle \langle w \rangle)}{\partial z} + f \langle u \rangle + \frac{1}{\rho_0}\frac{\partial \langle p \rangle}{\partial y}$$
$$= \frac{\partial}{\partial x}\left(\nu\frac{\partial \langle v \rangle}{\partial x} - \langle u'v' \rangle \right) + \frac{\partial}{\partial y}\left(\nu\frac{\partial \langle v \rangle}{\partial y} - \langle v'v' \rangle \right) + \frac{\partial}{\partial z}\left(\nu\frac{\partial \langle v \rangle}{\partial z} - \langle v'w' \rangle \right)$$
$$(4.5)$$

$$\frac{\partial \langle w \rangle}{\partial t} + \frac{\partial (\langle u \rangle \langle w \rangle)}{\partial x} + \frac{\partial (\langle v \rangle \langle w \rangle)}{\partial y} + \frac{\partial (\langle w \rangle \langle w \rangle)}{\partial z} - f_* \langle u \rangle + \frac{1}{\rho_0} \frac{\partial \langle p \rangle}{\partial z} = -g \langle \rho \rangle$$

$$+ \frac{\partial}{\partial x} \left(\nu \frac{\partial \langle w \rangle}{\partial x} - \langle u'w' \rangle \right) + \frac{\partial}{\partial y} \left(\nu \frac{\partial \langle w \rangle}{\partial y} - \langle v'w' \rangle \right) + \frac{\partial}{\partial z} \left(\nu \frac{\partial \langle w \rangle}{\partial z} - \langle w'w' \rangle \right).$$

$$(4.6)$$

4.2   EDDY COEFFICIENTS

Computer models of geophysical fluid systems are limited in their spatial resolution. They are therefore incapable of resolving all but the largest turbulent fluctuations, and all motions of lengths shorter than the mesh size. In one way or another, we must state something about these unresolved turbulent and subgrid scale motions in order to incorporate their aggregate effect on the larger, resolved flow. This process is called *subgrid-scale parameterization*. Here, we present the simplest of all schemes. More sophisticated parameterizations will follow in later sections of the book, particularly Chapter 14.

The primary effect of fluid turbulence and of motions at subgrid scales (small eddies and billows) is dissipation. It is therefore tempting to represent both the Reynolds stress and the effect of unresolved motions as some form of super viscosity. This is done summarily by replacing the molecular viscosity ν of the fluid by a much larger *eddy viscosity* to be defined in terms of turbulence and grid properties. This rather crude approach was first proposed by Boussinesq.

However, the parameterization recognizes one essential property: the anisotropy of the flow field and its modeling grid. Horizontal and vertical directions are treated differently by assigning two distinct eddy viscosities, $\mathcal{A}$ in the horizontal and ν_E in the vertical. Because turbulent motions and mesh size cover longer distances in the horizontal than in the vertical, $\mathcal{A}$ covers a much larger span of unresolved motions and needs to be significantly larger than ν_E. Furthermore, as they ought to depend in some elementary way on flow properties and grid dimensions, each of which may vary from place to place, eddy viscosities should be expected to exhibit some spatial variations. Returning to the basic manner by which the momentum budget was established, with stress differentials among forces on the right-hand sides, we are led to retain these eddy coefficients inside the first derivatives as follows:

$$\frac{\partial u}{\partial t} + u \frac{\partial u}{\partial x} + v \frac{\partial u}{\partial y} + w \frac{\partial u}{\partial z} + f_* w - f v$$

$$= -\frac{1}{\rho_0} \frac{\partial p}{\partial x} + \frac{\partial}{\partial x} \left(\mathcal{A} \frac{\partial u}{\partial x} \right) + \frac{\partial}{\partial y} \left(\mathcal{A} \frac{\partial u}{\partial y} \right) + \frac{\partial}{\partial z} \left(\nu_E \frac{\partial u}{\partial z} \right), \qquad (4.7a)$$

$$\frac{\partial v}{\partial t} + u \frac{\partial v}{\partial x} + v \frac{\partial v}{\partial y} + w \frac{\partial v}{\partial z} + f u$$

$$= -\frac{1}{\rho_0} \frac{\partial p}{\partial y} + \frac{\partial}{\partial x} \left(\mathcal{A} \frac{\partial v}{\partial x} \right) + \frac{\partial}{\partial y} \left(\mathcal{A} \frac{\partial v}{\partial y} \right) + \frac{\partial}{\partial z} \left(\nu_E \frac{\partial v}{\partial z} \right), \qquad (4.7b)$$

$$\frac{\partial w}{\partial t} + u\frac{\partial w}{\partial x} + v\frac{\partial w}{\partial y} + w\frac{\partial w}{\partial z} - f_* u$$

$$= -\frac{1}{\rho_0}\frac{\partial p}{\partial z} - \frac{g\rho}{\rho_0} + \frac{\partial}{\partial x}\left(\mathcal{A}\frac{\partial w}{\partial x}\right) + \frac{\partial}{\partial y}\left(\mathcal{A}\frac{\partial w}{\partial y}\right) + \frac{\partial}{\partial z}\left(\nu_E\frac{\partial w}{\partial z}\right), \quad (4.7c)$$

Since we will work exclusively with averaged equations in the rest of the book (unless otherwise specified), there is no longer any need to denote averaged quantities with brackets. Consequently, $\langle u \rangle$ has been replaced by u and similarly for all other variables.

In the energy (density) equation, heat and salt molecular diffusion needs likewise to be superseded by the dispersing effect of unresolved turbulent motions and subgrid-scale processes. Using the same horizontal eddy viscosity $\mathcal{A}$ for energy as for momentum is generally adequate, because the larger turbulent motions and subgrid processes act to disperse heat and salt as effectively as momentum. However, in the vertical, the practice is usually to distinguish dispersion of energy from that of momentum by introducing a vertical *eddy diffusivity* κ_E that differs from the vertical eddy viscosity ν_E. This difference stems from the specific turbulent behavior of each state variable and will be further discussed in Section 14.3. The energy (density) equation then becomes

$$\frac{\partial \rho}{\partial t} + u\frac{\partial \rho}{\partial x} + v\frac{\partial \rho}{\partial y} + w\frac{\partial \rho}{\partial z}$$

$$= \frac{\partial}{\partial x}\left(\mathcal{A}\frac{\partial \rho}{\partial x}\right) + \frac{\partial}{\partial y}\left(\mathcal{A}\frac{\partial \rho}{\partial y}\right) + \frac{\partial}{\partial z}\left(\kappa_E\frac{\partial \rho}{\partial z}\right). \quad (4.8)$$

The linear continuity equation is not subjected to any such adaptation and remains unchanged:

$$\frac{\partial u}{\partial x} + \frac{\partial v}{\partial y} + \frac{\partial w}{\partial z} = 0. \quad (4.9)$$

For more details on eddy viscosity and diffusivity and some schemes to make those depend on flow properties, the reader is referred to textbooks on turbulence, such as Tennekes and Lumley (1972) or Pope (2000). A widely used method to incorporate subgrid-scale processes in the horizontal eddy viscosity is that proposed by Smagorinsky (1963):

$$\mathcal{A} = \Delta x \Delta y \sqrt{\left(\frac{\partial u}{\partial x}\right)^2 + \left(\frac{\partial v}{\partial y}\right)^2 + \frac{1}{2}\left(\frac{\partial u}{\partial y} + \frac{\partial v}{\partial x}\right)^2}, \quad (4.10)$$

in which Δx and Δy are the local grid dimensions. Because the horizontal eddy viscosity is meant to represent physical processes, it ought to obey certain symmetry properties, notably invariance with respect to rotation of the coordinate system in the horizontal plane. We leave it to the reader to verify that the preceding formulation for $\mathcal{A}$ does indeed meet this requirement.

4.3   SCALES OF MOTION

Simplifications of the equations established in the preceding section are possible beyond the Boussinesq approximation and averaging over turbulent fluctuations. However, these require a preliminary discussion of orders of magnitude. Accordingly, let us introduce a scale for every variable, as we already did in a limited way in Section 1.10. By *scale*, we mean a dimensional constant of dimensions identical to that of the variable and having a numerical value representative of the values of that same variable. Table 4.1 provides illustrative scales for the variables of interest in geophysical fluid flow. Obviously, scale values do vary with every application, and the values listed in Table 4.1 are only suggestive. Even so, the conclusions drawn from the use of these particular values stand in the vast majority of cases. If doubt arises in a specific situation, the following scale analysis can always be redone.

In the construction of Table 4.1, we were careful to satisfy the criteria of geophysical fluid dynamics outlined in Sections 1.5 and 1.6,

$$T \gtrsim \frac{1}{\Omega},$$ (4.11)

for the time scale and

$$\frac{U}{L} \lesssim \Omega,$$ (4.12)

for the velocity and length scales. It is generally not required to discriminate between the two horizontal directions, and we assign the same length scale L to both coordinates and the same velocity scale U to both velocity components. However, the same cannot be said of the vertical direction. Geophysical flows are typically confined to domains that are much wider than they are thick, and the *aspect ratio* H/L is small. The atmospheric layer that determines our weather is only about 10 km thick, yet cyclones and anticyclones spread over thousands of kilometers. Similarly, ocean currents are generally confined to the upper hundred meters of the water column but extend over tens of kilometers or more, up

---

TABLE 4.1 Typical Scales of Atmospheric and Oceanic Flows

Variable	Scale	Unit	Atmospheric Value	Oceanic Value
x, y	L	m	100 km = 10^5 m	10 km = 10^4 m
z	H	m	1 km = 10^3 m	100 m = 10^2 m
t	T	s	$\geq \frac{1}{2}$ day $\simeq 4 \times 10^4$ s	≥ 1 day $\simeq 9 \times 10^4$ s
u, v	U	m/s	10 m/s	0.1 m/s
w	W	m/s		
p	P	$\mathrm{kg\,m^{-1}\,s^{-2}}$	variable	
ρ	$\Delta\rho$	kg/m^3		

to the width of the ocean basin. It follows that for large-scale motions,

$$H \ll L, \tag{4.13}$$

and we expect W to be vastly different from U.

The continuity equation in its reduced form (4.9) contains three terms of respective orders of magnitude:

$$\frac{U}{L}, \quad \frac{U}{L}, \quad \frac{W}{H}.$$

We ought to examine three cases: W/H is much less than, on the order of, or much greater than U/L. The third case must be ruled out. Indeed, if $W/H \gg U/L$, the equation reduces in first approximation to $\partial w/\partial z = 0$, which implies that w is constant in the vertical; because of a bottom somewhere, that flow must be supplied by lateral convergence (see later section 4.6.1), and we deduce that the terms $\partial u/\partial x$ and/or $\partial v/\partial y$ may not be both neglected at the same time. In sum, w must be much smaller than initially thought.

In the first case, the leading balance is two dimensional, $\partial u/\partial x + \partial v/\partial y = 0$, which implies that convergence in one horizontal direction must be compensated by divergence in the other horizontal direction. This is very possible. The intermediate case, with W/H on the order of U/L, implies a three-way balance, which is also acceptable. In summary, the vertical-velocity scale must be constrained by

$$W \lesssim \frac{H}{L} U \tag{4.14}$$

and, by virtue of Eq. (4.13),

$$W \ll U. \tag{4.15}$$

In other words, large-scale geophysical flows are shallow ($H \ll L$) and nearly two-dimensional ($W \ll U$).

Let us now consider the x-momentum equation in its Boussinesq and turbulence-averaged form (4.7a). Its various terms scale sequentially as

$$\frac{U}{T}, \quad \frac{U^2}{L}, \quad \frac{U^2}{L}, \quad \frac{WU}{H}, \quad \Omega W, \quad \Omega U, \quad \frac{P}{\rho_0 L}, \quad \frac{AU}{L^2}, \quad \frac{AU}{L^2}, \quad \frac{\nu_E U}{H^2}.$$

The previous remark immediately shows that the fifth term (ΩW) is always much smaller than the sixth (ΩU) and can be safely neglected.[1]

---

[1] Note, however, that near the equator, where f goes to zero while f_* reaches its maximum, the simplification may be invalidated. If this is the case, a reexamination of the scales is warranted. The fifth term is likely to remain much smaller than some other terms, such as the pressure gradient, but there may be instances when the f_* term must be retained. Because such a situation is exceptional, we will dispense with the f_* term here.

Because of the fundamental importance of the rotation terms in geophysical fluid dynamics, we can anticipate that the pressure-gradient term (the driving force) will scale as the Coriolis terms, that is,

$$\frac{P}{\rho_0 L} = \Omega U \quad \rightarrow \quad P = \rho_0 \Omega L U. \tag{4.16}$$

For typical geophysical flows, this dynamic pressure is much smaller than the basic hydrostatic pressure due to the weight of the fluid.

Although horizontal and vertical dissipations due to turbulent and subgrid-scale processes is retained in the equation (its last three terms), it cannot dominate the Coriolis force in geophysical flows, which ought to remain among the dominant terms. This implies

$$\frac{AU}{L^2} \quad \text{and} \quad \frac{\nu_E U}{H^2} \lesssim \Omega U. \tag{4.17}$$

Similar considerations apply to the y-momentum equation (4.7b). But the vertical momentum equation (4.7c) may be subjected to additional simplifications. Its various terms scale sequentially as

$$\frac{W}{T}, \quad \frac{UW}{L}, \quad \frac{UW}{L}, \quad \frac{W^2}{H}, \quad \Omega U, \quad \frac{P}{\rho_0 H}, \quad \frac{g\Delta\rho}{\rho_0}, \quad \frac{AW}{L^2}, \quad \frac{AW}{L^2}, \quad \frac{\nu_E W}{H^2}.$$

The first term (W/T) cannot exceed ΩW, which is itself much less than ΩU, by virtue of Eqs. (4.11) and (4.15). The next three terms are also much smaller than ΩU; this time because of Eqs. (4.12), (4.14), and (4.15). Thus, the first four terms may all be neglected compared to the fifth. But this fifth term is itself quite small. Its ratio to the first term on the right-hand side is

$$\frac{\rho_0 \Omega H U}{P} \sim \frac{H}{L},$$

which according to Eqs. (4.16) and (4.13) is much less than 1.

Finally, the last three terms are small. When W is substituted for U in Eq. (4.17), we have

$$\frac{AW}{L^2} \quad \text{and} \quad \frac{\nu_E W}{H^2} \lesssim \Omega W \ll \Omega U. \tag{4.18}$$

Thus, the last three terms on the right-hand side of the equation are much less than the fifth term on the left, which was already found to be very small. In summary, only two terms remain, and the vertical-momentum balance reduces to the simple *hydrostatic balance*:

$$0 = -\frac{1}{\rho_0}\frac{\partial p}{\partial z} - \frac{g\rho}{\rho_0}. \tag{4.19}$$

In the absence of stratification (density perturbation ρ nil), the next term in line that should be considered a possible balance to the pressure gradient

$(1/\rho_0)(\partial p/\partial z)$ is $f_* u$. However, under such balance, the vertical variation of the pressure p would be given by the vertical integration of $\rho_0 f_* u$ and its scale be $\rho_0 \Omega H U$. Since this is much less than the already established pressure scale (4.16), it is negligible, and we conclude that the vertical variation of p is very weak. In other words, p is nearly z-independent in the absence of stratification:

$$0 = -\frac{1}{\rho_0}\frac{\partial p}{\partial z}. \tag{4.20}$$

So, the hydrostatic balance (4.19) continues to hold in the limit $\rho \to 0$.

Since the pressure p is already a small perturbation to a much larger pressure, itself in hydrostatic balance, we conclude that geophysical flows tend to be fully hydrostatic even in the presence of substantial motions.[2] Looking back, we note that the main reason behind this reduction is the strong geometric disparity of geophysical flows $(H \ll L)$.

In rare instances when this disparity between horizontal and vertical scales does not exist, such as in convection plumes and short internal waves, the hydrostatic approximation ceases to hold and the vertical-momentum balance includes a three-way balance between vertical acceleration, pressure gradient, and buoyancy.

4.4 RECAPITULATION OF EQUATIONS GOVERNING GEOPHYSICAL FLOWS

The Boussinesq approximation performed in the previous chapter and the preceding developments have greatly simplified the equations. We recapitulate them here.

x-momentum:
$$\frac{\partial u}{\partial t} + u\frac{\partial u}{\partial x} + v\frac{\partial u}{\partial y} + w\frac{\partial u}{\partial z} - fv$$
$$= -\frac{1}{\rho_0}\frac{\partial p}{\partial x} + \frac{\partial}{\partial x}\left(A\frac{\partial u}{\partial x}\right) + \frac{\partial}{\partial y}\left(A\frac{\partial u}{\partial y}\right) + \frac{\partial}{\partial z}\left(\nu_E\frac{\partial u}{\partial z}\right) \tag{4.21a}$$

y-momentum:
$$\frac{\partial v}{\partial t} + u\frac{\partial v}{\partial x} + v\frac{\partial v}{\partial y} + w\frac{\partial v}{\partial z} + fu$$
$$= -\frac{1}{\rho_0}\frac{\partial p}{\partial y} + \frac{\partial}{\partial x}\left(A\frac{\partial v}{\partial x}\right) + \frac{\partial}{\partial y}\left(A\frac{\partial v}{\partial y}\right) + \frac{\partial}{\partial z}\left(\nu_E\frac{\partial v}{\partial z}\right) \tag{4.21b}$$

---

[2] According to Nebeker (1995, page 51), the scientist deserving credit for the hydrostatic balance in geophysical flows is Alexis Clairaut (1713–1765).

$$z\text{-momentum: } 0 = -\frac{\partial p}{\partial z} - \rho g \tag{4.21c}$$

$$\text{continuity: } \frac{\partial u}{\partial x} + \frac{\partial v}{\partial y} + \frac{\partial w}{\partial z} = 0 \tag{4.21d}$$

$$\text{energy: } \frac{\partial \rho}{\partial t} + u\frac{\partial \rho}{\partial x} + v\frac{\partial \rho}{\partial y} + w\frac{\partial \rho}{\partial z}$$

$$= \frac{\partial}{\partial x}\left(\mathcal{A}\frac{\partial \rho}{\partial x}\right) + \frac{\partial}{\partial y}\left(\mathcal{A}\frac{\partial \rho}{\partial y}\right) + \frac{\partial}{\partial z}\left(\kappa_E\frac{\partial \rho}{\partial z}\right), \tag{4.21e}$$

where the reference density ρ_0 and the gravitational acceleration g are constant coefficients, the Coriolis parameter $f = 2\Omega\sin\varphi$ is dependent on latitude or taken as a constant, and the eddy viscosity and diffusivity coefficients $\mathcal{A}$, ν_E, and κ_E may be taken as constants or functions of flow variables and grid parameters. These five equations for the five variables u, v, w, p, and ρ form a closed set of equations, the cornerstone of geophysical fluid dynamics, sometimes called *primitive equations*.

Using the continuity equation (4.21d), the horizontal-momentum and density equations can be written in *conservative form*:

$$\frac{\partial u}{\partial t} + \frac{\partial (uu)}{\partial x} + \frac{\partial (vu)}{\partial y} + \frac{\partial (wu)}{\partial z} - fv$$

$$= -\frac{1}{\rho_0}\frac{\partial p}{\partial x} + \frac{\partial}{\partial x}\left(\mathcal{A}\frac{\partial u}{\partial x}\right) + \frac{\partial}{\partial y}\left(\mathcal{A}\frac{\partial u}{\partial y}\right) + \frac{\partial}{\partial z}\left(\nu_E\frac{\partial u}{\partial z}\right) \tag{4.22a}$$

$$\frac{\partial v}{\partial t} + \frac{\partial (uv)}{\partial x} + \frac{\partial (vv)}{\partial y} + \frac{\partial (wv)}{\partial z} + fu$$

$$= -\frac{1}{\rho_0}\frac{\partial p}{\partial y} + \frac{\partial}{\partial x}\left(\mathcal{A}\frac{\partial v}{\partial x}\right) + \frac{\partial}{\partial y}\left(\mathcal{A}\frac{\partial v}{\partial y}\right) + \frac{\partial}{\partial z}\left(\nu_E\frac{\partial v}{\partial z}\right) \tag{4.22b}$$

$$\frac{\partial \rho}{\partial t} + \frac{\partial (u\rho)}{\partial x} + \frac{\partial (v\rho)}{\partial y} + \frac{\partial (w\rho)}{\partial z}$$

$$= \frac{\partial}{\partial x}\left(\mathcal{A}\frac{\partial \rho}{\partial x}\right) + \frac{\partial}{\partial y}\left(\mathcal{A}\frac{\partial \rho}{\partial y}\right) + \frac{\partial}{\partial z}\left(\kappa_E\frac{\partial \rho}{\partial z}\right), \tag{4.22c}$$

These will be found useful in numerical discretization.

4.5 IMPORTANT DIMENSIONLESS NUMBERS

The scaling analysis of Section 4.3 was developed to justify the neglect of some small terms. But this does not necessarily imply that the remaining terms are equally large. We now wish to estimate the relative sizes of those terms that have been retained.

The terms of the horizontal momentum equations in their last form (4.21a) and (4.21b) scale sequentially as

$$\frac{U}{T}, \quad \frac{U^2}{L}, \quad \frac{U^2}{L}, \quad \frac{WU}{H}, \quad \Omega U, \quad \frac{P}{\rho_0 L}, \quad \frac{AU}{L^2}, \quad \frac{\nu_E U}{H^2}.$$

By definition, geophysical fluid dynamics treats those motions in which rotation is an important factor. Thus, the term ΩU is central to the preceding sequence. A division by ΩU, to measure the importance of all other terms relative to the Coriolis term, yields the following sequence of dimensionless ratios:

$$\frac{1}{\Omega T}, \quad \frac{U}{\Omega L}, \quad \frac{U}{\Omega L}, \quad \frac{WL}{UH} \cdot \frac{U}{\Omega L}, \quad 1, \quad \frac{P}{\rho_0 \Omega L U}, \quad \frac{A}{\Omega L^2}, \quad \frac{\nu_E}{\Omega H^2}.$$

The first ratio,

$$Ro_T = \frac{1}{\Omega T}, \tag{4.23}$$

is called the *temporal Rossby number*. It compares the local time rate of change of the velocity to the Coriolis force and is on the order of unity or less as has been repeatedly stated, see Eq. (4.11). The next number,

$$Ro = \frac{U}{\Omega L}, \tag{4.24}$$

which compares advection to Coriolis force, is called the *Rossby number*[3] and is fundamental in geophysical fluid dynamics. Like its temporal analogue Ro_T, it is at most on the order of unity by virtue of Eq. (4.12). As a general rule, the characteristics of geophysical flows vary greatly with the values of the Rossby numbers.

The next number is the product of the Rossby number by WL/UH, which is on the order of 1 or less by virtue of Eq. (4.14). It will be shown in Section 11.5 that the ratio WL/UH is generally on the order of the Rossby number itself. The next ratio, $P/\rho_0 \Omega L U$, is on the order of unity by virtue of Eq. (4.16).

The last two ratios measure the relative importance of horizontal and vertical frictions. Of the two, only the latter bears a name:

$$Ek = \frac{\nu_E}{\Omega H^2}, \tag{4.25}$$

is called the *Ekman number*. For geophysical flows, this number is small. For example, with an eddy viscosity ν_E as large as $10^{-2} m^2/s$, $\Omega = 7.3 \times 10^{-5}\, s^{-1}$ and $H = 100$ m, $Ek = 1.4 \times 10^{-2}$. The Ekman number is even smaller in laboratory experiments where the viscosity reverts to its molecular value and the height scale H is much more modest. [Typical experimental values are

---

[3] See biographic note at the end of this chapter.

$\Omega = 4$ s^{-1}, $H = 20$ cm, and ν(water) $= 10^{-6}$ m^2/s, yielding $Ek = 6 \times 10^{-6}$.]
Although the Ekman number is small, indicating that the dissipative terms in
the momentum equation may be negligible, these need to be retained. The reason will become clear in Chapter 8, when it is shown that vertical friction creates
a very important boundary layer.

In nonrotating fluid dynamics, it is customary to compare inertial and frictional forces by defining the *Reynolds number*, Re. In the preceding scaling,
inertial and frictional forces were not compared to each other, but each was
instead compared to the Coriolis force, yielding the Rossby and Ekman numbers. There exists a simple relationship between the three numbers and the
aspect ratio H/L:

$$Re = \frac{UL}{\nu_E} = \frac{U}{\Omega L} \cdot \frac{\Omega H^2}{\nu_E} \cdot \frac{L^2}{H^2} = \frac{Ro}{Ek} \left(\frac{L}{H} \right)^2. \tag{4.26}$$

Since the Rossby number is on the order of unity or slightly less, but the
Ekman number and the aspect ratio H/L are both much smaller than unity,
the Reynolds number of geophysical flows is extremely large, even after the
molecular viscosity has been replaced by a much larger eddy viscosity.

With Eq. (4.16), the two terms in the hydrostatic equation (4.21c) scale,
respectively, as

$$\frac{P}{H}, \quad g \Delta \rho$$

and the ratio of the latter over the former is

$$\frac{gH\Delta\rho}{P} = \frac{gH\Delta\rho}{\rho_0 \Omega L U} = \frac{U}{\Omega L} \cdot \frac{gH\Delta\rho}{\rho_0 U^2} = Ro \cdot \frac{gH\Delta\rho}{\rho_0 U^2}.$$

This leads to the additional dimensionless ratio

$$Ri = \frac{gH\Delta\rho}{\rho_0 U^2}, \tag{4.27}$$

which we already encountered in Section 1.6. It is called the *Richardson number*.[4] For geophysical flows, this number may be much less than, on the order
of, or much greater than unity, depending on whether stratification effects are
negligible, important, or dominant.

4.6 BOUNDARY CONDITIONS

The equations of Section 4.4 governing geophysical flows form a closed set of
equations, with the number of unknown functions being equal to the number

---

[4] See biography at the end of Chapter 14

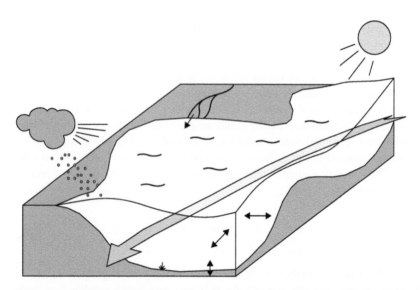

FIGURE 4.1 Schematic representation of possible exchanges between a coastal system under investigation and the surrounding environment. Boundary conditions must specify the influence of this outside world on the evolution within the domain. Exchanges may take place at the air–sea interface, in bottom layers, along coasts, and/or at any other boundary of the domain.

of available independent equations. However, the solution of those equations is uniquely defined only when additional specifications are provided. Those auxiliary conditions concern information on the initial state and geographical boundaries of the system (Fig. 4.1).

Because the governing equations (4.21) contain first-order time derivatives of u, v, and ρ, *initial conditions* are required, one for each of these three-dimensional fields. Because the respective equations, (4.21a), (4.21b), and (4.21e), provide *tendencies* for these variables in order to calculate future *values*, it is necessary to specify from where to start. The variables for which such initial conditions are required are called *state variables*. The remaining variables w and p, which have no time derivative in the equations, are called *diagnostic* variables, that is, they are variables that can be determined at any moment from the knowledge of the other variables at the same moment. Note that if a nonhy-drostatic formalism is retained, the time derivative of the vertical velocity arises [see (4.7c)], and w passes from being a diagnostic variable to a state variable, and an initial condition becomes necessary for it, too.

The determination of pressure needs special care depending on whether the hydrostatic approximation is applied and on the manner in which sea surface height is modeled. Since the pressure gradient is a major force in geophysical flows, the handling of pressure is a central question in the development of GFD models. This point deserves a detailed analysis, which we postpone to Section 7.6.

The conditions to impose at *spatial* boundaries of the domain are more difficult to ascertain than initial conditions. The mathematical theory of partial differential equations teaches us that the number and type of required boundary conditions depend on the nature of the partial differential equations. Standard classification (e.g., Durran, 1999) of second-order partial differential equations makes the distinction between *hyperbolic, parabolic,* and *elliptic* equations. This classification is based on the concept of *characteristics*, which are lines along which information propagates. The geometry of these lines constrains where information is propagated from the boundary into the domain or from the domain outward across the boundary and therefore prescribes along which portion of the domain's boundary information needs to be specified in order to define uniquely the solution within the domain.

A major problem with the GFD governing equations is that their classification cannot be established once and for all. First, the coupled set of equations (4.21) is more complicated than a single second-order equation for which standard classification can be performed; second, the equation type can change with the solution itself. Indeed, propagation of information is mostly accomplished by a combination of flow advection and wave propagation, and these may at various times leave and enter through the same boundary segment. Thus, the number and type of required boundary conditions is susceptible to change over time with the solution of the problem, which is obviously not known a priori. It is far from a trivial task to establish the mathematically correct set of boundary conditions, and the reader is referred to specialized literature for further information (e.g., Blayo & Debreu, 2005; Durran, 1999). The imposition of boundary conditions during analytical studies in this book will be guided by purely physical arguments, and the well-behaved nature of the subsequent solution will serve as a posteriori verification.

For the many situations when no analytical solution is available, not only is a posteriori verification out of the question but the problem is further complicated by the fact that numerical discretization solves modified equations with truncation errors, rather than the original equations. The equations may demand fewer or more boundary and initial conditions. If the numerical scheme asks for more conditions than those provided by the original equations, these conditions must be related to the truncation error in such a way that they disappear when the grid size (or time step) vanishes: We demand that all boundary and initial conditions be consistent.

Let us, for example, revisit the initialization problem of the leapfrog scheme from this point of view. As we have seen (Section 2.9), the leapfrog discretization $\partial u / \partial t = Q \rightarrow \tilde{u}^{n+1} = \tilde{u}^{n-1} + 2\Delta t Q^n$ needs two values, $\tilde{u}^0$ and $\tilde{u}^1$, to start the time stepping. However, the original problem indicates that only one initial condition, $\tilde{u}^0$, may be imposed, the value of which is dictated by the physics of the problem. The second condition, $\tilde{u}^1$, must then be such that its influence disappears in the limit $\Delta t \rightarrow 0$. This will be the case with the explicit Euler scheme $\tilde{u}^1 = \tilde{u}^0 + \Delta t Q$ [where $Q(t^0, \tilde{u}^0)$ stands for the other terms in the equation at time t^0]. Indeed, $\tilde{u}^1$ tends to the actual initial value $\tilde{u}^0$ and the first leapfrog step yields $\tilde{u}^2 = \tilde{u}^0 + 2\Delta t Q(t^1, \tilde{u}^0 + \mathcal{O}(\Delta t))$, which is consistent with a finite difference over a $2\Delta t$ time step.

Leaving for later sections the complexity of the additional conditions that may be required by virtue of the discretization schemes, the following sections present the boundary conditions that are most commonly encountered in GFD problems. They stem from basic physical requirements.

4.6.1 Kinematic Conditions

A most important condition, independent of any physical property or subgrid-scale parameterization, is that air and water flows do not penetrate land.[5] To translate this impermeability requirement into a mathematical boundary condition, we simply express that the velocity must be tangent to the land boundary, that is, the gradient vector of the boundary surface and the velocity vector are orthogonal to each other.

Consider the solid bottom of the domain. With this boundary defined as $z - b(x, y) = 0$, the gradient vector is given by $[\partial(z-b)/\partial x,\ \partial(z-b)/\partial y,\ \partial(z-b)/\partial z] = [-\partial b/\partial x,\ -\partial b/\partial y,\ 1]$, the boundary condition is

$$w = u\frac{\partial b}{\partial x} + v\frac{\partial b}{\partial y} \quad \text{at the bottom.} \tag{4.28}$$

We can interpret this condition in terms of a fluid budget at the bottom (Fig. 4.2) or alternatively as the condition that the bottom is a material surface of the fluid, not crossed by the flow and immobile. Expressing that the bottom is a material surface indeed demands

$$\frac{\mathrm{d}}{\mathrm{d}t}(z - b) = 0, \tag{4.29}$$

which is equivalent to Eq. (4.28) since $\mathrm{d}z/\mathrm{d}t = w$ and $\partial b/\partial t = 0$.

At a *free surface*, the situation is similar to the bottom except for the fact that the boundary is moving with the fluid. If we exclude overturning waves, the position of the surface is uniquely defined at every horizontal point by its vertical position η (Fig. 4.3), and $z - \eta = 0$ is the equation of the boundary. We then express that it is a material surface[6]:

$$\frac{\mathrm{d}}{\mathrm{d}t}(z - \eta) = 0 \quad \text{at the free surface} \tag{4.30}$$

---

[5] There is no appreciable penetration of land by water and air at geophysical scales. For ground flows, known to have a strong impact on geochemical behaviors of coastal systems, an appropriate flux can always be imposed if necessary.

[6] Exceptions are evaporation and precipitation at the air–sea interface. When important, these may be accommodated in a straightforward manner.

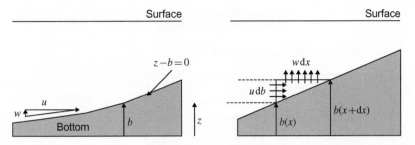

FIGURE 4.2 Notation and two physical interpretations of the bottom boundary condition illustrated here in a (x, z) plane for a topography independent of y. The impermeability of the bottom imposes that the velocity be tangent to the bottom defined by $z - b = 0$. In terms of the fluid budget, which can be extended to a finite-volume approach, expressing that the horizontal inflow matches the vertical outflow requires $u\,(b(x + dx) - b(x)) = w\,dx$, which for $dx \to 0$ leads to Eq. (4.28). Note that the velocity ratio w/u is equal to the topographic slope db/dx, which scales like the ratio of vertical to horizontal length scales, i.e., the aspect ratio.

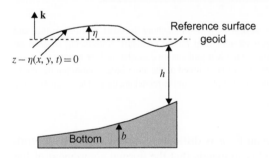

FIGURE 4.3 Notation for the surface boundary condition. Expressing impermeability of the moving surface $z = \eta$ results in boundary condition (4.31). The elevation of the sea surface height η is exaggerated compared to h for the purpose of illustration.

and obtain the surface boundary condition

$$w = \frac{\partial \eta}{\partial t} + u\frac{\partial \eta}{\partial x} + v\frac{\partial \eta}{\partial y} \quad \text{at} \quad z = \eta. \tag{4.31}$$

Particularly simple cases are those of a flat bottom and of a free surface of which the vertical displacements are neglected (such as small water waves on the surface of the deep sea)—called the *rigid-lid* approximation, which will be scrutinized in Section 7.6. In such cases, the vertical velocity is simply zero at the corresponding boundary.

A difficulty with the free surface boundary arises because the boundary condition is imposed at $z = \eta$, that is, at a location changing over time, depending on the flow itself. Such a problem is called a *moving boundary problem*, a topic which is a discipline unto itself in computational fluid dynamics (CFD) (e.g., Crank, 1987).

In oceanic models, lateral walls are introduced in addition to bottom and top boundaries so that the water depth remains nonzero all the way to the edge (Fig. 4.4). This is because watering and dewatering of land that would otherwise

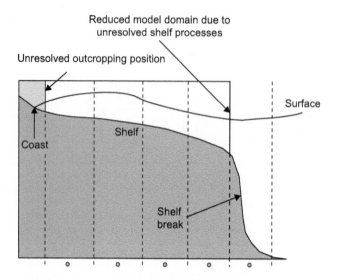

FIGURE 4.4 Vertical section across an oceanic domain reaching the coast. Besides surface and bottom boundaries, the coast introduces an additional *lateral* boundary. Introducing an artificial vertical wall is necessary because a fixed numerical grid cannot describe well the exact position of the water's edge. Occasionally, a vertical wall is assumed at the shelf break, removing the entire shelf area from the domain, because the reduced physics of the model are incapable of representing some processes on the shelf.

occur at the outcrop of the ocean floor is difficult to model with a fixed grid. At a vertical wall, impermeability demands that the normal component of the horizontal velocity be zero.

4.6.2 Dynamic Conditions

The previous impermeability conditions are purely kinematic, involving only velocity components. Dynamical conditions, implicating forces, are sometimes also necessary, for example, when requiring continuity of pressure at the air–sea interface.

Ignoring the effect of surface tension, which is important only for very short water waves (*capillary* waves, with wavelengths no longer than a few centimeters), the pressure p_{atm} exerted by the atmosphere on the sea must equal the total pressure p_{sea} exerted by the ocean onto the atmosphere:

$$p_{atm} = p_{sea} \quad \text{at air–sea interface.} \tag{4.32}$$

If the sea surface elevation is η and pressure is hydrostatic below, it follows that continuity of pressure at the actual surface $z = \eta$ implies

$$p_{sea}(z=0) = p_{atm \text{ at sea level}} + \rho_0 g \eta \tag{4.33}$$

at the more convenient reference sea level $z = 0$.

Another dynamical boundary condition depends on whether the fluid is considered inviscid or viscous. In reality, all fluids are subject to internal friction so that, in principle, a fluid particle next to fixed boundary must adhere to that boundary and its velocity be zero. However, the distance over which the velocity falls to zero near a boundary is usually short because viscosity is weak. This short distance restricts the influence of friction to a narrow band of fluid along the boundary, called a *boundary layer*. If the extent of this boundary layer is negligible compared to the length scale of interest, and generally it is, it is permissible to neglect friction altogether in the momentum equations. In this case, slip between the fluid and the boundary must be allowed, and the only boundary condition to be applied is the impermeability condition.

However, if viscosity is taken into account, zero velocity must be imposed at a fixed boundary, whereas along a moving boundary between two fluids, continuity of both velocity and tangential stress is required. From the oceanic point of view, this requires

$$\rho_0 \nu_E \left(\frac{\partial u}{\partial z} \right) \bigg|_{\text{at surface}} = \tau^x, \quad \rho_0 \nu_E \left(\frac{\partial v}{\partial z} \right) \bigg|_{\text{at surface}} = \tau^y \qquad (4.34)$$

where τ^x and τ^y are the components of the wind stress exerted by the atmosphere onto the sea. These are usually taken as quadratic functions of the wind velocity $\mathbf{u}_{10}$ 10 meters above the sea and parameterized using a drag coefficient:

$$\tau^x = C_d \, \rho_{\text{air}} \, U_{10} u_{10}, \quad \tau^y = C_d \, \rho_{\text{air}} \, U_{10} v_{10}, \qquad (4.35)$$

where u_{10} and v_{10} are the x and y components of the wind vector $\mathbf{u}_{10}$, $U_{10} = \sqrt{u_{10}^2 + v_{10}^2}$ is the wind speed, and C_d is a drag coefficient with approximate value of 0.0015 for wind over the sea.

Finally, an edge of the model may be an *open boundary*, by which we mean that the model domain is terminated at some location that cuts across a broader natural domain. Such a situation arises because computer resources or data availability restrict the attention to a portion of a broader system. Examples are regional meteorological models and coastal ocean models (Fig. 4.5). Ideally, the influence of the outside system onto the system of interest should be specified along the open boundary, but this is most often impossible in practice, for the obvious reason that the unmodeled part of the system is not known. However, certain conditions can be applied. For example, waves may be allowed to exit but not enter through the open boundary, or flow properties may be specified where the flow enters the domain but not where it leaves the domain. In oceanic tidal models, the sea surface may be imposed as a periodic function of time.

With increased computer power over the last decade, it has become common nowadays to *nest* models into one another, that is, the regionally limited model of interest is embedded in another model of lower spatial resolution but larger size, which itself may be embedded in a yet larger model of yet lower resolution, all the way to a model that has no open boundary (entire ocean basin or globe for the atmosphere). A good example is regional weather forecasting over a

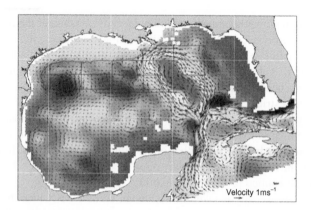

FIGURE 4.5   Open boundaries are common in regional modeling. Conditions at open boundaries are generally difficult to impose. In particular, the nature of the condition depends on whether the flow enters the domain (carrying unknown information from the exterior) or leaves it (exporting known information). (*Courtesy of the HYCOM Consortium on Data-Assimilative Modeling*). A color version can be found online at http://booksite.academicpress.com/9780120887590/

particular country: A grid covering this country and a few surrounding areas is nested into a grid that covers the continent, which itself is nested inside a grid that covers the entire globe.

4.6.3   Heat, Salt, and Tracer Boundary Conditions

For equations similar to those governing the evolution of temperature, salt, or density, that is, including advection and diffusion terms, we have the choice of imposing the value of the variable, its gradient, or a mixture of both. Prescribing the value of the variable (*Dirichlet condition*) is natural in situations where it is known from observations (sea surface temperature from satellite data, for example). Setting the gradient (*Neumann condition*) is done to impose the diffusive flux of a quantity (e.g., heat flux) and is therefore often associated with the prescription of turbulent air–sea exchanges. A mixed condition (*Cauchy condition, Robin condition*) is typically used to prescribe a total, advective plus diffusive, flux. For a 1D heat flux, for example, one sets the value of $uT - \kappa_T \partial T/\partial x$ at the boundary. For an insulating boundary, this flux is simply zero.

To choose the value of the variable or its gradient at the boundary, either observations are invoked or exchange laws prescribed. The most complex exchange laws are those for the air–sea interface, which involve calculation of fluxes depending on the sea surface water temperature T_{sea} (often called SST), air temperature T_{air}, wind speed $\mathbf{u}_{10}$ at 10 m above the sea, cloudiness, moisture, etc, Formally,

$$-\kappa_T \left.\frac{\partial T}{\partial z}\right|_{z=\eta} = F(T_{\mathrm{sea}}, T_{\mathrm{air}}, \mathbf{u}_{10}, \text{cloudiness, moisture, ...}). \qquad (4.36)$$

For heat fluxes, imposing the condition at $z = 0$ rather than at the actual position $z = \eta$ of the sea surface introduces an error much below the error in the heat flux estimate itself and is a welcomed simplification.

If the density equation is used as a combination of both salinity and temperature equations by invoking the linearized state equation, $\rho = -\alpha T + \beta S$, and if it can be reasonably assumed that all are dispersed with the same turbulent diffusivity, the boundary condition on density can be formulated as a weighted sum of prescribed temperature and salt fluxes:

$$\kappa_E \frac{\partial \rho}{\partial z} = -\alpha \kappa_E \frac{\partial T}{\partial z} + \beta \kappa_E \frac{\partial S}{\partial z}. \tag{4.37}$$

For any tracer (a quantity advected and dispersed by the flow), a condition similar to those on temperature and salinity can be imposed, and in particular, a zero total flux is common when there is no tracer input at the boundary.

4.7 NUMERICAL IMPLEMENTATION OF BOUNDARY CONDITIONS

Once mathematical boundary conditions are specified and values assigned at the boundaries, we can tackle the task of implementing the boundary condition numerically. We illustrate the process again with temperature as an example.

In addition to nodes forming the grid covering the domain being modeled, other nodes may be placed exactly at or slightly beyond the boundaries (Fig. 4.6). These additional nodes are introduced to facilitate the implementation of the boundary condition. If the condition is to specify the value T_b of the numerical variable $\tilde{T}$, it is most natural to place a node at the boundary (Fig. 4.6 right side) so that

$$\tilde{T}_m = T_b \tag{4.38}$$

requires no interpolation and forms an exact implementation.

If the boundary condition is in the form of a flux, it is more practical to have two grid nodes straddling the boundary, with one slightly outside the domain and the other slightly inside (Fig. 4.6 left side). In this manner, the derivative of the variable is more precisely formulated at the location of the boundary. With the index notation of Fig. 4.6,

$$\frac{\tilde{T}_2 - \tilde{T}_1}{\Delta x} \simeq \left.\frac{\partial T}{\partial x}\right|_{x_b} + \frac{\Delta x^2}{24} \left.\frac{\partial^3 T}{\partial x^3}\right|_{x_b} \tag{4.39}$$

yields a second-order approximation, and the flux boundary condition $-\kappa_T(\partial T/\partial x) = q_b$ turns into

$$\tilde{T}_1 = \tilde{T}_2 + \Delta x \frac{q_b}{\kappa_T}. \tag{4.40}$$

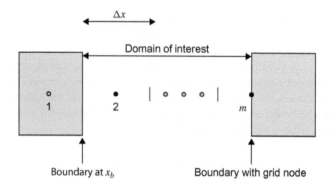

FIGURE 4.6   Grid nodes cover the interior of the domain of interest. Additional nodes may be placed *beyond* a boundary as illustrated on the left side or placed *on* the boundary as illustrated on the right. The numerical implementation of the boundary condition depends on the arrangement selected.

However, there are cases when the situation is less ideal. This occurs when a total, advective plus diffusive, flux boundary condition is specified $(uT - \kappa_T(\partial T/\partial x) = q_b)$. Either the ending node is *at* the boundary, complicating the discretization of the derivative, or it is placed *beyond* the boundary, and the value of T must be extrapolated. In the latter case, extrapolation is performed with second-order accuracy,

$$\frac{\tilde{T}_1 + \tilde{T}_2}{2} \simeq T(x_b) + \frac{\Delta x^2}{8} \left. \frac{\partial^2 T}{\partial x^2} \right|_{x_b} , \qquad (4.41)$$

and the total flux condition becomes

$$u_b \frac{\tilde{T}_1 + \tilde{T}_2}{2} - \kappa_T \frac{\tilde{T}_2 - \tilde{T}_1}{\Delta x} = q_b \qquad (4.42)$$

yielding the following condition on the end value $\tilde{T}_1$:

$$\tilde{T}_1 = \frac{2 q_b \Delta x + (2\kappa_T - u_b \Delta x)\tilde{T}_2}{2\kappa_T + u_b \Delta x}. \qquad (4.43)$$

In the former case, when the ending grid node lies exactly at the boundary, the straightfoward difference

$$\frac{\partial T}{\partial x} \simeq \frac{\tilde{T}_m - \tilde{T}_{m-1}}{\Delta x} \qquad (4.44)$$

provides only first-order accuracy at point x_m, and to recover second-order accuracy with this node placement, we need a numerical stencil that extends further into the domain (see Numerical Exercise 4.8). Therefore, to implement a flux condition, the preferred placement of the ending node is half a grid step beyond the boundary. With this configuration, the accuracy is greater than with the ending point placed at the boundary itself. The same conclusion is reached for the finite-volume approach, since imposing a flux condition consists of replacing the flux calculation at the boundary by the imposed value. We immediately realize that in this case, the natural placement of the boundary is at the interface between grid points because it is the location where fluxes are calculated in the finite-volume approach.

The question that comes to mind at this point is whether or not the level of truncation error in the boundary-condition implementation is adequate. To answer the question, we have to compare this truncation error to other errors, particularly the truncation error within the domain. Since there is no advantage in having a more accurate method at the relatively few boundary points than at the many interior points, the sensible choice is to use the same truncation order at the boundary as within the domain. The model then possesses a uniform level of approximation. Sometimes, however, a lower order near the boundary may be tolerated because there are many fewer boundary points than interior points, and a locally higher error level should not penalize the overall accuracy of the solution. In the limit of $\Delta x \to 0$, the ratio of boundary points to the total number of grid points tends to zero, and the effect of less-accurate approximations at the boundaries disappears.

In Eq. (4.43), we used the boundary condition to calculate a value at a point outside of the domain so that when applying the numerical scheme at the first interior point, the boundary condition is automatically satisfied. The same approach can also be used to implement the artificial boundary conditions that are sometimes required by the numerical scheme. Consider, for example, the fourth-order discretization (1.26) now applied to spatial derivatives in the domain interior coupled with the need to impose a single boundary condition at

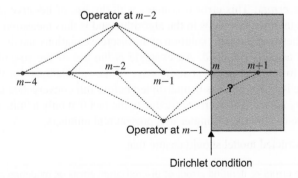

FIGURE 4.7   An operator spanning two points on each side of the calculation point can be applied only up to $m-2$ if a single Dirichlet condition is prescribed. When applying the same operator at $m-1$, we face the problem that the value at $m+1$ does not exist.

x_m of Dirichlet type. The discrete operator in the interior

$$\left.\frac{\partial T}{\partial x}\right|_{x_i} \simeq \frac{4}{3}\left(\frac{\tilde{T}_{i+1}-\tilde{T}_{i-1}}{2\Delta x}\right) - \frac{1}{3}\left(\frac{\tilde{T}_{i+2}-\tilde{T}_{i-2}}{4\Delta x}\right) \tag{4.45}$$

can be applied up to $i=m-2$. At $i=m-1$, the formula can no longer be applied, unless we provide a value at a virtual point $\tilde{T}_{m+1}$ (Fig. 4.7). This can be accomplished by requiring that a skewed fourth-order discretization near the boundary have the same effect as the centered version using the virtual value.

4.8   ACCURACY AND ERRORS

Errors in a numerical model can be of several types. Following Ferziger and Perić (1999), we classify them according to their origin.

- *Modeling errors*: This error is caused by the imperfections of the mathematical model in representing the physical system. It is thus the difference between the evolution of the real system and that of the exact solution of its mathematical representation. Earlier in this chapter, we introduced simplifications to the equations and added parameterizations of unresolved processes, which all introduce errors of representation. Furthermore, even if the model formulation had been ideal, coefficients remain imperfectly known. Uncertainties in the accompanying boundary conditions also contribute to modeling errors.
- *Discretization errors*: This error is introduced when the original equations are approximated to transform them into a computer code. It is thus the difference between the exact solution of the continuous problem and the exact numerical solution of the discretized equations. Examples are the replacement of derivatives by finite differences and the use of guesses in predictor-corrector schemes.
- *Iteration errors*: This error originates with the use of iterative methods to perform intermediate steps in the algorithm and is thus measured as the difference between the exact solution of the discrete equations and the numerical solution actually obtained. An example is the use of the so-called Jacobi method to invert a matrix at some stage of the calculations: for the sake of time, the iterative process is interrupted before full convergence is reached.
- *Rounding errors*: These errors are due to the fact that only a finite number of digits are used in the computer to represent real numbers.

A well constructed model should ensure that

> rounding errors $\ll$ iteration errors $\ll$ discretization errors $\ll$ modeling errors.

The order of these inequalities is easily understood: If the discretization error were larger than the modeling error, there would be no way to tell whether the mathematical model is an adequate approximation of the physical system we are trying to describe. If the iteration error were larger than the discretization error, the claim could not be made that the algorithm generates a numerical solution that satisfies the discretized equations, etc.

In the following, we will deal neither with rounding errors (generally controlled by appropriate compiler options, loop arrangements, and double-precision instructions), nor with iteration errors (generally controlled by sensitivity analysis or a priori knowledge of acceptable error levels for the convergence of the iterations). Modeling errors are discussed when performing scale analysis and additional modeling hypotheses or simplifications (see, e.g., the Boussinesq and hydrostatic approximations) so that we may restrict our attention here to the discretization error associated with the transformation of a continuous mathematical model into a discrete numerical scheme.

The concepts of consistency, convergence, and stability mentioned in Chapter 1 only provide information on the discretization error behavior when Δt tends to zero. In practice, however, time steps (and spatial steps as well) are never tending toward zero but are kept at fixed values, and the question arises about how accurate is the numerical solution compared to the exact solution. In that case, convergence is only marginally interesting, and even inconsistent schemes, if clever, may be able to provide results that cause lower actual errors than consistent and convergent methods.

By definition, the discretization error ϵ_u on a variable u is the difference between the exact numerical solution $\tilde{u}$ of the discretized equation and the mathematical solution u of the continuous equation:

$$\epsilon_u = \tilde{u} - u. \tag{4.46}$$

4.8.1   Discretization Error Estimates

In the case of explicit discretization (2.24) of inertial oscillations, we can obtain differential equations for the errors by subtracting the modified equations (2.28) from the exact continuous equation (2.23), to the leading order:

$$\frac{d\epsilon_u}{dt} - f\epsilon_v = f^2 \frac{\Delta t}{2} \tilde{u} + \mathcal{O}(\Delta t^2)$$

$$\frac{d\epsilon_v}{dt} + f\epsilon_u = f^2 \frac{\Delta t}{2} \tilde{v} + \mathcal{O}(\Delta t^2).$$

Obviously, we are not going to solve these equations to calculate the error because it would be tantamount to solving the exact problem directly. However, what we notice is that the error equations have source terms on the order of Δt (which vanish as $\Delta t \to 0$ because the scheme is consistent) and we anticipate that these will give rise to a proportional solution for ϵ_u and ϵ_v. The truncation error of the solution should therefore be of first order:

$$\epsilon_u = \mathcal{O}(\Delta t) \sim \frac{f\Delta t}{2} \|\tilde{\mathbf{u}}\|. \tag{4.47}$$

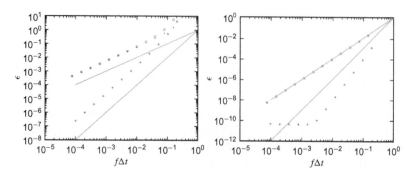

FIGURE 4.8 Relative discretization error $\epsilon = \epsilon_u / \|\tilde{u}\|$ as a function of the dimensionless variable $f\Delta t$ in the case of inertial oscillations. The log–log graphs show the real errors (dots) and estimated values of the error (circles) for an explicit scheme (left panel) and semiexplicit scheme (right panel). The slope of the theoretical convergence rates ($m=1$ on the left panel and $m=2$ on the right panel) are shown, as well as the next order $m+1$. Actual errors after a Richardson extrapolation (crosses) prove that the order is increased by 1 after extrapolation.

We can verify that the actual error is indeed divided by a factor 2 when the time step is halved (Fig. 4.8).

This is to be expected, since the equivalence theorem also states that for a linear problem, the numerical solution and its truncation error share the same order, say m. The difficulty with this approach is that for nonlinear problems, no guarantee can be made that this property continues to hold or that the actual error can be estimated by inspection of the modified equation.

To quantify the discretization error in nonlinear systems, we can resort to a sensitivity analysis. Suppose the leading error of the solution is

$$\epsilon_u = \tilde{u} - u = a\Delta t^m, \tag{4.48}$$

where the coefficient a is unknown and the order m may or may not be known. If m is known, the parameter a can be determined by comparing the solution $\tilde{u}_{2\Delta t}$ obtained with a double step $2\Delta t$ with the solution $\tilde{u}_{\Delta t}$ obtained with the original time step Δt:

$$\tilde{u}_{2\Delta t} - u = a2^m \Delta t^m, \quad \tilde{u}_{\Delta t} - u = a\Delta t^m \tag{4.49}$$

from which[7] falls the value of a:

$$a = \frac{\tilde{u}_{2\Delta t} - \tilde{u}_{\Delta t}}{(2^m - 1)\Delta t^m}. \tag{4.50}$$

---

[7] Notice that the difference must be done *at the same moment t*, not at the same time step n.

The error estimate associated with the higher resolution solution $\tilde{u}_{\Delta t}$ is

$$\epsilon_u = \tilde{u}_{\Delta t} - u = a\Delta t^m = \frac{\tilde{u}_{2\Delta t} - \tilde{u}_{\Delta t}}{(2^m - 1)}, \tag{4.51}$$

with which we can improve our solution by using Eq. (4.48)

$$\tilde{u} = \tilde{u}_{\Delta t} - \frac{\tilde{u}_{2\Delta t} - \tilde{u}_{\Delta t}}{(2^m - 1)}. \tag{4.52}$$

This suggests that the two-time-step approach may yield the exact answer because it determines the error. Unfortunately, this cannot be the case because we are working with a discrete representation of a continuous function. The paradox is resolved by realizing that, by using Eq. (4.50), we discarded higher-order terms and therefore did not calculate the exact value of a but only an estimate of it. What our manipulation accomplished was the elimination of the leading error term. This procedure is called a *Richardson extrapolation*:

$$u = \tilde{u}_{\Delta t} - \frac{\tilde{u}_{2\Delta t} - \tilde{u}_{\Delta t}}{(2^m - 1)} + \mathcal{O}\left(\Delta t^{m+1}\right). \tag{4.53}$$

Numerical calculations of the real error and error estimates according to Eq. (4.51) show good performance of the estimators in the context of inertial oscillations (Fig. 4.8). Also, the Richardson extrapolation increases the order by 1, except for the semi-implicit scheme at high resolution, when no gain is achieved because saturation occurs (Fig. 4.8, right panel). This asymptote corresponds to the inevitable rounding errors, and we can claim to have solved the discrete equations "exactly."

When considering the error estimate (4.51), we observe that the error estimate of a first-order scheme ($m = 1$) is simply the difference between two solutions obtained with different time steps. This is the basic justification for performing resolution sensitivity analysis on more complicated models: Differences in model results due to a variation in resolution may be taken as estimates of the discretization error. By extension, performing multiple simulations with different model parameter values leads to differences that are indicators of modeling errors.

When the truncation order m is not known, a third evaluation of the numerical solution, with a quadruple time step $4\Delta t$, yields an estimate of both the order m and the coefficient a of the discretization error:

$$m = \frac{1}{\log 2} \log\left(\frac{\tilde{u}_{4\Delta t} - \tilde{u}_{2\Delta t}}{\tilde{u}_{2\Delta t} - \tilde{u}_{\Delta t}}\right) \tag{4.54}$$

$$a = -\frac{\tilde{u}_{2\Delta t} - \tilde{u}_{\Delta t}}{(2^m - 1)\Delta t^m}.$$

As we can see in practice (Fig. 4.9), this estimate provides a good estimate of m when resolution is sufficiently fine. This method can thus be used to determine the truncation order of discretizations numerically, which can be useful to assess convergence rates of nonlinear discretized systems or to verify the proper numerical implementation of a discretization (for which the value of m is known). In the latter case, if a method should be of second order but the numerical estimate of m according to Eq. (4.54) reveals only first-order convergence on well behaved problems, a programming or implementation error is very likely to blame.

Having access now to an error estimate, we can think of choosing the time step so as to keep discretization errors below a prescribed level. If the time step is prescribed a priori, the error estimate allows us to verify that the solution remains within error bounds. The use of a fixed time step is common but might not be the most appropriate choice when the process exhibits a mix of slower and faster processes (Fig. 4.10). Then, it may be preferable that the time step be

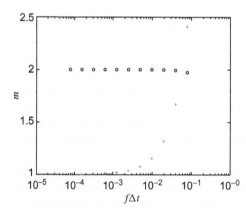

FIGURE 4.9 Estimator of m for explicit ($+$, tending to 1 for small time steps) and semi-implicit discretization (o, tending to 2) as function of $f\Delta t$.

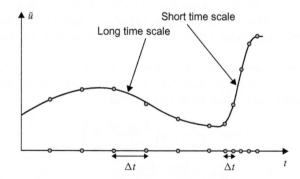

FIGURE 4.10 Use of different time steps Δt in function of the local error and time scales. The time step is decreased until the local error estimate is smaller than a prescribed value. When estimated errors are much smaller than allowed, the time step is increased.

adjusted over time so as to follow the time scale of the system. In this case, we speak about *adaptive time stepping*.

Adaptive time stepping can be implemented by decreasing the time step whenever the error estimate begins to be excessive. Vice versa, when the error estimate indicates an unnecessarily short time step, it should be allowed to increase again. Adaptive time stepping seems appealing, but the additional work required to track the error estimate (doubling/halving the time step and recomputing the solution) can exceed the gain obtained by maintaining a fixed time step, which is occasionally too short. Also, multistep methods are not easily generalized to adaptive time steps.

ANALYTICAL PROBLEMS

4.1. From the weather chart in today's edition of your newspaper, identify the horizontal extent of a major atmospheric feature and find the forecast wind speed. From these numbers, estimate the Rossby number of the weather pattern. What do you conclude about the importance of the Coriolis force? (*Hint*: When converting latitudinal and longitudinal differences in kilometers, use the Earth's mean radius, 6371 km.)

4.2. Using the scale given in Eq. (4.16), compare the dynamic pressure induced by the Gulf Stream (speed = 1 m/s, width = 40 km, and depth = 500 m) to the main hydrostatic pressure due to the weight of the same water depth. Also, convert the dynamic pressure scale in equivalent height of hydrostatic pressure (head). What can you infer about the possibility of measuring oceanic dynamic pressures by a pressure gauge?

4.3. Consider a two-dimensional periodic fluctuation of the type ($u' = U\sin(\phi+\alpha_u)$, $v' = V\sin(\phi+\alpha_v)$, $w' = 0$) with $\phi(x, y, t) = k_x x + k_y y - \omega t$ and all other quantities constant. Calculate the Reynolds stresses, such as $-\langle u'v'\rangle$, by taking the average over a 2π-period of the phase ϕ. Show that these stresses are not zero in general (proving that traveling waves may exert a finite stress and therefore accelerate or slow down a background flow on which they are superimposed). Under which relation between α_u and α_v does the shear stress $-\langle u'v'\rangle$ vanish?

4.4. Show that the horizontal eddy viscosity defined in Eq. (4.10) vanishes for a vortex flow with velocity components ($u = -\Omega y$, $v = +\Omega x$) with Ω being a constant. Is this a desirable property?

4.5. Why do we need to know the surface pressure distribution when using the hydrostatic approximation?

4.6. Theory tells us that in a pure advection problem for temperature T, a single boundary condition should be imposed at the inflow and none at the outflow, but when diffusion is present, a boundary condition must be imposed at both ends. What do you expect to happen at the outflow boundary when diffusion is very small? How would you measure the "smallness" of diffusion?

4.7. In forming energy budgets, the momentum equations are multiplied by their respective velocity components (i.e., the $\partial u/\partial t$ equation is multiplied by u and so forth), and the results are added. Show that in this manipulation, the Coriolis terms in f and f_* cancel one another out. What would be your reaction if someone presented you with a model in which the $f_* w$ term were dropped from Eq. (4.7a) because w is small compared to u and the term $f_* u$ were retained in Eq. (4.7c) for the same reason?

NUMERICAL EXERCISES

4.1. When air and sea surface temperatures, T_{air} and T_{sea}, are close to each other, it is acceptable to use a linearized form to express the heat flux across the air–sea heat interface, such as

$$-\kappa_T \frac{\partial T_{\text{sea}}}{\partial z} = \frac{h}{\rho_0 C_v} (T_{\text{sea}} - T_{\text{air}}), \qquad (4.55)$$

where h is an exchange coefficient in $(\text{W m}^{-2}\,\text{K}^{-1})$. The coefficient multiplying the temperature difference $T_{\text{sea}} - T_{\text{air}}$ has the units of a velocity and, for this reason, is sometimes called *piston velocity* in the context of gas exchange between air and water. Implement this boundary condition for a finite-volume ocean model. How would you calculate T_{sea} involved in the flux in order to maintain second-order accuracy of the standard second derivative within the ocean domain?

4.2. In some cases, particularly analytical and theoretical studies, the unknown field can be assumed to be periodic in space. How can periodic boundary conditions be implemented in a numerical one-dimensional model, for which the discretization scheme uses one point on each side of every calculation point? How would you adapt the scheme if the interior discretization needs two points on each side instead? Can you imagine what the expression *halo* used in this context refers to?

4.3. How do you generalize periodic boundary conditions (see preceding problem) to two dimensions? Is there an efficient algorithmic scheme that ensures periodicity without particular treatment of corner points? (*Hint:* Think about a method/order of copying rows/columns that ensure proper values in corners.)

4.4. Assume you implemented a Dirichlet condition for temperature along a boundary on which a grid node exists but would like to diagnose the heat flux across the boundary. How would you determine the turbulent flux at that point with third-order accuracy?

4.5. Models can be used on parallel machines by distributing work among different processors. One of the possibilities is the so-called *domain decomposition* in which each processor is dedicated to a portion of the total domain. The model of each subdomain can be interpreted as an open-boundary model. Assuming that the numerical scheme for a single variable uses q points on each side of the local node, how would you subdivide a one-dimensional domain into subdomains and design data exchange between these subdomains to avoid the introduction of new errors? Can you imagine the problems you are likely to encounter in two dimensions? (*Hint*: Think how periodic boundary conditions were handled in the halo approach of the preceding two problems.)

4.6. Develop a MATLAB™ program to automatically calculate finite-difference weighting coefficients a_i for an arbitrary derivative of order p using l points to the left and m points to the right of the point of interest:

$$\frac{d^p \tilde{u}}{dt^p}\bigg|_{t_n} \simeq a_{-l} \tilde{u}^{n-l} + \cdots + a_{-1} \tilde{u}^{n-1} + a_0 \tilde{u}^n + a_1 \tilde{u}^{n+1} + \cdots + a_m \tilde{u}^{n+m}.$$

(4.56)

The step is taken constant. Test your program on the fourth-order approximation of the first derivative. (*Hint*: Construct the linear system to be solved by observing that Δt should cancel out in all terms except for the relevant derivative so that $\Delta t^p a_i$ can be chosen as the unknowns.)

4.7. Imagine that you perform a series of simulations of the same model with time steps of $8\Delta t$, $4\Delta t$, $2\Delta t$, and Δt. The numerical discretization scheme is of order m. Which combination of the different solutions would best approximate the exact solution, and what truncation order would the combined solution have?

4.8. For a grid node placed on the boundary, show that using the value

$$\tilde{T}_m = \frac{2}{3}\left(-\Delta x \frac{q_b}{\kappa_T} - \frac{1}{2}\tilde{T}_{m-2} + 2\tilde{T}_{m-1}\right)$$

(4.57)

allows us to impose a flux condition at node m with second-order accuracy.

Osborne Reynolds
1842–1912

Osborne Reynolds was taught mathematics and mechanics by his father. While a teenager, he worked as an apprentice in the workshop of a mechanical engineer and inventor, where he realized that mathematics was essential for the explanation of certain mechanical phenomena. This motivated him to study mathematics at Cambridge, where he brilliantly graduated in 1867. Later, as a professor of engineering at the University of Manchester, his teaching philosophy was to subject engineering to mathematical description while also stressing the contribution of engineering to human welfare. His best known work is that on fluid turbulence, famous for the idea of separating flow fluctuations from the mean velocity and for his study of the transition from laminar to turbulent flow, leading to the dimensionless ratio that now bears his name. He made other significant contributions to lubrication, friction, heat transfer, and hydraulic modeling. Books on fluid mechanics are peppered with the expressions Reynolds number, Reynolds equations, Reynolds stress, and Reynolds analogy. (*Photo courtesy of Manchester School of Engineering*)

Carl-Gustaf Arvid Rossby
1898–1957

A Swedish meteorologist, Carl-Gustav Rossby is credited with most of the fundamental principles on which geophysical fluid dynamics rests. Among other contributions, he left us the concepts of radius of deformation (Section 9.2), planetary waves (Section 9.4), and geostrophic adjustment (Section 15.2). However, the dimensionless number that bears his name was first introduced by the Soviet scientist I. A. Kibel' in 1940.

Inspiring to young scientists, whose company he constantly sought, Rossby viewed scientific research as an adventure and a challenge. His accomplishments are marked by a broad scope and what he liked to call the *heuristic approach*, that is, the search for a useful answer without unnecessary complications. During a number of years spent in the United States, he established the meteorology departments at MIT and the University of Chicago. He later returned to his native Sweden to become the director of the Institute of Meteorology in Stockholm. (*Photo courtesy of Harriet Woodcock*)

Chapter 5

Diffusive Processes

ABSTRACT

All geophysical motions are diffusive because of turbulence. Here, we consider a relatively crude way of representing turbulent diffusion, by means of an eddy diffusivity. Although the theory is straightforward, numerical handling of diffusion terms requires care, and the main objective of this chapter is to treat the related numerical issues, leading to the fundamental concept of numerical stability.

5.1 ISOTROPIC, HOMOGENEOUS TURBULENCE

It was mentioned in Sections 3.4 and 3.5 that fluid properties such as heat, salt, and humidity diffuse, that is, they are exchanged between neighboring particles. In laminar flow, this is accomplished by random (so-called Brownian) motion of the colliding molecules, but in large-scale geophysical systems, turbulent eddies accomplish a similar effect far more efficiently. The situation is analogous to mixing milk in coffee or tea: Left alone, the milk diffuses very slowly through the beverage, but the action of a stirrer generates turbulent eddies that mix the two liquids far more effectively and create a homogeneous mixture in a short time. The difference is that eddying in geophysical fluids is generally not induced by a stirring mechanism but is self-generated by hydrodynamic instabilities.

In Section 4.1, we introduced turbulent fluctuations without saying anything specific about them; we now begin to elucidate some of their properties. At a very basic level, turbulent motion can be interpreted as a population of many eddies (vortices), of different sizes and strengths, embedded within one another and forever changing, giving a random appearance to the flow (Fig. 5.1). Two variables then play a fundamental role: d, the characteristic diameter of the eddies and $\mathring{u}$, their characteristic orbital velocity. Since the turbulent flow consists of many eddies, of varying sizes and speeds, $\mathring{u}$ and d do not each assume a single value but vary within a certain range. In stationary, homogeneous, and isotropic turbulence, that is, a turbulent flow that statistically appears unchanging in time, uniform in space, and without preferential direction, all eddies of a given size (same d) behave more or less in the same way and can be

FIGURE 5.1 Drawing of a turbulent flow by Leonardo da Vinci circa
1507–1509, who recognized that turbulence involves a multitude of eddies
at various scales.

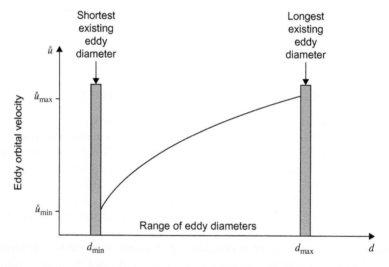

FIGURE 5.2 Eddy orbital velocity versus eddy length scale in homogeneous and isotropic
turbulence. The largest eddies have the largest orbital velocity.

assumed to share the same characteristic velocity $\mathring{u}$. In other words, we make
the assumption that $\mathring{u}$ is a function of d (Fig. 5.2).

5.1.1 Length and Velocity Scales

In the view of Kolmogorov (1941), turbulent motions span a wide range of
scales, from a macroscale at which the energy is supplied to a microscale at
which energy is dissipated by viscosity. The interaction among the eddies of

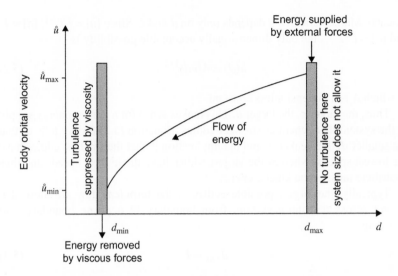

FIGURE 5.3 The turbulent energy cascade. According to this theory, the energy fed by external forces excites the largest possible eddies and is gradually passed to over smaller eddies, all the way to a minimum scale where this energy is ultimately dissipated by viscosity.

various scales passes energy gradually from the larger eddies to the smaller ones. This process is known as the *turbulent energy cascade* (Fig. 5.3).

If the state of turbulence is statistically steady (statistically unchanging turbulence intensity), then the rate of energy transfer from one scale to the next must be the same for all scales, so that no group of eddies sharing the same scale sees its total energy level increase or decrease over time. It follows that the rate at which energy is supplied at the largest possible scale (d_{max}) is equal to that dissipated at the shortest scale (d_{min}). Let us denote by ϵ this rate of energy supply/dissipation, per unit mass of fluid:

ϵ = energy supplied to fluid per unit mass and time

   = energy cascading from scale to scale, per unit mass and time

   = energy dissipated by viscosity, per unit mass and time.

The dimensions of ϵ are

$$[\epsilon] = \frac{ML^2T^{-2}}{MT} = L^2T^{-3}. \tag{5.1}$$

With Kolmogorov, we further assume that the characteristics of the turbulent eddies of scale d depend solely on d and on the energy cascade rate ϵ. This is to say that the eddies know how large they are, at what rate energy is supplied to them, and at what rate they must supply it to the next smaller eddies in the cascade. Mathematically, $\mathring{u}$ depends only on d and ϵ. Since $[\mathring{u}] = LT^{-1}$, $[d] = L$, and $[\epsilon] = L^2 T^{-3}$, the only dimensionally acceptable possibility is

$$\mathring{u}(d) = A(\epsilon d)^{1/3}, \tag{5.2}$$

in which A is a dimensionless constant.

Thus, the larger ϵ, the larger $\mathring{u}$. This makes sense for a greater energy supply to the system that generates stronger eddies. Equation (5.2) further tells us that the smaller d, the weaker $\mathring{u}$, and the implication is that the smallest eddies have the lowest speeds, whereas the largest eddies have the highest speeds and thus contribute most of the kinetic energy.

Typically, the largest possible eddies in the turbulent flow are those that extend across the entire system, from boundary to opposite boundary, and therefore

$$d_{max} = L, \tag{5.3}$$

where L is the geometrical dimension of the system (such as the width of the domain or the cubic root of its volume). In geophysical flows, there is a noticeable scale disparity between the short vertical extent (depth, height) and the long horizontal extent (distance, length) of the system. We must therefore clearly distinguish eddies that rotate in the vertical plane (about a horizontal axis) from those that rotate horizontally (about a vertical axis). The nearly two-dimensional character of the latter gives rise to a special form of turbulence, called geostrophic turbulence, which will be discussed in Section 18.3. In this chapter, we restrict our attention to three-dimensional isotropic turbulence.

The shortest eddy scale is set by viscosity and can be defined as the length scale at which molecular viscosity becomes dominant. Molecular viscosity, denoted by ν, has dimensions[1]:

$$[\nu] = L^2 T^{-1}.$$

If we assume that d_{min} depends only on ϵ, the rate at which energy is supplied to that scale, and on ν because these eddies feel viscosity, then the only dimensionally acceptable relation is

$$d_{min} \sim \nu^{3/4} \epsilon^{-1/4}. \tag{5.4}$$

The quantity $\nu^{3/4} \epsilon^{-1/4}$, called the *Kolmogorov scale*, is typically on the order of a few millimeters or shorter. We leave it to the reader to verify that at this length scale, the corresponding Reynolds number is on the order of unity.

---

[1] Values for ambient air and water are $\nu_{air} = 1.51 \times 10^{-5}$ m^2/s and $\nu_{water} = 1.01 \times 10^{-6}$ m^2/s.

The span of length scales in a turbulent flow is related to its Reynolds number. Indeed, in terms of the largest velocity scale, which is the orbital velocity of the largest eddies, $U = \mathring{u}(d_{max}) = A(\epsilon L)^{1/3}$, the energy supply/dissipation rate is

$$\epsilon = \frac{U^3}{A^3 L} \sim \frac{U^3}{L}, \tag{5.5}$$

and the length scale ratio can be expressed as

$$\frac{L}{d_{min}} \sim \frac{L}{\nu^{3/4}\epsilon^{-1/4}}$$

$$\sim \frac{LU^{3/4}}{\nu^{3/4}L^{1/4}}$$

$$\sim Re^{3/4}, \tag{5.6}$$

where $Re = UL/\nu$ is the Reynolds number of the flow. As we could have expected, a flow with a higher Reynolds number contains a broader range of eddies.

5.1.2   Energy Spectrum

In turbulence theory, it is customary to consider the so-called *power spectrum*, which is the distribution of kinetic energy per mass across the various length scales. For this, we need to define a wavenumber. Because velocity reverses across the diameter of an eddy, the eddy diameter should properly be considered as half of the wavelength:

$$k = \frac{2\pi}{\text{wavelength}} = \frac{\pi}{d}. \tag{5.7}$$

The lowest and highest wavenumber values are $k_{min} = \pi/L$ and $k_{max} \sim \epsilon^{1/4}\nu^{-3/4}$.

The kinetic energy E per mass of fluid has dimensions $ML^2T^{-2}/M = L^2T^{-2}$. The portion dE contained in the eddies with wavenumbers ranging from k to $k+dk$ is defined as

$$dE = E_k(k)\,dk.$$

It follows that the dimension of E_k is L^3T^{-2}, and dimensional analysis prescribes

$$E_k(k) = B\,\epsilon^{2/3}\,k^{-5/3}, \tag{5.8}$$

where B is a second dimensionless constant. It can be related to A of Eq. (5.2) because the integration of $E_k(k)$ from $k_{min} = \pi/L$ to $k_{max} \sim \infty$ is the total energy

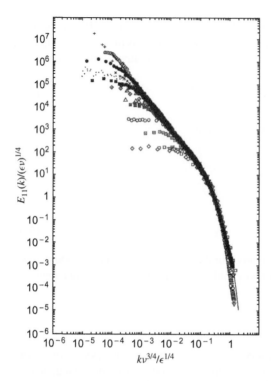

FIGURE 5.4 Longitudinal power spectrum of turbulence calculated from numerous observations taken outdoors and in the laboratory. (*From Saddoughi and Veeravalli, 1994*).

per mass in the system, which in good approximation is that contained in the largest eddies, namely $U^2/2$. Thus,

$$\int_{k_{\min}}^{\infty} E_k(k)\, \mathrm{d}k = \frac{U^2}{2}, \tag{5.9}$$

from which follows

$$\frac{3}{2\pi^{2/3}}\, B = \frac{1}{2}\, A^2. \tag{5.10}$$

The value of B has been determined experimentally and found to be about 1.5 (Pope, 2000, page 231). From this, we estimate A to be 1.45.

The $-5/3$ power law of the energy spectrum has been observed to hold well in the *inertial range*, that is, for those intermediate eddy diameters that are remote from both largest and shortest scales. Figure 5.4 shows the superposition of a large number of longitudinal power spectra.[2] The straight line where most data overlap in the range $10^{-4} < k\nu^{3/4}/\epsilon^{1/4} < 10^{-1}$ corresponds to the $-5/3$

---

[2]The longitudinal power spectrum is the spectrum of the kinetic energy associated with the velocity component in the direction of the wavenumber.

decay law predicted by the Kolmogorov turbulent cascade theory. The higher the Reynolds number of the flow, the broader the span of wavenumbers over which the $-5/3$ law holds. Several crosses visible at the top of the plot, which extend from a set of crosses buried in the accumulation of data below, correspond to data in a tidal channel (Grant, Stewart & Moilliet, 1962) for which the Reynolds number was the highest.

There is, however, some controversy over the $-5/3$ power law for E_k. Some investigators (Long, 1997, 2003; Saffman, 1968) have proposed alternative theories that predict a -2 power law.

5.2   TURBULENT DIFFUSION

Our concern here is not to pursue the study of turbulence but to arrive at a heuristic way to represent the dispersive effect of turbulence on those scales too short to be resolved in a numerical model.

Turbulent diffusion or *dispersion* is the process by which a substance is moved from one place to another under the action of random turbulent fluctuations in the flow. Given the complex nature of these fluctuations, it is impossible to describe the dispersion process in an exact manner but some general remarks can be made that lead to a useful parameterization.

Consider the two adjacent cells of Fig. 5.5 exchanging fluid between each other. The fluid in the left cell contains a concentration (mass per volume) c_1 of some substance, whereas the fluid in the right cell contains a different concentration c_2. Think of c_1 being less than c_2, although this does not necessarily have to be the case. Further assume, in order to focus exclusively on diffusion, that there is no net flow from one cell to the other but that the only exchange velocity is due to a single eddy moving fluid at velocity $\mathring{u}$ on one flank and at velocity $-\mathring{u}$ on its opposite flank. The amount of substance carried per unit area perpendicular to the x-axis and per time, called the *flux*, is equal to the product of the concentration and the velocity, $c_1\mathring{u}$ from left to right and $c_2\mathring{u}$ in the opposite direction. The net flux q in the x-direction is the flux from 1 to 2 minus the flux from 2 to 1:

$$q = c_1 \mathring{u} - c_2 \mathring{u}$$
$$= -\mathring{u}\,\Delta c,$$

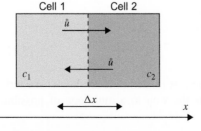

FIGURE 5.5 Exchange between two adjacent cells illustrating turbulent diffusion. Because of the difference between concentrations, the exchange between cells is uneven. The cell with the least concentration loses less than it receives.

where $\Delta c = c_2 - c_1$ is the concentration difference. Multiplying and dividing by the distance Δx between cell centers, we may write:

$$q = -(\mathring{u}\Delta x)\frac{\Delta c}{\Delta x}.$$

When considering the variation of c over larger scales, those for which the eddy-size Δx appears to be small, we may approximate the previous equation to

$$q = -D\frac{dc}{dx}, \qquad (5.11)$$

where D is equal to the product $\mathring{u}\Delta x$ and is called the turbulent diffusion coefficient or *diffusivity*. Its dimension is $[D] = L^2 T^{-1}$.

The diffusive flux is proportional to the gradient of the concentration of the substance. In retrospect, this makes sense; if there were no difference in concentrations between cells, the flux from one into the other would be exactly compensated by the flux in the opposite direction. It is the concentration difference (the gradient) that matters.

Diffusion is "down-gradient," that is, the transport is from high to low concentrations, just as heat conduction moves heat from the warmer side to the colder side. (In the preceding example with $c_1 < c_2$, q is negative, and the net flux is from cell 2 to cell 1.) This implies that the concentration increases on the low side and decreases on the high side, and the two concentrations gradually become closer to each other. Once they are equal ($dc/dx = 0$), diffusion stops, although turbulent fluctuations never do. Diffusion acts to homogenize the substance across the system.

The pace at which diffusion proceeds depends critically on the value of the diffusion coefficient D. This coefficient is inherently the product of two quantities, a velocity ($\mathring{u}$) and a length scale (Δx), representing respectively the magnitude of fluctuating motions and their range. Since the numerical model resolves scales down to the grid scale Δx, the turbulent diffusion that remains to represent is that due to the all shorter scales, starting with $d = \Delta x$. As seen in the previous section, for shorter scales, d correspond to slower eddy velocities $\mathring{u}$ and thus lower diffusivities. It follows that diffusion is chiefly accomplished by eddies at the largest unresolved scale, Δx, because these generate the greatest value of $\mathring{u}\Delta x$:

$$
\begin{aligned}
D &= \mathring{u}(\Delta x)\,\Delta x \\
&= A\,\epsilon^{1/3}\,\Delta x^{4/3}. \qquad (5.12)
\end{aligned}
$$

The manner by which the dissipation rate ϵ is related to local flow characteristics, such as a velocity gradient, opens the way to a multitude of possible parameterizations.

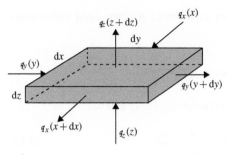

FIGURE 5.6 An infinitesimal piece of fluid for the local budget of a substance of concentration c in the fluid.

The preceding considerations in one dimension were generic in the sense that the direction x could stand for any of the three directions of space, x, y, or z. Because of the typical disparity in mesh size between the horizontal and vertical directions in GFD models ($\Delta x \approx \Delta y \gg \Delta z$), care must be taken to use two distinct diffusivities, which we denote $\mathcal{A}$ for the horizontal directions and κ for the vertical direction.[3] While κ must be constructed from the length scale Δz, $\mathcal{A}$ must be formed from a length scale that is hybrid between Δx and Δy. The Smagorinsky formulation presented in (4.10) is a good example.

The components of the three-dimensional flux vector are

$$q_x = -\mathcal{A}\frac{\partial c}{\partial x} \tag{5.13a}$$

$$q_y = -\mathcal{A}\frac{\partial c}{\partial y} \tag{5.13b}$$

$$q_z = -\kappa\frac{\partial c}{\partial z}. \tag{5.13c}$$

And, we are in a position to write a budget for the concentration $c(x, y, z, t)$ of the substance in the flow, by taking an elementary volume of fluid of size dx, dy, and dz, as illustrated in Fig. 5.6. The net import in the x-direction is the difference in x-fluxes times the area $dy\,dz$ they cross, that is, $[q_x(x, y, z) - q_x(x + dx, y, z)]\,dy\,dz$, and similarly in the y-and z-directions. The net import from all directions is then

$$\text{Net import in } dx\,dy\,dz = [q_x(x, y, z) - q_x(x + dx, y, z)]\,dy\,dz$$
$$+ [q_y(x, y, z) - q_y(x, y + dy, z)]\,dx\,dz$$
$$+ [q_z(x, y, z) - q_z(x, y, z + dz)]\,dx\,dy,$$

---

[3]GFD models generally use the same horizontal diffusivity for all variables, including momentum and density—see (4.21)—but distinguish between various diffusivities in the vertical.

on a per-time basis. This net import contributes to increasing the amount $c\,dx\,dy\,dz$ inside the volume:

$$\frac{d}{dt}(c\,dx\,dy\,dz) = \text{Net import}.$$

In the limit of an infinitesimal volume (vanishing dx, dy and dz), we have

$$\frac{\partial c}{\partial t} = -\frac{\partial q_x}{\partial x} - \frac{\partial q_y}{\partial y} - \frac{\partial q_z}{\partial z}, \tag{5.14}$$

and, after replacement of the flux components by their expressions (5.13),

$$\frac{\partial c}{\partial t} = \frac{\partial}{\partial x}\left(\mathcal{A}\frac{\partial c}{\partial x}\right) + \frac{\partial}{\partial y}\left(\mathcal{A}\frac{\partial c}{\partial y}\right) + \frac{\partial}{\partial z}\left(\kappa\frac{\partial c}{\partial z}\right), \tag{5.15}$$

where $\mathcal{A}$ and κ are, respectively, the horizontal and vertical eddy diffusivities. Note the similarity with the dissipation terms in the momentum and energy equations (4.21) of the previous chapter.

For a comprehensive exposition of diffusion and some of its applications, the reader is referred to Ito (1992) and Okubo and Levin (2002).

5.3   ONE-DIMENSIONAL NUMERICAL SCHEME

We now illustrate discretization methods for the diffusion equation and begin with a protypical one-dimensional system, representing a horizontally homogeneous piece of ocean or atmosphere, containing a certain substance, such as a pollutant or tracer, which is not exchanged across either bottom or top boundaries. To simplify the analysis further, we begin by taking the vertical diffusivity κ as constant until further notice. We then have to solve the following equation:

$$\frac{\partial c}{\partial t} = \kappa\frac{\partial^2 c}{\partial z^2}, \tag{5.16}$$

with no-flux boundary conditions at both bottom and top:

$$q_z = -\kappa\frac{\partial c}{\partial z} = 0 \quad \text{at} \quad z = 0 \text{ and } z = h, \tag{5.17}$$

where h is the thickness of the domain.

To complete the problem, we also prescribe an initial condition. Suppose for now that this initial condition is a constant C_0 plus a cosine function of amplitude C_1 ($C_1 \leq C_0$):

$$c(z, t = 0) = C_0 + C_1 \cos\left(j\pi\frac{z}{h}\right), \tag{5.18}$$

with j being an integer. Then, it is easily verified that

$$c = C_0 + C_1 \cos\left(j\pi \frac{z}{h}\right) \exp\left(-j^2\pi^2 \frac{\kappa t}{h^2}\right) \qquad (5.19)$$

satisfies the partial differential equation (5.16), both boundary conditions (5.17), and initial condition (5.18). It is thus the exact solution of the problem. As we can expect from the dissipative nature of diffusion, this solution represents a temporal attenuation of the nonuniform portion of c, which is more rapid under stronger diffusion (greater κ) and shorter scales (higher j).

Let us now design a numerical method to solve the problem and check its solution against the preceding, exact solution. First, we discretize the spatial derivative by applying a standard finite-difference technique. With a Neumann boundary condition applied at each end, we locate the end grid points not at, but surrounding the boundaries (see Section 4.7) and place the grid nodes at the following locations:

$$z_k = \left(k - \tfrac{3}{2}\right)\Delta z \quad \text{for} \quad k = 1, 2, \dots, m, \qquad (5.20)$$

with $\Delta z = h/(m-2)$ so that we use m grid points, among which the first and last are ghost points lying a distance $\Delta z/2$ beyond the boundaries (Fig. 5.7).

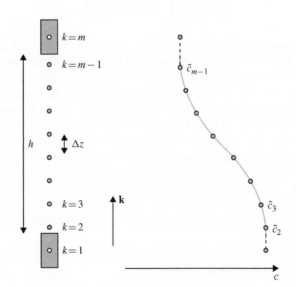

FIGURE 5.7 Gridding of a vertical domain with m nodes, of which the first and last lie beyond the bottom and top boundaries, respectively. Such points are called *ghost points*. With m nodes and $m-1$ intervals between nodes among which two are only half long, it follows that $(m-2)$ segments cover the domain, and the grid spacing is thus $\Delta z = h/(m-2)$. Neumann conditions (zero derivatives) at both boundaries are implemented by assigning the values $\tilde{c}_1 = \tilde{c}_2$ and $\tilde{c}_m = \tilde{c}_{m-1}$ to the end points, which implies zero derivatives in the middle of the first and last intervals. The calculations using the discretized form of the equation then proceed from $k=2$ to $k=m-1$.

Discretizing the second spatial derivative with a three-point centered scheme and before performing time discretization, we have

$$\frac{d\tilde{c}_k}{dt} = \frac{\kappa}{\Delta z^2} (\tilde{c}_{k+1} - 2\tilde{c}_k + \tilde{c}_{k-1}) \quad \text{for} \quad k = 2, \ldots, m-1. \tag{5.21}$$

We thus have $m-2$ ordinary, coupled, differential equations for the $m-2$ unknown time dependent functions $\tilde{c}_k$. We can determine the numerical error introduced in this *semidiscrete* set of equations by trying a solution similar to the exact solution:

$$\tilde{c}_k = C_0 + C_1 \cos\left(j\pi \frac{z_k}{h}\right) a(t). \tag{5.22}$$

Trigonometric formulas provide the following equation for the temporal evolution of the amplitude $a(t)$:

$$\frac{da}{dt} = -4a \frac{\kappa}{\Delta z^2} \sin^2 \phi \quad \text{with} \quad \phi = j\pi \frac{\Delta z}{2h}, \tag{5.23}$$

of which the solution is

$$a(t) = \exp\left(-4 \sin^2 \phi \frac{\kappa t}{\Delta z^2}\right). \tag{5.24}$$

With this spatial discretization, we thus obtain an exponential decrease of amplitude a, like in the exact equation (5.19) but with a different damping rate. The ratio τ of the numerical damping rate $4\kappa \sin^2 \phi / \Delta z^2$ to the true damping rate $j^2 \pi^2 \kappa / h^2$ is $\tau = \phi^{-2} \sin^2 \phi$. For small Δz compared with the length scale h/j of the c distribution, ϕ is small, and the correct damping is nearly obtained with the semidiscrete numerical scheme. Nothing anomalous is therefore expected from the approach thus far as long as the discretization of the domain is sufficiently dense to capture adequately the spatial variations in c. Also, the boundary conditions cause no problem because the mathematical requirement of one boundary condition on each side of the domain matches exactly what we need to calculate the discrete values $\tilde{c}_k$ for $k = 2, \ldots, m-1$. An initial condition is also needed at each node to start the time integration. This is all consistent with the mathematical problem.

We now proceed with the time discretization. First, let us try the simplest of all methods, the explicit Euler scheme:

$$\frac{\tilde{c}_k^{n+1} - \tilde{c}_k^n}{\Delta t} = \frac{\kappa}{\Delta z^2} \left(\tilde{c}_{k+1}^n - 2\tilde{c}_k^n + \tilde{c}_{k-1}^n\right) \quad \text{for} \quad k = 2, \ldots, m-1 \tag{5.25}$$

in which $n \geq 1$ stands for the time level. For convenience, we define a dimensionless number that will play a central role in the discretization and solution:

$$D = \frac{\kappa \Delta t}{\Delta z^2}. \tag{5.26}$$

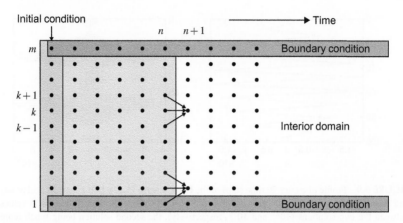

FIGURE 5.8   Initialized for each grid point, algorithm (5.27) advances the value at node k to the next time step (from n to $n+1$) using the previous values on stencil spanning points $k-1$, k, and $k+1$. A boundary condition is thus needed on each side of the domain, as the original mathematical problem requires. The calculations for the discretized governing equations proceed from $k=2$ to $k=m-1$.

This definition allows us to write the discretized equation more conveniently as

$$\tilde{c}_k^{n+1} = \tilde{c}_k^n + \mathrm{D}\left(\tilde{c}_{k+1}^n - 2\tilde{c}_k^n + \tilde{c}_{k-1}^n\right) \quad \text{for} \quad k=2,\ldots,m-1. \tag{5.27}$$

The scheme updates the discrete $\tilde{c}_k$ values from their initial values and with the aforementioned boundary conditions (Fig. 5.8). Obviously, the algorithm is easily programmed (e.g., `firstdiffusion.m`) and can be tested rapidly.

For simplicity, we start with a gentle profile ($j=1$, half a wavelength across the domain) and, equipped with our insight in scale analysis, we use a sufficiently small grid spacing $\Delta z \ll h$ to resolve the cosine function well. To be sure, we take 20 grid points. For the time scale T of the physical process, we use the scale provided by the original equation:

$$\frac{\partial c}{\partial t} = \kappa \frac{\partial^2 c}{\partial z^2}$$

$$\frac{\Delta c}{T} \quad \kappa \frac{\Delta c}{h^2}$$

to find $T = h^2/\kappa$. Dividing this time scale in 20 steps, we take $\Delta t = T/20 = h^2/(20\kappa)$ and begin to march algorithm (5.27) forward.

Surprisingly, it is not working. After only 20 time steps, the $\tilde{c}_k$ values do not show attenuation but have instead increased by a factor 10^{20}! Furthermore, increasing the spatial resolution to 100 points and reducing the time step proportionally does not help but worsens the situation (Fig. 5.9). Yet, there has been no programming error in `firstdiffusion.m`. The problem is more serious: We have stumbled on a crucial aspect of numerical integration, by falling prey to *numerical instability*. The symptoms of numerical instability are explosive

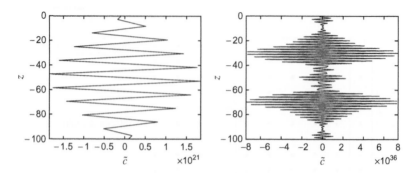

FIGURE 5.9   Profile of $\tilde{c}$ after 20 time steps of the Euler scheme (5.27). Left panel: 20 grid points and $\Delta t = T/20$. Right panel: 100 grid points and $\Delta t = T/100$. Note the vast difference in values between the two solutions (10^{21} and 10^{36}, respectively), the second solution being much more explosive than the first. Conclusion: Increasing resolution worsens the problem.

behavior and worsening of the problem with increased spatial resolution. At best, the scheme is used outside of a certain domain of validity or, at worst, it is hopeless and in need of replacement by a better, stable scheme. What makes a scheme stable and another unstable is the objective of numerical stability analysis.

5.4   NUMERICAL STABILITY ANALYSIS

The most widely used method to investigate the stability of a given numerical scheme is due to John von Neumann.[4] The basic idea of the method is to consider the temporal evolution of simple numerical solutions. As continuous signals and distributions can be expressed as Fourier series of sines and cosines, discrete functions can, too, be decomposed in elementary functions. If one or several of these elementary functions increase without bound over time ("explode"), the reconstructed solution, too, will increase without bound, and the scheme is unstable. Put the other way: A scheme is stable if none among all elementary functions grows without bound over time.

As for Fourier series and simple wave propagation, the elementary functions are periodic. In analogy with the continuous function

$$c(z, t) = A\, e^{i\,(k_z z - \omega t)}, \tag{5.28}$$

we use the discrete function $\tilde{c}_k^n$ formed by replacing z by $k\Delta z$ and t by $n\Delta t$:

$$\tilde{c}_k^n = A\, e^{i\,(k_z\, k\Delta z - \omega\, n\Delta t)}, \tag{5.29}$$

---

[4] See biography at the end of this chapter.

where k_z is a vertical wavenumber and ω a frequency. To consider periodic behavior in space and possibly explosive behavior in time, k_z is restricted to be real positive, whereas $\omega = \omega_r + i\omega_i$ may be complex. Growth without bound occurs if $\omega_i > 0$. (If $\omega_i < 0$, the function decreases exponentially and raises no concern). The origins of z and t do not matter, for they can be adjusted by changing the complex amplitude A.

The range of k_z values is restricted. The lowest value is $k_z = 0$, corresponding to the constant component in Eq. (5.22). At the other extreme, the shortest wave is the "$2\Delta x \, mode$" or "saw-tooth" ($+1, -1, +1, -1$, etc.) with $k_z = \pi/\Delta z$. It is most often with this last value that trouble occurs, as seen in the rapidly oscillating values generated by the ill-fated Euler scheme (Fig. 5.9) and, earlier, aliasing (Section 1.12).

The elementary function, or trial solution, can be recast in the following form to distinguish the temporal growth (or decay) from the propagating part:

$$\tilde{c}_k^n = A \, e^{+\omega_i \Delta t n} e^{i(k_z \Delta z k - \omega_r \Delta t n)}. \tag{5.30}$$

An alternative way of expressing the elementary function is by introducing a complex number ϱ called the *amplification factor* such that

$$\tilde{c}_k^n = A \, \varrho^n e^{i(k_z \Delta z)k} \tag{5.31a}$$

$$\varrho = |\varrho| e^{i \, \arg(\varrho)} \tag{5.31b}$$

$$\omega_i = \frac{1}{\Delta t} \ln |\varrho|, \quad \omega_r = -\frac{1}{\Delta t} \arg(\varrho). \tag{5.31c}$$

The choice of expression among Eqs. (5.29), (5.30), and (5.31a) is a matter of ease and convenience.

Stability requires a nongrowing numerical solution, with $\omega_i \leq 0$ or equivalently $|\varrho| \leq 1$. Allowing for *physical* exponential growth—such as the growth of a physically unstable wave—we should entertain the possibility that $c(t)$ may grow as $\exp(\omega_i t)$, in which case $c(t + \Delta t) = c(t) \exp(\omega_i \Delta t) = c(t)[1 + \mathcal{O}(\Delta t)]$ and $\varrho = 1 + \mathcal{O}(\Delta t)$. In other words, instead of $|\varrho| \leq 1$, we should adopt the slightly less demanding criterion

$$|\varrho| \leq 1 + \mathcal{O}(\Delta t). \tag{5.32}$$

Since there is no exponential growth associated with diffusion, the criterion $|\varrho| \leq 1$ applies here.

We can now try Eq. (5.31a) as a solution of the discretized diffusion equation (5.27). After division by the factor $A\varrho^n \exp[i(k_z \Delta z)k]$ common to all terms, the discretized equation reduces to

$$\varrho = 1 + D\left[e^{+ik_z \Delta z} - 2 + e^{-ik_z \Delta z}\right], \tag{5.33}$$

which is satisfied when the amplification factor equals

$$\varrho = 1 - 2D[1 - \cos(k_z\Delta z)]$$

$$= 1 - 4D\sin^2\left(\frac{k_z\Delta z}{2}\right). \tag{5.34}$$

Since in this case ϱ happens to be real, the stability criterion stipulates $-1 \le \varrho \le 1$, that is, $4D\sin^2(k_z\Delta z/2) \le 2$, for all possible k_z values. The most dangerous value of k_z is the one that makes $\sin^2(k_z\Delta z/2) = 1$, which is $k_z = \pi/\Delta z$, the wavenumber of the saw-tooth mode. For this mode, ϱ violates $-1 \le \varrho$ unless

$$D = \frac{\kappa\Delta t}{\Delta z^2} \le \frac{1}{2}. \tag{5.35}$$

In other words, the Euler scheme is stable only if the time step is shorter than $\Delta z^2/(2\kappa)$. We are in the presence of *conditional stability*, and Eq. (5.35) is called the *stability condition* of the scheme.

Generally, criterion (5.35) or a similar one in another case is neither necessary nor sufficient since it neglects any effect due to boundary conditions, which can either stabilize an unstable mode or destabilize a stable one. In most situations, however, the criterion obtained by this method turns out to be a necessary condition since it is unlikely that in the middle of the domain, boundaries could stabilize an unstable solution, especially the shorter waves that are most prone to instability. On the other hand, boundaries can occasionally destabilize a stable mode in their vicinity. For the preceding scheme applied to the diffusion equation, this is not the case, and criterion (5.35) is both necessary and sufficient.

In addition to stability information, the amplification-factor method also enables a comparison between a numerical property and its physical counterpart. In the case of the diffusion equation, it is the damping rate, but, should the initial equation have described wave propagation, it would have been the dispersion relation. The general solution (5.19) of the exact equation (5.16) leads to the relation

$$\omega_i = -\kappa\, k_z^2, \tag{5.36}$$

which we can compare with the numerical damping rate

$$\tilde{\omega}_i = \frac{1}{\Delta t}\ln|\varrho|$$

$$= \frac{1}{\Delta t}\ln\left|1 - 4D\sin^2\left(\frac{k_z\Delta z}{2}\right)\right|. \tag{5.37}$$

The ratio τ of the numerical damping to the actual damping rate is then given by

$$\tau = \frac{\tilde{\omega}_i}{\omega_i} = -\frac{\ln|1 - 4D\sin^2(k_z\Delta z/2)|}{Dk_z^2\Delta z^2}, \tag{5.38}$$

which for small $k_z\Delta z$, that is, numerically well-resolved modes, behaves as

$$\tau = 1 + \left(2D - \frac{1}{3}\right)\left(\frac{k_z\Delta z}{2}\right)^2 + \mathcal{O}\left(k_z^4\Delta z^4\right). \tag{5.39}$$

For $D < 1/6$, the numerical scheme dampens less fast than the physical process ($\tau < 1$), whereas for larger values $1/6 < D < 1/2$ (i.e., relatively large but still stable time steps), overdamping occurs ($\tau > 1$). In practice, when $D > 1/4$ (leading to $\varrho < 0$ for the higher k_z values), this overdamping can be unrealistically large and unphysical. The shortest wave resolved by the spatial grid with $k_z\Delta z = \pi$ exhibits not only a saw-tooth pattern in space (as it should) but also a flip-flop behavior in time. This is because, for real negative ϱ, the sequence $\varrho^1, \varrho^2, \varrho^3, \ldots$ alternates in sign. For $-1 < \varrho < 0$, the solution vanishes not by monotonically decreasing toward zero but instead by oscillating around zero. Though the scheme is stable, the numerical solution behaves unlike the exact solution, and this should be avoided. It is therefore prudent to keep $D \le 1/4$ to guarantee a realistic solution.

Let us now give a physical interpretation of the stability condition $2\Delta t \le \Delta z^2/\kappa$. First, we observe that the instability appears most strongly for the component with the largest wavenumber according to Eq. (5.34). Since the length scale of this signal is Δz, the associated diffusion time scale is $\Delta z^2/\kappa$, and the stability criterion expresses the requirement that Δt be set shorter than a fraction of this time scale. It is equivalent to ensuring that the time step provides an adequate representation of the *shortest* component resolved by the spatial grid. Even when this shortest component is absent from the mathematical solution (in our initial problem only a single length scale, h, was present), it does occur in the numerical solution because of computer round-off errors, and stability is thus conditioned by the *possible* presence of the shortest resolved component. The stability condition ensures that all possible solution components are treated with an adequate time step.

As the preceding simple example shows, the amplification-factor method is easily applied and provides a stability condition, as well as other properties of the numerical solution. In practice, however, nonconstant coefficients (such as a spatially variable diffusivity κ) or nonuniform spacing of grid points may render its application difficult. Since nonuniform grids may be interpreted as a coordinate transformation, stretching and compressing grid node positions (see also Section 20.6.1), a nonuniform grid is equivalent to introducing nonconstant coefficients into the equation. The procedure is to "freeze" the coefficients at some value before applying the amplification-factor method and then repeat the analysis with different frozen values within the allotted ranges. Generally, this provides quite accurate estimates of permissible time steps. For nonlinear problems, the approach is to perform a preliminary linearization of the equation, but the quality of the stability condition is not always reliable. Finally, it is important to remember that the amplification-factor method does not deal with boundary conditions. To treat accurately cases with variable coefficients and nonuniform grids and to take boundary conditions into account, the so-called *matrix method* is available (e.g., Kreiss, 1962; Richtmyer & Morton, 1967).

We now have some tools to guarantee stability. Since our diffusion scheme is also consistent, we anticipate convergence by virtue of the Lax–Richtmyer Theorem (see Section 2.7). Let us then verify numerically whether the scheme leads to a linear decrease of the error with decreasing time step. Leaning on the exact solution (5.19) for comparison, we observe (Fig. 5.10) that the numerical solution does indeed exhibit a decrease of the error with decreasing time step, but only up to a point (for D decreasing from the stability limit of 0.5 to 1/6). The error increases again for smaller Δt. What happens?

The fact is that two sources of errors (space and time discretization) are simultaneously present and what we are measuring is the *sum* of these errors,

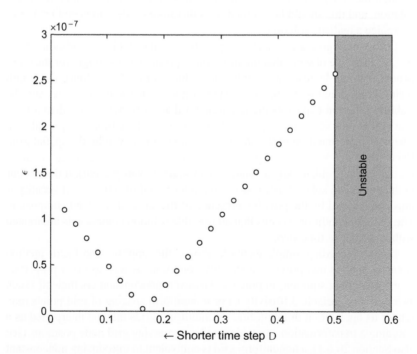

FIGURE 5.10 Root mean square of error $c - \tilde{c}$ scaled by the initial variation Δc at time $T = h^2/\kappa$ for a fixed space grid ($m = 50$) and decreasing time step (going from right to left). Above $D = 1/2$, the scheme is unstable and the error is extremely large (not plotted). For shorter time steps, the scheme is stable and the error first decreases linearly with D. Below $D = 1/6 = 0.167$, the error increases again.

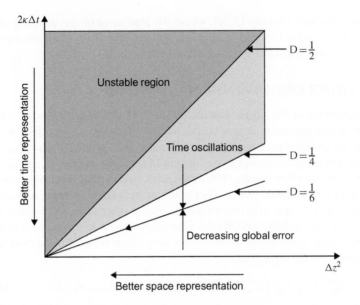

FIGURE 5.11   Different paths to convergence in the $(\Delta z^2, \Delta t)$ plane for the explicit scheme. For excessive values of Δt, $D \geq 1/2$, the scheme is unstable. Convergence can only be obtained by remaining within the stability region. When Δt alone is reduced (progressing vertically downward in the graph), the error decreases and then increases again. If Δz alone is decreased (progressing horizontally to the left in the graph), the error similarly decreases first and then increases, until the scheme becomes unstable. Reducing both Δt and Δz simultaneously at fixed D within the stability sector leads to monotonic convergence. The convergence rate is highest along the line $D = 1/6$ because the scheme then happens to be fourth-order accurate.

not the temporal error in isolation. This can be shown by looking at the modified equation obtained by using a Taylor-series expansion of discrete values $\tilde{c}_{k+1}^n$ etc. around $\tilde{c}_k^n$ in the difference equation (5.21). Some algebra leads to

$$\frac{\partial \tilde{c}}{\partial t} = \kappa \frac{\partial^2 \tilde{c}}{\partial z^2} + \frac{\kappa \Delta z^2 (1 - 6D)}{12} \frac{\partial^4 \tilde{c}}{\partial z^4} + \mathcal{O}\left(\Delta t^2, \Delta z^4, \Delta t \Delta z^2\right), \tag{5.40}$$

which shows that the scheme is of first order in time (through D) and second order in space. The rebounding error exhibited in Fig. 5.10 when Δt is gradually reduced (changing D alone) is readily explained in view of Eq. (5.40).

To check on convergence, we should consider the case when both parameters Δt and Δz are reduced simultaneously (Fig. 5.11). This is most naturally performed by keeping fixed the stability parameter D, which is a combination of both according to Eq. (5.26). The leading error (second term on the right) decreases as Δz^2, except when $D = 1/6$ in which case the scheme is then of fourth order. It can be shown[5] that in that case the error is on the order of Δz^4.

_____

[5] To show this, consider that for $D = 1/6$, $\Delta t = \Delta z^2 / 6\kappa$ and all contributions to the error term become proportional to Δz^4.

This is consistent with Eq. (5.39), where the least error on the damping rate is obtained with $2D = 1/3$, that is, $D = 1/6$, and with Fig. 5.10, where the error for fixed Δz is smallest when the time step corresponds to $D = 1/6$.

5.5 OTHER ONE-DIMENSIONAL SCHEMES

A disadvantage of the simple scheme (5.25) is its fast increase in cost when a higher spatial resolution is sought. For stability reasons Δt decreases as Δz^2, forcing us to calculate values at not only more grid points but also more frequently. For integration over a fixed length of time, the number of calculations grows as m^3. In other words, 1000 times more calculations must be performed if the grid size is divided by 10. Because this penalizing increase is rooted in the stability condition, it is imperative to explore other schemes that may have more attractive stability conditions. One such avenue is to consider implicit schemes. With a *fully implicit scheme*, the new values are used in the discretized derivative, and the algorithm is

$$\tilde{c}_k^{n+1} = \tilde{c}_k^n + D\left(\tilde{c}_{k+1}^{n+1} - 2\tilde{c}_k^{n+1} + \tilde{c}_{k-1}^{n+1}\right) \quad k = 2, \ldots, m-1. \tag{5.41}$$

The application of the stability analysis provides an amplification factor ϱ given implicitly by

$$\varrho = 1 - \varrho 2D\left[1 - \cos(k_z \Delta z)\right],$$

of which the solution is

$$\varrho = \frac{1}{1 + 4D \sin^2(k_z \Delta z/2)} \leq 1. \tag{5.42}$$

Because this amplification factor is always real and less than unity, there is no stability condition to be met, and the scheme is stable for any time step. This is called *unconditional stability*. The implicit scheme therefore allows us in principle to use a time step as large as we wish. We immediately sense, of course, that a large time step cannot be acceptable. Should the time step be too large, the calculated values would not "explode" but would provide a very inaccurate approximation to the true solution. This is confirmed by comparing the damping of the numerical scheme against its true value:

$$\tau = \frac{\tilde{\omega}_i}{\omega_i} = \frac{\ln|1 + 4D \sin^2(k_z \Delta z/2)|}{4D\,(k_z \Delta z/2)^2}. \tag{5.43}$$

For small D, the scheme behaves reasonably well, but for larger D, even for scales ten times larger than the grid spacing, the error on the damping rate is similar to the damping rate itself (Fig. 5.12).

Setting aside the accuracy restriction, we still have another obstacle to overcome. To calculate the left-hand side of Eq. (5.41) at grid node k, we have to know the values of the still unknown $\tilde{c}_{k+1}^{n+1}$ and $\tilde{c}_{k-1}^{n+1}$, which in turn depend on the

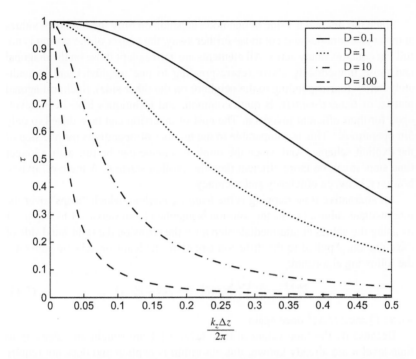

FIGURE 5.12   Ratio $\tau = \tilde{\omega}_i/\omega_i$ of the numerical damping rate of the implicit scheme to the exact value, as function of $k_z \Delta z/2$ for different values of D. For increasing time steps (increasing value of D), the numerical damping deteriorates rapidly even for relatively well-resolved solution components, and it is prudent to use a short time step, if not for stability, at least for accuracy.

unknown values at their adjacent nodes. This creates a circular dependency. It is, however, a linear dependency, and all we need to do is to formulate the problem as a set of simultaneous linear equations, that is, to frame the problem as a matrix to be inverted, once at each time step. Standard numerical techniques are available for such problem, most of them based on the so-called *Gaussian elimination* or *lower-upper decomposition* (e.g., Riley, Hobson & Bence, 1977). These methods are the most efficient ones for inverting arbitrary matrices of dimension N, and their computational cost increases as N^3. For the one-dimensional case with $N \sim m$, the matrix inversion requires m^3 operations to be performed.[6] Even if we executed only a single time step, the cost would be the same as for the execution of the explicit scheme during the full simulation. We may wonder: Is there some law of conservation of difficulty? Apparently there is, but we can exploit the particular form of the system to reduce the cost.

---

[6] If we anticipate generalization to three dimensions with $N \sim 10^6 - 10^7$ unknowns, a matrix inversion would demand a number of operations proportional to N^3 (at each time step!) and cannot be seriously considered as a viable approach.

Since the unknown value at one node depends on only the unknown values at the adjacent nodes and not those further away, the matrix of the system is not full but contains many zeros. All elements are zero except those on the diagonal and those immediately above (corresponding to one neighbor) and immediately below (corresponding to the neighbor on the other side). Such tridiagonal matrix, or *banded matrix*, is quite common, and techniques have been developed for their efficient inversion. The cost of inversion can be reduced to only 5m operations.[7] This is comparable to the number of operations for one step of the explicit scheme. And, since the implicit scheme can be run with a longer time step, it can be more efficient than the explicit scheme. A trade-off exists, however, between efficiency and accuracy.

An alternative time stepping is the *leapfrog method*, which "leaps" over the intermediate values, that is, the solution is marched from step $n-1$ to step $n+1$ by using the values at intermediate step n for the terms on the right-hand side of the equation. Applied to the diffusion equation, the leapfrog scheme generates the following algorithm:

$$\tilde{c}_k^{n+1} = \tilde{c}_k^{n-1} + 2D \left(\tilde{c}_{k+1}^n - 2\tilde{c}_k^n + \tilde{c}_{k-1}^n \right). \tag{5.44}$$

where $D = \kappa \Delta t / \Delta z^2$ once again.

Because by the time values at time level $n+1$ are sought all values up to time level n are already known, this algorithm is explicit and does not require any matrix inversion. We can analyze its stability by considering, as before, a single Fourier mode of the type (5.29). The usual substitution into the discrete equation, this time (5.44), application of trigonometric formulas, and division by the Fourier mode itself then lead to the following equation for the amplification factor ϱ of the leapfrog scheme:

$$\varrho = \frac{1}{\varrho} - 8D \sin^2 \left(\frac{k_z \Delta z}{2} \right). \tag{5.45}$$

This equation is quadratic and has therefore two solutions for ϱ, corresponding to two temporal modes. Only a single mode was expected because the original equation had only a first-order time derivative in time, but obviously, the scheme has introduced a second, *spurious mode*. With $b = 4D \sin^2(k_z \Delta z/2)$, the two solutions are

$$\varrho = -b \pm \sqrt{b^2 + 1}. \tag{5.46}$$

The physical mode is $\varrho = -b + \sqrt{b^2 + 1}$ because for well- resolved components ($k_z \Delta z \ll 1$ and thus $b \ll 1$) it is approximately $\varrho \simeq 1 - b \simeq 1 - Dk_z^2 \Delta z^2$, as it should be [see Eq. (5.34)]. Its value is always less than one, and the physical

---

[7] See Appendix C for the formulation of the algorithm.

mode is numerically stable. The other solution, $\varrho = -b - \sqrt{b^2 + 1}$, corresponds to the spurious mode and, unfortunately, has a magnitude always larger than one, jeopardizing the overall stability of the scheme. This is an example of *unconditional instability*. Note, however, that although unstable in the diffusion case, the leapfrog scheme will be found to be stable when applied to other equations.

The spurious mode causes numerical instability and must therefore be suppressed. One basic method is *filtering* (see Section 10.6) Because numerical instability is manifested by flip-flop in time (due to the negative ϱ value), averaging over two consecutive time steps or taking some kind of running average, called filtering, eliminates the flip-flop mode. Filtering, unfortunately, also alters the physical mode, and, as a rule, it is always prudent not to have a large flip-flop mode in the first place. Its elimination should be done a priori, not a posteriori. In the case of models using leapfrog for the sake of other terms in the equation, such as advection terms which it handles in a stable manner, the diffusion term is generally discretized at time level $n - 1$ rather than n, rendering the scheme as far as the diffusion part is concerned equivalent to the explicit Euler scheme with time step of $2\Delta t$.

Finally, we can illustrate the *finite-volume technique* in the more general case of nonuniform diffusion and variable grid spacing. In analogy with Eq. (3.35), we integrate the diffusion equation over an interval between two consecutive cell boundaries and over one time step to obtain the grid-cell averages $\bar{c}$ (Fig. 5.13)

$$\frac{\bar{c}_k^{n+1} - \bar{c}_k^n}{\Delta t_n} + \frac{\hat{q}_{k+1/2} - \hat{q}_{k-1/2}}{\Delta z_k} = 0, \tag{5.47}$$

assuming that the time-averaged flux at the interface between cells

$$\hat{q} = \frac{1}{\Delta t_n} \int_{t^n}^{t^{n+1}} -\kappa \frac{\partial c}{\partial z} \mathrm{d}t \tag{5.48}$$

is somehow known. Up to this point, the equations are exact. The variable c appearing in the expression of the flux is the actual function, including all its subgrid-scale variations, whereas Eq. (5.47) deals only with space-time averages. Discretization enters the formulation as we relate the time-averaged flux

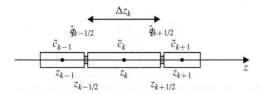

FIGURE 5.13 Arrangement of cells and interfaces for the finite-volume technique. Concentration values are defined at cell centers, whereas flux values are defined between cells. Cell lengths do not have to be uniform.

to the space-averaged function $\bar{c}$ to close the problem. We can, for example, estimate the flux using a factor α of implicitness and a gradient approximation:

$$\hat{q}_{k-1/2} \simeq -(1-\alpha)\,\kappa_{k-1/2}\frac{\tilde{c}_k^n - \tilde{c}_{k-1}^n}{z_k - z_{k-1}} - \alpha\,\kappa_{k-1/2}\frac{\tilde{c}_k^{n+1} - \tilde{c}_{k-1}^{n+1}}{z_k - z_{k-1}}, \tag{5.49}$$

where $\tilde{c}$ is now interpreted as the numerical estimate of the spatial averages. The numerical scheme reads

$$\tilde{c}_k^{n+1} = \tilde{c}_k^n + (1-\alpha)\frac{\kappa_{k+1/2}\Delta t_n}{\Delta z_k}\frac{\tilde{c}_{k+1}^n - \tilde{c}_k^n}{z_{k+1} - z_k} - (1-\alpha)\frac{\kappa_{k-1/2}\Delta t_n}{\Delta z_k}\frac{\tilde{c}_k^n - \tilde{c}_{k-1}^n}{z_k - z_{k-1}}$$
$$+ \alpha\frac{\kappa_{k+1/2}\Delta t_n}{\Delta z_k}\frac{\tilde{c}_{k+1}^{n+1} - \tilde{c}_k^{n+1}}{z_{k+1} - z_k} - \alpha\frac{\kappa_{k-1/2}\Delta t_n}{\Delta z_k}\frac{\tilde{c}_k^{n+1} - \tilde{c}_{k-1}^{n+1}}{z_k - z_{k-1}}. \tag{5.50}$$

With uniform grid spacing, κ constant, and $\alpha = 0$, we recover Eq. (5.25). Since the present finite-volume scheme is by construction conservative (see Section 3.9), we have incidentally proven that (5.25) is conservative in the case of a uniform grid and constant diffusivity, a property that can be verified numerically with `firstdiffusion.m` even in the unstable case.

In practice, it is expedient to program the calculations with the flux values defined and stored alongside the concentration values. The computations then entail two stages in every step: first the computation of the flux values from the concentration values at the same time level and then the updation of the concentration values from these most recent flux values. In this manner, it is clear how to take into account variable parameters such as the local value of the diffusivity κ (at cell edges rather than at cell centers), local cell length, and momentary time step. The approach is also naturally suited for the implementation of flux boundary conditions.

5.6   MULTI-DIMENSIONAL NUMERICAL SCHEMES

Explicit schemes are readily generalized to two and three dimensions[8] with indices i, j, and k being grid positions in the respective directions x, y, and z:

$$\tilde{c}^{n+1} = \tilde{c}^n + \frac{A\Delta t}{\Delta x^2}\left(\tilde{c}_{i+1}^n - 2\tilde{c}^n + \tilde{c}_{i-1}^n\right)$$
$$+ \frac{A\Delta t}{\Delta y^2}\left(\tilde{c}_{j+1}^n - 2\tilde{c}^n + \tilde{c}_{j-1}^n\right)$$
$$+ \frac{\kappa\Delta t}{\Delta z^2}\left(\tilde{c}_{k+1}^n - 2\tilde{c}^n + \tilde{c}_{k-1}^n\right). \tag{5.51}$$

---

[8] In order not to overload the notation, indices are written here only if they differ from the local grid point index. Therefore, $\tilde{c}(t^n, x_i, y_j, z_k)$ is written $\tilde{c}^n$, whereas $\tilde{c}_{j+1}^n$ stands for $\tilde{c}(t^n, x_i, y_{j+1}, z_k)$.

The stability condition is readily obtained by using the amplification-factor analysis. Substituting the Fourier mode

$$\tilde{c}^n = A \varrho^n e^{i(i\,k_x \Delta x)} e^{i(j\,k_y \Delta y)} e^{i(k\,k_z \Delta z)} \tag{5.52}$$

in the discrete equation, we obtain the following generalization of Eq. (5.35):

$$\frac{\mathcal{A}\Delta t}{\Delta x^2} + \frac{\mathcal{A}\Delta t}{\Delta y^2} + \frac{\kappa \Delta t}{\Delta z^2} \le \frac{1}{2}. \tag{5.53}$$

The implicit formulation of the scheme is not much more complicated and is, again, unconditionally stable. The associated matrix, however, is no longer tridiagonal but has a slightly more complicated structure (Fig. 5.14). Unfortunately, there exists no direct solver for which the cost remains proportional to the size of the problem. Several strategies can be developed to keep the method "implicit" with affordable costs.

In any case, a direct solver is in some way an overkill. It inverts the matrix exactly up to rounding errors, and such precision is not necessary in view of the much larger errors associated with the discretization (see Section 4.8). We can therefore afford to invert the matrix only approximately, and this can be accomplished by the use of iterative methods, which deliver solutions to any degree of approximation depending on the number of iterations performed. A small number of iterations usually yields an acceptable solution because the starting guess values may be taken as the values computed at the preceding time step. Two popular *iterative solvers* of linear systems are the Gauss–Seidel method and the Jacobi method, but there exist many other iterative solvers, more or less optimized for different kinds of problems and computers (e.g., Dongarra, Duffy, Sorensen & van der Vorst, 1998). In general, most software libraries offer

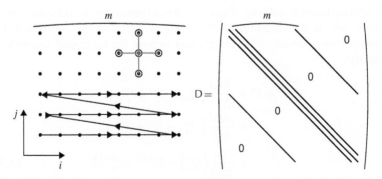

FIGURE 5.14 If the numerical state vector is constructed row by row in two dimensions, $\tilde{c}_{i,j}$ is the element $(j-1)m+i$ of $\mathbf{x}$. Since the diffusion operator at point i, j involves $\tilde{c}_{i,j}$, $\tilde{c}_{i+1,j}$, $\tilde{c}_{i-1,j}$, $\tilde{c}_{i,j-1}$, and $\tilde{c}_{i,j+1}$, the matrix to be inverted has zero elements everywhere, except on the diagonal (the point itself), the superdiagonal (point $i+1, j$), the subdiagonal (point $i-1, j$), and two lines situated $\pm m$ away from the diagonal (point $i, j \pm 1$).

a vast catalogue of methods, and we will only mention a few general approaches, giving more detail on specific methods later when we need to solve a Poisson equation for a pressure or streamfunction (Section 7.6).

Any linear system of simultaneous equations can be cast as

$$\mathbf{A}\mathbf{x} = \mathbf{b} \tag{5.54}$$

where the matrix $\mathbf{A}$ gathers all the coefficients, the vector $\mathbf{x}$ all the unknowns, and the vector $\mathbf{b}$ the boundary values and external forcing terms, if any. The objective of an iterative method is to solve this system by generating a sequence $\mathbf{x}^{(p)}$ that starts from a guess vector $\mathbf{x}^0$ and gradually converges toward the solution. The algorithm is a repeated application of

$$\mathbf{B}\mathbf{x}^{(p+1)} = \mathbf{C}\mathbf{x}^{(p)} + \mathbf{b} \tag{5.55}$$

where $\mathbf{B}$ must be easy to invert, otherwise there is no gain, and is typically a diagonal or triangular matrix (nonzero elements only on the diagonal or on the diagonal and one side of it). At convergence, $\mathbf{x}^{(p+1)} = \mathbf{x}^{(p)}$ and we must therefore have $\mathbf{B} - \mathbf{C} = \mathbf{A}$ to have solved Eq. (5.54). The closer $\mathbf{B}$ is to $\mathbf{A}$, the faster the convergence since at the limit of $\mathbf{B} = \mathbf{A}$, a single iteration would yield the exact answer. Using $\mathbf{C} = \mathbf{B} - \mathbf{A}$, we can rewrite the iterative step as

$$\mathbf{x}^{(p+1)} = \mathbf{x}^{(p)} + \mathbf{B}^{-1}\left(\mathbf{b} - \mathbf{A}\mathbf{x}^{(p)}\right) \tag{5.56}$$

which is reminiscent of a time stepping method. Here, $\mathbf{B}^{-1}$ denotes the inverse of $\mathbf{B}$. The Jacobi method uses a diagonal matrix $\mathbf{B}$, whereas the Gauss–Seidel method uses a triangular matrix $\mathbf{B}$. More advanced methods exist that converge faster than either of these. Those will be outlined in Section 7.6.

In GFD applications, diffusion is rarely dominant (except for vertical diffusion in strongly turbulent regime), and stability restrictions associated with diffusion are rarely penalizing. Therefore, it is advantageous to make the scheme implicit only in the direction of the strongest diffusion (or largest variability of diffusion), usually the vertical, and to treat the horizontal components explicitly:

$$\begin{aligned}
\tilde{c}^{n+1} = \tilde{c}^n &+ \frac{A\Delta t}{\Delta x^2}\left(\tilde{c}_{i+1}^n - 2\tilde{c}^n + \tilde{c}_{i-1}^n\right) \\
&+ \frac{A\Delta t}{\Delta y^2}\left(\tilde{c}_{j+1}^n - 2\tilde{c}^n + \tilde{c}_{j-1}^n\right) \\
&+ \frac{\kappa\Delta t}{\Delta z^2}\left(\tilde{c}_{k+1}^{n+1} - 2\tilde{c}^{n+1} + \tilde{c}_{k-1}^{n+1}\right).
\end{aligned} \tag{5.57}$$

Then, instead of inverting a matrix with multiple bands of nonzero elements, we only need to invert a tridiagonal matrix at each point of the horizontal grid. *Alternating direction implicit* (ADI) methods use the same approach but change the direction of the implicit sweep through the matrix at every time step. This helps when stability of the horizontal diffusion discretization is a concern.

The biggest challenge associated with diffusion in GFD models is, however, not their numerical stability but rather their physical basis because diffusion is often introduced as a parameterization of unresolved processes. Occasionally, the unphysical behavior of the discretization may create a problem (e.g., Beckers, Burchard, Deleersnijder & Mathieu, 2000).

ANALYTICAL PROBLEMS

5.1. What would the energy spectrum $E_k(k)$ be in a turbulent flow where all length scales were contributing equally to dissipation? Is this spectrum realistic?

5.2. Knowing that the average atmospheric pressure on Earth's surface is 1.013×10^5 N/m^2 and that Earth's average radius is 6371 km, deduce the mass of the atmosphere. Then, using this and the fact that the earth receives 1.75×10^{17} W from the sun globally, and assuming that half of the energy received from the sun is being dissipated in the atmosphere, estimate the rate of dissipation ϵ in the atmosphere. Assuming finally that turbulence in the atmosphere obeys the Kolmogorov theory, estimate the smallest eddy scale in the air, its ratio to the largest scale (the earth's radius), and the large-scale wind velocity. Is this velocity scale realistic?

5.3. In a 15-m coastal zone, the water density is 1032 kg/m^3 and the horizontal velocity scale is 0.80 m/s. What are the Reynolds number and the diameter of the shortest eddies? Approximately how many watts are dissipated per square meter of the ocean?

5.4. If you have to simulate the coastal ocean of the previous problem with a numerical model that includes 20 grid points over the vertical, what would be a reasonable value for the vertical eddy diffusivity?

5.5. Estimate the time it takes to reduce by a factor 2 a salinity variation in an ocean of depth $H = 1000$ m in the presence of salt diffusion, with a diffusion coefficient κ. Compare two solutions, one using the molecular diffusion ($\kappa = 10^{-9}$ m^2/s) and the other a turbulent diffusion typical of the deep ocean ($\kappa = 10^{-4}$ m^2/s).

5.6. A deposition at the sea surface of a tracer (normalized and without units) can be modeled by a constant flux $q = -10^{-4}$ m/s. At depth $z = -99$ m, a strong current is present and flushes the vertically diffused tracer so that $c = 0$ is maintained at that level. Assuming the diffusion coefficient has the profile of Fig. 5.15, calculate the steady solution for the tracer distribution.

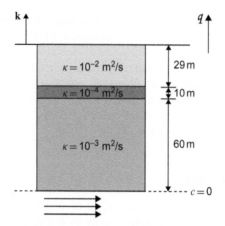

FIGURE 5.15 Values of a nonuniform eddy diffusion for Analytical Problem 5.6. A flux condition is imposed at the surface while $c = 0$ at the base of the domain.

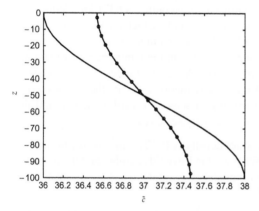

FIGURE 5.16 With a time step such that $D = 0.1$, the initial condition (single line) of the 1D diffusion problem has been damped after 500 time steps and the numerical solution of the explicit scheme (open circles) is almost indistinguishable from the exact solution (shown as a line crossing the circles), even with only 30 grid points across the domain.

5.7. Verify the assertion made below Eq. (5.4) that the Reynolds number corresponding to the Kolmogorov scale is on the order of unity.

NUMERICAL EXERCISES

5.1. Cure the unstable version `firstdiffusion.m` by adapting the time step and verify that below the limit (5.35) the scheme is indeed stable and provides accurate solutions (Fig. 5.16).

5.2. For a 1D Euler scheme with implicit factor α, constant grid size and constant diffusion coefficient, prove that the stability condition is $(1 - 2\alpha)D \leq 1/2$.

5.3. Implement periodic boundary conditions in the 1D diffusion problem (i.e., $c_{top} = c_{bottom}$ and $q_{top} = q_{bottom}$). Then, search the internet for a tridiagonal matrix inversion algorithm adapted to periodic boundary conditions and implement it.

5.4. Implement the 1D finite-volume method with an implicit factor α and variable diffusion coefficient. Set the problem with the same initial and boundary conditions as in the beginning of Section 5.3. Verify your solution against the exact solution (5.19).

5.5. Apply the code developed in Section 5.6 to the Analytical Problem 5.6. Start with an arbitrary initial condition and march in time until the solution becomes stationary. Estimate a priori the permitted time step and the minimum total number of time steps, depending on the implicit factor. Take $\Delta z = 2$, track convergence during the calculations and compare your final solution with the exact solution. Also try to implement the naive discretization

$$
\begin{aligned}
\frac{\partial}{\partial z}\left(\kappa \frac{\partial c}{\partial z}\right)\bigg|_{z_k} &= \kappa \frac{\partial^2 c}{\partial z^2}\bigg|_{z_k} + \frac{\partial \kappa}{\partial z}\bigg|_{z_k} \frac{\partial c}{\partial z}\bigg|_{z_k} \\
&\sim \frac{\kappa_k\left(\tilde{c}_{k+1} - 2\tilde{c}_k + \tilde{c}_{k-1}\right)}{\Delta z^2} \\
&\quad + \frac{\left(\kappa_{k+1} - \kappa_{k-1}\right)}{2\Delta z} \frac{\left(\tilde{c}_{k+1} - \tilde{c}_{k-1}\right)}{2\Delta z}.
\end{aligned} \tag{5.58}
$$

5.6. The Dufort–Frankel scheme approximates the diffusion equation by

$$
\tilde{c}_k^{n+1} = \tilde{c}_k^{n-1} + 2D\left[\tilde{c}_{k+1}^n - \left(\tilde{c}_k^{n+1} + \tilde{c}_k^{n-1}\right) + \tilde{c}_{k-1}^n\right]. \tag{5.59}
$$

Verify the consistency of this scheme. What relation must be imposed between Δt and Δz when each approaches zero to ensure consistency? Then, analyze numerical stability using the amplification-factor method.

Andrey Nikolaevich Kolmogorov
1903–1987

Andrey Kolmogorov was attracted to mathematics from an early age and, at the time of his studies at Moscow State University, sought the company of the most outstanding mathematicians. While still an undergraduate student, he began research and published several papers of international importance, chiefly on set theory. He had already 18 publications by the time he completed his doctorate in 1929. Kolmogororov's contributions to mathematics spanned a variety of topics, and he is perhaps best known for his work on probability theory and stochastic processes.

Research in stochastic processes led to a study of turbulent flow from a jet engine and, from there, to two famous papers on isotropic turbulence in 1941. It has been remarked that these two papers rank among the most important ones since Osborne Reynolds in the long and unfinished history of turbulence theory.

Kolmogorov found much inspiration for his work during nature walks in the outskirts of Moscow accompanied by colleagues and students. The brainstorming that had occurred during the walk often concluded in serious work around the dinner table upon return home. (*Photo from American Mathematical Society*)

John Louis von Neumann
1903–1957

John von Neumann was a child prodigy. At age six, he could mentally divide eight-digit numbers and memorize the entire page of a telephone book in a matter of minutes, to the amazement of his parents' guests at home. Shortly after obtaining his doctorate in 1928, he left his native Hungary to take an appointment at Princeton University (USA). When the Institute for Advanced Studies was founded there in 1933, he was named one of the original Professors of Mathematics.

Besides seminal contributions to ergodic theory, group theory, and quantum mechanics, his work included the application of electronic computers to applied mathematics. Together with Jule Charney (see biography at end of Chapter 16) in the 1940s, he selected weather forecasting as the first challenge for the emerging electronic computers, which he helped assemble. Unlike Lewis Richardson before them, von Neumann and Charney started with a single equation, the barotropic vorticity equation. The results exceeded expectations, and scientific computing was launched.

A famous quote attributed to von Neumann is: "If people do not believe that mathematics is simple, it is only because they do not realize how complicated life is." (*Photo from Virginia Polytechnic Institute and State University*)

Chapter 6

Transport and Fate

ABSTRACT

In this Chapter, we augment the diffusion equation of the preceding chapter to include the effects of advection (transport by the moving fluid) and fate (diffusion, plus possible source, and decay along the way). The numerical section begins with the design of schemes for advection in a fixed (Eulerian) framework and then extends those to include the discretization of diffusion and source/decay terms. Most of the developments are presented in one dimension before generalization to multiple dimensions.

6.1 COMBINATION OF ADVECTION AND DIFFUSION

When considering the heat (3.23), salt (3.14), humidity (3.15), or density (4.8) equations of Geophysical Fluid Dynamics, we note that they each include three types of terms. The first, a time derivative, tells how the variable is changing over time. The second is a group of three terms with velocity components and spatial derivatives, sometimes hidden in the material derivative d/dt. They represent the transport of the substance with the flow and are collectively called *advection*. Finally, the last group of terms, on the right-hand side, includes an assortment of diffusivities and second-order spatial derivatives. In the light of Chapter 5, we identify these with *diffusion*. They represent the spreading of the substance along and across the flow. Using a generic formulation, we are brought to study an equation of the type

$$\frac{\partial c}{\partial t} + u\frac{\partial c}{\partial x} + v\frac{\partial c}{\partial y} + w\frac{\partial c}{\partial z} = \frac{\partial}{\partial x}\left(\mathcal{A}\frac{\partial c}{\partial x}\right) + \frac{\partial}{\partial y}\left(\mathcal{A}\frac{\partial c}{\partial y}\right) + \frac{\partial}{\partial z}\left(\kappa\frac{\partial c}{\partial z}\right), \quad (6.1)$$

where the variable c may stand for any of the aforementioned variables or represent a substance imbedded in the fluid, such as a pollutant in the atmosphere or in the sea. Note the anisotropy between the horizontal and vertical directions ($\mathcal{A}$ generally $\gg \kappa$).

The examples in the following figures illustrate the combined effects of advection and diffusion. Figure 6.1 shows the fate of the Rhône River waters as

FIGURE 6.1   Rhône River plume discharging in the Gulf of Lions (circa 43°N) and carrying sediments into the Mediterranean Sea. This satellite picture was taken on 26 February 1999. (*Satellite image provided by the SeaWiFS Project, NASA/Goddard Space Flight Center*). A color version can be found online at http://booksite.academicpress.com/9780120887590/

they enter the Mediterranean Sea. Advection pulls the plume offshore, whereas diffusion dilutes it. Figure 6.2 is a remarkable satellite picture, showing wind advection of sand from the Sahara Desert westward from Africa to Cape Verde (white band across the lower part of the picture) at the same time as, and independently from, marine transport of suspended matter southwestward from the Cape Verde islands (von Kármán vortices in left of middle of the picture). Although sand is being blown quickly and without much diffusion in the air, the sediments follow convoluted paths in the water, pointing to a disparity between the relative effects of advection and diffusion in the atmosphere and ocean.

Often, the substance being carried by the fluid is not simply moved and diffused by the flow. It may also be created or lost along the way. Such is the case of particle matter, which tends to settle at the bottom. Chemical species can be produced by reaction between parent chemicals and be lost by participating in other reactions. An example of this is sulfuric acid (H_2SO_4) in the atmosphere: It is produced by reaction of sulfur dioxide (SO_2) from combustion and lost by precipitation (acid rain or snow). Tritium, a naturally radioactive form of hydrogen enters the ocean by contact with air at the surface and disintegrates along its oceanic journey to become Helium. Dissolved oxygen in the sea is consumed by biological activity and is replenished by contact with air at the surface.

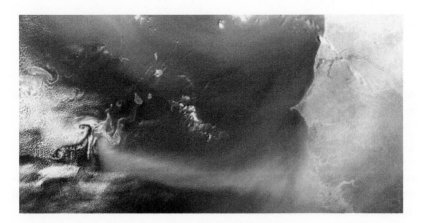

FIGURE 6.2   Sahara dust blown by the wind from the African continent over the ocean toward Cape Verde islands (15–17°N), while suspended matter in the water is being transported southwestward by a series of von Kármán vortices in the wake of the islands. Note in passing how these vortices in the water relate to the overlying cloud patterns. (*Jacques Descloitres, MODIS Land Science Team*)

To incorporate these processes, we augment the advection–diffusion equation (6.1) by adding source and sink terms in the right-hand side:

$$\frac{\partial c}{\partial t} + u\frac{\partial c}{\partial x} + v\frac{\partial c}{\partial y} + w\frac{\partial c}{\partial z}$$
$$= \frac{\partial}{\partial x}\left(\mathcal{A}\frac{\partial c}{\partial x}\right) + \frac{\partial}{\partial y}\left(\mathcal{A}\frac{\partial c}{\partial y}\right) + \frac{\partial}{\partial z}\left(\kappa\frac{\partial c}{\partial z}\right) + S - Kc, \qquad (6.2)$$

where the term S stands for the source, the formulation of which depends on the particular process of formation of the substance, and K is a coefficient of decay, which affects how quickly (large K) or slowly (small K) the substance is being lost.

At one dimension, say in the x-direction, and with constant diffusivity $\mathcal{A}$, the equation reduces to:

$$\frac{\partial c}{\partial t} + u\frac{\partial c}{\partial x} = \mathcal{A}\frac{\partial^2 c}{\partial x^2} + S - Kc. \qquad (6.3)$$

Several properties of the advection–diffusion equation are worth noting because they bear on the numerical procedures that follow: In the absence of source and sink, the total amount of the substance is conserved, and, in the further absence of diffusion, the variance of the concentration distribution, too, is conserved over time.

When we integrate Eq. (6.2) over the domain volume $\mathcal{V}$, we can readily integrate the diffusion terms and, if the flux is zero at all boundaries, these vanish, and we obtain the following:

$$\frac{d}{dt}\int_V c\,dV = -\int_V \left(u\frac{\partial c}{\partial x} + v\frac{\partial c}{\partial y} + w\frac{\partial c}{\partial z} \right)dV + \int_V S\,dV - \int_V Kc\,dV.$$

After an integration by parts, the first set of terms on the right can be rewritten as

$$\frac{d}{dt}\int_V c\,dV = +\int_V c\left(\frac{\partial u}{\partial x} + \frac{\partial v}{\partial y} + \frac{\partial w}{\partial z} \right)dV + \int_V S\,dV - \int_V Kc\,dV,$$

as long as there is no flux or no advection at all boundaries. Invoking the continuity equation (4.21d) reduces the first term on the right to zero, and we obtain simply:

$$\frac{d}{dt}\int_V c\,dV = \int_V S\,dV - \int_V Kc\,dV. \tag{6.4}$$

Since the concentration c represents the amount of the substance per volume, its integral over the volume is its total amount. Equation (6.4) simply states that this amount remains constant over time when there is no source ($S = 0$) or sink ($K = 0$). Put another way, the substance is moved around but conserved.

Now, if we multiply Eq. (6.2) by c and then integrate over the domain, we can integrate the diffusion terms by parts and, if the flux is again zero at all boundaries, we have the following:

$$\frac{1}{2}\frac{d}{dt}\int_V c^2\,dV = -\int_V \left[A\left(\frac{\partial c}{\partial x}\right)^2 + A\left(\frac{\partial c}{\partial y}\right)^2 + \kappa\left(\frac{\partial c}{\partial z}\right)^2 \right]dV$$
$$+ \int_V Sc\,dV - \int_V Kc^2\,dV. \tag{6.5}$$

With no diffusion, source, or sink, the right-hand side is zero, and *variance* is conserved in time. Diffusion and decay tend to reduce variance, whereas a (positive) source tends to increase it.

This conservation property can be extended, still in the absence of diffusion, source, and sink, to any power c^p of c, by multiplying the equation by c^{p-1} before integration. The conservation property even holds for any function $F(c)$. It goes without saying that numerical methods cannot conserve all these quantities, but it is highly desirable that they conserve at least the first two (total amount and variance).

There is one more property worth mentioning, which we will state without demonstration but justify in a few words. Because diffusion acts to smooth the distribution of c, it removes the substance from the areas of higher concentration and brings it into regions of lower concentration. Hence, due to diffusion alone, the maximum of c can only diminish and its minimum can only increase. Advection, by contrast, redistributes existing values, thus not changing either minimum and maximum. In the absence of source and sink, therefore, no future value of c can fall outside the initial range of values. This is called the *max-min property*. Exceptions are the presence of a source or sink, and the import through one of the boundaries of a concentration outside the initial range.

We call a numerical scheme that maintains the max-min property a *monotonic scheme* or *monotonicity preserving* scheme[1]. Alternatively, the property of *boundedness* is often used to describe a physical solution that does not generate new extrema. If c is initially positive everywhere, as it should be, the absence of new extrema keeps the variable positive at all future times, another property called *positiveness*. A monotonic scheme is thus positive but the reverse is not necessarily true.

6.2   RELATIVE IMPORTANCE OF ADVECTION: THE PECLET NUMBER

Since the preceding equations combine the effects of both advection and diffusion, it is important to compare the relative importance of one with the other. In a specific situation, could it be that one dominates over the other or that both impact concentration values to the same extent? To answer this question, we turn once again to scales. Introducing a length scale L, velocity scale U, diffusivity scale D, and a scale Δc to measure concentration differences, we note that advection scales like $U\Delta c/L$ and diffusion like $D\Delta c/L^2$. We can then compare the two processes by forming the ratio of their scales:

$$\frac{\text{advection}}{\text{diffusion}} = \frac{U\Delta c/L}{D\Delta c/L^2} = \frac{UL}{D}.$$

This ratio is by construction dimensionless. It bears the name of Peclet number[2] and is denoted by Pe:

$$Pe = \frac{UL}{D}, \tag{6.6}$$

where the scales U, L, and D may stand for the scales of either horizontal (u, v, x, y, and $\mathcal{A}$) or vertical (w, z, and κ) variables but not a mix of them. The Peclet number leads to an immediate criterion, as follows.

If $Pe \ll 1$ (in practice, if $Pe < 0.1$), the advection term is significantly smaller than the diffusion term. Physically, diffusion dominates, and advection is negligible. Diffusive spreading occurs almost symmetrically despite the directional bias of the weak flow. If we wish to simplify the problem, we may drop the

---

[1] Some computational fluid dynamicists do make a difference between these two labels, but this minor point lies beyond our present text.
[2] In honor of Jean Claude Eugène Péclet (1793–1857), French physicist who wrote a treatise on heat transfer.

advection term [$u \partial c / \partial x$ in (6.3)], as if u were zero. The relative error committed in the solution is expected to be on the order of the Peclet number, and
the smaller Pe, the smaller the error. The methods developed in the preceding
chapter were based on such simplification and thus apply whenever $Pe \ll 1$.

   If $Pe \gg 1$ (in practice, if $Pe > 10$), it is the reverse: the advection term is now
significantly larger than the diffusion term. Physically, advection dominates, and
diffusion is negligible. Spreading is weak, and the patch of substance is mostly
moved along, and possibly distorted by, the flow. If we wish to simplify the
problem, we may drop the diffusion term [$\mathcal{A} \partial^2 c / \partial x^2$ in (6.3)], as if $\mathcal{A}$ were
zero. The relative error committed in the solution by doing so is expected to be
on the order of the inverse of the Peclet number ($1/Pe$), and the larger Pe, the
smaller the error.

6.3 HIGHLY ADVECTIVE SITUATIONS

When a system is highly advective in one direction (high Pe number based on
scales U, L, and D corresponding to that direction), diffusion is negligible *in
that same direction*. This is not to say that it is also negligible in the other directions. For example, high advection in the horizontal does not preclude vertical
diffusion, as this is often the case in the lower atmosphere. In such a case, the
governing equation is

$$\frac{\partial c}{\partial t} + u \frac{\partial c}{\partial x} + v \frac{\partial c}{\partial y} = \frac{\partial}{\partial z} \left(\kappa \frac{\partial c}{\partial z} \right) + S - Kc. \tag{6.7}$$

   Because the diffusion terms are of higher-order (second derivatives) than
those of advection (first derivatives), the neglect of a diffusion term reduces
the order of the equation and, therefore, also reduces the need of boundary
conditions by one in the respective direction. The boundary condition at the
downstream end of the domain must be dropped: The concentration and flux
there are whatever the flow brings to that point. A problem occurs when the
situation is highly advective, but the small diffusion term is not dropped. In that
case, because the order of the equation is not reduced, a boundary condition is
enforced at the downstream end, and a locally high gradient of concentration
may occur.

   To see this, consider the one-dimensional, steady situation with no
source and sink, with constant velocity and diffusivity in the x-direction. The
equation is

$$u \frac{dc}{dx} = \mathcal{A} \frac{d^2 c}{dx^2}, \tag{6.8}$$

and its most general solution is

$$c(x) = C_0 + C_1 e^{ux/\mathcal{A}}. \tag{6.9}$$

For $u > 0$, the downstream end is to the right of the domain, and the solution increases exponentially towards the right boundary. Rather, it could be said that the solution decays away from this boundary as x decreases away from it. In other words, a boundary layer exists at the downstream end. The e-folding length of this boundary layer is $\mathcal{A}/u$, and it can be very short in a highly advective situation (large u and small $\mathcal{A}$). Put another way, the Peclet number is the ratio of the domain length to this boundary-layer thickness, and the larger the Peclet number, the smaller the fraction of the domain occupied by the boundary layer. Why do we need to worry about this? Because in a numerical model it may happen that the boundary-layer thickness falls below the grid size. It is therefore important to check the Peclet number in relation to the spatial resolution. Should the ratio of the grid size to the length scale of the system be comparable with, or larger than, the inverse of the Peclet number, diffusion must be neglected in that direction, or, if it must be retained for some reason, special care must be taken at the downstream boundary.

6.4 CENTERED AND UPWIND ADVECTION SCHEMES

In GFD, advection is generally dominant compared with diffusion, and we therefore begin with the case of pure advection of a tracer concentration $c(x, t)$ along the x-direction. The aim is to solve numerically the following equation:

$$\frac{\partial c}{\partial t} + u \frac{\partial c}{\partial x} = 0. \qquad (6.10)$$

For simplicity, we further take the velocity u as constant and positive so that advection carries c in the positive x-direction. The exact solution of this equation is

$$c(x, t) = c_0(x - ut), \qquad (6.11)$$

where $c_0(x)$ is the initial concentration distribution (at $t = 0$).

A spatial integration from $x_{i-1/2}$ to $x_{i+1/2}$ across a grid cell (Fig. 6.3) leads to the following budget

$$\frac{d\bar{c}_i}{dt} + \frac{q_{i+1/2} - q_{i-1/2}}{\Delta x} = 0, \quad q_{i-1/2} = uc|_{i-1/2}, \qquad (6.12)$$

which is the basis for the finite-volume technique, as in (3.33). To close the system, we need to relate the local fluxes q to the cell-average concentrations $\bar{c}$. To do so, we must introduce an approximation, because we do not know the actual value of c at the interfaces between cells, but only the average value in the cell on each side of it. It appears reasonable to use the following, consistent, numerical interpolation for the flux:

$$\tilde{q}_{i-1/2} = u \left(\frac{\bar{c}_i + \bar{c}_{i-1}}{2} \right), \qquad (6.13)$$

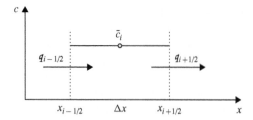

FIGURE 6.3 One-dimensional finite-volume approach with fluxes at the interfaces between grid cells for a straightforward budget calculation.

which is tantamount to assume that the local tracer concentration at the interface is equal to the mean of the surrounding cell averages. Before proceeding with time discretization, we can show that this centered approximation conserves not only the total amount of substance but also its variance, $\sum_i \bar{c}_i$ and $\sum_i (\bar{c}_i)^2$. Substitution of the flux approximation into (6.12) leads to the following semidiscrete equation for cell-averaged concentrations:

$$\frac{d\bar{c}_i}{dt} = -u \frac{\bar{c}_{i+1} - \bar{c}_{i-1}}{2\Delta x}. \tag{6.14}$$

Sum over index i leads to cancellation of terms by pairs on the right, leaving only the first and last $\bar{c}$ values. Then, multiplication of the same equation by c_i followed by the sum over the domain provides the time-evolution equation of the discretized variance:

$$\frac{d}{dt} \left(\sum_i (\bar{c}_i)^2 \right) = -\frac{u}{\Delta x} \sum_i \bar{c}_i \bar{c}_{i+1} + \frac{u}{\Delta x} \sum_i \bar{c}_i \bar{c}_{i-1},$$

where the sum covers all grid cells. By shifting the index of the last term from i to $i+1$, we note again cancellation of terms by pairs, leaving only contributions from the first and last grid points. Thus, except for possible contributions from the boundaries, the numerical scheme conserves both total amount and variance as the original equation does.

However, the conservation of global variance only holds for the semidiscrete equations. When time discretization is introduced as it must eventually be, conservation properties are often lost. In the literature, it is not always clearly stated whether conservation properties hold for the semidiscrete or fully discretized equations. The distinction, however, is important: The centered-space differencing conserves the variance of the semidiscrete solution, but a simple explicit time discretization renders the scheme unconditionally unstable and certainly does not conserve the variance. On the contrary, the latter quantity rapidly increases. Only a scheme that is both stable and consistent leads in the limit of vanishing time step to a solution that satisfies (6.12) and ensures conservation of the variance.

We might wonder why place emphasis on such conservation properties of semidiscrete equations, since by the time the algorithm is keyed into the computer it will always rely on fully discretized numerical approximations in both space and time. A reason to look at semidiscrete conservation properties is that some special time discretizations maintain the property in the fully discretized case. We now show that the trapezoidal time discretization conserves variance.

Consider the more general linear equation

$$\frac{d\tilde{c}_i}{dt} + \mathcal{L}(\tilde{c}_i) = 0, \tag{6.15}$$

where $\mathcal{L}$ stands for a linear discretization operator applied to the discrete field $\tilde{c}_i$. For our centered advection, the operator is $\mathcal{L}(\tilde{c}_i) = u(\tilde{c}_{i+1} - \tilde{c}_{i-1})/(2\Delta x)$. Suppose that the operator is designed to satisfy conservation of variance, which demands that at any moment t and for any discrete field $\tilde{c}_i$ the following relation holds

$$\sum_i \tilde{c}_i \mathcal{L}(\tilde{c}_i) = 0, \tag{6.16}$$

because only then does $\sum_i \tilde{c}_i \, d\tilde{c}_i/dt$ vanish according to (6.15) and (6.16). The trapezoidal time discretization applied to (6.15) is

$$\frac{\tilde{c}_i^{n+1} - \tilde{c}_i^n}{\Delta t} = -\frac{\mathcal{L}\left(\tilde{c}_i^{n+1}\right) + \mathcal{L}\left(\tilde{c}_i^n\right)}{2} = -\frac{1}{2}\mathcal{L}\left(\tilde{c}_i^{n+1} + \tilde{c}_i^n\right), \tag{6.17}$$

where the last equality follows from the linearity of operator $\mathcal{L}$. Multiplying this equation by $\left(\tilde{c}_i^{n+1} + \tilde{c}_i^n\right)$ and summing over the domain then yields

$$\sum_i \frac{\left(\tilde{c}_i^{n+1}\right)^2 - \left(\tilde{c}_i^n\right)^2}{\Delta t} = -\frac{1}{2}\sum_i \left(\tilde{c}_i^{n+1} + \tilde{c}_i^n\right)\mathcal{L}\left(\tilde{c}_i^{n+1} + \tilde{c}_i^n\right). \tag{6.18}$$

The term on the right is zero by virtue of (6.16). Therefore, any spatial discretization scheme that conserves variance continues to conserve variance if the trapezoidal scheme is used for the time discretization. As an additional benefit, the resulting scheme is also unconditionally stable. This does not mean, however, that the scheme is satisfactory, as Numerical Exercise 6.9 shows for the advection of the top-hat signal. Furthermore, there is a price to pay for stability because a system of simultaneous linear equations needs to be solved at each time step if the operator $\mathcal{L}$ uses several neighbors of the local grid point i.

To avoid solving simultaneous equations, alternative methods must be sought for time differencing. Let us explore the leapfrog scheme on the finite-volume approach. Time integration of (6.12) from t^{n-1} to t^{n+1} yields

$$\tilde{c}_i^{n+1} = \tilde{c}_i^{n-1} - 2\frac{\Delta t}{\Delta x}\left(\hat{q}_{i+1/2} - \hat{q}_{i-1/2}\right), \tag{6.19}$$

where $\hat{\tilde{q}}_{i-1/2}$ is the time-average advective flux uc across the cell interfaces between cells $i-1$ and i during the time interval from t^{n-1} to t^{n+1}. Using centered operators, this flux can be estimated as

$$\hat{\tilde{q}}_{i-1/2} = \frac{1}{2\Delta t} \int\limits_{t^{n-1}}^{t^{n+1}} uc|_{i-1/2}\, dt \quad \rightarrow \quad \tilde{q}_{i-1/2} = u\left(\frac{\tilde{c}_i^n + \tilde{c}_{i-1}^n}{2}\right), \tag{6.20}$$

so that the ultimate scheme is as follows:

$$\tilde{c}_i^{n+1} = \tilde{c}_i^{n-1} - C\left(\tilde{c}_{i+1}^n - \tilde{c}_{i-1}^n\right), \tag{6.21}$$

where the coefficient C is defined as

$$C = \frac{u\Delta t}{\Delta x}. \tag{6.22}$$

The same discretization could have been obtained by straightforward finite differencing of (6.10).

The parameter C is a dimensionless ratio central to the numerical discretization of advective problems called the *Courant number* or *CFL parameter* (Courant, Friedrichs & Lewy, 1928). It compares the displacement $u\Delta t$ made by the fluid during one time step to the grid size Δx. More generally, the Courant number for a process involving a propagation speed (such as a wave speed) is defined as the ratio of the distance of propagation during one time step to the grid spacing.

To use (6.21), two initial conditions are needed, one of which is physical and the other artificial. The latter must be consistent with the former. As usual, an explicit Euler step may be used to start from the single initial condition $\tilde{c}_i^0$:

$$\tilde{c}_i^1 = \tilde{c}_i^0 - \frac{C}{2}\left(\tilde{c}_{i+1}^0 - \tilde{c}_{i-1}^0\right). \tag{6.23}$$

In considering boundary conditions, we first observe that the exact solution of (6.10) obeys the simple law

$$c(x - ut) = \text{constant}. \tag{6.24}$$

By virtue of this property, a specified value of c somewhere along the line $x - ut = a$, called a *characteristic*, determines the value of c everywhere along that line. It is then easily seen (Fig. 6.4) that, in order to obtain a uniquely defined solution within the domain, a boundary condition must be provided at the upstream boundary, but no boundary condition is required at the outflow boundary. The centered space differencing, however, needs a value of $\tilde{c}$ given at each boundary. When discussing artificial boundary conditions (Section 4.7), we argued that these are acceptable as long as they are consistent with the mathematically correct boundary condition. But then, what requirement should the artificial boundary condition at the outflow obey with, since there is no physical

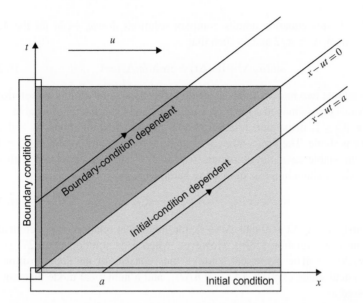

FIGURE 6.4 The characteristic line $x - ut = a$ propagates information from the initial condition or boundary condition into the domain. If the boundary is located at $x = 0$, and the initial condition given at $t = 0$, the line $x = ut$ divides the space-time frame into two distinct regions: For $x \leq ut$ the boundary condition defines the solution, whereas for $x \geq ut$ the initial condition defines the solution.

boundary condition for it to be consistent with? In practice, a one-sided space differencing is used at the outflow for the last calculation point $i = m$, so that its value is consistent with the local evolution equation:

$$\tilde{c}_m^{n+1} = \tilde{c}_m^n - C\left(\tilde{c}_m^n - \tilde{c}_{m-1}^n\right). \tag{6.25}$$

This provides the necessary value at the last grid cell.

For the inflow condition, the physical boundary condition is imposed, and algorithm (6.21) can be used starting from $n = 1$ and marching in time over all points $i = 2, ..., m - 1$. Numerically, we thus have sufficient information to calculate a solution that will be second-order accurate in both space and time, except near the initial condition and at the outflow boundary. In order to avoid any bad surprise when implementing the method, a stability analysis is advised.

For convenience, we use the von Neumann method written in Fourier-mode formalism (5.31)

$$\tilde{c}_i^n = A \mathrm{e}^{\mathrm{i}(k_x i \Delta x - \tilde{\omega} n \Delta t)}, \tag{6.26}$$

where the frequency $\tilde{\omega}$ may be complex. Substitution in the difference equation (6.21) provides the numerical dispersion relation

$$\sin(\tilde{\omega}\Delta t) = C \sin(k_x \Delta x). \tag{6.27}$$

If $|C| > 1$, this equation admits complex solutions $\tilde{\omega} = \tilde{\omega}_r + i\tilde{\omega}_i$ for the $4\Delta x$ wave with $\tilde{\omega}_r \Delta t = \pi/2$ and $\tilde{\omega}_i$ such that

$$\sin(\tilde{\omega}_r \Delta t + i\tilde{\omega}_i \Delta t) = \cosh(\tilde{\omega}_i \Delta t) = C, \tag{6.28}$$

which admits two real solutions $\tilde{\omega}_i$ of opposite signs. One of the two solutions, therefore, corresponds to a growing amplitude, and the scheme is unstable.

For $|C| \leq 1$, dispersion relation (6.27) has two real solutions $\tilde{\omega}$, and the scheme is stable. Therefore, numerical stability requires the condition $|C| \leq 1$.

In the stable case, the numerical frequency $\tilde{\omega}$ may be compared with the exact value written in terms of discrete parameters

$$\omega = uk_x \quad \rightarrow \quad \omega\Delta t = Ck_x \Delta x. \tag{6.29}$$

Obviously, for $k_x \Delta x \rightarrow 0$ and $\Delta t \rightarrow 0$, the numerical relation (6.27) coincides with the exact relation (6.29). However, when $\tilde{\omega}$ is solution of (6.27) so is also $\pi/\Delta t - \tilde{\omega}$. The numerical solution thus consists of the superposition of the physical mode $\exp[i(k_x i\Delta x - \tilde{\omega} n\Delta t)]$, and a numerical mode that can be expressed as

$$\tilde{c}_i^n = A e^{i(k_x i\Delta x + \tilde{\omega} n\Delta t)} e^{in\pi} \tag{6.30}$$

which, by virtue of $e^{in\pi} = (-1)^n$, flip-flops in time, irrespectively of how small the time step is or how well the spatial scale is resolved. This second component of the numerical solution, called *spurious mode* or *computational mode*, is traveling upstream, as indicated by the change of sign in front of the frequency. For the linear case discussed here, this spurious mode can be controlled by careful initialization (see Numerical Exercise 6.2), but for nonlinear equations, the mode may still be generated despite careful initialization and boundary conditions. In this case, it might be necessary to use time-filtering (see Section 10.6) to eliminate unwanted signals even if the spurious mode is stable for $|C| \leq 1$.

The leapfrog scheme is thus conditionally stable. The stability condition $|C| \leq 1$ was given a clear physical interpretation by Courant, Friedrichs and Lewy in their seminal 1928 paper. It is based on the fact that algorithm (6.21) defines a domain of dependence: Calculation of the value at point i and moment n (at the top of the gray pyramid in Fig. 6.5) implicates neighbor points $i \pm 1$ at time $n-1$ and the cell value i at time $n-2$. Those values in turn depend on their two neighboring and past values so that a network of points can be constructed that affect the value at grid point i and moment n. This network is the domain of numerical dependence. Physically, however, the solution at point i and time n depends only on the value along the characteristic $x - ut = x_i - ut^n$ according to (6.24). It is clear that, if this line does not fall inside the domain of dependence, there is trouble, for an attempt is made to determine a value from an irrelevant set of other values. Numerical instability is the symptom of this unsound approach. It is therefore necessary that the characteristic line passing through (i, n) be included in the domain of numerical dependence.

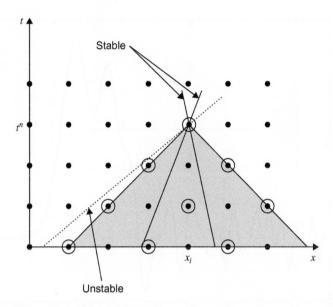

FIGURE 6.5 Domain of numerical dependence of the leapfrog scheme (in gray) covered by the points (circled dots) that influence the calculation at point i, n. This network of points is constructed recursively by identifying the grid points involved in prior calculations. The physical solution depends only on values along the characteristic. If the characteristic lies within the domain of numerical dependence (one of the solid lines, for example), this value can be captured by the calculations. On the contrary, when the physical characteristic lies outside the domain of numerical dependence (dashed line, for example), the numerical scheme relies on information that is physically unrelated to the advection process, and this flaw is manifested by numerical instability. Note also that the leapfrog scheme divides the grid points in two sets according to a checkerboard pattern (circled and noncircled dots). Unless some smoothing is performed, this risks to generate two numerically independent sets of values.

Except for the undesirable spurious mode, the leapfrog scheme has desirable features, because it is stable for $|C| \leq 1$, conserves variance for sufficiently small time steps, and leads to the correct dispersion relation for well-resolved spatial scales. But, is it sufficient to ensure a well-behaved solution? A standard test for advection schemes is the translation of a "top-hat" signal. In this case, the use of Eq. (6.21) leads to the result shown in Fig. 6.6, which is somewhat disappointing. The odd behavior can be explained: In terms of Fourier modes, the solution consists of a series of sine/cosine signals of different wavelengths, each of which by virtue of the numerical dispersion relation (6.27) travels at its own speed, thus unraveling the signal overtime. This also explains the unphysical appearance of both negative values and values in excess of the initial maximum. The scheme does not possess the monotonicity property but creates new extrema.

The cause of the poor performance of the leapfrog scheme is evident: The actual integration should be performed using upstream information exclusively,

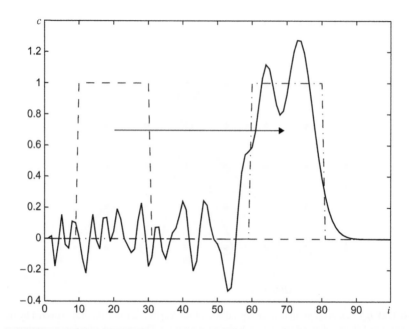

FIGURE 6.6 Leapfrog scheme applied to the advection of a "top-hat" signal with $C=0.5$ for 100 times steps. The exact solution is a mere translation from the initial position (dashed curve on the left) by 50 grid points downstream (dash-dotted curve on the right). The numerical method generates a solution that is roughly similar to the exact solution, with the solution varying around the correct value.

whereas the scheme uses a central average that disregards the origin of the information. In other words, it ignores the physical bias of advection.

To remedy the situation, we now try to take into account the directional information of advection and introduce the so-called *upwind* or *donor cell* scheme. A simple Euler scheme over a single time step Δt is chosen, and fluxes are integrated over this time interval. The essence of this scheme is to calculate the inflow based solely on the average value across the grid cell from where the flow arrives (the donor cell). For positive velocity and a time integration from t^n to t^{n+1}, we obtain

$$\bar{c}_i^{n+1} = \bar{c}_i^n - \frac{\Delta t}{\Delta x}\left(\hat{q}_{i+1/2} - \hat{q}_{i-1/2}\right) \tag{6.31}$$

with

$$\hat{q}_{i-1/2} = \frac{1}{\Delta t}\int_{t^n}^{t^{n+1}} q_{i-1/2}\mathrm{d}t \simeq u\tilde{c}_{i-1}^n, \tag{6.32}$$

so that the scheme is

$$\tilde{c}_i^{n+1} = \tilde{c}_i^n - C\left(\tilde{c}_i^n - \tilde{c}_{i-1}^n\right). \tag{6.33}$$

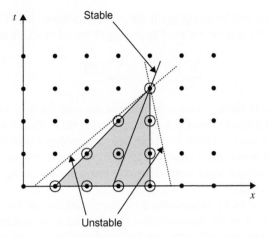

FIGURE 6.7   Domain of dependence of the upwind scheme. If the characteristic lies outside the numerical domain of dependence (dashed lines), unphysical behavior will be manifested as numerical instability. The necessary CFL stability condition therefore requires $0 \leq C \leq 1$ so that the characteristic lies within the numerical domain of dependence (solid lines). One initial condition and one upstream boundary condition are sufficient to determine the numerical solution.

Interestingly enough, the scheme can be used without need of artificial boundary conditions or special initialization, as we can see from algorithm (6.33) or the domain of numerical dependence (Fig. 6.7). The CFL condition $0 \leq C \leq 1$ provides the necessary condition for stability.

The stability of the scheme could be analyzed with the von Neumann method, but the simplicity of the scheme permits another approach, the so-called *energy method*. The energy method considers the sum of squares of $\tilde{c}$ and determines whether it remains bounded over time, providing a sufficient condition for stability. We start with (6.33), square it, and sum over the domain:

$$\sum_i \left(\tilde{c}_i^{n+1}\right)^2 = \sum_i (1-C)^2 \left(\tilde{c}_i^n\right)^2 + \sum_i 2C(1-C)\tilde{c}_i^n \tilde{c}_{i-1}^n + \sum_i C^2 \left(\tilde{c}_{i-1}^n\right)^2.$$

(6.34)

The first and last terms on the right can be grouped by shifting the index i in the last sum and invoking cyclic boundary conditions so that

$$\sum_i \left(\tilde{c}_i^{n+1}\right)^2 = \sum_i [(1-C)^2 + C^2]\left(\tilde{c}_i^n\right)^2 + \sum_i 2C(1-C)\tilde{c}_i^n \tilde{c}_{i-1}^n.$$ (6.35)

We can find an upper bound for the last term by using the following inequality:

$$0 \leq \sum_i \left(\tilde{c}_i^n - \tilde{c}_{i-1}^n\right)^2 = 2\sum_i \left(\tilde{c}_i^n\right)^2 - 2\sum_i \tilde{c}_i^n \tilde{c}_{i-1}^n,$$ (6.36)

which can be proved by using again the cyclic condition. If $C(1-C) > 0$, the last term in (6.35) may be replaced by the upper bound of (6.36) so that

$$\sum_i \left(\tilde{c}_i^{n+1}\right)^2 \leq \sum_i \left(\tilde{c}_i^{n}\right)^2, \tag{6.37}$$

and the scheme is stable because the norm of the solution does not increase in time. Although it is not related to a physical energy, the method derives its name from its reliance on a quadratic form that bears resemblance with kinetic energy. Methods which prove that a quadratic form is conserved or bounded over time are similar to energy-budget methods used to prove that the energy of a physical system is conserved.

The energy method provides only a sufficient stability condition because the upper bounds used in the demonstration do not need to be reached. But, since in the present case the sufficient stability condition was found to be identical to the necessary CFL condition, the condition $0 \leq C \leq 1$ is both necessary and sufficient to guarantee the stability of the upwind scheme.

Testing the upwind scheme on the "top-hat" problem (Fig. 6.8), we observe that, unlike leapfrog, the scheme does not create new minima or maxima, but somehow diffuses the distribution by reducing its gradients. The fact that the scheme is monotonic is readily understood by examining (6.33): The new value at point i is a linear interpolation of previous values found at i and $i-1$ so that no new value can ever fall outside the range of these previous values as long as the condition $0 \leq C \leq 1$ is satisfied.

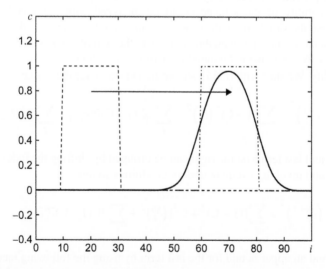

FIGURE 6.8   Upwind scheme with $C=0.5$ applied to the advection of a "top-hat" signal after 100 times steps. Ideally the signal should be translated without change in shape by 50 grid points, but the solution is characterized by a certain diffusion and a reduction in gradient.

The diffusive behavior can be explained by analyzing the modified equation associated with (6.33). A Taylor expansion of the discrete solution around point (i, n) in (6.33) provides the equation that the numerical scheme actually solves the following:

$$\frac{\partial \tilde{c}}{\partial t} + \frac{\Delta t}{2} \frac{\partial^2 \tilde{c}}{\partial t^2} + \mathcal{O}\left(\Delta t^2\right) + u\left(\frac{\partial \tilde{c}}{\partial x} - \frac{\Delta x}{2} \frac{\partial^2 \tilde{c}}{\partial x^2} + \mathcal{O}\left(\Delta x^2\right)\right) = 0. \qquad (6.38)$$

The scheme is only of first order as can be expected from the use of a one-sided finite difference. To give a physical interpretation to the equation, the second time derivative should be replaced by a spatial derivative. Taking the derivative of the modified equation with respect to t provides an equation for the second time derivative, which we would like to eliminate, but it involves a cross derivative[3]. This cross derivative can be obtained by differentiating the modified equation with respect to x. Some algebra ultimately provides

$$\frac{\partial^2 \tilde{c}}{\partial t^2} = u^2 \frac{\partial^2 \tilde{c}}{\partial x^2} + \mathcal{O}\left(\Delta t, \Delta x^2\right),$$

which can finally be introduced into (6.38) to yield the following equation

$$\frac{\partial \tilde{c}}{\partial t} + u\frac{\partial \tilde{c}}{\partial x} = \frac{u\Delta x}{2}(1 - C)\frac{\partial^2 \tilde{c}}{\partial x^2} + \mathcal{O}\left(\Delta t^2, \Delta x^2\right). \qquad (6.39)$$

This is the equation that the upwind scheme actually solves.

Up to $\mathcal{O}\left(\Delta t^2, \Delta x^2\right)$, therefore, the numerical scheme solves an advection–diffusion equation instead of the pure advection equation, with diffusivity equal to $(1 - C)u\Delta x/2$. For obvious reasons, this is called an artificial diffusion or *numerical diffusion*. The effect is readily seen in Fig. 6.8. To decide whether this level of artificial diffusion is acceptable or not, we must compare its size with that of physical diffusion. For a diffusivity coefficient $\mathcal{A}$, the ratio of numerical to physical diffusion is $(1 - C)u\Delta x/(2\mathcal{A})$. Since it would be an aberration to have numerical diffusion equal or exceed physical diffusion (recall the error analysis of Section 4.8: Discretization errors should not be larger than modeling errors), the *grid Peclet number* $U\Delta x/\mathcal{A}$ may not exceed a value of $\mathcal{O}(1)$ for the upwind scheme to be valid.

When no physical diffusion is present, we must require that the *numerical* diffusion term be small compared with the *physical* advection term, a condition

---

[3] Note that using the *original* equations, the physical solution satisfies $\partial^2 c/\partial t^2 = u^2 \partial^2 c/\partial x^2$, which is sometimes used as a shortcut to eliminate the second time derivative from the modified equation. This is, however, incorrect because $\tilde{c}$ does not solve the original equation. In practice, this kind of shortcuts can lead to correct leading truncation errors, but without being sure that no essential term is overlooked.

that can be associated with another grid Peclet number:

$$\tilde{P}e = \frac{UL}{U\Delta x(1-C)/2} \sim \frac{L}{\Delta x} \gg 1, \qquad (6.40)$$

where L stands for the length scale of any solution component worth resolving. Even for well-resolved signals in GFD flows, this Peclet number associated with numerical diffusion is often insufficiently large, and numerical diffusion is a problem that plagues the upwind scheme.

The observation that the scheme introduces artificial diffusion is interesting and annoying, and the question now is to identify its origin in order to reduce it. Compared with the centered scheme, which is symmetric and of second order, the upwind scheme uses exclusive information from the upstream side, the donor cell, and is only of first order. Numerical diffusion must, therefore, be associated with the asymmetry in the flux calculation, and to reduce numerical diffusion, we must somehow take into account values of $\tilde{c}$ on both sides of the interface to calculate the flux and thereby seek a scheme that is second-order accurate.

This can be accomplished with the *Lax-Wendroff scheme*, which estimates the flux at the cell interface by assuming that the function is not constant within the cell but varies linearly across it:

$$\begin{aligned} \tilde{q}_{i-1/2} &= u\left[\frac{\tilde{c}_i^n + \tilde{c}_{i-1}^n}{2} - \frac{C}{2}\left(\tilde{c}_i^n - \tilde{c}_{i-1}^n\right) \right] \\ &= u\tilde{c}_{i-1}^n + \underbrace{(1-C)\frac{u\Delta x}{2}\frac{\tilde{c}_i^n - \tilde{c}_{i-1}^n}{\Delta x}}_{\simeq (1-C)\frac{u\Delta x}{2}\frac{\partial \tilde{c}}{\partial x}}. \end{aligned} \qquad (6.41)$$

The last term, in addition to the upwind flux $u\tilde{c}_{i-1}^n$, is designed to oppose numerical diffusion. Substitution of this flux into the finite-volume scheme leads to the following scheme:

$$\tilde{c}_i^{n+1} = \tilde{c}_i^n - C\left(\tilde{c}_i^n - \tilde{c}_{i-1}^n\right) - \frac{\Delta t}{\Delta x^2}(1-C)\frac{u\Delta x}{2}\left(\tilde{c}_{i+1}^n - 2\tilde{c}_i^n + \tilde{c}_{i-1}^n\right) \qquad (6.42)$$

which, compared with the upwind scheme, includes an additional antidiffusion term with coefficient constructed to negate the numerical diffusion of the upwind scheme. The effect of this higher-order approach on the solution of our test case is a reduced overall error but the appearance of dispersion (Fig. 6.9). This is due to the fact that we eliminated the truncation error proportional to the second spatial derivative (an even derivative associated with dissipation) and now have a truncation error proportional to the third spatial derivative (an odd derivative associated with dispersion).

The same dispersive behavior is observed with the *Beam-Warming scheme*, in which the antidiffusion term is shifted upstream so as to anticipate the

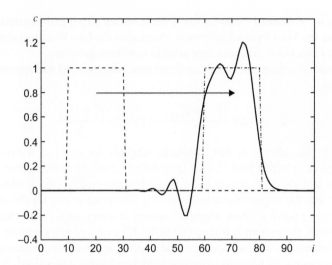

FIGURE 6.9   Second-order Lax-Wendroff scheme applied to the advection of a "top-hat" signal with $C = 0.5$ after 100 times steps. Dispersion and nonmonotonic behavior are noted.

gradient that will arrive later at the interface:

$$\tilde{q}_{i-1/2} = u\tilde{c}_{i-1}^{n} + (1-C)\frac{u}{2}\left(\tilde{c}_{i-1}^{n} - \tilde{c}_{i-2}^{n}\right). \tag{6.43}$$

This scheme is still of second order, since the correction term is only shifted upstream by Δx. The effect of anticipating the incoming gradients enhances stability but does not reduce dispersion (see Numerical Exercise 6.8).

Other methods spanning more grid points can be constructed to obtain higher-order integration of fluxes, implicit methods to increase stability, predictor-corrector methods, or combinations of all these schemes. Here, we only outline some of the approaches and refer the reader to more specialized literature for details (e.g., Chung, 2002; Durran, 1999).

A popular *predictor-corrector method* is the second-order MacCormack scheme: The predictor uses a forward spatial difference (antidiffusion),

$$\tilde{c}_{i}^{\star} = \tilde{c}_{i}^{n} - C\left(\tilde{c}_{i+1}^{n} - \tilde{c}_{i}^{n}\right) \tag{6.44}$$

and the corrector a backward spatial difference on the predicted field (diffusion):

$$\tilde{c}_{i}^{n+1} = \tilde{c}_{i}^{n+1/2} - \frac{C}{2}\left(\tilde{c}_{i}^{\star} - \tilde{c}_{i-1}^{\star}\right) \quad \text{with} \quad \tilde{c}_{i}^{n+1/2} = \frac{\tilde{c}_{i}^{\star} + \tilde{c}_{i}^{n}}{2}. \tag{6.45}$$

The elimination of the intermediate value $\tilde{c}_{i}^{n+1/2}$ from which starts the corrector step provides the expanded corrector step:

$$\tilde{c}_{i}^{n+1} = \frac{1}{2}\left[\tilde{c}_{i}^{n} + \tilde{c}_{i}^{\star} - C\left(\tilde{c}_{i}^{\star} - \tilde{c}_{i-1}^{\star}\right)\right], \tag{6.46}$$

assuming $u > 0$ as usual. Substitution of the predictor step into the corrector step shows that the MacCormack scheme is identical to the Lax-Wendroff scheme in the linear case, but differences may arise in nonlinear problems.

An *implicit scheme* can handle centered space differencing and approximates the flux as

$$\tilde{q}_{i-1/2} = \alpha u \frac{\tilde{c}_i^{n+1} + \tilde{c}_{i-1}^{n+1}}{2} + (1-\alpha)u \frac{\tilde{c}_i^n + \tilde{c}_{i-1}^n}{2}. \tag{6.47}$$

For $\alpha = 1$, the scheme is fully implicit, whereas for $\alpha = 1/2$ it becomes a semi-implicit or trapezoidal scheme (also called *Crank-Nicholson scheme*). The latter has already been shown to be unconditionaly stable [see variance conservation and the trapezoidal scheme (6.17)]. The price to pay for this stability is the need to solve a linear algebraic system at every step. As for the diffusion problem, the system is tridiagonal in the 1D case and more complicated in higher dimensions. The advantage of the implicit approach is a robust scheme when C occasionally happens to exceed unity in a known dimension[4]. It should be noted, however, that for too large a Courant number accuracy degrades.

All of the previous schemes can be mixed in a *linear combination*, as long as the sum of the weights attributed to each scheme is unity for the sake of consistency. An example of combining two schemes consists in averaging the flux calculated with a lower-order scheme $\tilde{q}_{i-1/2}^L$ with that of a higher-order scheme $\tilde{q}_{i-1/2}^H$:

$$\tilde{q}_{i-1/2} = (1-\Phi)\,\tilde{q}_{i-1/2}^L + \Phi\,\tilde{q}_{i-1/2}^H,$$

in which the weight Φ ($0 \leq \Phi \leq 1$) acts as a trade-off between the undesirable numerical diffusion of the lower-order scheme and numerical dispersion and loss of monotonicity of the higher-order scheme.

All these methods lead to sufficiently accurate solutions, but none except the upwind scheme ensures monotonic behavior. The reason for this disappointing fact can be found in the frustrating theorem by Godunov (1959) regarding the discretized advection equation:

A consistent linear numerical scheme that is monotonic can at most be first-order accurate.

Therefore, the upwind scheme is the inevitable choice if no overshoot or undershoot is permitted. To circumvent the Godunov theorem, state-of-the-art advection schemes relax the linear nature of the discretization and adjust the parameter Φ locally, depending on the behavior of the solution. The function

---

[4] Typically the vertical Courant number may be so variable that it becomes difficult to ensure that the local vertical C value remains below one. In particular, it is prudent to use an implicit scheme in the vertical when the model has nonuniform grid spacing, and when the vertical velocity is weak except on rare occasions.

that defines the way Φ is adapted locally is called a *limiter*. Such an approach is able to capture large gradients (fronts). Because of its advanced nature, we delay its presentation until Section 15.7. An example of a nonlinear scheme called TVD, however, is already included in the computer codes provided for the analysis of advection schemes in several dimensions.

6.5 ADVECTION–DIFFUSION WITH SOURCES AND SINKS

Having considered separately advection schemes (this chapter), diffusion schemes (Chapter 5) and time discretizations with arbitrary forcing terms (Chapter 2), we can now combine them to tackle the general advection–diffusion equation with sources and sinks. For a linear sink, the 1D equation to be discretized is

$$\frac{\partial c}{\partial t} + u \frac{\partial c}{\partial x} = -K c + \frac{\partial}{\partial x}\left(A\frac{\partial c}{\partial x}\right). \tag{6.48}$$

Since we already have a series of discretization possibilities for each individual process, the combination of these provides an even greater number of possible schemes which we cannot describe exhaustively here. We simply show one example to illustrate two important facts that should not be forgotten when combining schemes: The properties of the combined scheme are neither simply the sum of the properties of the individual schemes, nor is its stability condition the most stringent condition of the separate schemes.

To prove the first statement, we consider (6.48) without diffusion ($A=0$) and solve by applying the second-order Lax-Wendroff advection scheme with the second-order trapezoidal scheme applied to the decay term. The discretization, after some rearrangement of terms, is as follows:

$$\tilde{c}_i^{n+1} = \tilde{c}_i^n - \frac{B}{2}\left(\tilde{c}_i^n + \tilde{c}_i^{n+1}\right) - \frac{C}{2}\left(\tilde{c}_{i+1}^n - \tilde{c}_{i-1}^n\right) + \frac{C^2}{2}\left(\tilde{c}_{i+1}^n - 2\tilde{c}_i^n + \tilde{c}_{i-1}^n\right), \tag{6.49}$$

where $B = K\Delta t$ and $C = u\Delta t/\Delta x$. This scheme actually solves the following modified equation:

$$\frac{\partial \tilde{c}}{\partial t} + u \frac{\partial \tilde{c}}{\partial x} + K \tilde{c} = -\frac{\Delta t}{2}\frac{\partial^2 \tilde{c}}{\partial t^2} - K\frac{\Delta t}{2}\frac{\partial \tilde{c}}{\partial t} + \frac{u^2\Delta t}{2}\frac{\partial^2 \tilde{c}}{\partial x^2} + \mathcal{O}(\Delta t^2, \Delta x^2)$$

$$= -\frac{uK\Delta t}{2}\frac{\partial \tilde{c}}{\partial x} + \mathcal{O}(\Delta t^2, \Delta x^2) \tag{6.50}$$

where the last equality was obtained by a similar procedure as the one used to find the modified equation (6.39). It is not possible to cancel the leading term on the right unless $K=0$ or $u=0$, in which case we recover the second-order Lax-Wendroff or the second-order trapezoidal scheme. Thus, in the combined advection-decay case, what was expected to be a second-order scheme degenerates into a first-order scheme.

For the purpose of illustrating the second statement on stability, we combine the second-order Lax-Wendroff advection scheme (stability condition $|C| \leq 1$) with the explicit Euler diffusion scheme (stability condition $0 \leq D \leq 1/2$) and an explicit scheme for the sink term with rate K (stability condition $B \leq 2$). The discretization, after some rearrangement of the terms, is as follows:

$$\tilde{c}_i^{n+1} = \tilde{c}_i^n - B\,\tilde{c}_i^n - \frac{C}{2}\,(\tilde{c}_{i+1}^n - \tilde{c}_{i-1}^n) + \left(D + \frac{C^2}{2}\right)(\tilde{c}_{i+1}^n - 2\tilde{c}_i^n + \tilde{c}_{i-1}^n), \quad (6.51)$$

where $D = \mathcal{A}\Delta t/\Delta x^2$. Application of the von Neumann stability analysis yields the following amplification factor

$$\varrho = 1 - B - 4\left(D + \frac{C^2}{2}\right)\sin^2\theta - i\,2C\sin\theta\cos\theta, \quad (6.52)$$

where $\theta = k_x\Delta x/2$, so that

$$|\varrho|^2 = \left[1 - B - 4\left(D + \frac{C^2}{2}\right)\xi\right]^2 + 4C^2\xi(1 - \xi), \quad (6.53)$$

where $0 \leq \xi = \sin^2\theta \leq 1$. For $\xi \simeq 0$ (long waves), we obtain the necessary stability condition $B \leq 2$, corresponding to the stability condition of the sink term alone. For $\xi \simeq 1$ (short waves), we find the more demanding necessary stability condition

$$B + 2C^2 + 4D \leq 2. \quad (6.54)$$

We can show that the latter condition is also sufficient (Numerical Exercise 6.13), which proves that the stability condition of the combined schemes is more stringent than each stability condition taken individually. Only when two processes are negligible does the stability condition revert to the stability condition of the single remaining process. This seems evident but is not always the case. In some situations, adding even an infinitesimally-small stable process can require a discontinuous reduction in time step (e.g., Beckers & Deleersnijder, 1993).

In other situations, adding a process can stabilize an otherwise unconditionally unstable scheme (Numerical Exercise 6.14). Therefore, in theory, it is not enough to consider the stability of each piece of the scheme separately, but the stability of the full scheme must be investigated. In practice, however, if a complete scheme is too difficult to analyze, subschemes (i.e., including only a few processes) are isolated with the hope that the full scheme does not demand a drastically shorter time step than the one required by the most stringent stability condition of each elementary scheme taken separately.

Stability is an important property of any scheme as is, at least for tracers, monotonic behavior. If we assume the scheme to be explicit, linear and covering a stencil spanning p grid points upstream and q downstream (for a total of

$p+q+1$ points), it can be written in the general form:

$$\tilde{c}_i^{n+1} = a_{-p}\tilde{c}_{i-p}^n + \cdots + a_{-1}\tilde{c}_{i-1}^n + a_0\tilde{c}_i^n + a_1\tilde{c}_{i+1}^n + \cdots + a_q\tilde{c}_{i+q}^n. \tag{6.55}$$

To be consistent with (6.48), we need at least to ensure $a_{-p} + \cdots + a_{-1} + a_0 + a_1 + \cdots + a_q = 1 - B$, otherwise, not even a spatially uniform field would be represented correctly.

If there is a negative coefficient a_k, the scheme will not be monotonic, for indeed, if the function is positive at point $i+k$ but zero everywhere else, it will take on a negative value $\tilde{c}_i^{n+1}$. However, if all coefficients are positive, the sum of the total weights is obviously positive but less than one because it is equal to $(1-B)$. The scheme thus interpolates while damping, in agreement with physical decay. And, since damping does not create new extrema, we conclude that positive coefficients ensure a monotonic behavior in all situations. For our example (6.51), this demands $B + C^2 + 2D \leq 1$ and $C \leq C^2 + 2D$. The former condition is a slightly more constraining version of the stability condition (6.54), whereas the latter condition imposes a constraint on the grid Peclet number:

$$Pe_{\Delta x} = \frac{u\Delta x}{\mathcal{A}} = \frac{C}{D} \leq \frac{u^2\Delta t}{\mathcal{A}} + 2. \tag{6.56}$$

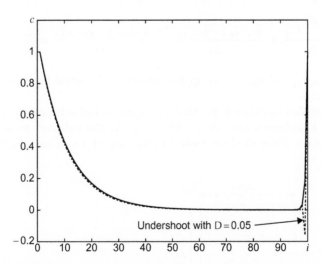

FIGURE 6.10 Simulation using (6.51) with $B = 0.05$, $C = 0.5$, and $D = 0.25$, after convergence to a stationary solution (solid line). With decreasing diffusion, the scheme eventually fails to resolve adequately the outflow boundary layer, and undershoot appears ($D = 0.05$, dot-dash line). This corresponds to a situation in which one of the coefficients in the numerical scheme has become negative. The program `advdiffsource.m` can be used to test other combinations of the parameter values.

For short time steps, this imposes a maximum value of 2 on the grid Peclet number. It is not a stability condition but a necessary condition for monotonic behavior.

The scheme is now tested on a physical problem. Because of the second derivative, we impose boundary conditions at both upstream *and* downstream and for simplicity hold $\tilde{c} = 1$ steady at these locations. We then iterate from a zero initial condition until the scheme converges to a stationary solution. This solution (Fig. 6.10) exhibits a boundary layer at the downstream end because of weak diffusion, in agreement with the remark made in Section 6.3. For weak diffusion, the grid Peclet number $Pe_{\Delta x}$ is too large and violates (6.56). Undershooting appears, although the solution remains stable. In conclusion, besides the parameters B, C, and D that control stability, the grid Peclet number C/D appears as a parameter controlling monotonic behavior.

6.6 MULTIDIMENSIONAL APPROACH

In addition to the various combinations already encountered in the 1D case, generalization to more dimensions allows further choices and different methods. Here, we concentrate on the 2D advection case because generalizations to 3D do not generally cause more fundamental complications.

The finite-volume approach can be easily extended to a 2D grid cell with the fluxes perpendicular to the interfaces (Fig. 6.11):

$$\frac{\tilde{c}_{i,j}^{n+1} - \tilde{c}_{i,j}^n}{\Delta t} + \frac{\tilde{q}_{x,i+1/2,j} - \tilde{q}_{x,i-1/2,j}}{\Delta x} + \frac{\tilde{q}_{y,i,j+1/2} - \tilde{q}_{y,i,j-1/2}}{\Delta y} = 0, \qquad (6.57)$$

where $\tilde{q}_{x,i\pm1/2,j}$ and $\tilde{q}_{y,i,j\pm1/2}$ are approximations of the actual fluxes uc and vc, respectively.

For any flux calculation, the least we require is that it be able to represent correctly a uniform tracer field $\mathcal{C}$. All of our 1D flux calculations do so and should do so. When (6.57) is applied to the case of a uniform concentration

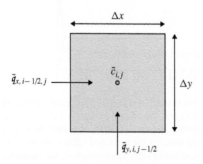

FIGURE 6.11 Finite volume in 2D with fluxes at the interfaces. The budget involves the total balance of inflowing and outflowing fluxes during one time step.

distribution $\tilde{c} = C$, we obtain

$$\frac{\tilde{c}_{i,j}^{n+1} - C}{\Delta t} + \frac{\tilde{u}_{i+1/2,j} - \tilde{u}_{i-1/2,j}}{\Delta x} C + \frac{\tilde{v}_{i,j+1/2} - \tilde{v}_{i,j-1/2}}{\Delta y} C = 0.$$

This can only lead to $\tilde{c}^{n+1} = C$ at the next time step if the discrete velocity field satisfies the condition

$$\frac{\tilde{u}_{i+1/2,j} - \tilde{u}_{i-1/2,j}}{\Delta x} + \frac{\tilde{v}_{i,j+1/2} - \tilde{v}_{i,j-1/2}}{\Delta y} = 0. \tag{6.58}$$

Since this requirement is an obvious discretization of $\partial u/\partial x + \partial v/\partial y = 0$, the 2D form of the continuity equation (4.21d), it follows that a prerequisite to solve the concentration equation by the finite-volume approach is a nondivergent flow field *in its discretized form*. Ensuring that (6.58) holds is the role of the discretization of the dynamical equations, those governing velocity and pressure.

Here, in order to test numerical advection schemes, we take the flow field as known and obeying (6.58). We can easily generate such a discrete velocity distribution by invoking a discretized *streamfunction* ψ:

$$\tilde{u}_{i-1/2,j} = -\frac{\psi_{i-1/2,j+1/2} - \psi_{i-1/2,j-1/2}}{\Delta y} \tag{6.59}$$

$$\tilde{v}_{i,j-1/2} = \frac{\psi_{i+1/2,j-1/2} - \psi_{i-1/2,j-1/2}}{\Delta x}. \tag{6.60}$$

It is straightforward to show that these $\tilde{u}$ and $\tilde{v}$ values satisfy (6.58) for any set of streamfunction values.

Note that if we had discretized directly the continuous equation

$$\frac{\partial c}{\partial t} + u \frac{\partial c}{\partial x} + v \frac{\partial c}{\partial y} = 0, \tag{6.61}$$

we would have obtained

$$\frac{\partial \tilde{c}_{i,j}}{\partial t} + u_{i,j} \left.\frac{\partial \tilde{c}}{\partial x}\right|_{i,j} + v_{i,j} \left.\frac{\partial \tilde{c}}{\partial y}\right|_{i,j} = 0,$$

which guarantees that an initially uniform tracer distribution remains uniform at all later times regardless of the structure of the discretized velocity distribution as long as the discretized form of the spatial derivatives return zeros for a uniform distribution (a mere requirement of consistency). Such a scheme could appear to offer a distinct advantage, but it is easy to show that it has a major drawback. It looses important conservation properties, including conservation of the quantity of tracer (heat for temperature, salt for salinity, etc.).

Assuming the discrete velocity field to be divergence-free in the sense of
(6.58), the first method that comes to mind is to calculate the flux components
using the discretizations developed in 1D along each coordinate line separately
(Fig. 6.12). The upwind scheme is then easily generalized as follows:

$$\tilde{q}_{x,i-1/2,j} = \tilde{u}_{i-1/2,j}\, \tilde{c}_{i-1,j}^{n} \quad \text{if} \quad \tilde{u}_{i-1/2,j} > 0, \quad \tilde{u}_{i-1/2,j}\, \tilde{c}_{i,j}^{n} \quad \text{otherwise} \quad (6.62a)$$

$$\tilde{q}_{y,i,j-1/2} = \tilde{v}_{i,j-1/2}\, \tilde{c}_{i,j-1}^{n} \quad \text{if} \quad \tilde{v}_{i,j-1/2} > 0, \quad \tilde{v}_{i,j-1/2}\, \tilde{c}_{i,j}^{n} \quad \text{otherwise.} \quad (6.62b)$$

The other 1D schemes can be generalized similarly. Applying such schemes
to the advection of an initially cone-shaped distribution (single peak with same
linear drop in all radial directions) embedded in a uniform flow field crossing the
domain at 45°, we observe that the upwing scheme is plagued by a very strong
numerical diffusion (left panel of Fig. 6.13). Using the TVD scheme keeps the
signal to a higher amplitude but strongly distorts the distribution (right panel of
Fig. 6.13).

This distortion is readily understood in terms of the advection process: The
information should be carried by the oblique flow, but the flux calculation relies
on information strictly along the x or y axes. In the case of a flow oriented at
45° from the $x-$axis, this ignores that grid point (i,j) is primarily influenced
by point $(i-1,j-1)$ whereas points $(i-1,j)$ and $(i,j-1)$ are used in the flux
calculations. In conclusion, the double 1D approach is unsatisfactory and rarely
used.

The *Corner Transport Upstream* (CTU) scheme (e.g., Colella, 1990) takes
into account the different contributions of the four grid cells involved in the
displacement (Fig. 6.14). Assuming uniform positive velocities to illustrate the
approach, the following discretization ensures that a diagonal flow brings to

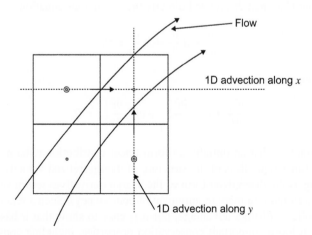

FIGURE 6.12 Naive 2D generalization using 1D methods along each coordinate line to approximate the advection operator as the sum of $\partial(uc)/\partial x$ and $\partial(vc)/\partial y$.

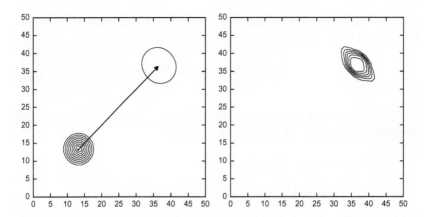

FIGURE 6.13   Oblique advection of a cone-shaped distribution using the upwind scheme generalized to 2D (left panel) and a TVD scheme (right panel) with $C_x = C_y = 0.12$. The upwind scheme severely dampens the signal, to less then 20% of its initial amplitude, whereas the TVD scheme used as a double 1D problem greatly distorts the solution.

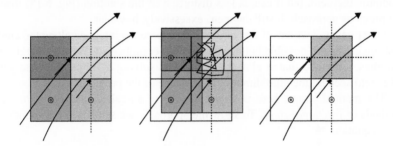

FIGURE 6.14   2D generalization designed to advect the field obliquely along streamlines. The associated numerical diffusion can then be interpreted as the necessary grid averaging (i.e., mixing) used in the finite-volume technique after displacement of the donor cells. The flux calculations (thick arrows) need to integrate the inflow of c along the flow instead of along the grid lines.

the interface a correct mixing of two donor cells:

$$\tilde{q}_{x,i-1/2,j} = \left(1 - \frac{C_y}{2}\right) \tilde{u}\, \tilde{c}^n_{i-1,j} + \frac{C_y}{2}\, \tilde{u}\, \tilde{c}^n_{i-1,j-1} \tag{6.63a}$$

$$\tilde{q}_{y,i,j-1/2} = \left(1 - \frac{C_x}{2}\right) \tilde{v}\, \tilde{c}^n_{i,j-1} + \frac{C_x}{2}\, \tilde{v}\, \tilde{c}^n_{i-1,j-1}, \tag{6.63b}$$

leading to the expanded scheme

$$\begin{aligned}
\tilde{c}^{n+1}_{i,j} = \tilde{c}^n_{i,j} &- C_x\left(\tilde{c}^n_{i,j} - \tilde{c}^n_{i-1,j}\right) - C_y\left(\tilde{c}^n_{i,j} - \tilde{c}^n_{i,j-1}\right) \\
&+ C_x C_y\left(\tilde{c}^n_{i,j} - \tilde{c}^n_{i-1,j} - \tilde{c}^n_{i,j-1} + \tilde{c}^n_{i-1,j-1}\right),
\end{aligned} \tag{6.64}$$

where the last term is an additional term compared with the double 1D approach. Two distinct Courant numbers arise, one for each direction:

$$C_x = \frac{u\Delta t}{\Delta x}, \qquad C_y = \frac{v\Delta t}{\Delta y}. \tag{6.65}$$

For $C_x = C_y = 1$, the scheme provides $\tilde{c}_{i,j}^{n+1} = \tilde{c}_{i-1,j-1}^{n}$, with obvious physical interpretation. The scheme may also be written as

$$\begin{aligned}
\tilde{c}_{i,j}^{n+1} = {} & (1-C_x)(1-C_y)\,\tilde{c}_{i,j}^{n} \\
& + (1-C_y)C_x\,\tilde{c}_{i-1,j}^{n} + (1-C_x)C_y\,\tilde{c}_{i,j-1}^{n} + C_x C_y\,\tilde{c}_{i-1,j-1}^{n}
\end{aligned} \tag{6.66}$$

highlighting the relative weights attached to the four grid points involved in the calculation (Fig. 6.14). This expression proves that the method is monotonic for Courant numbers smaller than one (ensuring that all coefficients on the right-hand side are positive). The method is only of first order according to the Godunov theorem, but it causes less distortion of the solution (Fig. 6.15) than the previous approach. It still dampens excessively, however.

Other generalizations of the various 1D schemes to integrate along the current directions are possible but become rapidly complicated. We will therefore introduce a method that is almost as simple as solving a 1D problem but yet takes into account the multidimensional essence of the problem.

The method shown here is a special case of so-called *operator splitting* methods or *fractional steps*. To show the approach, we start from the semidiscrete equation

$$\frac{d\tilde{c}_i}{dt} + \mathcal{L}_1(\tilde{c}_i) + \mathcal{L}_2(\tilde{c}_i) = 0, \tag{6.67}$$

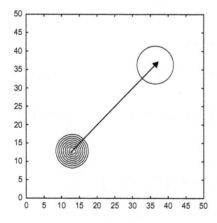

FIGURE 6.15 2D oblique advection using the CTU scheme (6.66). The solution's asymmetric deformation is reduced, but numerical diffusion still reduces the amplitude significantly.

where $\mathcal{L}_1$ and $\mathcal{L}_2$ are two discrete operators, which in the present case are advection operators along x and y. Temporal discretization by time splitting executes the following:

$$\frac{\tilde{c}_i^\star - \tilde{c}_i^n}{\Delta t} + \mathcal{L}_1(\tilde{c}_i^n) = 0 \qquad (6.68a)$$

$$\frac{\tilde{c}_i^{n+1} - \tilde{c}_i^\star}{\Delta t} + \mathcal{L}_2(\tilde{c}_i^\star) = 0, \qquad (6.68b)$$

where the second operator is marched forward with a value already updated by the first operator.

In this manner, we solve two sequential one-dimensional problems, which is not more complicated than what we just did and is a major improvement compared with the naive double 1D approach used in (6.62): The initial (predictor) step creates a field that is already advected in the direction of the $\mathcal{L}_1$ operator, and the second (corrector) step advects in the remaining direction the partially displaced field. In this way, point (i,j) is influenced by the upstream value, point $(i-1,j-1)$ in the case of positive velocities (Fig. 6.16).

To verify this, we can see how the splitting works with the 1D upwind scheme for positive and uniform velocities:

$$\tilde{c}_{i,j}^\star = \tilde{c}_{i,j}^n - C_x\left(\tilde{c}_{i,j}^n - \tilde{c}_{i-1,j}^n\right) \qquad (6.69a)$$

$$\tilde{c}_{i,j}^{n+1} = \tilde{c}_{i,j}^\star - C_y\left(\tilde{c}_{i,j}^\star - \tilde{c}_{i,j-1}^\star\right). \qquad (6.69b)$$

Substitution of the intermediate values $\tilde{c}_{i,j}^\star$ into the final step then proves that the scheme is identical to the CTU scheme (6.64) for uniform velocities. Such elimination, however, is not done in practice, and the sequence (6.69) is used. This is particularly convenient because, in the computer program, $\tilde{c}^\star$ may be stored during the first step in the future place of $\tilde{c}^{n+1}$ and then moved to the place of $\tilde{c}^n$ for the second step; $\tilde{c}^{n+1}$ can then be calculated and stored without need of additional storage.

Using always the same operator with current values and the other with "predicted" values is unsatisfactory because it breaks the symmetry between the two spatial dimensions. Hence, it is recommended to alternate the order in which operators are applied, depending on whether the time step is even or odd. Following a time step using (6.68) we then switch the order of operators by performing

$$\frac{\tilde{c}_i^\star - \tilde{c}_i^{n+1}}{\Delta t} + \mathcal{L}_2(\tilde{c}_i^{n+1}) = 0 \qquad (6.70a)$$

$$\frac{\tilde{c}_i^{n+2} - \tilde{c}_i^\star}{\Delta t} + \mathcal{L}_1(\tilde{c}_i^\star) = 0. \qquad (6.70b)$$

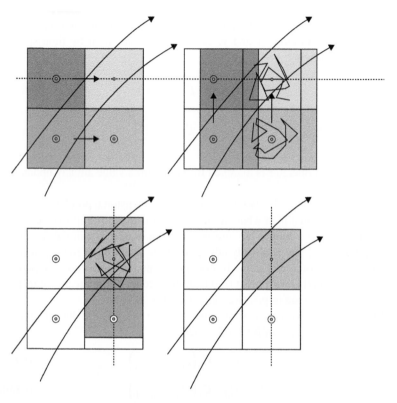

FIGURE 6.16 The splitting method uses two sequential 1D advection schemes. First, the signal is transported along the x-direction, and then the intermediate solution is advected along the y-direction. In this way, the information at upstream point $(i-1, j-1)$ is involved in the evolution of the value at (i, j) (case of positive velocity components).

This approach, alternating the order of the directional splitting, is a special case of the more general *Strang splitting* method designed to maintain second-order time accuracy when using time splitting (Strang, 1968).

The splitting approach thus seems attractive. It is not more complicated than applying two successive one-dimensional schemes. In the general case, however, attention must be given to the direction of the local flow so that "upwinding" consistently draws the information from upstream, whatever that direction may be. There is no other complication, and we now proceed with a test of the method with the TVD scheme. As Fig. 6.17 reveals, the result is a significant improvement with no increase of computational burden.

A more complicated case of advection can now be tried. For this, we choose an initially square distribution of tracer and place it in a narrow sheared flow (Fig. 6.18). We expect that the distribution will be distorted by the shear flow. To assess the quality of the advection scheme, we could try to obtain an analytical

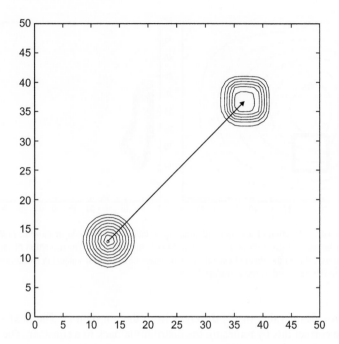

FIGURE 6.17   Advection at 45° of an initially conical distribution using the splitting method and the TVD scheme.

solution by calculating trajectories from the known velocity field, but a much simpler approach is to flip the sign of the velocity field after some time and continue the integration for an equal amount time. If the scheme were perfect, the patch would return to its original position and shape (without diffusion, the system is reversible and trajectories integrated forward and then backward should bring all particles back to their original position), but this won't be the case, and the difference between initial and final states is a measure of the error. Because some of the error generated during the flow in one direction may be negated during the return flow, we also need to consider the result at the moment of current reversal, i.e., the moment of farthest displacement.

For the method developed up to now, some degradation occurs, and bizarre results happen, even in regions of almost uniform flow (see Numerical Exercise 6.15). To discern the cause of this degradation, we first have to realize that the oblique advection test case is special in the sense that during a 1D step, the corresponding velocity is uniform. In the present case, the velocity during a substep is no longer uniform, and application of the first substep on a uniform field $\tilde{c} = C$ yields

$$\frac{\tilde{c}_{i,j}^{\star} - C}{\Delta t} + \frac{\tilde{u}_{i+1/2,j} - \tilde{u}_{i-1/2,j}}{\Delta x} C = 0,$$

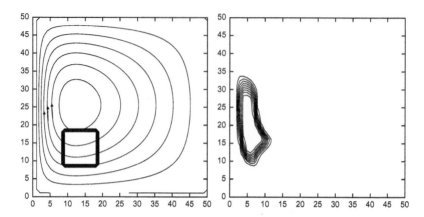

FIGURE 6.18 Advection of a square signal along a sheared boundary-layer current. *Left panel*: Initial distribution and streamlines. *Right panel*: After some advection. The distortion of the distribution is mostly due to the sheared current, which causes cross-stream squeezing and downstream stretching as the tracer enters the boundary layer.

which provides a value of $\tilde{c}^{\star}_{i,j}$ different from the constant $\mathcal{C}$. The next step is unable to correct this by returning the distribution back to a constant. The problem arises because the substep is not characterized by zero divergence of the 1D velocity field, and conservation of the tracer is not met. Conservation in the presence of a converging/diverging velocity in 1D is, however, encountered in another, physical problem: compressible flow. Mimicking this problem, we introduce a *pseudo-compressible approach*, which introduces a density-like variable ρ, to calculate the pseudo-mass conservation written as

$$\frac{\partial}{\partial t}(\rho) + \frac{\partial}{\partial x}(\rho u) + \frac{\partial}{\partial y}(\rho v) = 0 \qquad (6.71)$$

and the tracer budget as

$$\frac{\partial}{\partial t}(\rho c) + \frac{\partial}{\partial x}(\rho u c) + \frac{\partial}{\partial y}(\rho v c) = 0. \qquad (6.72)$$

The splitting method starts with a constant ρ during the first substep and yields

$$\frac{\rho^{\star} - \rho}{\Delta t} + \frac{\tilde{u}_{i+1/2,j} - \tilde{u}_{i-1/2,j}}{\Delta x}\, \rho = 0$$

for the pseudo-mass equation and

$$\frac{\rho^{\star}\tilde{c}^{\star}_i - \rho\tilde{c}^n_i}{\Delta t} + \rho \mathcal{L}_1(\tilde{c}^n_i) = 0. \qquad (6.73)$$

for the tracer concentration. In each calculation, the constant ρ is a multiplicative constant, which can be taken out of the advection operator $\mathcal{L}_1$. The second

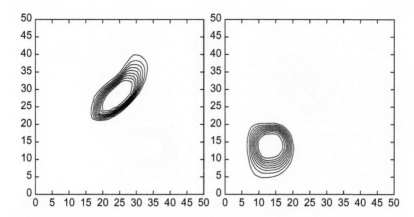

FIGURE 6.19 Advection with TVD scheme, Strang splitting, and pseudo-compressibility. The initial condition is as shown in the left panel of Fig. 6.18. *Left panel*: The patch of tracer at its furthest distance from the point of release, at the time of flow reversal. Its deformation is mostly physical and should ideally be undone during the return travel. *Right panel*: End state after return travel. The patch has nearly returned to its original location and shape, indicative of the scheme's good level of performance.

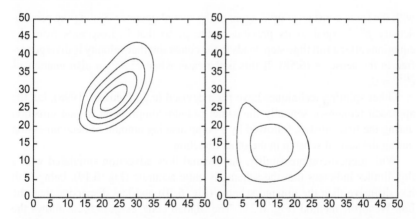

FIGURE 6.20 Same as Fig. 6.19 but with advection by the upwind scheme, using Strang splitting and including pseudo-compressibility. The situation at time of flow reversal (left panel) and after return (right panel) shows that numerical diffusion is clearly stronger than with the TVD scheme. The final distribution is hardly identifiable with the initial condition. The contour values are the same as in Fig. 6.19.

substep similarly follows with

$$\frac{\rho^{n+1} - \rho^{\star}}{\Delta t} + \frac{\tilde{v}_{i,j+1/2} - \tilde{v}_{i,j-1/2}}{\Delta y}\,\rho = 0,$$

$$\frac{\rho^{n+1}\tilde{c}_i^{n+1} - \rho^{\star}\tilde{c}_i^{\star}}{\Delta t} + \rho\mathcal{L}_2(\tilde{c}_i^{\star}) = 0, \tag{6.74}$$

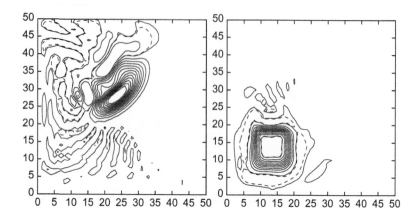

FIGURE 6.21   Same as Fig. 6.19 but with advection by the Lax-Wendroff scheme, using Strang splitting and including pseudo-compressibility. The situation at flow reversal (left panel) shows much dispersion, which is partly undone during the return travel (right panel). At the end, the distribution has been fairly well reconstructed but there is some undershooting around the edges and overshooting in the center. Dashed lines indicate values outside the initial range.

with the same constant ρ used again in the spatial operator. Setting the pseudo-density ρ^{n+1} equal to its previous value ρ, so that it disappears from the equations after a full time step, leads to the constraint that velocity is divergence-free in the sense of (6.58). If this is the case, when $\tilde{c}^n = \mathcal{C}$, it also guarantees $\tilde{c}^{n+1} = \mathcal{C}$.

Other splitting techniques have been devised (e.g., Pietrzak, 1998), but the approach remains essentially the same: pseudo-compression in one direction during the first substep, followed by a compensating amount of decompression during the second substep in the other direction.

With pseudo-compressibility, the sheared flow advection simulated with a flux limiter indicates that the scheme is quite accurate (Fig. 6.19), being both less diffusive than the upwind scheme (Fig. 6.20) and less dispersive than the Lax-Wendroff method (Fig. 6.21). The Matlab code `tvdadv2D.m` allows the reader to experiment with various strategies, by turning pseudo-compressibility on or off, enabling and disabling time splitting, and using different limiters in sheared and unsheared flow fields (Numerical Exercise 6.15).

ANALYTICAL PROBLEMS

6.1.   Show that

$$c(x, y, t) = \frac{M}{4\pi \mathcal{A}t}\, e^{-[(x-ut)^2 + (y-vt)^2]/(4\mathcal{A}t)} \tag{6.75}$$

is the solution of the two-dimensional advection-diffusion equation with uniform velocity components u and v. Plot the solution for decreasing values of t and infer the type of physical problem the initial condition is supposed to represent. Provide an interpretation of M. (*Hint*: Integrate over the infinite domain.)

6.2. Extend solution (6.75) to a radioactive tracer with decay constant K. (*Hint*: Look for a solution of similar structure but with one more exponential factor.)

6.3. Assuming a highly advective situation (high Peclet number), construct the 2D solution corresponding to the continuous release of a substance (S, in mass per time) from a punctual source (located at $x=y=0$) in the presence of velocity u in the x-direction and diffusion $\mathcal{A}$ in the y-direction.

6.4. An unreported ship accident results in an instantaneous release of a conservative pollutant. This substance floats along the sea surface and disperses for some time until it is eventually detected and measured. The maximum concentration, then equal to $c=0.1$ $\mu g/m^2$, is found just West of the Azores at 38°30′N 30°00′W. A month later, the maximum concentration has decreased to 0.05 $\mu g/m^2$ and is located 200 km further South. Assuming a fixed diffusivity $\mathcal{A}=1000$ m^2/s and uniform steady flow, can you infer the amount of substance that was released from the ship, and the time and location of the accident? Finally, how long will it be before the concentration no longer exceeds 0.01 $\mu g/m^2$ anywhere?

6.5. Study the dispersion relation of the equation

$$\frac{\partial c}{\partial t} = \kappa \, \frac{\partial^p c}{\partial x^p} \tag{6.76}$$

where p is a positive integer. Distinguish between even and odd values of p. What should be the sign of the coefficient κ for the solution to be well behaved? Then, compare the cases $p=2$ (standard diffusion) and $p=4$ (biharmonic diffusion). Show that the latter generates a more scale-selective damping behavior than the former.

6.6. Explain the behavior found in Fig. 6.10 by finding the analytical solution of the corresponding physical problem (6.48).

6.7. In the interior of the Pacific Ocean, a slow upwelling compensating the deep convection of the high latitudes creates an average upward motion of about 5 m/year between depths of 4 km to 1 km. The average background turbulent diffusion in this region is estimated to be on the order of 10^{-4} m^2/s. From the deep region, Radium ^{226}Ra found in the sediments, is brought up, while Tritium ^{3}H of atmospheric origin diffuses downward from the surface. Radium has a half-life (time for 50% decay) of 1620 years, and Tritium has a half-life of 12.43 years. Determine the steady-state solution using a one-dimensional vertical advection-diffusion model, assuming fixed and unit value of Tritium at the surface and zero at 4 km depth. For Radium, assume a unit value at depth of 4 km and zero value at the surface. Compare solutions with and without advection. Which tracer is most influenced by advection? Analyze the relative importance of advective and diffusive fluxes for each tracer at 4 km depth and 1 km depth.

6.8. If you intend to use a numerical scheme with an upwind advection to solve the preceding problem for Carbon-14 ^{14}C (half-life of 5730 years), what vertical resolution would be needed so that the numerical calculation does not introduce an excessively large numerical diffusion?

NUMERICAL EXERCISES

6.1. Prove the assertion that a forward-in-time, central-in-space approximation to the advection equation is unconditionaly unstable.

6.2. Use `advleap.m` with different initialization techniques for the first time step of the leapfrog scheme. What happens if an inconsistent approach is used (e.g., zero values)? Can you eliminate the spurious mode totally by a clever initialization of the auxiliary initial condition c^1 when a pure sinusoidal signal is being advected?

6.3. Use the stability analysis under the form (5.31) using an amplification factor. Verify that the stability condition is $|C| \leq 1$.

6.4. Verify numerically that the leapfrog scheme conserves variance of the concentration distribution when $\Delta t \rightarrow 0$. Compare with the Lax-Wendroff scheme behavior for the same time steps.

6.5. Analyze the numerical phase speed of the upwind scheme. What happens for $C = 1/2$? Which particular behavior is observed when $C = 1$?

6.6. Design a fourth-order spatial difference and explicit time stepping for the 1D advection problem. What is the CFL condition of this scheme? Compare with the von Neumann stability condition. Simulate the standard advection test case.

6.7. Design a higher-order finite-volume approach by using higher-order polynomials to calculate the flux integrals. Instead of a linear interpolation as in the Lax-Wendroff scheme, use a parabolic interpolation.

6.8. Show that the von Neumann stability condition of the Beam-Warming scheme is $0 \leq C \leq 2$.

6.9. Implement the trapezoidal scheme with centered space difference using the tridiagonal algorithm `thomas.m`. Apply it to the standard problem of the top-hat signal advection and verify that you find the result shown in Fig. 6.22. Provide an interpretation of the result in terms of the numerical

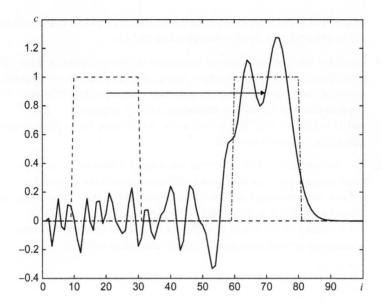

FIGURE 6.22   Standard test case with the trapezoidal scheme and centered spatial derivatives.

dispersion relation. Verify numerically that the variance is conserved exactly.

6.10. Show that the higher-order method for flux calculation at an interface using a linear interpolation on a nonuniform grid with spacing Δx_i between interfaces of cell i leads to the following flux, irrespective of the sign of the velocity

$$\tilde{q}_{i-1/2} = u \, \frac{(\Delta x_{i-1} - u\Delta t)\tilde{c}_i^n + (\Delta x_i + u\Delta t)\tilde{c}_{i-1}^n}{\Delta x_i + \Delta x_{i-1}}. \tag{6.77}$$

6.11. Use a leapfrog centered scheme for advection with diffusion. Apply it to the standard top-hat for different values of the diffusion parameter and interpret your results.

6.12. Find an explanation for why the $2\Delta x$ mode is stationary in all discretizations of advection. (*Hint*: Use a sinusoidal signal of wavelength $2\Delta x$ and zero phase, then sample it. Change the phase (corresponding to a displacement) by different values less than π and resample. What do you observe?)

6.13. Prove that (6.54) is the sufficient stability condition of scheme (6.51). (*Hint*: Rewrite $|\varrho|^2$ as $(\phi - 2C^2\xi)^2 + 4C^2\xi(1 - \xi)$ and observe that as a function of ϕ the amplification factor reaches its maxima at the locations of the extrema of ϕ, itself constrained by (6.54).)

6.14. Consider the one-dimensional advection-diffusion equation with Euler time discretization. For advection, use a centered difference with implicit factor α, and for diffusion the standard second-order difference with implicit factor β. Show that numerical stability requires $(1 - 2\alpha)C^2 \leq 2D$ and $(1 - 2\beta)D \leq 1/2$. Verify that, without diffusion, the explicit centered advection scheme is unstable.

6.15. Use `tvdadv2D.m` with different parameters (splitting or not, pseudo-mass conservation or not) under different conditions (sheared velocity field or solid rotation) and initial conditions (smooth field or strong gradients) with different flux limiters (upwind, Lax-Wendroff, TVD, etc.) to get a feeling of the range of different numerical solutions an advection scheme can provide compared with the analytical solution.

Richard Courant
1888–1972

Born in Upper Silesia, now in Poland but then part of Germany, Richard Courant was a precocious child and, because of economic difficulties at home, started to support himself by tutoring at an early age. His talents in mathematics led him to study in Göttingen, a magnet of mathematicians at the time, and Courant studied under David Hilbert with whom he eventually published in 1924 a famous treatise on methods of mathematical physics. In the foreword, Courant insists on the need for mathematics to be related to physical problems and warns against the trend of that time to loosen that link.

In 1928, well before the invention of computers, Richard Courant published with Kurt Friedrichs and Hans Lewy a most famous paper on the solution of partial difference equations, in which the now-called CFL stability condition was derived for the first time.

Courant left Germany for the United States, where he was offered a position at New York University. The Courant Institute of Mathematical Sciences at that institution is named after him. (*Photo from the MacTutor History of Mathematics archive at the University of St. Andrews*)

Peter David Lax
1926–

Born in Budapest (Hungary), Peter Lax quickly attracted attention for his mathematical prowess. His parents and he had barely moved to the United States, in December 1941 with the last ship from Lisbon during the war, and Peter was still in high school when he was visited in his home by John von Neumann (see biography at end of Chapter 5), who had heard about this outstanding Hungarian mathematician. After working on the top-secret Manhattan atomic bomb project in 1945–1946, he completed his first university degree in 1947 and obtained his doctorate in 1949, both at New York University (NYU). In his own words, "these were years of explosive growth in computing." Lax quickly gained a reputation for his work in numerical analysis.

Lax served as director of Courant Institute at NYU from 1972–1980 and was instrumental in getting the US Government to provide supercomputers for scientific research. (*Photo from the MacTutor History of Mathematics archive at the University of St. Andrews*)

Rotation Effects

Part II

Rotation Effects

Chapter 7

Geostrophic Flows and Vorticity Dynamics

ABSTRACT

This chapter treats homogeneous flows with small Rossby and Ekman numbers. It is shown that such flows have a tendency to display vertical rigidity. The concept of potential vorticity is then introduced. The solution of vertically homogeneous flows often involves a Poisson equation for the pressure distribution, and numerical techniques are presented for this purpose.

7.1 HOMOGENEOUS GEOSTROPHIC FLOWS

Let us consider rapidly rotating fluids by restricting our attention to situations where the Coriolis acceleration strongly dominates the various acceleration terms. Let us further consider homogeneous fluids and ignore frictional effects, by assuming

$$Ro_T \ll 1, \quad Ro \ll 1, \quad Ek \ll 1, \tag{7.1}$$

together with $\rho = 0$ (no density variation). The lowest-order equations governing such homogeneous, frictionless, rapidly rotating fluids are the following simplified forms of equations of motion, Eq. (4.21):

$$-fv = -\frac{1}{\rho_0} \frac{\partial p}{\partial x} \tag{7.2a}$$

$$+fu = -\frac{1}{\rho_0} \frac{\partial p}{\partial y} \tag{7.2b}$$

$$0 = -\frac{1}{\rho_0} \frac{\partial p}{\partial z} \tag{7.2c}$$

$$\frac{\partial u}{\partial x} + \frac{\partial v}{\partial y} + \frac{\partial w}{\partial z} = 0, \tag{7.2d}$$

where f is the Coriolis parameter.

This reduced set of equations has a number of surprising properties. First, if we take the vertical derivative of the first equation, (7.2a), we obtain, successively,

$$-f\frac{\partial v}{\partial z} = -\frac{1}{\rho_0}\frac{\partial}{\partial z}\left(\frac{\partial p}{\partial x}\right) = -\frac{1}{\rho_0}\frac{\partial}{\partial x}\left(\frac{\partial p}{\partial z}\right) = 0,$$

where the right-hand side vanishes because of Eq. (7.2c). The other horizontal momentum equation, (7.2b), succumbs to the same fate, bringing us to conclude that the vertical derivative of the horizontal velocity must be identically zero:

$$\frac{\partial u}{\partial z} = \frac{\partial v}{\partial z} = 0. \tag{7.3}$$

This result is known as the *Taylor–Proudman theorem* (Proudman, 1953; Taylor, 1923). Physically, it means that the horizontal velocity field has no vertical shear and that all particles on the same vertical move in concert. Such vertical rigidity is a fundamental property of rotating homogeneous fluids.

Next, let us solve the momentum equations in terms of the velocity components, a trivial task:

$$u = \frac{-1}{\rho_0 f}\frac{\partial p}{\partial y}, \qquad v = \frac{+1}{\rho_0 f}\frac{\partial p}{\partial x}, \tag{7.4}$$

with the corollary that the vector velocity (u, v) is perpendicular to the vector $(\partial p/\partial x, \partial p/\partial y)$. Since the latter vector is none other than the pressure gradient, we conclude that the flow is not down-gradient but rather across-gradient. The fluid particles are not cascading from high to low pressures, as they would in a nonrotating viscous flow but, instead, are navigating along lines of constant pressure, called *isobars* (Fig. 7.1). The flow is said to be isobaric, and isobars are streamlines. It also implies that no pressure work is performed either on the fluid or by the fluid. Hence, once initiated, the flow can persist without a continuous source of energy.

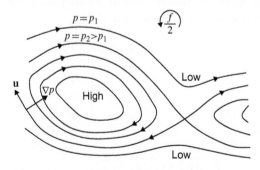

FIGURE 7.1   Example of geostrophic flow. The velocity vector is everywhere parallel to the lines of equal pressure. Thus, pressure contours act as streamlines. In the northern hemisphere (as pictured here), the fluid circulates with the high pressure on its right. The opposite holds for the southern hemisphere.

Such a flow field, where a balance is struck between the Coriolis and pressure forces, is called *geostrophic* (from the Greek, $\gamma\eta$ = Earth and $\sigma\tau\rho o\varphi\eta$ = turning). The property is called *geostrophy*. Hence, by definition, all geostrophic flows are isobaric.

A remaining question concerns the direction of flow along pressure lines. A quick examination of the signs in expressions (7.4) reveals that where f is positive (northern hemisphere, counterclockwise ambient rotation), the currents/winds flow with the high pressures on their right. Where f is negative (southern hemisphere, clockwise ambient rotation), they flow with the high pressures on their left. Physically, the pressure force is directed from the high pressure toward the low pressure initiating a flow in that direction, but on the rotating planet, this flow is deflected to the right (left) in the northern (southern) hemisphere. Figure 7.2 provides a meteorological example from the northern hemisphere.

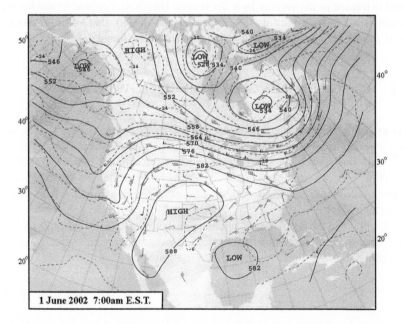

FIGURE 7.2 A meteorological example showing the high degree of parallelism between wind velocities and pressure contours (isobars), indicative of geostrophic balance. The solid lines are actually height contours of a given pressure (500 mb in this case) and not pressure at a given height. However, because atmospheric pressure variations are large in the vertical and weak in the horizontal, the two sets of contours are nearly identical by virtue of the hydrostatic balance. According to meteorological convention, wind vectors are depicted by arrows with flags and barbs: on each tail, a flag indicates a speed of 50 knots, a barb 10 knots, and a half-barb 5 knots (1 knot = 1 nautical mile per hour = 0.5144 m/s). The wind is directed toward the bare end of the arrow because meteorologists emphasize where the wind comes from, not where it is blowing. The dashed lines are isotherms. (*Chart by the National Weather Service, Department of Commerce, Washington, D.C.*)

If the flow field extends over a meridional span that is not too wide, the variation of the Coriolis parameter with latitude is negligible, and f can be taken as a constant. The frame of reference is then called the *f-plane*. In this case, the horizontal divergence of the geostrophic flow vanishes:

$$\frac{\partial u}{\partial x} + \frac{\partial v}{\partial y} = -\frac{\partial}{\partial x}\left(\frac{1}{\rho_0 f}\frac{\partial p}{\partial y}\right) + \frac{\partial}{\partial y}\left(\frac{1}{\rho_0 f}\frac{\partial p}{\partial x}\right) = 0. \tag{7.5}$$

Hence, geostrophic flows are naturally nondivergent on the *f*-plane. This leaves no room for vertical convergence or divergence, as the continuity equation (7.2d) implies:

$$\frac{\partial w}{\partial z} = 0. \tag{7.6}$$

A corollary is that the vertical velocity, too, is independent of height. If the fluid is limited in the vertical by a flat bottom (horizontal ground or sea for the atmosphere) or by a flat lid (sea surface for the ocean), this vertical velocity must simply vanish, and the flow is strictly two-dimensional.

7.2 HOMOGENEOUS GEOSTROPHIC FLOWS OVER AN IRREGULAR BOTTOM

Let us still consider a rapidly rotating fluid so that the flow is geostrophic, but now over an irregular bottom. We neglect the possible surface displacements, assuming that they remain modest in comparison with the bottom irregularities (Fig. 7.3). An example would be the flow in a shallow sea (homogeneous waters) with depth ranging from 20 to 50 m and under surface waves a few centimeters high.

As shown in the development of kinematic boundary conditions (4.28), if the flow were to climb up or down the bottom, it would undergo a vertical velocity proportional to the slope:

$$w = u\frac{\partial b}{\partial x} + v\frac{\partial b}{\partial y}, \tag{7.7}$$

where b is the bottom elevation above the reference level. The analysis of the previous section implies that the vertical velocity is constant across the entire

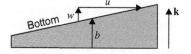

FIGURE 7.3 Schematic view of a flow over a sloping bottom. A vertical velocity must accompany flow across isobaths.

depth of the fluid. Since it must be zero at the top, it must be so at the bottom as well; that is,

$$u\frac{\partial b}{\partial x} + v\frac{\partial b}{\partial y} = 0, \tag{7.8}$$

and the flow is prevented from climbing up or down the bottom slope. This property has profound implications. In particular, if the topography consists of an isolated bump (or dip) in an otherwise flat bottom, the fluid on the flat bottom cannot rise onto the bump, even partially, but must instead go around it. Because of the vertical rigidity of the flow, the fluid parcels at all levels—including levels above the bump elevation—must likewise go around. Similarly, the fluid over the bump cannot leave the bump but must remain there. Such permanent tubes of fluids trapped above bumps or cavities are called *Taylor columns* (Taylor, 1923).

   In flat-bottomed regions, a geostrophic flow can assume arbitrary patterns, and the actual pattern reflects the initial conditions. But over a bottom where the slope is nonzero almost everywhere (Fig. 7.4), the geostrophic flow has no choice but to follow the depth contours (called *isobaths*). Pressure contours are then aligned with topographic contours, and isobars coincide with isobaths. These lines are sometimes also called *geostrophic contours*. Note that a relation between pressure and fluid thickness exists but cannot be determined without additional information on the flow.

   Open isobaths that start and end on a side boundary cannot support any flow, otherwise fluid would be required to enter or leave through lateral boundaries. The flow is simply blocked along the entire length of these lines. In other words, geostrophic flow can occur only along closed isobaths.

   The preceding conclusions hold true as long as the upper boundary is horizontal. If this is not the case, it can then be shown that geostrophic flows are

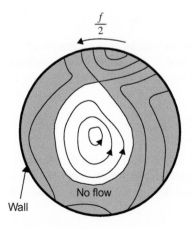

FIGURE 7.4 Geostrophic flow in a closed domain and over irregular topography. Solid lines are isobaths (contours of equal depth). Flow is permitted only along closed isobaths.

constrained to be directed along lines of constant fluid depth. (See Analytical Problem 7.3.) Thus, the fluid is allowed to move up and down, but only as long as it is not being vertically squeezed or stretched. This property is a direct consequence of the inability of geostrophic flows to undergo any two-dimensional divergence.

7.3 GENERALIZATION TO NONGEOSTROPHIC FLOWS

Let us now suppose that the fluid is not rotating as rapidly so that the Coriolis acceleration no longer dwarfs other acceleration terms. We still continue to suppose that the fluid is homogeneous and frictionless. The momentum equations are now augmented to include the relative acceleration terms:

$$\frac{\partial u}{\partial t} + u\frac{\partial u}{\partial x} + v\frac{\partial u}{\partial y} + w\frac{\partial u}{\partial z} - fv = -\frac{1}{\rho_0}\frac{\partial p}{\partial x} \tag{7.9a}$$

$$\frac{\partial v}{\partial t} + u\frac{\partial v}{\partial x} + v\frac{\partial v}{\partial y} + w\frac{\partial v}{\partial z} + fu = -\frac{1}{\rho_0}\frac{\partial p}{\partial y}. \tag{7.9b}$$

Pressure still obeys (7.2c), and continuity equation (7.2d) has not changed.

If the horizontal flow field is initially independent of depth, it will remain so at all future times. Indeed, the nonlinear advection terms and the Coriolis terms are initially z-independent, and the pressure terms are, too, z-independent by virtue of Eq. (7.2c). Thus, $\partial u/\partial t$ and $\partial v/\partial t$ must be z-independent, which implies that u and v tend not to become depth varying and thus remain z-independent at all subsequent times. Let us restrict our attention to such flows, which in the jargon of geophysical fluid dynamics are called *barotropic*. Equations (7.9) then reduce to

$$\frac{\partial u}{\partial t} + u\frac{\partial u}{\partial x} + v\frac{\partial u}{\partial y} - fv = -\frac{1}{\rho_0}\frac{\partial p}{\partial x} \tag{7.10a}$$

$$\frac{\partial v}{\partial t} + u\frac{\partial v}{\partial x} + v\frac{\partial v}{\partial y} + fu = -\frac{1}{\rho_0}\frac{\partial p}{\partial y}. \tag{7.10b}$$

Although the flow has no vertical structure, the similarity to geostrophic flow ends here. In particular, the flow is not required to be aligned with the isobars or it is devoid of vertical velocity. To determine the vertical velocity, we turn to continuity equation (7.2d),

$$\frac{\partial u}{\partial x} + \frac{\partial v}{\partial y} + \frac{\partial w}{\partial z} = 0,$$

in which we note that the first two terms are independent of z but do not necessarily add up to zero. A vertical velocity varying linearly with depth can exist, enabling the flow to support two-dimensional divergence and thus allowing a flow across isobaths.

An integration of the preceding equation over the entire fluid depth yields

$$\left(\frac{\partial u}{\partial x}+\frac{\partial v}{\partial y}\right)\int_{b}^{b+h} dz + [w]_{b}^{b+h} = 0, \tag{7.11}$$

where b is the bottom elevation above a reference level, and h is the local and instantaneous fluid layer thickness (Fig. 7.5). Because fluid particles on the surface cannot leave the surface and particles on the bottom cannot penetrate through the bottom, the vertical velocities at these levels are given by Eqs. (4.28) and (4.31)

$$w(z=b+h) = \frac{\partial}{\partial t}(b+h)+u\frac{\partial}{\partial x}(b+h)+v\frac{\partial}{\partial y}(b+h) \tag{7.12}$$

$$= \frac{\partial \eta}{\partial t}+u\frac{\partial \eta}{\partial x}+v\frac{\partial \eta}{\partial y}$$

$$w(z=b) = u\frac{\partial b}{\partial x}+v\frac{\partial b}{\partial y}. \tag{7.13}$$

Equation (7.11) then becomes, using the surface elevation $\eta = b+h-H$:

$$\frac{\partial \eta}{\partial t}+\frac{\partial}{\partial x}(hu)+\frac{\partial}{\partial y}(hv) = 0, \tag{7.14}$$

which supersedes Eq. (7.2d) and eliminates the vertical velocity from the formalism.

Finally, since the fluid is homogeneous, the dynamic pressure, p, is independent of depth. In the absence of a pressure variation above the fluid surface (e.g., uniform atmospheric pressure over the ocean), this dynamic pressure is

$$p = \rho_0 g \eta, \tag{7.15}$$

where g is the gravitational acceleration according to (4.33). With p replaced by the preceding expression, Eqs. (7.10) and (7.14) form a 3-by-3 system for the variables u, v, and η. The vertical variable no longer appears, and the

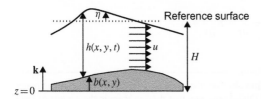

FIGURE 7.5 Schematic diagram of unsteady flow of a homogeneous fluid over an irregular bottom and the attending notation.

independent variables are x, y, and t. This system is

$$\frac{\partial u}{\partial t} + u\frac{\partial u}{\partial x} + v\frac{\partial u}{\partial y} - fv = -g\frac{\partial \eta}{\partial x} \tag{7.16a}$$

$$\frac{\partial v}{\partial t} + u\frac{\partial v}{\partial x} + v\frac{\partial v}{\partial y} + fu = -g\frac{\partial \eta}{\partial y} \tag{7.16b}$$

$$\frac{\partial \eta}{\partial t} + \frac{\partial}{\partial x}(hu) + \frac{\partial}{\partial y}(hv) = 0. \tag{7.16c}$$

Although this system of equations is applied as frequently to the atmosphere as to the ocean, it bears the name *shallow-water model*.[1] If the bottom is flat, the equations become

$$\frac{\partial u}{\partial t} + u\frac{\partial u}{\partial x} + v\frac{\partial u}{\partial y} - fv = -g\frac{\partial h}{\partial x} \tag{7.17a}$$

$$\frac{\partial v}{\partial t} + u\frac{\partial v}{\partial x} + v\frac{\partial v}{\partial y} + fu = -g\frac{\partial h}{\partial y} \tag{7.17b}$$

$$\frac{\partial h}{\partial t} + \frac{\partial}{\partial x}(hu) + \frac{\partial}{\partial y}(hv) = 0. \tag{7.17c}$$

This is a formulation that we will encounter in layered models (Chapter 12).

7.4   VORTICITY DYNAMICS

In the study of geostrophic flows (Section 7.1), it was noted that the pressure terms cancel in the expression of the two-dimensional divergence. Let us now repeat this operation while keeping the added acceleration terms by subtracting the y-derivative of Eq. (7.10a) from the x-derivative of Eq. (7.10b). After some manipulations, the result can be cast as follows:

$$\frac{d}{dt}\left(f + \frac{\partial v}{\partial x} - \frac{\partial u}{\partial y}\right) + \left(\frac{\partial u}{\partial x} + \frac{\partial v}{\partial y}\right)\left(f + \frac{\partial v}{\partial x} - \frac{\partial u}{\partial y}\right) = 0, \tag{7.18}$$

where the material time derivative is defined as

$$\frac{d}{dt} = \frac{\partial}{\partial t} + u\frac{\partial}{\partial x} + v\frac{\partial}{\partial y}.$$

---

[1] In the absence of rotation, these equations also bear the name of *Saint-Venant equations*, in honor of Jean Claude Barré de Saint-Venant (1797–1886) who first derived them in the context of river hydraulics.

In the derivation, care was taken to allow for the possibility of a variable Coriolis parameter (which on a sphere varies with latitude and thus with position). The grouping

$$f + \frac{\partial v}{\partial x} - \frac{\partial u}{\partial y} = f + \zeta \qquad (7.19)$$

is interpreted as the sum of the ambient vorticity (f) with the relative vorticity ($\zeta = \partial v/\partial x - \partial u/\partial y$). To be precise, the vorticity is a vector, but since the horizontal flow field has no depth dependence, there is no vertical shear and no eddies with horizontal axes. The vorticity vector is strictly vertical, and the preceding expression merely shows that vertical component.

Similarly, terms in the continuity equation, (7.14), can be regrouped as

$$\frac{d}{dt}h + \left(\frac{\partial u}{\partial x} + \frac{\partial v}{\partial y}\right)h = 0. \qquad (7.20)$$

If we now consider a narrow fluid column of horizontal cross-section ds, its volume is $h\,ds$, and by virtue of conservation of volume in an incompressible fluid, the following equation holds:

$$\frac{d}{dt}(h\,ds) = 0. \qquad (7.21)$$

This implies, as intuition suggests, that if the parcel is squeezed vertically (decreasing h), it stretches horizontally (increasing ds), and vice versa (Fig. 7.6). Combining Eq. (7.20) for h with Eq. (7.21) for $h\,ds$ yields an equation for ds:

$$\frac{d}{dt}ds = \left(\frac{\partial u}{\partial x} + \frac{\partial v}{\partial y}\right)ds, \qquad (7.22)$$

which simply says that horizontal divergence ($\partial u/\partial x + \partial v/\partial y > 0$) causes widening of the cross-sectional area ds, and convergence ($\partial u/\partial x + \partial v/\partial y < 0$) a

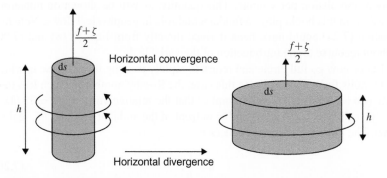

FIGURE 7.6 Conservation of volume and circulation of a fluid parcel undergoing vertical squeezing or stretching. The products $h\,ds$ and $(f+\zeta)\,ds$ are conserved during the transformation. As a corollary, the ratio $(f+\zeta)/h$, called the potential vorticity, is also conserved.

narrowing of the cross-section. It could have been derived from first principles (see Analytical Problem 7.4).

Now, combining Eqs. (7.18) and (7.22) yields

$$\frac{d}{dt}[(f+\zeta)ds]=0 \tag{7.23}$$

and implies that the product $(f+\zeta)ds$ is conserved by the fluid parcel. This product can be interpreted as the vorticity flux (vorticity integrated over the cross section) and is therefore the *circulation* of the parcel. Equation (7.23) is the particular expression for rotating, two-dimensional flows of Kelvin's theorem, which guarantees conservation of circulation in inviscid fluids (Kundu, 1990, pages 124–128).

This conservation principle is akin to that of angular momentum for an isolated system. The best example is that of a ballerina spinning on her toes; with her arms stretched out, she spins slowly, but with her arms brought against her body, she spins more rapidly. Likewise in homogeneous geophysical flows, when a parcel of fluid is squeezed laterally (ds decreasing), its vorticity must increase ($f+\zeta$ increasing) to conserve circulation.

Now, if both circulation and volume are conserved, so is their ratio. This ratio is particularly helpful, for it eliminates the parcel's cross section and thus depends only on local variables of the flow field:

$$\frac{d}{dt}\left(\frac{f+\zeta}{h}\right)=0, \tag{7.24}$$

where

$$q=\frac{f+\zeta}{h}=\frac{f+\partial v/\partial x-\partial u/\partial y}{h} \tag{7.25}$$

is called the *potential vorticity*. The preceding analysis interprets potential vorticity as circulation per volume. This quantity, as will be shown on numerous occasions in this book, plays a fundamental role in geophysical flows. Note that equation (7.24) could have been derived directly from Eqs. (7.18) and (7.20) without recourse to the introduction of the variable ds.

Let us now go full circle and return to rapidly rotating flows, those in which the Coriolis force dominates. In this case, the Rossby number is much less than unity ($Ro=U/\Omega L\ll1$), which implies that the relative vorticity ($\zeta=\partial v/\partial x-\partial u/\partial y$, scaling as U/L) is negligible in front of the ambient vorticity (f, scaling as Ω). The potential vorticity reduces to

$$q=\frac{f}{h}, \tag{7.26}$$

which, if f is constant—such as in a rotating laboratory tank or for geophysical patterns of modest meridional extent—implies that each fluid column must conserve its height h. In particular, if the upper boundary is horizontal, fluid parcels must follow isobaths, consistent with the existence of Taylor columns (Section 7.2). If f is variable (see also Section 9.4) and topography flat, the same constraint (7.26) tells us that the flow cannot cross latitudinal circles, while in the general case, the flow must follow lines of constant f/h.

Before closing this section, let us derive a germane result, which will be useful later. Consider the dimensionless expression

$$\sigma = \frac{z - b}{h}, \tag{7.27}$$

which is the fraction of the local height above the bottom to the full depth of the fluid, or, in short, the relative height above bottom ($0 \le \sigma \le 1$). This expression will later be defined as the so-called σ-coordinate (see Section 20.6.1). Its material time derivative is

$$\frac{d\sigma}{dt} = \frac{1}{h}\frac{d}{dt}(z - b) - \frac{z - b}{h^2}\frac{dh}{dt}. \tag{7.28}$$

Since $dz/dt = w$ by definition of the vertical velocity and because w varies linearly from db/dt at the bottom ($z = b$) to $d(b+h)/dt$ at the top ($z = b+h$), we have

$$\frac{dz}{dt} = w = \frac{db}{dt} + \frac{z - b}{h}\frac{dh}{dt}. \tag{7.29}$$

Use of this last expression to eliminate dz/dt from Eq. (7.28) cancels all terms on the right, leaving only

$$\frac{d\sigma}{dt} = 0. \tag{7.30}$$

Thus, a fluid parcel retains its relative position within the fluid column. Even if there is a vertical velocity, the structure of the velocity field is such that layers of fluid remain invariably stacked on each other. Hence, there is no internal overturning, and layers are simply squeezed or stretched.

7.5   RIGID-LID APPROXIMATION

Except in the case when fast surface waves are of interest (Section 9.1), we can exploit the fact that large-scale motions in the ocean are relatively slow and introduce the so-called rigid-lid approximation. Large-scale movements with small Rossby numbers are close to geostrophic equilibrium, and their dynamic pressure thus scales as $p \sim \rho_0 \Omega U L$ (see (4.16)), and since $p = \rho_0 g \eta$ in a homogeneous fluid, the scale ΔH of surface-height displacements is $\Delta H \sim \Omega U L/g$. Using the latter in the vertically integrated volume-conservation equation, we can then compare the sizes of the different terms. Assuming that the timescale is not shorter than the inertial timescale $1/\Omega$, we have

$$\frac{\partial \eta}{\partial t} + \frac{\partial}{\partial x}(hu) + \frac{\partial}{\partial y}(hv) = 0$$

$$\Omega \Delta H \qquad \frac{HU}{L} \qquad \frac{HU}{L},$$

in which $\Omega \Delta H \sim \Omega^2 UL/g$, and the scale ratio of the first term to the other terms is $\Omega^2 L^2/gH$. In many situations, this ratio is very small,

$$\frac{\Omega^2 L^2}{gH} \ll 1, \tag{7.31}$$

and the time derivative in the volume-conservation equation may be neglected:

$$\frac{\partial}{\partial x}(hu) + \frac{\partial}{\partial y}(hv) = 0. \tag{7.32}$$

This is called the *rigid-lid approximation* (Fig. 7.7).

However, this approximation has a major implication when we solve the equations numerically because now, instead of using the time derivative of the continuity equation to march η forward in time and determine the hydrostatic pressure p from it, we somehow need to find a pressure field that ensures that at any moment the *transport* field $(U, V) = (hu, hv)$ is nondivergent.

The momentum equations of the shallow-water model can be recast in transport form:

$$\frac{\partial}{\partial t}(hu) = -\frac{h}{\rho_0}\frac{\partial p}{\partial x} + F_x \tag{7.33a}$$

$$\text{with } F_x = -\frac{\partial}{\partial x}(huu) - \frac{\partial}{\partial y}(hvu) + f\,hv$$

$$\frac{\partial}{\partial t}(hv) = -\frac{h}{\rho_0}\frac{\partial p}{\partial y} + F_y \tag{7.33b}$$

$$\text{with } F_y = -\frac{\partial}{\partial x}(huv) - \frac{\partial}{\partial y}(hvv) - f\,hu.$$

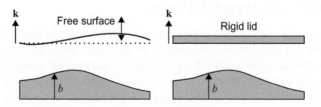

FIGURE 7.7 A free-surface formulation (left panel) allows the surface to move with the flow, whereas a rigid-lid formulation assumes a fixed surface, under which pressure is not uniform because the "lid" resists any local upward or downward force.

Since we have neglected the variation η in surface elevation, we can take in the preceding equations $h = H - b$, a known function of the coordinates x and y. The task ahead of us is to find a way to calculate from the preceding two equations (7.33a) and (7.33b), a pressure field p that leads to satisfaction of constraint (7.32). To do so, we have two approaches at our disposal. The first one is based on a diagnostic equation for pressure (Section 7.6), and the second one on a streamfunction formulation (Section 7.7).

7.6   NUMERICAL SOLUTION OF THE RIGID-LID PRESSURE EQUATION

The pressure method uses Eqs. (7.33a) and (7.33b) to construct an equation for pressure while enforcing the no-divergence constraint. This is accomplished by adding the x-derivative of Eq. (7.33a) to the y-derivative of Eq. (7.33b) and exploiting Eq. (7.32) to eliminate the time derivatives. Placing the pressure terms on the left then yields

$$\frac{\partial}{\partial x}\left(\frac{h}{\rho_0}\frac{\partial p}{\partial x}\right) + \frac{\partial}{\partial y}\left(\frac{h}{\rho_0}\frac{\partial p}{\partial y}\right) = \frac{\partial F_x}{\partial x} + \frac{\partial F_y}{\partial y} \qquad (7.34)$$
$$= Q.$$

This equation for pressure is the archetype of a so-called *elliptic equation*.

To complement it, appropriate boundary conditions must be provided. These pressure conditions are deduced from the impermeability of solid lateral boundaries or from the inflow/outflow conditions at open boundaries (see Section 4.6). For example, if the boundary is parallel to the y-axis (say $x = x_0$) and is impermeable, we need to impose $hu = 0$ (no normal transport), and the x-momentum equation in transport form reduces there to

$$\frac{h}{\rho_0}\frac{\partial p}{\partial x} = F_x, \qquad (7.35)$$

while along an impermeable boundary parallel to the x-axis (say $y = y_0$), we need to impose $hv = 0$ and obtain from the y-momentum equation

$$\frac{h}{\rho_0}\frac{\partial p}{\partial y} = F_y. \qquad (7.36)$$

In other words, the normal pressure gradient is given along impermeable boundaries. At inflow/ouflow boundaries, the expression is more complicated but it is still the normal pressure gradient that is imposed. An elliptic equation with the normal derivative prescribed all along the perimeter of the domain is called a Neumann problem.[2]

---

[2]If the pressure itself had been imposed all along the perimeter of the domain, the problem would have been called a Dirichlet problem.

One and only one condition at every point along all boundaries of the domain is necessary and sufficient to determine the solution of the elliptic equation (7.34). Since the pressure appears only through its derivatives in both the elliptic equation (7.34) and the boundary condition (7.35) and (7.36), the solution is only defined within an additional arbitrary constant, the value of which may be chosen freely without affecting the resulting velocity field. However, there is a natural choice, which is to select the constant so that the pressure has a zero average over the domain. By virtue of $p = \rho_0 g\eta$, this corresponds to stating that η has a zero average over the domain.

Numerically, the solution can be sought by discretizing the elliptic equation for pressure across a rectangular box:

$$
\frac{1}{\Delta x}\left(h_{i+1/2}\,\frac{\tilde{p}_{i+1,j}-\tilde{p}_{i,j}}{\Delta x}-h_{i-1/2}\,\frac{\tilde{p}_{i,j}-\tilde{p}_{i-1,j}}{\Delta x}\right)
$$
$$
+\frac{1}{\Delta y}\left(h_{j+1/2}\,\frac{\tilde{p}_{i,j+1}-\tilde{p}_{i,j}}{\Delta y}-h_{j-1/2}\,\frac{\tilde{p}_{i,j}-\tilde{p}_{i,j-1}}{\Delta y}\right)=\rho_0 Q_{ij}.
\tag{7.37}
$$

This forms a set of linear equations for the $\tilde{p}_{i,j}$ values across the grid, connecting five unknowns at each grid point (Fig. 7.8), a situation already encountered in

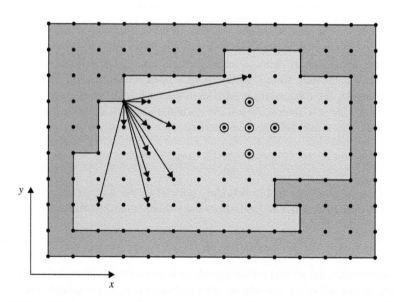

FIGURE 7.8 Discretization of the two-dimensional elliptic equation. The stencil is a five-point array consisting of the point where the calculation is performed and its four neighbors. These neighboring points are in turn dependent on their respective neighbors, and so on until boundary points are reached. In other words, the value at every point inside the domain is influenced by all other interior and boundary values. Simple accounting indicates that one and only one boundary condition is needed at all boundary points.

the treatment of two-dimensional implicit diffusion (Section 5.6). But there is another circular dependence: The right-hand side $\rho_0 Q_{ij}$ is not known until the velocity components are determined and the determination of these requires the knowledge of the pressure gradient. Because the momentum equations are nonlinear, this is a nonlinear dependence, and the method for constructing and solving a linear system cannot be applied.

The natural way to proceed is to progress incrementally. If we assume that at time level n, we have a divergent-free velocity field $(\tilde{u}^n, \tilde{v}^n)$, we can use Eq. (7.37) to calculate the pressure at the same time level n and use its gradient in the momentum equations to update the velocity components for time level $n+1$. But this offers no guarantee that the updated velocity components will be divergence free, despite the fact that the pressure distribution corresponds to a divergent-free flow field at the previous time level.

Once again, we face a situation in which discretized equations do not inherit certain mathematical properties of the continuous equations. In this case, we used properties of divergence and gradient operators to build a diagnostic pressure equation from the original equations, but these properties are not transferable to the numerical space unless special care is taken.

However, the design of adequate discrete equations can be inspired by the mathematical operations used to reach the pressure equation (7.34): We started with the velocity equation and applied the divergence operator to make appear the divergence of the transport that we then set to zero, and we should perform the same operations in the discrete domain to ensure that at any moment the discrete transport field is nondivergent in a finite volume. This is expressed by discrete volume conservation as

$$\frac{h_{i+1/2}\tilde{u}_{i+1/2} - h_{i-1/2}\tilde{u}_{i-1/2}}{\Delta x} + \frac{h_{j+1/2}\tilde{v}_{j+1/2} - h_{j-1/2}\tilde{v}_{j-1/2}}{\Delta y} = 0. \tag{7.38}$$

Anticipating a *staggered grid* configuration (Fig. 7.9), we realize that it would be natural to calculate for each cell the velocity $\tilde{u}$ at the middle of the left and right interfaces $(i\pm 1/2, j)$ and the other velocity component $\tilde{v}$ at the middle of the top and bottom interfaces $(i, j\pm 1/2)$ so that the divergence may be calculated most naturally in Eq. (7.38). In contrast, $\tilde{p}$ values are calculated at cell centers. Leapfrog time discretization applied to (7.33a) and (7.33b) then

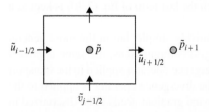

FIGURE 7.9 Arrangement of numerical unknowns for easy enforcement of numerical volume conservation and pressure-gradient calculations.

provides

$$h_{i+1/2}\tilde{u}_{i+1/2}^{n+1} = h_{i+1/2}\tilde{u}_{i+1/2}^{n-1} + 2\Delta t F_{xi+1/2} - 2\Delta t h_{i+1/2} \frac{\tilde{p}_{i+1} - \tilde{p}}{\rho_0 \Delta x} \qquad (7.39a)$$

$$h_{j+1/2}\tilde{v}_{j+1/2}^{n+1} = h_{j+1/2}\tilde{v}_{j+1/2}^{n-1} + 2\Delta t F_{yj+1/2} - 2\Delta t h_{j+1/2} \frac{\tilde{p}_{j+1} - \tilde{p}}{\rho_0 \Delta y}, \qquad (7.39b)$$

in which we omitted for clarity the obvious indices i, j, and n.

Requesting now that the discretized version (7.38) of the nondivergence constraint hold at time level $n+1$, we can eliminate the velocity values at that time level by combining the equations of (7.39) so that these terms cancel out. The result is the sought-after discretized equation for pressure:

$$\frac{1}{\Delta x}\left(h_{i+1/2}\frac{\tilde{p}_{i+1} - \tilde{p}}{\Delta x} - h_{i-1/2}\frac{\tilde{p} - \tilde{p}_{i-1}}{\Delta x}\right) + \frac{1}{\Delta y}\left(h_{j+1/2}\frac{\tilde{p}_{j+1} - \tilde{p}}{\Delta y} - h_{j-1/2}\frac{\tilde{p} - \tilde{p}_{j-1}}{\Delta y}\right)$$

$$= \rho_0\left(\frac{F_{xi+1/2} - F_{xi-1/2}}{\Delta x} + \frac{F_{yj+1/2} - F_{yj-1/2}}{\Delta y}\right)$$

$$+ \frac{\rho_0}{2\Delta t}\left(\frac{h_{i+1/2}\tilde{u}_{i+1/2}^{n-1} - h_{i-1/2}\tilde{u}_{i-1/2}^{n-1}}{\Delta x} + \frac{h_{j+1/2}\tilde{v}_{j+1/2}^{n-1} - h_{j-1/2}\tilde{u}_{j-1/2}^{n-1}}{\Delta y}\right),$$

$$(7.40)$$

where once again the obvious indices have been omitted.

It is clear that, up to the last term, this equation is a discrete version of Eq. (7.34) and resembles Eq. (7.37). The difference lies in the last term, which would vanish if the transport field were divergence free at time level $n-1$. We kept that term should the numerical solution of the discrete equation not be exact. Keeping the nonzero discrete divergence at $n-1$ in the equation is a way of applying an automatic correction to the discrete equation in order to insure the nondivergence of the transport at the new time level $n+1$. Neglecting this correction term would result in a gradual accumulation of errors and thus an eventually divergent transport field.

To summarize, the algorithm works as follows: Knowing velocity values at time levels n and $n-1$, we solve Eq. (7.40) iteratively for pressure, which is then used to advance velocity in time using Eqs. (7.39a) and (7.39b). For quickly converging iterations, the pressure calculations can be initialized with the values from the previous time step. This iterative procedure is one of the sources of numerical errors against which the last term of Eq. (7.40) is kept as a precaution.

The discretization shown here is relatively simple, but in the more general case of higher-order methods or other grid configurations, the same approach can be used. We must ensure that the divergence operator applied to the transport field is discretized in the same way as the divergence operator is applied to the pressure gradient. Furthermore, the pressure gradient needs to be discretized in the same way in both the velocity equation and the elliptic pressure equation.

In summary, the derivatives similarly labeled in the equations below must be discretized in identical ways to generate a mathematically coherent scheme:

$$\underbrace{\frac{\partial}{\partial x}\left(\frac{h}{\rho_0}\frac{\partial p}{\partial x}\right)}_{(1)\qquad(3)} + \underbrace{\frac{\partial}{\partial y}\left(\frac{h}{\rho_0}\frac{\partial p}{\partial y}\right)}_{(2)\qquad(4)} = \underbrace{\frac{\partial}{\partial x}F_x}_{(1)} + \underbrace{\frac{\partial}{\partial y}F_y}_{(2)}$$

$$\frac{\partial}{\partial t}(hu) = -\underbrace{\frac{h}{\rho_0}\frac{\partial p}{\partial x}}_{(3)} + F_x$$

$$\frac{\partial}{\partial t}(hv) = -\underbrace{\frac{h}{\rho_0}\frac{\partial p}{\partial y}}_{(4)} + F_y$$

$$\underbrace{\frac{\partial}{\partial x}(hu)}_{(1)} + \underbrace{\frac{\partial}{\partial y}(hv)}_{(2)} = 0.$$

It also means one can generally not resort to a "black box" elliptic-equation solver to obtain a pressure field that is used in "hand-made" discrete velocity equations.

7.7 NUMERICAL SOLUTION OF THE STREAMFUNCTION EQUATION

Instead of calculating pressure, a second method in use with the rigid-lid approximation is a generalization of the velocity streamfunction ψ to the *volume-transport streamfunction* Ψ:

$$hu = -\frac{\partial(h\psi)}{\partial y} = -\frac{\partial\Psi}{\partial y} \tag{7.42a}$$

$$hv = +\frac{\partial(h\psi)}{\partial x} = +\frac{\partial\Psi}{\partial x}. \tag{7.42b}$$

The difference between two isolines of Ψ can be interpreted as the volume transport between those lines, directed with the higher Ψ values to its right.

When transport components are calculated according to Eq. (7.42), volume conservation (7.32) is automatically satisfied, and as shown in Section 6.6, the numerical counterpart can also be divergence free. We may therefore discretize the equation governing Ψ without hesitation, sure that its discrete solution will lead to a well-bahaved discrete velocity field.

To obtain a mathematical equation for the streamfunction, all we have to do is to eliminate the pressure from the momentum equations. This is accomplished by dividing Eqs. (7.33a) and (7.33b) by h, differentiating the former by y and the latter by x, and finally subtracting one from the other. Replacement of the transport components hu and hv in terms of the streamfunction then yields

$$\frac{\partial}{\partial t}\left[\frac{\partial}{\partial x}\left(\frac{1}{h}\frac{\partial \Psi}{\partial x}\right)+\frac{\partial}{\partial y}\left(\frac{1}{h}\frac{\partial \Psi}{\partial y}\right)\right]=\frac{\partial}{\partial x}\left(\frac{F_y}{h}\right)-\frac{\partial}{\partial y}\left(\frac{F_x}{h}\right) \quad (7.43)$$

$$=Q.$$

The right-hand side could be further expanded in terms of the streamfunction, but for the sake of the following discussion it is sufficient to lump all its terms into a single "forcing" term Q.

We now consider a leapfrog time discretization or any other time discretization that allows us to write

$$\left[\frac{\partial}{\partial x}\left(\frac{1}{h}\frac{\partial \tilde{\Psi}}{\partial x}\right)^{n+1}+\frac{\partial}{\partial y}\left(\frac{1}{h}\frac{\partial \tilde{\Psi}}{\partial y}\right)^{n+1}\right]=F(\tilde{\Psi}^n,\tilde{\Psi}^{n-1},\ldots). \quad (7.44)$$

In the case of a leapfrog discretization, the right-hand side is

$$F(\tilde{\Psi}^n,\tilde{\Psi}^{n-1},\ldots)=\left[\frac{\partial}{\partial x}\left(\frac{1}{h}\frac{\partial \tilde{\Psi}}{\partial x}\right)^{n-1}+\frac{\partial}{\partial y}\left(\frac{1}{h}\frac{\partial \tilde{\Psi}}{\partial y}\right)^{n-1}\right]$$

$$+2\Delta t\, Q^n, \quad (7.45)$$

which can be evaluated numerically knowing $\tilde{\Psi}^n$ and $\tilde{\Psi}^{n-1}$. The problem then amounts to solving Eq. (7.44) for $\tilde{\Psi}^{n+1}$. Again, an elliptic equation must be solved, as for the pressure equation in the previous section, and the same method can be applied.

Differences, other than the terms in the right-hand side, are noteworthy. First, instead of h appearing inside the derivatives, $1/h$ is involved, which increases the role played by the streamfunction derivatives in shallow regions ($h \to 0$, usually near boundaries), possibly amplifying errors on boundary conditions. This is in contrast to the pressure formulation, in which the influence of the vertically integrated pressure gradient decreases in shallow regions. Applications indeed reveal that the solution of the Poisson equation (7.40) is better conditioned and converges better than Eq. (7.44).

A second difference is related to the formulation of boundary conditions. While in the pressure approach imposing zero normal velocity leads to a condition on the normal derivative of pressure, the streamfunction formulation has the apparent advantage of only demanding that the streamfunction be constant along a solid boundary, a Dirichlet condition. A problem arises for ocean models when islands are present within the domain (Fig. 7.10). Knowing that the streamfunction is constant on an impermeable boundary does not tell us what the value of the constant ought to be. This is no small matter because the difference of streamfunction values across a channel defines the volume transport in that

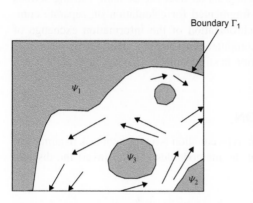

FIGURE 7.10 Boundary conditions on the streamfunction in an ocean model with islands. The streamfunction value must be prescribed constant along impermeable boundaries. Setting Ψ_1 and Ψ_2 for the outer boundaries is reasonable and amounts to imposing the total flow across the domain, but setting a priori the value of Ψ_3 along the perimeter of an island is in principle not permitted because the flow around the island should depend on the interior solution and its temporal evolution. Clearly, a prognostic equation for the streamfunction value on islands is needed.

channel. Such volume transport should be determined by the dynamics of the flow and not by the modeler's choice.

The streamfunction equation being linear with known right-hand side allows superposition of solutions, and we take one island at a time:

$$\frac{\partial}{\partial x}\left(\frac{1}{h}\frac{\partial \psi_k}{\partial x}\right) + \frac{\partial}{\partial y}\left(\frac{1}{h}\frac{\partial \psi_k}{\partial y}\right) = 0 \qquad (7.46)$$

with ψ_k set to zero on all boundaries except $\psi_k = 1$ on the boundary for the kth island. Each island thus engenders a dimensionless streamfunction $\psi_k(x, y)$ that can be used to construct the overall solution

$$\Psi(x, y, t) = \Psi_f(x, y, t) + \sum_k \Psi_k(t)\psi_k(x, y), \qquad (7.47)$$

where Ψ_f is the particular solution of Eq. (7.44) with streamfunction set to zero along all island boundaries and prescribed values along the outer boundaries and wherever the volume flow is known (e.g., at the inflow boundary as depicted in Fig. 7.10). The $\Psi_k(t)$ coefficients are the time-dependent factors by which the island contributions must be multiplied to construct the full solution. What should these factors be is the question.

One possibility is to project the momentum equations onto the direction locally tangent to the island boundary, similarly to what was done to determine the boundary conditions in the pressure formulation. Invoking Stokes theorem on the closed contour formed by the perimeter of the kth island then provides an equation including the time derivative $d\Psi_k/dt$. Repeating the procedure for each island leads to a linear set of N equations, where N is the number of islands. These equations can then be integrated in time (e.g., Bryan & Cox, 1972). This approach has become less popular over the years for several reasons, among which is the nonlocal nature of the equations. Indeed, each island equation involves both area and contour integrals all over the domain, causing serious difficulties when the domain is fragmented for calculation on separate computers working in parallel. Synchronization of the information exchange of different integral pieces across computers can be very challenging. Nevertheless, the streamfunction formulation is still available in most large-scale ocean models.

7.8   LAPLACIAN INVERSION

Because the inversion of a Poisson-type equation is a recurrent task in numerical models, we now outline some of the methods designed to invert the discrete Poisson equation

$$\frac{\tilde{\psi}_{i+1,j} - 2\tilde{\psi}_{i,j} + \tilde{\psi}_{i-1,j}}{\Delta x^2} + \frac{\tilde{\psi}_{i,j+1} - 2\tilde{\psi}_{i,j} + \tilde{\psi}_{i,j-1}}{\Delta y^2} = \tilde{q}_{i,j}, \qquad (7.48)$$

where the right-hand side is given and $\tilde{\psi}$ is the unknown field.[3] Iterative methods outlined in Section 5.6 using pseudo-time iterations were the first methods used to solve a linear system for $\tilde{\psi}_{i,j}$. The Jacobi method with over-relaxation reads

$$\tilde{\psi}_{i,j}^{(k+1)} = \tilde{\psi}_{i,j}^{(k)} + \omega\,\epsilon_{i,j}^{(k)}$$

$$\left(\frac{2}{\Delta x^2} + \frac{2}{\Delta y^2}\right)\epsilon_{i,j}^{(k)} =$$

$$\frac{\tilde{\psi}_{i+1,j}^{(k)} - 2\tilde{\psi}_{i,j}^{(k)} + \tilde{\psi}_{i-1,j}^{(k)}}{\Delta x^2} + \frac{\tilde{\psi}_{i,j+1}^{(k)} - 2\tilde{\psi}_{i,j}^{(k)} + \tilde{\psi}_{i,j-1}^{(k)}}{\Delta y^2} - \tilde{q}_{i,j}, \qquad (7.49)$$

in which the residual ϵ is used to correct the previous estimate at iteration (k). Taking the relaxation parameter $\omega > 1$ (i.e., performing over-relaxation) accelerates convergence toward the solution, at the risk of instability. By considering iterations as evolution in pseudo-time, we can assimilate the parameter ω to a pseudo-time step and perform a numerical stability analysis. The outcome is that iterations are stable (i.e., they converge) provided $0 \leq \omega < 2$. In terms of the general iterative solvers of Section 5.6, matrix **B** in (5.56) is diagonal. The algorithm requires at least as many iterations to propagate the information once through the domain as they are grid points across the domain. If M is the total number of grid points in the 2D model, then $\sqrt{M}$ is an estimate of the "width" of the grid, and it takes $\sqrt{M}$ iterations to propagate information once from side

---

[3]Generalization to equations with variable coefficients such as h or $1/h$, as encountered in the preceding two sections for example, is relatively straightforward, and we keep the notation simple here by assuming constant coefficients.

to side. Usually, M iterations are needed for convergence, and the cost rapidly becomes prohibitive with increased resolution.

The finite speed at which information is propagated during numerical iterations does not reflect the actual nature of elliptic equation, the interconnectness of which theoretically implies instantaneous adjustment to any change anywhere, and we sense that we should be able to do better. Because in practice the iterations are only necessary to arrive at the converged solution, we do not need to mimic the process of a time-dependent equation and can tamper with the pseudo-time.

The Gauss-Seidel method with over-relaxation calculates the residual instead as

$$
\begin{aligned}
&\left(\frac{2}{\Delta x^2} + \frac{2}{\Delta y^2}\right) \epsilon_{i,j}^{(k)} \\
&= \frac{\tilde{\psi}_{i+1,j}^{(k)} - 2\tilde{\psi}_{i,j}^{(k)} + \tilde{\psi}_{i-1,j}^{(k+1)}}{\Delta x^2} + \frac{\tilde{\psi}_{i,j+1}^{(k)} - 2\tilde{\psi}_{i,j}^{(k)} + \tilde{\psi}_{i,j-1}^{(k+1)}}{\Delta y^2} - \tilde{q}_{i,j}
\end{aligned}
\tag{7.50}
$$

in which the updated values at the previous neighbors $(i-1, j)$ and $(i, j-1)$ are immediately used (assuming that we loop across the domain with increasing i and j). In other words, the algorithm (7.50) does not delay using the most updated values. With this time saving also comes a saving of storage as old values can be replaced by new values as soon as these are calculated. Matrix **B** of equation (5.56) is triangular, and the Gauss–Seidel loop (7.50) is the matrix inversion performed by backward substitution. The method is called *SOR*, successive over-relaxation.

The use of the most recent $\tilde{\psi}$ values during the iterations accelerates convergence but not in a drastic way. Only when the relaxation parameter ω is set at a very particular value can the number of iterations be reduced significantly, from $\mathcal{O}(M)$ down to $\mathcal{O}(\sqrt{M})$ (see Numerical Exercise 7.6). Unfortunately, the optimal value of ω depends on the geometry and type of boundary conditions, and a small departure from the optimal value quickly deteriorates the convergence rate. As a guideline, the optimal value behaves as

$$
\omega \sim 2 - \alpha \frac{2\pi}{m}
\tag{7.51}
$$

for a square and isotropic grid with m grid points in each direction, and with parameter $\alpha = \mathcal{O}(1)$ depending on the nature of the boundary conditions.

Because of its easy implementation, the SOR method was very popular in the early days of numerical modeling, but when vector and, later, parallel computers appeared, some adaptation was required. The recurrence relationships that appear in the loops do not allow to calculate $\tilde{\psi}_{i,j}^{(k+1)}$ before the calculations of $\tilde{\psi}_{i-1,j}^{(k+1)}$ and $\tilde{\psi}_{i,j-1}^{(k+1)}$ are finished, and this prevents independent calculations on parallel processors or vector machines. In response, the so-called *red-black*

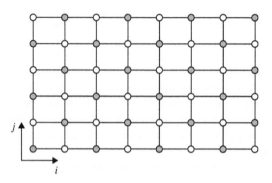

FIGURE 7.11 To avoid recurrence relationships, the discrete domain is swept by two loops, working on white and gray dots separately. During the loop updating white nodes, only values of the gray nodes are used so that all white nodes can be updated independently and immediately. The reverse holds for the gray nodes in the second loop, and all calculations may be performed in parallel. Because the original algorithm is directly related to the so-called red-black partitioning of trees (e.g., Hageman & Young, 2004), the nodes can be colored accordingly, and "red-black" is the name given to the two-stage sweep mechanism.

methods were developed. These perform two Jacobi iterations on two interlaced grids, nicknamed "red" and "black" (Fig. 7.11).

If we want to reduce further the computational burden associated with the inversion of the Poisson equation, we must exploit the very special nature of (7.48) and the resulting linear system to be solved. For the discrete version (7.48) of the Poisson equation, the matrix $\mathbf{A}$ relating the unknowns, now stored in an array $\mathbf{x}$, is symmetric and positive definite (Numerical Exercise 7.10). In this case, the solution of $\mathbf{A}\mathbf{x} = \mathbf{b}$ is equivalent to solving the minimization of

$$J = \frac{1}{2}\mathbf{x}^{\mathsf{T}}\mathbf{A}\mathbf{x} - \mathbf{x}^{\mathsf{T}}\mathbf{b} \qquad (7.52)$$

$$\nabla_x J = \mathbf{A}\mathbf{x} - \mathbf{b} \qquad (7.53)$$

with respect to $\mathbf{x}$. We then have to search for minima rather than to solve a linear equation and, though apparently more complicated, the task can also be tackled by iterative methods. The minimum of J is reached when the gradient with respect to $\mathbf{x}$ is zero: $\nabla_x J = 0$. This is the case when the residual $\mathbf{r} = \mathbf{A}\mathbf{x} - \mathbf{b}$ is zero, that is, when the linear problem is solved.

The use of a minimization approach (e.g., Golub & Van Loan, 1990) instead of a linear-system solver relies on the possibility of using efficient minimization methods. The gradient of J, the residual, is easily calculated and only takes $4M$ operations for the matrix $\mathbf{A}$ arising from the discrete Poisson equation. The value of J, if desired, is also readily obtained by calculating two scalar products involving the already available gradient. A standard minimization method used in optimization problems is to minimize the residual J by following its gradient. In this method, called the *steepest descent* method, a better estimate of $\mathbf{x}$ is sought in the direction in which J decreases fastest. Starting from $\mathbf{x}_0$ and associated residual $\mathbf{r} = \mathbf{A}\mathbf{x}_0 - \mathbf{b}$, a better estimate of $\mathbf{x}$ is sought as

$$\mathbf{x} = \mathbf{x}_0 - \alpha\mathbf{r}, \tag{7.54}$$

which is reminiscent of a relaxation method. The parameter α is then chosen to minimize J. Because the form is quadratic in α, this can be achieved easily (see Numerical Exercise 7.7) by taking

$$\alpha = \frac{\mathbf{r}^T\mathbf{r}}{\mathbf{r}^T\mathbf{A}\mathbf{r}} \tag{7.55}$$

which, because $\mathbf{A}$ is positive definite, can always be calculated as long as the residual is nonzero. If the residual vanishes, iterations can be stopped because the solution has been found. Otherwise, from the new estimate $\mathbf{x}$, a new residual and gradient are computed, and iterations proceed:

> Initialize by first guess $\mathbf{x}^{(0)} = \mathbf{x}_0$
> Loop on increasing k until the residual $\mathbf{r}$ is small enough
> $\quad \mathbf{r} = \mathbf{A}\mathbf{x}^{(k)} - \mathbf{b}$
> $\quad \alpha = \dfrac{\mathbf{r}^T\mathbf{r}}{\mathbf{r}^T\mathbf{A}\mathbf{r}}$
> $\quad \mathbf{x}^{(k+1)} = \mathbf{x}^{(k)} - \alpha\mathbf{r}$
> End of loop on k.

where residual and optimal descent parameter α change at each iteration. It is interesting to note that the residuals of two successive iterations are orthogonal to each other (Numerical Exercise 7.7).

Although very natural, the approach does not converge rapidly, and *conjugate gradient* methods have been developed to provide better convergence rates. In these methods, the direction of progress is no longer the direction of the steepest descent but is prescribed from among a set, noted $\mathbf{e}_i$. We then look for the minimum along these possible directions:

$$\mathbf{x} = \mathbf{x}_0 - \alpha_1\mathbf{e}_1 - \alpha_2\mathbf{e}_2 - \alpha_3\mathbf{e}_3 - \cdots - \alpha_M\mathbf{e}_M. \tag{7.56}$$

If there are M vectors $\mathbf{e}_i$, chosen to be linearly independent, minimization with respect to the M parameters α_i will yield the exact minimum of J. So, instead of searching for the M components of $\mathbf{x}$, we search for the M parameters α_i leading to the optimal state $\mathbf{x}$. This solves the linear system exactly. A simplification in the calculations arises if we choose

$$\mathbf{e}_i^T\mathbf{A}\mathbf{e}_j = 0 \quad \text{when} \quad i \neq j \tag{7.57}$$

because in this case the quadratic form J takes the form

$$J = \frac{1}{2}\mathbf{x}_0^\top \mathbf{A}\mathbf{x}_0 - \mathbf{x}_0^\top \mathbf{b}$$
$$+ \frac{\alpha_1^2}{2}\mathbf{e}_1^\top \mathbf{A}\mathbf{e}_1 - \alpha_1 \mathbf{e}_1^\top (\mathbf{A}\mathbf{x}_0 - \mathbf{b})$$
$$+ \frac{\alpha_2^2}{2}\mathbf{e}_2^\top \mathbf{A}\mathbf{e}_2 - \alpha_2 \mathbf{e}_2^\top (\mathbf{A}\mathbf{x}_0 - \mathbf{b})$$
$$+ \cdots$$
$$+ \frac{\alpha_M^2}{2}\mathbf{e}_M^\top \mathbf{A}\mathbf{e}_M - \alpha_M \mathbf{e}_M^\top (\mathbf{A}\mathbf{x}_0 - \mathbf{b}). \tag{7.58}$$

This expression is readily minimized with respect to each parameter α_k and yields, with $\mathbf{r}_0 = \mathbf{A}\mathbf{x}_0 - \mathbf{b}$,

$$\alpha_k = \frac{\mathbf{e}_k^\top \mathbf{r}_0}{\mathbf{e}_k^\top \mathbf{A}\mathbf{e}_k}, \quad k = 1, \ldots, M \tag{7.59}$$

In other words, to reach the global minimum and thus the solution of the linear system, we simply have to minimize each term individually. However, the difficulty is that the construction of the set of directions $\mathbf{e}_k$ is complicated. Hence, the idea is to proceed step by step and construct the directions as we iterate, with the plan of stopping iterations when residuals have become small enough. We can start with a first arbitrary direction, typically the steepest descent $\mathbf{e}_1 = \mathbf{A}\mathbf{x}_0 - \mathbf{b}$. Then, once we have a set of k directions that satisfy Eq. (7.57), we only minimize along direction $\mathbf{e}_k$ by Eq. (7.59):

$$\mathbf{x}^{(k)} = \mathbf{x}^{(k-1)} - \alpha_k \mathbf{e}_k. \tag{7.60}$$

This leads to a new residual $\mathbf{r}_k = \mathbf{A}\mathbf{x}^{(k)} - \mathbf{b}$

$$\mathbf{r}_k = \mathbf{r}_{k-1} - \alpha_k \mathbf{A}\mathbf{e}_k$$
$$= \mathbf{r}_0 - \alpha_1 \mathbf{A}\mathbf{e}_1 - \alpha_2 \mathbf{A}\mathbf{e}_2 - \cdots - \alpha_k \mathbf{A}\mathbf{e}_k. \tag{7.61}$$

This shows that, instead of calculating α_k according to Eq. (7.59), we can use

$$\alpha_k = \frac{\mathbf{e}_k^\top \mathbf{r}_{k-1}}{\mathbf{e}_k^\top \mathbf{A}\mathbf{e}_k} \tag{7.62}$$

because of property (7.57). We can interpret this result together with (7.61) by showing that the successive residuals are orthogonal to all previous search directions $\mathbf{e}_i$ so that no new search in those directions is needed. Expression (7.62) is also more practical because it requires the storage of only the residual calculated at the previous iteration. The construction of the next direction $\mathbf{e}_{k+1}$ is then performed by a variation of the Gram–Schmidt orthogonalization process of a series of linearly independent vectors. The conjugate gradient method chooses for this set of vectors the residuals already calculated, which can be shown to be orthogonal to one another and hence linearly independent. When applying the Gram–Schmidt orthogonalization in the sense of Eq. (7.57), it turns out that the new direction e_{k+1} is surprisingly easy to calculate in terms of the last residuals and search direction (e.g., Golub & Van Loan, 1990):

$$\mathbf{e}_{k+1} = \mathbf{r}_k + \frac{\|\mathbf{r}_k\|^2}{\|\mathbf{r}_{k-1}\|^2} \mathbf{e}_k \tag{7.63}$$

from which we can proceed to the next step. The algorithm is therefore only slightly more complicated than the steepest-descent method, and we note that we no longer need to store all residuals or search directions, not even intermediate values of $\mathbf{x}$. Only the most recent one needs to be stored at any moment for the following algorithm:

Initialize by first guess
$$\mathbf{x}^{(0)} = \mathbf{x}_0, \quad \mathbf{r}_0 = \mathbf{A}\mathbf{x}_0 - \mathbf{b}, \quad \mathbf{e}_1 = \mathbf{r}_0, \quad s_0 = \|\mathbf{r}_0\|^2$$
Loop on increasing k until the residual $\mathbf{r}$ is small enough
$$\alpha_k = \frac{\mathbf{e}_k^T \mathbf{r}_{k-1}}{\mathbf{e}_k^T \mathbf{A}\mathbf{e}_k}$$
$$\mathbf{x}^{(k)} = \mathbf{x}^{(k-1)} - \alpha_k \mathbf{e}_k$$
$$\mathbf{r}_k = \mathbf{r}_{k-1} - \alpha_k \mathbf{A}\mathbf{e}_k$$
$$s_k = \|\mathbf{r}_k\|^2$$
$$\mathbf{e}_{k+1} = \mathbf{r}_k + \frac{s_k}{s_{k-1}} \mathbf{e}_k$$
End of loop on k.

Because we minimize independent terms, we are sure to reach the minimum of J in M steps, within rounding errors. For the conjugate-gradient method, the exact solution is therefore obtained within M iterations, and the overall cost of our special sparse matrix inversion arising from the two-dimensional discrete Poisson equation behaves as M^2. However, there is no need to find the exact minimum, and in practice, only a certain number of successive minimizations are necessary, and convergence is generally obtained within $M^{3/2}$ operations. This does not seem an improvement over the optimal over-relaxation, but the conjugate-gradient method is generally robust and has no need of an over-relaxation parameter. If in addition a proper preconditioning is applied, it can lead to spectacular convergence rates.

Preconditioning needs to preserve the symmetry of the problem and is performed by introducing a sparse triangular matrix $\mathbf{L}$ and writing the original problem as

$$\mathbf{L}^{-1}\mathbf{A}\mathbf{L}^{-T}\mathbf{L}^T\mathbf{x} = \mathbf{L}^{-1}\mathbf{b}, \tag{7.64}$$

so that we now work with the new unknown $L^T x$ and modified matrix $L^{-1} A L^{-T}$. This matrix is symmetric, and, if L is chosen correctly, also positive definite if A is. The resulting algorithm, involving the modified matrix and unknown, is very close to the original conjugate-gradient algorithm after clever rearrangement of the matrix-vector products. The only difference is the appearance of $M^{-1} r$, where $M^{-1} = L^{-T} L^T$. Because $u = M^{-1} r$ is the solution of $M u = r$, the sparseness and triangular nature of L allows us to perform this operation quite efficiently. If $L = C$ is obtained from a Cholesky decomposition of the symmetric positive-definite matrix $A = C C^T$, where C is a triangular matrix, a single step of the conjugate-gradient method would suffice because the inversion of A would be directly available. For this reason, L is often constructed by the Cholesky decomposition but with incomplete and cost-effective calculations, imposing on L a given sparse pattern. This leads to the incomplete Cholesky preconditioning[4] and reduces the cost of the decomposition but increases the number of iterations needed compared to a situation in which the full Cholesky decomposition is available. On the other hand, it generally reduces the number of iterations compared to the version without preconditioning. An optimum is therefore to be found in the amount of preconditioning, and the particular choice of preconditioning is problem dependent. Stability of the iterations might be an occasional problem.

Most linear-algebra packages contain conjugate-gradient methods including generalizations to solve nonsymmetric problems. In this case, we can consider the augmented (double) problem

$$\begin{pmatrix} 0 & A \\ A^T & 0 \end{pmatrix} \begin{pmatrix} y \\ x \end{pmatrix} = \begin{pmatrix} b \\ c \end{pmatrix} \tag{7.65}$$

which is symmetric and possesses the same solution x.

More efficient solution methods for special linear systems, such as our Poisson equation, exist and exhibit a close relationship with Fast Fourier Transforms (FFT, see Appendix C). The cyclic block reduction methods (e.g., Ferziger & Perić, 1999), for example, can be applied when the discretization constants are uniform and boundary conditions simple. But in such a case, we could also use a spectral method coupled with FFT for immediate inversion of the Laplacian operator (see Section 18.4). In these methods, costs can be reduced down to $M \log M$.

Finally, the most efficient methods for very large problems are *multigrid* methods. These start from the observation that the pseudo-evolution approach mimics diffusion, which generally acts more efficiently at smaller scales [see damping rates of discrete diffusion operators (5.34)], leaving larger scales to converge more slowly. But these larger scales can be made to appear as

---

[4]More generally, an incomplete LU decomposition approximates any matrix A by the product of lower and upper sparse triangular matrices.

relatively shorter scales on a grid with wider grid spacing so that their convergence can be accelerated (using, incidentally, a larger pseudo-time step). Thus, introducing a hierarchy of grids, as a *multigrid* method does, accelerates convergence by iterating on different grids for different length scales. Typically, the method begins with a very coarse grid, on which a few iterations lead to a good estimate of the broad shape of the solution. This solution is then interpolated onto a finer grid on which several more iterations are performed, and so on down to the ultimate resolution of interest. The iterations may also be redone on the coarser grids after some averaging to estimate the broad solution from the finer grid. Multigrid methods, therefore, cycle through different grids (e.g., Hackbusch, 1985), and the art is to perform the right number of iterations on each grid and to choose wisely the next grid on which to iterate (finer or coarser). For well-chosen strategies, the number of operations required for convergence behaves asymptotically as M, and multigrid methods are therefore the most effective ones for very large problems. Iterations on each of the grids may be of red-black type with over-relaxation or any other method with appropriate convergence properties.

We only scratched here the surface of the problem of solving large and sparse linear algebraic systems to give a flavor of the possible approaches, and the reader should be aware that there is a large number of numerical solvers available for specific problems. Since these are optimized for specific computer hardware, the practical and operational task of large-system inversion of the discrete Poisson equation should be left to libraries provided with the computing system available. Only the choice of when to stop the iterations and the proper preconditioning strategy should be left to the modeler.

ANALYTICAL PROBLEMS

7.1. A laboratory experiment is conducted in a cylindrical tank 20 cm in diameter, filled with homogeneous water (15 cm deep at the center) and rotating at 30 rpm. A steady flow field with maximum velocity of 1 cm/s is generated by a source-sink device. The water viscosity is 10^{-6} m^2/s. Verify that this flow field meets the conditions of geostrophy.

7.2. (Generalization of the Taylor–Proudman theorem) By reinstating the f_*–terms of equations (3.19) and (3.22) into (7.2a) and (7.2c) show that motions in fluids rotating rapidly around an axis not parallel to gravity exhibit columnar behavior in the direction of the axis of rotation.

7.3. Demonstrate the assertion made at the end of Section 7.2, namely, that geostrophic flows between irregular bottom and top boundaries are constrained to be directed along lines of constant fluid depth.

7.4. Establish Eq. (7.22) for the evolution of a parcel's horizontal cross section from first principles.

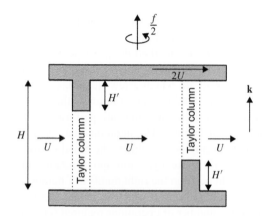

FIGURE 7.12   Schematic view of a hypothetical system as described in Analytical Problem 7.5.

7.5.   In a fluid of depth H rapidly rotating at the rate $f/2$ (Fig. 7.12), there exists a uniform flow U. Along the bottom (fixed), there is an obstacle of height H' ($<H/2$), around which the flow is locally deflected, leaving a quiescent Taylor column. A rigid lid, translating in the direction of the flow at speed $2U$, has a protrusion identical to the bottom obstacle, also locally deflecting the otherwise uniform flow and entraining another quiescent Taylor column. The two obstacles are aligned with the direction of motion so that there will be a time when both are superimposed. Assuming that the fluid is homogeneous and frictionless, what do you think will happen to the Taylor columns?

7.6.   As depicted in Fig. 7.13, a vertically uniform but laterally sheared coastal current must climb a bottom escarpment. Assuming that the jet velocity still vanishes offshore, determine the velocity profile and the width of the jet downstream of the escarpment using $H_1 = 200\,$m, $H_2 = 160\,$m, $U_1 = 0.5\,$m/s, $L_1 = 10\,$km, and $f = 10^{-4}\,$s^{-1}. What would happen if the downstream depth were only $100\,$m?

7.7.   What are the differences in dynamic pressure across the coastal jet of Problem 7.6 upstream and downstream of the escarpment? Take $H_2 = 160\,$m and $\rho_0 = 1022\,$kg/m^3.

7.8.   In Utopia, a narrow 200-m-deep channel empties in a broad bay of varying bottom topography (Fig. 7.14). Trace the path to the sea and the velocity profile of the channel outflow. Take $f = 10^{-4}\,$s^{-1}. Solve only for straight stretches of the flow and ignore corners.

7.9.   A steady ocean current of uniform potential vorticity $q = 5 \times 10^{-7}\,$m^{-1}s^{-1} and volume flux $Q = 4 \times 10^5\,$m^3/s flows along isobaths of a uniformly sloping bottom (with bottom slope $S = 1\,$m/km). Show that the velocity profile across the current is parabolic. What are the width

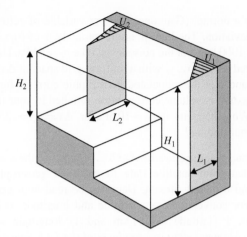

FIGURE 7.13   A sheared coastal jet negotiating a bottom escarpment (Problem 7.6).

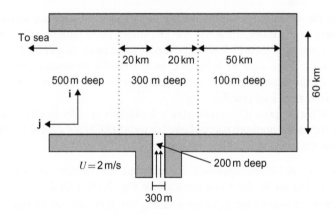

FIGURE 7.14   Geometry of the idealized bay and channel mentioned in Analytical Problem 7.8.

of the current and the depth of the location of maximum velocity? Take $f = 7 \times 10^{-5}\,\text{s}^{-1}$.

7.10. Show that the rigid-lid approximation can also be obtained by assuming that the vertical velocity at top is much smaller than at the bottom. Establish the necessary scaling conditions that support your assumptions.

NUMERICAL EXERCISES

7.1. An atmospheric pressure field p over a flat bottom is given on a rectangular grid according to

$$p_{i,j} = P_H \exp(-r^2/L^2) + p_\epsilon \, \xi_{i,j} \quad r^2 = (x_i - x_c)^2 + (y_j - y_c)^2 \qquad (7.66)$$

where ξ is a normal (Gaussian) random variable of zero mean and unit standard deviation. The high pressure anomaly is of $P_H = 40$ hPa and its radius $L = 1000$ km. For the noise level, take $p_\epsilon = 5$ hPa. Use a rectangular grid centered around x_c, y_c with a uniform grid spacing $\Delta x = \Delta y = 50$ km. Calculate and plot the associated geostrophic currents for $f = 10^{-4}$ s^{-1}. To which extent is volume conservation satisfied in your finite-difference scheme? What happens if $p_\epsilon = 10$ hPa or $\Delta x = \Delta y = 25$ km? Can you interpret your finding?

7.2. Open file `madt_oer_merged_h_18861.nc` and use the sea surface height reconstructed from satellite data to calculate geostrophic ocean currents around the Gulf Stream. Data can be read with `topexcircula-tion.m`. For conversion from latitude and longitude to local Cartesian coordinates, $1°$ latitude $= 111$ km and $1°$ longitude $= 111$ km $\times \cos$ (latitude). (*Altimeter data are products of the CLS Space Oceanography Division; see also Ducet, Le Traon & Reverdin, 2000*).

7.3. Use the meteorological pressure field at sea level to calculate geostrophic winds over Europe. First use the December 2000 monthly average sea-level pressure, then look at daily variations. `Era40.m` will help you read the data. Take care of using the local Coriolis parameter value. Geographical distances can be calculated from the conversion factors given in Numerical Exercise 7.2.

What happens if you redo your calculations in order to plan your sailing trip in the northern part of Lake Victoria? (*ECMWF ERA-40 data were obtained from the ECMWF data server*)

7.4. Use the red-black approach to calculate the numerical solution of Eq. (7.48) inside the basin depicted in Fig. 7.15, with $\tilde{q}_{i,j} = -1$ on the right-hand side of the equation and $\tilde{\psi}_{i,j} = 0$ along all boundaries. Implement a stopping criterion based on a relative measure of the residual compared to **b**.

7.5. Use the conjugate-gradient implementation called in `testpcg.m` to solve the problem of Numerical Exercise 7.4 with improved convergence.

7.6. Redo Numerical Exercise 7.4 with the Gauss–Seidel approach using over-relaxation and several values of ω between 0.7 and 1.999. For each value of ω, start from zero and converge until reaching a preset threshold for the residual. Plot the required number of iterations until convergence as a function of ω. Design a numerical tool to find the optimal value of ω numerically. Then repeat the problem by varying the spatial resolution, taking successively 20, 40, 60, 80, and 100 grid points in each direction. Look at the number of iterations and the optimal value of ω as functions of resolution.

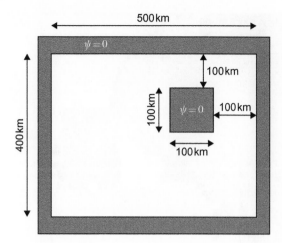

FIGURE 7.15 Geometry of the idealized basin mentioned in Numerical Exercise 7.4.

7.7. Prove that the parameter α given by Eq. (7.55) leads to a minimum of J defined in Eq. (7.52), for a given starting point and fixed gradient **r**. Also prove that at the next iteration of the steepest-descent method, the new residual is orthogonal to the previous residual. Implement the steepest-descent algorithm to find the minimum of Eq. (7.52) with

$$\mathbf{A} = \begin{pmatrix} 3 & 1 \\ 1 & 1 \end{pmatrix}, \quad \mathbf{b} = \begin{pmatrix} 6 \\ 2 \end{pmatrix} \tag{7.67}$$

starting at the origin of the axes. Observe the successive approximations obtained by the method. (*Hint*: In the plane defined by the two unknowns, plot isolines of J and plot the line connecting the successive approximations to the solution. Make several zooms near the solution point.)

7.8. Write a general solver for the Poisson equation as a Matlab^{TM} function. Provide for masked grids and a variable rectangular grid such that Δx depends on i and Δy on j. Also permit variable coefficients in the Laplacian operator, as found in Eq. (7.44) and apply to the following situation.

In shallow, wind-driven basins, such as small lakes and lagoons, the flow often strikes a balance between the forces of surface wind, pressure gradient, and bottom friction. On defining a streamfunction ψ and eliminating the pressure gradient, one obtains for steady flow (Mathieu, Deleersnijder, Cushman-Roisin, Beckers & Bolding, 2002):

$$\frac{\partial}{\partial x}\left(\frac{2\nu_E}{h^3}\frac{\partial \psi}{\partial x}\right) + \frac{\partial}{\partial y}\left(\frac{2\nu_E}{h^3}\frac{\partial \psi}{\partial y}\right) = \frac{\partial}{\partial y}\left(\frac{\tau^x}{\rho_0 h}\right) - \frac{\partial}{\partial x}\left(\frac{\tau^y}{\rho_0 h}\right), \tag{7.68}$$

where ν_E is the vertical eddy viscosity, $h(x, y)$ is the local bottom depth, and (τ^x, τ^y) the components of the surface wind stress. In the application,

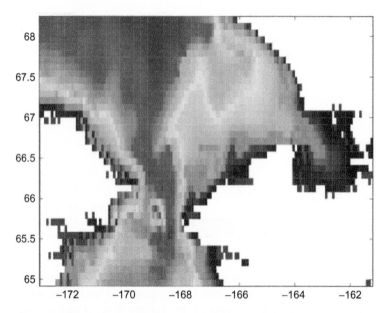

FIGURE 7.16 Model of the Bering Sea for Numerical Exercise 7.9. For the calculation of the streamfunction, assume that West of $-169°$ longitude all land points have a prescribed streamfunction $\Psi_1 = 0$ and those to the East $\Psi_2 = 0.8$ Sv $(1$ Sv $= 10^6$ m^3/s$)$. For convenience, you may consider closing the western and eastern boundary completely and imposing a zero normal derivative of Ψ along the open boundaries.

take $\nu_E = 10^{-2}$ m^2/s, $\rho_0 = 1000$ kg/m^3, $h(x, y) = 50 - (x^2 + 4y^2/10)$ (in m, with x and y in km), $\tau^x = 0.1$ N/m^2, and $\tau^y = 0$ within the elliptical domain $x^2 + 4y^2 \leq 400$ km^2.

7.9. Use the tool developed in Numerical Exercise 7.8 to simulate the stationary flow across the Bering Sea, assuming the right-hand side of Eq. (7.68) is zero. Use `beringtopo.m` to read the topography of Fig. 7.16. To pass from latitude and longitude to Cartesian coordinates, use the conversion factors given in Numerical Exercise 7.2 but with cos(latitude) taken as cos(66.5°N) to obtain a rectangular grid. Compare your solution to the case of uniform average depth in place of the real topography, maintaining the same land mask.

7.10. Prove that matrix **A** arising from the discretization of Eq. (7.48) is symmetric and positive definite if we change the sign of each side. Show also that the latter property ensures $\mathbf{z}^T\mathbf{A}\mathbf{z} > 0$ for any $\mathbf{z} \neq 0$.

7.11. Calculate the amplification factor of Gauss–Seidel iterations including over-relaxation. Can you infer the optimal over-relaxation coefficient for Dirichlet conditions? (*Hint*: The optimal parameter will ensure that the slowest damping is accomplished as fast as possible.)

Geoffrey Ingram Taylor
1886–1975

Considered one of the great physicists of the twentieth century, Sir Geoffrey Taylor contributed enormously to our understanding of fluid dynamics. Although he did not envision the birth and development of geophysical fluid dynamics, his research on rotating fluids laid the foundation for the discipline. His numerous contributions to science also include seminal work on turbulence, aeronautics, and solid mechanics. With a staff consisting of a single assistant engineer, he maintained a very modest laboratory, constantly preferring to undertake entirely new problems and to work alone. (*Photo courtesy of Cambridge University Press*)

James Cyrus McWilliams
1946–

A student of George Carrier at Harvard University, James McWilliams is a pioneer in the synthesis of mathematical theory and computational simulation in geophysical fluid dynamics. A central theme of his research is how advection produces the peculiar combinations of global order and local chaos—and vice versa—evident in oceanic currents, as well as analogous phenomena in atmospheric and astrophysical flows. His contributions span a formidable variety of topics across the disciplines of rotating and stratified flows, waves, turbulence, boundary layers, oceanic general circulation, and computational methods.

McWilliams' scientific style is the pursuit of phenomenological discovery in the virtual reality of simulations, leading, "on good days," to dynamical understanding and explanation and to confirmation in nature. (*Photo credit: J. C. McWilliams*)

Chapter 8

The Ekman Layer

ABSTRACT

Frictional forces, neglected in the previous chapter, are now investigated. Their main effect is to create horizontal boundary layers that support a flow transverse to the main flow of the fluid. The numerical treatment of the velocity profiles dominated by friction is illustrated with a spectral approach.

8.1 SHEAR TURBULENCE

Because most geophysical fluid systems are much shallower than they are wide, their vertical confinement forces the flow to be primarily horizontal. Unavoidable in such a situation is friction between the main horizontal motion and the bottom boundary. Friction acts to reduce the velocity in the vicinity of the bottom, thus creating a vertical shear. Mathematically, if u is the velocity component in one of the horizontal directions and z the elevation above the bottom, then u is a function of z, at least for small z values. The function $u(z)$ is called the *velocity profile* and its derivative du/dz, the *velocity shear*.

Geophysical flows are invariably turbulent (high Reynolds number), and this greatly complicates the search for the velocity profile. As a consequence, much of what we know is derived from observations of actual flows, either in the laboratory or in the nature.

The turbulent nature of the shear flow along a flat or rough surface includes variability at short time and length scales, and the best observational techniques for the detailed measurements of these have been developed for the laboratory rather than outdoor situations. Laboratory measurements of nonrotating turbulent flows along smooth straight surfaces have led to the conclusion that the velocity varies solely with the stress τ_b exerted against the bottom, the fluid molecular viscosity ν, the fluid density ρ and, of course, the distance z above the bottom. Thus,

$$u(z) = F(\tau_b, \nu, \rho, z).$$

Dimensional analysis permits the elimination of the mass dimension shared by τ_b and ρ but not present in u, v, and z, and we may write more simply:

$$u(z) = F\left(\frac{\tau_b}{\rho}, v, z\right).$$

The ratio τ_b/ρ has the same dimension as the square of a velocity, and for this reason, it is customary to define

$$u_* = \sqrt{\frac{\tau_b}{\rho}}, \tag{8.1}$$

which is called the *friction velocity* or *turbulent velocity*. Physically, its value is related to the orbital velocity of the vortices that create the cross-flow exchange of particles and the momentum transfer.

The velocity structure thus obeys a relation of the form $u(z) = F(u_*, v, z)$, and further use of dimensional analysis reduces it to a function of a single variable:

$$\frac{u(z)}{u_*} = F\left(\frac{u_* z}{v}\right). \tag{8.2}$$

In the presence of rotation, the Coriolis parameter enters the formalism, and the preceding function depends on two variables:

$$\frac{u(z)}{u_*} = F\left(\frac{u_* z}{v}, \frac{fz}{u_*}\right). \tag{8.3}$$

8.1.1   Logarithmic Profile

The observational determination of the function F in the absence of rotation has been repeated countless times, yielding the same results every time, and it suffices here to provide a single report (Fig. 8.1). When the velocity ratio u/u_* is plotted versus the logarithm of the dimensionless distance $u_* z/v$, not only do all the points coalesce onto a single curve, confirming that there is indeed no other variable to be invoked, but the curve also behaves as a straight line over a range of two orders of magnitude (from $u_* z/v$ between 10^1 and 10^3).

If the velocity is linearly dependent on the logarithm of the distance, then we can write for this portion of the velocity profile:

$$\frac{u(z)}{u_*} = A \ln \frac{u_* z}{v} + B.$$

Numerous experimental determinations of the constants A and B provide $A = 2.44$ and $B = 5.2$ within a 5% error (Pope, 2000). Tradition has it to write

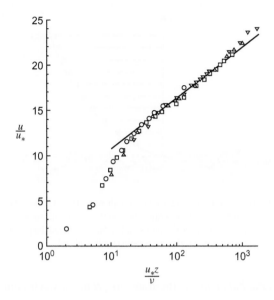

FIGURE 8.1   Mean velocity profiles in fully developed turbulent channel flow measured by Wei and Willmarth (1989) at various Reynolds numbers: circles $Re = 2970$, squares $Re = 14914$, upright triangles $Re = 22776$, and downright triangles $Re = 39582$. The straight line on this log-linear plot corresponds to the logarithmic profile of Eq. (8.2). (*From Pope, 2000*)

the function as:

$$u(z) = \frac{u_*}{\mathcal{K}} \ln \frac{u_* z}{\nu} + 5.2 u_*, \qquad (8.4)$$

where $\mathcal{K} = 1/A = 0.41$ is called the *von Kármán constant*[1]

The portion of the curve closer to the wall, where the logarithmic law fails, may be approximated by the laminar solution. Constant laminar stress $\nu du/dz = \tau_b/\rho = u_*^2$ implies $u(z) = u_*^2 z/\nu$ there. Ignoring the region of transition in which the velocity profile gradually changes from one solution to the other, we can attempt to connect the two. Doing so yields $u_* z/\nu = 11$. This sets the thickness of the laminar boundary layer δ as the value of z for which $u_* z/\nu = 11$, that is,

$$\delta = 11 \frac{\nu}{u_*}. \qquad (8.5)$$

Most textbooks (e.g., Kundu, 1990) give $\delta = 5\nu/u_*$, for the region in which the velocity profile is strictly laminar, and label the region between $5\nu/u_*$

---

[1] In honor of Theodore von Kármán (1881–1963), Hungarian-born physicist and engineer who made significant contributions to fluid mechanics while working in Germany and who first introduced this notation.

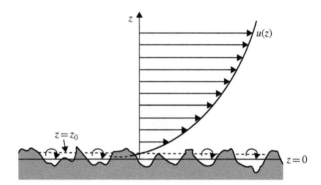

FIGURE 8.2   Velocity profile in the vicinity of a rough wall. The roughness height z_0 is smaller than the averaged height of the surface asperities. So, the velocity u falls to zero somewhere within the asperities, where local flow degenerates into small vortices between the peaks, and the negative values predicted by the logarithmic profile are not physically realized.

and $30\nu/u_*$ as the *buffer layer*, the transition zone between laminar and fully turbulent flow.

For water in ambient conditions, the molecular viscosity ν is equal to 1.0×10^{-6} m^2/s, whereas the friction velocity in the ocean rarely falls below 1 mm/s. This implies that δ hardly exceeds a centimeter in the ocean and is almost always smaller than the height of the cobbles, ripples, and other asperities that typically line the bottom of the ocean basin. Similarly for the atmosphere, the air viscosity at ambient temperature and pressure is about 1.5×10^{-5} m^2/s, and u_* rarely falls below 1 cm/s, giving $\delta < 5$ cm, smaller than most irregularities on land and wave heights at sea.

When this is the case, the velocity profile above the bottom asperities no longer depends on the molecular viscosity of the fluid but on the so-called *roughness height* z_0, such that

$$u(z) = \frac{u_*}{\mathcal{K}} \ln \frac{z}{z_0}, \tag{8.6}$$

as depicted in Fig. 8.2. It is important to note that the roughness height is not the average height of bumps on the surface but is a small fraction of it, about one tenth (Garratt, 1992, page 87).

8.1.2   Eddy Viscosity

We have already mentioned in Section 5.2 what an eddy diffusivity or viscosity is and how it can be formulated in the case of a homogeneous turbulence field, that is, away from boundaries. Near a boundary, the turbulence ceases to be isotropic and an alternate formulation needs to be developed.

In analogy with Newton's law for viscous fluids, which has the tangential stress τ proportional to the velocity shear du/dz with the coefficient of proportionality being the molecular viscosity v, we write for turbulent flow:

$$\tau = \rho_0 v_E \frac{du}{dz}, \tag{8.7}$$

where the turbulent viscosity v_E supersedes the molecular viscosity v. For the logarithmic profile (8.6) of a flow along a rough surface, the velocity shear is $du/dz = u_*/\mathcal{K}z$ and the stress τ is uniform across the flow (at least in the vicinity of the boundary for lack of other significant forces): $\tau = \tau_b = \rho u_*^2$, giving

$$\rho_0 u_*^2 = \rho_0 v_E \frac{u_*}{\mathcal{K}z}$$

and thus

$$v_E = \mathcal{K}z u_*. \tag{8.8}$$

Note that unlike the molecular viscosity, the turbulent viscosity is not constant in space, for it is not a property of the fluid but of the flow, including its structure. From its dimension $(L^2 T^{-1})$, we verify that Eq. (8.8) is dimensionally correct and note that it can be expressed as the product of a length by the friction velocity:

$$v_E = l_m u_*, \tag{8.9}$$

with the *mixing length* l_m defined as

$$l_m = \mathcal{K}z. \tag{8.10}$$

This parameterization is occasionally used for cases other than boundary layers (see Chapter 14).

The preceding considerations ignored the effect of rotation. When rotation is present, the character of the boundary layer changes dramatically.

8.2   FRICTION AND ROTATION

After the development of the equations governing geophysical motions (Sections 4.1 to 4.4), a scale analysis was performed to evaluate the relative importance of the various terms (Section 4.5). In the horizontal momentum equations [(4.21a) and (4.21b)], each term was compared with the Coriolis term, and a corresponding dimensionless ratio was defined. For vertical friction, the dimensionless ratio was the *Ekman number*:

$$Ek = \frac{v_E}{\Omega H^2}, \tag{8.11}$$

where v_E is the eddy viscosity, Ω the ambient rotation rate, and H the height (depth) scale of the motion (the total thickness if the fluid is homogeneous).

Typical geophysical flows, as well as laboratory experiments, are characterized by very small Ekman numbers. For example, in the ocean at midlatitudes ($\Omega \simeq 10^{-4}$ s^{-1}), motions modeled with an eddy-intensified viscosity $\nu_E = 10^{-2}$ m^2/s (much larger than the molecular viscosity of water, equal to 1.0×10^{-6} m^2/s) and extending over a depth of about 1000 m have an Ekman number of about 10^{-4}.

The smallness of the Ekman number indicates that vertical friction plays a very minor role in the balance of forces and may, consequently, be omitted from the equations. This is usually done and with great success. However, something is then lost. The frictional terms happen to be those with the highest order of derivatives among all terms of the momentum equations. Thus, when friction is neglected, the order of the set of differential equations is reduced, and not all boundary conditions can be applied simultaneously. Usually, slipping along the bottom must be accepted.

Since Ludwig Prandtl[2] and his general theory of boundary layers, we know that in such a circumstance, the fluid system exhibits two distinct behaviors: At some distance from the boundaries, in what is called the *interior*, friction is usually negligible, whereas, near a boundary (wall) and across a short distance, called the *boundary layer*, friction acts to bring the finite interior velocity to zero at the wall.

The thickness, d, of this thin layer is such that the Ekman number is on the order of one at that scale, allowing friction to be a dominant force:

$$\frac{\nu_E}{\Omega d^2} \sim 1,$$

which leads to

$$d \sim \sqrt{\frac{\nu_E}{\Omega}}. \tag{8.12}$$

Obviously, d is much less than H, and the boundary layer occupies a very small portion of the flow domain. For the oceanic values cited above ($\nu_E = 10^{-2}$ m^2/s and $\Omega = 10^{-4}$ s^{-1}), d is about 10 m.

Because of the Coriolis effect, the frictional boundary layer of geophysical flows, called the *Ekman layer*, differs greatly from the boundary layer in nonrotating fluids. Although, the traditional boundary layer has no particular thickness and grows either downstream or with time, the existence of the depth scale d in rotating fluids suggests that the Ekman layer can be characterized by a fixed thickness. (Note that as the rotational effects disappear ($\Omega \to 0$), d tends to infinity, exemplifying this essential difference between rotating and nonrotating fluids.) Rotation not only imparts a fixed length scale to the boundary layer, but we will now show that it also changes the direction of the velocity vector when approaching the boundary, leading to transverse currents.

---

[2] See biography at the end of this chapter.

8.3   THE BOTTOM EKMAN LAYER

Let us consider a uniform, geostrophic flow in a homogeneous fluid over a flat bottom (Fig. 8.3). This bottom exerts a frictional stress against the flow, bringing the velocity gradually to zero within a thin layer above the bottom. We now solve for the structure of this layer.

In the absence of horizontal gradients (the interior flow is said to be uniform) and of temporal variations, continuity equation (4.21d) yields $\partial w/\partial z = 0$ and thus $w = 0$ in the thin layer near the bottom. The remaining equations are the following reduced forms of Eq. (4.21a) through Eq. (4.21c):

$$-fv = -\frac{1}{\rho_0}\frac{\partial p}{\partial x} + \nu_E\frac{\partial^2 u}{\partial z^2} \tag{8.13a}$$

$$+fu = -\frac{1}{\rho_0}\frac{\partial p}{\partial y} + \nu_E\frac{\partial^2 v}{\partial z^2} \tag{8.13b}$$

$$0 = -\frac{1}{\rho_0}\frac{\partial p}{\partial z}, \tag{8.13c}$$

where f is the Coriolis parameter (taken as a constant here), ρ_0 is the fluid density, and ν_E is the eddy viscosity (taken as a constant for simplicity). The horizontal gradient of the pressure p is retained because a uniform flow requires a uniformly varying pressure (Section 7.1). For convenience, we align the x-axis with the direction of the interior flow, which is of velocity $\bar{u}$. The boundary conditions are then

$$\text{Bottom } (z=0): \quad u=0, \quad v=0, \tag{8.14a}$$

$$\text{Toward the interior } (z \gg d): \quad u=\bar{u}, \quad v=0, \quad p=\bar{p}(x,y). \tag{8.14b}$$

By virtue of Eq. (8.13c), the dynamic pressure p is the same at all depths; thus, $p = \bar{p}(x, y)$ in the interior flow as well as throughout the boundary layer.

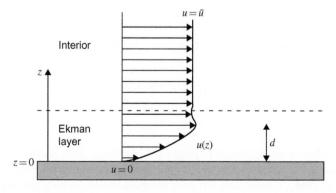

FIGURE 8.3   Frictional influence of a flat bottom on a uniform flow in a rotating framework.

In the interior flow ($z \gg d$, mathematically equivalent to $z \to \infty$), Eqs. (8.13a) and (8.13b) relate the velocity to the pressure gradient:

$$0 = -\frac{1}{\rho_0}\frac{\partial \bar{p}}{\partial x},$$

$$f\bar{u} = -\frac{1}{\rho_0}\frac{\partial \bar{p}}{\partial y} = \text{independent of } z.$$

Substitution of these derivatives in the same equations, which are now taken at any depth, yields

$$-fv = \nu_E \frac{\mathrm{d}^2 u}{\mathrm{d}z^2} \tag{8.15a}$$

$$f(u - \bar{u}) = \nu_E \frac{\mathrm{d}^2 v}{\mathrm{d}z^2}. \tag{8.15b}$$

Seeking a solution of the type $u = \bar{u} + A\exp(\lambda z)$ and $v = B\exp(\lambda z)$, we find that λ obeys $\nu^2\lambda^4 + f^2 = 0$; that is,

$$\lambda = \pm(1 \pm i)\frac{1}{d}$$

where the distance d is defined by

$$d = \sqrt{\frac{2\nu_E}{f}}. \tag{8.16}$$

Here, we have restricted ourselves to cases with positive f (northern hemisphere). Note the similarity to Eq. (8.12). Boundary conditions (8.14b) rule out the exponentially growing solutions, leaving

$$u = \bar{u} + e^{-z/d}\left(A\cos\frac{z}{d} + B\sin\frac{z}{d}\right) \tag{8.17a}$$

$$v = e^{-z/d}\left(B\cos\frac{z}{d} - A\sin\frac{z}{d}\right), \tag{8.17b}$$

and the application of the remaining boundary conditions (8.14a) yields $A = -\bar{u}$, $B = 0$, or

$$u = \bar{u}\left(1 - e^{-z/d}\cos\frac{z}{d}\right) \tag{8.18a}$$

$$v = \bar{u}e^{-z/d}\sin\frac{z}{d}. \tag{8.18b}$$

This solution has a number of important properties. First and foremost, we notice that the distance over which it approaches the interior solution is on the order of d. Thus, expression (8.16) gives the thickness of the boundary layer. For

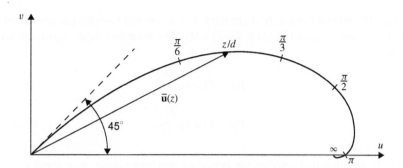

FIGURE 8.4 The velocity spiral in the bottom Ekman layer. The figure is drawn for the northern hemisphere ($f > 0$), and the deflection is to the left of the current above the layer. The reverse holds for the southern hemisphere.

this reason, d is called the *Ekman depth*. A comparison with Eq. (8.12) confirms the earlier argument that the boundary-layer thickness is the one corresponding to a local Ekman number near unity.

The preceding solution also tells us that there is, in the boundary layer, a flow transverse to the interior flow ($v \neq 0$). Very near the bottom ($z \to 0$), this component is equal to the downstream velocity ($u \sim v \sim \bar{u}z/d$), thus implying that the near-bottom velocity is at 45 degrees to the left of the interior velocity (Fig. 8.4). (The boundary flow is to the right of the interior flow for $f < 0$.) Further up, where u reaches a first maximum ($z = 3\pi d/4$), the velocity in the direction of the flow is greater than in the interior ($u = 1.07\bar{u}$). (Viscosity can occasionally fool us!)

It is instructive to calculate the net transport of fluid transverse to the main flow:

$$V = \int_0^\infty v\,dz = \frac{\bar{u}d}{2},$$ (8.19)

which is proportional to the interior velocity and the Ekman depth.

8.4 GENERALIZATION TO NONUNIFORM CURRENTS

Let us now consider a more complex interior flow, namely, a spatially nonuniform flow that is varying on a scale sufficiently large to be in geostrophic equilibrium (low Rossby number, as in Section 7.1). Thus,

$$-f\bar{v} = -\frac{1}{\rho_0}\frac{\partial \bar{p}}{\partial x}, \quad f\bar{u} = -\frac{1}{\rho_0}\frac{\partial \bar{p}}{\partial y},$$

where the pressure $\bar{p}(x, y, t)$ is arbitrary. For a constant Coriolis parameter, this flow is nondivergent ($\partial\bar{u}/\partial x + \partial\bar{v}/\partial y = 0$). The boundary-layer equations are now

$$-f(v - \bar{v}) = \nu_E \frac{\partial^2 u}{\partial z^2} \tag{8.20a}$$

$$f(u - \bar{u}) = \nu_E \frac{\partial^2 v}{\partial z^2}, \tag{8.20b}$$

and the solution that satisfies the boundary conditions aloft ($u \to \bar{u}$ and $v \to \bar{v}$ for $z \to \infty$) is

$$u = \bar{u} + e^{-z/d}\left(A\cos\frac{z}{d} + B\sin\frac{z}{d}\right) \tag{8.21}$$

$$v = \bar{v} + e^{-z/d}\left(B\cos\frac{z}{d} - A\sin\frac{z}{d}\right). \tag{8.22}$$

Here, the "constants" of integration A and B are independent of z but will be dependent on x and y through $\bar{u}$ and $\bar{v}$. Imposing $u = v = 0$ along the bottom ($z = 0$) sets their values, and the solution is

$$u = \bar{u}\left(1 - e^{-z/d}\cos\frac{z}{d}\right) - \bar{v}e^{-z/d}\sin\frac{z}{d} \tag{8.23a}$$

$$v = \bar{u}e^{-z/d}\sin\frac{z}{d} + \bar{v}\left(1 - e^{-z/d}\cos\frac{z}{d}\right). \tag{8.23b}$$

The transport attributed to the boundary-layer flow has components given by

$$U = \int_0^\infty (u - \bar{u})\,dz = -\frac{d}{2}(\bar{u} + \bar{v}) \tag{8.24a}$$

$$V = \int_0^\infty (v - \bar{v})\,dz = \frac{d}{2}(\bar{u} - \bar{v}). \tag{8.24b}$$

Since this transport is not necessarily parallel to the interior flow, it is likely to have a nonzero divergence. Indeed,

$$\frac{\partial U}{\partial x} + \frac{\partial V}{\partial y} = \int_0^\infty \left(\frac{\partial u}{\partial x} + \frac{\partial v}{\partial y}\right)dz = -\frac{d}{2}\left(\frac{\partial\bar{v}}{\partial x} - \frac{\partial\bar{u}}{\partial y}\right)$$

$$= -\frac{d}{2\rho_0 f}\nabla^2\bar{p}. \tag{8.25}$$

The flow in the boundary layer converges or diverges if the interior flow has a relative vorticity. The situation is depicted in Fig. 8.5. The question is: From where does the fluid come, or where does it go, to meet this convergence

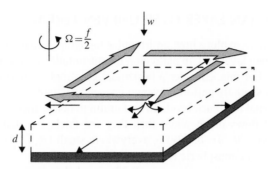

FIGURE 8.5 Divergence in the bottom Ekman layer and compensating downwelling in the interior. Such a situation arises in the presence of an anticyclonic gyre in the interior, as depicted by the large horizontal arrows. Similarly, interior cyclonic motion causes convergence in the Ekman layer and upwelling in the interior.

or divergence? Because of the presence of a solid bottom, the only possibility is that it be supplied from the interior by means of a vertical velocity. But, remember (Section 7.1) that geostrophic flows must be characterized by

$$\frac{\partial \bar{w}}{\partial z} = 0, \tag{8.26}$$

that is, the vertical velocity must occur throughout the depth of the fluid. Of course, since the divergence of the flow in the Ekman layer is proportional to the Ekman depth, d, which is very small, this vertical velocity is weak.

The vertical velocity in the interior, called *Ekman pumping*, can be evaluated by a vertical integration of the continuity equation (4.21d), using $w(z=0)=0$ and $w(z\rightarrow\infty)=\bar{w}$:

$$\bar{w} = -\int_0^\infty \left(\frac{\partial u}{\partial x} + \frac{\partial v}{\partial y} \right) dz = \frac{d}{2}\left(\frac{\partial \bar{v}}{\partial x} - \frac{\partial \bar{u}}{\partial y} \right)$$

$$= \frac{d}{2\rho_0 f}\nabla^2 \bar{p} = \frac{1}{\rho_0}\sqrt{\frac{\nu_E}{2f^3}}\nabla^2 \bar{p}. \tag{8.27}$$

So, the greater the vorticity of the mean flow, the greater the upwelling or downwelling. Also, the effect increases toward the equator (decreasing $f = 2\Omega\sin\varphi$ and increasing d). The direction of the vertical velocity is upward in a cyclonic flow (counterclockwise in the northern hemisphere) and downward in an anticyclonic flow (clockwise in the northern hemisphere).

In the southern hemisphere, where $f < 0$, the Ekman layer thickness d must be redefined with the absolute value of f: $d = \sqrt{2\nu_E/|f|}$, but the previous rule remains: the vertical velocity is upward in a cyclonic flow and downward in an anticyclonic flow. The difference is that cyclonic flow is clockwise and anticyclonic flow is counterclockwise.

8.5   THE EKMAN LAYER OVER UNEVEN TERRAIN

It is noteworthy to explore how an irregular topography may affect the structure of the Ekman layer and, in particular, the magnitude of the vertical velocity in the interior. For this, consider a horizontal geostrophic interior flow $(\bar{u}, \bar{v})$, not necessarily spatially uniform, over an uneven terrain of elevation $z = b(x, y)$ above a horizontal reference level. To be faithful to our restriction (Section 4.3) to geophysical flows much wider than they are thick, we shall assume that the bottom slope $(\partial b / \partial x, \partial b / \partial y)$ is everywhere small $(\ll 1)$. This is hardly a restriction in most atmospheric and oceanic situations.

Our governing equations are again (8.20), coupled to the continuity equation (4.21d), but the boundary conditions are now

$$\text{Bottom } (z = b): \quad u = 0, \quad v = 0, \quad w = 0, \tag{8.28}$$

$$\text{Toward the interior } (z \gg b + d): \quad u = \bar{u}, \quad v = \bar{v}. \tag{8.29}$$

The solution is the previous solution (8.23) with z replaced by $z - b$:

$$u = \bar{u} - e^{(b-z)/d} \left(\bar{u} \cos \frac{z-b}{d} + \bar{v} \sin \frac{z-b}{d} \right) \tag{8.30a}$$

$$v = \bar{v} + e^{(b-z)/d} \left(\bar{u} \sin \frac{z-b}{d} - \bar{v} \cos \frac{z-b}{d} \right). \tag{8.30b}$$

We note that the vertical thickness of the boundary layer is still measured by $d = \sqrt{2 \nu_E / f}$. However, the boundary layer is now oblique, and its true thickness, measured perpendicularly to the bottom, is slightly reduced by the cosine of the small bottom slope.

The vertical velocity is then determined from the continuity equation:

$$
\begin{aligned}
\frac{\partial w}{\partial z} &= -\frac{\partial u}{\partial x} - \frac{\partial v}{\partial y} \\
&= e^{(b-z)/d} \left\{ \left(\frac{\partial \bar{v}}{\partial x} - \frac{\partial \bar{u}}{\partial y} \right) \sin \frac{z-b}{d} \right. \\
&\quad + \frac{1}{d} \frac{\partial b}{\partial x} \left[(\bar{u} - \bar{v}) \cos \frac{z-b}{d} + (\bar{u} + \bar{v}) \sin \frac{z-b}{d} \right] \\
&\quad \left. + \frac{1}{d} \frac{\partial b}{\partial y} \left[(\bar{u} + \bar{v}) \cos \frac{z-b}{d} - (\bar{u} - \bar{v}) \sin \frac{z-b}{d} \right] \right\},
\end{aligned}
$$

where use has been made of the fact that the interior geostrophic flow has no divergence $[\partial \bar{u} / \partial x + \partial \bar{v} / \partial y = 0 -$ See (7.5)]. A vertical integration from the bottom $(z = b)$, where the vertical velocity vanishes $(w = 0$ because u and v are also zero there) into the interior $(z \rightarrow +\infty)$ where the vertical velocity assumes a vertically uniform value $(w = \bar{w})$, yields

$$\bar{w} = \left(\bar{u} \frac{\partial b}{\partial x} + \bar{v} \frac{\partial b}{\partial y} \right) + \frac{d}{2} \left(\frac{\partial \bar{v}}{\partial x} - \frac{\partial \bar{u}}{\partial y} \right). \tag{8.31}$$

The interior vertical velocity thus consists of two parts: a component that ensures no normal flow to the bottom [see (7.7)] and an Ekman-pumping contribution, as if the bottom were horizontally flat [see (8.27)].

The vanishing of the flow component perpendicular to the bottom must be met by the inviscid dynamics of the interior, giving rise to the first contribution to $\bar{w}$. The role of the boundary layer is to bring the tangential velocity to zero at the bottom. This explains the second contribution to $\bar{w}$. Note that the Ekman pumping is not affected by the bottom slope.

The preceding solution can also be applied to the lower portion of the atmospheric boundary layer. This was first done by Akerblom (1908), and matching between the logarithmic layer close to the ground (Section 8.1.1) and the Ekman layer further aloft was performed by Van Dyke (1975). Oftentimes, however, the lower atmosphere is in a stable (stratified) or unstable (convecting) state, and the neutral state during which Ekman dynamics prevail is more the exception than the rule.

8.6   THE SURFACE EKMAN LAYER

An Ekman layer occurs not only along bottom surfaces, but wherever there is a horizontal frictional stress. This is the case, for example, along the ocean surface, where waters are subject to a wind stress. In fact, this is precisely the situation first examined by Vagn Walfrid Ekman.[3] Fridtjof Nansen[4] had noticed during his cruises to northern latitudes that icebergs drift not downwind but systematically at some angle to the right of the wind. Ekman, his student at the time, reasoned that the cause of this bias was the earth's rotation and subsequently developed the mathematical representation that now bears his name. The solution was originally published in his 1902 doctoral thesis and again, in a more complete article, three years later (Ekman, 1905). In a subsequent article (Ekman, 1906), he mentioned the relevance of his theory to the lower atmosphere, where the wind approaches a geostrophic value with increasing height.

Let us consider the situation depicted in Fig. 8.6, where an ocean region with interior flow field $(\bar{u}, \bar{v})$ is subjected to a wind stress (τ^x, τ^y) along its surface. Again, assuming steady conditions, a homogeneous fluid, and a geostrophic interior, we obtain the following equations and boundary conditions for the flow field (u, v) in the surface Ekman layer:

$$-f(v - \bar{v}) = \nu_E \frac{\partial^2 u}{\partial z^2} \tag{8.32a}$$

$$+f(u - \bar{u}) = \nu_E \frac{\partial^2 v}{\partial z^2} \tag{8.32b}$$

---

[3] See biography at the end of this chapter.

[4] Fridtjof Nansen (1861–1930), Norwegian oceanographer famous for his Arctic expeditions and Nobel Peace Prize laureate (1922).

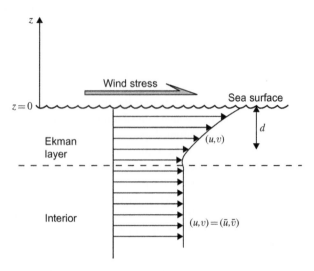

FIGURE 8.6   The surface Ekman layer generated by a wind stress on the ocean.

$$\text{Surface } (z = 0): \quad \rho_0 \nu_E \frac{\partial u}{\partial z} = \tau^x, \quad \rho_0 \nu_E \frac{\partial v}{\partial z} = \tau^y \qquad (8.32\text{c})$$

$$\text{Toward interior } (z \to -\infty): \quad u = \bar{u}, \quad v = \bar{v}. \qquad (8.32\text{d})$$

The solution to this problem is

$$u = \bar{u} + \frac{\sqrt{2}}{\rho_0 f d} e^{z/d} \left[\tau^x \cos\left(\frac{z}{d} - \frac{\pi}{4}\right) - \tau^y \sin\left(\frac{z}{d} - \frac{\pi}{4}\right) \right] \qquad (8.33\text{a})$$

$$v = \bar{v} + \frac{\sqrt{2}}{\rho_0 f d} e^{z/d} \left[\tau^x \sin\left(\frac{z}{d} - \frac{\pi}{4}\right) + \tau^y \cos\left(\frac{z}{d} - \frac{\pi}{4}\right) \right], \qquad (8.33\text{b})$$

in which we note that the departure from the interior flow $(\bar{u}, \bar{v})$ is exclusively due to the wind stress. In other words, it does not depend on the interior flow. Moreover, this wind-driven flow component is inversely proportional to the Ekman-layer depth, d, and may be very large. Physically, if the fluid is almost inviscid (small ν_E, hence short d), a moderate surface stress can generate large drift velocities.

The wind-driven horizontal transport in the surface Ekman layer has components given by

$$U = \int_{-\infty}^{0} (u - \bar{u})\,\mathrm{d}z = \frac{1}{\rho_0 f} \tau^y \qquad (8.34\text{a})$$

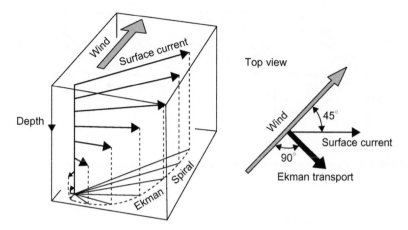

FIGURE 8.7 Structure of the surface Ekman layer. The figure is drawn for the northern hemisphere ($f > 0$), and the deflection is to the right of the surface stress. The reverse holds for the southern hemisphere.

$$V = \int_{-\infty}^{0} (v - \bar{v})\,dz = \frac{-1}{\rho_0 f}\tau^x. \tag{8.34b}$$

Surprisingly, it is oriented perpendicular to the wind stress (Fig. 8.7), to the right in the northern hemisphere and to the left in the southern hemisphere. This fact explains why icebergs, which float mostly underwater, systematically drift to the right of the wind in the North Atlantic, as observed by Fridtjof Nansen.

As for the bottom Ekman layer, let us determine the divergence of the flow, integrated over the boundary layer:

$$\int_{-\infty}^{0} \left(\frac{\partial u}{\partial x} + \frac{\partial v}{\partial y} \right) dz = \frac{1}{\rho_0} \left[\frac{\partial}{\partial x}\left(\frac{\tau^y}{f} \right) - \frac{\partial}{\partial y}\left(\frac{\tau^x}{f} \right) \right]. \tag{8.35}$$

At constant f, the contribution is entirely due to the wind stress since the interior geostrophic flow is nondivergent. It is proportional to the wind-stress curl and, most importantly, it is independent of the value of the viscosity. It can be shown furthermore that this property continues to hold even when the turbulent eddy viscosity varies spatially (see Analytical Problem 8.7).

If the wind stress has a nonzero curl, the divergence of the Ekman transport must be provided by a vertical velocity throughout the interior. A vertical integration of the continuity equation, (4.21d), across the Ekman layer with $w(z = 0)$

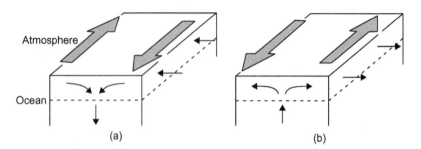

FIGURE 8.8 Ekman pumping in an ocean subject to sheared winds (case of northern hemisphere).

and $w(z \to -\infty) = \bar{w}$ yields

$$\bar{w} = + \int_{-\infty}^{0} \left(\frac{\partial u}{\partial x} + \frac{\partial v}{\partial y} \right) dz$$

$$= \frac{1}{\rho_0} \left[\frac{\partial}{\partial x} \left(\frac{\tau^y}{f} \right) - \frac{\partial}{\partial y} \left(\frac{\tau^x}{f} \right) \right] = w_{Ek}.$$

(8.36)

This vertical velocity is called *Ekman pumping*. In the northern hemisphere ($f > 0$), a clockwise wind pattern (negative curl) generates a downwelling (Fig. 8.8a), whereas a counterclockwise wind pattern causes upwelling (Fig. 8.8b). The directions are opposite in the southern hemisphere. Ekman pumping is a very effective mechanism by which winds drive subsurface ocean currents (Pedlosky, 1996; see also Chapter 20).

8.7 THE EKMAN LAYER IN REAL GEOPHYSICAL FLOWS

The preceding models of bottom and surface Ekman layers are highly ideal-ized, and we do not expect their solutions to match actual atmospheric and oceanic observations closely (except in some cases; see Fig. 8.9). Three factors, among others, account for substantial differences: turbulence, stratification, and horizontal gradients.

It was noted at the end of Chapter 4 that geophysical flows have large Reynolds numbers and are therefore in a state of turbulence. Replacing the molecular viscosity of the fluid by a much greater eddy viscosity, as performed in Section 4.2, is a first attempt to recognize the enhanced transfer of momen-tum in a turbulent flow. However, in a shear flow such as in an Ekman layer, the turbulence is not homogeneous, being more vigorous where the shear is greater and also partially suppressed in the proximity of the boundary where the size of turbulent eddies is restricted. In the absence of an exact theory of

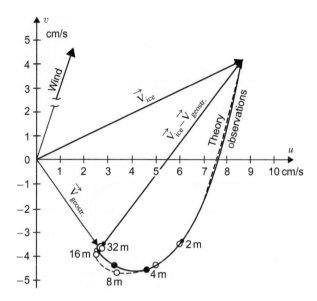

FIGURE 8.9 Comparison between observed currents below a drifting ice floe at 84.3°N and the-oretical predictions based on an eddy viscosity $\nu_E = 2.4 \times 10^{-3}$ m^2/s. (*Reprinted from Deep-Sea Research, 13, Kenneth Hunkins, Ekman drift currents in the Arctic Ocean, p. 614, ©1966, with kind permission from Pergamon Press Ltd, Headington Hill Hall, Oxford OX3 0BW, UK*)

turbulence, several schemes have been proposed. For the bottom layer, the eddy viscosity has been made to vary in the vertical (Madsen, 1977) and to depend on the bottom stress (Cushman-Roisin & Malačič, 1997). Other schemes have been formulated (see Section 4.2), with varying degrees of success. Despite numerous disagreements among models and with field observations, two results nonetheless stand out as quite general. The first is that the angle between the near-boundary velocity and that in the interior or that of the surface stress (depending on the type of Ekman layer) is always substantially less than the theoretical value of 45° and is found to range between 5° and 20° (Fig. 8.10). See also Stacey, Pond and LeBlond (1986).

The second result is a formula for the vertical scale of the Ekman-layer thickness:

$$d \simeq 0.4 \, \frac{u_*}{f}, \tag{8.37}$$

where u_* is the turbulent friction velocity defined in Eq. (8.1). The numerical factor is derived from observations (Garratt, 1992, Appendix 3). Although 0.4 is the most commonly accepted value, there is evidence that certain oceanic conditions call for a somewhat smaller value (Mofjeld & Lavelle, 1984; Stigebrandt, 1985).

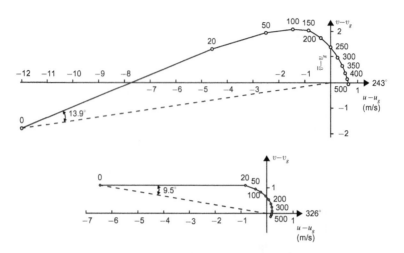

FIGURE 8.10 Wind vectors minus geostrophic wind as a function of height (in meters) in the maritime friction layer near the Scilly Isles. Top diagram: Case of warm air over cold water. Bottom diagram: Case of cold air over warm water. (*Adapted from Roll, 1965*)

Taking u_* as the turbulent velocity and the (unknown) Ekman-layer depth scale, d, as the size of the largest turbulent eddies, we write

$$\nu_E \sim u_* d. \tag{8.38}$$

Then, using rule (8.12) to determine the boundary-layer thickness, we obtain

$$1 \sim \frac{\nu_E}{fd^2} \sim \frac{u_*}{fd},$$

which immediately leads to Eq. (8.37).

The other major element missing from the Ekman-layer formulations of the previous sections is the presence of vertical density stratification. Although the effects of stratification are not discussed in detail until Chapter 11, it can be anticipated here that the gradual change of density with height (lighter fluid above heavier fluid) hinders vertical movements, thereby reducing vertical mixing of momentum by turbulence; it also allows the motions at separate levels to act less coherently and to generate internal gravity waves. As a consequence, stratification reduces the thickness of the Ekman layer and increases the veering of the velocity vector with height (Garratt, 1992, Section 6.2). For a study of the oceanic wind-driven Ekman layer in the presence of density stratification, the reader is referred to Price and Sundermeyer (1999).

The surface atmospheric layer during daytime over land and above warm currents at sea is frequently in a state of convection because of heating from below. In such situations, the Ekman dynamics give way to convective motions, and a controlling factor, besides the geostrophic wind aloft, is the intensity of the surface heat flux. An elementary model is presented later (Section 14.7).

Because Ekman dynamics then play a secondary role, the layer is simply called the *atmospheric boundary layer*. The interested reader is referred to books on the subject by Stull (1988), Sorbjan (1989), Zilitinkevich (1991), or Garratt (1992).

8.8   NUMERICAL SIMULATION OF SHALLOW FLOWS

The theory presented till now largely relies on the assumption of a constant turbulent viscosity. For real flows, however, turbulence is rarely uniform, and eddy diffusion profiles must be considered. Such complexity renders the analytical treatment tedious or even impossible, and numerical methods need to be employed.

To illustrate the approach, we reinstate nonstationary terms and assume a vertically varying eddy viscosity (Fig. 8.11) but retain the hydrostatic approximation (8.13c) and continue to consider a fluid of homogeneous density. The governing equations for u and v are

$$\frac{\partial u}{\partial t} - fv = -\frac{1}{\rho_0}\frac{\partial p}{\partial x} + \frac{\partial}{\partial z}\left(\nu_E(z)\frac{\partial u}{\partial z}\right) \tag{8.39a}$$

$$\frac{\partial v}{\partial t} + fu = -\frac{1}{\rho_0}\frac{\partial p}{\partial y} + \frac{\partial}{\partial z}\left(\nu_E(z)\frac{\partial v}{\partial z}\right) \tag{8.39b}$$

$$0 = -\frac{1}{\rho_0}\frac{\partial p}{\partial z}. \tag{8.39c}$$

From the last equation, it is clear that the horizontal pressure gradient is independent of z.

A standard finite-volume approach could be applied to the equations, but since we already used the approach several times, its implementation is left here as an exercise (see Numerical Problem 8.5). Instead, we introduce another

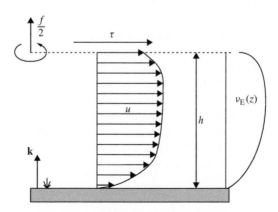

FIGURE 8.11   A vertically confined fluid flow, with bottom and top Ekman layers bracketing a nonuniform velocity profile. The vertical structure can be calculated by a one-dimensional model spanning the entire fluid column eventhough the turbulent viscosity $\nu_E(z)$ may vary in the vertical.

numerical method, which consists in expanding the solution in terms of preselected functions ϕ_j, called *basis functions*. A finite set of N basis functions is
used to construct a *trial solution*:

$$\tilde{u}(z,t) = a_1(t)\phi_1(z) + a_2(t)\phi_2(z) + \cdots + a_N(t)\phi_N(z)$$

$$= \sum_{j=1}^{N} a_j(t)\,\phi_j(z) \tag{8.40a}$$

$$\tilde{v}(z,t) = b_1(t)\phi_1(z) + b_2(t)\phi_2(z) + \cdots + b_N(t)\phi_N(z)$$

$$= \sum_{j=1}^{N} b_j(t)\,\phi_j(z). \tag{8.40b}$$

The problem then reduces to finding a way to calculate the unknown coefficients
$a_j(t)$ and $b_j(t)$ for $j = 1$ to N such that the trial solution is as close as possible to
the exact solution. In other words, we demand that the residual r_u obtained by
substituting the trial solution $\tilde{u}$ into the differential x-momentum equation,

$$\frac{\partial \tilde{u}}{\partial t} - f\tilde{v} + \frac{1}{\rho_0}\frac{\partial p}{\partial x} - \frac{\partial}{\partial z}\left(\nu_E \frac{\partial \tilde{u}}{\partial z}\right) = r_u, \tag{8.41}$$

be as small as possible, and similarly with the residual r_v of the y-momentum
equation. The residuals r_u and r_v quantify the truncation error of the trial
solution, and the objective is to minimize them.

Collocation methods require that the residuals be zero at a finite number
of locations z_k across the domain. If each of the two series contains N terms,
then taking also N points where the two residuals r_u and r_v are forced to vanish
provides $2N$ constraints for the $2N$ unknowns a_j and b_j. With a little chance,
these constraints will be necessary and sufficient to determine the time evolution
of the coefficients $a_j(t)$ and $b_j(t)$. In the present case, the situation is almost
certain because the equations are linear, and the temporal derivatives $da_j(t)/dt$
and $db_j(t)/dt$ appear in linear differential equations, a relatively straightforward
problem to be solved numerically, though the matrices involved may have few
zeros. In some cases, however, the equations may be ill-conditioned because of
inadequate choices of the collocation points z_k (e.g., Gottlieb & Orszag, 1977).

An alternative to requiring zero residuals at selected points is to minimize a
global measure of the error. For example, we can multiply the equations by
N different weighting functions $w_i(z)$ and integrate over the domain before
requiring that the weighted-average error vanish:

$$\int_0^h w_i r_u \, dz = 0, \tag{8.42}$$

and similarly for the companion equation. Note that we require Eq. (8.42) to hold only for a *finite set* of functions w_i, $i = 1, \dots, N$. Had we asked instead that the integral be zero for *any* function w, the trial solution would be the exact solution of the equation since r_u and r_v would then be zero everywhere, but this not possible because we have only $2N$ and not an infinity of degrees of freedom at our disposal. The weighted residuals (8.42) must satisfy

$$\int_0^h w_i \left[\frac{\partial \tilde{u}}{\partial t} - f\tilde{v} + \frac{1}{\rho_0} \frac{\partial p}{\partial x} - \frac{\partial}{\partial z}\left(\nu_E \frac{\partial \tilde{u}}{\partial z} \right) \right] dz = 0, \tag{8.43}$$

for every value of the index i, which leads to

$$\sum_{j=1}^N \int_0^h \left(w_i \phi_j \, dz \right) \frac{da_j}{dt} - f \sum_{j=1}^N \int_0^h \left(w_i \phi_j \, dz \right) b_j + \frac{1}{\rho_0} \frac{\partial p}{\partial x} \int_0^h (w_i \, dz)$$

$$-w_i(h) \, \nu_E \frac{\partial u}{\partial z}\bigg|_h + w_i(0) \, \nu_E \frac{\partial u}{\partial z}\bigg|_0 + \sum_{j=1}^N \int_0^h \left(\nu_E \frac{dw_i}{dz} \frac{d\phi_j}{dz} \, dz \right) a_j = 0, \tag{8.44}$$

and similarly for the y-momentum equation, with the as replaced by bs, bs by $-a$s, x by y, and u by v. Note that use was made of the fact that the pressure gradient is independent of z. The top and bottom stresses (first and second terms of the last line) can be replaced by their value, if known (such as a wind stress on the sea surface).

As already mentioned, if Eq. (8.44) holds for any weighting function, an exact solution is obtained, but if it only holds for a finite series of weighting functions, an approximate solution is found for which the residual is not zero everywhere but is orthogonal to every weighting function.[5] If N different weights are used and the weighted residuals of each of the 2 equations are required to be zero, we obtain $2N$ ordinary differential equations for the $2N$ unknowns a_j and b_j. To write the sets of equations in compact form, we define square matrices $\mathbf{M}$ and $\mathbf{K}$ and column vector $\mathbf{s}$ by

$$M_{ij} = \int_0^h w_i \phi_j \, dz, \quad K_{ij} = \int_0^h \nu_E \frac{dw_i}{dz} \frac{d\phi_j}{dz} \, dz, \quad s_i = \int_0^h w_i \, dz. \tag{8.45}$$

We then group the coefficients a_j and b_j into column vectors $\mathbf{a}$ and $\mathbf{b}$ and the functions $w_i(z)$ into a column vector $\mathbf{w}(z)$. The weighted-residual equations can

---

[5] Orthogonality of two functions is understood here as the property that the product of the two functions integrated over the domain is zero. In the present case, the residual is orthogonal to all weighting functions.

then be written in matrix notation as

$$\mathbf{M}\frac{d\mathbf{a}}{dt} = +f\mathbf{M}\mathbf{b} - \mathbf{K}\mathbf{a} - \frac{1}{\rho_0}\frac{\partial p}{\partial x}\mathbf{s} + \frac{\tau^x}{\rho_0}\mathbf{w}(h) - \frac{\tau_b^x}{\rho_0}\mathbf{w}(0) \qquad (8.46a)$$

$$\mathbf{M}\frac{d\mathbf{b}}{dt} = -f\mathbf{M}\mathbf{a} - \mathbf{K}\mathbf{b} - \frac{1}{\rho_0}\frac{\partial p}{\partial y}\mathbf{s} + \frac{\tau^y}{\rho_0}\mathbf{w}(h) - \frac{\tau_b^y}{\rho_0}\mathbf{w}(0). \qquad (8.46b)$$

This set of ordinary differential equations can be solved by any of the time-integration methods of Chapter 2 as long as the temporal evolution of the surface stress, bottom stress, and pressure gradient is known.

There remains to provide an initial condition on the coefficients a_j and b_j, which must be deduced from the initial flow condition (see Numerical Exercise 8.1). After solving Eq. (8.46), the known values of the coefficients $a_j(t)$ and $b_j(t)$ permit the reconstruction of the solution by means of expansion (8.40). This method is called the *weighted-residual method*.

A word of caution is necessary with respect to boundary conditions. Top and bottom conditions on the shear stress are automatically taken into account, since the stress appears explicitly in the discrete formulation. A Neumann boundary condition, called a *natural condition*, is thus easily applied. No special demand is placed on the basis functions, and weights are simply required to be different from zero at boundaries where a stress condition is applied. There are situations, however, when the stress on the boundary is not known. This is generally the case at a solid boundary along which a no-slip boundary condition $u = v = 0$ is enforced. The integration method does not make the boundary values of u and v appear, and the basis functions must be chosen carefully to be compatible with the boundary condition. In the case of a no-slip condition, it is required that $\phi_j = 0$ at the concerned boundary, so that the velocity is made to vanish there. This is called an *essential boundary condition*.

Till now, both basis functions ϕ_j and weights w_i were arbitrary, except for the aforementioned boundary-related constraints, and smart choices can lead to effective methods. The *Galerkin method* makes the rather natural choice of taking weights equal to the basis functions used in the expansion. The error is then orthogonal to the basis functions. With a well-chosen set of functions ϕ_j, an increasing number of functions can be made to lead ultimately to the exact solution. With the Galerkin method, the matrices $\mathbf{M}$ and $\mathbf{K}$ are symmetric, with components[6]:

$$M_{ij} = \int_0^h \phi_i\phi_j\,dz, \qquad K_{ij} = \int_0^h \nu_E \frac{d\phi_i}{dz}\frac{d\phi_j}{dz}\,dz, \qquad s_i = \int_0^h \phi_i\,dz. \qquad (8.47)$$

The basis functions ϕ_j do not need to span the entire domain but may be chosen to be zero everywhere, except in finite subdomains. The solution can

---

[6] In finite-element jargon, $\mathbf{M}$ and $\mathbf{K}$ are called, respectively, the mass matrix and stiffness matrix.

then be interpreted as the superposition of elementary local solutions. For this, the numerical domain is divided into subdomains called *finite elements*, which are linear segments in 1D, triangles in 2D, and on each of which only a few basis functions differ from zero. This greatly reduces the calculations of the matrices $\mathbf{M}$ and $\mathbf{K}$. The finite-element method is one of the most advanced and flexible methods available for the solution of partial differential equations but is also one of the most difficult to implement correctly (see, e.g., Hanert, Legat & Deleersnijder, 2003 for the implementation of a 2D ocean model). The interested reader is referred to the specialized literature: Buchanan (1995) and Zienkiewicz and Taylor (2000) for an introduction to general finite-element methods, and Zienkiewicz, Taylor and Nithiarasu (2005) for the application of finite elements to fluid dynamics.

For the bottom boundary condition $u = v = 0$, one takes $w_j(0) = \phi_j(0) = 0$, and Eq. (8.46) are unchanged, except for the fact that the term including the bottom stress disappears. The method involves matrices coupling all unknowns a_j and b_j, demanding a preliminary matrix inversion (N^3 operations) and then matrix-vector multiplications (N^2 operations) at every time step.

For the 1D Ekman layer, the problem can be further simplified (e.g., Davies, 1987; Heaps, 1987) by choosing special basis functions that are designed to obey

$$\frac{\partial}{\partial z}\left(\nu_E(z)\, \frac{\partial \phi_j}{\partial z}\right) = -\varrho_j\, \phi_j(z) \tag{8.48a}$$

$$\phi_i(0) = 0, \qquad \left.\frac{\partial \phi_j}{\partial z}\right|_{z=h} = 0. \tag{8.48b}$$

In other words, ϕ_j are chosen as the eigenfunctions of the diffusion operator (8.48a), with ϱ_j as the eigenvalues. Multiplication of Eq. (8.48a) by ϕ_i and subsequent integration by parts in the left-hand side and use of the boundary conditions (8.48b) yield

$$\int_0^h \nu_E\, \frac{d\phi_i}{dz}\frac{d\phi_j}{dz}\, dz = \varrho_j \int_0^h \phi_i\phi_j\, dz. \tag{8.49}$$

Note that for $i = j$, this relationship proves the eigenvalues to be positive for positive diffusion coefficients, since all other terms involved are quadratic and thus positive. Switching the indices i and j, we also have

$$\int_0^h \nu_E\, \frac{d\phi_j}{dz}\frac{d\phi_i}{dz}\, dz = \varrho_i \int_0^h \phi_j\phi_i\, dz, \tag{8.50}$$

and subtracting this equation from the preceding one, we obtain

$$(\varrho_i - \varrho_j) \int_0^h \phi_i \phi_j \, dz = 0,$$ (8.51)

showing that for nonequal eigenvalues, the basis functions ϕ_i and ϕ_j are orthogonal in the sense that

$$\int_0^h \phi_i(z) \, \phi_j(z) \, dz = 0 \quad \text{if} \quad i \neq j.$$ (8.52)

Finally, since the basis functions are defined within an arbitrary multiplicative factor, we may normalize them such that

$$\int_0^h \phi_i(z) \, \phi_j(z) \, dz = \delta_{ij} = \begin{cases} 0 & \text{if } i \neq j \\ 1 & \text{if } i = j \end{cases}$$ (8.53)

When eigenfunctions are used as basis functions in the Galerkin method, a so-called *spectral method* is obtained. It is a very elegant method because the equations for the coefficients are greatly reduced. The orthonormality (8.53) of the eigenfunctions yields $\mathbf{M} = \mathbf{I}$, the identity matrix, and Eq. (8.49) in matrix form reduces to $\mathbf{K} = \varrho \mathbf{M} = \varrho$, where ϱ is a diagonal matrix formed with the eigenvalues ϱ_j. Finally, the equations for components j of $\mathbf{a}$ and $\mathbf{b}$ become

$$\frac{da_j}{dt} = +f b_j - \varrho_j a_j - \frac{1}{\rho_0} \frac{\partial p}{\partial x} s_j + \frac{\tau^x}{\rho_0} \phi_j(h),$$ (8.54a)

$$\frac{db_j}{dt} = -f a_j - \varrho_j b_j - \frac{1}{\rho_0} \frac{\partial p}{\partial y} s_j + \frac{\tau^y}{\rho_0} \phi_j(h).$$ (8.54b)

Note that, since the eigenvalues are positive, the second term on the right corresponds to a damping of the amplitudes a_j and b_j, consistent with physical damping by diffusion.

Because of the decoupling[7] achieved by a set of orthogonal basis functions, we no longer solve a system of $2N$ equations but N systems of 2 equations. This leads to a significant reduction in the number of operations to be performed: The standard Galerkin method requires, at every time step, one inversion of a matrix of size $2N \times 2N$ and a matrix multiplication of cost $4N^2$, whereas the spectral method demands solving N times a 2×2 system, with cost proportional to N. For a large number of time steps, the computational burden is roughly reduced by a factor N. With typically $10^2 - 10^3$ basis functions retained, the savings are

---

[7] Only when equations are linear.

very significant, and the use of a spectral method generates important gains in computing time. It is well worth the preliminary search of eigenfunctions.

In principle, for well-behaved $v_E(z)$, there exist an infinite but countable number of eigenvalues ϱ_j, and the full set of eigenfunctions ϕ_j allows the decomposition of any function. An approximate solution can thus be obtained by retaining only a finite number of eigenfunctions, and the questions that naturally come to mind are how many functions should be retained and which ones. To know which to retain, we can assume a constant viscosity v_E, in which case the solution to the eigenproblem is (for $j = 1, 2, \ldots$)

$$\phi_j = \sqrt{\frac{2}{h}} \sin\left[(2j-1)\frac{\pi}{2}\frac{z}{h}\right]$$

$$\varrho_j = (2j-1)^2 \frac{\pi^2 v_E}{4h^2}$$

$$s_j = \frac{2\sqrt{2h}}{\pi(2j-1)}$$

$$\phi_j(h) = (-1)^{(j+1)}\sqrt{\frac{2}{h}}$$

in which the scaling factor $\sqrt{2/h}$ was introduced to satisfy the normalization requirement (8.53). The name *spectral method* is now readily understood in view of the type of eigenfunctions used in the expansion. The sine functions are indeed nothing else than those used in Fourier series to decompose periodic functions into different wavelengths. The coefficients a_j and b_j are directly interpretable in terms of modal amplitudes or, in other words, the energy associated with the corresponding Fourier modes. The sets of a_j and b_j then provide an insight into the spectrum of the solution.

We further observe that, the larger the eigenvalue ϱ_j, the more rapidly the function oscillates in space, allowing the capture of finer structures. Thus, a higher resolution is achieved by retaining more eigenfunctions in the expansion, just as adding grid points in finite differencing is done to obtain higher resolution. The number N of functions being retained is a matter of scales to be resolved. For a finite-difference representation with N degrees of freedom, the domain is covered with a uniform grid with spacing $\Delta z = h/N$, and the shortest scale that can be resolved has wavenumber $k_z = \pi/\Delta z$ (see Section 1.12). In the spectral method, the highest mode retained corresponds to wavenumber $k_z = N\pi/h$, which is identical to the one resolved in the finite-difference approach. Both methods are thus able to represent the same spectrum of wavenumbers with an identical number of unknowns. Also, the cost of both methods is directly proportional to the number of unknowns. So, where is the advantage of using a spectral method?

Except for the straightforward interpretation of the coefficients a_j and b_j in terms of Fourier components, the essential advantage of spectral methods resides in their rapid convergence as the number N of basis functions is increased, for sufficiently gentle solutions and boundary conditions (e.g., Canuto, Hussaini, Quarteroni & Zang, 1988). To illustrate this claim, we calculate the stationary solution of a geostrophic current without surface stress by dropping $\mathrm{d}a_j/\mathrm{d}t$ and $\mathrm{d}b_j/\mathrm{d}t$ from the matrix equations and solve for **a** and **b** before recombining the solution. Even with only five basis functions (i.e., equivalent to using five grid points), the behavior of the solution is well captured (Fig. 8.12). Furthermore, since the equations for different a_j coefficients are decoupled, increasing the value of N does not modify the values of the previously calculated coefficients but simply adds more terms, each one bringing additional resolution. The amplitude of the new terms is directly proportional to the value of the coefficients a_j and b_j, and their rapid decrease as a function of index j (Fig. 8.13—note the logarithmic scales) explains why fast convergence can be expected.

In order not to miss the most important parts of the solution, it is imperative to use eigenfunctions in their order, that is, without skipping any in the series, up to the preselected number N. In the limit of large N, it can be shown

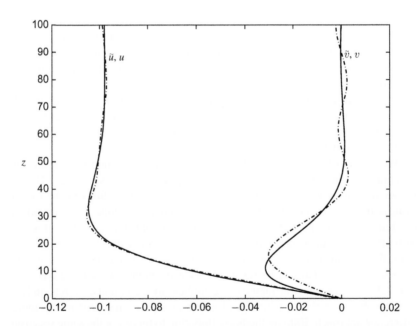

FIGURE 8.12 Velocity profile forced by a pressure gradient directed along the y-axis above a no-slip bottom and below a stress-free surface. The geostrophic flow aloft has components $u = 0.1$ and $v = 0$. Solid lines represent the exact solution, whereas dash-dotted lines depict the numerical solution obtained by the spectral method with only the first five modes. Note the excellent agreement. Oscillations appearing in the numerical solution give a hint of the sine functions used in expanding the solution.

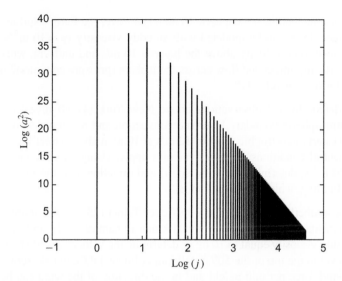

FIGURE 8.13   Sequential values of the coefficients a_j (scaled by an arbitrary coefficient) obtained with the spectral method applied to the problem of Fig. 8.12. Scales are logarithmic, showing the rapid decrease of amplitudes with increasing number of modes in the expansion. Convergence toward the exact solution is fast.

that convergence for a relatively smooth solution is faster than with any finite-difference method of any order (e.g., Gottlieb & Orszag, 1977). This is the distinct advantage of the spectral method, which explains why it is often used in cases when nearly exact numerical solutions are sought.

An alternative to the Galerkin spectral approach is to force the error to vanish at particular grid points, leading to so-called *pseudospectral methods* (e.g., Fornberg, 1988). As for all collocation methods, these do not require evaluation of integrals over the domain.

In concluding the presentation of the function-expansion approach, we insist on the fundamental aspect that the numerical approximation is very different from the point-value sampling used in finite-difference methods. In space, the basis functions ϕ_j are continuous and can therefore be differentiated or manipulated mathematically without approximation. The numerical error arises only due to the fact that a finite number of basis functions is used to represent the solution.

ANALYTICAL PROBLEMS

8.1.   It is observed that fragments of tea leaves at the bottom of a stirred tea cup conglomerate toward the center. Explain this phenomenon with Ekman-layer dynamics. Also explain why the tea leaves go to the center irrespectively of the direction of stirring (clockwise or counterclockwise).

8.2. Assume that the atmospheric Ekman layer over the Earth's surface at latitude 45°N can be modeled with an eddy viscosity $\nu_E = 10$ m²/s. If the geostrophic velocity above the layer is 10 m/s and uniform, what is the vertically integrated flow across the isobars (pressure contours)? Is there any vertical velocity?

8.3. Meteorological observations above New York (41°N) reveal a neutral atmospheric boundary layer (no convection and no stratification) and a westerly geostrophic wind of 12 m/s at 1000 m above street level. Under neutral conditions, Ekman dynamics apply. Using an eddy viscosity of 10 m²/s, determine the wind speed and direction atop the Empire State Building, which stands 381 m tall.

8.4. A southerly wind blows at 9 m/s over Taipei (25°N). Assuming neutral atmospheric conditions so that Ekman dynamics apply and taking the eddy viscosity equal to 10 m²/s, determine the velocity profile from street level to the top of the 509 m tall Taipei Financial Center skyscraper. The wind force per unit height and in the direction of the wind can be taken as $F = 0.93\rho L V^2$, where $\rho = 1.20$ kg/m³ is the standard air density, $L = 25$ m is the building width, and $V(z) = (u^2 + v^2)^{1/2}$ is the wind speed at the height considered. With this, determine the total wind force on the southern facade of the Taipei Financial Center.

8.5. Show that although $\bar{w}$ may not be zero in the presence of horizontal gradients, the vertical advection terms $w\partial u/\partial z$ and $w\partial v/\partial z$ of the momentum equations are still negligible, even if the short distance d is taken as the vertical length scale.

8.6. You are working for a company that plans to deposit high-level radioactive wastes on the bottom of the ocean, at a depth of 3000 m. This site (latitude: 33°N) is known to be at the center of a permanent counterclockwise vortex. Locally, the vortex flow can be assimilated to a solid-body rotation with angular speed equal to 10^{-5} s⁻¹. Assuming a homogeneous ocean and a steady, geostrophic flow, estimate the upwelling rate at the vortex center. How many years will it take for the radioactive wastes to arrive at the surface? Take $f = 8 \times 10^{-5}$ s⁻¹ and $\nu_E = 10^{-2}$ m²/s.

8.7. Derive Eq. (8.36) more simply not by starting from solution (8.33) as done in the text but by vertical integration of the momentum equations (8.32). Consider also the case of nonuniform eddy viscosity, in which case ν_E must be kept inside the vertical derivative on the right-hand side of the equations, as in the original governing equations (4.21a) and (4.21b).

8.8. Between 15°N and 45°N, the winds over the North Pacific Ocean consist mostly of the easterly trades (15°N to 30°N) and the midlatitude westerlies (30°N to 45°N). An adequate representation is

$$\tau^x = \tau_0 \sin\left(\frac{\pi y}{2L}\right), \qquad \tau^y = 0 \quad \text{for} \quad -L \le y \le L,$$

with $\tau_0 = 0.15$ N/m^2 (maximum wind stress) and $L = 1670$ km. Taking $\rho_0 = 1028$ kg/m^3 and the value of the Coriolis parameter corresponding to 30°N, calculate the Ekman pumping. Which way is it directed? Calculate the vertical volume flux over the entire 15°–45°N strip of the North Pacific (width = 8700 km). Express your answer in sverdrup units (1 sverdrup = 1 Sv = 10^6 m^3/s).

8.9. The variation of the Coriolis parameter with latitude can be approximated as $f = f_0 + \beta_0 y$, where y is the northward coordinate (beta-plane approximation, see Section 9.4). Using this, show that the vertical velocity below the surface Ekman layer of the ocean is given by

$$\bar{w}(z) = \frac{1}{\rho_0}\left[\frac{\partial}{\partial x}\left(\frac{\tau^y}{f}\right) - \frac{\partial}{\partial y}\left(\frac{\tau^x}{f}\right)\right] - \frac{\beta_0}{f}\int_z^0 \bar{v}\,dz, \qquad (8.55)$$

where τ^x and τ^y are the zonal and meridional wind-stress components, respectively, and $\bar{v}$ is the meridional velocity in the geostrophic interior below the Ekman layer.

8.10. Determine the vertical distribution of horizontal velocity in a 4-m deep lagoon subject to a northerly wind stress of 0.2 N/m^2. The density of the brackish water in the lagoon is 1020 kg/m^3. Take $f = 10^{-4}$ s^{-1} and $\nu_E = 10^{-2}$ m^2/s. In which direction is the net transport in this brackish layer?

8.11. Redo Problem 8.10 with $f = 0$ and compare the two solutions. What can you conclude about the role of the Coriolis force in this case?

8.12. Find the stationary solution of (8.13a)–(8.13c) for constant viscosity, a uniform pressure gradient in the y-direction in a domain of finite depth h with no stress at the top and no slip at the bottom. Study the behavior of the solution as h/d varies and compare with the solution in the infinite domain. Then, derive the stationary solution without pressure gradient but with a top stress in the y-direction.

NUMERICAL EXERCISES

8.1. How can we obtain initial conditions for a_j and b_j in the expansions (8.40) from initial conditions on the physical variables $u = u_0(z)$ and $v = v_0(z)$?

(*Hint*: Investigate a least-square approach and an approach in which the initial error is forced to vanish in the sense of Eq. (8.42). When do the two approaches lead to the same result?)

8.2. Use `spectralekman.m` to calculate numerically the stationary solutions of Analytical Problem 8.12. Compare the exact and numerical solutions for $h/d = 4$ and assess the convergence rate as a function of $1/N$, where N is the number of eigenfunctions retained in the trial solution. Compare the convergence rate of both cases (with and without stress at the top) and comment.

8.3. Use `spectralekman.m` to explore how the solution changes as a function of the ratio h/d and how the number N of modes affects your resolution of the boundary layers.

8.4. Modify `spectralekman.m` to allow for time evolution, but maintaining constant wind stress and pressure gradient. Use a trapezoidal method for time integration. Start from rest and observe the temporal evolution. What do you observe?

8.5. Use a finite-volume approach with time splitting for the Coriolis terms and an explicit Euler method to discretize diffusion in Eqs. (8.39a) and (8.39b). Verify your program in the case of uniform eddy viscosity by comparing with the steady analytical solution. Then, use the viscosity profile $\nu_E(z) = \mathcal{K}z(1 - z/h)u_*$. In this case, can you find the eigenfunctions of the diffusion operator and outline the Galerkin method? (*Hint*: Look for Legendre polynomials and their properties.)

8.6. Assume that your vertical grid spacing in a finite-difference scheme is large compared with the roughness length z_0 and that your first point for velocity calculation is found at the distance $\Delta z/2$ above the bottom. Use the logarithmic profile to deduce the bottom stress as a function of the computed velocity at level $\Delta z/2$. Then, use this expression in the finite-volume approach of Numerical Exercise 8.5 to replace the no-slip condition by a stress condition at the lowest level of the grid.

Vagn Walfrid Ekman
1874–1954

Born in Sweden, Ekman spent his formative years under the tutelage of Vilhelm Bjerknes and Fridtjof Nansen in Norway. One day, Nansen asked Bjerknes to let one of his students make a theoretical study of the influence of the earth's rotation on wind-driven currents, on the basis of Nansen's observations during his polar expedition that ice drifts with ocean currents to the right of the wind. Ekman was chosen and later presented a solution in his doctoral thesis of 1902.

As professor of mechanics and mathematical physics at the University of Lund in Sweden, Ekman became the most famous oceanographer of his generation. The distinguished theoretician also proved to be a skilled experimentalist. He designed a current meter, which bears his name and which has been used extensively. Ekman was also the one who explained the phenomenon of dead water by a celebrated laboratory experiment (see Fig. 1.4). (*Photo courtesy of Pierre Welander*)

Ludwig Prandtl
1875–1953

A German engineer, Ludwig Prandtl was attracted by fluid phenomena and their mathematical representation. He became professor of mechanics at the University of Hannover in 1901, where he established a world renowned institute for aerodynamics and hydrodynamics. It was while working on wing theory in 1904 and studying friction drag in particular that he developed the concept of boundary layers and the attending mathematical technique. His central idea was to recognize that frictional effects are confined to a thin layer in the vicinity of the boundary, allowing the modeler to treat rest of the flow as inviscid.

Prandtl also made noteworthy advances in the study of elasticity, supersonic flows, and turbulence, particularly shear turbulence in the vicinity of a boundary. A mixing length and a dimensionless ratio are named after him.

It has been remarked that Prandtl's keen perception of physical phenomena was balanced by a limited mathematical ability and that this shortcoming prompted him to seek ways of reducing the mathematical description of his objects of study. Thus perhaps, the boundary-layer technique was an invention born out of necessity. (*Photo courtesy of the Emilio Segrè Visual Archives, American Institute of Physics*)

Chapter 9

Barotropic Waves

ABSTRACT

The aim of this chapter is to describe an assortment of waves that can be supported by an inviscid, homogeneous fluid in rotation and to analyze numerical grid arrangements that facilitate the simulation of wave propagation, in particular for the prediction of tides and storm surges.

9.1 LINEAR WAVE DYNAMICS

Chiefly because linear equations are most amenable to methods of solution, it is wise to gain insight into geophysical fluid dynamics by elucidating the possible linear processes and investigating their properties before exploring more intricate, nonlinear dynamics. The governing equations of the previous section are essentially nonlinear; consequently, their linearization can proceed only by imposing restrictions on the flows under consideration.

The Coriolis acceleration terms present in the momentum equations [(4.21a) and (4.21b)] are, by nature, linear and need not be subjected to any approximation. This situation is extremely fortunate because these are the central terms of geophysical fluid dynamics. In contrast, the so-called advective terms (or convective terms) are quadratic and undesirable at this moment. Hence, our considerations will be restricted to low-Rossby-number situations:

$$Ro = \frac{U}{\Omega L} \ll 1. \tag{9.1}$$

This is usually accomplished by restricting the attention to relatively weak flows (small U), large scales (large L), or, in the laboratory, fast rotation (large Ω). The terms expressing the local time rate of change of the velocity ($\partial u/\partial t$ and $\partial v/\partial t$) are linear and are retained here in order to permit the investigation of unsteady flows. Thus, the temporal Rossby number is assumed to be on the order of unity:

$$Ro_T = \frac{1}{\Omega T} \sim 1. \tag{9.2}$$

Contrasting conditions (9.1) and (9.2), we conclude that we are about to consider slow flow fields that evolve relatively fast. Aren't we asking for the impossible? Not at all, for rapidly moving disturbances do not necessarily require large velocities. In other words, information may travel faster than material particles, and when this is the case, the flow takes the aspect of a wave field. A typical example is the spreading of concentric ripples on the surface of a pond after throwing a stone; energy radiates but there is no appreciable water movement across the pond. In keeping with the foregoing quantities, a scale C for the wave speed (or celerity) can be defined as the velocity of a signal covering the distance L of the flow during the nominal evolution time T, and, by virtue of restrictions (9.1) and (9.2), it can be compared with the flow velocity:

$$C = \frac{L}{T} \sim \Omega L \gg U. \tag{9.3}$$

Thus, our present objective is to consider wave phenomena.

To shed the best possible light on the mechanisms of the basic wave processes typical in geophysical flows, we further restrict our attention to homogeneous and inviscid flows, for which the shallow-water model (Section 7.3) is adequate. With all the preceding restrictions, the horizontal momentum equations (7.9a) and (7.9b) reduce to

$$\frac{\partial u}{\partial t} - fv = - g \frac{\partial \eta}{\partial x} \tag{9.4a}$$

$$\frac{\partial v}{\partial t} + fu = - g \frac{\partial \eta}{\partial y}, \tag{9.4b}$$

where f is the Coriolis parameter, g is the gravitational acceleration, u and v are the velocity components in the x- and y-directions, respectively, and η is the surface displacement (equal to $\eta = h - H$, the total fluid depth h minus the mean fluid thickness H). The independent variables are x, y, and t; the vertical coordinate is absent, for the flow is vertically homogeneous (Section 7.3).

In terms of surface height, η, the continuity equation (7.14) can be expanded in several groups of terms:

$$\frac{\partial \eta}{\partial t} + \left(u \frac{\partial \eta}{\partial x} + v \frac{\partial \eta}{\partial y} \right) + H \left(\frac{\partial u}{\partial x} + \frac{\partial v}{\partial y} \right) + \eta \left(\frac{\partial u}{\partial x} + \frac{\partial v}{\partial y} \right) = 0$$

if the mean depth H is constant (flat bottom). Introducing the scale ΔH for the vertical displacement η of the surface, we note that the four groups of terms in the preceding equation are, sequentially, on the order of

$$\frac{\Delta H}{T}, \quad U \frac{\Delta H}{L}, \quad H \frac{U}{L}, \quad \Delta H \frac{U}{L}.$$

According to Eq. (9.3) L/T is much larger than U, and the second and fourth groups of terms may be neglected compared with the first term, leaving us with the linearized equation

$$\frac{\partial \eta}{\partial t} + H \left(\frac{\partial u}{\partial x} + \frac{\partial v}{\partial y} \right) = 0, \tag{9.5}$$

the balance of which requires $\Delta H/T$ to be on the order of UH/L or, again by virtue of (9.3),

$$\Delta H \ll H. \tag{9.6}$$

We are thus restricted to waves of small amplitudes.

The system of Eqs. (9.4a) through (9.5) governs the linear wave dynamics of inviscid, homogeneous fluids under rotation. For the sake of simple notation, we will perform the mathematical derivations only for positive values of the Coriolis parameter f and then state the conclusions for both positive and negative values of f. The derivations with negative values of f are left as exercises. Before proceeding with the separate studies of geophysical fluid waves, the reader not familiar with the concepts of phase speed, wavenumber vector, dispersion relation, and group velocity is directed to Appendix B. A comprehensive account of geophysical waves can be found in the book by LeBlond and Mysak (1978), with additional considerations on nonlinearities in Pedlosky (2003).

9.2 THE KELVIN WAVE

The Kelvin wave is a traveling disturbance that requires the support of a lateral boundary. Therefore, it most often occurs in the ocean where it can travel along coastlines. For convenience, we use oceanic terminology such as coast and offshore.

As a simple model, consider a layer of fluid bounded below by a horizontal bottom, above by a free surface, and on one side (say, the y-axis) by a vertical wall (Fig. 9.1). Along this wall ($x = 0$, the coast), the normal velocity

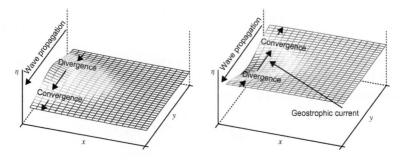

FIGURE 9.1 Upwelling and downwelling Kelvin waves. In the northern hemisphere, both waves travel with the coast on their right, but the accompanying currents differ. Geostrophic equilibrium in the x−momentum equation leads to a velocity v that is maximum at the bulge and directed as the geostrophic equilibrium requires. Because of the different geostrophic velocities at the bulge and further away, convergence and divergence patterns create a lifting or lowering of the surface. The lifting and lowering are such that the wave propagates toward negative y in either case (positive or negative bulge).

must vanish ($u=0$), but the absence of viscosity allows a nonzero tangential velocity.

As he recounted in his presentation to the Royal Society of Edinburgh in 1879, Sir William Thomson (later to become Lord Kelvin) thought that the vanishing of the velocity component normal to the wall suggested the possibility that it be zero everywhere. So, let us state, in anticipation,

$$u=0 \tag{9.7}$$

throughout the domain and investigate the consequences. Although Eq. (9.4a) contains a remaining derivative with respect to x, Eqs. (9.4b) and (9.5) contain only derivatives with respect to y and time. Elimination of the surface elevation leads to a single equation for the alongshore velocity:

$$\frac{\partial^2 v}{\partial t^2}=c^2\frac{\partial^2 v}{\partial y^2}, \tag{9.8}$$

where

$$c=\sqrt{gH} \tag{9.9}$$

is identified as the speed of surface gravity waves in nonrotating shallow waters.

The preceding equation governs the propagation of one-dimensional nondispersive waves and possesses the general solution

$$v=V_1(x,\ y+ct)+V_2(x,\ y-ct), \tag{9.10}$$

which consists of two waves, one traveling toward decreasing y and the other in the opposite direction. Returning to either Eq. (9.4b) or (9.5) where u is set to zero, we easily determine the surface displacement:

$$\eta=-\sqrt{\frac{H}{g}}\ V_1(x,\ y+ct)+\sqrt{\frac{H}{g}}\ V_2(x,\ y-ct). \tag{9.11}$$

(Any additive constant can be eliminated by a proper redefinition of the mean depth H.) The structure of the functions V_1 and V_2 is then determined by the use of the remaining equation, i.e., (9.4a):

$$\frac{\partial V_1}{\partial x}=-\frac{f}{\sqrt{gH}}V_1, \quad \frac{\partial V_2}{\partial x}=+\frac{f}{\sqrt{gH}}V_2$$

or

$$V_1=V_{10}(y+ct)\ e^{-x/R}, \quad V_2=V_{20}(y-ct)\ e^{+x/R},$$

where the length R, defined as

$$R=\frac{\sqrt{gH}}{f}=\frac{c}{f}, \tag{9.12}$$

combines all three constants of the problem. Within a numerical factor, it is the distance covered by a wave, such as the present one, traveling at the speed c during one inertial period $(2\pi/f)$. For reasons that will become apparent later, this quantity is called the *Rossby radius of deformation* or, more simply, the radius of deformation.

Of the two independent solutions, the second increases exponentially with distance from shore and is physically unfit. This leaves the other as the most general solution:

$$u = 0 \tag{9.13a}$$

$$v = \sqrt{gH} F(y+ct)\, e^{-x/R} \tag{9.13b}$$

$$\eta = -HF(y+ct)\, e^{-x/R}, \tag{9.13c}$$

where F is an arbitrary function of its variable.

Because of the exponential decay away from the boundary, the Kelvin wave is said to be trapped. Without the boundary, it is unbounded at large distances and thus cannot exist; the length R is a measure of the trapping distance. In the longshore direction, the wave travels without distortion at the speed of surface gravity waves. In the northern hemisphere ($f > 0$, as in the preceding analysis), the wave travels with the coast on its right; in the southern hemisphere, the wave travels with the coast on its left. Note that, although the direction of wave propagation is unique, the sign of the longshore velocity is arbitrary: An upwelling wave (i.e., a surface bulge with $\eta > 0$) has a current flowing in the direction of the wave, whereas a downwelling wave (i.e., a surface trough with $\eta < 0$) is accompanied by a current flowing in the direction opposite to that of the wave (Fig. 9.1).

In the limit of no rotation ($f \to 0$), the trapping distance increases without bound, and the wave reduces to a simple gravity wave with crests and troughs oriented perpendicularly to the coast.

Surface Kelvin waves (as described previously, and to be distinguished from internal Kelvin waves, which require a stratification, see the end of Chapter 13) are generated by the ocean tides and by local wind effects in coastal areas. For example, a storm off the northeast coast of Great Britain can send a Kelvin wave that follows the shores of the North Sea in a counterclockwise direction and eventually reaches the west coast of Norway. Traveling in approximately 40 m of water and over a distance of 2200 km, it accomplishes its journey in about 31 h.

The decay of the Kelvin wave amplitude away from the coast is clearly manifested in the English Channel. The North Atlantic tide enters the Channel from the west and travels eastward toward the North Sea (Fig. 9.2). Being essentially a surface wave in a rotating fluid bounded by a coast, the tide assumes the character of a Kelvin wave and propagates while leaning against a coast on its right, namely, France. This partly explains why tides are noticeably higher along the French coast than along the British coast a few tens of kilometers across (Fig. 9.2).

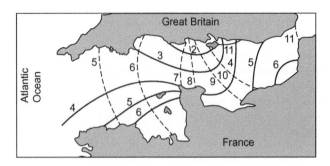

FIGURE 9.2 Cotidal lines (dashed) with time in lunar hours for the M2 tide in the English Channel showing the eastward progression of the tide from the North Atlantic Ocean. Lines of equal tidal range (solid, with value in meters) reveal larger amplitudes along the French coast, namely to the right of the wave progression in accordance with Kelvin waves. (*From Proudman, 1953, as adapted by Gill, 1982*)

9.3 INERTIA-GRAVITY WAVES (POINCARÉ WAVES)

Let us now do away with the lateral boundary and relax the stipulation $u = 0$. The system of Eqs. (9.4a) through (9.5) is kept in its entirety. With f constant and in the presence of a flat bottom, all coefficients are constant, and a Fourier-mode solution can be sought. With u, v, and η taken as constant factors times a periodic function

$$\begin{pmatrix} \eta \\ u \\ v \end{pmatrix} = \Re \begin{pmatrix} A \\ U \\ V \end{pmatrix} e^{i\,(k_x x + k_y y - \omega t)} \tag{9.14}$$

where the symbol $\Re$ indicates the real part of what follows, k_x and k_y are the wavenumbers in the x-and y-directions, respectively, and ω is a frequency, the system of equations becomes algebraic:

$$-i\omega U - fV = -igk_x A \tag{9.15a}$$

$$-i\omega V + fU = -igk_y A \tag{9.15b}$$

$$-i\omega A + H(ik_x U + ik_y V) = 0. \tag{9.15c}$$

This system admits the trivial solution $U = V = A = 0$ unless its determinant vanishes. Thus, waves occur only when the following condition is met:

$$\omega[\omega^2 - f^2 - gH\,(k_x^2 + k_y^2)] = 0. \tag{9.16}$$

This condition, called the *dispersion relation*, provides the wave frequency in terms of the wavenumber magnitude $k = (k_x^2 + k_y^2)^{1/2}$ and the constants of the problem. The first root, $\omega = 0$, corresponds to a steady geostrophic state. Returning to Eqs. (9.4a) through (9.5) with the time derivatives set to zero, we recognize the equations governing the geostrophic flow described in Section 7.1.

In other words, geostrophic flows can be interpreted as arrested waves of any wavelength. If the bottom were not flat, these "waves" would cease to exist and be replaced by Taylor columns.

The remaining two roots,

$$\omega = \sqrt{f^2 + gHk^2} \qquad (9.17)$$

and its opposite, correspond to bona fide traveling waves, called *Poincaré waves*, whose frequency is always superinertial. In the limit of no rotation $(f = 0)$, the frequency is $\omega = k\sqrt{gH}$, and the phase speed is $c = \omega/k = \sqrt{gH}$. The waves become classical shallow-water gravity waves. The same limit also occurs at large wavenumbers $[k^2 \gg f^2/gH$, i.e., wavelengths much shorter than the deformation radius defined in Eq. (9.12)]. This is not too surprising, since such waves are too short and too fast to feel the rotation of the earth.

At the opposite extreme of low wavenumbers $(k^2 \ll f^2/gH$, i.e., wavelengths much longer than the deformation radius), the rotation effect dominates, yielding $\omega \simeq f$. At this limit, the flow pattern is virtually laterally uniform, and all fluid particles move in unison, each describing a circular inertial oscillation, as described in Section 2.3. For intermediate wavenumbers, the frequency (Fig. 9.3) is always greater than f, and the waves are said to be *superinertial*. Since Poincaré waves exhibit a mixed behavior between gravity waves and inertial oscillations, they are often called *inertia-gravity waves*.

Because the phase speed $c = \omega/k$ depends on the wavenumber, wave components of different wavelengths travel at different speeds, and the wave is said to be *dispersive*. This is in contrast with the nondispersive Kelvin wave, the signal

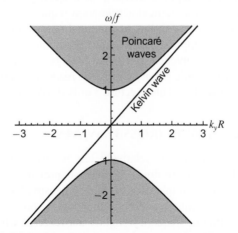

FIGURE 9.3   Recapitulation of the dispersion relation of Kelvin and Poincaré waves on the f-plane and on a flat bottom. Although Poincaré waves (gray shades) can travel in all directions and occupy therefore a continuous spectrum in terms of k_y, the Kelvin wave (diagonal line) propagates only along a boundary.

of which travels without distortion, irrespective of its profile. See Appendix B for additional information on these notions.

Seiches, tides, and tsunamis are examples of barotropic gravity waves. A *seiche* is a standing wave, formed by the superposition of two waves of equal wavelength and propagating in opposite directions due to reflection on lateral boundaries. Seiches occur in confined water bodies, such as lakes, gulfs, and semi-enclosed seas. In the Adriatic Sea, the untimely superposition of a wind-generated seiche with high tide can cause flooding in Venice (Robinson, Tomasin & Artegiani, 1973).

A *tsunami* is a wave triggered by an underwater earthquake. With wavelengths ranging from tens to hundreds of kilometers, tsunamis are barotropic waves, but their relatively high frequency (period of a few minutes) makes them only slightly affected by the Coriolis force. What makes tsunamis disastrous is the gradual amplification of their amplitude as they enter shallower waters, so that what may begin as an innocuous 1m wave in the middle of the ocean, which a ship hardly notices, can turn into a catastrophic multimeter surge on the beach. Disastrous tsunamis occurred in the Pacific Ocean on 22 May 1960, in the Indian Ocean on 26 December 2004, and in the Pacific Ocean off Japan on 11 March 2011. Tsunami propagation is relatively easy to forecast with computer models. The key to an effective warning system is the early detection of the originating earthquake to track the rapid propagation (at speed $\sqrt{gH}$) of the tsunami from point of origin to the coastline on time to issue a warning before the high wave strikes.

Before concluding this section, a note is in order to warn about the possibility of violating the hydrostatic assumption. Indeed, at short wavelengths (on the order of the fluid depth or shorter), the frequency is high (period much shorter than $2\pi/f$), and the vertical acceleration (equal to $\partial^2\eta/\partial t^2$ at the surface) becomes comparable with the gravitational acceleration g. When this is the case, the hydrostatic approximation breaks down, the assumption of vertical rigidity may no longer be invoked, and the problem becomes three dimensional. For a study of nonhydrostatic gravity waves, the reader is referred to Section 10 of LeBlond and Mysak (1978) and Lecture 3 of Pedlosky (2003).

9.4   PLANETARY WAVES (ROSSBY WAVES)

Kelvin and Poincaré waves are relatively fast waves, and we may wonder whether rotating, homogeneous fluids could not support another breed of slower waves. Could it be, for example, that the steady geostrophic flows, those corresponding to the zero frequency solution found in the preceding section, may develop a slow evolution (frequency slightly above zero) when the system is modified somewhat? The answer is yes, and one such class consists of planetary waves, in which the time evolution originates in the weak but important *planetary effect*.

As we may recall from Section 2.5, on a spherical earth (or planet or star, in general), the Coriolis parameter, f, is proportional to the rotation rate, Ω, times the sine of the latitude, φ:

$$f = 2\Omega \sin\varphi.$$

Large wave formations such as alternating cyclones and anticyclones contribute to our daily weather and, to a lesser extent, Gulf Stream meanders span several degrees of latitude; for them, it is necessary to consider the meridional change in the Coriolis parameter. If the coordinate y is directed northward and is measured from a reference latitude φ_0 (say, a latitude somewhere in the middle of the wave under consideration), then $\varphi = \varphi_0 + y/a$, where a is the earth's radius (6371 km). Considering y/a as a small perturbation, the Coriolis parameter can be expanded in a Taylor series:

$$f = 2\Omega \sin\varphi_0 + 2\Omega \frac{y}{a}\cos\varphi_0 + \cdots \tag{9.18}$$

Retaining only the first two terms, we write

$$f = f_0 + \beta_0 y, \tag{9.19}$$

where $f_0 = 2\Omega \sin\varphi_0$ is the reference Coriolis parameter, and $\beta_0 = 2(\Omega/a)\cos\varphi_0$ is the so-called *beta parameter*. Typical midlatitude values on earth are $f_0 = 8 \times 10^{-5}$ s^{-1} and $\beta_0 = 2 \times 10^{-11}$ m^{-1}s^{-1}. The Cartesian framework where the beta term is not retained is called the *f-plane*, and that where it is retained is called the *beta plane*. The next step in order of accuracy is to retain the full spherical geometry (which we avoid throughout this book). Rigorous justifications of the beta-plane approximation can be found in Veronis (1963, 1981), Pedlosky (1987), and Verkley (1990).

Note that the beta-plane representation is validated at midlatitudes only if the $\beta_0 y$ term is small compared with the leading f_0 term. For the motion's meridional length scale L, this implies

$$\beta = \frac{\beta_0 L}{f_0} \ll 1, \tag{9.20}$$

where the dimensionless ratio β can be called the *planetary number*.

The governing equations, having become

$$\frac{\partial u}{\partial t} - (f_0 + \beta_0 y)v = -g\frac{\partial \eta}{\partial x} \tag{9.21a}$$

$$\frac{\partial v}{\partial t} + (f_0 + \beta_0 y)u = -g\frac{\partial \eta}{\partial y} \tag{9.21b}$$

$$\frac{\partial \eta}{\partial t} + H\left(\frac{\partial u}{\partial x} + \frac{\partial v}{\partial y}\right) = 0, \tag{9.21c}$$

are now mixtures of small and large terms. The larger ones (f_0, g, and H terms) comprise the otherwise steady, f-plane geostrophic dynamics; the smaller ones (time derivatives and β_0 terms) come as perturbations, which, although small, will govern the wave evolution. In first approximation, the large terms dominate, and thus $u \simeq -(g/f_0)\partial\eta/\partial y$ and $v \simeq +(g/f_0)\partial\eta/\partial x$. Use of this first approximation in the small terms of Eqs. (9.21a) and (9.21b) yields

$$-\frac{g}{f_0}\frac{\partial^2\eta}{\partial y\partial t} - f_0 v - \frac{\beta_0 g}{f_0}y\frac{\partial\eta}{\partial x} = -g\frac{\partial\eta}{\partial x} \tag{9.22}$$

$$+\frac{g}{f_0}\frac{\partial^2\eta}{\partial x\partial t} + f_0 u - \frac{\beta_0 g}{f_0}y\frac{\partial\eta}{\partial y} = -g\frac{\partial\eta}{\partial y}. \tag{9.23}$$

These equations are trivial to solve with respect to u and v:

$$u = -\frac{g}{f_0}\frac{\partial\eta}{\partial y} - \frac{g}{f_0^2}\frac{\partial^2\eta}{\partial x\partial t} + \frac{\beta_0 g}{f_0^2}y\frac{\partial\eta}{\partial y} \tag{9.24}$$

$$v = +\frac{g}{f_0}\frac{\partial\eta}{\partial x} - \frac{g}{f_0^2}\frac{\partial^2\eta}{\partial y\partial t} - \frac{\beta_0 g}{f_0^2}y\frac{\partial\eta}{\partial x}. \tag{9.25}$$

These last expressions can be interpreted as consisting of the leading and first-correction terms in a regular perturbation series of the velocity field. We identify the first term of each expansion as the geostrophic velocity. By contrast, the next and smaller terms are called *ageostrophic*.

Substitution in continuity equation (9.21c) leads to a single equation for the surface displacement:

$$\frac{\partial\eta}{\partial t} - R^2\frac{\partial}{\partial t}\nabla^2\eta - \beta_0 R^2\frac{\partial\eta}{\partial x} = 0, \tag{9.26}$$

where ∇^2 is the two-dimensional Laplace operator, and $R = \sqrt{gH}/f_0$ is the deformation radius, defined in Eq. (9.12) but now suitably amended to be a constant. Unlike the original set of equations, this last equation has constant coefficients and a solution of the Fourier type, $\cos(k_x x + k_y y - \omega t)$, can be sought. The dispersion relation follows:

$$\omega = -\beta_0 R^2\frac{k_x}{1 + R^2(k_x^2 + k_y^2)}, \tag{9.27}$$

providing the frequency ω as a function of the wavenumber components k_x and k_y. The waves are called *planetary waves* or *Rossby waves*, in honor of Carl-Gustaf Rossby, who first proposed this wave theory to explain the systematic movement of midlatitude weather patterns. We note immediately that if the beta corrections had not been retained ($\beta_0 = 0$), the frequency would have been nil. This is the $\omega = 0$ solution of Section 9.3, which corresponds to a steady geostrophic flow on the f-plane. The absence of the other two roots is explained by our approximation. Indeed, treating the time derivatives as small terms (i.e., having in effect assumed a very small temporal Rossby number, $Ro_T \ll 1$), we have retained only the low frequency, the one much less than f_0. In the parlance of wave dynamics, this is called *filtering*.

That the frequency given by Eq. (9.27) is indeed small can be verified easily. With L ($\sim 1/k_x \sim 1/k_y$) as a measure of the wavelength, two cases can arise either $L \lesssim R$ or $L \gtrsim R$; the frequency scale is then given by

$$\text{Shorter waves:} \quad L \lesssim R, \quad \omega \sim \beta_0 L \qquad (9.28)$$

$$\text{Longer waves:} \quad L \gtrsim R, \quad \omega \sim \frac{\beta_0 R^2}{L} \lesssim \beta_0 L. \qquad (9.29)$$

In either case, our premise (9.20) that $\beta_0 L$ is much less than f_0 implies that ω is much smaller than f_0 (subinertial wave), as we anticipated.

Let us now explore other properties of planetary waves. First and foremost, the zonal phase speed

$$c_x = \frac{\omega}{k_x} = \frac{-\beta_0 R^2}{1 + R^2 \, (k_x^2 + k_y^2)} \qquad (9.30)$$

is always negative, implying a phase propagation to the west (Fig. 9.4). The sign of the meridional phase speed $c_y = \omega/k_y$ is undetermined, since the wavenumber k_y may have either sign. Thus, planetary waves can propagate only northwestward, westward, or southwestward. Second, very long waves ($1/k_x$ and $1/k_y$ both much larger than R) propagate always westward and at the speed

$$c = -\beta_0 R^2, \qquad (9.31)$$

which is the largest wave speed allowed.

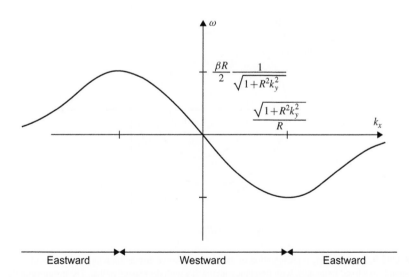

FIGURE 9.4 Dispersion relation of planetary (Rossby) waves. The frequency ω is plotted against the zonal wavenumber k_x at constant meridional wavenumber k_y. As the slope of the curve reverses, so does the direction of zonal propagation of energy.

Lines of constant frequency ω in the (k_x, k_y) wavenumber space are circles defined by

$$\left(k_x + \frac{\beta_0}{2\omega}\right)^2 + k_y^2 = \left(\frac{\beta_0^2}{4\omega^2} - \frac{1}{R^2}\right), \tag{9.32}$$

and are illustrated in Fig. 9.5. Such circles exist only if their radius is a real number, that is, if $\beta_0^2 > 4\omega^2/R^2$. This implies the existence of a maximum frequency

$$|\omega|_{\max} = \frac{\beta_0 R}{2}, \tag{9.33}$$

beyond which planetary waves do not exist.

The group velocity, at which the energy of a wave packet propagates, defined as the vector $(\partial\omega/\partial k_x, \partial\omega/\partial k_y)$, is the gradient of the function ω in the (k_x, k_y) wavenumber plane (see Appendix B). It is thus perpendicular to the circles of constant ω. A little algebra reveals that the group-velocity vector is directed inward, toward the center of the circle. Therefore, long waves (small k_x and k_y, points near the origin in Fig. 9.5) have westward group velocities, whereas energy is carried eastward by the shorter waves (larger k_x and k_y, points on the

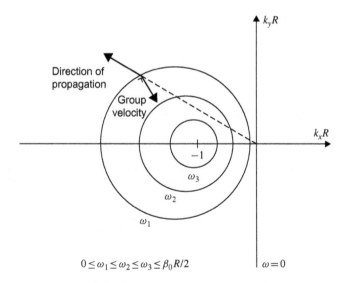

FIGURE 9.5   Geometric representation of the planetary-wave dispersion relation. Each circle corresponds to a fixed frequency, with frequency increasing with decreasing radius. The group velocity of the (k_x, k_y) wave is a vector perpendicular to the circle at point (k_x, k_y) and directed toward its center.

opposite side of the circle). This dichotomy is also apparent in Fig. 9.4, which exhibits reversals in slope ($\partial\omega/\partial k_x$ changing sign).

9.5 TOPOGRAPHIC WAVES

Just as small variations in the Coriolis parameter can turn a steady geostrophic flow into slowly moving planetary waves, so can a weak bottom irregularity. Admittedly, topographic variations can come in a great variety of sizes and shapes, but for the sake of illustrating the wave process in its simplest form, we limit ourselves here to the case of a weak and uniform bottom slope. We also return to the use of a constant Coriolis parameter. This latter choice allows us to choose convenient directions for the reference axes, and, in anticipation of an analogy with planetary waves, we align the y-axis with the direction of the topographic gradient. We thus express the depth of the fluid at rest as:

$$H = H_0 + \alpha_0 y, \tag{9.34}$$

where H_0 is a mean reference depth, and α_0 is the bottom slope, which is required to be gentle so that

$$\alpha = \frac{\alpha_0 L}{H_0} \ll 1, \tag{9.35}$$

where L is the horizontal length scale of the motion. The topographic parameter α plays a role similar to the planetary number, defined in Eq. (9.20).

The bottom slope gives rise to new terms in the continuity equation. Starting with the continuity equation (7.14) for shallow water and expressing the instantaneous fluid layer depth as (Fig. 9.6)

$$h(x, y, t) = H_0 + \alpha_0 y + \eta(x, y, t), \tag{9.36}$$

we obtain

$$\frac{\partial \eta}{\partial t} + \left(u\frac{\partial \eta}{\partial x} + v\frac{\partial \eta}{\partial y} \right) + (H_0 + \alpha_0 y)\left(\frac{\partial u}{\partial x} + \frac{\partial v}{\partial y} \right)$$
$$+ \eta\left(\frac{\partial u}{\partial x} + \frac{\partial v}{\partial y} \right) + \alpha_0 v = 0.$$

Once again, we strike out the nonlinear terms by invoking a very small Rossby number (much smaller than the temporal Rossby number) for the sake of linear dynamics. The term $\alpha_0 y$ can also be dropped next to H_0 by virtue of Eq. (9.35).

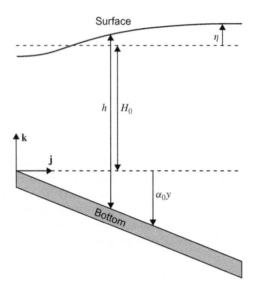

FIGURE 9.6 A layer of homogeneous fluid over a sloping bottom and the attending notation.

With the momentum equations (9.4a) and (9.4b), our present set of equations is

$$\frac{\partial u}{\partial t} - fv = -g\frac{\partial \eta}{\partial x} \tag{9.37a}$$

$$\frac{\partial v}{\partial t} + fu = -g\frac{\partial \eta}{\partial y} \tag{9.37b}$$

$$\frac{\partial \eta}{\partial t} + H_0\left(\frac{\partial u}{\partial x} + \frac{\partial v}{\partial y}\right) + \alpha_0 v = 0. \tag{9.37c}$$

In analogy with the system of equations governing planetary waves, the preceding set contains both small and large terms. The large ones (terms including f, g, and H_0) comprise the otherwise steady geostrophic dynamics, which correspond to a zero frequency. But, in the presence of the small α_0 term in the last equation, the geostrophic flow cannot remain steady, and the time-derivative terms come into play. We naturally expect them to be small and, compared with the large terms, on the order of α. In other words, the temporal Rossby number, $Ro_T = 1/\Omega T$, is expected to be comparable with α, leading to wave frequencies

$$\omega \sim \frac{1}{T} \sim \alpha\Omega \sim \alpha f \ll f$$

that are very subinertial, just as in the case of planetary waves, for which $\omega \sim \beta f_0$.

Capitalizing on the smallness of the time-derivative terms, we take in first approximation the large geostrophic terms: $u \simeq -(g/f)\partial \eta/\partial y$, $v \simeq +(g/f)\partial \eta/\partial x$.

Substitution of these expressions in the small time derivatives yields to the next degree of approximation:

$$u = -\frac{g}{f}\frac{\partial \eta}{\partial y} - \frac{g}{f^2}\frac{\partial^2 \eta}{\partial x \partial t} \tag{9.38a}$$

$$v = +\frac{g}{f}\frac{\partial \eta}{\partial x} - \frac{g}{f^2}\frac{\partial^2 \eta}{\partial y \partial t}. \tag{9.38b}$$

The relative error is only on the order of α^2. Replacement of the velocity components, u and v, by their last expressions (9.38a) and (9.38b) in the continuity equation (9.37c) provides a single equation for the surface displacement η, which to the leading order is

$$\frac{\partial \eta}{\partial t} - R^2 \frac{\partial}{\partial t}\nabla^2 \eta + \frac{\alpha_0 g}{f}\frac{\partial \eta}{\partial x} = 0. \tag{9.39}$$

(The ageostrophic component of v is dropped from the $\alpha_0 v$ term for being on the order of α^2, whereas all other terms are on the order of α.) Note the analogy with Eq. (9.26) that governs the planetary waves: It is identical, except for the substitution of $\alpha_0 g/f$ for $-\beta_0 R^2$. Here, the deformation radius is defined as

$$R = \frac{\sqrt{gH_0}}{f}, \tag{9.40}$$

that is, the closest constant to the original definition (9.12). A wave solution of the type $\cos(k_x x + k_y y - \omega t)$ immediately provides the dispersion relation:

$$\omega = \frac{\alpha_0 g}{f} \frac{k_x}{1 + R^2(k_x^2 + k_y^2)}, \tag{9.41}$$

the topographic analog of Eq. (9.27). Again, we note that if the additional ingredient, here the bottom slope α_0, had not been present, the frequency would have been nil, and the flow would have been steady and geostrophic. Because they owe their existence to the bottom slope, these waves are called *topographic waves*.

The discussion of their direction of propagation, phase speed, and maximum possible frequency follows that of planetary waves. The phase speed in the x-direction—that is, along the isobaths—is given by

$$c_x = \frac{\omega}{k_x} = \frac{\alpha_0 g}{f} \frac{1}{1 + R^2\,(k_x^2 + k_y^2)} \tag{9.42}$$

and has the sign of $\alpha_0 f$. Thus, topographic waves propagate in the northern hemisphere with the shallower side on their right. Because planetary waves propagate westward, i.e., with the north to their right, the analogy between the two kinds of waves is "shallow–north" and "deep–south". (In the southern hemisphere, topographic waves propagate with the shallower side on their left, and the analogy is "shallow–south," "deep–north.")

The phase speed of topographic waves varies with the wavenumber; they are thus dispersive. The maximum possible wave speed along the isobaths is

$$c = \frac{\alpha_0 g}{f},$$ (9.43)

which is the speed of the very long waves $\left(k_x^2 + k_y^2 \to 0\right)$. With (9.41) cast in the form

$$\left(k_x - \frac{\alpha_0 g}{2f\omega R^2}\right)^2 + k_y^2 = \left(\frac{\alpha_0^2 g^2}{4f^2 R^4 \omega^2} - \frac{1}{R^2}\right),$$ (9.44)

we note that there exists a maximum frequency:

$$|\omega|_{max} = \frac{|\alpha_0| g}{2|f| R}.$$ (9.45)

The implication is that a forcing at a frequency higher than the preceding threshold cannot generate topographic waves. The forcing then generates either a disturbance that is unable to propagate or higher frequency waves, such as inertia-gravity waves. However, such a situation is rare because, unless the bottom slope is very weak, the maximum frequency given by (9.45) approaches or exceeds the inertial frequency f, and the theory fails before (9.45) can be applied.

As an example, let us take the West Florida Shelf, which is in the eastern Gulf of Mexico. There the ocean depth increases gradually offshore to 200 m over 200 km ($\alpha_0 = 10^{-3}$) and the latitude (27°N) yields $f = 6.6 \times 10^{-5}$ s^{-1}. Using an average depth $H_0 = 100$ m, we obtain $R = 475$ km and $\omega_{max} = 1.6 \times 10^{-4}$ s^{-1}. This maximum frequency, corresponding to a minimum period of 11 min, is larger than f, violates the condition of subinertial motions and is thus meaningless. The wave theory, however, applies to waves whose frequencies are much less than the maximum value. For example, a wavelength of 150 km along the isobaths ($k_x = 4.2 \times 10^{-5}$ m^{-1}, $k_y = 0$) yields $\omega = 1.6 \times 10^{-5}$ s^{-1} (period of 4.6 days) and a wave speed of $c_x = 0.38$ m/s.

Where the topographic slope is confined between a coastal wall and a flat-bottom abyss, such as for a continental shelf, topographic waves can be trapped, not unlike the Kelvin wave. Mathematically, the solution is not periodic in the offshore, cross-isobath direction but assumes one of several possible profiles (eigenmodes). Each mode has a corresponding frequency (eigenvalue). Such waves are called *continental shelf waves*. The interested reader can find an exposition of these waves in LeBlond and Mysak (1978) and Gill (1982, pages 408–415).

9.6   ANALOGY BETWEEN PLANETARY AND TOPOGRAPHIC WAVES

We have already discussed some of the mathematical similarities between the two kinds of low-frequency waves. The objective of this section is to go to the root of the analogy and to compare the physical processes at work in both kinds of waves.

Let us turn to the quantity called potential vorticity and defined in Eq. (7.25). On the beta plane and over a sloping bottom (oriented meridionally for convenience), the expression of the potential vorticity becomes

$$q = \frac{f_0 + \beta_0 y + \partial v/\partial x - \partial u/\partial y}{H_0 + \alpha_0 y + \eta} \, . \tag{9.46}$$

Our assumptions of a small beta effect and a small Rossby number imply that the numerator is dominated by f_0, all other terms being comparatively very small. Likewise, H_0 is the leading term in the denominator because both bottom slope and surface displacements are weak. A Taylor expansion of the fraction yields

$$q = \frac{1}{H_0} \left(f_0 + \beta_0 y - \frac{\alpha_0 f_0}{H_0} y + \frac{\partial v}{\partial x} - \frac{\partial u}{\partial y} - \frac{f_0}{H_0} \eta \right). \tag{9.47}$$

In this form, it is immediately apparent that the planetary and topographic terms (β_0 and α_0 terms, respectively) play identical roles. The analogy between the coefficients β_0 and $-\alpha_0 f_0/H_0$ is identical to the one noted earlier between $-\beta_0 R^2$ of Eq. (9.20) and $\alpha_0 g/f$ of Eq. (9.35), since now $R = (gH_0)^{1/2}/f_0$. The physical significance is the following: Just as the planetary effect imposes a potential-vorticity gradient, with higher values toward the north, so the topographic effect, too, imposes a potential-vorticity gradient, with higher values toward the shallower side.

The presence of an ambient gradient of potential vorticity is what provides the *bouncing* effect necessary to the existence of the waves. Indeed, consider Fig. 9.7, where the first panel represents a north-hemispheric fluid (seen from the top) at rest in a potential-vorticity gradient and think of the fluid as consisting of bands tagged by various potential-vorticity values. The next two panels show the same fluid bands after a wavy disturbance has been applied in the presence of either the planetary or the topographic effect.

Under the planetary effect (middle panel), fluid parcels caught in crests have been displaced northward and have seen their ambient vorticity, $f_0 + \beta_0 y$, increase. To compensate and conserve their initial potential vorticity, they must develop some negative relative vorticity, that is, a clockwise spin. This is

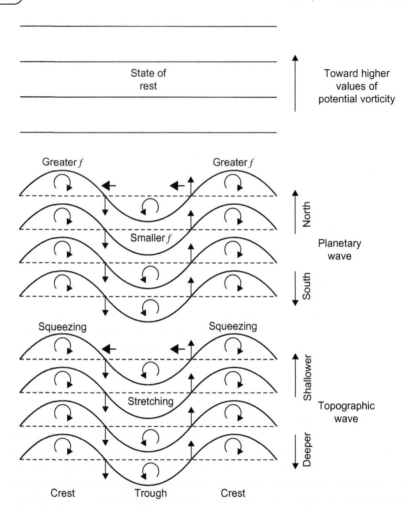

FIGURE 9.7 Comparison of the physical mechanisms that propel planetary and topographic waves. Displaced fluid parcels react to their new location by developing either clockwise or counterclockwise vorticity. Intermediate parcels are entrained by neighboring vortices, and the wave progresses forward.

indicated by curved arrows. Similarly, fluid parcels in troughs have been displaced southward, and the decrease of their ambient vorticity is met with an increase of relative vorticity, that is, a counterclockwise spin. Focus now on those intermediate parcels that have not been displaced so far. They are sandwiched between two counterrotating vortex patches, and, like an unfortunate finger caught between two gears or the newspaper zipping through the rolling press, they are entrained by the swirling motions and begin to move in the meridional direction. From left to right on the figure, the displacements are southward from crest to trough and northward from trough to crest. Southward displacements set up new troughs, whereas northward displacements generate new crests. The net effect is a westward drift of the pattern. This explains why planetary waves propagate westward.

In the third panel of Fig. 9.7, the preceding exercise is repeated in the case of an ambient potential-vorticity gradient due to a topographic slope. In a crest, a fluid parcel is moved into a shallower environment. The vertical squeezing causes a widening of the parcel's horizontal cross-section (see Section 7.4), which in turn is accompanied by a decrease in relative vorticity. Similarly, parcels in troughs undergo vertical stretching, a lateral narrowing, and an increase in relative vorticity. From there on, the story is identical to that of planetary waves. The net effect is a propagation of the trough-crest pattern with the shallow side on the right.

The analogy between the planetary and topographic effects has been found to be extremely useful in the design of laboratory experiments. A sloping bottom in rotating tanks can substitute for the beta effect, which would otherwise be impossible to model experimentally. Caution must be exercised, however, for the substitution is acceptable so long as the analogy holds. The following three conditions must be met: absence of stratification, gentle slope, and slow motion. If stratification is present, the sloping bottom affects preferentially the fluid motions near the bottom, whereas the true beta effect operates evenly at all levels. And, if the slope is not gentle, and the motions are not weak, the expression of potential vorticity cannot be linearized as in Eq. (9.47), and the analogy is invalidated.

9.7   ARAKAWA'S GRIDS

The preceding developments had for aim to explain the basic physical mechanisms responsible for shallow-water wave propagation, by simplifying the governing equations down to their simplest, yet meaningful ingredients. Numerical models help us do better in cases where such simplifications are questionable, or when it is necessary to calculate wave motions in more realistic geometries. For the sake of clarity, broadly applicable numerical techniques will be illustrated on simplified cases. The simplest situation arises with inertia-gravity waves, for which the core mechanisms are rotation and gravity (see Section 9.3). In a one-dimensional domain of uniform fluid depth H, the linearized governing equations are

$$\frac{\partial \eta}{\partial t} + H \frac{\partial u}{\partial x} = 0 \tag{9.48a}$$

$$\frac{\partial u}{\partial t} - fv = -g \frac{\partial \eta}{\partial x} \tag{9.48b}$$

$$\frac{\partial v}{\partial t} + fu = 0. \tag{9.48c}$$

A straightforward second-order central finite differencing in space yields the following:

$$\frac{d\tilde{\eta}_i}{dt} + H \frac{\tilde{u}_{i+1} - \tilde{u}_{i-1}}{2\Delta x} = 0 \qquad (9.49a)$$

$$\frac{d\tilde{u}_i}{dt} - f\tilde{v}_i = -g \frac{\tilde{\eta}_{i+1} - \tilde{\eta}_{i-1}}{2\Delta x} \qquad (9.49b)$$

$$\frac{d\tilde{v}_i}{dt} + f\tilde{u}_i = 0. \qquad (9.49c)$$

When analyzing this second-order method (upper part of Fig. 9.8), we observe that the effective grid size is $2\Delta x$ in the sense that all derivatives are taken over this distance. This is somehow unsatisfactory because the real grid size, i.e., the distance between adjacent grid nodes is only Δx. To improve the situation, we notice that the spatial derivatives of u are needed to calculate η, while the calculation of u requires the gradients of η. The most natural place to calculate a derivative of η is then at a point midway between two grid points of η, since the gradient approximation there is of second order while using a step of only Δx, and the most natural position to calculate the velocity u is therefore at mid-distance between η-grid nodes. Likewise, the most natural place to calculate the time evolution of η is at mid-distance between u nodes. It appears therefore that locating grid nodes for η and u in an interlaced fashion allows a second-order space differencing of *both* fields over a distance Δx (lower part of Fig. 9.8).

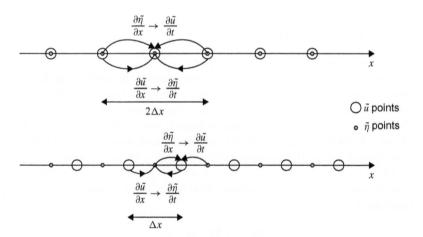

FIGURE 9.8 For variables $\tilde{\eta}$ and $\tilde{u}$ defined at the same grid points (upper panel), the discretization of inertia-gravity waves demands that approximations of spatial derivatives be made over distances $2\Delta x$, even if the underlying grid has a resolution of Δx. However, if variables $\tilde{\eta}$ and $\tilde{u}$ are defined on two different grids (lower panel), shifted one with respect to the other by $\Delta x/2$, the spatial derivatives can conveniently be discretized over the grid spacing Δx. The origin of each arrow indicates which variable influences the time evolution of the node where the arrow ends.

Formally, the discretization on such a *staggered grid* takes the form[1]:

$$\frac{d\tilde{\eta}_i}{dt} + H\frac{\tilde{u}_{i+1/2} - \tilde{u}_{i-1/2}}{\Delta x} = 0 \tag{9.50a}$$

$$\frac{d\tilde{u}_{i+1/2}}{dt} - f\tilde{v}_{i+1/2} = -g\frac{\tilde{\eta}_{i+1} - \tilde{\eta}_i}{\Delta x} \tag{9.50b}$$

$$\frac{d\tilde{v}_{i+1/2}}{dt} + f\tilde{u}_{i+1/2} = 0. \tag{9.50c}$$

Thus, spatial differencing can be performed over a distance Δx instead of $2\Delta x$. Since the scheme is centered in space, it is of second order, and its discretization error is reduced by a factor 4 without any additional calculation.[2] This advantage of a staggered grid over the elementary collocated grid is a prime example of optimization of numerical methods at fixed cost.

But, as we will now show, the performance gain is not the sole advantage of the staggered-grid approach. In the case of a negligible Coriolis force ($f \to 0$), elimination of velocity from Eq. (9.48) leads to a single wave equation for η:

$$\frac{\partial^2 \eta}{\partial t^2} = c^2 \frac{\partial^2 \eta}{\partial x^2}, \tag{9.51}$$

where $c^2 = gH$. This equation is the archetype of a hyperbolic equation, which possesses a general solution of the form

$$\eta = E_1(x+ct) + E_2(x-ct), \tag{9.52}$$

where the functions E_1 and E_2 are set by initial and boundary conditions. The general solution is therefore the combination of two signals, travelling in opposite directions at speeds $\pm c$ (Fig. 9.9). The lines of constant $x+ct$ and $x-ct$ define the characteristics along which the solution is propagated. For $t=0$, it is then readily seen that two initial conditions are needed, one for η and the other on its time derivative, i.e., the velocity field, before one can determine the two functions E_1 and E_2 at the conclusion of the first step. Later on, when characteristics no longer originate from the initial conditions but have their root in the boundaries, the solution becomes influenced first by the most proximate boundary condition and ultimately by both. If the boundary is impermeable, the condition is $u=0$, which can be translated into a zero-gradient condition on η. For discretization (9.50), the necessary numerical boundary conditions are consistent with the analytical conditions, whereas the nonstaggered version (9.53) needs additional conditions to reach the near boundary points. We have already learned (Section 4.7) how to deal with artificial conditions.

---

[1] We arbitrarily choose to place $\tilde{\eta}$ at integer grid indices and $\tilde{u}$ at half indices. We could have chosen the reverse.

[2] Both approaches use the same number of grid points to cover a given domain, and both schemes demand the same number of operations.

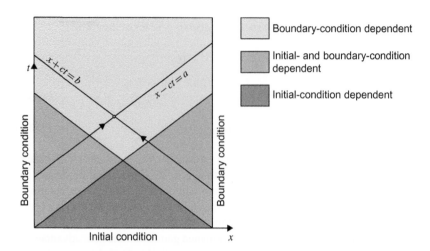

FIGURE 9.9 Characteristics $x + ct$ and $x - ct$ of wave equation (9.51). Information propagates along these lines from two initial conditions and one boundary condition on each side to set in a unique way the value of the solution at any point (x, t) of the domain.

We now turn our attention to the remaining problem of the nonstaggered grid, which is the appearance of spurious, stationary, and decoupled modes within the domain. To illustrate the issue, we use a standard leapfrog time discretization with zero Coriolis force so that the nonstaggered discretization becomes

$$\frac{\tilde{\eta}_i^{n+1} - \tilde{\eta}_i^{n-1}}{2\Delta t} = -H \frac{\tilde{u}_{i+1}^n - \tilde{u}_{i-1}^n}{2\Delta x} \tag{9.53a}$$

$$\frac{\tilde{u}_i^{n+1} - \tilde{u}_i^{n-1}}{2\Delta t} = -g \frac{\tilde{\eta}_{i+1}^n - \tilde{\eta}_{i-1}^n}{2\Delta x}. \tag{9.53b}$$

For the grid-staggered version, we can also introduce a form of time staggering by using a forward-backward approach in time:

$$\frac{\tilde{\eta}_i^{n+1} - \tilde{\eta}_i^n}{\Delta t} = -H \frac{\tilde{u}_{i+1/2}^n - \tilde{u}_{i-1/2}^n}{\Delta x} \tag{9.54a}$$

$$\frac{\tilde{u}_{i+1/2}^{n+1} - \tilde{u}_{i+1/2}^n}{\Delta t} = -g \frac{\tilde{\eta}_{i+1}^{n+1} - \tilde{\eta}_i^{n+1}}{\Delta x}. \tag{9.54b}$$

Although it may first appear that we are dealing with an implicit scheme because of the presence of $\tilde{\eta}^{n+1}$ on the right of the second equation, it is noted that this quantity has just been calculated when marching the preceding equation one step forward in time. The scheme is thus explicit in the sense that we solve the first equation to get $\tilde{\eta}^{n+1}$ everywhere in the domain, and then use it immediately in the second equation to calculate $\tilde{u}^{n+1}$ without having to invert any matrix.

As for all forms of leapfrogging and staggering, we should be concerned by spurious modes. These can be sought here rather simply by eliminating the discrete field $\tilde{u}$ from each set of equations. This elimination can be performed by taking a finite time difference of the first equation and a finite space difference of the second equation.[3] This is the direct analog of the mathematical differentiation used in eliminating the velocity between the two governing equations (9.48) to obtain (9.51). For the nonstaggered and staggered grids, we obtain

$$\tilde{\eta}_i^{n+2} - 2\tilde{\eta}_i^n + \tilde{\eta}_i^{n-2} = \frac{c^2 \Delta t^2}{\Delta x^2} \left(\tilde{\eta}_{i+2}^n - 2\tilde{\eta}_i^n + \tilde{\eta}_{i-2}^n \right) \qquad (9.55)$$

$$\tilde{\eta}_i^{n+1} - 2\tilde{\eta}_i^n + \tilde{\eta}_i^{n-1} = \frac{c^2 \Delta t^2}{\Delta x^2} \left(\tilde{\eta}_{i+1}^n - 2\tilde{\eta}_i^n + \tilde{\eta}_{i-1}^n \right). \qquad (9.56)$$

These equations are straightforward second-order discretizations of the wave equation (9.51), the first one with spatial and temporal steps twice as large as for the second one. The CFL criterion is $|c|\Delta t/\Delta x \leq 1$ in each case, since the propagation speeds of the hyperbolic equation are $\pm c$, and the corresponding characteristics must lie in the numerical domain of dependence.

The discretization (9.55) shows that the nonstaggered grid is prone to decoupled modes. Indeed, for even values of n and i, all grid indexes involved are even, and the evolution is completely independent of that on points with odd values of n or i, which are nonetheless proximate in both time and space. In fact, there are four different solutions evolving independently, with their only link being through the initial and boundary conditions (left panel of Fig. 9.10). Although theoretically acceptable, such decoupled modes typically increase their "distance" from one another in the course of the simulation and induce

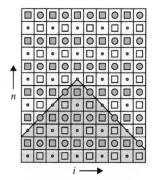
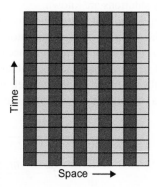

FIGURE 9.10 Four different solutions, each identified by a different symbol, evolve independently on the nonstaggered grid. The numerical domain of dependence is shown as the shaded region (left panel). A spurious stationary mode alternating between two constants (right panel) is incompatible with the original governing equations.

---

[3]For the nonstaggered version, we make the following formal elimination: Δx [(9.53a)$^{n+1}$ - (9.53a)$^{n-1}$] - $\Delta t H$[(9.53b)$_{i+1}$ - (9.53b)$_{i-1}$]

undesirable space-time oscillations in the solution. Occasionally, this can lead to stationary solutions that are simply unphysical (Fig. 9.10), such as a solution with zero velocity and $\tilde{\eta}$ alternating in space between two different constants. Solutions of this type are clearly spurious. In contrast, the only stationary solution produced by the staggered equation (9.56) is the physical one. This is a desirable property.

More generally, the spurious stationary solutions of a space discretization can be analyzed in terms of the state variable vector $\mathbf{x}$, space discretization operator $\mathbf{D}$, and the semi-discrete equation

$$\frac{d\mathbf{x}}{dt} + \mathbf{D}\mathbf{x} = 0 \tag{9.57}$$

so that spurious stationary modes can be found among the nonzero solutions of

$$\mathbf{D}\mathbf{x} = 0. \tag{9.58}$$

In the jargon of matrix calculation (linear algebra), spurious modes lie within the *null-space* of matrix $\mathbf{D}$. In the case of the wave equation, the solution depicted in the right panel of Fig. 9.10 is certainly not a physically valid solution but satisfies Eq. (9.58).

All nonzero stationary solutions (members of the null-space), however, do not need to be spurious, and a physically admissible nonzero stationary solution is possible in the presence of the Coriolis terms, namely the geostrophic equilibrium. In that case, the discretized geostrophic equilibrium solution is also part of the null-space (9.58). It is therefore worthwhile sometimes to analyze the null-space of discretization operators for which the corresponding physical stationary solutions are known.

Having found that staggering has advantages in one dimension, we can now explore the situation in two dimensions but immediately realize that there is no single way to generalize the approach. Indeed, we have three state variables, u, v, and η, which can each be calculated on a different grid. The collocated version, the so-called A-grid model, is readily defined, and discretization[4] of Eqs. (9.4a) through (9.5) with uniform fluid thickness leads to:

$$\frac{d\tilde{\eta}}{dt} = -H \frac{\tilde{u}_{i+1} - \tilde{u}_{i-1}}{2\Delta x} - H \frac{\tilde{v}_{j+1} - \tilde{v}_{j-1}}{2\Delta y} \tag{9.59a}$$

$$\frac{d\tilde{u}}{dt} = +f\tilde{v} - g \frac{\tilde{\eta}_{i+1} - \tilde{\eta}_{i-1}}{2\Delta x} \tag{9.59b}$$

$$\frac{d\tilde{v}}{dt} = -f\tilde{u} - g \frac{\tilde{\eta}_{j+1} - \tilde{\eta}_{j-1}}{2\Delta y}. \tag{9.59c}$$

---

[4]As before, we only write indices that differ from i and j.

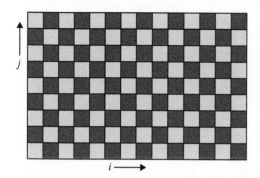

FIGURE 9.11 A spurious stationary $\tilde{\eta}$ mode alternating between two constants (depicted by two different gray levels) on the A-grid. This mode is called for obvious reasons the *checkerboard mode*.

Clearly, a spurious stationary solution exists, again with zero velocity ($\tilde{u} = \tilde{v} = 0$) and $\tilde{\eta}$ alternating between two constants on the spatial grid (Fig. 9.11).

Starting from the collocated grid, we can distribute the variables with respect to one another in different ways and create various staggered grids. These are named Arakawa's grids in honor of Akio Arakawa[5] and bear the letters A, B, C, D, and E depending on where the state variables are located across the mesh (Fig. 9.12; Arakawa & Lamb, 1977). For the linear system of equations considered here, it can be shown (e.g., Mesinger & Arakawa, 1976) that the E-Grid is a rotated B-grid so that we do not need to analyze it further.

A two-dimensional staggered grid we already encountered is the so-called C-grid (bottom left of Fig. 9.12). For advection [recall Eq. (6.58)] and the rigid-lid pressure formulation [recall Eq. (7.38)], we tacitly assumed that the velocity u was being calculated halfway between pressure nodes $(i+1,j)$ and (i,j) and v halfway between nodes (i,j) and $(i,j+1)$. In the present wave problem, this approach yields a straightforward second-order discretization of both divergence and pressure gradient terms

$$\left(\frac{\partial u}{\partial x} + \frac{\partial v}{\partial y}\right)_{i,j} \simeq \frac{\tilde{u}_{i+1/2} - \tilde{u}_{i-1/2}}{\Delta x} + \frac{\tilde{v}_{j+1/2} - \tilde{v}_{j-1/2}}{\Delta y} + \mathcal{O}(\Delta x^2, \Delta y^2) \qquad (9.60)$$

$$\left.\frac{\partial \tilde{\eta}}{\partial x}\right|_{i+1/2,j} \simeq \frac{\tilde{\eta}_{i+1} - \tilde{\eta}}{\Delta x} + \mathcal{O}(\Delta x^2) \qquad (9.61a)$$

$$\left.\frac{\partial \tilde{\eta}}{\partial y}\right|_{i,j+1/2} \simeq \frac{\tilde{\eta}_{j+1} - \tilde{\eta}}{\Delta y} + \mathcal{O}(\Delta y^2) \qquad (9.61b)$$

exactly as in the advection and surface pressure problems (Fig. 9.13). But if we proceed with the discretization of the Coriolis term, a problem arises for the C-grid because the velocity components are not defined at the same points. The

---

[5] See biography at the end of this Chapter.

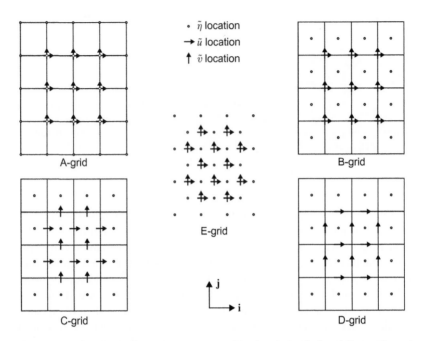

FIGURE 9.12 The five Arakawa grids. On the A-grid, the variables $\tilde{\eta}$, $\tilde{u}$, and $\tilde{v}$ are collocated, and staggered on other grids, called B-, C-, D-and E-grids. Note that the E-grid (center) has a higher grid-point density than the other grids for the same distance between adjacent nodes.

integration of the du/dt equation at the u grid node $(i+1/2,j)$ requires knowledge of the velocity v, which is only available at node $(i,j\pm1/2)$. Therefore, an interpolation is necessary. The simplest scheme takes an average of surrounding values:

$$v|_{i+1/2,j} \simeq \frac{\tilde{v}_{j+1/2} + \tilde{v}_{i+1,j+1/2} + \tilde{v}_{j-1/2} + \tilde{v}_{i+1,j-1/2}}{4} \; , \qquad (9.62)$$

where the right-hand side can now be calculated from the available values of $\tilde{v}$. Similar averaging to estimate variables at locations where they are not defined can be used to discretize the equations on the other staggered grids. For example, the B-grid is a grid where η is defined on integer grid indices whereas velocity components are defined at corner points $(i\pm1/2,j\pm1/2)$. For this grid, the Coriolis term does not require any averaging, since both velocity components are collocated, but the grid arrangement requires the derivative of η in the x-direction at location $(i+1/2,j+1/2)$. We approximate such a term by the appropriate average

$$\frac{\partial \eta}{\partial x}\bigg|_{i+1/2,j+1/2} \simeq \frac{\frac{\tilde{\eta}_{i+1,j+1} + \tilde{\eta}_{i+1}}{2} - \frac{\tilde{\eta}_{j+1} + \tilde{\eta}}{2}}{\Delta x} \; , \qquad (9.63)$$

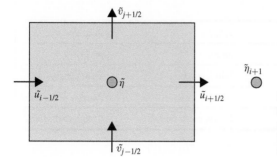

FIGURE 9.13 Discretization on the C-grid. The divergence operator is discretized most naturally by (9.60) while pressure gradients are calculated with (9.61).

and similarly for $\partial u / \partial x$, which is needed at (i,j). The full spatial discretization on each grid can be achieved in similar manner, and the derivation is left as an exercise (Numerical Exercise 9.2).

We can investigate the wave propagation properties on the various grids by a Fourier analysis, for which we take

$$\begin{pmatrix} \tilde{\eta} \\ \tilde{u} \\ \tilde{v} \end{pmatrix} = \Re \begin{pmatrix} A \\ U \\ V \end{pmatrix} e^{i\left(ik_x \Delta x + jk_y \Delta y - \tilde{\omega} t\right)}. \tag{9.64}$$

Insertion of this type of solution in the various finite-difference schemes and division by the common exponential factor provide the following equations

$$-i\tilde{\omega}U - f\alpha V = -ig\alpha_x k_x A \tag{9.65a}$$

$$-i\tilde{\omega}V + f\alpha U = -ig\alpha_y k_y A \tag{9.65b}$$

$$-i\tilde{\omega}A + H\left(i\alpha_x k_x U + i\alpha_y k_y V\right) = 0, \tag{9.65c}$$

where the coefficients α, α_x, and α_y vary with the type of grid and are given in Table 9.1.

As for the physical solution, a nonzero solution is only possible when the determinant of the system vanishes, and this provides the dispersion relation of the discretized wave physics:

$$\tilde{\omega}\left[\tilde{\omega}^2 - \alpha^2 f^2 - gH\left(\alpha_x^2 k_x^2 + \alpha_y^2 k_y^2\right)\right] = 0, \tag{9.66}$$

which is the discrete analog of (9.16).

For small wavenumber values ($k_x \Delta x \ll 1$ and $k_y \Delta y \ll 1$), i.e., long, well-resolved waves, we recover the physical dispersion relation because α, α_x, and α_y all tend towards unity. For shorter waves, the numerical dispersion relation

TABLE 9.1 Definition of the Parameters Involved in the Discrete Dispersion Relation for A-, B-, C-and D-grids with $2\theta_x = k_x \Delta x$ and $2\theta_y = k_y \Delta y$.

Grid	α	$\alpha_x k_x \Delta x$	$\alpha_y k_y \Delta y$
A	1	$\sin 2\theta_x$	$\sin 2\theta_y$
B	1	$2\sin\theta_x \cos\theta_y$	$2\sin\theta_y \cos\theta_x$
C	$\cos\theta_x \cos\theta_y$	$2\sin\theta_x$	$2\sin\theta_y$
D	$\cos\theta_x \cos\theta_y$	$2\cos\theta_x \cos\theta_y \sin\theta_x$	$2\cos\theta_x \cos\theta_y \sin\theta_y$

can be analyzed in detail through the error estimate

$$\frac{\omega^2 - \tilde{\omega}^2}{\omega^2} = \frac{(1-\alpha^2)+R^2\left[(1-\alpha_x^2)k_x^2+(1-\alpha_y^2)k_y^2\right]}{1+R^2\left(k_x^2+k_y^2\right)} \geq 0. \qquad (9.67)$$

Except for the simple statement $\tilde{\omega}^2 \leq \omega^2$, the analysis of the error is rather complex because it involves five length scales[6]: R, $1/k_x$, $1/k_y$, Δx, and Δy. For simplicity, we take $\Delta x \sim \Delta y$ and $k_x \sim k_y$ to reduce the problem. We then define the length scale $L \sim 1/k_x \sim 1/k_y$ of the wave under consideration. In this case $\alpha_x \sim \alpha_y$, and we can distinguish two situations:

$$\text{Shorter waves:} \quad L \lesssim R, \quad \omega^2 \sim \frac{gH}{L^2} \qquad (9.68)$$

$$\text{Longer waves:} \quad L \gtrsim R, \quad \omega^2 \sim f^2. \qquad (9.69)$$

The shorter waves are dominated by gravity, and the relative error on ω^2 behaves as

$$\frac{\omega^2 - \tilde{\omega}^2}{gH/L^2} \sim (1-\alpha_x^2). \qquad (9.70)$$

If the wave is well resolved ($\Delta x \ll L$), the error tends toward zero for all four grids because $\alpha_x \to 1$. For the barely resolved waves ($\Delta x \sim L$), the errors are largest for the discretizations in which α_x and α_y depart most from unity. In this sense, the A-, B-, and D-grids have larger errors than the C-grid (see Table 9.1).

---

[6]Note how the discretization has added two length scales, Δx and Δy, to the discussion.

The longer waves are dominated by rotation, and the relative error on ω^2 behaves as

$$\frac{\omega^2 - \tilde{\omega}^2}{f^2} \sim (1 - \alpha^2).$$

(9.71)

Again, α should remain close to unity for all wavenumbers so that the B-grid outperforms both the C- and D-grids. For details on the errors, an exploration with abcdgrid.m in parameter space provides relative error fields as those depicted in Fig. 9.14 for various resolution levels $R/\Delta x$, etc. Errors can be further investigated through the analysis of the group-velocity behavior (Numerical Exercise 9.5) and in the context of generalized dynamics, including planetary waves, with a clear distinction between zonal and meridional wave behaviors (Dukowicz, 1995; Haidvogel & Beckmann, 1999).

Because the A-grid suffers from spurious modes, and the D-grid is always penalized in terms of accuracy, the B-and C-grids are the most interesting ones among the four types. Since wavelengths up to $\Delta x \sim L$ are to be resolved in a

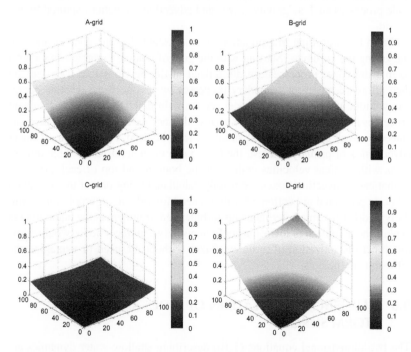

FIGURE 9.14   For medium resolution compared with the deformation radius ($R/\Delta x = R/\Delta y = 2$), the frequency error (9.67) is depicted as a function of $k_x \Delta x$ and $k_y \Delta y$. Waves with wavenumber higher than $k_x \Delta x = \pi/2$ are not shown, and the x- and y-axes are labeled in percents of $\pi/2$. The D-grid clearly exhibits the worst behavior. The B-grid keeps the error low for well-resolved waves, whereas the C-grid creates lower errors for shorter waves. A color version can be found online at http://booksite.academicpress.com/9780120887590/

significant way, the C-grid is the better choice as long as $\Delta x \ll R$, whereas for $R \ll \Delta x$ the B-grid should be preferred on the ground that the error in Eqs. (9.70) and (9.71) is less. This confirms more detailed error analyses of the semidiscrete equations on staggered grids (e.g., Haidvogel & Beckman, 1999; Mesinger & Arakawa, 1976), although additional time discretization or boundary condition implementation can introduce stability problems (e.g., Beckers, 1999; Beckers & Deleersnijder, 1993). Also, time discretization further complicates the error analysis and may sometimes inverse the error behavior (Beckers, 2002). Nevertheless, the choice of the B-grid for larger grid spacing and the C-grid for finer grid spacing is justified by the fact that for large grid spacing we only capture large-scale movements, which are nearly geostrophic. Since the Coriolis force is dominant in this case, its discretization is crucial. Because the B-grid does not require a spatial average of the velocity components, contrary to the C-grid, its use should be advantageous. The pressure gradient, which is the other dominant force, could arguably be better represented on the C-grid. If the grid is very fine, averaging the large-scale geostrophic equilibrium over four closely spaced nodes does not deteriorate the geostrophic solution, whereas smaller-scale processes such as gravity waves and advection are better captured by the C-grid.

From the preceding interpretation, we can establish some general rules for staggering the variables. Starting with the goal of placing variables on the grid so that dominant processes are discretized in the best possible way, we can then afford to represent secondary processes by less accurate discrete operators without affecting overall model accuracy. In practice, however, dominant processes may change in time and space so that no single approach can be guaranteed to work uniformly, but it should at least be tried. For example, if tracer advection is of primary interest, the C-grid can be generalized to three dimensions with vertical velocities defined at the bottom and top of each grid cell. In that case, advection fluxes are readily calculated using one of the advection schemes presented in Section 6.6 without need for velocity interpolations. Similarly, if diffusion in a heterogeneous turbulent environment is the main process at play, the definition of diffusion coefficients between tracer nodes would allow the direct discretization of turbulent fluxes without the need of averaging the diffusion coefficients.

9.8 NUMERICAL SIMULATION OF TIDES AND STORM SURGES

The two-dimensional equations (7.16) describing shallow-water dynamics are the basic equations from which storm-surge models of vertically mixed coastal seas have been developed. The prediction of rising sea level (surge) along a coast depends on remotely generated waves that propagate from the stormy area toward the shore. Because shallow-water equations describe well the propagation of such waves, their prediction is indeed feasible, although a few additional processes must be taken into account. Among these other processes are the surface wind stress, through which waves are generated, and bottom friction, which causes attenuation during travel. To include these last two stresses, we can start from the observation that the shallow-water equations assume that the flow is independent of the vertical coordinate. If this is not the case, we can at least try to predict the evolution of the depth-averaged velocity,

$$\bar{u} = \frac{1}{h} \int_b^{b+h} u \, dz \quad \bar{v} = \frac{1}{h} \int_b^{b+h} v \, dz \tag{9.72}$$

where $z = b$ is the bottom level and h the water depth. We can derive a governing equation for $\bar{u}$ by integrating vertically the three-dimensional governing equations including the x-momentum equation:

$$\frac{\partial u}{\partial t} = \frac{\partial}{\partial z}\left(\nu_E \frac{\partial u}{\partial z}\right) + F(u), \tag{9.73}$$

where the term $F(u)$ gathers all terms other than the time derivative and vertical diffusion. We can then integrate vertically to obtain

$$\frac{1}{h} \int_b^{b+h} \frac{\partial u}{\partial t} \, dz = \frac{\tau^x}{\rho_0 h} - \frac{\tau_b^x}{\rho_0 h} + \overline{F(u)}, \tag{9.74}$$

where boundary conditions similar to (4.34) have been used for the surface wind stress τ and bottom stress τ_b, respectively. Physically, these stresses appear here as body forces applied to the layer h of fluid moving as a slab with the depth-averaged velocity (Fig. 9.15).

Two difficulties arise, however, during the integration. The first is that the elevation of the surface is time dependent and does not allow a simple permu-

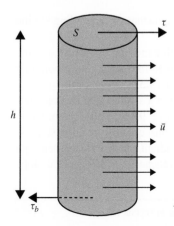

FIGURE 9.15 For a fluid column of volume hS and moving with the average velocity $\bar{u}$, Newton's second law in the absence of lateral friction and pressure force implicates the forces associated with the surface stress τ and bottom friction: $\rho_0 h S \, d\bar{u}/dt = (\tau - \tau_b)S$.

tation of the integration with the time derivative in the left-hand side of (9.74).[7] The second and more fundamental difficulty is due to the nonlinearities of the equations, which prevent us from equating the average of $F(u)$ with $F(\bar{u})$, i.e.,

$$\overline{F(u)} \neq F(\bar{u}) \tag{9.75}$$

so that we cannot express the right-hand side of (9.74) as a function of only the average velocity. The integrated equation then requires some form of parameterization. Hence, shallow-water models include additional parameterization of the horizontal diffusion type.

For simplicity, the governing equations are written as if depth averaging had not taken place, and the overbar ($\bar{\ }$) operator is ignored. The outcome is that it is sufficient to add the $\tau/(\rho_0 h)$ and $-\tau_b/(\rho_0 h)$ terms to the right-hand side of the two-dimensional momentum equations (7.9a) and (7.9b), and then to include a parameterization of the nonlinear effects. The wind stress τ appears as an externally imposed source term in the equations, whereas the bottom stress is depending on the flow itself, i.e., $\tau_b = \tau_b(u, v)$. A difficulty arises here because the bottom stress depends on the velocity profile near the bottom, whereas the governing equations provide only the vertical average of the velocity. A parameterization is needed here, too. The simplest version is linear bottom friction, in which the frictional term is made linear in, and opposite to, velocity:

$$\tau_b^x = -r\rho_0 u, \quad \tau_b^y = -r\rho_0 v, \tag{9.76}$$

where r is a coefficient with dimensions of velocity (LT^{-1}). The linear formulation is particularly advantageous in analytical studies or with spectral methods. They fail, however, to take into account the turbulent nature of the bottom boundary layer, with stress better expressed as a quadratic function of velocity (see Chapter 14):

$$\tau_b^x = -\rho_0 C_d \sqrt{u^2 + v^2}\, u, \quad \tau_b^y = -\rho_0 C_d \sqrt{u^2 + v^2}\, v, \tag{9.77}$$

with a dimensionless *drag coefficient* C_d either constant or depending on the flow itself.

Finally, the direct driving force associated with a moving disturbance of the atmospheric pressure $p_{atm}(x, y, t)$ can be easily taken into account by including it in the pressure boundary condition at the surface (4.32), $p = \rho_0 g \eta + p_{atm}$.

We can now estimate the wind-induced surge in a shallow sea by considering how the storm piles up water near the coast (Fig. 9.16). This accumulation of water creates a surface elevation (surge) and, consequently, an adverse pressure gradient. Eventually, this adverse pressure gradient can grow strong enough to cancel the wind stress. When this balance is reached, the sea surface slope

---

[7]This problem can be overcome by using Leibniz rule as done in Section 15.6.

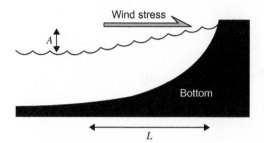

FIGURE 9.16   The piling up of water by a storm near a coast an adverse pressure-gradient force, and an equilibrium can exist if it cancels the wind-stress force.

caused by the wind stress is governed by

$$\frac{\partial \eta}{\partial x} \simeq \frac{\tau}{\rho_0 g h} \,. \tag{9.78}$$

This relation provides an estimate of the storm-surge amplitude A as a function of the distance L over which the wind blows

$$A \simeq \frac{L\tau}{\rho_0 g h} \,. \tag{9.79}$$

Note that the shallower the water, the stronger the effect. In other words, storm surges intensify near the coast where the water is shallower.

Storm surges can become dramatic when superimposed to the tide, and it is therefore important to know how to calculate tidal elevations, too. Tides are forced gravity waves caused by the gravitational attraction of the moon and sun. The following development is also valid for the atmosphere, but the velocities associated with atmospheric tides are much smaller than the wind speed due to atmospheric disturbances. Tides, therefore, are generally negligible in the atmosphere, whereas tidal currents in the ocean can be an order of magnitude larger than other currents.

To quantify the net effect of the gravitational acceleration of the moon and sun, we have to realize that the whole system is moving. Therefore, Newton's law cannot simply be written with respect to axes fixed at the earth center as we did in Chapter 2. Instead, using Newton's law in absolute axes $\mathbf{I}$, $\mathbf{J}$, and $\mathbf{K}$ of the solar system, we can calculate the absolute acceleration $\mathbf{A}$ of the fluid parcel and of the earth $\mathbf{A_e}$ under the moon's attraction[8]:

$$\rho \mathbf{A} = \rho \boldsymbol{\gamma} + \rho \mathbf{f} \tag{9.80}$$

$$M_e \mathbf{A_e} = M_e \boldsymbol{\gamma}_e. \tag{9.81}$$

---

[8]The sun's influence can be studied in an analogous way.

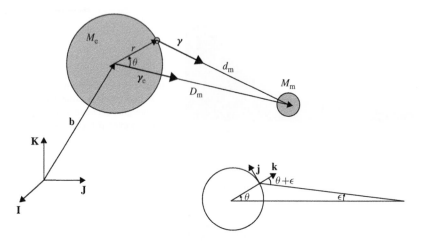

FIGURE 9.17　The moon acts simultaneously on a fluid parcel lying on the earth's surface as well as on the entire earth. The tidal force results from the mutual attraction between earth and moon and from the local gravitational attraction on earth.

We regrouped under the force $\rho\mathbf{f}$ all forces acting on the fluid parcel other than the moon's attraction. The gravitational forces[9] $\rho\boldsymbol{\gamma}$ and $M_e\boldsymbol{\gamma}_e$ involve the gravitational constant $G = 6.67 \times 10^{-11}$ N m^2/kg^2, the earth's mass $M_e = 5.9736 \times 10^{24}$ kg, the moon's mass $M_m = 7.349 \times 10^{22}$ kg, the distance $D_m \sim 385000$ km between earth and moon, and the actual distance d_m of the point under consideration to the moon (Fig. 9.17). The two gravitational accelerations are directed towards the center of the moon and of magnitude

$$\gamma = \frac{GM_m}{d_m^2}, \qquad \gamma_e = \frac{GM_m}{D_m^2}. \tag{9.82}$$

We are not so much interested in the movement of the earth *per se*, but we need its acceleration to subtract it from the fluid parcel's absolute acceleration, which is $\mathbf{A} = \mathbf{A}_e + d^2\mathbf{r}/dt^2$. The fluid's parcel's relative acceleration with respect to the earth is thus

$$\frac{d^2\mathbf{r}}{dt^2} = \mathbf{f} + (\boldsymbol{\gamma} - \boldsymbol{\gamma}_e). \tag{9.83}$$

Without the astronomical force, the equation would have been $d^2\mathbf{r}/dt^2 = \mathbf{f}$, and so we note that its effect is the addition of a so-called *tidal force* (per unit volume):

$$\rho\mathbf{f}_t = \rho(\boldsymbol{\gamma} - \boldsymbol{\gamma}_e). \tag{9.84}$$

---

[9] It should be clear that for a fluid parcel, forces are per unit volume, whereas for the earth, we speak about full forces.

We notice that this force is the difference between two almost identical forces. Its locally vertical component is as follows:

$$f_\uparrow = \gamma \cos(\theta + \epsilon) - \gamma_e \cos\theta. \qquad (9.85)$$

The expression can be simplified because the angle ϵ is extremely small (Fig. 9.17, diagram in lower right). By expanding $\cos(\theta + \epsilon)$ and using

$$\cos\epsilon \simeq 1, \qquad D_m \sin\epsilon \simeq r\sin\theta, \qquad (9.86)$$

we obtain

$$f_\uparrow = \frac{GM_m}{D_m^2}\left[\left(\frac{D_m^2}{d_m^2} - 1\right)\cos\theta - \frac{D_m^2}{d_m^2}\frac{r\sin\theta}{D_m}\sin\theta\right]. \qquad (9.87)$$

The use of d_m in the formulation of the tidal forcing is not very practical. (Can you tell at any moment the precise distance of your position with respect to the moon?) So, we use the identity $r\cos\theta + d_m\cos\epsilon = D_m$ and the smallness of ϵ to obtain

$$d_m \simeq D_m\left(1 - \frac{r}{D_m}\cos\theta\right). \qquad (9.88)$$

For the same reason that ϵ is small, the ratio r/D_m is also considered small,[10] and we drop higher-order terms in r/D_m:

$$\frac{D_m^2}{d_m^2} \simeq 1 + 2\frac{r}{D_m}\cos\theta, \qquad (9.89)$$

so that the vertical component of the tidal force can be reduced to

$$f_\uparrow \simeq \frac{GM_m}{D_m^3}r\left(3\cos^2\theta - 1\right). \qquad (9.90)$$

To compare its magnitude to $g = GM_e/r^2 = 9.81$ m/s², the gravitational acceleration of the earth on its surface, we form the ratio $\delta = f_\uparrow/g$ and find it to be on the order of

$$\delta \sim \frac{r^3 M_m}{D_m^3 M_e} \sim \mathcal{O}(10^{-7}). \qquad (9.91)$$

It appears therefore that the tidal force associated with the moon is completely negligible, not only compared with gravity g but also to any of the typical forces acting along the vertical direction. So does it mean tidal forces are not responsible for the observed tides? Of course they are, but not through the local

---

[10]In the case of the earth–moon system, its value is about 6400/385000 $\sim$ 0.017.

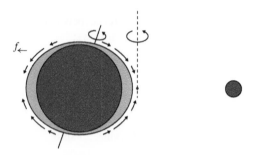

FIGURE 9.18 The movement of earth and moon around their common center of mass creates a centrifugal force that is weaker than the moon's gravitational attraction for earth points facing the moon. The resulting horizontal tidal force $f_\leftarrow$ has a tendency to create a bulge toward the moon. On the earth's face opposite to the moon, the centrifugal force is larger than the moon's gravitational attraction, and the horizontal force creates a second bulge facing away from the moon. Since the earth rotates around its own South–North axis, the two bulges move with respect to the continents.

vertical attraction as sometimes erroneously thought, but through the horizontal component, which we now proceed to calculate.

The component of the tidal force along the local northward axis ($\mathbf{j}$ in Fig. 2.9) is, after several simplifications similar to those made above,

$$f_\leftarrow \simeq -\frac{GM_m}{D_m^3} 3r \cos\theta \sin\theta. \qquad (9.92)$$

The order of magnitude of this force component is the same as that of the vertical one, but since all horizontal forces are much smaller than gravity, the horizontal tidal force is *not* negligible and acts to make the fluid converge or diverge. The spatial distribution of this force along the earth's surface is such that it tends to create a bulge in the region of the earth facing the moon and a second bulge at the diametrically *opposite* place. The explanation is that, for a point closer to the moon than D_m, the gravitational pull of the moon exceeds the centrifugal force associated with the earth–moon corotation, while on the opposite side of the earth the inverse is true; the centrifugal force of the earth–moon corotation exceeds the gravitational pull of the moon. This is the essential mechanism of lunar tides (Fig. 9.18). Solar tides are similar, with the sun taking the place of the moon but being much larger and much further away.

The angle θ involved in our formula is constantly changing in time because of the terrestrial rotation and lunar motion, and it must be determined through astronomical calculations (e.g., Doodson, 1921). These calculations also take into account variations in the earth–moon distance D_m, which induce slow modulations of the tidal force. Trigonometric calculations reveal different periods of motion, the most noticeable one being due to the corotation of the earth and moon, giving rise to an apparent rotation of the moon over a given point on the earth every 24 h and 50 min (24 h of terrestrial rotation and a delay caused by the moon rotating around the sun in the same direction). But, because there are two bulges half an earth's circumference apart from each other [mathematically because of the product $\cos\theta\sin\theta$ in Eq. (9.92)], the period of the main lunar tide is only half of that, i.e., 12 h 25 min.

For practical purposes, it is worth noting that the tidal force can be derived from the so-called *tidal potential* (see Analytical Problem 9.8). In the local Cartesian coordinate system, the tidal force can be expressed as

$$\mathbf{f}_t = -\left(\frac{\partial V}{\partial x}, \frac{\partial V}{\partial y}, \frac{\partial V}{\partial z}\right) \quad \text{with} \quad V = -\frac{GM_m}{D_m^3}\frac{r^2}{2}(3\cos^2\theta - 1). \tag{9.93}$$

All we have to do then is to calculate the local tidal potential, take its local derivatives, and introduce these as tidal forces in the shallow-water equations.

The tidal potential can also be used to estimate tidal amplitudes. Since the tidal force, i.e., the gradient of the potential, has a form similar to the pressure-gradient force associated with the sea surface height, we can ask which distribution of η, denoted η_e, would cancel the tidal force so that no motion would result. Obviously, this is the case when

$$\eta_e = -\frac{V}{g} = \frac{GM_m}{D_m^3}\frac{r^2}{2g}(3\cos^2\theta - 1)$$

$$= \mathcal{O}\left(\frac{GM_m}{D_m^3}\frac{r^2}{g}\right) \sim 0.36 \text{ m}. \tag{9.94}$$

This defines the so-called *equilibrium tide*, first derived by Isaac Newton. It would be the tidal elevation if the fluid could follow the tidal force in order to remain in equilibrium with the pressure gradient generated by the bulges. In reality, however, continents and topographic features in the ocean do not allow sea water to stay at the equilibrium. Not only is the equilibrium tide never reached, but the tidal potential is also in need of further adaptation to take into account the solid earth deformation due to tides and the self-attraction of tides (e.g., Hendershott, 1972).

In the same way as we defined the equilibrium tide, we can determine the sea surface height that would exactly cancel the effect of an atmospheric pressure disturbance p_{atm}:

$$\eta = -\frac{p_{atm}}{\rho_0 g}, \tag{9.95}$$

which can be used as a first approximation to estimate the effect of atmospheric pressure on measurements of η and is called the *inverse barometric response*.

For actual tidal predictions, we resort to numerical methods. For this, we gather all terms previously mentioned in this chapter and add the components of the tidal force. The governing equations used in a shallow-water model to predict both tides and storm surges are

$$
\frac{\partial u}{\partial t}+u\frac{\partial u}{\partial x}+v\frac{\partial u}{\partial y}-fv=-\frac{1}{\rho_0}\frac{\partial p}{\partial x}+\frac{\tau^x}{\rho_0 h}-\frac{\tau_b^x}{\rho_0 h}-\frac{\partial V}{\partial x}
$$
$$
+\frac{1}{h}\frac{\partial}{\partial x}\left(A\frac{\partial hu}{\partial x}\right)+\frac{1}{h}\frac{\partial}{\partial y}\left(A\frac{\partial hu}{\partial y}\right) \tag{9.96a}
$$

$$
\frac{\partial v}{\partial t}+u\frac{\partial v}{\partial x}+v\frac{\partial v}{\partial y}+fu=-\frac{1}{\rho_0}\frac{\partial p}{\partial y}+\frac{\tau^y}{\rho_0 h}-\frac{\tau_b^y}{\rho_0 h}-\frac{\partial V}{\partial y}
$$
$$
+\frac{1}{h}\frac{\partial}{\partial x}\left(A\frac{\partial hv}{\partial x}\right)+\frac{1}{h}\frac{\partial}{\partial y}\left(A\frac{\partial hv}{\partial y}\right) \tag{9.96b}
$$

$$
p=\rho_0 g\eta+p_{\mathrm{atm}} \tag{9.96c}
$$

together with Eqs. (7.14) and (9.93). Note that the driving forces of wind and tide act very differently. While the wind stress acts as a surface force and therefore appears with a factor $1/h$, the tidal force is a body force acting over the whole water column. Consequently, the tidal force is more important in the deeper parts of the ocean. This might surprise us since we are used to observe the highest tides near the coasts, where h is small! In most cases, tides are generated in the deeper parts of the oceans, where the tidal force acts on a thick layer of water, creates a pattern of convergence/divergence and locally modifies the sea surface height. The sea surface elevation is then propagated as a set of Kelvin and inertia-gravity waves into shelf seas and coastal regions, where the reduced depth increases their amplitudes (Fig. 9.2).

Some shelf models can provide tidal predictions by imposing tidal elevations at distant open boundaries and propagating the waves into the domain while discarding the local tidal force. This is consistent with the idea that, in shallow seas, the wind stress is the dominant local forcing. Indeed, in a 10,000-m-deep basin, the tidal force is equivalent to the surface friction of a 75 m/s wind, whereas in a shallow sea of 100 m, a wind of 7.5 m/s already matches the local tidal force. An example of a tidal calculation in which the tides are imposed along an open boundary is given in Fig. 9.19. In this figure, we note in passing the presence of nodes where the tidal amplitude is nil and the phase undefined. Each such node, called an *amphidromic point*, is a place where the various wave components cancel each other (destructive interference).

The numerical implementation of the model we have just developed is readily feasible since we have already encountered all its ingredients: time stepping, advection, Coriolis term, pressure gradient, diffusion, which were all treated in detail in previous sections. The only remaining term is that including the bottom stress, and for it we suggest to discretize it with the Patankar technique (to be discussed in Section 14.6) if the quadratic relationship is selected:

$$
\tau_b^x=-\rho_0 C_d\sqrt{(u^n)^2+(v^n)^2}\ u^{n+1},\quad \tau_b^y=-\rho_0 C_d\sqrt{(u^n)^2+(v^n)^2}\ v^{n+1}. \tag{9.97}
$$

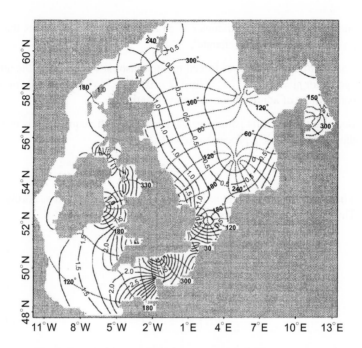

FIGURE 9.19 Tidal amplitudes (full lines) and phases (dotted lines) over the Northwestern European continental shelf, generated by the moon. (*Eric Delhez*)

Since several methods are available for each process, the combination of the various processes leads to a very wide array of possible numerical implementations, all at relatively low cost with two spatial dimensions. This explains the large number of two-dimensional numerical models that were developed relatively early in geophysical fluid modeling (e.g., Backhaus, 1983; Heaps, 1987; Nihoul, 1975).

ANALYTICAL PROBLEMS

9.1.   Prove that Kelvin waves propagate with the coast on their left in the southern hemisphere.

9.2.   The Yellow Sea between China and Korea (mean latitude: 37°N) has an average depth of 50 m and a coastal perimeter of 2600 km. How long does it take for a Kelvin wave to go around the shores of the Yellow Sea?

9.3.   Prove that at extremely large wavelengths, inertia-gravity waves degenerate into a flow field where particles describe circular inertial oscillations.

9.4.   An oceanic channel is modeled by a flat-bottom strip of ocean between two vertical walls. Assume that the fluid is homogeneous and inviscid, and that the Coriolis parameter is constant. Describe all waves that can propagate along such a channel.

9.5. Consider planetary waves forced by the seasonal variations of the annual cycle. For $f_0 = 8 \times 10^{-5}\,\text{s}^{-1}$, $\beta_0 = 2 \times 10^{-11}\,\text{m}^{-1}\text{s}^{-1}$, $R = 1000\,\text{km}$, what is the range of admissible zonal wavelengths?

9.6. Because the Coriolis parameter vanishes along the equator, it is usual in the study of tropical processes to write

$$f = \beta_0 y,$$

where y is the distance measured from the equator (positive northward). The linear wave equations then take the form

$$\frac{\partial u}{\partial t} - \beta_0 y v = -g \frac{\partial \eta}{\partial x} \tag{9.98}$$

$$\frac{\partial v}{\partial t} + \beta_0 y u = -g \frac{\partial \eta}{\partial y} \tag{9.99}$$

$$\frac{\partial \eta}{\partial t} + H\left(\frac{\partial u}{\partial x} + \frac{\partial v}{\partial y}\right) = 0, \tag{9.100}$$

where u and v are the zonal and meridional velocity components, η is the surface displacement, g is gravity, and H is the ocean depth at rest. Explore the possibility of a wave traveling zonally with no meridional velocity. At which speed does this wave travel and in which direction? Is it trapped along the equator? If so, what is the trapping distance? Does this wave bear any resemblance to a midlatitude wave (f_0 not zero)?

9.7. Seek wave solutions to the nonhydrostatic system of equations with nonstrictly vertical rotation vector:

$$\frac{\partial u}{\partial t} - f v + f_* w = -\frac{1}{\rho_0} \frac{\partial p}{\partial x} \tag{9.101a}$$

$$\frac{\partial v}{\partial t} + f u = -\frac{1}{\rho_0} \frac{\partial p}{\partial y} \tag{9.101b}$$

$$\frac{\partial w}{\partial t} - f_* u = -\frac{1}{\rho_0} \frac{\partial p}{\partial z} \tag{9.101c}$$

$$\frac{\partial u}{\partial x} + \frac{\partial v}{\partial y} + \frac{\partial w}{\partial z} = 0. \tag{9.101d}$$

The fluid is homogeneous ($\rho = 0$), inviscid ($\nu = 0$) and infinitely deep. Consider in particular the equivalent of the Kelvin wave ($u = 0$ at $x = 0$) and Poincaré waves.

9.8. Prove by using a local polar coordinate system that tidal forces derive from the tidal potential (9.93).

9.9. Estimate the average travel time for a gravity wave to circle the earth along the equator, assuming that there are no continents, and that the average depth of the ocean is 3800 m. Compare with the tidal period.

9.10. Based on the mass of the sun and its distance to the earth, how intense do you expect solar tides to be compared with lunar tides? At what period do the combined forces give rise to the strongest tides?

9.11. Knowing that a hurricane approaching Florida has a diameter of 100 km and wind-speeds U of 150 km/h, which storm surge height do you expect in a 10-m-deep coastal sea? Use the following wind-stress formula: $\tau = 10^{-6} \rho_0 U^2$.

9.12. Assuming the earthquake near Indonesia's Sumatra Island on 26 December 2004 generated a surface wave (tsunami) by an upward motion of the sea floor during 10 min, estimate the wavelength of the wave. For simplicity, assume a uniform depth $h = 4$ km. Estimate also the time available between the detection of the earthquake and the moment the tsunami reaches a coastline 4000 km away. If instead of a uniform depth, you use the depth profile $h(x)$ provided in sumatra.m, how would you estimate the travel time? Investigate under which conditions you can use the local wave speed of gravity waves over uneven topography. (*Hint*: Compare the wavelength with the length scale of topographic variations.)

9.13. In order to avoid the problem in Section 9.5 of an infinitely deep layer at large distances, assume now that the flow takes place in a channel of width L. How are the topographic waves modified by the presence of the lateral boundaries?

9.14. Consider an inertia-gravity wave of wavelength $\lambda = 2\pi/k$ on the f-plane and align the x-axis with the direction of propagation (i.e., $k_x = k$ and $k_y = 0$). Write the partial differential equations and solve them for u and η proportional to $\cos(kx - \omega t)$ and v proportional to $\sin(kx - \omega t)$. Then, calculate the kinetic and potential energies per unit horizontal area, defined as

$$KE = \frac{1}{\lambda} \int_0^\lambda \frac{1}{2} \rho_0 (u^2 + v^2) H \, dx \qquad (9.102a)$$

$$PE = \frac{1}{\lambda} \int_0^\lambda \frac{1}{2} \rho_0 g \eta^2 \, dx, \qquad (9.102b)$$

each in terms of the amplitude of η and show that the kinetic energy is always greater than the potential energy, except in the case $f = 0$ (pure gravity waves), in which case there is equipartition of energy.

9.15. Take a hurricane or typhoon from last summer season and note the pressure anomaly in its eye as it approached the coast. Determine the inverse barometric response at that time.

NUMERICAL EXERCISES

9.1. Establish the numerical stability condition of schemes (9.53) and (9.54). Can you provide an interpretation for the parameter $c\Delta t/\Delta x$? Compare with the CFL criterion.

9.2. Spell out the spatial discretization on the B-, C-, and D-grids of Eqs. (9.4a) through (9.5).

9.3. Implement the C-grid in Matlab for Eqs. (9.4a) through (9.5) with a variable fluid thickness given on a grid at the same location as η. Use a time discretization as in Eq. (9.54) and a fractional-step approach for the Coriolis term. Then, use your code to simulate a pure Kelvin wave for different values of $\Delta x/R$ and $k_x^2 R^2$ by initializing with the exact solution. (*Hint*: Use a periodic domain in the x-direction and a second impermeable boundary in the y-direction, to be justified, at $y = 10R$. Start from shallow.m.)

9.4. Analyze the way geostrophic equilibrium is represented in discrete Fourier modes on the B- and C-grids.

9.5. Investigate group-velocity errors for the different Arakawa grids using the numerical dispersion relation given in Eq. (9.66). Use $\Delta x = \Delta y$ and distinguish two types of waves: $k_x \neq 0$, $k_y = 0$ and $k_x = k_y$. Vary the resolution by taking $R/\Delta x = 0.2, 1, 5$, where R is the deformation radius.

9.6. Design the ideal staggering strategy for a model in which the eddy viscosity ν_E is chosen proportional to $|\partial u/\partial z| l_m^2$, where l_m is a specified mixing length and velocity u is determined numerically from a governing equation that includes vertical turbulent diffusion.

9.7. Assume you need to calculate the vertical component of relative vorticity from a discrete velocity field provided on the two-dimensional C-grid. Where is the most natural node to calculate the relative vorticity? Can you see an advantage to using a D-grid here?

9.8. Can you think of possibilities to include bottom topographic variations as those inducing tsunamis in shallow-water equations?

9.9. Take the variable depth implementation of Numerical Exercise 9.3 and apply it to the following topography

$$h = H_0 + \Delta H \left[1 + \tanh\left(\frac{x - \frac{L}{2}}{D}\right) \right]$$

with $H_0 = 50\,\text{m}$, $D = L/8$, and $\Delta H = 5H_0$. Use a solid wall at $x = 0$ and $x = L = 100\,\text{km}$ and periodic boundary conditions in the y-direction across a domain of length $5L$. Start with zero velocities and a Gaussian sea surface elevation of width $L/4$ and height of $1\,\text{m}$ in the center of the basin. Use linear bottom friction with friction coefficient $r = 10^{-4}\,\text{m/s}$. Trace the evolution of the sea surface elevation for $f = 10^{-4}\,\text{s}^{-1}$.

9.10. Perform a storm-surge simulation with the implementation of Numerical Exercise 9.3 by using a uniform wind stress over a square basin with a uniform topography and then with the topography given in Numerical Exercise 9.9. Use the quadratic law (9.77) for bottom friction.

William Thomson, Lord Kelvin
1824–1907
(Standing at right, in laboratory of Lord Rayleigh, left)

Named professor of natural philosophy at the University of Glasgow, Scotland, at age 22, William Thomson became quickly regarded as the leading inventor and scientist of his time. In 1892, he was named Baron Kelvin of Largs for his technological and theoretical contributions leading to the successful laying of a transatlantic cable. A friend of James P. Joule, he helped establish a firm theory of thermodynamics and first defined the absolute scale of temperature. He also made major contributions to the study of heat engines. With Hermann von Helmholtz, he estimated the ages of the earth and sun and ventured into fluid mechanics. His theory of the so-called Kelvin wave was published in 1879 (under the name William Thomson). His more than 300 original papers left hardly any aspect of science untouched. He is quoted as saying that he could understand nothing of which he could not make a model. (*Photo by A.G. Webster*)

Akio Arakawa
1927–

Akio Arakawa entered the Japanese Meteorological Agency in 1950 and received his doctorate at the University of Tokyo in 1961. He then went to the University of California in Los Angeles (UCLA) to pursue research, at a time when the atmospheric circulation computer models could reproduce weather-like motion but not for long. Beyond a two-week simulation, the computed patterns no longer looked like weather, and Arakawa's work demonstrated that the problem lied in the artificial generation of energy by inadequate numerical procedures. He also found the remedy. This remedy consisted of enforcing conservation of energy and of enstrophy (the square of vorticity) at the grid level.

The grids, which he proposed and later came to bear his name, were developed in the context of a study (Arakawa & Lamb, 1977) on the effects of grid topology on the dispersion of inertia-gravity waves. Arakawa's legacy to the science of weather prediction by computer modeling is significant and enduring. (*Photo credit: Akio Arakawa*)

Chapter 10

Barotropic Instability

ABSTRACT

The waves explored in the previous chapter evolve in a fluid otherwise at rest, propagating without either growth or decay. Here, we investigate waves riding on an existing current and find that, under certain conditions, they may grow at the expense of the energy contained in the mean current while respecting conservation of vorticity. The numerical section exposes the method of contour dynamics, designed specifically for applications in which conservation of vorticity is important.

10.1 WHAT MAKES A WAVE GROW UNSTABLE?

The planetary and topographic waves described in the previous chapter (Sections 9.4 through 9.5) owe their existence to the presence of an ambient potential-vorticity gradient. In the case of planetary waves, the cause is the sphericity of the planet, whereas for topographic waves the gradient results from the bottom slope. We may naturally wonder whether a sheared current that possesses a gradient of relative vorticity, would, too, be able to sustain similar low-frequency waves.

However, the situation is quite different for several reasons. First, the current would not only create the required ambient potential-vorticity gradient but would also transport the wave pattern; because of the current shear, this translation would be differential, and the wave pattern would be rapidly distorted. Moreover, there is likely to be a place within the current where the speed of the wave matches the velocity of the current; such a location, termed a *critical level*, typically permits a vigorous transfer of energy between the current and the wave. As a consequence, the wave may draw energy from the current and grow in time. If this happens, insignificant little wiggles may turn into very large perturbations, and the initial flow can become highly contorted, to the point of becoming unrecognizable. The flow is said to be unstable. To distinguish this situation from other instabilities occurring in baroclinic fluids (i.e., those possessing a stratification; see Chapters 14 and 17), the preceding process is generally known as *barotropic instability*.

The stability theory of homogeneous shear flows is a well-developed chapter in fluid mechanics (see, e.g., Kundu, 1990, Section 11.9; Lindzen, 1988). Here,

Introduction to Geophysical Fluid Dynamics

we address the problem with the inclusion of the Coriolis force but limit our investigation to establishing general properties and solving one particular case.

10.2   WAVES ON A SHEAR FLOW

To investigate the behavior of waves on an existing current in a relatively clear and tractable formalism, it is customary to make the following assumptions: The fluid is homogeneous and inviscid, and the bottom and the surface are flat and horizontal. However, the Coriolis parameter is allowed to vary (i.e., the beta effect is retained). The governing equations are (Section 4.4)

$$\frac{\partial u}{\partial t} + u\frac{\partial u}{\partial x} + v\frac{\partial u}{\partial y} + w\frac{\partial u}{\partial z} - fv = -\frac{1}{\rho_0}\frac{\partial p}{\partial x} \qquad (10.1a)$$

$$\frac{\partial v}{\partial t} + u\frac{\partial v}{\partial x} + v\frac{\partial v}{\partial y} + w\frac{\partial v}{\partial z} + fu = -\frac{1}{\rho_0}\frac{\partial p}{\partial y} \qquad (10.1b)$$

$$0 = -\frac{\partial p}{\partial z} \qquad (10.1c)$$

$$\frac{\partial u}{\partial x} + \frac{\partial v}{\partial y} + \frac{\partial w}{\partial z} = 0, \qquad (10.1d)$$

where the Coriolis parameter $f = f_0 + \beta_0 y$ varies with the northward coordinate y (Section 9.4). As demonstrated in Section 7.3, a horizontal flow that is initially uniform in the vertical will, in the absence of vertical friction, remain so at all times. In GFD parlance, this is what is called a *barotropic flow*, and we consider such a case. Consequently, we drop the terms $w\partial u/\partial z$ and $w\partial v/\partial z$ in Eqs. (10.1a) and (10.1b), respectively. According to (10.1d), $\partial w/\partial z$ must be z-independent, too, which implies that w is linear in z. But because the vertical velocity vanishes at both top and bottom, it must be zero everywhere ($w = 0$). The continuity equation reduces to

$$\frac{\partial u}{\partial x} + \frac{\partial v}{\partial y} = 0. \qquad (10.2)$$

For the basic state, we choose a zonal current with arbitrary meridional profile: $u = \bar{u}(y)$, $v = 0$. This is an exact solution to the nonlinear equations as long as the pressure distribution, $p = \bar{p}(y)$, satisfies the geostrophic balance

$$(f_0 + \beta_0 y)\,\bar{u}(y) = -\frac{1}{\rho_0}\frac{d\bar{p}}{dy}. \qquad (10.3)$$

Next, we add a small perturbation, meant to represent an arbitrary wave of weak amplitude. We write

$$u = \bar{u}(y) + u'(x, y, t) \qquad (10.4a)$$
$$v = v'(x, y, t) \qquad (10.4b)$$
$$p = \bar{p}(y) + p'(x, y, t), \qquad (10.4c)$$

where the perturbations u', v', and p' are taken to be much smaller that the corresponding variables of the basic flow (i.e., u' and v' much less than $\bar{u}$, and p' much less than $\bar{p}$). Substitution in Eqs. (10.1a), (10.1b), and (10.2) and subsequent linearization to take advantage of the smallness of the perturbation yield:

$$\frac{\partial u'}{\partial t} + \bar{u}\frac{\partial u'}{\partial x} + v'\frac{d\bar{u}}{dy} - (f_0 + \beta_0 y)v' = -\frac{1}{\rho_0}\frac{\partial p'}{\partial x} \tag{10.5a}$$

$$\frac{\partial v'}{\partial t} + \bar{u}\frac{\partial v'}{\partial x} + (f_0 + \beta_0 y)u' = -\frac{1}{\rho_0}\frac{\partial p'}{\partial y} \tag{10.5b}$$

$$\frac{\partial u'}{\partial x} + \frac{\partial v'}{\partial y} = 0. \tag{10.5c}$$

The last equation admits the streamfunction ψ, defined as

$$u' = -\frac{\partial \psi}{\partial y}, \quad v' = +\frac{\partial \psi}{\partial x}. \tag{10.6}$$

The choice of signs corresponds to a flow along streamlines with the higher streamfunction values on the right.

A cross-differentiation of the momentum equations (10.5a) and (10.5b) and the elimination of the velocity components leads to a single equation for the streamfunction:

$$\left(\frac{\partial}{\partial t} + \bar{u}\frac{\partial}{\partial x}\right)\nabla^2\psi + \left(\beta_0 - \frac{d^2\bar{u}}{dy^2}\right)\frac{\partial \psi}{\partial x} = 0. \tag{10.7}$$

This equation has coefficients that depend on $\bar{u}$ and, therefore, on the meridional coordinate y only. A sinusoidal wave in the zonal direction is then a solution:

$$\psi(x, y, t) = \phi(y)e^{i(kx-\omega t)}. \tag{10.8}$$

Substitution provides the following second-order ordinary differential equation for the amplitude $\phi(y)$:

$$\frac{d^2\phi}{dy^2} - k^2\phi + \frac{\beta_0 - d^2\bar{u}/dy^2}{\bar{u}(y) - c}\phi = 0, \tag{10.9}$$

where $c = \omega/k$ is the zonal speed of propagation. An equation of this type is called a *Rayleigh equation* (Rayleigh, 1880). Its key features are the nonconstant coefficient in the third term and the fact that its denominator may be zero, creating a singularity.

For boundary conditions, let us assume for simplicity that the fluid is contained between two walls, at $y = 0$ and L. We are thus considering waves on a zonal flow in a zonal channel. Obviously, there is no such zonal channel in either the atmosphere or ocean, but wavy zonal flows of limited meridional extent abound. The atmospheric jet stream in the upper troposphere, the Gulf Stream after its seaward turn off Cape Hatteras (36°N), and the Antarctic Circumpolar Current are all good examples. Also, the atmosphere on Jupiter, with the exception of the Great Red Spot and other vortices, consists almost entirely of zonal bands of alternating winds, called *belts* or *stripes* (see Fig. 1.5).

If the boundaries prevent fluid from entering and leaving the channel, v' is zero there, and Eq. (10.6) implies that the streamfunction must be a constant along each wall. In other words, walls are streamlines. This is possible only if the wave amplitude obeys

$$\phi(y=0) = \phi(y=L) = 0. \tag{10.10}$$

The second-order, homogeneous problem of Eqs. (10.9) and (10.10) can be viewed as an eigenvalue problem: The solution is trivial ($\phi=0$), unless the phase velocity assumes a specific value (eigenvalue), in which case a nonzero function ϕ (eigenfunction) can be determined within an arbitrary multiplicative constant.

In general, the eigenvalues c may be complex. If c admits the function ϕ, then the complex conjugate c^* admits the complex conjugate function ϕ^* and is thus another eigenvalue. This can be readily verified by taking the complex conjugate of Eq. (10.9). Hence, complex eigenvalues come in pairs.

Decomposing the eigenvalue into its real and imaginary components,

$$c = c_r + ic_i, \tag{10.11}$$

we note that the streamfunction ψ has an exponential factor of the form $\exp(kc_i t)$, which grows or decays according to the sign of c_i. Because the eigenvalues come in pairs to any decaying mode will correspond a growing mode. Therefore, the presence of a nonzero imaginary part in the phase velocity c automatically guarantees the existence of a growing disturbance and thus the instability of the basic flow. The product kc_i is then called the *growth rate*. Conversely, for the basic flow to be stable, it is necessary that the phase speed c be purely real.

Because mathematical difficulties prevent a general determination of the c values for an arbitrary velocity profile $\bar{u}(y)$ (the analysis is difficult even for idealized but nontrivial profiles), we shall not attempt to solve the problems (10.9)–(10.10) exactly but will instead establish some of its integral properties and, in so doing, reach weaker stability criteria.

When we multiply Eq. (10.9) by ϕ^* and then integrate across the domain, we obtain

$$-\int_0^L \left(\left| \frac{d\phi}{dy} \right|^2 + k^2 |\phi|^2 \right) dy + \int_0^L \frac{\beta_0 - d^2\bar{u}/dy^2}{\bar{u} - c} |\phi|^2 \, dy = 0, \tag{10.12}$$

after an integration by parts. The imaginary part of this expression is

$$c_i \int_0^L \left(\beta_0 - \frac{d^2\bar{u}}{dy^2} \right) \frac{|\phi|^2}{|\bar{u} - c|^2} \, dy = 0. \tag{10.13}$$

Two cases are possible: Either c_i vanishes or the integral does. If c_i is zero, the basic flow admits no growing disturbance and is stable. But, if c_i is not zero, then the integral must vanish, which requires that the quantity

$$\beta_0 - \frac{d^2\bar{u}}{dy^2} = \frac{d}{dy}\left(f_0 + \beta_0 y - \frac{d\bar{u}}{dy}\right) \qquad (10.14)$$

must change sign at least once within the confines of the domain. Summing up, we conclude that a necessary condition for instability is that expression (10.14) vanish somewhere inside the domain. Conversely, a sufficient condition for stability is that expression (10.14) not vanish anywhere within the domain (on the boundaries maybe, but not inside the domain). Physically, the total vorticity of the basic flow, $f_0 + \beta_0 y - d\bar{u}/dy$, must reach an extremum within the domain to cause instabilities. This result was first derived by Kuo (1949).

This first criterion can be strengthened by considering next the real part of Eq. (10.12), which takes the form

$$\int_0^L (\bar{u} - c_r)\left(\beta_0 - \frac{d^2\bar{u}}{dy^2}\right)\frac{|\phi|^2}{|\bar{u}-c|^2}\,dy = \int_0^L \left(\left|\frac{d\phi}{dy}\right|^2 + k^2|\phi|^2\right)dy. \qquad (10.15)$$

In the event of instability, the integral in Eq. (10.13) vanishes. Multiplying it by $(c_r - \bar{u}_0)$, where $\bar{u}_0$ is any real constant, adding the result to Eq. (10.15), and noting that the right-hand side of Eq. (10.15) is always positive for nonzero perturbations, we obtain

$$\int_0^L (\bar{u} - \bar{u}_0)\left(\beta_0 - \frac{d^2\bar{u}}{dy^2}\right)\frac{|\phi|^2}{|\bar{u}-c|^2}\,dy > 0. \qquad (10.16)$$

This inequality demands that the expression

$$(\bar{u} - \bar{u}_0)\left(\beta_0 - \frac{d^2\bar{u}}{dy^2}\right) \qquad (10.17)$$

be positive in at least some finite portion of the domain. Because this must hold true for any constant $\bar{u}_0$, it must be true in particular if $\bar{u}_0$ is the value of $\bar{u}(y)$ where $\beta_0 - d^2\bar{u}/dy^2$ vanishes. Hence, a stronger criterion is: Necessary conditions for instability are that $\beta_0 - d^2\bar{u}/dy^2$ vanish at least once within the domain *and* that $(\bar{u} - \bar{u}_0)(\beta_0 - d^2\bar{u}/dy^2)$, where $\bar{u}_0$ is the value of $\bar{u}(y)$ at which the first expression vanishes, be positive in at least some finite portion of the domain. Although this stronger criterion still offers no sufficient condition for instability, it is generally quite useful.

10.3   BOUNDS ON WAVE SPEEDS AND GROWTH RATES

The preceding analysis taught us that instabilities may occur when certain conditions are met. A question then naturally arises: If the flow is unstable, how fast will perturbations grow? In the general case of an arbitrary shear flow $\bar{u}(y)$, a precise determination of the growth rate of unstable perturbations is not possible. However, an upper bound can be derived relatively easily, and in the process, we can also determine lower and upper bounds on the phase speed of the perturbations. For simplicity, we will restrict our attention to the f-plane ($\beta_0 = 0$), in which case the derivation is due to Howard (1961). Afterwards, we will cite, without demonstration, the result for the beta plane.

The analysis begins by a change of variable[1]:

$$\phi = (\bar{u} - c)\, a, \tag{10.18}$$

which transforms Eq. (10.9) into

$$\frac{d}{dy}\left[(\bar{u} - c)^2 \frac{da}{dy}\right] - k^2 (\bar{u} - c)^2\, a = 0, \tag{10.19}$$

with β_0 set to zero. Because of Eq. (10.18), the boundary conditions on a are identical to those on ϕ, namely, $a(0) = a(L) = 0$.

We consider the case of an unstable wave. In this case, c has a nonzero imaginary part, and a is nonzero and complex. Multiplying by the complex conjugate a^* and integrating across the domain, we obtain an expression whose real and imaginary parts are

$$\text{Real part:} \quad \int_0^L [(\bar{u} - c_r)^2 - c_i^2]P\, dy = 0 \tag{10.20}$$

$$\text{Imaginary part:} \quad \int_0^L (\bar{u} - c_r)P\, dy = 0, \tag{10.21}$$

where $P = |da/dy|^2 + k^2|a|^2$ is a nonzero positive quantity. With Eq. (10.21), Eq. (10.20) can also be recast as

$$\int_0^L [\bar{u}^2 - (c_r^2 + c_i^2)]P\, dy = 0. \tag{10.22}$$

---

[1] It can be shown that the new variable a is the meridional displacement, the material time derivative of which is the v component of velocity.

It immediately follows from Eq. (10.21) that $(\bar{u} - c_r)$ must vanish somewhere in the domain, implying that the phase speed c_r lies between the minimum and maximum values of $\bar{u}(y)$:

$$U_{\min} < c_r < U_{\max}. \tag{10.23}$$

Physically, the wavy perturbation, if unstable, must travel with a speed that matches that of the entraining flow, in at least one location. In other words, there will always be a place in the domain where the wave does not drift with respect to the ambient flow and grows in place. It is precisely this local coupling between wave and flow that allows the wave to extract energy from the flow and to grow at its expense. The location where the phase speed is equal to the flow velocity is called a *critical level*.

Armed with bounds for the real part of c, we now seek bounds on its imaginary part. To do so, we introduce the obvious inequality

$$\int_0^L (\bar{u} - U_{\min}) \, (U_{\max} - \bar{u}) P \, dy \geq 0 \tag{10.24}$$

and then expand the expression, replace all linear terms in $\bar{u}$ using Eq. (10.21), and replace the quadratic term using Eq. (10.22) to arrive at

$$\left[\left(c_r - \frac{U_{\min} + U_{\max}}{2} \right)^2 + c_i^2 - \left(\frac{U_{\max} - U_{\min}}{2} \right)^2 \right] \int_0^L P \, dy \leq 0. \tag{10.25}$$

Because the integral of P can only be positive, the preceding bracketed quantity must be negative:

$$\left(c_r - \frac{U_{\min} + U_{\max}}{2} \right)^2 + c_i^2 \leq \left(\frac{U_{\max} - U_{\min}}{2} \right)^2. \tag{10.26}$$

This inequality implies that, in the complex plane, the number $c_r + ic_i$ must lie within the circle centered at $[(U_{\min} + U_{\max})/2, 0]$ and of radius $(U_{\max} - U_{\min})/2$. Since we are interested in modes that grow in time, c_i is positive, and only the upper half of that circle is relevant (Fig. 10.1). This result is called *Howard's semicircle theorem*.

It is readily evident from inequality (10.26) or Fig. 10.1 that c_i is bounded above by

$$c_i \leq \frac{U_{\max} - U_{\min}}{2}. \tag{10.27}$$

The perturbation's growth rate kc_i is thus likewise bounded above.

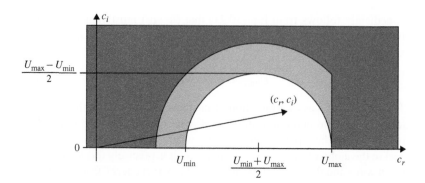

c_i

$\dfrac{U_{max} - U_{min}}{2}$

(c_r, c_i)

0

U_{min} $\dfrac{U_{min} + U_{max}}{2}$ U_{max} c_r

FIGURE 10.1 The semicircle theorem. Growing perturbations of wavenumber k must have phase speeds c_r and growth rates kc_i such that the tip of the vector (c_r, c_i) falls within the half-circle constructed from the minimum and maximum velocities of the ambient shear flow $\bar{u}(y)$, as depicted in the figure. When the beta effect is taken into account, the tip of the vector must lie in the slightly enlarged domain that includes the semicircle and the light gray area.

On the beta plane, the treatment of integrals and inequalities is somewhat more elaborate but still feasible. Pedlosky (1987, Section 7.5) showed that the preceding inequalities on c_r and c_i must be modified to

$$U_{min} - \frac{\beta_0 L^2}{2(\pi^2 + k^2 L^2)} < c_r < U_{max} \qquad (10.28)$$

$$\left(c_r - \frac{U_{min} + U_{max}}{2}\right)^2 + c_i^2 \leq \left(\frac{U_{max} - U_{min}}{2}\right)^2 + \frac{\beta_0 L^2 (U_{max} - U_{min})}{2(\pi^2 + k^2 L^2)}, \qquad (10.29)$$

where L is the domain's meridional width and k the zonal wavenumber (Fig. 10.1). The westward velocity shift on the left side of Eq. (10.28) is related to the existence of planetary waves [see the zonal phase speed, (9.30)]. The last inequality readily leads to an upper bound for the growth rate kc_i. Knowing bounds for the phase speed c_r and growth rate kc_i is useful in the numerical search of stability threshold in specific applications (Proehl, 1996).

10.4 A SIMPLE EXAMPLE

The preceding considerations on the existence of instabilities and their properties are rather abstract. So, let us work out an example to illustrate the concepts. For simplicity, we restrict ourselves to the f-plane ($\beta_0 = 0$) and take a shear flow that is piecewise linear (Fig. 10.2):

$$y < -L: \quad \bar{u} = -U, \quad \frac{d\bar{u}}{dy} = 0, \quad \frac{d^2\bar{u}}{dy^2} = 0 \qquad (10.30)$$

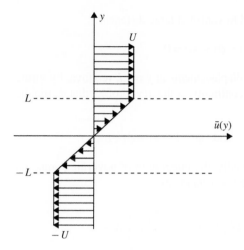

FIGURE 10.2 An idealized shear-flow profile that lends itself to analytic treatment. This profile meets both necessary conditions for instability and is found to be unstable to long waves.

$$-L < y < +L: \quad \bar{u} = \frac{U}{L}\, y, \quad \frac{d\bar{u}}{dy} = \frac{U}{L}, \quad \frac{d^2\bar{u}}{dy^2} = 0 \tag{10.31}$$

$$+L < y: \quad \bar{u} = +U, \quad \frac{d\bar{u}}{dy} = 0, \quad \frac{d^2\bar{u}}{dy^2} = 0, \tag{10.32}$$

where U is a positive constant, and the domain width is now infinity. Although the second derivative vanishes within each of the three segments of the domain, it is nonzero at their junctions. As y increases, the first derivative $d\bar{u}/dy$ changes from zero to a positive value and back to zero, so it can be said that the second derivative is positive at the first junction $(y = -L)$ and negative at the second $(y = +L)$. Thus, $d^2\bar{u}/dy^2$ changes sign in the domain, and this satisfies the first condition for the existence of instabilities. The second condition, that expression (10.17), now reduced to

$$-\bar{u}\, \frac{d^2\bar{u}}{dy^2},$$

be positive in some portion of the domain, is also satisfied because $d^2\bar{u}/dy^2$ has the sign opposite to $\bar{u}$ at each junction of the profile. Thus, the necessary conditions for instability are met, and although instabilities are not guaranteed to exist, we ought to expect them.

We now proceed with the solution. In each of the three domain segments, governing Eq. (10.9) reduces to

$$\frac{d^2\phi}{dy^2} - k^2\, \phi = 0, \tag{10.33}$$

and admits solutions of the type $\exp(+ky)$ and $\exp(-ky)$. This introduces two constants of integration per domain segment, for a total of six. Six conditions are then applied. First, ϕ is required to vanish at large distances:

$$\phi(-\infty) = \phi(+\infty) = 0.$$

Next, continuity of the meridional displacements at $y = \pm L$ requires, by virtue of Eq. (10.19) and by virtue of the continuity of the $\bar{u}(y)$ profile, that ϕ, too, be continuous there:

$$\phi(-L-\epsilon) = \phi(-L+\epsilon) \quad \text{and} \quad \phi(+L-\epsilon) = \phi(+L+\epsilon),$$

for arbitrarily small values of ϵ. Finally, the integration of governing equation (10.9) across the lines joining the domain segments

$$\int_{\pm L-\epsilon}^{\pm L+\epsilon} \left[(\bar{u}-c) \frac{d^2\phi}{dy^2} - k^2 (\bar{u}-c) \phi - \frac{d^2\bar{u}}{dy^2} \phi \right] dy = 0,$$

followed by an integration by parts, implies that

$$(\bar{u}-c) \frac{d\phi}{dy} - \frac{d\bar{u}}{dy} \phi$$

must be continuous at both $y = -L$ and $y = +L$. An alternative way of obtaining this result is to integrate Eq. (10.19), which is in conservative form, across a discontinuity.

Applying these six conditions leads to a homogeneous system of equations for the six constants of integration. Nonzero perturbations exist when this system admits a nontrivial solution—that is, when its determinant vanishes. Some tedious algebra yields

$$\frac{c^2}{U^2} = \frac{(1-2kL)^2 - e^{-4kL}}{(2kL)^2}. \tag{10.34}$$

Equation (10.34) is the dispersion relation, providing the wave velocity c in terms of the wavenumber k and the flow parameters L and U. It yields a unique and real c^2, either positive or negative. If it is positive, c is real and the perturbation behaves as a non-amplifying wave. But, if c^2 is negative, c is imaginary and one of the two solutions yields an exponentially growing mode [a proportional to $\exp(kc_i t)$]. Obviously, the instability threshold is $c^2 = 0$, in which case the dispersion relation (10.34) yields $kL = 0.639$. There thus exists a critical wavenumber $k = 0.639/L$ or critical wavelength $2\pi/k = 9.829L$ separating stable from unstable waves (Fig. 10.3). It can be shown by inspection of the same dispersion relation that shorter waves ($kL > 0.639$) travel without growth (because $c_i = 0$), whereas longer waves ($kL < 0.639$) grow exponentially without propagation (because $c_r = 0$). In sum, the basic shear flow is unstable to long-wave disturbances.

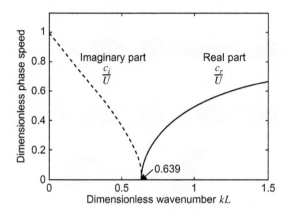

FIGURE 10.3 Plot of the dispersion relation (10.34) for waves riding on the shear flow depicted in Fig. 10.2. The lower wavenumbers k for which c_i is nonzero correspond to growing waves.

An interesting quest is the search for the fastest growing wave because this is the dominant wave, at least until finite-amplitude effects become important and the preceding theory loses its validity. For this, we look for the value of kL that maximizes kc_i, where c_i is the positive imaginary root of Eq. (10.34). The answer is $kL = 0.398$, from which follows the wavelength of the fastest growing mode:

$$\lambda_{\text{fastest growth}} = \frac{2\pi}{k} = 15.77\, L = 7.89\,(2L). \tag{10.35}$$

This means that the wavelength of the perturbation that dominates the early stage of instability is about eight times the width of the shear zone. Its growth rate is

$$(kc_i)_{\text{max}} = 0.201\,\frac{U}{L}, \tag{10.36}$$

corresponding to $c_i = 0.505 U$. It is left to the reader as an exercise to verify the preceding numerical values.

At this point, it is instructive to unravel the physical mechanism responsible for the growth of long-wave disturbances. Figure 10.4 displays the basic flow field, on which is superimposed a wavy disturbance. The phase shift between the two lines of discontinuity is that propitious to wave amplification. As the middle fluid, endowed with clockwise vorticity, intrudes in either neighboring strip where the vorticity is nonexistent, it produces local vorticity anomalies, which can be viewed as vortices. These vortices generate clockwise rotating flows in their vicinity, and if the wavelength is sufficiently long, the interval between the two lines of discontinuity appears relatively short and the vortices from each side interact with those on the other side. Under a proper phase difference, such as the one depicted in Fig. 10.4, the vortices entrain one another further into the regions of no vorticity, thereby amplifying the crests and troughs of the wave. The wave amplifies, and the basic shear flow cannot persist. As the wave grows, nonlinear terms are no longer negligible, and some level of saturation is

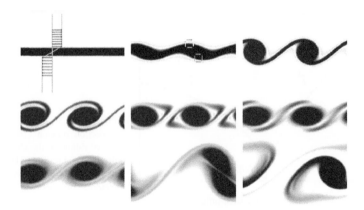

FIGURE 10.4 Finite-amplitude development of the instability of the shear flow depicted in Fig. 10.2. The troughs and crests of the wave induce a vortex field, which, in turn, amplifies those troughs and crests. The wave does not travel but amplifies with time. (The sequence of figures shown here were generated with shearedflow.m developed in Chapter 16.)

reached. The ultimate state (Fig. 10.4) is that of a series of clockwise vortices embedded in a weakened ambient shear flow (Dritschel, 1989; Zabusky, Hughes & Roberts, 1979).

Lindzen (1988) offers an alternative mechanism for the instability based on the fact that there are two special locations across the system. The first is the critical level y_c, where the wave speed matches the velocity of the basic flow $[c_r = \bar{u}(y_c)]$ and the other is y_0, where the vorticity of the basic flow reaches an extremum [where expression (10.14) changes sign]. A wave traveling in the direction of y_0 to y_c undergoes overreflection, that is, on entering the $[y_0, y_c]$ interval, it is being reflected toward its region of origin with a greater amplitude than on arrival. If there is a boundary or other place where the wave can be (simply) reflected, then it returns toward the region of overreflection, and on it goes. The successive overreflections of the echoing wave lead to exponential growth.

10.5   NONLINEARITIES

From Section 10.2, we note that the nonlinear advection term is responsible for the instability of the basic flow $\bar{u}(y)$. We analyzed the stability by linearizing the equations around the steady-state solution and replaced terms such as $u \partial u / \partial x$ by $\bar{u} \partial u' / \partial x$ and this led to linear equations and wave-like solutions, yet retaining the advection by the basic current $\bar{u}$. When instability occurs, the velocity perturbations grow in time, and after an initial phase during which linearization holds, they eventually reach such an intensity that $u' \partial u' / \partial x$ may no longer be neglected. We enter a nonlinear regime requiring numerical simulation. In an inviscid problem such as the present one, we then face a serious problem, already mentioned in the Introduction, namely the aliasing of short waves into longer waves.

As shown in Section 1.12 for time series, sampling (read: discretization) sets some limits on the frequencies that can be resolved. In space instead of time, the same analysis applies, and waves of wavenumbers k_x and $k_x + 2\pi/\Delta x$ cannot be distinguished from each other in a discretization with space interval Δx. If a wavenumber k_x larger than $\pi/\Delta x$ exists, it is mistaken by the discretization as being of smaller wavenumber $k_x - 2\pi/\Delta x$ or $2\pi/\Delta x - k_x$. This misinterpretation of too rapidly varying waves can be depicted (Fig. 10.5) as a reflection of the wavenumber about the cutoff value $\pi/\Delta x$. Any wave can be decomposed in its spectral components, and let us suppose that the spectrum of a set waves (wave packet) takes the form shown in Fig. 10.6. Since waves of higher wavenumbers are reflected around the cutoff wavenumber into the resolved range, the associated spectral energy will also be transferred from the shorter unresolved waves to the longer resolved waves. If the energy level decreases with decreasing wavenumber, the spectrum alteration will be strongest near the cutoff value. In other words, the energy content of marginally resolved waves is the one most influenced by aliasing, and the manifestation is an unwanted excess of energy among barely resolved waves. This is one reason why model results ought generally to viewed as suspect at scales comparable to the grid spacing. But there is more to the problem.

The aliasing problem is particularly irksome when nonlinear advection comes into play because the quadratic term in the equation creates wave harmonics: If the velocity field is resulting from the superposition of two waves, one of wavenumber k_1 and another of wavenumber k_2 of equal amplitude and phase,

$$u = u_1 + u_2 \quad \text{with} \quad u_1 = A e^{ik_1 x} \quad \text{and} \quad u_2 = A e^{ik_2 x}, \tag{10.37}$$

then the advection term $u \partial u / \partial x$ generates a contribution of the form

$$A^2 i (k_1 + k_2) e^{i(k_1 + k_2)x} \tag{10.38}$$

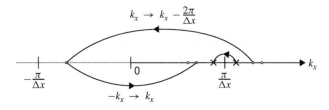

FIGURE 10.5 Transformation of a unresolved short wave of wavenumber $k_x > \pi/\Delta x$ into a resolved wavenumber $|k_x - 2\pi/\Delta x|$, corresponding to a reflection of wavenumber about the cutoff value $\pi/\Delta x$ as indicated for three particular values of the wavenumber identified by an open circle, a cross, and a gray dot to the right of $\pi/\Delta x$ and the wavenumbers into which they are aliased (all to the left of $\pi/\Delta x$).

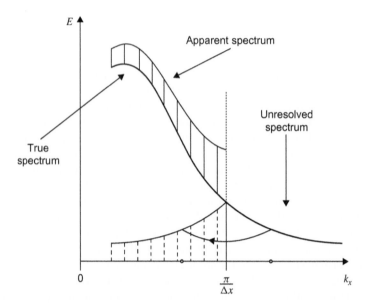

FIGURE 10.6 Spectrum alteration by aliasing, which effectively folds the numerically unresolved part of the spectrum ($k_x \geq \pi/\Delta x$) into resolved scales. The steeper the energy spectrum decrease around the cutoff wavenumber, the less aliasing is a problem. The spectrum alteration is then limited to the vicinity of the shortest resolved waves.

which introduces a new spectral component of higher wavenumber $k_1 + k_2$. Even if the two original waves are resolved by the grid, the newly created and shorter wave may be aliased and mistaken for a longer wave. This happens when $k_1 + k_2 > \pi/\Delta x$. The nonlinear advection thus creates an aliasing problem, which can seriously handicap calculations, especially if the aliasing is such that the newly created waves have a wavenumber identical to one of the original waves, k_1 for example. In this case, we have a feedback loop in which component u_1 interacts with another one and, instead of generating a shorter wave as it ought, increases its own amplitude. The process is self-repeating, and before long, the amplitude of the self-amplifying wave will reach an untolerable level. This is known as *nonlinear numerical instability*, which was first identified by Phillips (1956).

Such a self-amplification occurs when the interaction of k_1 and k_2 satisfies the aliasing condition and the new wave is aliased back into one of the original wavenumbers (here k_1):

$$(k_1 + k_2) \geq \frac{\pi}{\Delta x} \quad \text{and} \quad \frac{2\pi}{\Delta x} - (k_1 + k_2) = k_1. \tag{10.39}$$

To avoid such a situation for any wavenumber resolved by the grid (i.e., for all admissible values for k_2 varying between 0 and $\pi/\Delta x$), k_1 should not be allowed to take values in the interval $\pi/(2\Delta x)$ to $\pi/\Delta x$. This requirement is a little bit too strong since if k_1 is not allowed to exceed $\pi/(2\Delta x)$, k_2, too, should not be allowed to do so. The highest permitted value for either k_1 or k_2 is then found by letting $k_1 = k_2$ in (10.39), and this yields $k_{max} = 2\pi/(3\Delta x)$. In other words, if we are able to avoid all waves of wavelength shorter than $2\pi/k_{max} = 3\Delta x$, nonlinear instability by aliasing can be prevented. However, disallowing these waves from the initial condition is not enough because sooner or later, they will be generated by nonlinear interaction among the longer waves. The remedy is to eliminate the shorter waves as they are being generated, and this is accomplished by *filtering*.

Filtering is a form of dissipation that mimics physical dissipation but is designed to remove preferentially the undesirable waves, that is, only those on the shortest scales resolved by the numerical grid. This can be accomplished by the filters discussed in the following section. Other methods to address aliasing and nonlinear numerical instability related to the advection term will be encountered later in Section 16.7. Before concluding this chapter, we will also describe an entirely different approach, which avoids aliasing altogether by not using a grid at all. This method, known as *contour dynamics*, follows fluid parcels along their path of motion, thus absorbing the advection terms in the material time derivative.

10.6   FILTERING

We saw earlier that the leapfrog method generates spurious modes (flip-flop in time), and we just realized that spatial modes near the $2\Delta x$ cutoff ("saw-tooth" structure in space) are poorly reproduced and prone to aliasing. We further showed how nonlinearities can create aliasing problems around the $2\Delta x$ mode. Naturally, we would now like to remove these unwanted oscillations from the numerical solution. For the spatial saw-tooth structure, we already have at our disposal a method for eliminating shorter waves: physical diffusion. However, physical diffusion in the model may not always be sufficient to suppress or even control the $2\Delta x$ mode, and additional dissipation, of a numerical nature, becomes necessary. This is called *filtering*.

In this section, we concentrate on explicit filtering, designed to damp short waves. Let us start with a discrete filter inspired by the physical diffusion operator:

$$\widehat{c}_i^{\,n} = \tilde{c}_i^{\,n} + \varkappa \underbrace{\left(\tilde{c}_{i+1}^{\,n} - 2\tilde{c}_i^{\,n} + \tilde{c}_{i-1}^{\,n} \right)}_{\simeq \Delta x^2 \frac{\partial^2 c}{\partial x^2}}, \tag{10.40}$$

in which the new (filtered) value $\widehat{c}_i^{\,n}$ is henceforth replacing the original (unfiltered) value $\tilde{c}_i^{\,n}$. The preceding formulation is equivalent to introducing a diffusion term with diffusivity $\varkappa \Delta x^2/\Delta t$, which enhances physical diffusion, if any.

The behavior of this filter can be analyzed with the aid of Fourier modes [$\exp(i k_x i \Delta x)$], thus providing the "amplification" factor, which in this case is actually a damping factor:

$$\varrho = 1 - 4\varkappa \sin^2 \left(\frac{k_x \Delta x}{2} \right). \tag{10.41}$$

For well-resolved waves, the amplification factor is close to unity (no change of amplitude), whereas for the $2\Delta x$ wave ($k_x = 2\pi/2\Delta x$), its value is $1 - 4\varkappa$. The value $\varkappa = 1/4$ therefore eliminates the shortest wave in a single pass of the filter, but intermediate wavelengths are partially reduced, too, and a smaller value of $\varkappa$ is generally used in order not to dampen unnecessarily the intermediate scales of the solution. Therefore, a compromise needs to be reached between our desire to eliminate the $2\Delta x$ component while least affecting the rest of the solution.

To alleviate such compromise, more selective filters can be implemented. These are of the biharmonic type and require a wider stencil (more grid points). For example,

$$\widehat{c}_i^n = \tilde{c}_i^n + \frac{\varkappa}{4} \underbrace{\left(-\tilde{c}_{i+2}^n + 4\tilde{c}_{i-1}^n - 6\tilde{c}_i^n + 4\tilde{c}_{i+1}^n - \tilde{c}_{i+2}^n \right)}_{\simeq -\Delta x^4 \frac{\partial^4 \tilde{c}}{\partial x^4}} \tag{10.42}$$

leads to the more scale-selective damping factor

$$\varrho = 1 - \varkappa \left[4\sin^2 \left(\frac{k_x \Delta x}{2} \right) - \sin^2 \left(\frac{2k_x \Delta x}{2} \right) \right]. \tag{10.43}$$

The difference in the case $\varkappa = 1/4$ is illustrated in Fig. 10.7. Both diffusion-like and biharmonic filters [Eqs. (10.40) and (10.42), respectively] eliminate the $2\Delta x$ mode with the same value of $\varkappa$. Figure 10.7 also shows that components of intermediate scales are less affected by the biharmonic filter than by the diffusion-like filter. However, the biharmonic filter may introduce nonmonotonic behavior because there are negative coefficients in its stencil (10.42).

As for the diffusion-like filter, the biharmonic filter is sometimes made explicit in the undiscretized model equations by an additional term of the form $-\mathcal{B} \partial^4 \tilde{c}/\partial x^4$, with $\mathcal{B} = \varkappa \Delta x^4/(4\Delta t)$. The approach can, of course, be extended to ever larger stencils with increased scale selectivity but at the cost of additional computations.

It should be noted that the coefficients used in the filters are depending on the grid spacing and time step, whereas physical parameters do not, unless they parameterize subgrid-scale effects. In the latter case, the grid size can be involved in the parameterization, as seen in Section 4.2. However, we should not confuse the different concepts: The physical molecular diffusion, the standard

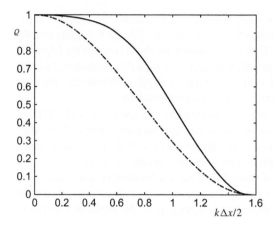

FIGURE 10.7 Damping factor as a function of wavelength for two different filters: regular diffusion [Eq. (10.40), dashed-dotted line] and biharmonic operator [Eq. (10.42), solid line], each for $\varkappa = 1/4$. Both filters eliminate the $2\Delta x$ mode completely ($\varrho = 0$), but the biharmonic filter is more scale selective in the sense that it damps less the intermediate-scale components (ϱ closer to unity for these).

microturbulent (eddy) diffusion, subgrid-scale diffusion introduced to parameterize mixing at scales longer than turbulent motions yet shorter than the grid spacing, diffusion associated with explicit filtering (the subject of the present section), and finally, numerical diffusion caused by the numerical scheme (totally uncoded). It is unfortunately not always clearly stated in model applications which type of diffusion is being meant when the authors mention their model's diffusion parameters.

For filtering in time, we can adopt the same filtering technique. Because the spatial filter replaces the model values by a filtered version obtained via Eq. (10.40), one way of eliminating the flip-flop mode is

$$\widehat{c}^{\,n} = \tilde{c}^{\,n} + \varkappa \left(\tilde{c}^{\,n+1} - 2\tilde{c}^{\,n} + \tilde{c}^{\,n-1} \right). \tag{10.44}$$

However, this is not very practical since it requires that past values of $\tilde{c}$ be stored for later filtering. Note also how filtering at time level n must wait until values have been computed at time level $n+1$. This does not avoid the nonlinear interactions of the spurious mode with the physical modes. It is better, therefore, to blend the filtering with time stepping and replace the unfiltered solution by the filtered one as soon as it becomes available. Suppose for example that we have a new value of $\tilde{c}^{\,n+1}$ obtained with the leapfrog scheme,

$$\tilde{c}^{\,n+1} = \widehat{c}^{\,n-1} + 2\Delta t\, Q\left(t, \tilde{c}^{\,n} \right), \tag{10.45}$$

with the usual source term Q regrouping all spatial operators. We can then filter $\tilde{c}^{\,n}$ with

$$\widehat{c}^{\,n} = \tilde{c}^{\,n} + \varkappa \left(\tilde{c}^{\,n+1} - 2\tilde{c}^{\,n} + \widehat{c}^{\,n-1} \right), \tag{10.46}$$

and immediately store it in the array holding $\tilde{c}^n$. Note how $\widehat{c}^{n-1}$ appears in the filtering and leapfrog step instead of $\tilde{c}^{n-1}$ because the filtered value has already superseded the original one. This filter, known as the *Asselin filter* (Asselin, 1972), is commonly used in models with leapfrog time discretization. In order not to filter excessively, small values of $\varkappa$ can be used. Alternatively, the filter can be applied only intermittently or with varying intensity $\varkappa$.

More selective filters in time can be inspired by the spatial filter (10.42), but these would require the storage of additional intermediate values of the state vector because the filter involves more time levels (five in the biharmonic case), while the leapfrog scheme requires that only three levels be stored.

Other filters exist, some of them based on intermittent re-initialization of the leapfrog time integration by simple Euler steps, but all of them should be applied with caution because they always filter part of the physical solution or alter the truncation error.

10.7 CONTOUR DYNAMICS

The preceding stability analysis and aliasing problem gives us a nice opportunity to introduce yet another numerical method, the family of so-called boundary element methods. This method was first applied to vortex calculations by Norman Zabusky[2] (Zabusky et al., 1979). To illustrate the approach, we start from the simple task of retrieving the velocity field from a known vorticity distribution in two dimensions. The vorticity ω is related to the velocity components u and v by

$$\frac{\partial v}{\partial x} - \frac{\partial u}{\partial y} = \omega, \tag{10.47}$$

and it follows by inversion of this definition that the velocity accompanying a localized vortex patch of area ds and uniform vorticity ω in the absence of boundary conditions (i.e., for an infinite domain) is given by

$$2\pi r \, dv_\theta = \omega \, ds, \tag{10.48}$$

where $r = \sqrt{(x-x')^2 + (y-y')^2}$, and v_θ is the velocity component perpendicular to the line joining the vortex patch to the point under consideration (left side of Fig. 10.8). The result follows from a straightforward application of Stokes theorem, which states that the circulation of the velocity along a contour (here circle of radius r), that is $2\pi r \, dv_\theta$, is equal to the integration of the vorticity within that contour (here ωds).

---

[2] See his biography at the end of this chapter.

In vectorial notation, the infinitesimal velocity associated with a differential patch ds of vorticity ω is

$$d\mathbf{u} = \frac{\omega}{2\pi r}\frac{\mathbf{k} \times (\mathbf{x} - \mathbf{x}')}{r}\, ds, \qquad (10.49)$$

which can be integrated over space for a nonuniform distribution $\omega(x, y)$ over a finite area (right side of Fig. 10.8). We obtain

$$\mathbf{u}(x, y) = \frac{1}{2\pi}\int\int \omega(x', y')\,\frac{\mathbf{k} \times (\mathbf{x} - \mathbf{x}')}{r^2}\, dx'\, dy'. \qquad (10.50)$$

This provides the velocity field as a function of the vorticity distribution, up to an irrotational velocity field. In an infinite domain (i.e., with no boundary conditions), the latter is zero. Suppose for now that we have a single patch of constant vorticity so that

$$u(x, y) = \frac{\omega}{2\pi}\int\int \frac{-(y - y')}{(x - x')^2 + (y - y')^2}\, dx'\, dy' \qquad (10.51a)$$

$$v(x, y) = \frac{\omega}{2\pi}\int\int \frac{(x - x')}{(x - x')^2 + (y - y')^2}\, dx'\, dy', \qquad (10.51b)$$

where the integral is performed over the vorticity patch delimited by its contour $\mathcal{C}$ (Fig. 10.8). Noting that integrands are derivatives of the function

$$\phi = \ln\left[\frac{(x - x')^2 + (y - y')^2}{L^2}\right], \qquad (10.52)$$

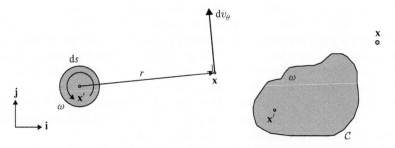

FIGURE 10.8   Element dv_θ of velocity associated with an infinitesimal vortex patch of area ds and vorticity ω (left panel). Integration over a finite patch of nonzero vorticity within contour $\mathcal{C}$ gives the associate velocity field (right panel). Note that the domain is infinitely wide outside the vortex patch.

we can rewrite the velocity components as

$$u(x,y) = \frac{\omega}{4\pi} \int\int \frac{\partial \phi}{\partial y'} \, dx' \, dy' = -\frac{\omega}{4\pi} \oint_C \phi \, dx' \qquad (10.53a)$$

$$v(x,y) = \frac{\omega}{4\pi} \int\int -\frac{\partial \phi}{\partial x'} \, dx' \, dy' = -\frac{\omega}{4\pi} \oint_C \phi \, dy', \qquad (10.53b)$$

for which we performed integration by parts to reduce the integral over the area of nonzero vorticity to a line integral along its perimeter. The symbol $\oint$ means that the integral is taken as x' and y' vary along the closed perimeter C with the patch on the left. Thus, we can express the velocity vector $\mathbf{u}$ at any point due to a patch of uniform vorticity ω as

$$\mathbf{u}(x,y) = -\frac{\omega}{4\pi} \oint_C \ln\left[\frac{(x-x')^2+(y-y')^2}{L^2}\right] dx'. \qquad (10.54)$$

When several vorticity patches are present, all we have to do is to add the contributions of the different patches. However, there is a slight difficulty when a patch is contained within another one. For example, in Fig. 10.9, the ω_3 vorticity lies entirely within the ω_2 vorticity patch. For the ω_2 patch, the contour integration breaks into two parts, one for the outer contour C_2 traveled counterclockwise (with vorticity ω_2 to its left) and the other for the inner contour C_3 traveled clockwise (again with vorticity ω_2 to its left). The latter contour integral needs to be repeated for the ω_3 patch, this time traveled counterclockwise and with ω_3 in its integrand. The addition of the last two integrals leads to a single integration along C_3 performed counterclockwise with the vorticity jump $\delta\omega_3 = \omega_3 - \omega_2$ in its integrand. For any number of contours, we have

$$\mathbf{u}(x,y) = -\frac{1}{4\pi} \sum_m \delta\omega_m \oint_{C_m} \ln\left[\frac{(x-x')^2+(y-y')^2}{L^2}\right] dx', \qquad (10.55)$$

where the sum is performed over all existing contours, and where $\delta\omega_m$ is the vorticity jump across contour C_m (inside value minus outside value).

Up to here we only established a diagnostic tool to retrieve the velocity field from a given distribution of vorticity patches. To predict the evolution of these patches, we now have to solve the governing equation for vorticity. In the absence of friction or any other vorticity-altering process, vorticity is conserved and simply advected by the flow. Thus, points within a given vortex patch will retain their vorticity and remain within their original patch. All we have to do is to predict the evolution of the *boundary* of each patch, that is, the contours, hence the name *contour dynamics* given to the method.

Points along the contours are physical fluid points and therefore move with the local flow velocity, that is, the velocity field of Eq. (10.55) taken at contour points. In practice, such integration can rarely be performed analytically, and numerical methods must be devised. The most natural discretization consists of dividing all contours into segments (Fig. 10.10), and the contour integrals then

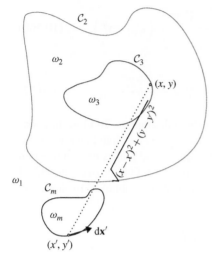

FIGURE 10.9 When several contours are involved in the velocity determination, contour integrals must be added to one another, and the relevant quantity along a contour is the vorticity jump across it. For the case depicted here, it is $\omega_3 - \omega_2$ for contour C_3.

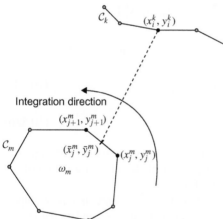

FIGURE 10.10 Discretization of contour integrals achieved by using a mid-point evaluation of the integrand along the mth contour at location $((x_j^m + x_{j+1}^m)/2, (y_j^m + y_{j+1}^m)/2)$. In this way, the singularity of the logarithm is avoided when the point (x_i^k, y_i^k) for which the integral is evaluated lies on the same contour along which the integral is performed.

reduce to sums of discrete contributions. The integral discretization has to deal with a singularity when the point (x, y) for which the velocity is computed lies on the same contour as where integration takes place and eventually coincides with point (x', y'). A simple way to avoid the problem is to use a staggered approach for the integration, that is, to evaluate the integrand at mid-distance between nodes j and $j+1$ (Fig. 10.10): For point (x_i^k, y_i^k) on contour k (with k possibly equal to m) where the velocity is being calculated, the pieces of the integral on contour C_m are approximated as

$$I_m(x_i^k, y_i^k) = \sum_{j=1}^{N} \ln\left[\frac{(x_i^k - \bar{x}_j^m)^2 + (y_i^k - \bar{y}_j^m)^2}{L^2}\right] \left(x_{j+1}^m - x_j^m\right) \qquad (10.56a)$$

$$J_m(x_i^k, y_i^k) = \sum_{j=1}^{N} \ln \left[\frac{(x_i^k - \bar{x}_j^m)^2 + (y_i^k - \bar{y}_j^m)^2}{L^2} \right] \left(y_{j+1}^m - y_j^m \right) \quad (10.56b)$$

$$\text{with} \quad \bar{x}_j^m = \frac{x_{j+1}^m + x_j^m}{2}, \quad \bar{y}_j^m = \frac{y_{j+1}^m + y_j^m}{2}, \quad (10.57)$$

where the sum covers the N segments[3] of the mth contour. To close the contour, we define for convenience $x_{N+1}^m = x_1$ and $y_{N+1}^m = y_1$. Note that there is no singularity because when m takes its turn to equal k and j takes its turn to equal i, the expression inside the logarithm remains nonzero. Finally, once individual integrals are calculated, the velocity components can be obtained by summing over all contour integrals:

$$u(x_i^k, y_i^k) = -\frac{1}{4\pi} \sum_m \delta\omega_m \, I_m\left(x_i^k, y_i^k\right)$$

$$v(x_i^k, y_i^k) = -\frac{1}{4\pi} \sum_m \delta\omega_m \, J_m\left(x_i^k, y_i^k\right),$$

and every node i on every contour k can be moved in time with the velocity:

$$\frac{dx_i^k}{dt} = u\left(x_i^k, y_i^k\right) \quad (10.58a)$$

$$\frac{dy_i^k}{dt} = v\left(x_i^k, y_i^k\right). \quad (10.58b)$$

The time integration can be performed by any method presented in Chapter 2. The Lagrangian (i.e., fluid following) displacements lead to deformation of the contours (see also Lagrangian approach of Section 12.8).

The simple numerical integration method outlined here is easily implemented (see, e.g., contourdyn.m). To try the method, we simulate the evolution of a narrow band of uniform vorticity (Fig. 10.11). Except for the curvature, this case is that of the shear layer instability seen in Section 10.4. Note the growing instabilities of the shear layer manifested as rolling waves.

FIGURE 10.11　Evolution of a narrow band of uniform vorticity simulated with contour dynamics.

---

[3] The number of segments per contour can, of course, be different for each contour, in which case $N = N_m$.

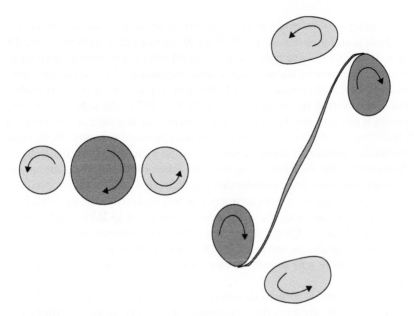

FIGURE 10.12   Two positive vorticity patches flanking a middle vortex of opposite vorticity and twice the area (left panel). Depending on the distances between the initial vortices, several outcomes are possible. One of them is vortex breakup and creation of vortex pairs (right panel).

The method can also be used to study the evolution and interaction of inviscid vortex patches in an infinite domain (Fig. 10.12), with the distinct advantage that no aliasing is present and that, in principle, no numerical dissipation needs to be added to stabilize the nonlinear advection. In reality, some dissipation is necessary because tearing and shearing of the eddies can generate filaments that become ever thinner, yet never disappear when there is no viscosity. Because the integration along one side of a thin filament almost nearly cancels that along the other side, the model makes unnecessary calculations, and it would be best if filaments could be severed.

Because the discretization uses only a finite number of fluid parcels on each contour, the contours cannot be tracked down to their shortest scales, and some special treatment becomes necessary when adjacent points are getting too close in some places and too distant in other places. Removing crowded points and inserting new ones in sparse areas is required. Procedures dealing with these problems are properly called *contour surgery* and have been optimized by Dritschel (1988). This eliminates some of the smallest structures and amounts to numerical dissipation.

To conclude the section, we observe that the method of contour dynamics cleverly replaces a two-dimensional problem of Eulerian vorticity evolution (i.e., on a 2D fixed array of points) with the problem of moving one-dimensional contours in a Lagrangian way (i.e., with points following the fluid). This reduces the complexity of the problem, but we must realize that the numerical cost of the methods is still proportional to $M^2 N^2$ for M contours of N segments, since for each of the MN discrete points, a sum over all other points must be performed. However, because of the one-dimensional distribution of the unknowns, resolution is increased compared to an Eulerian model in which a Poisson equation must be solved in two dimensions (see Section 16.7). The reduction of complexity is possible only because we exploited the fact that there are no boundary conditions and that vorticity remains constant between contours. To decrease further the number of computations, it can be noticed that the integrals are dominated by the contributions near the singularities. Hence, the contributions of points far away from singularities can be treated in a less precise manner without penalizing the overall accuracy. One way to do so is to group them. Such simplifications can bring the computational cost down to $MN \log(MN)$ operations (e.g., Vosbeek, Clercx & Mattheij, 2000). The accuracy of the numerical integration can also be enhanced by fitting a high-order analytical function to the contour points near singularities and then integrating the resulting integrand exactly.

For a continuous vorticity distribution without boundaries, we can still apply the approach by breaking the continuous vorticity into discrete vorticity levels (see Numerical Exercise 10.5). Generalization to stratified systems and more complicated governing equations is also possible (e.g., Mohebalhojeh & Dritschel, 2004).

ANALYTICAL PROBLEMS

10.1. Show that the variable a introduced in Eq. (10.18) is the amplitude of the meridional displacement, as claimed in the footnote.

10.2. What can you say of the stability properties of the following flow fields on the f-plane?

$$\bar{u}(y) = U\left(1 - \frac{y^2}{L^2}\right) \quad (-L \le y \le +L) \tag{10.59}$$

$$\bar{u}(y) = U \sin\frac{\pi y}{L} \quad (0 \le y \le L) \tag{10.60}$$

$$\bar{u}(y) = U \cos\frac{\pi y}{L} \quad (0 \le y \le L) \tag{10.61}$$

$$\bar{u}(y) = U \tanh\left(\frac{y}{L}\right) \quad (-\infty < y < +\infty). \tag{10.62}$$

10.3. A zonal shear flow with velocity profile

$$\bar{u}(y) = U\left(\frac{y}{L} - 3\frac{y^3}{L^3}\right)$$

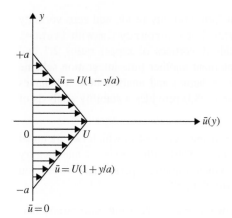

FIGURE 10.13   A jet-like profile (for Analytical Problem 10.6).

occupies the channel $-L \le y \le +L$ on the beta plane. Show that if $|U|$ is less than $\beta_0 L^2/12$, this flow is stable.

10.4. The atmospheric jet stream is a wandering zonal flow of the upper troposphere, which plays a central role in midlatitude weather. If we ignore the variations in air density, we can model the average jet stream as a purely zonal flow, independent of height and varying meridionally according to

$$\bar{u}(y) = U \exp\left(-\frac{y^2}{2L^2}\right),$$

in which the constants U and L, characteristics of the speed and width, are taken as $40\,\text{m/s}$ and $570\,\text{km}$, respectively. The jet center ($y=0$) is at $45°\text{N}$ where $\beta_0 = 1.61 \times 10^{-11}\,\text{m}^{-1}\text{s}^{-1}$. Is the jet stream unstable to zonally propagating waves?

10.5. Verify the semicircle theorem for the particular shear flow studied in Section 10.4. In other words, prove that $|c_r| < U$ for stable waves and $c_i < U$ for unstable waves. Also, prove that the wavelength leading to the highest growth rate, kc_i, is 15.77L, as stated in the text.

10.6. Derive the dispersion relation and establish a stability threshold for the jet-like profile of Fig. 10.13.

10.7. Redo Analytical Problem 10.6 in a channel between $y = -a$ and $y = a$.

NUMERICAL EXERCISES

10.1. Redo the analysis of nonlinear aliasing for a cubic term like $du/dt = -u^3$ in the governing equation for u. Why do you think aliasing is less of a concern in this particular case?

10.2. An elliptic vortex patch with uniform vorticity inside and zero vorticity outside is called a *Kirchhoff vortex*. Use `contourdyn.m` with `itest=1` to study the evolution of Kirchhoff vortices of aspect ratios 2:1 and 4:1. What do you observe? Implement another time-integration scheme (among which the explicit Euler scheme) and analyze how it behaves with a circular eddy. (*Hint*: Love (1983) provides a stability analysis of the Kirchhoff vortex.)

10.3. Experiment with `contourdyn.m` using `itest=4` in which two identical eddies are placed at various distances. Start with `dist=1.4` and then try the value `1.1`. What happens? Which numerical parameters would you adapt to improve the numerical simulation?

10.4. Simulate the eddy separation shown in Fig. 10.12 using `contourdyn.m`.

10.5. Discretize a circular eddy with vorticity varying linearly from zero at the rim to a maximum at the center by using M different vorticity values in concentric annuli. Then simulate its evolution with $M = 3$.

10.6. Verify your findings of Analytical Problem 10.6 by adapting `sheared-flow.m` to simulate the evolution of the most unstable periodic perturbation (for details on the numerical aspects, see Section 16.7).

10.7. Adapt `shearedflow.m` to investigate the so-called Bickley jet with profile given by

$$\bar{u}(y) = U \operatorname{sech}^2\left(\frac{y}{L}\right) \quad (-\infty < y < +\infty). \qquad (10.63)$$

Louis Norberg Howard
1929–

Applied mathematician and fluid dynamicist, Louis Norberg Howard has made numerous contributions to hydrodynamic stability and rotating flow. His famous semicircle theorem was published in 1961 as a short note extending some contemporary work by John Miles. Howard is also well known for his theoretical and experimental studies of natural convection. With Willem Malkus, he devised a simple waterwheel model of convection that, like real convection, can exhibit resting, steady, and periodic and chaotic behaviors. Howard has been a regular lecturer at the annual Geophysical Fluid Dynamics Summer Institute at the Woods Hole Oceanographic Institution, where his audiences have been much impressed by the breadth of his knowledge and the clarity of his explanations. (*Photo credit: L. N. Howard*)

Norman Julius Zabusky
1929–

Educated as an electrical engineer, Norman Zabusky spent the early part of his career working on plasma physics, and this led him to a lifetime pursuit of fluid turbulence by computational simulation. Vorticity dynamics lie at the center of his investigations. In the mid-1980s, he invented the method of contour dynamics (presented in this chapter) to investigate with greater precision the behavior of vorticity in two-dimensional flows in the absence of viscosity. Equipped with low-dissipation three-dimensional models of turbulent flows of his own design, Zabusky has been able to document in details the complicated processes of vortex tube deformation and reconnection. He firmly believes that progress in fluid turbulence demands a mathematical understanding of nonlinear coherent structures in weakly dissipative systems. In addition, Professor Zabusky has been fascinated by artistic renditions of waves and vortices in air and water, across ages and cultures. He wrote a book titled *From Art to Modern Science: Understanding Waves and Turbulence*. (*Photo credit: Rutgers University*)

Stratification Effects

Chapter 11

Stratification

ABSTRACT

After having studied the effects of rotation in homogeneous fluids, we now turn our attention toward the other distinctive feature of geophysical fluid dynamics, namely, stratification. A basic measure of stratification, the Brunt–Väisälä frequency, is introduced, and the accompanying dimensionless ratio, the Froude number, is defined and given a physical interpretation. The numerical part deals with the handling of unstable stratification in model simulations

11.1 INTRODUCTION

As Chapter 1 stated, problems in geophysical fluid dynamics concern fluid motions with one or both of two attributes, namely, ambient rotation and stratification. In the preceding chapters, attention was devoted exclusively to the effects of rotation, and stratification was avoided by the systematic assumption of a homogeneous fluid. We noted that rotation imparts to the fluid a strong tendency to behave in a columnar fashion—to be vertically rigid.

By contrast, a stratified fluid, consisting of fluid parcels of various densities, will tend under gravity to arrange itself so that the higher densities are found below lower densities. This vertical layering introduces an obvious gradient of properties in the vertical direction, which affects—among other things—the velocity field. Hence, the vertical rigidity induced by the effects of rotation will be attenuated by the presence of stratification. In return, the tendency of denser fluid to lie below lighter fluid imparts a horizontal rigidity to the system.

Because stratification induces a certain degree of decoupling between the various fluid masses (those of different densities), stratified systems typically contain more degrees of freedom than homogeneous systems, and we anticipate that the presence of stratification permits the existence of additional types of motions. When the stratification is mostly vertical (e.g., layers of various densities stacked on top of one another), gravity waves can be sustained internally (Chapter 13). When the stratification also has a horizontal component, additional waves can be permitted. These may lead to motion in equilibrium (Chapter 15), or, if they grow at the expense of the basic potential energy available in the system, may cause instabilities (Chapter 17).

11.2 STATIC STABILITY

Let us first consider a fluid in static equilibrium. Lack of motion can occur only in the absence of horizontal forces and thus in the presence of horizontal homogeneity. Stratification is then purely vertical (Fig. 11.1).

It is intuitively obvious that if the heavier fluid parcels are found below the lighter fluid parcels, the fluid is stable, whereas if heavier parcels lie above lighter ones, the system is apt to overturn, and the fluid is unstable. Let us now verify this intuition. Take a fluid parcel at a height z above a certain reference level, where the density is $\rho(z)$, and displace it vertically to the higher level $z+h$, where the ambient density is $\rho(z+h)$ (Fig. 11.1). If the fluid is incompressible, our displaced parcel retains its former density despite a slight pressure change, and that new level is subject to a net downward force equal to its own weight minus, by Archimedes' buoyancy principle, the weight of the displaced fluid, thus

$$g[\rho(z) - \rho(z+h)]V,$$

where V is the volume of the parcel. As it is written, this force is positive if it is directed downward. Newton's law (mass times acceleration equals upward force) yields

$$\rho(z)\, V \frac{d^2 h}{dt^2} = g\, [\rho(z+h) - \rho(z)]\, V. \tag{11.1}$$

Now, geophysical fluids are generally only weakly stratified; the density variations, although sufficient to drive or affect motions, are nonetheless relatively small compared with the average or reference density of the fluid. This remark was the essence of the Boussinesq approximation (Section 3.7). In the present case, this fact allows us to replace $\rho(z)$ on the left-hand side of Eq. (11.1) by the reference density ρ_0 and to use a Taylor expansion to approximate the density difference on the right by

$$\rho(z+h) - \rho(z) \simeq \frac{d\rho}{dz} h.$$

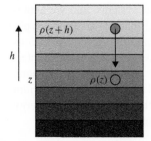

FIGURE 11.1 When an incompressible fluid parcel of density $\rho(z)$ is vertically displaced from level z to level $z+h$ in a stratified environment, a buoyancy force appears because of the density difference $\rho(z) - \rho(z+h)$ between the particle and the ambient fluid.

After a division by V, Eq. (11.1) reduces to

$$\frac{d^2h}{dt^2} - \frac{g}{\rho_0}\frac{d\rho}{dz}h = 0,\tag{11.2}$$

which shows that two cases can arise. The coefficient $-(g/\rho_0)d\rho/dz$ is either positive or negative. If it is positive ($d\rho/dz < 0$, corresponding to a fluid with the greater densities below the lesser densities), we can define the quantity N^2 as

$$N^2 = -\frac{g}{\rho_0}\frac{d\rho}{dz},\tag{11.3}$$

and the solution to the equation has an oscillatory character, with frequency N. Physically, this means that when displaced upward, the parcel is heavier than its surroundings, experiences a downward recalling force, falls down, and, in the process, acquires a vertical velocity; upon reaching its original level, the particle's inertia causes it to go further downward and to become surrounded by heavier fluid. The parcel, now buoyant, is recalled upward, and oscillations persist about the equilibrium level. The quantity N, defined by the square root of Eq. (11.3), provides the frequency of the oscillation and can thus be called the *stratification frequency*. It goes more commonly, however, by the name of Brunt–Väisälä frequency, in recognition of the two scientists who were the first to highlight the importance of this frequency in stratified fluids. (See their biographies at the end of this chapter.)

If the coefficient in Eq. (11.1) is negative (i.e., $d\rho/dz > 0$, corresponding to a top-heavy fluid configuration), the solution exhibits exponential growth, a sure sign of instability. The parcel displaced upward is surrounded by heavier fluid, finds itself buoyant, and moves farther and farther away from its initial position. Obviously, small perturbations will ensure not only that the single displaced parcel will depart from its initial position, but that all other fluid parcels will likewise participate in a general overturning of the fluid until it is finally stabilized, with the lighter fluid lying above the heavier fluid. If, however, a permanent destabilization is forced onto the fluid, such as by heating from below or cooling from above, the fluid will remain in constant agitation, a process called *convection*.

11.3   A NOTE ON ATMOSPHERIC STRATIFICATION

In a compressible fluid, such as the air of our planetary atmosphere, density can change in one of two ways: by pressure changes or by internal energy changes. In the first case, a pressure variation resulting in no heat exchange (i.e., an adiabatic compression or expansion) is accompanied by both density and temperature variations: All three quantities increase (or decrease) simultaneously, though not in equal proportions. If the fluid is made of fluid parcels all having the same heat content, the lower parcels, experiencing the weight of those above them, will be more compressed than those in the upper levels, and the system will appear stratified, with the denser and warmer fluid underlying the lighter and colder fluid. But such stratification cannot be dynamically relevant, for if parcels are interchanged adiabatically, they adjust their density and temperature according to the local pressure, and the system is left unchanged.

In contrast, internal energy changes are dynamically important. In the atmosphere, such variations occur because of a heat flux (such as heating in the tropics and cooling at high latitudes, or according to the diurnal cycle) or because of variations in air composition (such as water vapor). Such variations among fluid parcels do remain despite adiabatic compression or expansion and cause density differences that drive motions. It is thus imperative to distinguish, in a compressible fluid, the density variations that are dynamically relevant from those that are not. Such separation of density variations leads to the concept of potential density.

First, we consider a neutral (adiabatic) atmosphere—that is, one consisting of all air parcels having the same internal energy. Further, let us assume that the air, a mixture of various gases, behaves as a single ideal gas. Under these assumptions, we can write the equation of state and the adiabatic conservation law:

$$p = R\rho T, \tag{11.4}$$

$$\frac{p}{p_0} = \left(\frac{\rho}{\rho_0}\right)^{\gamma}, \tag{11.5}$$

where p, ρ, and T are, respectively, the pressure, density,[1] and absolute temperature; $R = C_p - C_v$ and $\gamma = C_p/C_v$ are the constants of an ideal gas.[2] Finally, p_0 and ρ_0 are reference pressure and density characterizing the level of internal energy of the fluid; the corresponding reference temperature T_0 is obtained from Eq. (11.4)—that is, $T_0 = p_0/R\rho_0$. Expressing both pressure and density in terms of the temperature, we obtain

$$\frac{p}{p_0} = \left(\frac{T}{T_0}\right)^{\gamma/(\gamma-1)} \tag{11.6a}$$

$$\frac{\rho}{\rho_0} = \left(\frac{T}{T_0}\right)^{1/(\gamma-1)}. \tag{11.6b}$$

Without motion, the atmosphere is in static equilibrium, which requires hydrostatic balance:

$$\frac{dp}{dz} = -\rho g. \tag{11.7}$$

---

[1] In contrast with preceding chapters, the variables p and ρ denote here the full pressure and density.

[2] For air, values are $C_p = 1005\,\mathrm{J\,kg^{-1}\,K^{-1}}$, $C_v = 718\,\mathrm{J\,kg^{-1}\,K^{-1}}$, $R = 287\,\mathrm{J\,kg^{-1}\,K^{-1}}$, and $\gamma = 1.40$.

Elimination of p and ρ by the use of Eqs. (11.6a) and (11.6b) yields a single equation for the temperature:

$$\frac{dT}{dz} = -\frac{\gamma-1}{\gamma}\frac{g}{R}$$
$$= -\frac{g}{C_p}. \tag{11.8}$$

In the derivation, it was assumed that p_0, ρ_0, and thus T_0 are not dependent on z, in agreement with our premise that the atmosphere is composed of parcels with identical internal energy contents. Equation (11.8) states that the temperature in such an atmosphere must decrease with increasing height at the uniform rate $g/C_p \simeq 10$ K/km. This gradient is called the *adiabatic lapse rate*. Physically, lower parcels are under greater pressure than higher parcels and thus have higher densities and temperatures. This explains why the air temperature is lower on mountain tops than in the valleys below.

It almost goes without saying that the departures from this adiabatic lapse rate—and not the actual temperature gradients—are to be considered in the study of atmospheric motions. We can demonstrate this clearly by redoing here, with a compressible fluid, the analysis of a vertical displacement performed in the previous section with an incompressible fluid. Consider a vertically stratified gas with pressure, density, and temperature, p, ρ, and T, varying with height z but not necessarily according to Eq. (11.8); that is, the heat content in the fluid is not uniform. The fluid is in static equilibrium so that Eq. (11.7) is satisfied. Consider now a parcel at height z; its properties are $p(z)$, $\rho(z)$, and $T(z)$. Imagine then that this fluid parcel is displaced adiabatically upward over a small distance h. According to the hydrostatic equation, this results in a pressure change $\delta p = -\rho g h$, which causes density and temperature changes given by the adiabatic constraints Eqs. (11.5) and (11.6a): $\delta\rho = -\rho g h/\gamma RT$ and $\delta T = -(\gamma-1)gh/\gamma R$. Thus, the new density is $\rho' = \rho + \delta\rho = \rho - \rho g h/\gamma RT$. But, at that new level, the ambient density is given by the stratification: $\rho(z+h) \simeq \rho(z) + (d\rho/dz)h$. The net force exerted on the parcel is the difference between its own weight and the weight of the displaced fluid at the new location (the buoyancy force), which per volume is

$$F = g\left[\rho_{\text{ambient}} - \rho_{\text{parcel}}\right]$$
$$= g\left[\rho(z+h) - \rho'\right]$$
$$\simeq g\left(\frac{d\rho}{dz} + \frac{\rho g}{\gamma RT}\right)h.$$

As the ideal gas law $(p = R\rho T)$ holds everywhere, the vertical gradients of pressure, density, and temperature are related by

$$\frac{dp}{dz} = RT\frac{d\rho}{dz} + R\rho\frac{dT}{dz}.$$

With the pressure gradient given by the hydrostatic balance (11.7), it follows that the density and temperature gradients are related by

$$\frac{1}{\rho}\frac{d\rho}{dz} + \frac{1}{T}\frac{dT}{dz} + \frac{g}{RT} = 0,$$

and the force on the fluid parcel can be expressed in terms of the temperature gradient

$$F \simeq -\frac{\rho g}{T}\left(\frac{dT}{dz} + \frac{g}{C_p}\right)h.$$

If

$$N^2 = -\frac{g}{\rho}\left(\frac{d\rho}{dz} + \frac{\rho g}{\gamma RT}\right) \tag{11.9a}$$

$$= +\frac{g}{T}\left(\frac{dT}{dz} + \frac{g}{C_p}\right) \tag{11.9b}$$

is a positive quantity, the force recalls the particle toward its initial level, and the stratification is stable. As we can clearly see, the relevant quantity is not the actual temperature gradient but its departure from the adiabatic gradient $-g/C_p$. As in the previous case of a stably stratified incompressible fluid, the quantity N is the frequency of vertical oscillations. It is called the stratification, or Brunt–Väisälä, frequency.

In order to avoid the systematic subtraction of the adiabatic gradient from the temperature gradient, the concept of potential temperature is introduced. The *potential temperature*, denoted by θ, is defined as the temperature that the parcel would have if it were brought adiabatically to a given reference pressure.[3] From Eq. (11.6a), we have

$$\frac{p}{p_0} = \left(\frac{T}{\theta}\right)^{\gamma/(\gamma-1)}$$

and hence

$$\theta = T\left(\frac{p}{p_0}\right)^{-(\gamma-1)/\gamma}. \tag{11.10}$$

The corresponding density is called the *potential density*, denoted by σ:

$$\sigma = \rho\left(\frac{p}{p_0}\right)^{-1/\gamma} = p_0/R\,\theta. \tag{11.11}$$

---

[3] In the atmosphere, this reference pressure is usually taken as the standard sea level pressure of 1013.25 millibars $= 1.01325 \times 10^5$ N/m^2.

The definition of the stratification frequency (11.9b) takes the more compact form:

$$N^2 = -\frac{g}{\sigma}\frac{\mathrm{d}\sigma}{\mathrm{d}z} = +\frac{g}{\theta}\frac{\mathrm{d}\theta}{\mathrm{d}z}. \tag{11.12}$$

Comparison with the earlier definition, (11.3), immediately shows that the substitution of potential density for density allows us to treat compressible fluids as incompressible.

During daytime and above land, the lower atmosphere is typically heated from below by the warmer ground and is in a state of turbulent convection. The convective layer not only covers the region where the time-averaged gradient of potential temperature is negative but also penetrates into the region above where it is positive (Fig. 11.2). Consequently, the sign of N^2 at a particular level is not

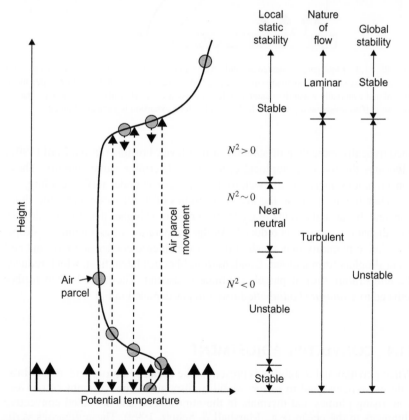

FIGURE 11.2   Typical profile of potential temperature in the lower atmosphere above warm ground. Heating from below destabilizes the air, generating convection and turbulence. Note how the convective layer extends not only over the region of negative N^2 but also slightly beyond, where N^2 is positive. Such a situation shows that a positive value of N^2 may not always be indicative of local stability. Global stability refers then to regions where even a finite amplitude displacement cannot destabilize the fluid parcel. (From *Stull, 1991*)

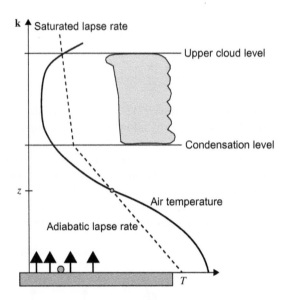

FIGURE 11.3   Fluid parcels located around level z amidst a temperature gradient (curved solid line) locally exceeding the adiabatic lapse rate (dashed line) are in an unstable situation. They move upward and eventually reach their saturation level, condensation takes place, and the lapse rate is decreased. If an inversion is present at higher levels, cloud extension is vertically limited.

unequivocally indicative of stability at that level. For this reason, Stull (1991) advocates the use of a nonlocal criterion to determine static stability. Those considerations apply equally well to the upper ocean under surface cooling.

When the air is moist, the thermodynamics of water vapor affect the situation, and, because the value of C_p for water vapor is higher than that for dry air, the adiabatic lapse rate is reduced. As the temperature of ascending air drops, the relative humidity may reach 100%, in which case condensation occurs and water droplets form a cloud. Condensation liberates latent heat, which reduces the temperature drop if parcels continue to ascend. The lapse is then further reduced to a *saturated adiabatic lapse rate*, as depicted in Fig. 11.3.

11.4 CONVECTIVE ADJUSTMENT

When gravitational instability is present in the ocean or atmosphere, nonhydrostatic movements tend to restore stability through narrow columns of convection, rising plumes and thermals in the atmosphere, and so-called convective chimneys in the ocean (e.g., Marshall & Schott, 1999). These vigorous vertical motions are not resolved by most computer models, and parameterizations called *convection schemes* are introduced to remove the instability and model the mixing associated with convection. Such parameterization can be achieved by additional terms in the governing equations, typically through a much increased eddy viscosity and diffusivity whenever $N^2 \leq 0$ (e.g., Cox, 1984; Marotzke, 1991). Other parameterizations are pieces of computer code of the type (see Fig. 11.4):

```
while there is any denser fluid being on top of lighter fluid
    loop over all layers
        if density of layer above > density of layer below
            mix properties of both layers, with a volume-weighted
                average
        end if
    end loop over all layers
end while
```

Oceanic circulation models (e.g., Bryan, 1969; Cox, 1984) were the first to use this type of parameterization.

The mixing accomplished by such scheme, however, is too strong in practice, because the model mixes fluid properties instantaneously over an entire horizontal grid cell of size $\Delta x \Delta y$, whereas physical convection operates at shorter scales and only partially mixes the physical properties at the spot. Therefore, numerical mixing should preferably be replaced by a mere swapping of fluid masses, under the assumption that convection carries part of the properties without alteration to their new level of equilibrium (e.g., Roussenov, Williams & Roether, 2001). It is clear that some arbitrariness remains and that every application demands its own calibration. Among other things, changing the time step clearly modifies the speed at which mixing takes place.

In atmospheric applications, the situation is more complicated as it may involve condensation, latent heat release, and precipitation during convective movement. Atmospheric convection parameterizations involve delicate adjustments of both temperature and moisture in the vertical (e.g., Betts, 1986; Kuo, 1974).

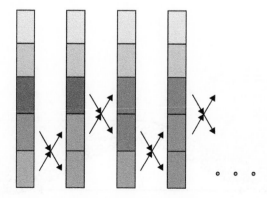

FIGURE 11.4   Illustration of convective adjustment within a fluid heated from below. Grid boxes below heavier neighbors are systematically mixed in pairs until the whole fluid column is rendered stable.

11.5 THE IMPORTANCE OF STRATIFICATION: THE FROUDE NUMBER

It was established in Section 1.5 that rotational effects are dynamically important when the Rossby number is on the order of unity or less. This number compares the distance traveled horizontally by a fluid parcel during one revolution ($\sim U/\Omega$) with the length scale over which the motions take place (L). Rotational effects are important when the former is less than the latter. By analogy, we may ask whether there exists a similar number measuring the importance of stratification. From the remarks in the preceding sections, we can anticipate that the stratification frequency, N, and the height scale, H, of a stratified fluid will play roles similar to those of Ω and L in rotating fluids.

To illustrate how such a dimensionless number can be derived, let us consider a stratified fluid of thickness H and stratification frequency N flowing horizontally at a speed U over an obstacle of length L and height Δz (Fig. 11.5). We can think of a wind in the lower atmosphere blowing over a mountain range. The presence of the obstacle forces some of the fluid to be displaced vertically and, hence, requires some supply of gravitational energy. Stratification will act to restrict or minimize such vertical displacements in some way, forcing the flow to pass around rather than over the obstacle. The greater the restriction, the greater the importance of stratification.

The time passed in the vicinity of the obstacle is approximately the time spent by a fluid parcel to cover the horizontal distance L at the speed U, that is, $T = L/U$. To climb a height of Δz, the fluid needs to acquire a vertical velocity on the order of

$$W = \frac{\Delta z}{T} = \frac{U \Delta z}{L}. \tag{11.13}$$

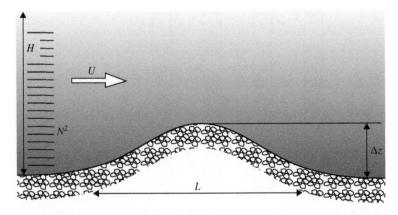

FIGURE 11.5 Situation in which a stratified flow encounters an obstacle, forcing some fluid parcels to move vertically against a buoyancy force.

The vertical displacement is on the order of the height of the obstacle and, in the presence of stratification $\rho(z)$, causes a density perturbation on the order of

$$\Delta\rho = \left|\frac{d\bar{\rho}}{dz}\right| \Delta z$$

$$= \frac{\rho_0 N^2}{g} \Delta z, \tag{11.14}$$

where $\bar{\rho}(z)$ is the fluid's vertical density profile upstream. In turn, this density variation gives rise to a pressure disturbance that scales, via the hydrostatic balance, as

$$\Delta P = gH\Delta\rho$$

$$= \rho_0 N^2 H \Delta z. \tag{11.15}$$

By virtue of the balance of forces in the horizontal, the pressure-gradient force must be accompanied by a change in fluid velocity $[u\partial u/\partial x + v\partial u/\partial y \sim (1/\rho_0)\partial p/\partial x]$:

$$\frac{U^2}{L} = \frac{\Delta P}{\rho_0 L} \implies U^2 = N^2 H \Delta z. \tag{11.16}$$

From this last expression, the ratio of vertical convergence, W/H, to horizontal divergence, U/L, is found to be

$$\frac{W/H}{U/L} = \frac{\Delta z}{H} = \frac{U^2}{N^2 H^2}. \tag{11.17}$$

We immediately note that if U is less than the product NH, W/H must be less than U/L, implying that convergence in the vertical cannot fully meet horizontal divergence. Consequently, the fluid is forced to be partially deflected horizontally so that the term $\partial u/\partial x$ can be met by $-\partial v/\partial y$ better than by $-\partial w/\partial z$. The stronger the stratification, the smaller is U compared with NH and, thus, W/H compared with U/L.

From this argument, we conclude that the ratio

$$Fr = \frac{U}{NH}, \tag{11.18}$$

called the *Froude number*, is a measure of the importance of stratification. The rule is as follows: If $Fr \lesssim 1$, stratification effects are important; the smaller Fr, the more important these effects are.

The analogy with the Rossby number of rotating fluids,

$$Ro = \frac{U}{\Omega L}, \tag{11.19}$$

where Ω is the angular rotation rate and L the horizontal scale, is immediate. Both Froude and Rossby numbers are ratios of the horizontal velocity scale by a product of frequency and length scale; for stratified fluids, the relevant frequency and length are naturally the stratification frequency and the height scale, whereas in rotating fluids they are, respectively, the rotation rate and the horizontal length scale.

The analogy can be pursued a little further. Just as the Froude number is a measure of the vertical velocity in a stratified fluid [via Eq. (11.17)], the Rossby number can be shown to be a measure of the vertical velocity in a rotating fluid. We saw (Section 7.2) that strongly rotating fluids (Ro nominally zero) allow no convergence of vertical velocity, even in the presence of topography. This results from the absence[4] of horizontal divergence in geostrophic flows. In reality, the Rossby number cannot be nil, and the flow cannot be purely geostrophic. The nonlinear terms, of relative importance measured by Ro, yield corrective terms to the geostrophic velocities of the same relative importance. Thus, the horizontal divergence, $\partial u/\partial x + \partial v/\partial y$, is not zero but is on the order of RoU/L. Since the divergence is matched by the vertical divergence, $-\partial w/\partial z$, on the order of W/H, we conclude that

$$\frac{W/H}{U/L} = Ro, \tag{11.20}$$

in rotating fluids. Contrasting Eqs. (11.17)–(11.20), we note that, with regard to vertical velocities, the square of the Froude number is the analogue of the Rossby number.

In continuation of the analogy, it is tempting to seek the stratified analogue of the Taylor column in rotating fluids. Recall that Taylor columns occur in rapidly rotating fluids ($Ro = U/\Omega L \ll 1$). Let us then ask what happens when a fluid is very stratified ($Fr = U/NH \ll 1$). By virtue of Eq. (11.17), the vertical displacements are severely restricted ($\Delta z \ll H$), implying that an obstacle causes the fluid at that level to be deflected almost purely horizontally. (In the absence of rotation, there is no tendency toward vertical rigidity, and parcels at levels above the obstacle can flow straight ahead without much disruption.) If the obstacle occupies the entire width of the domain, such a horizontal detour is not allowed, and the fluid at the level of the obstacle is blocked on both upstream and downstream sides. This horizontal blocking in stratified fluids is the analogue of the vertical Taylor columns in rotating fluids. Further analogies between homogeneous rotating fluids and stratified nonrotating fluids have been described by Veronis (1967).

11.6   COMBINATION OF ROTATION AND STRATIFICATION

In the light of the previous remarks, we are now in position to ask what happens when, as in actual geophysical fluids, the effects of rotation and stratification

---

[4]For the sake of the analogy, we rule out here an possible beta effect.

are simultaneously present. The preceding analysis remains unchanged, except that we now invoke the geostrophic balance [see Eq. (7.4)] in the horizontal momentum equation to obtain the horizontal velocity scale:

$$\Omega U = \frac{\Delta P}{\rho_0 L} \quad \Longrightarrow \quad U = \frac{N^2 H \Delta z}{\Omega L}. \tag{11.21}$$

The ratio of the vertical to horizontal convergence then becomes

$$\frac{W/H}{U/L} = \frac{\Delta z}{H} = \frac{\Omega L U}{N^2 H^2}$$

$$= \frac{Fr^2}{Ro}. \tag{11.22}$$

This is a particular case of great importance. According to our foregoing scaling analysis, the ratio of vertical convergence to horizontal divergence, $(W/H)/(U/L)$, is given by Fr^2, Fr^2/Ro, or Ro, depending on whether vertical motions are controlled by stratification, rotation, or both (Fig. 11.6). Thus, if Fr^2/Ro is less than Ro, stratification restricts vertical motions more than rotation and is the dominant process. The converse is true if Fr^2/Ro is greater than Ro.

Note that Ro is in the denominator of Eq. (11.22), which implies that the influence of rotation is to increase the scale for the vertical velocity when stratification is present. However, since vertical divergence cannot exist without horizontal convergence ($W/H \lesssim U/L$), the following inequality must hold:

$$Fr^2 \lesssim Ro, \tag{11.23}$$

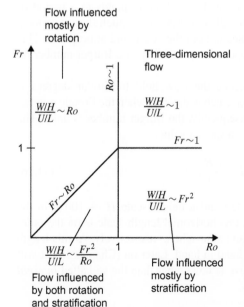

FIGURE 11.6 Recapitulation of the various scalings of the ratio of vertical convergence (divergence), W/H, to horizontal divergence (convergence), U/L, as a function of the Rossby number, $Ro = U/(\Omega L)$, and Froude number, $Fr = U/(NH)$.

that is,

$$\frac{U}{NH} \lesssim \frac{NH}{\Omega L}. \tag{11.24}$$

This sets an upper bound for the magnitude of the flow field in a fluid under given rotation (Ω) and of given stratification (N) in a domain of given dimensions (L, H). If the velocity is imposed externally (e.g., by an upstream condition), the inequality specifies either the horizontal or the vertical length scales of the possible disturbances. Finally, if the system is such that all quantities are externally imposed and that they do not meet Eq. (11.24), then special effects such as Taylor columns or blocking must occur.

Inequality Eq. (11.24) brings a new dimensionless number $NH/\Omega L$, namely, the ratio of the Rossby and Froude numbers. For historical reasons and also because it is more convenient in some dimensional analyses, the square of this quantity is usually defined:

$$Bu = \left(\frac{NH}{\Omega L}\right)^2 = \left(\frac{Ro}{Fr}\right)^2. \tag{11.25}$$

It bears the name of *Burger number*, in honor of Alewyn P. Burger (1927–2003), who contributed to our understanding of geostrophic scales of motions (Burger, 1958). In practice, the Burger number is a useful measure of stratification in the presence of rotation.

In typical geophysical fluids, the height scale is much less than the horizontal length scale $(H \ll L)$, but there is also a disparity between the two frequencies Ω and N. Although the rotation rate of the earth corresponds to a period of 24 h, the stratification frequency generally corresponds to much shorter periods, on the order of few to tens of minutes in both the ocean and atmosphere. This implies that generally $\Omega \ll N$ and opens the possibility of a Burger number on the order of unity.

Stratification and rotation influence the flow field to similar degrees if Fr^2/Ro and Ro are on the same order. Such is the case when the Froude number equals the Rossby number and, consequently, the Burger number is unity. The horizontal length scale then assumes a special value:

$$L = \frac{NH}{\Omega}. \tag{11.26}$$

For the values of Ω and N just cited and a height scale H of 100 m in the ocean and 1 km in the atmosphere, this horizontal length scale is on the order of 50 km and 500 km in the ocean and atmosphere, respectively. At this length scale, stratification and rotation go hand in hand. Later on (Chapter 15), it will be shown that the scale defined above is none other than the so-called *internal radius of deformation*.

ANALYTICAL PROBLEMS

11.1. Gulf Stream waters are characterized by surface temperatures around 22°C. At a depth of 800 m below the Gulf Stream, temperature is only 10°C. Using the value $2.1 \times 10^{-4} \, \text{K}^{-1}$ for the coefficient of thermal expansion, calculate the stratification frequency. What is the horizontal length at which both rotation and stratification play comparable roles? Compare this length scale to the width of the Gulf Stream.

11.2. An atmospheric inversion occurs when the temperature increases with altitude, in contrast to the normal situation when the temperature decays with height. This corresponds to a very stable stratification and, hence, to a lack of ventilation (smog, etc.). What is the stratification frequency when the inversion sets in $(dT/dz=0)$? Take $T=290 \, \text{K}$ and $C_p = 1005 \, \text{m}^2 \, \text{s}^{-2} \, \text{K}^{-1}$).

11.3. A meteorological balloon rises through the lower atmosphere, simultaneously measuring temperature and pressure. The reading, transmitted to the ground station where the temperature and pressure are, respectively, 17°C and 1028 millibars, reveals a gradient $\Delta T / \Delta p$ of 6°C per 100 millibars. Estimate the stratification frequency. If the atmosphere were neutral, what would the reading be?

11.4. Wind blowing from the sea at a speed of 10 m/s encounters Diamond Head, an extinct volcano on the southeastern coast of O'ahu Island in Hawai'i. This volcano is 232 m tall and 20 km wide. Stable air possesses a stratification frequency on the order of $0.02 \, \text{s}^{-1}$. How do vertical displacements compare to the height of the volcano? What does this imply about the importance of the stratification? Is the Coriolis force important in this case?

11.5. Redo Problem 11.4 with the same wind speed and stratification but with a mountain range 1000 m high and 500 km wide.

11.6. Vertical soundings of the atmosphere provided the temperature profiles displayed in Fig. 11.7. Analyze the stability of each profile.

NUMERICAL EXERCISES

11.1. Use `medprof.m` to read average Mediterranean temperature and salinity vertical profiles and calculate N^2 for various levels of vertical resolution (averaging data within cells). What do you conclude? (*Hint:* Use `ies80.m` for the state equation.)

11.2. Use the diffusion equation solver of Numerical Exercise 5.4 with a turbulent diffusion coefficient that changes from 10^{-4} to $10^{-2} \, \text{m}^2/\text{s}$ whenever N^2 is negative. Simulate the evolution of a 50-m high water column with

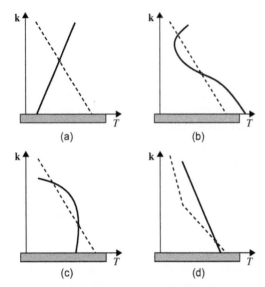

FIGURE 11.7 Various vertical profiles of temperature (solid lines) with
the lapse rate (dashed line) corresponding to a particular fluid parcel
(dot).

an initially stable vertical temperature gradient of 0.3°C/m subsequently
cooled at the surface by a heat loss of 100 W/m². Salinity is unchanged.
Study the effect of changes in Δz and Δt.

11.3. Implement the algorithm outlined in Section 11.4, to remove any gravitational instability instantaneously. Keep the turbulent diffusivity constant at 10^{-4} m²/s and simulate the same problem as in Numerical
Exercise 11.2.

David Brunt
1886–1965

As a bright young British mathematician, David Brunt began a career in astronomy, analyzing the statistics of celestial variables. Then, turning to meteorology during World War I, he became fascinated with weather forecasting and started to apply his statistical methods to atmospheric observations in the search for primary periodicities. By 1925, he had concluded that weather forecasting by extrapolation of cyclical behavior was not possible and turned his attention to the dynamic approach, which had been initiated in the late nineteenth century by William Ferrel and given new impetus by Vilhelm Bjerknes in recent years.

In 1926, he delivered a lecture at the Royal Meteorological Society on the vertical oscillations of particles in a stratified atmosphere. Lewis F. Richardson then led him to a paper published the preceding year by Finnish scientist Vilho Väisälä, in which the same oscillatory frequency was derived. This quantity is now jointly known as the Brunt–Väisälä frequency.

Continuing his efforts to explain observed phenomena by physical processes, Brunt contributed significantly to the theories of cyclones and anticyclones and of heat transfer in the atmosphere. His studies culminated in a textbook titled *Physical and Dynamical Meteorology* (1934) and confirmed him as a founder of modern meteorology. (*Photo credit: LaFayette, London*)

Vilho Väisälä
1889–1969

Altough he obtained his doctorate in mathematics (at the University of Helsinki, Finland), Vilho Väisälä found the subject rather uninspiring and became interested in meteorology. His positions at various Finnish institutes, including the Ilmala Meteorological Observation Station, required of him to develop instruments for atmospheric observations, which he did with much ingenuity. This eventually led him to establish in 1936 a commercial company for the manufacture of meteorological instrumentation, the Vaisala Company, a company now with branches on five continents and sales across the globe. In addition to his inventions and commercial activities, Väisälä retained an interest in the physics of the atmosphere, publishing over one hundred scientific papers, and mastered nine foreign languages. (*Photo credit: Vaisala Archives, Helsinki*)

Chapter 12

Layered Models

ABSTRACT

The assumption of density conservation by fluid parcels is advantageously used to change the vertical coordinate from depth to density. The new equations offer a clear discussion of potential-vorticity dynamics and lend themselves to discretization in the vertical. The result is a layered model. Splitting stratification in a series of layers may be interpreted as a vertical discretization in which the vertical grid is a material surface of the flow. This naturally leads to the presentation of Lagrangian approaches. *Note*: To avoid problems of terminology, we restrict ourselves here to the ocean. The case of the atmosphere follows with the replacement of depth by height and density by potential density.

12.1 FROM DEPTH TO DENSITY

Since a stable stratification requires a monotonic increase of density downward, density can be taken as a surrogate for depth and used as the vertical coordinate. If density is conserved by individual fluid parcels, as it is approximately the case for most geophysical flows, considerable mathematical simplification follows, and the new equations present a definite advantage in a number of situations. It is thus worth expounding on this change of variables at some length.

In the original Cartesian system of coordinates, z is an independent variable, and density $\rho(x, y, z, t)$ is a dependent variable, giving the water density at location (x, y), time t, and depth z. In the transformed coordinate system (x, y, ρ, t), density becomes an independent variable, and $z(x, y, \rho, t)$ has become the dependent variable giving the depth at which density ρ is found at location (x, y) and at time t. A surface along which density is constant is called an *isopycnal surface* or *isopycnic* for short.

From a differentiation of the expression $a = a(x, y, \rho(x, y, z, t), t)$, where a is any function, the rules for the change of are as follows:

$$\frac{\partial}{\partial x} \longrightarrow \frac{\partial a}{\partial x}\bigg|_z = \frac{\partial a}{\partial x}\bigg|_\rho + \frac{\partial a}{\partial \rho}\frac{\partial \rho}{\partial x}\bigg|_z$$

$$\frac{\partial}{\partial y} \longrightarrow \frac{\partial a}{\partial y}\bigg|_z = \frac{\partial a}{\partial y}\bigg|_\rho + \frac{\partial a}{\partial \rho}\frac{\partial \rho}{\partial y}\bigg|_z$$

$$\frac{\partial}{\partial z} \longrightarrow \frac{\partial a}{\partial z} = \frac{\partial a}{\partial \rho}\frac{\partial \rho}{\partial z}$$

$$\frac{\partial}{\partial t} \longrightarrow \frac{\partial a}{\partial t}\bigg|_z = \frac{\partial a}{\partial t}\bigg|_\rho + \frac{\partial a}{\partial \rho}\frac{\partial \rho}{\partial t}\bigg|_z.$$

Then, application to $a = z$ gives $0 = z_x + z_\rho \rho_x$, $1 = z_\rho \rho_z$, etc. (where a subscript indicates a derivative). This provides the rule to change the derivative of ρ at z constant to that of z at ρ constant. For a other than z, we can write

$$\frac{\partial a}{\partial x}\bigg|_z = \frac{\partial a}{\partial x}\bigg|_\rho - \frac{z_x}{z_\rho}\frac{\partial a}{\partial \rho}, \tag{12.1}$$

with similar expressions where x is replaced by y or t, and

$$\frac{\partial a}{\partial z} = \frac{1}{z_\rho}\frac{\partial a}{\partial \rho}. \tag{12.2}$$

Here, subscripts denote derivatives. Fig. 12.1 depicts a geometrical interpretation of rule (12.1).

The hydrostatic Eq. (4.19) readily becomes

$$\frac{\partial p}{\partial \rho} = -\rho g \frac{\partial z}{\partial \rho} \tag{12.3}$$

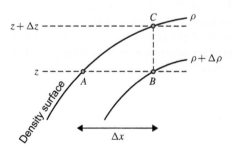

FIGURE 12.1 Geometrical interpretation of Eq. (12.1). The x-derivatives of any function a at constant depth z and at constant density ρ are $[a(B) - a(A)]/\Delta x$ and $[a(C) - a(A)]/\Delta x$, respectively. The difference between the two, $[a(C) - a(B)]/\Delta x$, represents the vertical derivative of a, $[a(C) - a(B)]/\Delta z$, times the slope of the density surface, $\Delta z/\Delta x$. Finally, the vertical derivative can be split as the ratio of the ρ-derivative of a, $[a(C) - a(B)]/\Delta \rho$, by $\Delta z/\Delta \rho$.

and leads to the following horizontal pressure gradient:

$$\left.\frac{\partial p}{\partial x}\right|_z = \left.\frac{\partial p}{\partial x}\right|_\rho - \frac{z_x}{z_\rho}\frac{\partial p}{\partial \rho} = \left.\frac{\partial p}{\partial x}\right|_\rho + \rho g\frac{\partial z}{\partial x} = \left.\frac{\partial P}{\partial x}\right|_\rho.$$

Similarly, $\partial p/\partial y$ at constant z becomes $\partial P/\partial y$ at constant ρ. The new function P, which plays the role of pressure in the density-coordinate system, is defined as

$$P = p + \rho g z \qquad (12.4)$$

and is called the *Montgomery potential*.[1] Later on, when there is no ambiguity, this potential may loosely be called pressure. With P replacing pressure, the hydrostatic balance, (12.3), now takes a more compact form:

$$\frac{\partial P}{\partial \rho} = g z, \qquad (12.5)$$

further indicating that P is the natural substitute for pressure when density is the vertical coordinate.

Beyond this point, all derivatives with respect to x, y, and time are meant to be taken at constant density, and the subscript ρ is no longer necessary.

With the use of Eqs. (12.1)–(12.3) and the obvious relation $\partial \rho/\partial x|_\rho = 0$, the density-conservation equation, (4.21e) in the absence of diffusion, can be solved for the vertical velocity

$$w = \frac{\partial z}{\partial t} + u\frac{\partial z}{\partial x} + v\frac{\partial z}{\partial y}. \qquad (12.6)$$

This last equation simply tells that the vertical velocity is that necessary for the particle to remain at all times on the same density surface in analogy with surface fluid particles having to remain on the surface [see Eq. (7.12)]. Armed with expression (12.6), we can now eliminate the vertical velocity throughout the set of governing equations. First, the material derivative (3.3) assumes a simplified, two-dimensional-like form

$$\frac{d}{dt} = \frac{\partial}{\partial t} + u\frac{\partial}{\partial x} + v\frac{\partial}{\partial y}, \qquad (12.7)$$

where the derivatives are now taken at constant ρ. The absence of an advective term in the third spatial direction results from the absence of motion across density surfaces.

---

[1] In honor of Raymond B. Montgomery who first introduced it in 1937. See his biography at the end of this chapter.

In the absence of friction and in the presence of rotation, the horizontal-momentum equations (4.21a) and (4.21b) become

$$\frac{du}{dt} - fv = -\frac{1}{\rho_0} \frac{\partial P}{\partial x} \tag{12.8a}$$

$$\frac{dv}{dt} + fu = -\frac{1}{\rho_0} \frac{\partial P}{\partial y}. \tag{12.8b}$$

We note that they are almost identical to their original versions. The differences are nonetheless important: The material derivative is now along density surfaces and expressed by Eq. (12.7), the pressure p has been replaced by the Montgomery potential P defined in Eq. (12.4), and all temporal and horizontal derivatives are taken at constant density. Note, however, that the components u and v are still the true horizontal velocity components and are not measured along sloping density surfaces. This property is important for the proper application of lateral boundary conditions.

To complete the set of equations, it remains to transform the continuity equation (4.21d) according to rules (12.1) and (12.2). Further elimination of the vertical velocity by using Eq. (12.6) leads to

$$\frac{\partial h}{\partial t} + \frac{\partial}{\partial x}(hu) + \frac{\partial}{\partial y}(hv) = 0, \tag{12.9}$$

where the quantity h introduced for convenience is proportional to $\partial z/\partial \rho$, the derivative of depth with respect to density. For practicality, we want h to have the dimension of height, and so we introduce an arbitrary but constant density difference, $\Delta\rho$, and define

$$h = -\Delta\rho \frac{\partial z}{\partial \rho}. \tag{12.10}$$

In this manner, h can be interpreted as the thickness of a fluid layer between the density ρ and $\rho + \Delta\rho$. At this point, the value of $\Delta\rho$ is arbitrary, but later, in the development of layered models, it will naturally be chosen as the density difference between adjacent layers.

The transformation of coordinates is now complete. The new set of governing equations consists of the two horizontal-momentum equations (12.8a) and (12.8b), the hydrostatic balance (12.5), the continuity equation (12.9), and the relation (12.10). It thus forms a closed 5-by-5 system for the dependent variables, u, v, P, z, and h. Once the solution is known, the pressure p and the vertical velocity w can be recovered from Eqs. (12.4) and (12.6).

The governing equations are accompanied by the relevant boundary and initial conditions of Section 4.6. We only have to evaluate the derivatives of the Cartesian coordinates according to Eqs. (12.1) and (12.2) in order to impose the auxiliary conditions in the new coordinate system. The fluxes of heat and mass, leading to buoyancy changes, are not easily incorporated because of the interplay with density, the new coordinate. Since processes that do not conserve density are neglected in most applications of isopycnal models, we will not investigate this point here but refer to Dewar (2001) for further details on the representation of mixed-layer dynamics in isopycnal models.

Since the aforementioned work of Montgomery (1937), the substitution of density as the vertical variable has been implemented in a number of applications, especially by Robinson (1965) in a study of inertial currents, by Hodnett (1978) and Huang (1989) in studies of the permanent oceanic thermocline, and by Sutyrin (1989) in a study of isolated eddies. A review in the meteorological context is provided by Hoskins, McIntyre and Robertson (1985).

12.2 LAYERED MODELS

A *layered model* is an idealization by which a stratified fluid flow is represented as a finite number of moving layers, stacked one upon another and each having a uniform density. Its evolution is governed by a discretized version of the system of equations in which density, taken as the vertical variable, is not varied continuously but in steps: density is restricted to assume a finite number of values. A layered model is the density analog of a *level model*, which is obtained after discretization of the vertical variable z.

Each layer ($k = 1$ to m, where m is the number of layers) is characterized by its density ρ_k (unchanging), thickness h_k, Montgomery potential P_k, and horizontal velocity components u_k and v_k. The surface marking the boundary between two adjacent layers is called an *interface* and is described by its elevation z_k, measured (negatively downward) from the mean surface level. The displaced surface level is denoted z_0 (Fig. 12.2a). The interfacial heights can be obtained recursively from the bottom[2]

$$z_m = b, \tag{12.11}$$

upward:

$$z_{k-1} = z_k + h_k, \quad k = m \text{ to } 1. \tag{12.12}$$

This geometrical relation can be regarded as the discretized version of Eq. (12.10) used to define h.

In a similar manner, the discretization of hydrostatic relation (12.5) provides another recursive relation, which can be used to evaluate the Montgomery potential P from the top,

$$P_1 = p_{atm} + \rho_0 g z_0, \tag{12.13}$$

---

[2] Note that contrary to our general approach of using indexes which increase with the Cartesian coordinate directions, we choose to increase the index k downward, in agreement with the traditional notation for isopycnal models and with the fact that our new vertical coordinate ρ is increasing downward, too.

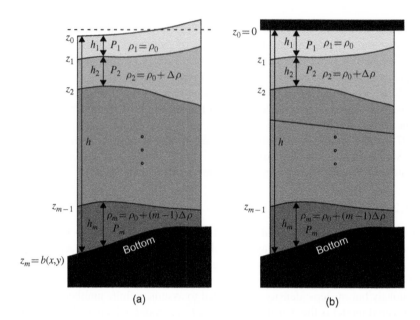

FIGURE 12.2   A layered model with m active layers: (a) with free surface, (b) with rigid lid.

downward:

$$P_{k+1} = P_k + \Delta\rho g z_k, \quad k = 1 \text{ to } m-1. \tag{12.14}$$

In writing (12.13), we have selected the uppermost density ρ_1 as the reference density ρ_0. Gradients of the atmospheric pressure p_{atm} rarely play a significant role, and the contribution of p_{atm} to P_1 is usually omitted. If the layered model is for the lower atmosphere, p_{atm} represents a pressure distribution aloft and may, too, be taken as an inactive constant.

When the *reduced gravity*,

$$g' = \frac{\Delta\rho}{\rho_0} g, \tag{12.15}$$

is introduced for convenience, the recursive relations (12.12) and (12.14) lead to simple expressions for the interfacial heights and Montgomery potentials. For up to three layers, these equations are summarized in Table 12.1.

   In certain applications, it is helpful to discard surface gravity waves because they travel much faster than internal waves and near-geostrophic disturbances. To do so, we eliminate the flexibility of the surface by imagining that the system is covered by a rigid lid (Fig. 12.2b). This is called the *rigid-lid approximation*, which has already been introduced in the study of barotropic motions in Section 7.5. In such a case, z_0 is set to zero, and there are only $(m-1)$ independent layer thicknesses. In return, one of the Montgomery potentials cannot be

---

TABLE 12.1 Layered Models

One Layer:
$z_0 = h_1 + b$ $P_1 = \rho_0 g(h_1 + b)$
$z_1 = b$

Two Layers:
$z_0 = h_1 + h_2 + b$ $P_1 = \rho_0 g(h_1 + h_2 + b)$
$z_1 = h_2 + b$ $P_2 = \rho_0 g h_1 + \rho_0 (g + g')(h_2 + b)$
$z_2 = b$

Three Layers:
$z_0 = h_1 + h_2 + h_3 + b$ $P_1 = \rho_0 g(h_1 + h_2 + h_3 + b)$
$z_1 = h_2 + h_3 + b$ $P_2 = \rho_0 g h_1 + \rho_0 (g + g')(h_2 + h_3 + b)$
$z_2 = h_3 + b$ $P_3 = \rho_0 g h_1 + \rho_0 (g + g') h_2$
$z_3 = b$ $+ \rho_0 (g + 2g')(h_3 + b)$

TABLE 12.2 Rigid-Lid Models

One Layer:
$z_1 = -h_1$ P_1 variable
$h_1 = h$, fixed

Two Layers:
$z_1 = -h_1$ $P_1 = P_2 + \rho_0 g' h_1$
$z_2 = -h_1 - h_2$ P_2 variable
$h_1 + h_2 = h$, fixed

Three Layers:
$z_1 = -h_1$ $P_1 = P_3 + \rho_0 g'(2h_1 + h_2)$
$z_2 = -h_1 - h_2$ $P_2 = P_3 + \rho_0 g'(h_1 + h_2)$
$z_3 = -h_1 - h_2 - h_3$ P_3 variable
$h_1 + h_2 + h_3 = h$, fixed

---

derived from the hydrostatic relation. If this potential is chosen as the one in the lowest layer, the recursive relations yield the equations of Table 12.2.

In some other instances, mainly in the investigation of upper-ocean processes, the lowest layer may be imagined to be infinitely deep and at rest (Fig. 12.3). Keeping m as the number of moving layers, we assign to this lowest (abyssal) layer the index $(m + 1)$. The absence of motion there implies a uniform Montgomery potential, the value of which may be set to zero without loss of generality: $P_{m+1} = 0$. For up to three active layers, the recursive relations provide equations of Table 12.3. Because these expressions do not involve the full gravity g but only its reduced value g', this type of model is known as a *reduced-gravity model*.

TABLE 12.3 Reduced Gravity Models

One Layer:

$z_1 = -h_1$ $P_1 = \rho_0 g' h_1$

Two Layers:

$z_1 = -h_1$ $P_1 = \rho_0 g'(2h_1 + h_2)$
$z_2 = -h_1 - h_2$ $P_2 = \rho_0 g'(h_1 + h_2)$

Three Layers:

$z_1 = -h_1$ $P_1 = \rho_0 g'(3h_1 + 2h_2 + h_3)$
$z_2 = -h_1 - h_2$ $P_2 = \rho_0 g'(2h_1 + 2h_2 + h_3)$
$z_3 = -h_1 - h_2 - h_3$ $P_3 = \rho_0 g'(h_1 + h_2 + h_3)$

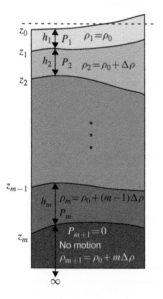

FIGURE 12.3 A reduced-gravity layered model. The assumption of a very deep ocean at rest can be justified by the need to keep the transport hu_{m+1} and kinetic energy hu_{m+1}^2 bounded so that velocities must vanish as the depth of the last layer increases to infinity. In this case, the pressure in the deeper layer tends towards a constant, which we may take as zero.

In this table, $z_1 = -h_1$ is an approximation that begs for an explanation. The free surface is not at $z = z_0 = 0$ but given by Eq. (12.13) when we arrive at the surface integrating upward. For a single layer this yields, in the absence of atmospheric pressure variation,

$$P_2 = 0 \rightarrow \quad P_1 = -\Delta\rho g z_1 = \rho_0 g z_0. \tag{12.16}$$

Hence, $g z_0 = -g' z_1$. Since $h_1 = z_0 - z_1$, we get $z_0 = -(g'/g)z_1$ and $h_1 = -(1 + g'/g)z_1 \simeq -z_1$ since $g' \ll g$. This implies a surface lifting over light-water lenses. Indeed, in order to preserve a uniform pressure in the lowest layer, a thickening of the light-water layer must be compensated by an addition of water above mean sea level.

Generalization to more than three moving layers is straightforward. When a configuration with few but physically relevant layers is desired, the preceding derivations may be extended to nonuniform density differences from layer to layer. Mathematically, this would correspond to a discretization of the vertical density coordinate in unevenly spaced gridpoints.

Once the layer thicknesses, interface depths, and layer pressures (more precisely, the Montgomery potentials) are all related, the system of governing equations is completed by gathering the horizontal-momentum equations (12.8a) and (12.8b) and the continuity equation (12.9), each written for every layer.

In Section 11.6, the length $L = NH/\Omega$ was derived as the horizontal scale at which rotation and stratification play equally important roles. It is noteworthy at this point to formulate the analog for a layered system. Introducing H as a typical layer thickness in the system (such as the maximum depth of the uppermost layer at some initial time) and $\Delta\rho$ as a density difference between two adjacent layers (such as the top two), an approximate expression of the stratification frequency squared is

$$N^2 = -\frac{g}{\rho_0}\frac{d\rho}{dz} \simeq \frac{g}{\rho_0}\frac{\Delta\rho}{H} = \frac{g'}{H}, \qquad (12.17)$$

where $g' = g\,\Delta\rho/\rho_0$ is the reduced gravity defined earlier. Substitution of Eq. (12.17) in the definition of L yields $L \simeq (g'H)^{1/2}/\Omega$. Finally, because the ambient rotation rate Ω enters the dynamics only through the Coriolis parameter f, it is more convenient to introduce the length scale

$$R = \frac{\sqrt{g'H}}{f}, \qquad (12.18)$$

called the *radius of deformation*. To distinguish this last scale from its cousin (9.12) derived for free-surface homogeneous rotating fluids (where the full gravitational acceleration g appears), it is customary in situations where ambiguity could arise to use the expressions *internal radius of deformation* and *external radius of deformation* for Eqs. (12.18) and (9.12), respectively. Because density differences within geophysical fluids are typically a percent or less of the average density, the internal radius is most often less than one-tenth the external radius.

When the model consists of a single moving layer above a motionless abyss, the governing equations reduce to

$$\frac{\partial u}{\partial t} + u\frac{\partial u}{\partial x} + v\frac{\partial u}{\partial y} - fv = -g'\frac{\partial h}{\partial x} \qquad (12.19a)$$

$$\frac{\partial v}{\partial t} + u\frac{\partial v}{\partial x} + v\frac{\partial v}{\partial y} + fu = -g'\frac{\partial h}{\partial y} \qquad (12.19b)$$

$$\frac{\partial h}{\partial t} + \frac{\partial}{\partial x}(hu) + \frac{\partial}{\partial y}(hv) = 0. \qquad (12.19c)$$

The subscripts indicating the layer have become superfluous and have been omitted. The coefficient $g' = g(\rho_2 - \rho_1)/\rho_0$ is called the *reduced gravity*. Except for the replacement of the full gravitational acceleration, g, by its reduced fraction, g', this system of equations is identical to that of the shallow-water model over a flat bottom [Eq. (7.17)] and is thus called the *shallow-water reduced-gravity model*. Because the vertical simplicity of this model permits the investigation of a number of horizontal processes with a minimum of mathematical complication, it will be used in some of the following chapters. Finally, recall that the Coriolis parameter, f, may be taken as either a constant (f-plane) or as a function of latitude ($f = f_0 + \beta_0 y$, beta plane).

12.3   POTENTIAL VORTICITY

For layered models, we can reproduce the vorticity analysis that we performed on the shallow-water model (Section 7.4). First, the relative vorticity ζ of the flow at any level is defined as

$$\zeta = \frac{\partial v}{\partial x} - \frac{\partial u}{\partial y}, \tag{12.20}$$

and the expression for potential vorticity is defined in analogy with (7.25):

$$\begin{aligned} q &= \frac{f + \zeta}{h} \\ &= \frac{f + \partial v/\partial x - \partial u/\partial y}{h}, \end{aligned} \tag{12.21}$$

which is identical to the expression for a barotropic fluid, except that the denominator is now a differential thickness given by Eq. (12.10) rather than the full thickness of the system. It can be shown that in the absence of friction, expression (12.21) is conserved by the flow (its material derivative is zero).

The interpretation of this conservation property follows that for a barotropic fluid: When the fluid layer between two consecutive density surface is squeezed (from left to right in Fig. 12.4), conservation of volume demands that it widens, and conservation of circulation in turn requires that it spins less fast; the net effect is that the vorticity $f + \zeta$ decreases in proportion to the thickness h of the fluid layer.

12.4   TWO-LAYER MODELS

For the representation of stratified systems with the simplest possible formalism, the two-layer model is often the tool of choice for it retains the effect of stratification in some basic way through the reduced gravity g' while keeping the number of equations and variables to a minimum. According to Table 12.1,

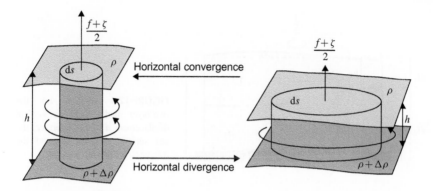

FIGURE 12.4   Conservation of volume and circulation in a fluid undergoing divergence (squeezing) or convergence (stretching). The products of hds and $(f+\zeta)ds$ are conserved during the transformation, implying conservation of $(f+\zeta)/h$, too.

the inviscid governing equations of the two-layer model are

$$\frac{\partial u_1}{\partial t}+u_1\frac{\partial u_1}{\partial x}+v_1\frac{\partial u_1}{\partial y}-fv_1=-g\frac{\partial(h_1+h_2+b)}{\partial x} \tag{12.22a}$$

$$\frac{\partial v_1}{\partial t}+u_1\frac{\partial v_1}{\partial x}+v_1\frac{\partial v_1}{\partial y}+fu_1=-g\frac{\partial(h_1+h_2+b)}{\partial y} \tag{12.22b}$$

$$\frac{\partial u_2}{\partial t}+u_2\frac{\partial u_2}{\partial x}+v_2\frac{\partial u_2}{\partial y}-fv_2=-g\frac{\partial h_1}{\partial x}-(g+g')\frac{\partial(h_2+b)}{\partial x} \tag{12.22c}$$

$$\frac{\partial v_2}{\partial t}+u_2\frac{\partial v_2}{\partial x}+v_2\frac{\partial v_2}{\partial y}+fu_2=-g\frac{\partial h_1}{\partial y}-(g+g')\frac{\partial(h_2+b)}{\partial y} \tag{12.22d}$$

$$\frac{\partial h_1}{\partial t}+\frac{\partial(h_1u_1)}{\partial x}+\frac{\partial(h_1v_1)}{\partial y}=0 \tag{12.22e}$$

$$\frac{\partial h_2}{\partial t}+\frac{\partial(h_2u_2)}{\partial x}+\frac{\partial(h_2v_2)}{\partial y}=0 \tag{12.22f}$$

for the six unknowns h_1, u_1, v_1, h_2, u_2, and v_2.

If we introduce the surface elevation η and the vertical displacement a of the interface between the two layers through $h_1+h_2+b=H+\eta$ and $h_2+b=H_2+a$, where H and H_2 are two constants representing, respectively, the mean surface level and the mean interface level, each measured from the reference datum from which the bottom elevation b, too, is measured (Fig. 12.5), the pressure terms in the x-direction can be rewritten as

$$-g\frac{\partial(h_1+h_2+b)}{\partial x}=-g\frac{\partial\eta}{\partial x} \tag{12.23a}$$

$$-g'\frac{\partial(h_2+b)}{\partial x}=-g'\frac{\partial a}{\partial x} \tag{12.23b}$$

and similarly for the pressure terms in the y-direction.

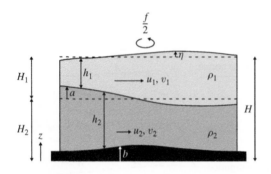

FIGURE 12.5 Notation for the two-layer model with the vertical displacement a of the interface, the sea surface elevation η, and the reference heights H_1 and H_2.

Oftentimes, the two-layer model is used in analytical studies in which case it is also linearized for added simplicity. The equations are then

$$\frac{\partial u_1}{\partial t} - f v_1 = -g \frac{\partial \eta}{\partial x} \tag{12.24a}$$

$$\frac{\partial v_1}{\partial t} + f u_1 = -g \frac{\partial \eta}{\partial y} \tag{12.24b}$$

$$\frac{\partial u_2}{\partial t} - f v_2 = -g \frac{\partial \eta}{\partial x} - g' \frac{\partial a}{\partial x} \tag{12.24c}$$

$$\frac{\partial v_2}{\partial t} + f u_2 = -g \frac{\partial \eta}{\partial y} - g' \frac{\partial a}{\partial y} \tag{12.24d}$$

$$\frac{\partial (\eta - a)}{\partial t} + \frac{\partial (H_1 u_1)}{\partial x} + \frac{\partial (H_1 v_1)}{\partial y} = 0 \tag{12.24e}$$

$$\frac{\partial a}{\partial t} + \frac{\partial [(H_2 - b) u_2]}{\partial x} + \frac{\partial [(H_2 - b) v_2]}{\partial y} = 0, \tag{12.24f}$$

where $H_1 = H - H_2$ is the mean thickness of the top layer.

In the case of flat bottom ($b = 0$), it is interesting to decompose this set of six coupled equations in two sets of three in order to facilitate the solution and clarify the dynamics. For this, we seek proportionality of the type $u_2 = \lambda u_1$, $v_2 = \lambda v_1$ and $\eta = \mu a$ between variables of one layer with those of the other layer. Momentum equations (12.24c)–(12.24d) become identical to (12.24a)–(12.24b) if

$$\frac{\lambda}{1} = \frac{g\mu + g'}{g\mu}, \tag{12.25}$$

while continuity equation (12.24f) replicates (12.24e) if

$$\frac{1}{\mu - 1} = \frac{H_2 \lambda}{H_1}. \tag{12.26}$$

Elimination of μ between the preceding two equations yields an equation for the proportionality coefficient λ:

$$H_2\lambda^2 + \left(H_1 - H_2 - \frac{g'}{g}H_2\right)\lambda - H_1 = 0. \tag{12.27}$$

Neglecting the small ratio $g'/g = \Delta\rho/\rho_0 \ll 1$, the pair of solutions is

$$\lambda = \frac{(H_2 - H_1) \pm (H_2 + H_1)}{2H_2}. \tag{12.28}$$

Selection of the $+$ sign gives $\lambda = 1$, implying a vertically uniform flow ($u_1 = u_2$ and $v_1 = v_2$). This is called the *barotropic mode*. The interfacial displacement a is related to the surface elevation η by $a = \eta/\mu = H_2\eta/H$ and is thus a vertically prorated fraction of the latter. This mode behaves as if the density difference were absent.

Selection of the $-$ sign in Eq. (12.28) provides the other mode, with $\lambda = -H_1/H_2$, $H_2u_2 = -H_1u_1$, and $H_2v_2 = -H_1v_1$. The vertically integrated transport is nil for this mode. Equation (12.25) then provides the ratio between vertical elevations, $\mu = -g'H_2/gH$, which is small because it is on the order of the relative density difference $\Delta\rho/\rho_0$. This means that the surface elevation η is weak compared with the interfacial displacement a. For this mode, therefore, the flow is vertically compensated, and its surface is nearly rigid. In other words, it is an internal mode called the *baroclinic mode*.

The equations governing each mode separately can be obtained as follows. For the barotropic mode, we define $u_T = u_1 = u_2$, $v_T = v_1 = v_2$, and put $a = H_2\eta/H$. Within an error on the order of $\Delta\rho/\rho_0$, the momentum equations reduce to a single pair,

$$\frac{\partial u_T}{\partial t} - fv_T = -g\frac{\partial \eta}{\partial x} \tag{12.29a}$$

$$\frac{\partial v_T}{\partial t} + fu_T = -g\frac{\partial \eta}{\partial y} \tag{12.29b}$$

while each continuity equation reduces to

$$\frac{\partial \eta}{\partial t} + H\frac{\partial u_T}{\partial x} + H\frac{\partial v_T}{\partial y} = 0. \tag{12.29c}$$

If we scale time by $1/f$, distances by L and the velocity components by U, the momentum equations tell us that the surface elevation is on the order of fLU/g. Substitution of these scales in the continuity equation then requires that $f(fLU/g) \sim HU/L$, which sets the square of the length scale to $L^2 \sim gH/f^2$. This leads us to define the barotropic (or external) radius of deformation:

$$R_{\text{external}} = \frac{\sqrt{gH}}{f}. \tag{12.30}$$

Similarly, the equations governing the baroclinic mode are obtained by defining $u_B = u_1 - u_2$ and $v_B = v_1 - v_2$, and setting $\eta = -(g'H_2/gH)a$. Subtraction of the momentum equations exploiting $H_1 u_1 = -H_2 u_2$ for the baroclinic mode yields

$$\frac{\partial u_B}{\partial t} - f v_B = +g'\frac{\partial a}{\partial x} \tag{12.31a}$$

$$\frac{\partial v_B}{\partial t} + f u_B = +g'\frac{\partial a}{\partial y} \tag{12.31b}$$

while subtraction of the continuity equations gives

$$-\frac{\partial a}{\partial t} + \frac{H_1 H_2}{H}\frac{\partial u_B}{\partial x} + \frac{H_1 H_2}{H}\frac{\partial v_B}{\partial y} = 0. \tag{12.31c}$$

To determine the corresponding radius of deformation, we scale time by $1/f$, distances by L' and the velocity components by U'. According to the momentum equations, the interfacial displacement scales like $fL'U'/g'$. Substitution of these scales in the continuity equation requires $f(fL'U'/g') \sim H_1 H_2 U'/HL'$, which sets the square of the length scale to $L'^2 \sim g'H_1 H_2/f^2 H$. This in turn leads us to define the baroclinic (or internal) radius of deformation:

$$R_{internal} = \frac{1}{f}\sqrt{\frac{g'H_1 H_2}{H_1 + H_2}}. \tag{12.32}$$

Note that $R_{internal}$ is significantly shorter than $R_{external}$ because the reduced gravity g' is much smaller than the full gravity g. Another interpretation of the internal radius of deformation is obtained by observing that (12.29) has the same form as (12.31) if η is replaced by $-a$, barotropic velocities by their baroclinic counterpart, g by g', and H by $\bar{h} = H_1 H_2/(H_1 + H_2)$. Hence, the role played by the deformation radius in the barotropic mode is now played by the internal radius for the internal mode. Also, the gravity-wave propagation speed is replaced by the propagation speed of internal gravity waves

$$c = \sqrt{\frac{g'H_1 H_2}{H_1 + H_2}} = \sqrt{g'\bar{h}}, \tag{12.33}$$

which is much lower than the external gravity-wave speed. Because equations (12.31) are structurally identical to (12.29), the solutions will also be, and all wave solutions of the shallow-water equations of Chapter 9 can be applied to the internal mode with the appropriate definitions of gravity and depth. When interpreting the solution we simply have to keep in mind the difference in vertical structure. While the barotropic mode is uniform in the vertical (left panel of Fig. 12.6), the baroclinic mode has zero transport and hence opposite velocities in each layer (right panel of Fig. 12.6).

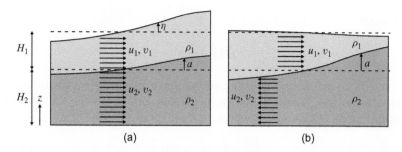

FIGURE 12.6   Barotropic mode (a) and baroclinic mode (b). For the barotropic mode, the interface moves in phase with the sea surface, and velocity is uniform over both layers. In the baroclinic mode, the surface displacement is very weak compared with the interfacial displacement and opposite to it. Velocities in the baroclinic mode are opposite to each other and create no net transport.

When more than two layers are present, the number of modes increases accordingly. For three layers, there will be three modes and so on. In the limit of an infinite number of levels representing continuous stratification, we therefore expect an infinite number of vertical modes, a situation we will encounter again in Section 13.4.

12.5   WIND-INDUCED SEICHES IN LAKES

An interesting application of the two-layer model is to the seiching of thermally stratified lakes. Most lakes of temperate latitudes undergo thermal stratification in summer, and by late summer when surface cooling begins, the water column is often divided into a relatively well mixed and warmer surface layer (called *epilimnion*) and a colder bottom layer (called *hypolimnion*), separated by a thin layer of rapid temperature variation (the *thermocline*). Waves can propagate along both surface and thermocline, and reflection of these waves at the lake's ends can create standing waves called *seiches*. A seiche is usually the response to a wind event: the wind blows for some time, dragging upper-layer water to the downwind side, thereby raising the water level and depressing the thermocline at the downwind end. When the wind relaxes, the situation is out-of-equilibrium, and the warm water begins to slush back and forth across the lake, creating the seiche.

The simplest seiche model assumes no rotational effect (because lakes are typically much smaller than the ocean), a flat bottom and the absence of friction (to simplify the analysis). To illustrate such a seiche model, we take the two-layer model without rotation ($f = 0$) for a domain confined between a flat bottom and two lateral boundaries.

First we concentrate on the barotropic mode with the system of Eqs. (12.29). If we differentiate (12.29a) with respect to x, (12.29b) with respect to y, and (12.29c) with respect to t, and then subtract the first two from the last one, we obtain

$$\frac{\partial^2 \eta}{\partial t^2} = gH \left(\frac{\partial^2 \eta}{\partial x^2} + \frac{\partial^2 \eta}{\partial y^2} \right), \tag{12.34}$$

We recognize the generic two-dimensional wave equation. Impermeability of lateral boundaries translates into zero normal derivative of η, as seen for example in Eq. (12.29a) when we set $f = 0$ and impose $u = 0$. Thus,

$$\frac{\partial \eta}{\partial x} = 0 \quad \text{at} \quad x = 0, L \tag{12.35}$$

for a basin of length L in the x-direction. Similarly in the y-direction,

$$\frac{\partial \eta}{\partial y} = 0 \quad \text{at} \quad y = 0, W \tag{12.36}$$

if the width is W.

For such a rectangular domain, it is easily verified that

$$\eta = A \cos(\omega t) \cos \left(\frac{m \pi x}{L} \right) \cos \left(\frac{n \pi y}{W} \right) \tag{12.37}$$

is the solution of Eq. (12.34) satisfing all four boundary conditions as long as m and n are integers and provided that

$$\omega^2 = gH \left(\frac{m^2 \pi^2}{L^2} + \frac{n^2 \pi^2}{W^2} \right) \tag{12.38}$$

The solution $m = n = 0$ corresponds to a situation of rest, and, for an elongated basin with $L \geq W$, the gravest mode is obtained for $m = 1$, $n = 0$. In this case, the seiche is a standing wave of frequency

$$\omega = \pi \frac{\sqrt{gH}}{L}. \tag{12.39}$$

The gravest mode is of special importance because it is the one with the smoothest structure and thus the one that is dissipated at the slowest rate. Also, wind forcing is more likely to generate the gravest mode because atmospheric forcing generally varies weakly over the length of a lake and can, at a first approximation, be considered uniform.

We can immediately extend the previous result to the baroclinic mode by replacing gH by $g'H_1H_2/(H_1 + H_2)$ and interpret the velocity oscillations in the light of the baroclinic mode of Fig. 12.6. The lowest-mode frequency becomes

$$\omega = \frac{\pi}{L} \sqrt{g' \frac{H_1 H_2}{H_1 + H_2}} \tag{12.40}$$

and corresponds to a much longer period of oscillation. Sloshing of the density interface is similar to an oscillation of the surface. In an internal seiche, however, the upper-layer velocity is opposite to that in the lower layer, so that water going to one side of the lake near the surface is compensated by flow near the bottom to the other side of the lake, with no appreciable change in surface elevation.

When rotation comes into play (f no longer zero), the standing wave pattern is no longer formed by the superposition of pure gravity waves but by the superposition of inertial-gravity waves, and the mathematical solution is more complicated. In particular, so-called *amphidromic points*, at which the amplitude is nil, can arise (see Section 9.8). The interested reader is referred to Taylor (1921).

Seiches occur not only in lakes but also in the coastal ocean. In the Adriatic Sea, a longitudinal seiche following an episode of sirocco wind can combine with a tidal elevation to create flooding in Venice (Cushman-Roisin, Gačić, Poulain & Artegiani, 2001). Internal seiches have also been observed in Scandinavian fjords (e.g., Arneborg & Liljebladh, 2001).

12.6   ENERGY CONSERVATION

Inspection of energetics in a layered model is insightful because it provides the formulation of the quantities that serve as kinetic and potential energies. To do this, we reinstate the nonlinear terms, the Coriolis acceleration and uneven bottom topography but restrict our attention to the two-layer system described by Eq. (12.22) with the pressure gradients given by Eq. (12.23).

In the absence of friction, no dissipation is present, and we expect conservation of total energy, sum of kinetic energy (*KE*), and potential energy (*PE*). These are defined respectively as:

$$
\begin{aligned}
KE &= \frac{\rho_0}{2} \iiint \left(u^2 + v^2 \right) dz \, dy \, dx \\
&= \frac{\rho_0}{2} \iint \left[\int_b^{H_2+a} \left(u_2^2 + v_2^2 \right) dz + \int_{H_2+a}^{H_1+H_2+\eta} \left(u_1^2 + v_1^2 \right) dz \right] dy \, dx \\
&= \frac{\rho_0}{2} \iint \left[h_2 \left(u_2^2 + v_2^2 \right) + h_1 \left(u_1^2 + v_1^2 \right) \right] dy \, dx.
\end{aligned}
\tag{12.41}
$$

$$
\begin{aligned}
PE &= \iiint \rho g z \, dz \, dy \, dx \\
&= \iint \int_b^{H_2+a} \rho_2 g z \, dz \, dy \, dx + \iint \int_{H_2+a}^{H_1+H2+\eta} \rho_1 g z \, dz \, dy \, dx
\end{aligned}
\tag{12.42}
$$

Note how we used the Boussinesq approximation: The actual density (ρ_1 or ρ_2) is used next to g in the potential energy but is replaced by the reference density (ρ_0) in the kinetic energy.

Mass conservation in a closed basin leads obviously to

$$\iint \eta \, dy \, dx = 0 \qquad \iint a \, dy \, dx = 0 \qquad (12.43)$$

so that, up to an additive constant, potential energy can be expressed as

$$PE = \frac{\rho_0}{2} \iint \left(g\eta^2 + g'a^2 \right) dy \, dx, \qquad (12.44)$$

which shows that vertical displacements a of the interface need to be much larger than surface elevations η to contribute equally to potential energy. With the previous definition, the reference state $\eta = 0$ and $a = 0$ has zero potential energy. Any departure from this situation leads to a positive amount of potential energy. This amount is thus called available potential energy (see also Section 16.4).

To construct the energy budget, we multiply (12.22a) by $h_1 u_1$ and exploit (12.22e) to obtain first:

$$\frac{\partial}{\partial t}\left(\frac{h_1 u_1^2}{2} \right) + \frac{\partial}{\partial x}\left(u_1 \frac{h_1 u_1^2}{2} \right) + \frac{\partial}{\partial y}\left(v_1 \frac{h_1 u_1^2}{2} \right)$$
$$- f h_1 u_1 v_1 h_1 = - \frac{\partial (g h_1 u_1 \eta)}{\partial x} + g\eta \frac{\partial (h_1 u_1)}{\partial x}. \qquad (12.45)$$

We then multiply (12.22b) by $h_1 v_1$, exploit (12.22e), add the result to Eq. (12.45), and integrate over a closed or periodic domain to obtain

$$\frac{d}{dt} \iint \left(h_1 \frac{u_1^2 + v_1^2}{2} \right) dy \, dx = \iint g\eta \left[\frac{\partial (h_1 u_1)}{\partial x} + \frac{\partial (h_1 v_1)}{\partial y} \right] dy \, dx, \qquad (12.46)$$

which shows that the amount of upper-layer kinetic energy can be altered by the divergence of the transport.

Similarly, we can obtain the equation governing the evolution of kinetic energy in the second layer:

$$\frac{d}{dt} \iint \left(h_2 \frac{u_2^2 + v_2^2}{2} \right) dy \, dx = \iint (g\eta + g'a) \left[\frac{\partial (h_2 u_2)}{\partial x} + \frac{\partial (h_2 v_2)}{\partial y} \right] dy \, dx. \qquad (12.47)$$

For the potential-energy budget, we multiply (12.22e) by $g\eta$ and integrate over the domain:

$$\frac{d}{dt} \iint g \frac{\eta^2}{2} dy \, dx = \iint \left\{ g\eta \frac{\partial a}{\partial t} - g\eta \left[\frac{\partial (h_1 u_1)}{\partial x} + \frac{\partial (h_1 v_1)}{\partial y} \right] \right\} dy \, dx. \qquad (12.48)$$

We then multiply (12.22f) by $g'a$, exploit the relation $\partial h_2/\partial t = \partial a/\partial t$, and integrate over the domain yields, to obtain

$$\frac{d}{dt} \iint g' \frac{a^2}{2} \, dy \, dx = \iint -g'a \left[\frac{\partial (h_2 u_2)}{\partial x} + \frac{\partial (h_2 v_2)}{\partial y} \right] dy \, dx. \tag{12.49}$$

Using Eq. (12.22f) to replace $\partial a/\partial t$ in the right-hand side of Eq. (12.48), we can identify similar terms on the right-hand side of Eqs. (12.46)–(12.49) but with opposite signs. These terms represent energy exchange between the different forms of potential and kinetic energy. Hence, by adding the four equations, these terms cancel one another out. After multiplying by ρ_0, we ultimately obtain the statement of energy conservation:

$$\frac{d}{dt} (PE + KE) = 0. \tag{12.50}$$

Besides certifying the expressions for kinetic and potential energy, this analysis has also identified for us the expressions for the exchange terms between the different forms of energy. The underlying mechanism is convergence/divergence of the horizontal flow (affecting kinetic energy) accompanied by piling/dropping of water (affecting potential energy in a compensating way).

12.7   NUMERICAL LAYERED MODELS

The development of numerical models based on the governing equations written in isopycnal coordinates is simplified by the fact that a discretization of the vertical coordinate is already performed through the layering (Section 12.2). Indeed, we arrived at governing equations for a set of m layers, in which the vertical coordinate no longer appears. In other words, we replaced a three-dimensional problem by m coupled two-dimensional problems.

Since it is straightforward to generalize the layering approach and use different values of $\Delta \rho$ across layers, we can easily define layers so as to follow physically meaningful water masses. Once the $\Delta \rho$ values are assigned to define the layers, the only discretization that remains to be done is the one related to the two-dimensional "horizontal" structure, a task we already performed in the context of the shallow-water equations (Section 9.7 and 9.8). Since the governing equations of each isopycnal layer are very similar to those of the inviscid shallow-water equations, all we have to do is to "repeat" the implementation of the shallow-water equations for each layer and adapt the pressure force. Here, we can notice how easily pressure can be calculated in the layered system once the layer thicknesses are known by simply integrating (12.14), or its straightforward generalization when density differences vary between layers. To calculate layer thicknesses, the volume-conservation equations are at our disposal and are also similar to those of the shallow-water system. Finally, as for shallow-water equations, additional processes neglected up to now can be reinstated. Bottom and top stresses can be taken into account at the lowest and uppermost layers by adding a frictional term as we did for the shallow-water equations. Also friction between layers can be accommodated by introducing terms that depend on the velocity difference across each interface. These internal friction terms must appear with opposite signs in the equations for the two layers scrubbing against each other, so that the momentum lost by one is gained by the other.

Finally, unresolved horizontal subgrid scale processes can be parameterized, for example using a lateral eddy viscosity (i.e., adding a Laplacian term). It is noteworthy to point out here that lateral diffusion formulated in the new coordinate system (with $\partial/\partial x$ derivatives taken at constant ρ) corresponds to mixing along isopycnals rather than in the horizontal plane (constant z) (see Section 20.6.2). This is generally considered advantageous if we consider that shorter-scale movements are more easily generated along a density surface because they do not implicate any buoyancy force. The parameterization of subgrid scale processes as diffusion along isopycnal surfaces is called *isopycnal diffusion* and is naturally included in the governing equations of layered models. In other words, no diapycnal (i.e., across-density) diffusion and erosion of stratification will take place, and water masses are conserved.

This is at the same time a major strength and weakness of the layer formulation. It is an advantage if the physical system prevents mixing, and the model simulates motions and oscillations without any numerical destruction of the density stratification, otherwise a common issue affecting three-dimensional models. However, if vertical mixing or vertical convection is significant in the physical system, the layered model requires an additional term to represent transfer (entrainment) of fluid from one layer into another. The danger is then that a layer may lose so much of its fluid and become so thin that is becomes dynamically irrelevant and should be removed, at least in some region of the domain. In addition, layer interfaces can intersect the bottom or top (Fig. 12.7). In other words, the region where each layer exists may not be the entire domain and may furthermore change over time. Tracking the edges of the layers is a problem that is far from trivial.

The problem is even worse when gravitational instabilities are present because the coordinate transformation then loses its validity. Problems associated with strong vertical mixing and gravitational instabilities explain why isopycnal models are rarely used for atmospheric simulations, where convection and associated gravitational instabilities are much more frequent than in the ocean.

Another difficulty with oceanic layered models is the fact that by construction, density is constant within each layer so that temperature and salinity cannot vary independently. Yet, physical boundary conditions are independent for temperature (heat flux) and salinity (evaporation, precipitation). Finally, isopycnal models are not easily applied when the same domain includes both the deep ocean and coastal areas because the variety of density structure does not lend itself to be represented with a single set of density layers.

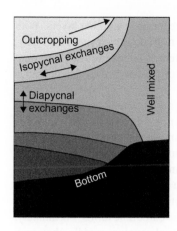

FIGURE 12.7 The application of a layered model needs some special care when isopycnals intersect the surface (outcropping), the bottom, or one another. Surface outcropping typically arises in the vicinity of fronts or following strong mixing events, while bottom intersection is likely to occur in regions of steep topography.

Numerical layered models offer a choice between the original model, the rigid-lid approximation, and the reduced-gravity version. This choice affects the numerical properties. For example, the reduced-gravity model has gravity g replaced by g' in all its equations, and as a result no longer allows propagation of surface gravity waves. This can be desirable from a numerical perspective. Indeed, numerical stability requirements of shallow-water models are typically of the type

$$\frac{\sqrt{gh}\,\Delta t}{\Delta x} \leq \mathcal{O}(1), \tag{12.51}$$

(see Section 9.7). With gravity replaced by reduced gravity, the numerical stability constraint becomes

$$\frac{\sqrt{g'h}\,\Delta t}{\Delta x} \leq \mathcal{O}(1), \tag{12.52}$$

which is much less stringent than (12.51) because $g' \ll g$. The full gravity g also disappears from the models using the rigid-lid approximation, and no stability condition of the type (12.51) applies. Longer time steps may be used in either case.

For the original version of the layered model, using neither rigid-lid approximation nor the reduced-gravity approach, surface gravity waves are possible, and stability condition (12.51) can be very constraining compared with other numerical stability conditions. Some optimization becomes necessary. A brute force approach would be implicit treatment of the terms responsible for the stability constraint. The velocity field in the equation governing layer thickness should then be treated implicitly together with the surface height term in the momentum equations. The latter appears in all momentum equations because the surface pressure is $P_1 = p_{\text{atm}} + \rho_0 g \eta$ according to Eq. (12.13), and the sum

(12.14) means that η becomes part of the pressure terms in all other layers. With an implicit scheme, this means that all equations must be solved simultaneously, forming a rather large, though sparse, linear system to be inverted at every time step. Using a long time step for the propagation of the fast waves also degrades the propagation properties of these waves.

A better approach is to recall that surface gravity waves are generated by surface displacements accompanied by divergence–convergence of the vertically integrated flow and to treat the corresponding subset of the dynamics with a shorter time step than the rest. This can be accomplished by averaging over the vertical in order to construct an equation governing the barotropic component of the flow. The nonlinear terms, however, cause a difficulty because the average of products is not equal to the product of the averages. The result is an equation governing the barotropic mode that contains some baroclinic terms. Fortunately, since those vary slowly whereas the barotropic mode evolves rapidly, they may be held frozen while marching the rest of the equation forward. This is called *mode splitting* (Fig. 12.8).

In such scheme, the surface elevation η is marched N times, whereas the rest of the dynamics is marched forward only once. In view of the typical values of g and g', the longer time step is typically an order of magnitude larger than the short time step required for surface waves, and since the solution of the shallow-water equations with the short time step involves only three instead of $3m$ equations, an order of magnitude can be gained in computational cost by the mode-splitting technique.

At the end of the N barotropic steps and the subsequent single baroclinic step, a problem may occur. The momentum equations of each layer calculated with the new elevation η^{n+1} from the barotropic equation will lead to

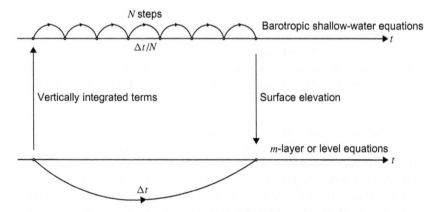

FIGURE 12.8 At a given time step, information from all layer thicknesses permit the calculation of the slow baroclinic terms, which can be held temporarily frozen. The barotropic component of the flow (shallow-water type equation) can then be marched forward in time over several short time steps, until the overall, longer time step Δt is covered. Then, the surface height η at the new time level can be used in all layers to forward in time the internal structure of the flow.

velocities at the new time level u_k^{n+1}, v_k^{n+1}. The transport obtained by summing up these velocities weighted by the corresponding layer thicknesses will, however, be different from the transport that would have been obtained after N substeps because of nonlinearities in the equations. If nothing is done to correct this mismatch, instabilities can occur (e.g., Killworth, Stainforth, Webb & Paterson, 1991). The problem can be avoided by correcting the velocity fields obtained after solving the individual layer equations so as to make sure that their weighted sum equals the predicted transport from the shallow-water equations. This approach can also be applied to level models or any 3D model with a free surface. The general idea remains the preliminary vertical integration of the governing equations in order to make the barotropic component explicit and to march forward in time the latter with a shorter time step than the rest of the equations.

Aside from the aforementioned difficulties, we can retain the distinct advantage of aligning coordinate lines (and hence numerical grids) with dynamically significant features. This explains the success of numerical layer models, and several widely used numerical isopycnal models are based on the successive developments of Hurlburt and Thompson (1980), Bleck, Rooth, Hu and Smith (1992), and Hallberg (1995). The modern tendency, however, is to go beyond pure layer models in favor of more general vertical-coordinate models, which we will encounter in Section 20.6.1.

12.8 LAGRANGIAN APPROACH

In a layered model, the conservation equation for ρ is vastly simplified because coordinate surfaces coincide with material surfaces of the flow. This is the hallmark of a Lagrangian approach. As opposed to a Eulerian representation of the flow, in which flow characteristics are assigned to fixed points, in a Lagrangian approach, characteristics of the flow are tracked following fluid parcels. A layered model, however, does this only in the vertical direction, not in all three directions of space. This motivates us nonetheless to explore here what fully Lagrangian models can offer.

A fully Lagrangian model tracks not merely material surfaces but individual fluid parcels, and just as the choice of density as the vertical coordinate in a layered model eliminates the vertical velocity from advection, a fully Lagrangian approach eliminates all advection terms and is therefore most interesting in problems associated with the discretization of advection terms. In Section 6.4, the astute reader may have already wondered why we went through all these complicated Eulerian schemes to find the solution to a pure-advection problem, which is stated so simply in terms of the material derivative

$$\frac{dc}{dt} = 0 \tag{12.53}$$

and has such a disarming solution $c = c^0$ for a parcel with initial value c^0. A single fluid parcel, however, does not provide the concentration everywhere across the domain but only at its particular location. To determine the concentration distribution, we first need to calculate the trajectories of an ensemble of parcels launched at different locations. For this, we need to return to the basic definition of velocity:

$$\frac{dx}{dt} = u[x(t), y(t), z(t), t] \tag{12.54a}$$

$$\frac{dy}{dt} = v[x(t), y(t), z(t), t] \tag{12.54b}$$

$$\frac{dz}{dt} = w[x(t), y(t), z(t), t], \tag{12.54c}$$

and integrate over time for each of, say, N fluid parcels. If the starting locations are (x_p^0, y_p^0, z_p^0) for $p = 1$ to N, integration of the previous equations provides the positions $[x_p(t), y_p(t), z_p(t)]$ of the same N parcels at time t. This is the core of a *Lagrangian approach*, and it obviously requires the knowledge of the velocity field at all times.

The value of c at any point in the domain can then be obtained by interpolation from the closest parcels or by averaging within grid cells (*binning*), depending on the number of parcels in the region. In either approach, for the method to work, the domain must at any moment be as uniformly covered as possible by a sufficiently dense number of parcels. If there are regions nearly void of parcels, concentrations cannot be inferred there, and this is the first problem with the Lagrangian approach: For an initially relatively uniform distribution of parcels, convergence and divergence of the flow can sooner or later concentrate parcels in some regions and depopulate other parts of the domain (Fig. 12.9). Algorithms must be designed to eliminate parcels in regions where they are redundant and to add new parcels in empty regions. Roughly, if L and H are, respectively, the horizontal and vertical length scales of the 3D flow with surface S and depth D, we need at least $DS/(L^2 H)$ parcels, which is about the required number of grid boxes, (1.17), needed in an Eulerian model of the same flow. However, because the Lagrangian parcels have a tendency to cluster and leave regions with lower coverage, 10–100 times more parcels are typically needed to resolve the same flow.

The time integration itself also imposes some constraints on the method: For accurate time integration of Eq. (12.54), we must be able to respect the spatial variations of the velocity field during a time step, and this requires

$$U \Delta t \leq L, \tag{12.55}$$

or the trajectory calculation will be inaccurate. For the same reason, we must also respect the time variations of the flow, $\Delta t \ll T$ if the flow varies with time scale T.

Another important aspect is related to the integration of the trajectories. In addition to the aforementioned accuracy requirement, two sources of errors

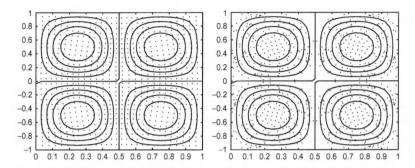

FIGURE 12.9   Calculation of particle displacement in a flow field with multiple circulation cells. Fluid parcels follow the stationary streamfunction (solid line) and deplete some regions. Use traj2D.m to view an animated version.

are encountered. First, the time discretization itself does not generally ensure a reversible calculation. If time or velocities are reversed, the numerical integration does not return parcels to their initial positions (see Numerical Exercise 12.5). Therefore, the time integration produces some dispersion. The second source of error is related to the knowledge of the velocity field itself calculated by a model and thus available only at discrete locations. For a velocity field given on a rectangular grid, calculations of a trajectory passing through the arbitrary location $[x(t), y(t), z(t)]$ require interpolation among adjacent grid points. The resulting interpolation error will then affect the subsequent calculation of the trajectory and induce some additional dispersion among parcels.

Except for those restrictions, the Lagrangian approach is easily implemented. Diffusion can be taken into account by simulating mixing through random displacement of particles, also called *random walk*, according to

$$x^{n+1} = x^n + \int_{t^n}^{t^{n+1}} \left[u(x(t), y(t), z(t), t) + \frac{\partial \mathcal{A}}{\partial x} \right] dt + \sqrt{2\Delta t \mathcal{A}} \, \xi \qquad (12.56)$$

where $\mathcal{A}$ is the desired diffusivity, and ξ is a random variable of Gaussian[3] distribution with zero mean and unit standard deviation (e.g., Gardiner, 1997).

It can be shown (e.g., Gardiner, 1997; Spagnol et al., 2002; Spivakovskaya, Heemink & Deleersnijder, 2007) that the previous stochastic equation leads to

---

[3] It should be noted that in most models the random variable does not obey a Gaussian but a uniform distribution. This is acceptable as long as time steps are short so that performing a succession of many time steps amounts to adding up a large number of random steps, and, by virtue of the *central limit theorem* (e.g., Riley, Hobson & Bence, 1977), the many-step process follows the Gaussian distribution.

particle distributions consistent with

$$\frac{\partial c}{\partial t} + u\frac{\partial c}{\partial x} = \frac{\partial}{\partial x}\left(\mathcal{A}\frac{\partial c}{\partial x}\right). \tag{12.57}$$

A large number of fluid parcels moved according to Eq. (12.56) then mimic the concentration evolution of Eq. (12.57). The role of the term $\partial \mathcal{A}/\partial x$ in the time integral can be easily understood and illustrated in a situation of a local minimum of $\mathcal{A}$ within the domain of interest, such as near a pycnocline (Numerical Exercise 12.8).

If momentum equations, too, are solved by a Lagrangian approach, the momentum of a particle is changed along its trajectory, mostly by the pressure-gradient force. An elegant method among others to obtain the pressure distribution corresponding to a given set of mass-endowed particles is the particle-in-cell method (Cushman-Roisin, Esenkov & Mathias, 2000; Esenkov & Cushman-Roisin, 1999; Pavia & Cushman-Roisin, 1988).

ANALYTICAL PROBLEMS

12.1. Generalize the theory of the coastal Kelvin wave (Section 9.2) to the two-layer system over a flat bottom and under a rigid lid. In particular, what are the wave speed and trapping scale?

12.2. In the case of the shallow-water reduced-gravity model, derive an energy-conservation principle. Then, separate the kinetic and potential energy contributions.

12.3. Show that a steady flow of the shallow-water reduced-gravity system conserves the Bernoulli function $B = g'h + (u^2 + v^2)/2$.

12.4. Establish the equations governing motions in a one-layer model above an uneven bottom and below a thick, motionless layer of slightly lesser density.

12.5. Seek a solution to the shallow-water reduced-gravity model of the type

$$h(x, t) = x^2 A(t) + 2xB(t) + C(t) \tag{12.58a}$$
$$u(x, t) = xU_1(t) + U_0(t) \tag{12.58b}$$
$$v(x, t) = xV_1(t) + V_0(t). \tag{12.58c}$$

To what type of motion does this solution correspond? What can you say of its temporal variability? (Take $f = $ constant.)

12.6. Using the rigid-lid approximation in the shallow-water equations (i.e., a single density layer), analyze the dispersion relation of waves on the beta plane. Show that there are no gravity waves and that the only dispersion relation that remains corresponds to planetary waves (9.27). How can the difference in dispersion relation be interpreted in terms of the hypotheses used in the rigid-lid approximation? (*Hint*: Look again at Section 7.5.)

12.7. In the western Mediterranean Sea, Atlantic waters flow along theAlgerian coast and are slightly lighter than the Mediterranean water. If the density difference is 1.0 kg/m^3, and the thickness of the Atlantic water layer is 150 m, how large is the sea surface displacement associated with the intrusion of the lighter water? (*Hint*: Assume the lower layer to be at rest.)

12.8. Prove the assertion made below Eq. (12.21) that the potential vorticity so defined is indeed conserved along the flow in the absence of friction. Specifically, show that its material derivative, using Eq. (12.7), vanishes.

NUMERICAL EXERCISES

12.1. Adapt the shallow-water equation model developed in Numerical Exercise 9.3 to simulate seiches in a lake with a reduced-gravity model.

12.2. Discretize the linearized two-layer model in time and with a horizontal Arakawa grid of your choice. Use a discretization that does not require the solution of a linear system. Provide a stability analysis neglecting the Coriolis force.

12.3. Implement the discretization of Numerical Exercise 12.2 and reproduce numerically the solution of Section 12.5.

12.4. Explore the implicit treatment of surface elevation in the barotropic component of the layer equations. Use an Arakawa C-grid and implicit treatment of both the divergence term in volume conservation and pressure gradient in the momentum equations. Neglect the Coriolis force. Eliminate the yet unknown velocity components from the equations to arrive at an equation for η^{n+1}. Compare the approach with the pressure calculation used in conjunction with the rigid-lid approximation of Section 7.6 and interpret what happens when you take very long time steps.

12.5. Use different time-integration techniques to calculate trajectories associated with the following 2D current field:

$$u = -\cos(\pi t)\, y, \qquad v = +\cos(\pi t)\, x \qquad (12.59)$$

from $t = 0$ to $t = 1$ and interpret the results. Prove that the trapezoidal scheme is reversible in time.

12.6. Implement a random walk into the calculation of the trajectories of Numerical Exercise 12.5 and verify the dispersive nature of the random walk. (*Hint*: Start with a dense cloud of particles in the middle of the domain.)

12.7. Use a large number of particles to advect the tracer field of Fig. 6.18 by distributing them initially on a regular grid. Look at `tvdadv2D.m` to see how the velocity field and initial concentration distribution are defined. Which problem do you face when you need to calculate the concentration at an arbitrary position at a later moment?

12.8. Implement Eq. (12.56) without advection in a periodic domain between $x = -10L$ and $x = 10L$ and with diffusion given by

$$A = A_0 \tanh^2\left(\frac{x}{L}\right). \tag{12.60}$$

Start with a uniform distribution of particles in the left half of the domain ($x < 0$) and no particles in the other half ($x > 0$). Simulate the evolution with and without the term $\partial A / \partial x$ and discuss your findings. Take $L = 1000$ km and $A_0 = 1000\,\text{m}^2/\text{s}$. (*Hint*: Periodicity of the domain can be ensured by proper repositioning of particles when they cross a boundary. Modulo is an interesting and useful function here.)

Raymond Braislin Montgomery
1910–1988

A student of Carl-Gustav Rossby, Raymond Braislin Montgomery earned a reputation as a brilliant descriptive physical oceanographer. Applying dynamic results derived by his mentor and other contemporary theoreticians to observations, he developed precise means of characterizing water masses and currents. By his choice of analyzing observations along density surfaces rather than along level surfaces, an approach that led him to formulate the potential now bearing his name, Montgomery was able to trace the flow of water masses across ocean basins and to arrive at a lucid picture of the general oceanic circulation. Montgomery's lectures and published works, marked by an unusual attention to clarity and accuracy, earned him great respect as a critic and reviewer. (*Photo by Hideo Akamatsu — courtesy of Mrs. R. B. Montgomery*)

James Joseph O'Brien
1935–

After a professional start as a chemist, Jim O'Brien decided to change career and turned to oceanography, a discipline in which he found greater intellectual challenges. This was around 1970, when computers were beginning to reveal their power for solving complicated dynamical systems, such as the ocean circulation. O'Brien quickly rose to become a leader in numerical modeling of physical oceanography and air–sea interactions. His early models of coastal upwelling taught us much about this important oceanic process. He was the first to show by numerical simulation that knowledge of wind over the tropical Pacific Ocean is essential to represent El Niño events. In the course of his numerical applications (many of them using layered models), he also discovered that the error made on the group velocity in a computer model is often more troublesome than the error made on the phase propagation speed.

Professor O'Brien has communicated his boundless enthusiasm for numerical modeling, oceanography, and air-sea interactions to a large number of graduate students and young researchers, who went on to occupy prominent positions across the United States and the world. (*Photo credit: Florida State University*)

Chapter 13

Internal Waves

ABSTRACT

This chapter presents internal gravity waves, which exist in the presence of vertical stratification. After the derivation of the dispersion relation and examination of wave properties, the chapter briefly considers mountain waves and nonlinear effects. Vertical-mode decomposition is introduced and treated numerically as an eigenvalue problem.

13.1 FROM SURFACE TO INTERNAL WAVES

Starting at an early age, everyone has seen, experienced, and wondered about surface waves. Sloshing of water in bathtubs and kitchen sinks, ripples on a pond, surf at the beach, and swell further offshore are all manifestations of surface water waves. Sometimes we look at them with disinterest, and sometimes they fascinate us. But whatever our reaction or interest, their mechanism relies on a simple balance between gravity and inertia. When the surface of the water is displaced upward, gravity pulls it back downward, the fluid develops a vertical velocity (potential energy turns into kinetic energy), and because of inertia, the surface penetrates below its level of equilibrium. An oscillation results. A change in the phase of the oscillations from place to place causes the wave to travel. Because surface waves carry energy and no volume, they naturally occur wherever there is agitation that causes no overall water displacements, such as the shaking of a half-full bottle, the throwing of a stone in a pond, or a storm at sea.

The gravitational force continuously strives to restore the water surface to a horizontal level because water density is greater than that of the air above. It goes almost without saying that the same mechanism is at work whenever two fluid densities differ. This frequently occurs in the atmosphere when warm air overlies cold air; waves may then be manifested by cloud undulations, which may at times be remarkably periodic (Fig. 13.1). An oceanic example, known as the phenomenon of dead water (Fig. 1.4), is the occurrence of waves at the interface between an upper layer of relatively light water and a denser lower layer. Those waves, although unseen from the surface, can cause a sizable drag on a sailing vessel (Section 1.3).

Introduction to Geophysical Fluid Dynamics

FIGURE 13.1   Evidence of internal waves in the atmosphere. The presence of moisture causes condensation in the rising air (wave crests), thus revealing the internal wave as a periodic succession of cloud bands. (Photo by one of the authors, February 2005, Tassili N'Ajjer, Algeria)

FIGURE 13.2   Surface manifestation of oceanic internal waves. The upward energy propagation of internal waves modifies the properties of surface waves rendering them visible from space. In this sunglint photograph taken from the space shuttle *Atlantis* on 19 November 1990 over Sibutu Passage in the Philippines (5°N, 119.5°E), a large group of tidally generated internal waves is seen to propagate northward into the Sulu Sea. (NASA Photo STS-38-084-060)

But the existence of such interfacial waves is not restricted to fluids with two distinct densities and a single interface. With three densities and two interfaces, two internal wave modes are possible; if the middle layer is relatively thin, the vertical excursions of the interfaces interact, letting energy pass from one level to the other. At the limit of a continuously stratified fluid, an infinite number of modes is possible, and wave propagation has both horizontal and vertical components (Fig. 13.2). Regardless of the level of apparent complexity in the wave pattern, the mechanism remains the same: There is a continuous interplay between gravity and inertia and a continuous exchange between potential and kinetic energies.

13.2 INTERNAL-WAVE THEORY

To study internal waves in their purest form, a few assumptions are necessary: There is no ambient rotation, the domain is infinite in all directions, there is no dissipative mechanism of any kind, and finally, the fluid motions and wave amplitudes are small. This last assumption is made to permit the linearization of the governing equations. However, we reinstate a term previously neglected, namely, the vertical acceleration term $\partial w/\partial t$ in the vertical momentum equation. We do so anticipating that vertical accelerations may play an important role in gravity waves. (Recall the discussion in Section 11.2 on the vertical oscillations of fluid parcels in a stratified fluid, which included the vertical acceleration). The inclusion of this term breaks the hydrostatic balance, but so be it! Finally, we decompose the fluid density as follows:

$$\text{Actual fluid density} = \rho_0 + \bar{\rho}(z) + \rho'(x, y, z, t), \tag{13.1}$$

where ρ_0 is the reference density (a pure constant), $\bar{\rho}(z)$ is the ambient equilibrium stratification, and $\rho'(x, y, z, t)$ is the density fluctuation induced by the wave (lifting and lowering of the ambient stratification). The inequality $|\bar{\rho}| \ll \rho_0$ is enforced to justify the Boussinesq approximation (Section 3.7), whereas the further inequality $|\rho'| \ll |\bar{\rho}|$ is required to linearize the wave problem. The total pressure field is decomposed in a similar manner. With the preceding assumptions, the governing equations become (Section 4.4)

$$\frac{\partial u}{\partial t} = -\frac{1}{\rho_0} \frac{\partial p'}{\partial x} \tag{13.2a}$$

$$\frac{\partial v}{\partial t} = -\frac{1}{\rho_0} \frac{\partial p'}{\partial y} \tag{13.2b}$$

$$\frac{\partial w}{\partial t} = -\frac{1}{\rho_0} \frac{\partial p'}{\partial z} - \frac{1}{\rho_0} g\rho' \tag{13.2c}$$

$$\frac{\partial u}{\partial x} + \frac{\partial v}{\partial y} + \frac{\partial w}{\partial z} = 0 \tag{13.2d}$$

$$\frac{\partial \rho'}{\partial t} + w \frac{d\bar{\rho}}{dz} = 0. \tag{13.2e}$$

The factor $d\bar{\rho}/dz$ in the last term can be transformed by introducing the stratification frequency (Brunt–Väisälä frequency) defined earlier in Eq. (11.3):

$$N^2 = -\frac{g}{\rho_0}\frac{d\bar{\rho}}{dz}. \tag{13.3}$$

For simplicity, we will assume it to be uniform over the extent of the fluid. This corresponds to a linear density variation in the vertical. Because all coefficients in the preceding linear equations are constant, a wave solution of the form

$$e^{i\,(k_x x + k_y y + k_z z - \omega t)}$$

is sought. Transformation of the derivatives into products (e.g., $\partial/\partial x$ becomes ik_x) leads to a 5-by-5 homogeneous algebraic problem. The solution is nonzero if the determinant vanishes, and this requires that the wave frequency ω be given by

$$\omega^2 = N^2\,\frac{k_x^2 + k_y^2}{k_x^2 + k_y^2 + k_z^2} \tag{13.4}$$

in terms of the wavenumbers, k_x, k_y, and k_z, and the stratification frequency, N. This is the dispersion relation of internal gravity waves.

A number of wave properties can be stated by examination of this relation. First and foremost, it is obvious that the numerator is always smaller than the denominator, meaning the wave frequency will never exceed the stratification frequency; that is,

$$\omega \le N \tag{13.5}$$

for positive frequencies. The reason for this upper bound can be traced back to the presence of the vertical acceleration term in Eq. (13.2c). Indeed, without that term the denominator in Eq. (13.4) reduces from $k_x^2 + k_y^2 + k_z^2$ to only k_z^2, implying that the nonhydrostatic term may be neglected as long as $k_z^2 + k_y^2 \ll k_z^2$. This occurs for waves with horizontal wavelengths much longer than their vertical wavelengths; the frequency of those waves is much less than N. For progressively shorter waves, the correction becomes increasingly important, the frequency rises but saturates at the value N. We may then ask what would happen if we agitate a stratified fluid at a frequency greater than its own stratification frequency. The answer is that with such short periods, fluid particles do not have the time to oscillate at their natural frequency and instead follow whatever displacements are forced on them; the disturbance turns into a local patch of turbulence, and no energy is carried away by waves. Using a neutrally buoyant float in the ocean, D'Asaro and Lien (2000) have shown that in stratified waters values of ω/N in the range 0.2–1 generally correspond to internal waves, whereas, at the same places, values above 1 ($1 < \omega/N < 50$) correspond to turbulent fluctuations.

Another important property derived from the dispersion relation (13.4) is that the frequency does not depend on the wavenumber magnitude (and thus on the wavelength) but only on its angle with respect to the horizontal plane. Indeed, with $k_x = k\cos\theta\cos\phi$, $k_y = k\cos\theta\sin\phi$, and $k_z = k\sin\theta$, where $k = \left(k_x^2 + k_y^2 + k_z^2\right)^{1/2}$ is the wavenumber magnitude, θ is its angle from the horizontal (positive or negative), and ϕ is the angle of its horizontal projection with the x–axis, we obtain

$$\omega = \pm N\cos\theta, \tag{13.6}$$

proving that the frequency depends only on the pitch of the wavenumber and, of course, the stratification frequency. The fact that two signs are allowed indicates that the wave can travel in one of two directions, upward or downward along the wavenumber direction. On the other hand, if the frequency is imposed (e.g., by tidal forcing), all waves regardless of wavelength propagate at fixed angles from the horizontal. The lower the frequency, the steeper the direction. At the limit of very low frequencies, the phase propagation is purely vertical ($\theta = 90°$).

13.3   STRUCTURE OF AN INTERNAL WAVE

Let us rotate the x and y axes so that the wavenumber vector is contained in the (x, z) vertical plane (i.e., $k_y = 0$, and there is no variation in the y-direction and no v velocity component). The expressions for the remaining two velocity components and the density fluctuation are

$$u = -A\,\frac{g\omega k_z}{N^2 k_x}\sin(k_x x + k_z z - \omega t) \tag{13.7a}$$

$$w = +A\,\frac{g\omega}{N^2}\sin(k_x x + k_z z - \omega t) \tag{13.7b}$$

$$p' = -A\,\frac{\rho_0 g k_z}{k_x^2 + k_z^2}\sin(k_x x + k_z z - \omega t) \tag{13.7c}$$

$$\rho' = +A\rho_0 \cos(k_x x + k_z z - \omega t). \tag{13.7d}$$

For k_x, k_z, and ω all positive, the structure of the wave is depicted on Fig. 13.3. The areas of upwelling (crests) and downwelling (troughs) alternate both horizontally and vertically, and lines of constant phase (e.g., following crests) tilt perpendicularly to the wavenumber vector. The trigonometric functions in solution (13.7) tell us that the phase $k_x x + k_z z - \omega t$ remains constant with time if one translates in the direction (k_x, k_z) of the wavenumber at the speed (see Appendix B):

$$c = \frac{\omega}{\sqrt{k_x^2 + k_z^2}}. \tag{13.8}$$

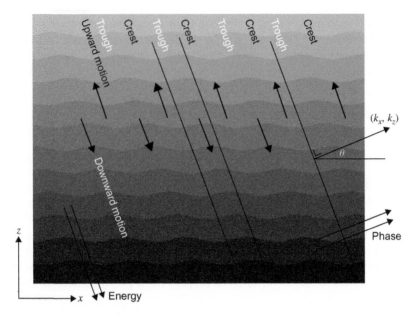

FIGURE 13.3   Vertical structure of an internal wave.

This is the phase speed, at which lines of crests and troughs translate. Because the velocity components, u and w, are in quadrature with the density fluctuations, the velocity is nil at the crests and troughs but is maximum a quarter of a wavelength away. The signs indicate that when one component is positive, the other is negative, implying downwelling to the right and upwelling to the left as indicated in Fig. 13.3. The ratio of velocities $(-k_x/k_z)$ further indicates that the flow is everywhere perpendicular to the wavenumber vector and thus parallel to the lines connecting crests and troughs. Internal waves are transverse waves. A comparison of the signs in the expressions of w and ρ' reveals that rising motions occur ahead of crests and sinking motions occur ahead of troughs, eventually forming the next crests and troughs, respectively. Thus, the wave moves forward and, because of the inclination of its wavenumber, also upward.

The propagation of the energy is given by the group velocity, which is the gradient of the frequency with respect to the wavenumber (Appendix B):

$$c_{gx} = \frac{\partial \omega}{\partial k_x} = + \frac{\omega k_z^2}{k_x \left(k_x^2 + k_z^2\right)} \tag{13.9}$$

$$c_{gz} = \frac{\partial \omega}{\partial k_z} = - \frac{\omega k_z}{\left(k_x^2 + k_z^2\right)}. \tag{13.10}$$

The direction is perpendicular to the wavenumber (k_x, k_z) and is downward. Thus, although crests and troughs appear to move upward, the energy actually sinks. The reader can verify that irrespective of the signs of the frequency and wavenumber components, the phase and energy always propagate in the same horizontal direction (though not at the same rates) and in opposite vertical directions.

Let us now turn our attention to the extreme cases. The first one is that of a purely horizontal wavenumber ($k_z = 0$, $\theta = 0$). The frequency is then N, and the phase speed is N/k_x. The absence of wave-like behavior in the vertical direction implies that all crests and troughs are vertically aligned. The motion is strictly vertical, and the group velocity vanishes, implying that the energy does not travel. The opposite extreme is that of a purely vertical wavenumber ($k_x = 0$, $\theta = 90°$). The frequency vanishes, implying a steady state. There is then no wave propagation. The velocity is purely horizontal and, of course, laterally uniform. The picture is that of a stack of horizontal sheets each moving, without distortion, with its own speed and in its own direction. If a boundary obstructs the flow at some depth, none of the fluid at that depth, however remote from the obstacle, is allowed to move. This phenomenon, occurring at very low frequencies in highly stratified fluids, is none other than the blocking phenomenon discussed at the end of Section 11.5 and presented as the stratified analogue of the Taylor column in rotating fluids.

In stratified and rotating fluids, the lowest possible internal-wave frequency is not zero but the inertial frequency f (see Analytical Problem 13.3). At that limit, the wave motion assumes the form of inertial oscillations, wherein fluid parcels execute horizontal circular trajectories (Section 2.3). Such limiting behavior is an attribute of inertia-gravity waves in homogeneous rotating fluids (Section 9.3) and is not surprising, since internal waves in stratified rotating fluids are the three-dimensional extensions of the inertia-gravity waves of homogeneous rotating fluids.

13.4   VERTICAL MODES AND EIGENVALUE PROBLEMS

Up to now, we considered internal waves in the rather schematic situation of an unbounded, nonrotating domain of uniform stratification. This is tantamount to considering only waves of wavelength much shorter than both the domain size and the length over which N^2 varies, and of frequency sufficiently high not to be influenced by the earth's rotation.

If we look at actual vertical profiles of density and their associated stratification frequencies (Fig. 13.4, for example), it is clear that stratification is far from uniform in the vertical, and we should question the validity of the preceding theory, except for very short vertical wavelengths. What happens then to internal waves with wavelengths comparable to the scale over which stratification changes?

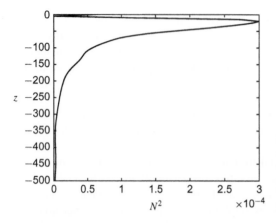

FIGURE 13.4 Stratification frequency squared (N^2 in s^{-2}) obtained from averaged climatological density profile in the Western Mediterranean Sea. (Data from Medar©)

To answer this question with a relatively simple analysis, we assume constant depth and eliminate surface waves by imposing a rigid lid (Sections 7.5 and 12.2). In return, we reinstate rotation by invoking the f-plane approximation. Because it is typically the case in nature, we assume that $f^2 < N^2(z)$ holds everywhere across the fluid column, that is,

$$N^2(z) = -\frac{g}{\rho_0}\frac{\mathrm{d}\bar{\rho}}{\mathrm{d}z} > f^2 > 0. \tag{13.11}$$

Within this framework, the equations governing small perturbations are now

$$\frac{\partial u}{\partial t} - fv = -\frac{1}{\rho_0}\frac{\partial p'}{\partial x} \tag{13.12a}$$

$$\frac{\partial v}{\partial t} + fu = -\frac{1}{\rho_0}\frac{\partial p'}{\partial y} \tag{13.12b}$$

$$\frac{\partial w}{\partial t} = -\frac{1}{\rho_0}\frac{\partial p'}{\partial z} - \frac{g}{\rho_0}\rho' \tag{13.12c}$$

$$\frac{\partial u}{\partial x} + \frac{\partial v}{\partial y} + \frac{\partial w}{\partial z} = 0 \tag{13.12d}$$

$$\frac{\partial \rho'}{\partial t} + w\frac{\mathrm{d}\bar{\rho}}{\mathrm{d}z} = 0. \tag{13.12e}$$

As for pure internal waves, we do not make use of the hydrostatic approximation. For a uniform topography, we can apply the technique of separation of variables and search for solutions of the type:

$$u = \mathcal{F}(z)\mathcal{U}(x, y)\, e^{-i\omega t} \tag{13.13a}$$

$$v = \mathcal{F}(z)\mathcal{V}(x, y)\, e^{-i\omega t} \tag{13.13b}$$

$$p' = \rho_0 \mathcal{F}(z)\mathcal{P}(x, y)\, e^{-i\omega t} \tag{13.13c}$$

$$w = i\omega \mathcal{W}(z)\mathcal{P}(x, y)\, e^{-i\omega t} \tag{13.13d}$$

$$\rho' = -N^2 \frac{\rho_0}{g}\mathcal{W}(z)\mathcal{P}(x, y)\, e^{-i\omega t}. \tag{13.13e}$$

Using these expressions in the governing equations (13.12), we realize that Eq. (13.12e) is already satisfied and that the four remaining equations reduce to

$$-i\omega\,\mathcal{U} = f\mathcal{V} - \frac{\partial \mathcal{P}}{\partial x} \tag{13.14a}$$

$$-i\omega\,\mathcal{V} = -f\mathcal{U} - \frac{\partial \mathcal{P}}{\partial y} \tag{13.14b}$$

$$\left(\omega^2 - N^2\right)\mathcal{W} = -\frac{d\mathcal{F}}{dz} \tag{13.14c}$$

$$\frac{1}{\mathcal{P}}\left(\frac{\partial \mathcal{U}}{\partial x} + \frac{\partial \mathcal{V}}{\partial y}\right) = -i\frac{\omega}{\mathcal{F}}\frac{d\mathcal{W}}{dz}. \tag{13.14d}$$

The first two equations do not depend on z, the third one depends on neither x or y, while the last equation has a left-hand side that does not depend on z and a right-hand side that does not depend on x and y. This last equation can only be met if both terms are constant. For dimensional reasons, we call this constant $i\omega/gh^{(j)}$, where $h^{(j)}$ has the dimension of a depth and is commonly called the *equivalent depth*. The reason for this label will become clear shortly.

Substitution of the right-hand side of Eq. (13.14d),

$$-i\frac{1}{\mathcal{F}}\frac{d\mathcal{W}}{dz} = \frac{i}{gh^{(j)}},$$

into the z-dependent equation (13.14c) leads to an equation governing the vertical mode $\mathcal{W}(z)$:

$$\frac{d^2\mathcal{W}}{dz^2} + \frac{\left(N^2 - \omega^2\right)}{gh^{(j)}}\,\mathcal{W} = 0, \tag{13.15}$$

while the horizontal structure $\mathcal{U}$, $\mathcal{V}$, $\mathcal{P}$ is solution of Eqs. (13.14a) and (13.14b) completed with the result of Eq. (13.14d):

$$\frac{\partial \mathcal{U}}{\partial x} + \frac{\partial \mathcal{V}}{\partial y} = \frac{i\omega}{gh^{(j)}}\,\mathcal{P}. \tag{13.16}$$

These three equations possess the same structure as the shallow-water wave equations with constant depth (Section 9.1). We observe that the surface elevation η is replaced by $\mathcal{P}/g$ and the depth h by $h^{(j)}$. Therefore, we can immediately recover the wave solutions of the shallow-water theory and verify that for a horizontal periodic solution of the type $(\mathcal{U}, \mathcal{V}, \mathcal{P}) = (U, V, P)e^{i(k_x x + k_y y)}$ with constant coefficients U, V and P, these waves obey the dispersion relation of inertial (Poincaré) waves:

$$\omega^2 = f^2 + gh^{(j)}\left(k_x^2 + k_y^2\right). \tag{13.17}$$

For horizontally finite domains, only discrete values of (k_x, k_y) pairs are allowed, but we will not explore this possibility here, preferring to consider the pair (k_x, k_y) as given.

13.4.1 Vertical Eigenvalue Problem

Now that we know the horizontal structure of the wave, we have to find its associated vertical structure. This can be done by substituting the horizontal dispersion relation (13.17) in the vertical-mode equation (13.15), which leads to the following problem:

$$\frac{d^2\mathcal{W}}{dz^2} + \left(k_x^2 + k_y^2\right)\frac{N^2(z) - \omega^2}{\omega^2 - f^2}\,\mathcal{W} = 0, \tag{13.18}$$

with boundary conditions

$$\mathcal{W} = 0 \quad \text{at } z = 0 \text{ and } z = H \tag{13.19}$$

corresponding to rigid lid on top and flat bottom below.

We are in the presence of a homogeneous differential equation with homogeneous boundary conditions, that is, the solution is trivially $\mathcal{W} = 0$ unless ω assumes a special value. As it turns out, there exists a whole series of special ω values for which $\mathcal{W}$ may be nonzero. These are called eigenvalues, and the corresponding $\mathcal{W}(z)$ solutions are called eigenfunctions or, more specifically in our case, *vertical modes*.

13.4.2 Bounds on Frequency

We anticipate that there should be some bounds on the wave frequencies as we are already aware of $\omega^2 < N^2$ from Section 13.2. In order to find these bounds, we apply an integral technique similar to that used to analyze stability of shear flows (Section 10.2).

When we multiply Eq. (13.18) by the complex conjugate $\mathcal{W}^*$, integrate vertically across the domain, perform an integration by part on the first term, and use boundary conditions (13.19), we obtain

$$\int_0^H \left|\frac{dW}{dz}\right|^2 dz = \left(k_x^2 + k_y^2\right) \int_0^H \frac{N^2 - \omega^2}{\omega^2 - f^2} |W|^2 dz. \tag{13.20}$$

For $f^2 < N^2$, the common situation, and as long as ω is real, it is clear that only values within the range

$$f^2 \leq \omega^2 \leq N_{max}^2 \tag{13.21}$$

are permitted, since outside this range the right-hand side of Eq. (13.20) would be negative and unable to match the positive left-hand side.

However, we can ask whether there could be complex values of ω. First, a purely imaginary solution is not possible since $\omega = i\omega_i$ would make the right-hand side of Eq. (13.20) always negative.[1] In the general case $\omega = \omega_r + i\omega_i$, it is sufficient to consider the imaginary part of the fraction in Eq. (13.20):

$$\Im\left(\frac{N^2 - \omega^2}{\omega^2 - f^2}\right) = -2\omega_r \omega_i \frac{N^2 - f^2}{(\omega_r^2 - \omega_i^2 - f^2)^2 + 4\omega_r^2 \omega_i^2} \tag{13.22}$$

and to realize that any $\omega_i \neq 0$ prevents Eq. (13.20) from being met because its real left-hand side cannot match its nonreal right-hand side. Therefore, it is true that Eq. (13.20) allows only real ω.

In conclusion, for $f^2 < N^2$, we find pure wave motions with frequencies in the range (13.21). Compared to Eq. (13.5), we see that one effect of rotation is to eliminate the lower frequencies. We also note that, as many times before, if a wave of frequency ω exists, so does one of frequency $-\omega$, corresponding to propagation in the opposite direction.

13.4.3 Simple Example of Constant N^2

We can solve analytically the case of uniform stratification in a rotating bounded domain. The eigenvalue problem has the following immediate solution

$$W(z) = \sin k_z z, \quad k_z = j\frac{\pi}{H}, \quad j = 1, 2, 3, \ldots \tag{13.23}$$

with the accompanying dispersion relation

$$\omega^2 = \frac{\left(k_x^2 + k_y^2\right) N^2 + k_z^2 f^2}{k_x^2 + k_y^2 + k_z^2}. \tag{13.24}$$

---

[1] We assumed from the beginning $0 \leq f^2 \leq N^2$, but the interested reader could analyze the case $N^2 < 0$, corresponding to gravitational instability. In this case, complex ω values are possible as we can expect on physical ground.

Due to the finiteness of the vertical domain size, the vertical wavenumber k_z now takes discrete values, and the corresponding functions $\mathcal{W}(z)$ form a discrete set of eigenfunctions. Yet, we have an infinite number of them as predicted in Section 12.4. We can verify that all frequencies fall in the range (13.21) and recognize that the spatial structures are the same as in the unbounded domain, except that only those wavelengths are permitted that satisfy the boundary conditions.

Likewise, the constant $gh^{(j)}$, which had been free sofar, may only take one among a discrete set of values:

$$gh^{(j)} = \frac{\omega^2 - f^2}{k_x^2 + k_y^2} = \frac{N^2 - f^2}{k_x^2 + k_y^2 + \left(j\frac{\pi}{H}\right)^2}. \tag{13.25}$$

Since $gh^{(j)}$ plays here the same role in the horizontal structure of each mode as gH did in a shallow-water system, we can form an analogous radius of deformation.

$$R_j = \frac{\sqrt{gh^{(j)}}}{f}. \tag{13.26}$$

This *internal* radius of deformation plays the same role as the *external* radius of deformation did in the shallow water system. Among other properties, it characterizes the horizontal scale at which both rotation and gravity, here through stratification, come into play, and, for example, the lateral trapping scale of an internal coastal Kelvin wave is the internal deformation radius (see Analytical Problem 13.9).

By virtue of $\omega^2/f^2 = 1 + (k_x^2 + k_y^2)\, R_j^2$, waves with a shorter wavelength than the deformation radius are influenced primarily by stratification, while those with longer scales are dominated by rotation. The radius of deformation is thus the scale at which rotation and stratification play equally significant roles. Note that it varies from mode to mode, with higher modes having shorter radii. Waves that vary rapidly in the vertical ($j \gg 1$) have a shorter radius of deformation, subjecting them to stronger rotational effects than waves with smoother vertical variation.

For small aspect ratios $k_x^2 + k_y^2 \ll k_z^2$ with strong stratification $f^2 \ll N^2$, the expression of the deformation radius reduces to

$$R_j \simeq \frac{NH}{j\pi f}, \quad j = 1, 2, 3, \ldots \tag{13.27}$$

The uniform-stratification application has the advantage of showing how rotation and finite domain influence the wave dispersion relation compared to Eq. (13.4), but it is inadequate to determine the eigenfrequencies of a system with highly nonuniform stratification, such as one with a localized pycnocline.

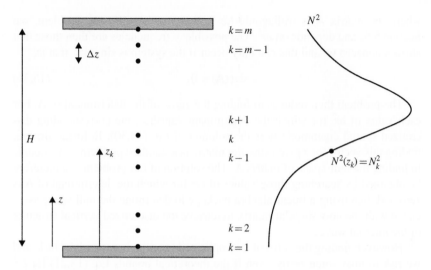

FIGURE 13.5   Notation for the discretization in search of vertical modes.

Although analytical asymptotic methods[2] exist to tackle such an eigenvalue problem, it is generally easier and more accurate to resort to numerical methods. This is even more true if the system to be analyzed is a concrete one, where the density profile was measured or obtained from climatological databases with data at discrete vertical levels.

13.4.4   Numerical Decomposition into Vertical Modes

The discretization chosen here is a straightforward finite-difference technique. For the sake of simplicity, uniform grid spacing Δz is assumed. The first and last grid points are chosen, respectively, on the flat bottom and at the rigid lid (Fig. 13.5), since boundary conditions on the unknown itself are imposed there. The discretized field w of the exact solution $\mathcal{W}$ at location z_k is noted w_k so that the discretization reads

$$w_{k+1} + w_{k-1} - 2w_k + \Delta z^2 \left(k_x^2 + k_y^2 \right) \frac{N^2(z_k) - \omega^2}{\omega^2 - f^2} \, w_k = 0, \quad k = 2, 3, \ldots, m-1$$

(13.28)

$$\text{with} \quad w_1 = 0, \quad w_m = 0.$$

(13.29)

This problem can be recast in matrix form by collecting all w_k into an array **w**:

$$\mathbf{A}(\omega^2)\mathbf{w} = 0,$$

(13.30)

[2]See WKB methods in Bender and Orszag (1978).

where the matrix $\mathbf{A}$ is tridiagonal (as it was for the diffusion problem, see Section 5.5) and depends on ω^2. The possible ω frequencies are then those that allow a nonzero $\mathbf{w}$, and this can only occur if the system is singular, that is,

$$\det(\mathbf{A}) = 0. \tag{13.31}$$

The problem then reduces to finding the zeros of the determinant of $\mathbf{A}$. For each value of ω^2 for which the determinant vanishes, the corresponding discretized spatial eigenmode $\mathbf{w}$ is the solution of Eq. (13.30). In linear algebra, finding this vector for a given singular matrix is a standard problem and amounts to finding the null-space of matrix $\mathbf{A}$. The solution of our problem can therefore be obtained by searching those values of ω^2 for which the determinant of $\mathbf{A}$ is zero and then using a linear algebra package to determine the null-space associated with the now singular matrix to retrieve the discretized vertical structure of the internal wave.

However, finding the zeros of a complicated function is not a trivial task, and we risk to miss some zeros, even if the theoretical bounds Eq. (13.21) for ω^2 can guide the search algorithm. We can neither be sure that numerical solutions of Eq. (13.31) fall into the same range, although those falling clearly outside should certainly qualify as nonphysical.

To isolate the variable ω^2, we reformulate the problem (13.30) by restating it as

$$-\frac{\omega^2}{f^2}\left[-w_{k+1} - w_{k-1} + (2+\epsilon)w_k\right] + \left[-w_{k+1} - w_{k-1} + \left(2 + \epsilon N_k^2 f^{-2}\right)w_k\right] = 0 \tag{13.32}$$

with $\epsilon = \Delta z^2\left(k_x^2 + k_y^2\right) > 0$. Values of k are limited to $2 \leq k \leq m-1$ since the boundary conditions are readily implemented. Equation (13.32) can be written as a linear problem:

$$\mathbf{Bw} = \lambda\mathbf{Cw}, \quad \lambda = \frac{\omega^2}{f^2} \tag{13.33}$$

where matrices $\mathbf{B}$ and $\mathbf{C}$ are tridiagonal and, most importantly, independent of ω^2. Both matrices also have -1 on their subdiagonal and superdiagonal lines, whereas the diagonal of $\mathbf{C}$ repeats $2+\epsilon$ and that of $\mathbf{B}$, $2+\epsilon N_k^2/f^2$. Further, $\mathbf{B}$ and $\mathbf{C}$ are symmetric positive definite since they are diagonally dominant.

Equation (13.33) is stated as a standard linear algebra problem (called generalized eigenvalue problem) for which solvers and theorems exist (e.g., the Rayleigh–Ritz inequalities that are the subject of Numerical Exercise 13-2).

We can recast the problem in an even more familiar form by noting that for any positive definite matrix, its inverse exists, and therefore, by defining $\tilde{\mathbf{w}} = \mathbf{Cw}$, we recover a standard eigenvalue problem

$$\mathbf{A}\tilde{\mathbf{w}} = \lambda\tilde{\mathbf{w}} \quad \text{with} \quad \mathbf{A} = \mathbf{BC}^{-1}. \tag{13.34}$$

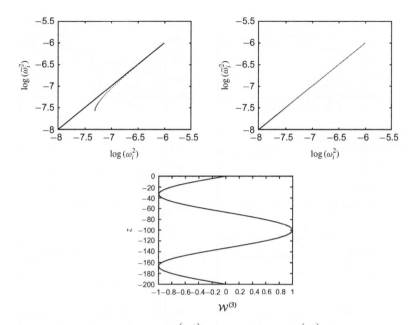

FIGURE 13.6   Numerical estimation $\log\left(\tilde{\omega}_j^2\right)$ versus exact value $\log\left(\omega_j^2\right)$ of internal-mode frequencies, using 50 (top left panel) and 450 (top right panel) discrete levels in the vertical. Bottom panel: Numerical profile of the third vertical mode, clearly corresponding to $\sin(3\pi z/H)$.

Once the eigenvalues λ and eigenvectors $\tilde{\mathbf{w}}$ are found, the discretized physical mode $\mathbf{w}$ can be reconstructed by means of $\mathbf{w} = \mathbf{C}^{-1}\tilde{\mathbf{w}}$.

We can be assured that the problem has only real solutions. Since both $\mathbf{B}$ and $\mathbf{C}$ are symmetric positive definite, multiplying Eq. (13.33) for a given eigenvalue λ_j and eigenvector $\mathbf{w}^j$ by the transposed complex conjugate $\mathbf{w}^{j*}$ reveals[3] that λ_j must be real for any j.

We note that, contrary to the analytical solution for which an infinite number of modes exist, only a finite number $(m-2)$ of eigenvalues and modes can be calculated in the discretized version.

To verify the numerical method outlined above, we calculate the numerical solution for the case of uniform N^2 and compare it with the known analytical solution Eq. (13.23)–(13.24). As expected, the largest eigenvalues, those corresponding the longest vertical wavelengths, are well represented, even for a moderate number of grid points (Fig. 13.6). However, the frequencies closer to f require finer resolution because of their shorter wavelengths. To represent mode j accurately, we indeed need a spacing $j\Delta z \ll H$.

---

[3] We use the fact that both $\mathbf{w}^{j*}\mathbf{B}\mathbf{w}^j$ and $\mathbf{w}^{j*}\mathbf{C}\mathbf{w}^j$ are real because of the positive definite nature of $\mathbf{B}$ and $\mathbf{C}$.

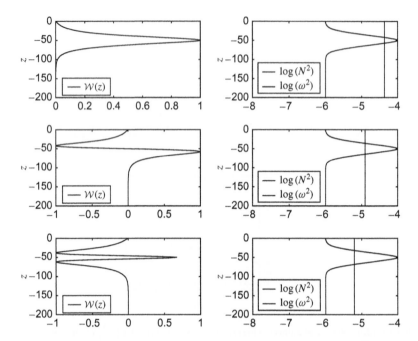

FIGURE 13.7   First three modes (left panel) of a nonuniform stratification with N^2 peak. The corresponding values of ω^2 are set against the N^2 profile in the right panel. Note that the ω^2 values lie between the extrema of N^2.

13.4.5   Waves Concentration at a Pycnocline

The numerical method shown above is now being used to analyze a situation in which the N^2 profile exhibits a maximum, which corresponds to a region of greater density variation, that is, a pycnocline. To do so, we take a schematic density profile (Figure 13.7), with a stratification frequency varying between N_0 and N_1 such that $f^2 < N_0^2 < N^2(z) < N_1^2$.

The first three modes (Fig. 13.7), correspond to the highest frequencies, and their vertical profiles (left panels) indicate a concentration of amplitude and highest gradients in the vicinity of the peak of $N^2(z)$, the pycnocline. This can be understood in the light of the sign of $(N^2 - \omega^2)/(f^2 - \omega^2)$ that appears in Eq. (13.18). Where this factor is positive, the eigenfunctions are oscillatory in nature and where it is negative, they exhibit exponential decay. Since the term changes sign within the domain because $N_0^2 < \omega^2 < N_1^2$ (see right panels of Fig. 13.7), the solution switches from oscillations in regions of $\omega^2 \leq N^2$ to exponential behavior elsewhere, decreasing toward zero at the boundaries of the domain. A point where $\omega^2 = N^2$ is called a *turning point*. With ω^2 cutting

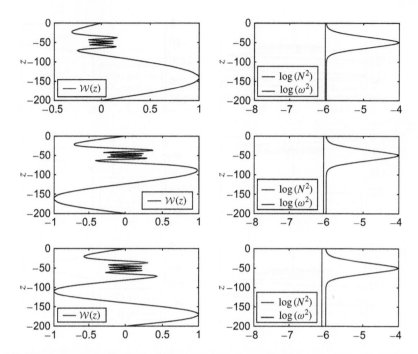

FIGURE 13.8   Modes 10–12 for the same nonuniform stratification as in the previous figure. Again, the left panel shows the vertical structure of the modes, whereas the right panel compares their ω^2 values to the N^2 values. Note the fine structure in the vicinity of the pycnocline and that the values of ω^2 fall below the minimum N^2 value.

twice across a peak in N^2, a pycnocline is accompanied by two turning points, one below and one above.

For higher modes (Fig. 13.8), ω^2 decreases and approaches the minimum of N^2. The turning points move away from the pycnocline until they disappear when, for high enough mode numbers, the corresponding ω^2 values fall below the N^2 minimum. Those higher modes have a structure that is oscillatory everywhere (Fig. 13.8). Surprisingly, amplitudes are now lowest near the pycnocline. This is due to the fact that the frequency difference is maximum ($\omega^2 - N^2$ largest) near the pycnocline, and resonant behavior is thus stronger away from the pycnocline, where the amplitude is consequently higher.

For even higher modes, the modal frequency ω approaches the inertial frequency f, and a new behavior emerges (Fig. 13.9). The regions above and below the pycnocline start to be decoupled, with one mode being almost entirely confined to one side of the pycnocline, the next mode to the other side, and so alternatively with mode number. The pycnocline appears to act as a barrier. In the limit of an extremely sharp pycnocline (very high N^2 peak), the

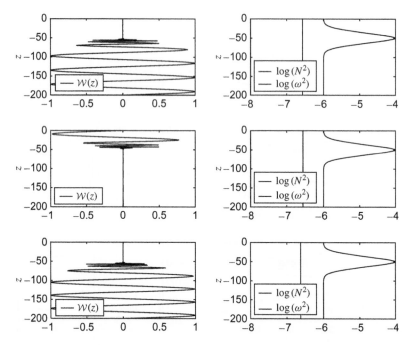

FIGURE 13.9 Modes 26–28 for the same nonuniform stratification as in the previous two figures. Again, the left panel shows the vertical structure of the modes, whereas the right panel compares their ω^2 values to the N^2 values. These modes, for which the frequencies approach the inertial frequency f, are almost entirely confined on one side or the other of the pycnocline. In other words, a pycnocline acts as a vertical barrier to near-inertial waves.

stratification effectively becomes a two-layer system, for which waves near the inertial frequency can exist in each layer independently of the other.

13.5   LEE WAVES

Internal waves in the atmosphere and ocean can be generated by a myriad of processes, almost wherever a source of energy has some temporal or spatial variability. Oceanic examples include the ocean tide over a sloping bottom, mixing processes in the upper ocean (especially during a hurricane), instabilities of shear flows, and the passage of a submarine. In the atmosphere, one particularly effective mechanism is the generation of internal waves by a wind blowing over an irregular terrain such as a mountain range or hilly countryside. We select the latter example to serve as an illustration of internal-wave theory because it has some meteorological importance and lends itself to a simple mathematical treatment.

To apply the previous linear-wave theory, we naturally restrict our attention to small-amplitude waves and, consequently, to small topographic irregularities.

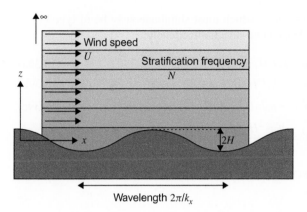

FIGURE 13.10 Stratified flow over a wavy terrain. The difference in elevation between crests and troughs is assumed small to justify a linear analysis. Then, the flow over any terrain configuration can be obtained from the superimposition of elementary wave solutions.

This restriction also permits us to study a single topographic wavelength, from which the principle of linear superposition should allow us to construct more general solutions. The model (Fig. 13.10) consists of a stratified air mass of uniform stratification frequency N flowing at speed U over a slightly wavy terrain. The ground elevation is taken as a sinusoidal function $b = H \cos k_x x$ of amplitude H (the trough-to-crest height difference is then $2H$) and wavenumber k_x (the wavelength is then $2\pi/k_x$). The wind direction (along the x-axis of the model) is chosen to be normal to the troughs and crests so that the problem is two-dimensional.

Because our theory has been developed for waves in the absence of a main flow, we translate the x-axis with the wind speed. The topography then appears to move at speed U in the negative x-direction:

$$z = b(x + Ut) = H \cos[k_x(x + Ut)]$$
$$= H \cos(k_x x - \omega t), \tag{13.35}$$

where the frequency is defined as

$$\omega = -k_x U \tag{13.36}$$

and is a negative quantity. Because a particle initially on the bottom must remain there at all times (no airflow through the ground), a boundary condition is

$$w = \frac{\partial b}{\partial t} + u \frac{\partial b}{\partial x} \quad \text{at} \quad z = b, \tag{13.37}$$

which can be immediately linearized to become

$$w = \frac{\partial b}{\partial t}$$
$$= H\omega \sin(k_x x - \omega t) \quad \text{at} \quad z = 0, \tag{13.38}$$

by virtue of our small-amplitude assumption.

The solution to the problem, which must simultaneously be of type (13.7) and meet condition (13.38), can be stated immediately:

$$u = k_z U H \sin(k_x x + k_z z - \omega t) \tag{13.39a}$$

$$w = -k_x U H \sin(k_x x + k_z z - \omega t) \tag{13.39b}$$

$$p' = -\rho_0 k_z U^2 H \sin(k_x x + k_z z - \omega t) \tag{13.39c}$$

$$\rho' = \frac{\rho_0 N^2 H}{g} \cos(k_x x + k_z z - \omega t), \tag{13.39d}$$

where the vertical wavenumber k_z is required to meet the dispersion relation (13.4):

$$k_z^2 = \frac{N^2}{U^2} - k_x^2. \tag{13.40}$$

The mathematical structure of this last expression shows that two cases must be distinguished: Either $N/U > k_x$ and k_z is real, or $N/U < k_x$ and k_z is imaginary. Note that solution (13.39) is formulated in the moving reference frame. To obtain the stationary solution in the fixed frame of the topography, one needs to add advection by the wind.

13.5.1   Radiating Waves

Let us first explore the former situation, which arises when the stratification is sufficiently strong ($N > k_x U$) or when the topographic wavelength is sufficiently long ($k_x < N/U$). Physically, the time $2\pi/k_x U$ taken by a particle traveling at the mean wind speed U to go from a trough to the next trough (i.e., up and down once) is longer than the natural oscillatory period $2\pi/N$, and internal waves can be excited. Solving Eq. (13.40) for k_z, we have two solutions at our disposal,

$$k_z = \pm\sqrt{\frac{N^2}{U^2} - k_x^2}, \tag{13.41}$$

but because the source of wave energy is at the bottom of the domain, only the wave with upward group velocity is physically relevant. According to Eqs. (13.10) and (13.36), we select the positive root.

The wave structure in the framework fixed with the topography (Fig. 13.11) is steady and such that all density surfaces undulate like the terrain, with no vertical attenuation but with an upwind phase tilt with height. The angle θ between the wave fronts (lines joining crests) and the vertical, which is also the angle between the wavenumber vector and the horizontal, is given by

$$\cos\theta = \frac{k_x U}{N}, \tag{13.42}$$

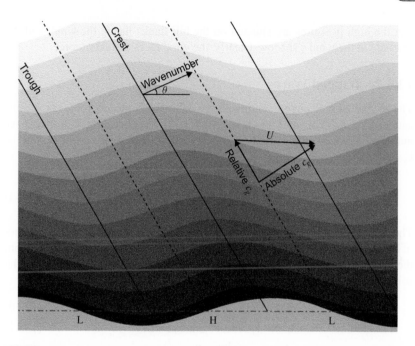

FIGURE 13.11 Structure of a mountain wave in the case of strong stratification or long wavelength ($k_x U < N$). Note the absence of vertical attenuation and the presence of a phase shift with height. The group velocity with respect to the ground is oriented upward and downwind. The pressure distribution, with highs on wind-facing slopes and lows on flanks in the wind's shadow, exerts a drag on the moving air mass.

so that $k_x = k\cos\theta$, $k_z = k\sin\theta$, with $k = \left(k_x^2 + k_z^2\right)^{1/2}$. The group velocity in the fixed frame is equal to the group velocity relative to the moving wind, given by Eq. (13.10) with $\omega = -k_x U$, plus the velocity U in the x-direction:

$$c_{gx} = -U\frac{k_z^2}{k^2} + U = U\cos^2\theta \tag{13.43}$$

$$c_{gz} = U\frac{k_x k_z}{k^2} = U\sin\theta\cos\theta. \tag{13.44}$$

It tilts upward as required, and its direction coincides with that of the wavenumber (Fig. 13.11). Energy is thus radiated upward and downwind. We shall not calculate the energy flux and will only show that the terrain exerts a drag force on the flowing air mass. The drag force, which is minus the Reynolds stress, is

$$\text{Drag force} = +\rho_0 \,\overline{uw}\,|_{z=0} = -\frac{1}{2}\,\rho_0 k_x k_z U^2 H^2,$$

where the overbar indicates an average over one wavelength. The minus sign indicates a retarding force. The existence of this force is also related to the fact that the high pressures are situated on the hill flanks facing the wind, and the lows are in the wind's shadow. Clearly, the wind faces a braking force.

13.5.2 Trapped Waves

The second case, leading to an imaginary value for k_z, occurs for weak stratifications ($N < k_x U$) or short waves ($k_x > N/U$). To avoid dealing with imaginary numbers, we define the quantity a as the positive imaginary part of k_z, that is, $k_z = \pm i a$ with

$$a = \sqrt{k_x^2 - \frac{N^2}{U^2}}. \tag{13.45}$$

The solution now contains exponential functions in z, and the physical nature of the problem dictates that we retain only the function that decays away from the ground. In the reference framework translating with the wind speed U, the solution is

$$u = aUHe^{-az} \cos(k_x x - \omega t) \tag{13.46}$$

$$w = -k_x UHe^{-az} \sin(k_x x - \omega t) \tag{13.47}$$

$$p' = -\rho_0 aU^2 He^{-az} \cos(k_x x - \omega t) \tag{13.48}$$

$$\rho' = \frac{\rho_0 N^2 H}{g} e^{-az} \cos(k_x x - \omega t). \tag{13.49}$$

The wave structure is depicted in Fig. 13.12. Density surfaces undulate at the same wavelength as the terrain, but the amplitude decays with height. There is no vertical phase shift. Because the waves are contained near the ground, in a boundary layer of thickness on the order of $1/a$, there is no upward energy radiation. The absence of such energy loss is corroborated by the absence of a drag force:

$$\text{Drag force} = +\rho_0 \overline{uw}\,|_{z=0} = 0.$$

The Reynolds stress vanishes because u and w are now in quadrature. Physically, the high pressures are in the valleys, the lows on the hilltops, and the pressure distribution causes no work against the wind.

13.6 NONLINEAR EFFECTS

All we have said thus far on internal waves is strictly applicable only if their amplitude is small, but internal-wave amplitudes can be quite large. For example, Liu et al. (2006) have observed internal waves in Luzon Strait (South China Sea) with amplitudes as high as 140 m.

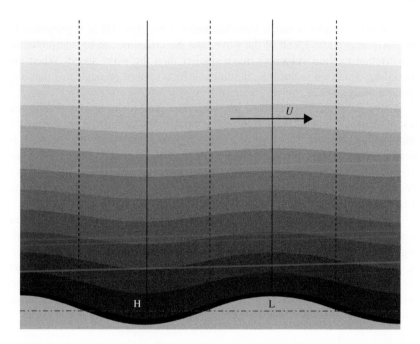

FIGURE 13.12   Structure of a mountain wave in the case of weak stratification or short wavelength ($N < k_x U$). Note the attenuation with height and the absence of vertical phase shift. The pressure distribution, with highs in valleys and lows on hill tops, causes no drag on the moving air mass.

The obvious question is: When do internal-wave dynamics become nonlinear? The answer lies in comparing the displacements of the particles caused by the wave to the wavelength: If those displacements are much shorter than the wavelength, then advective processes are unimportant, and the linear analysis is justified. The maximum horizontal displacement of fluid particles subject to an oscillatory horizontal velocity of the type $u = U\sin(k_x x + k_z z - \omega t)$ is U/ω, whereas the horizontal wavelength is $2\pi/k_x$. We thus require $U/\omega \ll 2\pi/k_x$, or because of Eq. (13.4),

$$U \ll \frac{2\pi N}{\sqrt{k_x^2 + k_z^2}} \leq \frac{2\pi N}{k_z}.$$

Since $2\pi/k_z$ is the vertical wavelength, which we can take as the depth scale of the motion and denote by H, the criterion becomes

$$Fr = \frac{U}{NH} \ll 1. \tag{13.50}$$

Thus, the preceding description of internal waves is applicable only to situations where the Froude number (based on the wave-induced velocities and the vertical wavelength) is much less than unity. Note that NH is approximately the horizontal phase speed of the wave, and the criterion can be interpreted as a restriction to fluid velocities much smaller than the wave speed. When the preceding condition is not met, nonlinear effects cannot be neglected, and the spectral analysis fails.

A first possible effect is wave breaking. The crest (or trough) overtakes the rest of the wave, and the wave rolls over not unlike the surf on the sea surface on approaching a beach. This type of instability, due to the wave motion itself, is termed *advective instability*. At lower energy levels, waves do not overturn but may nonetheless be sufficiently strong not to conform to the linear theory. Wave interactions create harmonics, and energy spreads over a continuous spectrum, usually spanning several decades in frequencies and wavenumbers.

Observed spectra in the deep ocean (i.e., in areas remote from important topographic influences) all show a striking resemblance (Munk, 1981), suggesting the existence of a universal spectrum. This observation led Christopher J. R. Garrett and Walter H. Munk to formulate in 1972 a prototypical spectrum for internal-wave energy. This model spectrum was subsequently modified and refined (Garrett & Munk, 1979; Munk, 1981) and has become known as the *Garrett–Munk spectrum*. The expression for the spectral energy density is

$$E(k, m) = \frac{3fNE\, m\, m_*^{3/2}}{\pi\, (m + m_*)^{5/2}\, (N^2 k^2 + f^2 m^2)}, \tag{13.51}$$

where $k = \sqrt{k_x^2 + k_y^2}$ is the horizontal wavenumber, $m = k_z$ the vertical wavenumber, m_* a reference wavenumber to be determined from observations, and E a dimensionless constant setting the overall energy level (see Fig. 13.13).

The Garrett–Munk spectrum is largely empirical in the sense that its formulation is based on observations, simple dimensional considerations, and elementary physics. Yet, it has been shown to conform to a large number of observations, prompting the conjecture (Munk, 1981) that the internal-wave climate in the deep ocean is somehow regulated by a saturation process rather than by external generation processes. Lvov and Tabak (2001) have advanced a theory that closely but not exactly reproduces the Garrett–Munk spectrum.

In coastal areas, where topographic irregularities play a dominant role in generating internal waves, it is not unusual to find coherent wave groups at a single (tidal) frequency. Under certain conditions, the dispersion effect (different wave speeds for different wavenumbers) can annihilate the nonlinear steepening effect (crests or troughs overtaking the rest of the wave), yielding a robust wave called *internal solitary wave* (Turner, 1973, Chapter 3). Figure 13.2 displays the surface signature of a train of internal solitary waves.

Theory, field observations, and laboratory simulations indicate that internal-wave characteristics are substantially altered in the presence of shear flows. Although a general theory is beyond our present scope, it is worth noting that, like waves in a laterally sheared flow of a homogeneous fluid, internal waves can

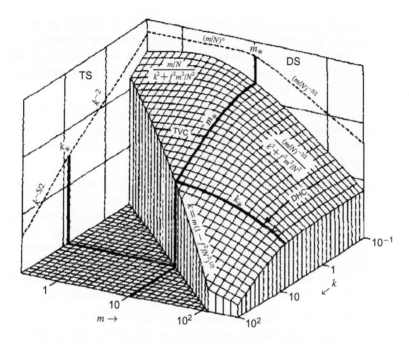

FIGURE 13.13   Universal spectrum of internal waves in the ocean, according to Garrett and Munk (1979). The spectral energy density E is plotted as a function of k and m, the horizontal and vertical wavenumbers, respectively. The two-dimensional spectrum can be integrated over one or the other of its variables to obtain the "towed spectrum" (TS) or the "dropped spectrum" (DS).

also encounter critical levels in a vertically sheared flow (wave speed equal to local flow velocity).

For weak-to-moderate shear flows ($du/dz < 2N$) in nonrotating (Booker & Bretherton, 1967) and rotating (Jones, 1967) stratified fluids, theoretical considerations show that on approaching a critical level, the internal-wave vertical wavenumber increases without limit and the group velocity becomes horizontal, thus aligning itself with the flow. The theory also shows that the time taken for the energy to reach the critical level is infinite, implying that dissipative effects become important. Physically, the energy is not focussed and amplified but absorbed and dissipated at the critical level.

In stronger shear flows ($du/dz > 2N$), instabilities develop. This *shear instability*, treated in the following chapter, is the vertical analogue of the barotropic instability in horizontally sheared flows (Chapter 10).

ANALYTICAL PROBLEMS

13.1.   In a coastal ocean, the water density varies from $1028 \, \text{kg/m}^3$ at the surface to $1030 \, \text{kg/m}^3$ at depth of $100 \, \text{m}$. What is the maximum internal-wave frequency? What is the corresponding period?

13.2. Internal waves are generated along the coast of Norway by the M_2 surface tide (period of 12.42 h). If the stratification frequency N is $2 \times 10^{-3}\,\text{s}^{-1}$, at which possible angles can the energy propagate with respect to the horizontal? (*Hint*: Energy propagates in the direction of the group velocity.)

13.3. Derive the dispersion relation of internal gravity waves in the presence of rotation, assuming $f < N$. Show that the frequency of these waves must always be higher than f but lower than N. Compare vertical phase speed to vertical group velocity.

13.4. A 10-m/s wind blows over a rugged terrain, and lee waves are generated. If the stratification frequency is equal to $0.03\,\text{s}^{-1}$ and if the topography is approximated to a sinusoidal pattern aligned perpendicularly to the wind, with a 25-km wavelength and a height difference from trough to crest of 500 m, calculate the vertical wavelength, the angle made by the wave fronts (surfaces of constant phase) with the horizontal, and the maximum horizontal velocity at the ground. Also, where is this maximum velocity observed (at crests, at troughs, or at the points of maximum slope)?

13.5. A 75-km/h gale wind blows over a hilly countryside. If the terrain elevation is approximated by a sinusoid of wavelength 4 km and amplitude of 40 m and if the stratification frequency of the air mass is $0.025\,\text{s}^{-1}$, what are the vertical displacements of the air particles at 1000 m and 2000 m above the mean ground level?

13.6. Determine the kinetic and potential energy densities of a pure internal wave.

13.7. Demonstrate that a single internal wave satisfying the dispersion relation is not only solution of the linearized perturbation equations but also of the fully nonlinear equations. What happens if two waves are present in the system?

13.8. Study internal waves in a stratified system of stratification frequency $N(z)$ in a vertically bounded domain using the hydrostatic approximation in Eq. (13.12). Show that the separation constant now appears as the eigenvalue to be calculated and that the associated radius of deformation no longer depends on the horizontal wavenumber squared $k_x^2 + k_y^2$.

13.9. Search for the existence of an internal Kelvin wave by using the separation-of-constants approach. Use the long-wave (or hydrostatic) assumption adapting Eq. (13.12) as necessary. Particularize to the case of uniform N^2.

13.10. Consider Skjomen Fjord near Narvik in northern Norway, with length L of 25 km, average depth H of about 110 m, and stratification frequency N

of about 2.0×10^{-3} s^{-1}. What are the lowest two frequencies of internal waves with wavelength equal to the fjord's length?

13.11. Despite the fact that system (13.2) has four time derivatives, we obtained a dispersion relation that yields only a pair of eigenfrequencies. Can you resolve the paradox?

NUMERICAL EXERCISES

13.1. Use the dispersion relation of pure internal waves (13.4) to illustrate the superposition of two waves by an animation in the (x, z) plane of size L_x, H. Can you see the group velocity when using two waves of equal amplitude and the two wavenumber vectors $(k_x = 35/L_x, k_z = 10/H)$ and $(k_x = 40/L_x, k_z = 12/H)$?

13.2. For the matrices and eigenvalues of problem (13.33), prove that

$$\lambda_{\min} \leq \frac{\mathbf{x}^\mathsf{T} \mathbf{B} \mathbf{x}}{\mathbf{x}^\mathsf{T} \mathbf{C} \mathbf{x}} \leq \lambda_{\max}. \tag{13.52}$$

To prove these so-called Rayleigh–Ritz inequalities Eq. (13.52) for symmetric positive definite matrices $\mathbf{B}$ and $\mathbf{C}$,

- assume that all eigenvalues are different,
- demonstrate that $\left(\mathbf{w}^i\right)^\mathsf{T} \mathbf{B} \mathbf{w}^j = \delta_{ij}$ and $\left(\mathbf{w}^i\right)^\mathsf{T} \mathbf{C} \mathbf{w}^j = \delta_{ij}$,
- prove that all eigenvectors $\mathbf{w}^i, i = 1, \ldots$ are linearly independent, and
- write any vector $\mathbf{x}$ as a weighted sum of those independent vectors, and use this expression in the Rayleigh–Ritz quotient to prove the Rayleigh–Ritz inequalities.

Adapt the code iwave.m to provide estimates of the upper and lower bound using a series of random vectors $\mathbf{x}$ to calculate the Rayleigh–Ritz estimator $(\mathbf{x}^\mathsf{T} \mathbf{B} \mathbf{x})/(\mathbf{x}^\mathsf{T} \mathbf{C} \mathbf{x})$ and store the minima and maxima of the calculated quotient. Seek how the bounds become increasingly precise as you choose additional random vectors.

13.3. By substitution of $\omega^2 = (1 + \tilde{\lambda}) f^2$ into Eq. (13.32) and redefining $\mathbf{B}$, show that all eigenvalues ω_i satisfy $\omega_i^2 \geq f^2$. Incidentally show that you can recast the problem into a standard eigenvalue problem.
By substitution of $\omega^2 = (1 - \tilde{\lambda}) N_{\max}^2$ into Eq. (13.32) and redefining $\mathbf{B}$, show that all eigenvalues ω_i satisfy $\omega_i^2 \leq N_{\max}^2$.

13.4. Adapt iwavemed.m to read temperature and salinity profiles from oceanographic data bases, such as the Levitus climatology,[4] and calcu-

---

[4] See http://www.cdc.noaa.gov/cdc/data.nodc.woa94.html.

late the radius of deformation for the Gulf Stream region. To read a climatological atlas, you can use `levitus.m` and select a location.

13.5. Assess numerically the convergence rate for eigenvalues and eigenfunctions in the case of uniform stratification by adapting `iwave.m`.

13.6. Discretize the eigenvalue problem of the sheared-flow instability (10.9) using the same techniques as in Section 13.4.4. What can you say about the positive-definite nature of the matrices involved? Try finding the numerical eigenvalues and growth rates of profiles you think are probably unstable in Analytical Problem 10.2.

Walter Heinrich Munk
1917–

Born in Austria and educated in the United States, Walter Heinrich Munk became interested in oceanography during a summer project under Harald Sverdrup at the Scripps Institution of Oceanography and quickly developed a fascination for ocean waves. This interest in waves arose partly because of the wartime need to predict sea and swell and also because Munk found wave research a challenge of intermediate complexity between simple periodic oscillations and hopeless chaos. As years went by, Munk eventually investigated all wavelengths, from the small capillary waves responsible for sun glitter to the ocean-wide tides. His studies of internal waves, in collaboration with Christopher Garrett, led him to propose a universal spectrum for the distribution of internal-wave energy in the deep ocean, now called the Garrett–Munk spectrum. More recently, pursuing an interest in acoustic waves, Munk initiated ocean tomography, a method for determining the large-scale temperature structure in the ocean from the measure of acoustic travel times. (*Photo by Jeff Cordia*)

Adrian Edmund Gill
1937–1986

Born in Australia, Adrien Edmund Gill pursued his career in Great Britain. His publications spanned a wide range of topics, including wind-forced currents, equatorially trapped ocean waves, tropical atmospheric circulation, and the El Niño–Southern Oscillation phenomenon, and culminated in his treatise *Atmosphere-Ocean Dynamics* (Gill, 1982). His greatest contributions relied on the formulation of simple yet illuminating models of geophysical flows. It has been said (only half jokingly) that he could reduce all problems to a simple ordinary differential equation with constant coefficients, with all the essential physics retained. Although he never held a professorship, Gill supervised numerous students at the Universities of Cambridge and Oxford. He is also remembered for his unassuming style and for the generosity with which he shared his ideas with students and colleagues. (*Photo credit: Gillman & Soame, Oxford*)

Chapter 14

Turbulence in Stratified Fluids

ABSTRACT

The previous chapter treated organized wave flows in stratified fluids, whereas the attention now turns to more complicated motions, such as vertical mixing, flow instability, forced turbulence, and convection. Because the study of such phenomena does not lend itself to analytical solutions, the emphasis is on budgets and scale analysis. The numerical section presents a few methods by which mixing and turbulence can be represented in numerical models.

14.1 MIXING OF STRATIFIED FLUIDS

Mixing by turbulence generates vertical motions and overturning. In a homogeneous fluid, the required energy is only that necessary to overcome mechanical friction (see Sections 5.1 and 8.1), but in a stratified fluid, work must also be performed to raise heavy fluid parcels and lower light parcels. Let us consider, for example, the system pictured in Fig. 14.1. Initially, it consists of two layers of equal thicknesses with fluids of different densities and horizontal velocities. After some time, mixing is assumed to have taken place, and the system consists of a single layer of average density flowing with the average velocity.[1] Because the heavier fluid (density ρ_2) lies initially below the lighter fluid (density ρ_1), the initial center of gravity is below mid-depth level, whereas in the final state, it is exactly at mid-depth. Thus, the center of gravity has been raised in the mixing process, and potential energy must have been provided to the system. Put another way, work has been performed against the buoyancy forces. With identical initial depths $H_1 = H_2 = H/2$, the average density is $\rho = (\rho_1 + \rho_2)/2$, and the potential energy gain is

$$PE\,\text{gain} = \int_0^H \rho_{\text{final}}\, gz \, \mathrm{d}z - \int_0^H \rho_{\text{initial}}\, gz \, \mathrm{d}z$$

---

[1] Credit for this illustrative example goes to Prof. William K. Dewar.

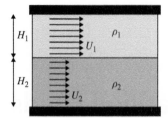 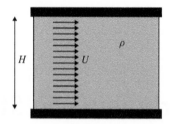

FIGURE 14.1 Mixing of a two-layer stratified fluid with velocity shear. Rising of dense fluid and lowering of light fluid both require work against buoyancy forces and thus lead to an increase in potential energy. Concomitantly, the kinetic energy of the system decreases during mixing. Only when the kinetic-energy drop exceeds the potential-energy rise can mixing proceed spontaneously.

$$= \frac{1}{2}\rho g H^2 - \left[\frac{1}{2}\rho_2 g \frac{H^2}{4} + \frac{1}{2}\rho_1 g \frac{3H^2}{4} \right]$$

$$= \frac{1}{8}(\rho_2 - \rho_1) g H^2. \tag{14.1}$$

The question arises as to the source of this energy increase. Because human intervention is ruled out in geophysical flows, a natural energy supply must exist or mixing would not take place. In this case, kinetic energy is released in the mixing process, as long as the initial velocity distribution is nonuniform. Conservation of momentum in the absence of external forces and in the context of the Boussinesq approximation ($\rho_1 \simeq \rho_2 \simeq \rho_0$) implies that the final, uniform velocity is the average of the initial velocities: $U = (U_1 + U_2)/2$. This indeed leads to a kinetic-energy loss

$$KE \text{ loss} = \int_0^H \frac{1}{2}\rho_0 u_{\text{initial}}^2 dz - \int_0^H \frac{1}{2}\rho_0 u_{\text{final}}^2 dz$$

$$= \frac{1}{2}\rho_0 U_2^2 \frac{H}{2} + \frac{1}{2}\rho_0 U_1^2 \frac{H}{2} - \frac{1}{2}\rho_0 U^2 H$$

$$= \frac{1}{8}\rho_0 (U_1 - U_2)^2 H. \tag{14.2}$$

Complete vertical mixing is naturally possible only if the kinetic-energy loss exceeds the potential-energy gain; that is,

$$\frac{(\rho_2 - \rho_1)gH}{\rho_0 (U_1 - U_2)^2} < 1. \tag{14.3}$$

Physically, the initial density difference should be sufficiently weak in order not to present an insurmountable gravitational barrier, or alternatively the initial velocity shear should be sufficiently large to supply the necessary amount of energy. When criterion (14.3) is not satisfied, mixing occurs only in the vicinity of the initial interface and cannot extend over the entire system. The determination of the characteristics of such localized mixing calls for a more detailed analysis.

For this purpose, let us now consider a two-fluid system of infinite extent (Fig. 14.2), with upper and lower densities and velocities denoted, respectively, by ρ_1, ρ_2 and U_1, U_2, and let us explore interfacial waves of infinitesimal

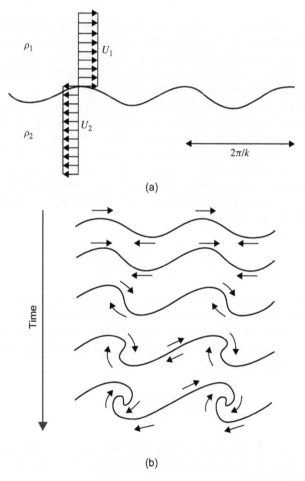

(a)

(b)

FIGURE 14.2   Kelvin–Helmholtz instability: (a) initial perturbation of wavenumber k, (b) temporal evolution of an unstable perturbation. The system is always unstable to short waves, which steepen, overturn, and ultimately cause mixing. As waves overturn, their vertical and lateral dimensions are comparable.

amplitudes. Mathematical derivations, not reproduced here, show that a sinusoidal perturbation of wavenumber k (corresponding to wavelength $2\pi/k$) is unstable if (Kundu, 1990, Section 11.6)

$$(\rho_2^2 - \rho_1^2)g < \rho_1\rho_2 k(U_1 - U_2)^2, \tag{14.4}$$

or for a Boussinesq fluid ($\rho_1 \simeq \rho_2 \simeq \rho_0$),

$$2(\rho_2 - \rho_1)g < \rho_0 k(U_1 - U_2)^2. \tag{14.5}$$

In a stability analysis, waves of all wavelengths must be considered, and we conclude that there will always be sufficiently short waves to cause instabilities. Therefore, a two-layer shear flow is always unstable. This is known as the *Kelvin–Helmholtz instability*. Among other instances, this instability plays a role in the generation of water waves by surface winds.

The details of the analysis leading to Eq. (14.5) reveal that the interfacial waves induce flow perturbations that extend on both sides of the interface across a height on the order of their wavelength. Thus, as unstable waves grow, they form rolls of height comparable to their width (Figs. 14.2, 14.3, and 14.4).

The rolling and breaking of waves induces turbulent mixing, and it is expected that the vertical extent of the mixing zone, which we denote by ΔH, scales like the wavelength of the longest unstable wave, that for which criterion (14.5) turns into an equality:

$$\Delta H \sim \frac{1}{k_{min}} = \frac{\rho_0(U_1 - U_2)^2}{2(\rho_2 - \rho_1)g}. \tag{14.6}$$

If the fluid system is of finite depth H, the preceding theory is no longer applicable, but we can anticipate, by virtue of dimensional analysis, that the results still hold, within some numerical factors. For a fluid depth H greater than ΔH, mixing must remain localized to a band of thickness ΔH, whereas for a fluid depth H less than ΔH, that is,

$$H \lesssim \frac{\rho_0(U_1 - U_2)^2}{(\rho_2 - \rho_1)g}, \tag{14.7}$$

mixing will engulf the entire system. Note the similarity between this last inequality, derived from a wave theory, and inequality (14.3) obtained from energy considerations.

Figures 14.5 and 14.6 show atmospheric instances of Kelvin–Helmholtz instabilities made visible by localized cloud formation. Kelvin–Helmholtz instabilities have also been observed to take place in the ocean (Woods, 1968).

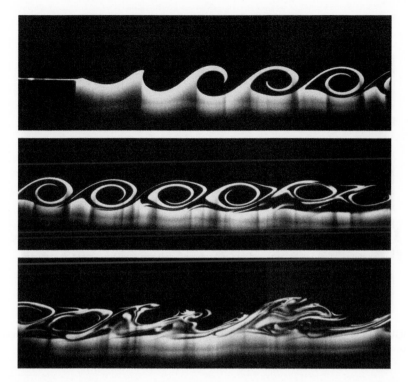

FIGURE 14.3   Development of a Kelvin–Helmholtz instability in the laboratory. Here, two layers flowing from left to right join downstream of a thin plate (visible on the left of the top photograph). The upper and faster moving layer is slightly less dense than the lower layer. Downstream distance (from left to right on each photograph and from top to bottom panel) plays the role of time. At first, waves form and overturn in a two-dimensional fashion (in the vertical plane of the photo), but eventually, three-dimensional motions appear that lead to turbulence and complete the mixing. (*Courtesy of Greg A. Lawrence. For more details on the laboratory experiment, see Lawrence, Browand & Redekopp, 1991*)

14.2   INSTABILITY OF A STRATIFIED SHEAR FLOW: THE RICHARDSON NUMBER

In the preceding section, we restricted our considerations to a discontinuity of the density and horizontal velocity, only to find that such a discontinuous stratification is always unstable. Instability causes mixing, and mixing will proceed until the velocity profile has been made stable. The question then is as follows: For a gradual density stratification, what is the critical velocity shear below which the system is stable and above which mixing occurs? To answer this question, we are led to study the stability of a stratified shear flow.

Let us consider a two-dimensional (x, z) inviscid and nondiffusive fluid with horizontal and vertical velocities (u, w), dynamic pressure p, and density

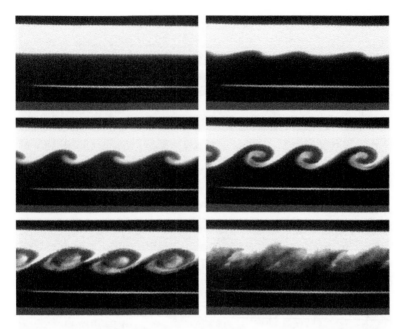

FIGURE 14.4   Kelvin–Helmholtz instability generated in a laboratory with fluids of two different densities and colors. (*Adapted from GFD-online, Satoshi Sakai, Isawo Iizawa, Eiji Aramaki*)

FIGURE 14.5   Kelvin–Helmholtz instability in the Algerian sky. (*Photo by one of the authors*). A color version can be found online at http://booksite.academicpress.com/9780120887590/

FIGURE 14.6 Kelvin–Helmholtz instability over the Sahara desert. (*Photo by one of the authors*)

anomaly ρ. In anticipation of the important role played by vertical motions, we reinstate the acceleration term in the vertical momentum equation (Section 4.3) and write

$$\frac{\partial u}{\partial t} + u\frac{\partial u}{\partial x} + w\frac{\partial u}{\partial z} = -\frac{1}{\rho_0}\frac{\partial p}{\partial x} \qquad (14.8a)$$

$$\frac{\partial w}{\partial t} + u\frac{\partial w}{\partial x} + w\frac{\partial w}{\partial z} = -\frac{1}{\rho_0}\frac{\partial p}{\partial z} - \frac{\rho g}{\rho_0} \qquad (14.8b)$$

$$\frac{\partial u}{\partial x} + \frac{\partial w}{\partial z} = 0 \qquad (14.8c)$$

$$\frac{\partial \rho}{\partial t} + u\frac{\partial \rho}{\partial x} + w\frac{\partial \rho}{\partial z} = 0. \qquad (14.8d)$$

Our basic state consists of a steady, sheared horizontal flow $[u = \bar{u}(z),\ w = 0]$ in a vertical density stratification $[\rho = \bar{\rho}(z)]$. The accompanying pressure field $\bar{p}(z)$ obeys $\mathrm{d}\bar{p}/\mathrm{d}z = -g\bar{\rho}(z)$. The addition of an infinitesimally small perturbation $(u = \bar{u}+u',\ w = w',\ p = \bar{p}+p',\ \rho = \bar{\rho}+\rho')$ and a subsequent linearization of the equations yield

$$\frac{\partial u'}{\partial t} + \bar{u}\frac{\partial u'}{\partial x} + w'\frac{\mathrm{d}\bar{u}}{\mathrm{d}z} = -\frac{1}{\rho_0}\frac{\partial p'}{\partial x} \qquad (14.9a)$$

$$\frac{\partial w'}{\partial t} + \bar{u}\frac{\partial w'}{\partial x} = -\frac{1}{\rho_0}\frac{\partial p'}{\partial z} - \frac{\rho' g}{\rho_0} \qquad (14.9b)$$

$$\frac{\partial u'}{\partial x} + \frac{\partial w'}{\partial z} = 0 \qquad (14.9c)$$

$$\frac{\partial \rho'}{\partial t} + \bar{u}\frac{\partial \rho'}{\partial x} + w'\frac{\mathrm{d}\bar{\rho}}{\mathrm{d}z} = 0. \qquad (14.9d)$$

Introducing the perturbation streamfunction ψ via $u' = +\partial\psi/\partial z$, $w' = -\partial\psi/\partial x$, the buoyancy frequency $N^2 = -(g/\rho_0)(d\bar{\rho}/dz)$, and the Fourier structure $\exp[i k (x - ct)]$ in the horizontal, we can reduce the problem to a single equation for ψ in terms of the remaining variable z:

$$(\bar{u} - c)\left(\frac{d^2\psi}{dz^2} - k^2\psi\right) + \left(\frac{N^2}{\bar{u} - c} - \frac{d^2\bar{u}}{dz^2}\right)\psi = 0. \tag{14.10}$$

This is called the *Taylor–Goldstein equation* (Goldstein, 1931; Taylor, 1931). It governs the vertical structure of a perturbation in a stratified parallel flow. Note the formal analogy with the Rayleigh equation (10.9) governing the structure of a perturbation on a horizontally sheared flow in the absence of stratification and in the presence of rotation. Therefore, the same analysis can be applied.

First, we state the boundary conditions. For a domain bounded vertically by two horizontal planes, at $z = 0$ and $z = H$, we impose a zero vertical velocity there, or, in terms of the streamfunction:

$$\psi(0) = \psi(H) = 0. \tag{14.11}$$

Then, we recognize that the equation and its accompanying boundary conditions form an eigenvalue problem: Unless the phase velocity c takes on a particular value (eigenvalue), the solution is trivial ($\psi = 0$). In general, the eigenvalues may be complex, but if c admits the function ψ, then its complex conjugate c^* admits the function ψ^* and is thus another eigenvalue. This can be easily verified by taking the complex conjugates of Eqs. (14.10) and (14.11). Hence, complex eigenvalues come in pairs. In each pair, one of the two eigenvalues will have a positive imaginary part and will correspond to an exponentially growing perturbation. The presence of a nonzero imaginary part to c automatically guarantees the existence of at least one unstable mode. Conversely, the basic flow is stable if and only if all possible phase speeds c are purely real.

Because it is impossible to solve problem (14.10) and (14.11) in the general case of an arbitrary shear flow $\bar{u}(z)$, we will limit ourselves, as in Section 10.2, to deriving integral constraints. A variety of such constraints can be established, but the most powerful one is obtained when the function ϕ, defined by

$$\psi = \sqrt{\bar{u} - c}\,\phi, \tag{14.12}$$

is used to replace ψ. Equation (14.10) and boundary conditions (14.11) become

$$\frac{d}{dz}\left[(\bar{u} - c)\frac{d\phi}{dz}\right] - \left[k^2(\bar{u} - c) + \frac{1}{2}\frac{d^2\bar{u}}{dz^2}\right.$$

$$\left. + \frac{1}{\bar{u} - c}\left(\frac{1}{4}\left(\frac{d\bar{u}}{dz}\right)^2 - N^2\right)\right]\phi = 0 \tag{14.13}$$

$$\phi(0) = \phi(H) = 0. \tag{14.14}$$

Multiplying Eq. (14.13) by the complex conjugate ϕ^*, integrating over the vertical extent of the domain, and utilizing conditions (14.14), we obtain

$$
\int_0^H \left[N^2 - \frac{1}{4} \left(\frac{d\bar{u}}{dz} \right)^2 \right] \frac{|\phi|^2}{\bar{u} - c} dz
$$

$$
= \int_0^H (\bar{u} - c) \left(\left| \frac{d\phi}{dz} \right|^2 + k^2 |\phi|^2 \right) dz + \frac{1}{2} \int_0^H \frac{d^2\bar{u}}{dz^2} |\phi|^2 dz, \tag{14.15}
$$

where vertical bars denote the absolute value of complex quantities. The imaginary part of this expression is

$$
c_i \int_0^H \left[N^2 - \frac{1}{4} \left(\frac{d\bar{u}}{dz} \right)^2 \right] \frac{|\phi|^2}{|\bar{u} - c|^2} dz = -c_i \int_0^H \left(\left| \frac{d\phi}{dz} \right|^2 + k^2 |\phi|^2 \right) dz, \tag{14.16}
$$

where c_i is the imaginary part of c. If the flow is such that $N^2 > \frac{1}{4} (d\bar{u}/dz)^2$ everywhere, then the preceding equality requires that c_i times a positive quantity equals c_i times a negative quantity and, consequently, that c_i must be zero. This leads us to define the *Richardson number*

$$
Ri = \frac{N^2}{(d\bar{u}/dz)^2} = \frac{N^2}{M^2}, \tag{14.17}
$$

with $M = |d\bar{u}/dz|$ (called the Prandtl frequency), and the criterion is that if the inequality

$$
Ri > \frac{1}{4} \tag{14.18}
$$

holds everywhere in the domain, the stratified shear flow is stable.

Note that the criterion does not imply that c_i must be nonzero if the Richardson number falls below 1/4 somewhere in the domain. Hence, inequality (14.18) is a sufficient condition for stability, whereas its converse is a necessary condition for instability. Atmospheric, oceanic, and laboratory data indicate, however, that the converse of (14.18) is generally a reliable predictor of instability.

If the shear flow is characterized by linear variations of velocity and density, with velocities and densities ranging from U_1 to U_2 and ρ_1 to ρ_2 ($\rho_2 > \rho_1$), respectively, over a depth H, then

$$
M = \left| \frac{d\bar{u}}{dz} \right| = \frac{|U_1 - U_2|}{H}, \quad N^2 = -\frac{g}{\rho_0} \frac{d\rho}{dz} = \frac{g}{\rho_0} \frac{\rho_2 - \rho_1}{H},
$$

and the Richardson criterion stated as the necessary condition for instability becomes

$$\frac{(\rho_2 - \rho_1)gH}{\rho_0(U_1 - U_2)^2} < \frac{1}{4}. \tag{14.19}$$

The similarity to Eq. (14.3) is not coincidental: Both conditions imply the possibility of large perturbations that could destroy the stratified shear flow. The difference in the numerical coefficients on the right-hand sides can be explained by the difference in the choice of the basic profile [discontinuous for Eq. (14.3), linear for Eq. (14.19)] and by the fact that the analysis leading to Eq. (14.3) did not make provision for a consumption of kinetic energy by vertical motions. The change from 1 in Eq. (14.3) to 1/4 in Eq. (14.19) is also consistent with the fact that condition (14.3) refers to complete mixing, whereas Eq. (14.19) is a condition for the onset of the instability.

More importantly, the similarity between Eqs. (14.3) and (14.19) imparts a physical meaning to the Richardson number: It is essentially a ratio between potential and kinetic energies, with the numerator being the potential-energy barrier that mixing must overcome if it is to occur and the denominator being the kinetic energy that the shear flow can supply when smoothed away. In fact, it was precisely by developing such energy considerations that British meteorologist Lewis Fry Richardson[2] first arrived, in 1920, to the dimensionless ratio that now rightfully bears his name. A first formal proof of criterion (14.18), however, did not come until four decades later (Miles, 1961).

In closing this section, it may be worth mentioning that bounds on the real and imaginary parts of the wave velocity c can be derived by inspection of certain integrals. This analysis, due to Louis N. Horward,[3] has already been applied to the study of barotropic instability (Section 10.3). Here, we summarize Howard's original derivation in the context of stratified shear flow. To begin, we introduce the vertical displacement a caused by the small wave perturbation, defined by

$$\frac{\partial a}{\partial t} + \bar{u}\frac{\partial a}{\partial x} = w$$

or

$$(\bar{u} - c)\, a = -\psi. \tag{14.20}$$

We then eliminate ψ from Eqs. (14.10) and (14.11) and obtain an equivalent problem for the variable a:

$$\frac{d}{dz}\left[(\bar{u} - c)^2 \frac{da}{dz}\right] + \left[N^2 - k^2(\bar{u} - c)^2\right]a = 0 \tag{14.21}$$

---

[2] See biography at the end of this chapter.
[3] See biography at end of Chapter 10.

$$a(0) = a(H) = 0. \tag{14.22}$$

A multiplication by the complex conjugate a^* followed by an integration over the domain and use of the boundary conditions yields

$$\int_0^H (\bar{u} - c)^2 P \, dz = \int_0^H N^2 |a|^2 dz, \tag{14.23}$$

where $P = |da/dz|^2 + k^2 |a|^2$ is a nonzero positive quantity. The imaginary part of this equation implies that if there is instability ($c_i \neq 0$), c_r must lie between the minimum and maximum values of $\bar{u}$, that is,

$$U_{min} < c_r < U_{max}. \tag{14.24}$$

Physically, the growing perturbation travels with the flow at some intermediate speed, and there exists at least one critical level in the domain where the perturbation is stationary with respect to the local flow. This local coupling between the wave and the flow is precisely what allows the wave to extract energy from the flow and to grow at its expense.

Now, the real part of Eq. (14.23),

$$\int_0^H \left[(\bar{u} - c_r)^2 - c_i^2 \right] P dz = \int_0^H N^2 |a|^2 dz \tag{14.25}$$

can be manipulated in a way similar to that used in Section 10.3 to obtain the following inequality:

$$\left(c_r - \frac{U_{min} + U_{max}}{2} \right)^2 + c_i^2 \leq \left(\frac{U_{max} - U_{min}}{2} \right)^2. \tag{14.26}$$

This implies that, in the complex plane, the number $c = c_r + ic_i$ must lie within the circle that has the range $\bar{u}$ as diameter on the real axis. Because instability requires a positive imaginary value c_i, the interest is restricted to the upper half of the circle (Fig. 10.1). This result is called the Howard semicircle theorem. In particular, it implies that c_i is bounded by $(U_{max} - U_{min})/2$, providing a useful upper bound on the growth rate of unstable perturbations:

$$kc_i \leq \frac{k}{2} (U_{max} - U_{min}). \tag{14.27}$$

14.3   TURBULENCE CLOSURE: k-MODELS

Reynolds averaging (Section 4.1) showed that small-scale processes, as those involved during turbulence and mixing, affect the mean flow through the so-called Reynolds stresses, which stem from the nonlinear advection terms in the momentum equations. Till now, Reynolds stresses were represented as diffusive fluxes and thus modeled with the help of an eddy viscosity. What value should be assigned to this eddy viscosity was not said. The fact is that setting a value to this parameter is far from trivial because it does not represent a unique fluid property, such as molecular viscosity, but rather reflects the level of turbulence in the particular flow under consideration. The value of the eddy viscosity should therefore not be expected to be a constant but ought to depend on characteristics of the flow conditions at the time and place of consideration. All we can hope for is that the local level of turbulence can be related to flow properties on the larger, resolved scale. Put another way, the determination of the eddy viscosity is in fact part of the problem. This forces us to consider how fluctuations actually behave.

A naive approach consists of calculating fluctuations such as u' by taking the original, nonaveraged equation for u and subtracting its Reynolds average (4.4) in order to obtain an equation for the fluctuation u', and similarly for the other variables. In principle, solving these "perturbation equations" should allow us to determine the fluctuations (such as u' and w'), from which we can then form products and obtain the Reynolds averages (such as $\langle u'w'\rangle$). This process, unfortunately, is not working.

To illustrate the nature of the problem, let us simplify the notation by introducing $\mathcal{L}$, an arbitrary linear operator, and start from a much reduced equation with quadratic nonlinearity of the type

$$\frac{\partial u}{\partial t} + \mathcal{L}(uu) = 0. \tag{14.28}$$

Its Reynolds average (see Section 4.1) is

$$\frac{\partial \langle u\rangle}{\partial t} + \mathcal{L}(\langle u\rangle\langle u\rangle) + \mathcal{L}(\langle u'u'\rangle) = 0, \tag{14.29}$$

from which we can obtain the equation governing the fluctuation u' by subtraction:

$$\frac{\partial u'}{\partial t} + 2\mathcal{L}(\langle u\rangle u') + \mathcal{L}(u'u') - \mathcal{L}(\langle u'u'\rangle) = 0. \tag{14.30}$$

Solving this equation should provide u' and allow us to calculate the Reynolds stress $\langle u'u'\rangle$, but it is clearly not realistic since we initially set out to separate the fluctuation so that we would not have to solve for it. What we want is only the average of a certain product and none of the details. With this in mind, let us start from the equation for the fluctuation (and not its solution) and seek an equation governing directly the desired average. To do so, we multiply

Eq. (14.30) by the fluctuation and then take the average of the product.[4] This yields a predictive equation for the desired quantity $\langle u'u' \rangle$:

$$\frac{1}{2}\frac{\partial \langle u'u' \rangle}{\partial t} + 2\langle u'\mathcal{L}(\langle u \rangle u') \rangle + \langle u'\mathcal{L}(u'u') \rangle = 0, \qquad (14.31)$$

in which we used $\langle u' \rangle = 0$ because a fluctuation has no average by definition.

From the last term of Eq. (14.31) emerges an annoying triple correlation, which no equation so far can provide. Should we try to establish a governing equation for this triplet (or third-order moment) by persevering with the same approach, it comes as no surprise that a fourth-order term arises, and so on endlessly. This means that we face a *closure problem*, and at some level of the process, we need to parameterize the unknown higher-order products in terms of those of lower order in a way that remains faithful to the physical phenomenology of turbulence and keeps modeling errors small.

It is generally accepted, more by intuition than by proof, that the higher the order at which truncation is performed and parameterization introduced, the lesser the modeling error. Stopping at second-order correlations is done almost universally in the context of field data and laboratory experiments (e.g., Gibson & Launder, 1978; Pope, 2000). The attending models are full *second-order closure* or *second-moment closure* schemes that calculate all Reynolds stresses involving products of variables by means of evolution equations relying on closure assumptions at the level of third-order correlations.

Here we restrict our attention to the two simplest versions of second-order schemes, in which only some of the second moments are determined by their evolution equations, whereas the others are governed by simpler equations containing no time derivative (so-called *diagnostic equations*). Such models continue to be called second-order closure schemes and are distinguished by explicitly naming the higher-order moments that are parameterized. An example is the k *model* that is presented below.

We begin by identifying key features of turbulence from which we can establish a practical closure scheme. The most obvious property of a turbulent flow is its ability to mix the fluid efficiently. This is why we stir our *café au lait* rather than wait for molecular diffusion to distribute the milk evenly across the black coffee. Turbulence-enhanced mixing is the reason why Reynolds stresses are most often expressed as diffusion terms. Shear in the mean flow generates instabilities, which are manifested by eddies at many scales. The larger eddies reflect the anisotropy of the mean flow, but rapid fluctuations at the shorter scales

---

[4]Here we assume that the averaging operation commutes with time and space derivatives. Also, the average of an average is the first average. Should our average not be an average over multiple realizations (so-called *ensemble average*) but rather an average over time or space, it is necessary that the temporal or spatial scales of the fluctuation be clearly separated from that of the mean flow (e.g., Burchard, 2002).

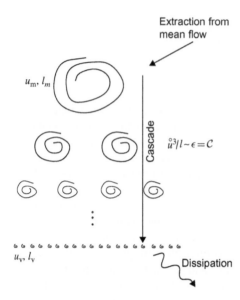

FIGURE 14.7 The oft-quoted lines of Lewis Fry Richardson *"the big whirls have little whirls that feed on their velocities, and little whirls have lesser whirls and so on to viscosity – in the molecular sense"* taken from his 1922 book aptly summarizes the idea used in turbulence modeling according to which turbulence effects a gradual transfer of energy from the broader and unstable flow to the smallest eddies that dissipate it.

appear erratic and isotropic. There is, however, no clear distinction between the two types of fluctuations, only a broad transition called the *energy cascade* (Kolmogorov, 1941; Fig. 14.7). For eddies of velocity scale $\mathring{u}$ and diameter d within the cascade, the Reynolds number is very large so that their evolution is fast compared with the decay time due to viscosity. Nonlinear advection is therefore dominant, rapidly breaking the eddies down into smaller ones. The energy continuously extracted from the mean flow by its instabilities is thus transferred gradually from the larger eddies to the smaller ones without appreciable loss to viscosity. This energy is eventually dissipated at the shortest scales. Because of the lack of dissipation through the cascade, the dissipation rate is thus conserved across the cascade. Since viscosity does not influence the dissipation rate within the cascade, the only parameters that can be related to the dissipation rate denoted as ϵ are the velocity scale $\mathring{u}$ and length scale d of the eddies, the time scale being determined by the turnover scale $d/\mathring{u}$, hence $\epsilon = \epsilon(\mathring{u}, d)$. The dimensionally correct relation is

$$\epsilon = \left(\frac{c_\mu^0}{4} \right)^{3/4} \frac{\mathring{u}^3}{d}, \qquad (14.32)$$

in which we introduced a calibration constant $c_\mu^0 \sim 0.1$ in a combination that will be useful later. We recover the result of Section 5.1. Also the value of ϵ is the one that is extracted from the mean flow at velocity scale u_m and length scale l_m, called *macro scales* and $\epsilon = \epsilon(\mathring{u}, d) = \epsilon(u_m, l_m)$.

Relation (14.32) shows that for a given turbulence cascade, $\mathring{u} \propto d^{1/3}$, stating that the smaller eddies contain less kinetic energy than the larger ones, so that the bulk of kinetic energy is attributable to the largest eddies of size l_m (Fig. 14.7). The kinetic-energy spectrum must therefore be decreasing with increasing wavenumber according to Eq. (5.8) of Section 5.1. The turbulent cascade may also be interpreted in terms of vorticity. In three dimensions, vortex tubes are twisted and stretched by other, adjacent or containing vortex tubes (in a manner similar to the two-dimensional straining encountered in Fig. 10.12), and because of incompressibility, stretching in one direction is accompanied by squeezing in another, and conservation of circulation demands an increase in vorticity. Vorticity, therefore, is increasing with decreasing eddy length scale.

The cascade cannot, of course, continue down to arbitrarily small scales, and at some small but finite scale, molecular viscosity comes into play. The scale l_v at which this occurs is the one for which viscous friction becomes a dominant term in the momentum equation, that is, the scale that renders the Reynolds number, ratio of inertia to friction, of order unity:

$$\frac{u \partial u / \partial x}{v \, \partial^2 u / \partial x^2} \sim \frac{u_v^2 / l_v}{v u_v / l_v^2} = \mathcal{O}(1). \tag{14.33}$$

The length and velocity scales, l_v and u_v, at which this *viscous sink* occurs are called *micro scales* or *viscous scales*. The preceding equation implies that they are linked by the following relation:

$$\frac{u_v l_v}{v} \sim 1. \tag{14.34}$$

Since Eq. (14.32) continues to hold down to that scale, we also have

$$\epsilon \propto \frac{u_v^3}{l_v} = \frac{u_m^3}{l_m}, \tag{14.35}$$

and we can determine the range of scales in the eddy cascade by eliminating u_v between Eqs. (14.35) and (14.34) and by expressing u_m in terms of the Reynolds number $u_m l_m / v$ of the macroscale:

$$\frac{l_m}{l_v} \sim Re^{3/4}. \tag{14.36}$$

High Reynolds-number flows, therefore, are characterized by broad eddy cascades.

Alternatively, we can express the scales at which dissipation takes place as functions of the dissipation rate and molecular viscosity by eliminating u_v from Eqs. (14.32) and (14.34):

$$l_v \sim \epsilon^{-1/4} v^{3/4}. \tag{14.37}$$

Thus, the more energy is fed into turbulence by the mean flow, the smaller the ultimate eddies are in order to dissipate that energy. Or, back to a more familiar situation, the more strongly we stir our coffee, the smaller the eddies, and the more efficient the mixing. It is worth insisting that it is molecular viscosity that is ultimately responsible for the diffusion. The turbulent cascade simply increases the shearing and tearing on fluid parcels, increasing contact between initially separated fluid parcels and increasing spatial gradients so that molecular diffusion can act more efficiently (Fig. 14.8).

Integrating the energy spectrum Eq. (5.8) from the longest to the shortest eddy sizes yields the total kinetic energy of the inertial range:

$$\int_{\pi/l_m}^{\pi/l_v} E_k dk = \frac{u_m^2}{2}\left(1 - \frac{1}{\sqrt{Re}}\right) \sim \frac{u_m^2}{2}, \qquad (14.38)$$

and hence the total turbulent kinetic energy in the flow does not differ much from that of the largest eddies.

Having now some idea on how energy is extracted from the mean flow, we can return to the challenge of representing the effect of this cascade in a model via an eddy viscosity. We limit our search to a parameterization in which the properties of the turbulent cascade are purely local (so-called *one-point closure model*) and do not involve remote parameters. For the mean flow, we suppose that l_m is the scale at which turbulence extracts energy, and since we do not resolve the cascade and its associated velocity fluctuations u' explicitly, the dissipation introduced by the eddy viscosity must extract the ϵ energy per time.

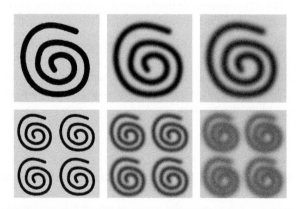

FIGURE 14.8   Effect of eddy size on diffusion over time. With time increasing from left to right, the top row illustrates the progressive action of diffusion on a structure of larger scale, whereas the bottom row shows that on a structure of shorter size. Note how diffusion acts more effectively at the shorter scale. For the same molecular diffusivity, mixing occurs more effectively at shorter scales because regions of different properties are in closer contact. (*Figure prepared at the suggestion of Hans Burchard*)

To accomplish this, the Reynolds number based on the eddy viscosity must be on the order of unity at the level of l_m:

$$\frac{u_m l_m}{\nu_E} \sim 1. \tag{14.39}$$

The eddy viscosity concept ensures that fluid parcels moving with the eddy velocity u_m over the distance l_m exchange momentum with other fluids parcels, as in molecular diffusion where momentum exchange between molecules occurs over the mean free path. Within this context, it is not surprising that the term *mixing length* was coined for l_m.

With Eq. (14.39), we ensure that some energy is extracted from the mean flow using an eddy viscosity approach. That this amount of energy extraction per unit time be equal to ϵ is an additional requirement. Assuming we know this dissipation rate, we require

$$\frac{u_m^3}{l_m} = \frac{\epsilon}{(c_\mu^0/4)^{3/4}}. \tag{14.40}$$

In summary, the formulation of the eddy viscosity demands that we know the scales u_m and l_m. If we also know the dissipation rate ϵ, we can use Eq. (14.40) to reduce the number of scales to be prescribed. If in addition, we know the kinetic energy k at the macroscale, this would add another relation that determines the velocity scale:

$$k = \frac{u_m^2}{2}. \tag{14.41}$$

This can also be used to calculate the dissipation rate (14.40) as follows:

$$\epsilon = \left(c_\mu^0\right)^{3/4} \frac{k^{3/2}}{l_m}. \tag{14.42}$$

Knowing k and ϵ would thus allow the calculation of the eddy viscosity ν_E and complete our closure scheme. At this point, the reader might object that calculating a macroscale length using a microscale dissipation rate sounds contradictory, but the paradox is easily explained by the fact that in the Kolmogorov theory, the dissipation rate at the microscale is equal to the energy input at the macroscale, hence the link. This justification and most of the previous reasoning rely on the idea of a statistical equilibrium state of turbulence in which, at each moment, energy input in the mean flow matches energy removal at the shortest scales.

One of the first successful attempts to quantify the eddy viscosity by means of a velocity and length scale is credited to Ludwig Prandtl[5] who considered

---

[5] See biography at the end of Chapter 8.

a vertically sheared horizontal flow $\langle u \rangle_{(z)}$ and assumed the velocity fluctuations to be statistically random except for a nonzero correlation between the horizontal and vertical velocity fluctuations, u' and w'. In such a case,

$$\langle u'w' \rangle = r\sqrt{\langle u'^2 \rangle}\sqrt{\langle w'^2 \rangle}, \qquad (14.43)$$

where r is the correlation coefficient between u' and w'. Assuming that each velocity fluctuation is proportional to the velocity scale u_m of the coherent structures causing the correlation, we can write

$$\langle u'w' \rangle = c_1 u_m^2 \qquad (14.44)$$

with a constant coefficient of proportionality c_1. Since the eddy viscosity is defined from

$$\langle u'w' \rangle = -\nu_E \frac{\partial \langle u \rangle}{\partial z}, \qquad \nu_E = u_m l_m, \qquad (14.45)$$

we obtain an expression for the eddy viscosity that depends only on mean flow quantities, thus yielding a first turbulence-closure model:

$$u_m = l_m \left| \frac{\partial \langle u \rangle}{\partial z} \right| \quad \text{and} \quad \nu_E = l_m^2 \left| \frac{\partial \langle u \rangle}{\partial z} \right|, \qquad (14.46)$$

Here, the absolute value is introduced to ensure a positive value for the eddy viscosity. We note that this model predicts increasing turbulent diffusion under increased shear, in accordance with our intuition that shear is destabilizing and the cause of turbulence.

The only remaining parameter that Prandtl needed to determine was the mixing length, in which all calibration constants come together. The determination of the mixing length depends on the particular situation, especially its geometry. For example, a flow along a rigid boundary is characterized by an eddy size increasing with distance from the boundary, and thus a larger l_m at a greater distance from the boundary.

Because $\langle u'w' \rangle$ must vanish at a rigid boundary, so must the Reynolds stress. Since it is now expressed as $-l_m^2 |\partial \langle u \rangle / \partial z|^2$ with the present closure scheme, it is clear that the mixing length, too, must go to zero at the boundary. As a result, the mixing length in the vicinity of a solid boundary is often expressed as $l_m = \mathcal{K}z$, in which z is the distance from the boundary and $\mathcal{K}$ the so-called *von Kármán constant* (e.g., Nezu & Nakagawa, 1993; see also Section 8.1).

The preceding closure scheme relying on the concept of a mixing length l_m may be modified in order to accommodate the stabilizing effect of stratification and also generalized to three dimensions. These modifications will be explored later when we show that the Prandtl model can be seen as a simplification of more complex turbulence closure schemes. Despite the advantage of its algebraic nature, which is easily implemented, Prandtl's scheme suffers from the fact that it assumes a turbulence level determined solely by the instantaneous flow. No memory effect is included, and as soon as mixing eradicates the shear, the eddy viscosity falls to zero, whereas in reality, turbulence never stops abruptly but undergoes gradual decay. This is one among several problems of a so-called *zero-equation turbulence model*. Clearly, more sophisticated schemes are often necessary.

To develop models with memory effect, we seek governing equations with time derivatives (*prognostic equations* as opposed to *diagnostic equations*) for some of the second moments. First, we note that the difference between volume conservation of the total flow, Eq. (3.17), and of the mean flow only, Eq. (4.9), provides a constraint on the velocity fluctuations:

$$\frac{\partial u'}{\partial x} + \frac{\partial v'}{\partial y} + \frac{\partial w'}{\partial z} = 0. \tag{14.47}$$

We can then use this constraint in our manipulation of the governing equations for the Reynolds stresses. The two most natural candidates for diagnostic equations are those for the turbulent kinetic energy k and dissipation rate ϵ because they together capture the primary characteristics of the turbulent environment. Their values also set the scales in formulating the eddy viscosity.

Here, we start by developing the so-called turbulent kinetic energy model in which k is defined[6] as

$$k = \frac{\langle u'^2 \rangle + \langle v'^2 \rangle + \langle w'^2 \rangle}{2}. \tag{14.48}$$

In view of Eqs. (14.32) and (14.38), the bulk of the turbulent kinetic energy is contained in the largest eddies so that we may use the velocity scale of the largest eddies as $\sqrt{2k}$. We are then in a position of establishing a governing equation for k by applying a closure approach.

Taking the difference between Eqs. (3.19) and (4.7a), we obtain the evolution equation for the fluctuation u', which we then multiply by u' itself. Doing the same with the equations for v' and w', and adding the three together, we obtain after some tedious algebra and the use of Eq. (14.47) the following evolution equation for k:

$$\frac{dk}{dt} = P_s + P_b - \epsilon - \left(\frac{\partial q_x}{\partial x} + \frac{\partial q_y}{\partial y} + \frac{\partial q_z}{\partial z} \right), \tag{14.49}$$

in which we arranged the various terms for a better understanding of the physics. First, the time derivative is the material derivative based on the mean flow,

---

[6]To be exact, k is the turbulent kinetic energy per unit of mass of fluid. By virtue of the Boussinesq approximation, however, the ratio between energy and energy per mass is the reference density ρ_0, which is a constant.

namely,

$$\frac{d}{dt} = \frac{\partial}{\partial t} + \langle u \rangle \frac{\partial}{\partial x} + \langle v \rangle \frac{\partial}{\partial y} + \langle w \rangle \frac{\partial}{\partial z},$$

and we note that the Coriolis terms have canceled one another out, which is not too surprising since the Coriolis force does not perform mechanical work and should therefore not affect a kinetic energy budget.

With no approximation invoked, the other terms[7] are

$$P_s = -\langle u'u' \rangle \frac{\partial \langle u \rangle}{\partial x} - \langle u'v' \rangle \frac{\partial \langle u \rangle}{\partial y} - \langle u'w' \rangle \frac{\partial \langle u \rangle}{\partial z}$$

$$- \langle v'u' \rangle \frac{\partial \langle v \rangle}{\partial x} - \langle v'v' \rangle \frac{\partial \langle v \rangle}{\partial y} - \langle v'w' \rangle \frac{\partial \langle v \rangle}{\partial z}$$

$$- \langle w'u' \rangle \frac{\partial \langle w \rangle}{\partial x} - \langle w'v' \rangle \frac{\partial \langle w \rangle}{\partial y} - \langle w'w' \rangle \frac{\partial \langle w \rangle}{\partial z} \qquad (14.50)$$

$$P_b = -\langle \rho'w' \rangle \frac{g}{\rho_0} \qquad (14.51)$$

$$\frac{\epsilon}{\nu} = \left\langle \frac{\partial u'}{\partial x} \frac{\partial u'}{\partial x} \right\rangle + \left\langle \frac{\partial u'}{\partial y} \frac{\partial u'}{\partial y} \right\rangle + \left\langle \frac{\partial u'}{\partial z} \frac{\partial u'}{\partial z} \right\rangle$$

$$+ \left\langle \frac{\partial v'}{\partial x} \frac{\partial v'}{\partial x} \right\rangle + \left\langle \frac{\partial v'}{\partial y} \frac{\partial v'}{\partial y} \right\rangle + \left\langle \frac{\partial v'}{\partial z} \frac{\partial v'}{\partial z} \right\rangle$$

$$+ \left\langle \frac{\partial w'}{\partial x} \frac{\partial w'}{\partial x} \right\rangle + \left\langle \frac{\partial w'}{\partial y} \frac{\partial w'}{\partial y} \right\rangle + \left\langle \frac{\partial w'}{\partial z} \frac{\partial w'}{\partial z} \right\rangle \qquad (14.52)$$

$$q_x = \frac{1}{\rho} \left\langle \left(p' + \frac{u'^2 + v'^2 + w'^2}{2} \right) u' \right\rangle - \nu \frac{\partial k}{\partial x} \qquad (14.53)$$

and similar expressions for q_y and q_z. All terms involve unknown averages for which we now need to make closure assumptions.

Because the quantity P_s involves both mean flow and turbulence, it stems from the interaction between the two. The presence of the shear of the large-scale flow suggests that we call it *shear production*. The second term, P_b, clearly involves the work performed by the turbulent buoyancy forces on the vertical stratification and is thus related to potential-energy changes. For obvious reasons we call it *buoyancy production*. The dissipation rate ϵ involves, as expected, the molecular viscous dissipation by turbulent motions. Finally, the vector (q_x, q_y, q_z) involves only turbulent fluctuations of pressure and velocity, and its divergence form in the turbulent kinetic energy budget (14.49) indicates

---

[7] Strictly speaking, the dissipation term should be $\epsilon = 2\nu \|\mathbf{D}\|^2$ in which the deformation tensor $\mathbf{D}$ of the fluctuations is similar to Eq. (14.54). The definition used here is often called pseudo dissipation.

that it represents a spatial redistribution of k by turbulence, not contributing to production or dissipation.

All those terms must now be modeled in terms of the state variables. For example, the Reynolds stresses appearing in the unknown terms are replaced by the eddy-viscosity parameterization already shown. By defining the deformation tensor (or strain-rate tensor)

$$
\mathbf{D} = \frac{1}{2}
\begin{pmatrix}
2\frac{\partial u}{\partial x} & \left(\frac{\partial u}{\partial y} + \frac{\partial v}{\partial x}\right) & \left(\frac{\partial u}{\partial z} + \frac{\partial w}{\partial x}\right) \\
\left(\frac{\partial u}{\partial y} + \frac{\partial v}{\partial x}\right) & 2\frac{\partial v}{\partial y} & \left(\frac{\partial v}{\partial z} + \frac{\partial w}{\partial y}\right) \\
\left(\frac{\partial u}{\partial z} + \frac{\partial w}{\partial y}\right) & \left(\frac{\partial v}{\partial z} + \frac{\partial w}{\partial x}\right) & 2\frac{\partial w}{\partial z}
\end{pmatrix}
\tag{14.54}
$$

and the Reynolds stress tensor

$$
\boldsymbol{\tau} =
\begin{pmatrix}
\langle u'u' \rangle & \langle u'v' \rangle & \langle u'w' \rangle \\
\langle u'v' \rangle & \langle v'v' \rangle & \langle v'w' \rangle \\
\langle u'w' \rangle & \langle v'w' \rangle & \langle w'w' \rangle
\end{pmatrix},
\tag{14.55}
$$

the eddy-viscosity model is

$$
\boldsymbol{\tau} = -2\nu_E \mathbf{D} + \frac{2k}{3}\mathbf{I}.
\tag{14.56}
$$

The first term on the right-hand side is a familiar expression relating the (turbulent) stress to the rate of strain. The appearance of the second term, proportional to k and involving the identity matrix $\mathbf{I}$, begs for an explanation. Without it the parameterization is faulty: The trace of the stress tensor on the left-hand side must be equal to the trace on the right-hand side, and this constraint justifies the presence of the second term. In practice, this term is not a dominant one, but it is readily calculated with a k model. Using Eq. (14.56) in Eq. (14.50) expresses the shear production P_s in terms of the mean-flow characteristics.

For the buoyancy production term, P_b, the velocity-density correlations are modeled with the help of the eddy-diffusivity approach

$$
\langle \rho'w' \rangle = -\kappa_E \frac{\partial \langle \rho \rangle}{\partial z} = \kappa_E \frac{\rho_0}{g} N^2
\tag{14.57}
$$

in which κ_E is a turbulent diffusivity that is part of the closure scheme. Aside from the latter, the term P_b does not require any further treatment.

Since the flux terms q_x, q_y, and q_z appear in divergence form in Eq. (14.49), they are only responsible for redistributing k in space. This is coupled to the fact that they involve turbulent quantities, suggesting that we model these terms as turbulent diffusion of k. In view of the velocity and pressure correlations involved,[8] the eddy viscosity ν_E is used for the flux calculation rather than the eddy diffusivity κ_E.

In summary, the various terms are parameterized as follows:

$$P_s = 2\nu_E \| \langle \mathbf{D} \rangle \| \tag{14.58}$$

$$P_b = -\kappa_E N^2 \tag{14.59}$$

$$q_x = -\nu_E \frac{\partial k}{\partial x} \quad q_y = -\nu_E \frac{\partial k}{\partial y} \quad q_z = -\nu_E \frac{\partial k}{\partial z}. \tag{14.60}$$

We notice that the sign of the modeled term P_s is consistent with the idea of turbulence extracting energy from the mean flow and transferring it to turbulence. Similarly, the sign of P_b indicates that stratification inhibits turbulence because of the increase of potential energy required by mixing a stably stratified system (Section 14.2).

With these parameterizations, Eq. (14.49) governing the evolution of turbulent kinetic energy becomes

$$\frac{dk}{dt} = P_s + P_b - \epsilon + \mathcal{D}(k), \tag{14.61}$$

$$\mathcal{D}(k) = \frac{\partial}{\partial z} \left(\nu_E \frac{\partial k}{\partial z} \right). \tag{14.62}$$

Note that the turbulent diffusion has been reduced to its the vertical component. The horizontal part is awaiting a subsequent parameterization of horizontal subgrid scales.

With appropriate boundary conditions, we can predict the evolution of k if we know how to calculate ϵ, the eddy viscosity ν_E, and the eddy diffusivity κ_E. With ϵ calculated using Eq. (14.42) and mixing length l_m prescribed, the turbulent closure scheme is called a k-model or *one-equation turbulence model*. The energy of the turbulent eddies gives a reliable estimate of their velocity and hence of the eddy viscosity via

$$\nu_E = \frac{c_\mu}{\left(c_\mu^0 \right)^{3/4}} \sqrt{k} \, l_m. \tag{14.63}$$

Note that in this scheme, the mixing length must be prescribed independently. This is usually done based on geometrical considerations of the flow. The value of ϵ is then obtained from Eq. (14.35) for use in Eq. (14.61) to predict k. The constant c_μ^0 is the same constant as in Eq. (14.42), whereas c_μ is a calibration parameter.

---

[8] In state-of-the art models, the eddy diffusivity for turbulent kinetic energy is the eddy diffusivity divided by the so-called *Schmidt number*.

The eddy diffusivity κ_E is obtained in a similar way:

$$\kappa_E = \frac{c'_\mu}{\left(c^0_\mu\right)^{3/4}} \sqrt{k}\, l_m, \tag{14.64}$$

where the calibration constant c'_μ differs from c_μ. The two parameters will be defined later as functions of shear and stratification.

For simplicity, we introduced only two different turbulent-diffusion coefficients, although each state variable could claim its own (e.g., Canuto, Howard, Cheng & Dubovikov, 2001). We utilize a unique κ_E for diffusion of all scalar fields because they are subjected to the same turbulent transport. Hence κ_E is used for the diffusion of density, salinity, temperature, moisture, or any other tracer concentration. In contrast, the eddy viscosity ν_E is used for the diffusion of momentum, k and ϵ, the dynamical variables.

Before exploring more advanced closure schemes, we can verify how the present model performs in a simple flow situation, such as a vertically sheared flow of uniform density. We align the x-axis with the mean flow and the z-axis with the shear. In this way, averaged fields are independent of x and y (Fig. 14.9), and the velocity field is simply $u = \langle u \rangle + u'$, $w = w'$. The mean flow depends only on z and obeys

$$\frac{\partial \langle u \rangle}{\partial t} = -\frac{1}{\rho_0}\frac{\partial \langle p \rangle}{\partial x} + \frac{\partial}{\partial z}\left(\nu\,\frac{\partial \langle u \rangle}{\partial z} - \langle u'w' \rangle\right). \tag{14.65}$$

The pressure gradient is uniform, and over a distance L along the x-axis, the pressure difference is $p_2 - p_1$. The kinetic energy $KE = \langle(u^2 + v^2 + w^2)\rangle/2$ of the flow can be split into mean and turbulent contributions:

$$KE = \frac{\langle u \rangle^2}{2} + k. \tag{14.66}$$

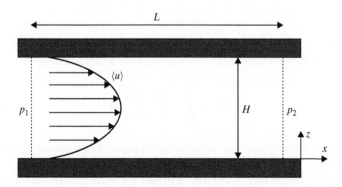

FIGURE 14.9   Flow between two horizontal planes forced by an external pressure gradient.

Multiplying Eq. (14.65) by $\langle u \rangle$, we can construct the governing equation for the kinetic energy of the mean flow. For the turbulent part, we can perform the same manipulation on the equations governing the fluctuations (not written down). With the simplifications pertinent to the present flow, the outcome is

$$\frac{\partial}{\partial t} \frac{\langle u \rangle^2}{2} = -\frac{1}{\rho_0} \frac{\partial \langle p \rangle \langle u \rangle}{\partial x} + \frac{\partial}{\partial z}\left[\langle u \rangle \left(\nu \frac{\partial \langle u \rangle}{\partial z} - \langle u'w' \rangle \right) \right]$$
$$- \nu \left(\frac{\partial \langle u \rangle}{\partial z} \right)^2 + \underline{\langle u'w' \rangle \frac{\partial \langle u \rangle}{\partial z}}$$

$$\frac{\partial k}{\partial t} = -\underline{\langle u'w' \rangle \frac{\partial \langle u \rangle}{\partial z}} - \epsilon + \frac{\partial q_z}{\partial z}. \tag{14.67}$$

The underlined term in the second equation is identified as the shear production of turbulence, and we note that it also appears in the first equation with opposite sign. Clearly, the shear production of turbulence is at the expense of mean-flow energy.

With the eddy-viscosity approach outlined earlier, the shear production term is expressed as

$$P_s = -\langle u'w' \rangle \frac{\partial \langle u \rangle}{\partial z} = \nu_E \left(\frac{\partial \langle u \rangle}{\partial z} \right)^2. \tag{14.68}$$

For positive eddy viscosity, energy is extracted from the mean flow and feeds the turbulence.

If we integrate both energy equations over a distance L and across the domain height, assume steady state (for averages), exploit the fact that the velocities (both mean and turbulent) are zero at bottom ($z = 0$) and top ($z = H$), and finally use the closure scheme, we obtain

$$\left(\frac{p_1 - p_2}{\rho_0} \right) \frac{UH}{L} = \int_0^H (\nu_E + \nu) \left(\frac{\partial \langle u \rangle}{\partial z} \right)^2 dz \tag{14.69}$$

in which $U = (1/H) \int_0^H \langle u \rangle \, dz$ is the average velocity over the inflow and outflow sections. For the turbulent kinetic energy, we have a similar budget

$$\int_0^H \epsilon \, dz = \int_0^H \nu_E \left(\frac{\partial \langle u \rangle}{\partial z} \right)^2 dz. \tag{14.70}$$

Let us now interpret these budgets. In the absence of turbulence, eddy viscosity is zero, and the energy equation (14.69) for the mean flow then shows that for an increased energy input by a higher pressure gradient, the flow must generate increasing shear so that molecular viscosity can dissipate this energy.

However, when shear increases, the flow is prone to instabilities and eventually becomes turbulent. We see that energy is now extracted from the mean flow at much lower values of the shear because of the presence of $\nu_E \gg \nu$ in the right-hand side. The energy budget (14.70) for the turbulence confirms that the energy extracted from the mean flow is dissipated in the viscous sink by ϵ. We verify that molecular viscosity remains the ultimate sink of energy but at the much shorter scales of turbulence.

14.4 OTHER CLOSURES: $k - \epsilon$ AND $k - k l_m$

To form a closed system of equations, the previous one-equation turbulence model needed the mixing length to be prescribed a priori. To avoid this, it is desirable to move from one governing equation (for k) to two governing equations (for k and l_m, or for k and ϵ). It is therefore no surprise to find a vast literature proposing governing equations for a combination of k and l_m, or equivalently a combination of k and ϵ. From the two calculated quantities, the third can always be determined from the algebraic relationship (14.42) and then the eddy viscosity via Eq. (14.63) or similar expression.

Since dissipation rates can be measured in the ocean by microprofilers (e.g., Lueck, Wolk & Yamazaki, 2002; Osborn, 1974), ϵ is an attractive candidate for a second equation in turbulence modeling. By manipulating the governing equations for velocity fluctuations in a similar way as for k, we can formulate a governing equation for ϵ of the type

$$\frac{d\epsilon}{dt} = Q, \tag{14.71}$$

in which the right-hand side contains a series of complicated expressions involving higher-order correlations (e.g., Burchard, 2002; Rodi, 1980). Unlike the k equation, however, these terms cannot be systematically modeled using the eddy-viscosity approach, and additional hypotheses, not to say educated guesses, are required. The most common approach is to use from the k equation the terms related to energy production and use them in linear combinations to close the energy dissipation source terms. Terms related to spatial redistribution of energy are, as usual, modeled by a turbulent diffusion. When all is said and done, the governing equation for ϵ is expressed as

$$\frac{d\epsilon}{dt} = \frac{\epsilon}{k}(c_1 P_s + c_3 P_b - c_2 \epsilon) + \mathcal{D}(\epsilon), \tag{14.72}$$

where the terms P_s and P_b are the same as in the k equation and the coefficients c_1, c_2, and c_3 are calibration constants: $c_1 \approx 1.44$, $c_2 \approx 1.92$, $-0.6 \lesssim c_3 \lesssim 0.3$. Because the two turbulent quantities that are calculated in this model are k and ϵ, it is useful to express the eddy viscosity as a function of these by eliminating the mixing length l_m between Eqs. (14.63) and (14.42) to obtain

$$\nu_E = c_\mu \frac{k^2}{\epsilon}. \tag{14.73}$$

This outlines a particular *two-equation turbulence model* within an array of possible other ones. Instead of formulating the governing equation for ϵ, an equation for l_m or a combination of l_m and k can also be established, with Eq. (14.42) providing a link between the three variables k, ϵ, and l_m.

A very popular scheme in the context of geophysical flows is the so-called $k - kl_m$ model of Mellor and Yamada (1982), of which the governing equation for kl_m is

$$\frac{dkl_m}{dt} = \frac{l_m}{2}\left[E_1 P_s + E_3 P_b - \left(1 - E_2\,\frac{l_m^2}{l_z^2}\right)\epsilon\right] + \mathcal{D}(kl_m), \tag{14.74}$$

where E_1, E_2, and E_3 are calibration constants. As for Eq. (14.72), the source term is a linear combination of P_s, P_b, and ϵ. In this closure scheme appears a new length scale l_z that needs to be prescribed to force l_m to vanish at solid boundaries. In the $k - \epsilon$ model, this is achieved "automatically" if the correct boundary conditions are applied (Burchard & Bolding, 2001). Except for this difference, the two formulations are structurally identical because, in the absence of spatial variations, Eqs. (14.72) and (14.74) are equivalent by virtue of Eq. (14.42). The difference lies in the quantity that is transported by the flow: dissipation in the $k - \epsilon$ model, kl_m in the Mellor–Yamada model.

In fact, it is possible to establish a generic evolution equation for $k^a \epsilon^b$ with two parameters a and b ($a = 0$, $b = 1$ to recover the $k - \epsilon$ model and $a = 5/2$, $b = -1$ to obtain the $k - kl_m$ model). Changing the values of a and b changes the nature of the second quantity that is transported by the flow (Umlauf & Burchard, 2003). Whatever combination is chosen, all such models fall under the label of *two-equation models* and, except for the background mixing length l_z, do not need additional prescribed spatial functions.

Leaving aside more complex closure schemes, we end our description of turbulence modeling with the observation that all closure schemes described here are based on local properties, that is, not using distant information to parameterize Reynolds stresses. These models are called *one-point closure* schemes. Schemes that use information from remote locations to infer local turbulence properties are referred to as two-point closure schemes (e.g., Stull, 1993).

We now return to the notation used in the rest of the book by no longer making a distinction between average flow properties and turbulent properties, so that from here on u stands again for the mean velocity.

14.5 MIXED-LAYER MODELING

The turbulence models presented in the previous two sections are applicable to three-dimensional flows in general. In geophysical fluid dynamics, models can be simplified by exploiting the small aspect ratio of the flows under investigation (e.g., Umlauf & Burchard, 2005). In particular, the strain-rate tensor can be reduced to

$$
\mathbf{D} = \frac{1}{2}\begin{pmatrix} \sim 0 & \sim 0 & \frac{\partial u}{\partial z} \\ \sim 0 & \sim 0 & \frac{\partial v}{\partial z} \\ \frac{\partial u}{\partial z} & \frac{\partial v}{\partial z} & \sim 0 \end{pmatrix}
$$

(14.75)

and the shear production to

$$
P_s = \nu_E M^2, \quad M^2 = \left(\frac{\partial u}{\partial z}\right)^2 + \left(\frac{\partial v}{\partial z}\right)^2,
$$

(14.76)

in which we took the opportunity to define the Prandtl frequency M. Furthermore, turbulent diffusion is chiefly acting in the vertical direction because of the shorter distances and larger gradients in that direction. On the other hand, the study of flows with small aspect ratios inevitably requires that a larger step size be taken in the horizontal and, consequently, that a series of horizontal subgrid scale processes be handled separately. The effects of these are generally modeled by a horizontal diffusion with diffusion coefficient $\mathcal{A}$:

$$
\mathcal{D}() = \frac{\partial}{\partial x}\left(\mathcal{A}\frac{\partial}{\partial x}\right) + \frac{\partial}{\partial y}\left(\mathcal{A}\frac{\partial}{\partial y}\right) + \frac{\partial}{\partial z}\left(\nu_E\frac{\partial}{\partial z}\right).
$$

(14.77)

It should be clear here that ν_E is intending to model actual turbulence, whereas $\mathcal{A}$ is an attempt to take into account processes unresolved in the horizontal, at scales longer than l_m but shorter than the horizontal grid step used in the model.

Assuming a Kolmogorov-type turbulent energy cascade in the horizontal, a possible closure is

$$
\mathcal{A} \sim (\Delta x)^{4/3}\epsilon_H^{1/3},
$$

(14.78)

directly inspired by $\nu_E \sim l_m^{4/3}\epsilon^{1/3}$ [deduced from Eqs. (14.63) and (14.42)]. For this estimation of $\mathcal{A}$, ϵ_H is the energy dissipation of the horizontally unresolved processes. According to Okubo (1971), this dissipation rate is relatively similar from case to case (Fig. 14.10).

Another subgrid scale parameterization of horizontal processes is directly inspired by the Prandtl model (e.g., Smagorinsky, 1963):

$$
\mathcal{A} \sim \Delta x \Delta y \left[\left(\frac{\partial u}{\partial x}\right)^2 + \left(\frac{\partial v}{\partial y}\right)^2 + \frac{1}{2}\left(\frac{\partial u}{\partial y} + \frac{\partial v}{\partial x}\right)^2\right]^{1/2}.
$$

(14.79)

In this formulation, the mixing length is replaced by the average grid spacing to ensure that all scales below the grid size are effectively treated as unresolved motions. In view of the factors $\Delta x \Delta y$ appearing in the front of Eq. (14.79)

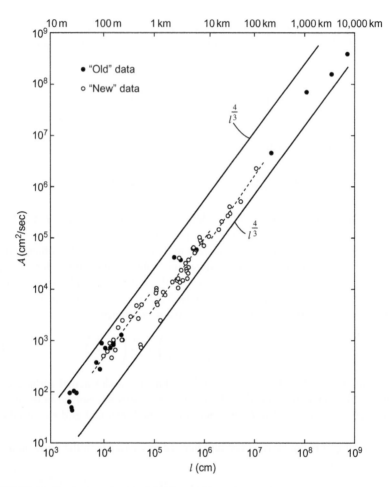

FIGURE 14.10   Horizontal eddy diffusivity as function of cutoff scales in typical geophysical flows (Okubo, 1971).

and the factors Δx^2 and Δy^2 arising in the denominators after the numerical discretization of the horizontal second derivative in the diffusion terms, we can interpret the Smagorinsky formulation as a numerical filter (Section 10.6). This filter acts at the grid resolution with the intensity of the filter cleverly made to depend on the local shear of the flow.

We now leave aside subgrid scale parameterizations because they are less well established than turbulence closure schemes and return to vertical turbulence modeling. In particular, we show that the Prandtl model can be recovered under the assumption of instantaneous and local equilibrium between shear production, buoyancy production, and dissipation (as in stationary and homogeneous turbulence, for example). In this case, the turbulent kinetic energy budget

(14.61) reduces to

$$P_s + P_b = \epsilon. \tag{14.80}$$

For a vertically sheared flow $\langle u \rangle$ in a stratified fluid of stratification frequency N, the equilibrium between production and dissipation yields, using Eqs. (14.63) and (14.42) for a given mixing length l_m,

$$k = \frac{c_\mu}{\left(c_\mu^0\right)^{3/2}} l_m^2 M^2 (1 - R_f), \tag{14.81}$$

in which the flux Richardson number R_f is defined as

$$R_f = \frac{-P_b}{P_s} = \frac{c_\mu'}{c_\mu} \frac{N^2}{M^2} = \frac{c_\mu'}{c_\mu} Ri. \tag{14.82}$$

The eddy viscosity follows:

$$\nu_E = \left(\frac{c_\mu}{c_\mu^0}\right)^{3/2} l_m^2 M \sqrt{1 - R_f}. \tag{14.83}$$

We recover the Prandtl closure (14.46) with eddy viscosity now taking into account the stabilizing effect of the stratification via the flux Richardson number. The simplest models are obtained as particular cases of the more complex models.

Further adaptations to mixed-layer flows can be made by introducing so-called *stability functions* into the parameterizations. The derivation of such formulations lies beyond the scope of this chapter, and we only outline here the general approach. The derivation begins with the governing equations for the individual components of the Reynolds stress tensor, obtained by multiplying the governing equations for the velocity fluctuations by other velocity fluctuations and taking their average. Again, higher-order terms demand simplifying hypotheses. Spatial and temporal variations are either neglected under an equilibrium hypothesis similar to Eq. (14.80) or rather heuristically described by an advection-diffusion equation. Depending on the nature of the closure hypotheses made along the way, the end results are so-called *algebraic Reynolds-stress models*. In these models, Reynolds stresses are often appearing in nonlinear algebraic systems that need to be solved to extract the individual stresses. This can be done with some additional approximations, eventually leading to expressions for the Reynolds stresses as functions of mean-flow characteristics. In all cases, formulations of the type (14.73) appear, in which the function c_μ may be quite complicated.

In all algebraic second-order turbulent closure schemes, Reynolds stresses depend on two dimensionless *stability parameters*:

$$\alpha_N = \frac{k^2}{\epsilon^2} N^2, \quad \alpha_M = \frac{k^2}{\epsilon^2} M^2. \tag{14.84}$$

Stability functions widely differ based on the various hypotheses used during their derivation (e.g., Canuto et al., 2001; Galperin, Kantha, Mellor & Rosati, 1989; Kantha & Clayson, 1994; Mellor & Yamada, 1982). If $P_s + P_b = \epsilon$ is assumed along the way, so-called *quasi-equilibrium versions* are obtained (Galperin, Kantha, Hassid & Rosati, 1988). These generally exhibit a more robust behavior than other formulations (see Deleersnijder, Hanert, Burchard & Dijkstra, 2008 for a discussion). An example of stability functions are those of Umlauf and Burchard (2005) depicted in Fig. 14.11 and given by

$$\nu_E = c_\mu \frac{k^2}{\epsilon} \tag{14.85}$$

$$\kappa_E = c'_\mu \frac{k^2}{\epsilon} \tag{14.86}$$

with the coefficients given by

$$c_\mu = \frac{s_0 + s_1 \alpha_N + s_2 \alpha_M}{1 + d_1 \alpha_N + d_2 \alpha_M + d_3 \alpha_N \alpha_M + d_4 \alpha_N^2 + d_5 \alpha_M^2}, \tag{14.87}$$

$$c'_\mu = \frac{s_4 + s_5 \alpha_N + s_6 \alpha_M}{1 + d_1 \alpha_N + d_2 \alpha_M + d_3 \alpha_N \alpha_M + d_4 \alpha_N^2 + d_5 \alpha_M^2}. \tag{14.88}$$

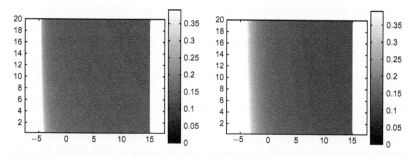

FIGURE 14.11 Stability functions c_μ and c'_μ of Umlauf and Burchard (2005) plotted as functions of α_N and α_M using parameter values given in Table 14.1.

TABLE 14.1 Parameters Used in the Closure Scheme of Canuto et al. (2001)

$s_0 =$	$s_1 =$	$s_2 =$	$s_4 =$	$s_5 =$	$s_6 =$
0.10666	0.01734	−0.00012	0.11204	0.00451	0.00088
$d_1 =$	$d_2 =$	$d_3 =$	$d_4 =$	$d_5 =$	$c_\mu^0 =$
0.2554	0.02871	0.00522	0.00867	−0.00003	0.0768

14.6 PATANKAR-TYPE DISCRETIZATIONS

A turbulent closure scheme does not lend itself to analytical solutions, and numerics are called to the rescue. Before dwelling into numerical discretization, however, it is worth insisting that the equations for turbulent variables are only models obtained after a long series of assumptions. If the derivation is not done with diligence, inconsistencies can arise. Turbulent kinetic energy, for example, must never be negative, but, if the flux Richardson number for some reason begins to exceed one, the equilibrium value of k becomes negative by virtue of Eq. (14.81) and the eddy viscosity Eq. (14.83) ceases to exist. A first constraint on any turbulence closure scheme is therefore that their answers make basic physical sense. For the k model, for example, it should be shown that the solution of Eq. (14.61) is always positive (see Analytical Problem 14.8).

Assuming that the turbulence closure scheme is respecting all physical and mathematical requirements, we should further ensure that subsequent numerical discretization respects these properties (Numerical Exercise 14.2). We may now appreciate why, during the treatment of advection problems (Section 6.4), much discussion was devoted to monotonic behavior. Negative values of a variable that should remain positive can have dramatic effects when nonlinearities are present. The values of the source terms in the equations governing turbulent kinetic energy and dissipation can hint at problems. Occasionally, well-defined mathematical and numerical operations may cause unexpected problems. For example, a quadratic sink for a tracer c with uniform spatial distribution,

$$\frac{dc}{dt} = -\mu c^2 \qquad (14.89)$$

starting from initial condition c^0 has for solution

$$c(t) = \frac{c^0}{1 + \mu t \, c^0}, \qquad (14.90)$$

which is well behaved if c^0 is positive but will eventually become unacceptable if c^0 is negative. In the presence of spatial variations, such a problem may be far more difficult to detect but is just as serious.

An implicit treatment of the nonlinear source or sink term would enhance numerical stability and therefore reduce over- or undershooting tendencies (e.g., avoid unphysical negative values), but the numerical cost is the need to invert or solve a nonlinear algebraic equation at each time step. This is tantamount to finding the zeros of a function and can be done with standard iterative methods, Picard, Regula Falsi, Newton–Raphson methods, for example (see Dahlquist & Björck, 1974; Stoer & Bulirsh, 2002). But since the procedure needs to be repeated at each grid point and at each time step, the approach can become quite burdensome. Also problems of robustness in finding roots are not uncommon in view of the large number of times the procedure needs to be repeated. There is always a risk of no convergence or of convergence toward an unphysical solution.

Patankar (1980) introduced a method that renders the discretization of a nonlinear source term somehow implicit without actually needing to solve a nonlinear algebraic equation. To present his idea, we start from the spatially discretized version of the governing equation for the vector state $\mathbf{x}$ (i.e., the vector consisting of the variables of the model):

$$\frac{\partial \mathbf{x}}{\partial t} = \mathbf{M}(\mathbf{x}) \tag{14.91}$$

where the right-hand side gathers all discretized spatial operators. Such a system can be discretized in time by one of the many methods already presented. The algorithm to update the numerical state variable $\mathbf{x}$ is then of the type

$$\mathbf{A}\mathbf{x}^{n+1} = \mathbf{B}\mathbf{x}^n + \mathbf{f}, \tag{14.92}$$

where $\mathbf{A}$ and $\mathbf{B}$ result from the chosen discretization and $\mathbf{f}$ may contain forcing terms, sources, sinks, and boundary conditions. If we now add a decay term to the governing equation

$$\frac{\partial \mathbf{x}}{\partial t} = \mathbf{M}(\mathbf{x}) - \mathbf{K}\mathbf{x} \tag{14.93}$$

with the matrix $\mathbf{K} = \mathrm{diag}(K_i)$ being a diagonal matrix with various decay rates K_i, one for each component of the state vector, an explicit discretization of the decay term then leads to the modified algorithm

$$\mathbf{A}\mathbf{x}^{n+1} = (\mathbf{B} - \mathbf{C})\mathbf{x}^n + \mathbf{f}, \tag{14.94}$$

in which $\mathbf{C} = \mathrm{diag}(K_i \Delta t)$ is, too, a diagonal matrix. Alternatively, an implicit treatment of the decay term would lead to

$$(\mathbf{A} + \mathbf{C})\mathbf{x}^{n+1} = \mathbf{B}\mathbf{x}^n + \mathbf{f}. \tag{14.95}$$

The only modification in the calculations is to invert $\mathbf{A} - \mathbf{C}$ instead of $\mathbf{A}$, which does not add much burden since only the diagonal is changed.

Patankar's simple yet powerful trick is to take a nonlinear sink written in a pseudo-linear fashion $-K(c)\,c$. As long as $K(c)$ remains bounded (and positive) for all c, we can always express any sink term in this way simply by defining K accordingly. The discretization then uses at each grid point

$$- K(\tilde{c}_i^n)\tilde{c}_i^{n+1}, \tag{14.96}$$

which is a consistent discretization. To calculate $\tilde{c}^{n+1}$, all that is required is to modify the system similarly to Eq. (14.95) by adding a term $K_i(\tilde{c}_i^n)\Delta t$ on the diagonal of $\mathbf{A}$.

The method is simple, but how does this trick help maintain positive values? The explicit discretization applied to a positive value $\tilde{c}^n$ does not ensure positiveness for arbitrary time steps because

$$\tilde{c}^{n+1} = \tilde{c}^n - K(\tilde{c}^n)\Delta t \tilde{c}^n = \left[1 - K(\tilde{c}^n)\Delta t\right]\tilde{c}^n$$

is negative whenever $K(\tilde{c}^n)\Delta t > 1$. In contrast, the Patankar trick replaces the explicit calculation by

$$\tilde{c}^{n+1} = \frac{\tilde{c}^n}{1 + K(\tilde{c}^n)\Delta t},$$

which remains positive at all times. The only requirement for the method to work is that $K(c)$ be bounded for $c \to 0$. Otherwise, upon approaching zero, overflows in the numerical code will occur.

But why not just enforce $K(\tilde{c}^n)\Delta t \leq 1$ by choosing a sufficiently short time step? In so-called *stiff problems*, K varies widely, and the time step being constrained by the largest value of K may have to be excessively short. Unless adaptive time-stepping is used to keep the instances of short time steps to a minimum, it is almost impossible to ensure a sufficiently small time step that keeps $\tilde{c}$ positive at all times without using excessively small time steps during most of the calculations. In coupled nonlinear equations, the stiffness is often difficult to gage, and ecosystem models among others are prone to nonpositive behavior whenever an explicit discretization is used. This is very frustrating because such a problem tends to occur only occasionally. The benefit of the Patankar method is to avoid the time-step penalty in the presence of quickly damped processes.

We now generalize the Patankar method slightly to take into account that sinks decrease values but sources increase them. For a single equation with a source (production term $P \geq 0$) and a sink (destruction term $-K(c)\, c \leq 0$) such that

$$\frac{dc}{dt} = P(c) - K(c)\, c, \tag{14.97}$$

a discretization *à la* Patankar would read

$$\tilde{c}^{n+1} = \tilde{c}^n + \Delta t \left\{ \frac{P^n}{\tilde{c}^n}\left[\alpha \tilde{c}^{n+1} + (1-\alpha)\tilde{c}^n\right] - K(\tilde{c}^n)\left[\beta \tilde{c}^{n+1} + (1-\beta)\tilde{c}^n\right] \right\} \tag{14.98}$$

where α and β are implicitness factors. This equation can directly be solved for $\tilde{c}^{n+1}$.

In some problems, the solution of Eq. (14.97) tends toward an equilibrium solution c^*, such that $P(c^*) = K(c^*)\, c^*$, without oscillating around this equilibrium. It is relatively easy to show that this is the case if

$$P(c) \gtreqless K(c)c \quad \text{for} \quad c \lesseqgtr c^*. \tag{14.99}$$

It is then possible to show (Numerical Exercise 14.8) that a numerical solution that keeps concentration values positive and converges toward the equilibrium value c^* without oscillation is guaranteed as long as

$$\frac{1}{\Delta t} \geq \frac{P-Kc}{c^*-c} + \frac{\alpha P - \beta Kc}{c}. \tag{14.100}$$

To obtain the least restrictive time step, the best choice is $\alpha = 0$, $\beta = 1$. For example with $P = c^r$ and $K = c^r$, the equilibrium is $c^* = 1$ for any value $r > 0$. If $\alpha = 0$ and $\beta = 1$, the Patankar scheme yields steady convergence toward this value for arbitrary large time steps.

The present example is not an academic one because the perceptive reader may have realized that $r = 1/2$ corresponds to the typical source/sink term in a turbulence closure scheme with fixed mixing length (Section 14.3). The case $r = 1$ arises with the logistic equation encountered in the modeling of biological processes.

The method outlined here has been adapted to a set of coupled equations, ensuring conservation between components and higher-order convergence than the Euler scheme shown here (Burchard, Deleersnijder & Meister, 2003, 2005). Such an approach is of special interest in ecosystem models that include transport.

14.7   WIND MIXING AND PENETRATIVE CONVECTION

Like mixing, turbulence in stratified fluids requires work against buoyancy forces, and stratification thus acts as a moderator of turbulence. This situation can be expressed quantitatively by applying to turbulence some of the concepts derived earlier, particularly the notion of mixing depth, as expressed by Eq. (14.6),

$$\Delta H = \frac{\rho_0 (U_1 - U_2)^2}{2g(\rho_2 - \rho_1)}. \tag{14.101}$$

An important measure of turbulence is the *friction velocity* u_*, a measure of the turbulent velocity fluctuations.[9] Thus, locally horizontal velocities are expected to differ by values on the order of u_*, and the numerator of Eq. (14.101) could be replaced by the dimensionally equivalent expression $\rho_0 u_*^2$. Likewise, the difference $(\rho_2 - \rho_1)$ can be interpreted as a local turbulent density fluctuation and the product $u_*(\rho_2 - \rho_1)$ as a measure of the vertical density flux $\overline{w'\rho'}$ (where primes denote turbulent fluctuation and an overbar indicates some average). The introduction of those quantities transforms Eq. (14.101) into a turbulent

---

[9] The attribute *friction* reflects the historical heritage of turbulent boundary-layer theory and does not imply that friction is of great importance here.

analogue:

$$L = \frac{\rho_0 u_*^3}{\mathcal{K} g \overline{w'\rho'}}. \tag{14.102}$$

This length scale represents the depth of fluid to which stratification confines eddies of strength u_*. It is called the *Monin–Obukhov length* in honor of the two Soviet oceanographers who, in 1954, first pointed to the importance of this scale in the study of stratified turbulence. In the denominator, the factor $\mathcal{K}$ is the von Kármán constant ($\mathcal{K} = 0.41$), which is traditionally introduced to facilitate mathematical development in boundary-layer applications and which was first encountered in Section 8.1.1.

If density variations are entirely due to temperature stratification, then the flux $\overline{w'\rho'}$ is equal to $-\alpha \rho_0 \overline{w'T'}$, where α is the coefficient of thermal expansion and T' is the temperature fluctuation. Because this is often the case, the Monin–Obukhov length is customarily defined as

$$L = \frac{u_*^3}{-\mathcal{K}\alpha g \,\overline{w'T'}}. \tag{14.103}$$

14.7.1 Wind Mixing

As an application, consider the development of a turbulent mixed layer in the upper ocean under the action of a wind stress (Fig. 14.12). Let us assume that, initially, the ocean stratification is characterized by a uniform stratification frequency N, so that the density increases linearly with depth according to

$$\rho = -\frac{\rho_0 N^2}{g}\, z, \tag{14.104}$$

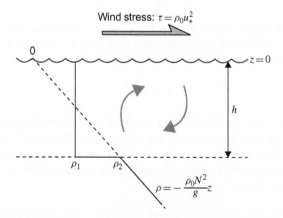

FIGURE 14.12 Development of a mixed layer in the ocean under the action of a wind stress.

where z is the vertical coordinate measured negatively downward ($z = 0$ is the surface) and ρ is the density departure from the reference density ρ_0, the initial surface density. After sometime t, this stratification has been partially eroded, and a mixed layer of depth h has developed (Fig. 14.12). In this layer, the density has been homogenized and, in the absence of surface heating, evaporation, and precipitation, has become the average density that initially existed over that depth:

$$\rho_1 = \frac{\rho_0 N^2 h}{2g}.$$

Below the mixed layer, density is still unchanged, $\rho_2 = \rho(z = -h) = \rho_0 N^2 h/g$, and there exists a density jump

$$\Delta\rho = \rho_2 - \rho_1 = \frac{\rho_0 N^2 h}{2g}. \tag{14.105}$$

Mixing has caused upwelling of denser waters and downwelling of lighter waters, thus raising the level of potential energy. The energy gain by time t is

$$PE = \int_{-h}^{0} \rho_1 gz \, dz - \int_{-h}^{0} \rho gz \, dz$$

$$= \frac{1}{12} \rho_0 N^2 h^3. \tag{14.106}$$

Therefore, potential energy increases at the rate

$$\frac{dPE}{dt} = \frac{1}{4} \rho_0 N^2 h^2 \frac{dh}{dt}. \tag{14.107}$$

The supply of energy is provided by the surface wind. If the wind stress is τ, the turbulent friction velocity u_* is given by Eq. (8.1) (see also Kundu, 1990, Section 12.11):

$$\tau = \rho_0 u_*^2, \tag{14.108}$$

and the rate of work performed by τ on fluid particles with typical velocities u_* is proportional to τu_* or $\rho_0 u_*^3$. Introducing a coefficient of proportionality m to account for the exact rate of work minus the portion diverted to kinetic-energy production (which eventually dissipates), we state: $dPE/dt = m\rho_0 u_*^3$, or by virtue of Eq. (14.107),

$$N^2 h^2 \frac{dh}{dt} = 4mu_*^3. \tag{14.109}$$

Observations and laboratory experiments suggest $m = 1.25$. This last equation can be readily integrated to obtain the instantaneous mixed-layer depth:

$$h = \left(\frac{12mu_*^3}{N^2} t \right)^{1/3}. \tag{14.110}$$

Of some interest here is the evaluation of the Monin–Obukhov length. As the layer erodes the underlying stratification at the rate dh/dt, turbulence must overcome the density jump $\Delta\rho$, causing a density flux at the base of the mixed layer of magnitude

$$\overline{w'\rho'} = \frac{dh}{dt} \Delta\rho$$
$$= \frac{2m\rho_0 u_*^3}{gh}, \tag{14.111}$$

by virtue of Eqs. (14.105) and (14.109). Based on this local flux value, the Monin–Obukhov length (14.102) is found to be

$$L = \frac{1}{2mK} h. \tag{14.112}$$

With the numerical values $K = 0.40$ and $m = 1.25$, L is exactly h. The exact identity between L and h is fortuitous (especially since K is closer to 0.41 than 0.40), but it remains that the depth of the turbulent mixed layer is on the order of the Monin–Obukhov length, thus imparting a direct physical meaning to the latter.

The preceding considerations illustrate but one aspect of the development of a mixed layer in the upper ocean. Much work has been done on this problem, and the reader desiring additional information is referred to the book edited by Kraus (1977) for a review and to the article by Pollard, Rhines and Thompson (1973) for a particularly clear discussion of Coriolis effects and of the relevance of the Richardson number. See also Section 8.7.

Considerations of mechanically induced mixing in the lower atmosphere and above the ocean floor can be found, respectively, in Sorbjan (1989, Section 4.4.1) and in Weatherly and Martin (1978). A review of laboratory experiments and associated theories is provided by Fernando (1991).

14.7.2   Penetrative Convection

Convection is defined as the process by which vertical motions modify the heat distribution in the system. In the example at the end of the previous subsection, the stirring of the upper ocean layer is caused by the mechanical action of the wind stress, and convection is said to be forced. Natural, or free, convection arises when the only source of energy is of thermal origin, such as an imposed temperature difference or an imposed heat flux, and the motions associated with the convective process derive their energy from the work generated by buoyancy forces as warm fluid rises and cold fluid sinks.

A common occurrence of natural convection in geophysical fluids is the development of an unstable atmospheric boundary layer (Sorbjan, 1989). During daytime, the solar radiation traverses the atmosphere and reaches the earth (ground or sea), where it is absorbed. The earth reemits this radiation in the infrared range, thus effectively heating the atmosphere from below. As a result, the lowest level of the atmosphere is usually an unstable, convective region, called the *atmospheric boundary layer*. The existence of this layer is very beneficial to humans because of the ventilation it causes. When the atmosphere is stably stratified down to the ground, a situation called *inversion*, the air is still and uncomfortable; moreover, if there is a source of pollution, this pollution stagnates and can become harmful. Such is the situation in Los Angeles (USA) when smog occurs (Stern, Boubel, Turner & Fox, 1984).

The intensity of stirring motions in natural convection depends, obviously, on the strength of the thermal forcing, as well as on the resistance of the fluid to move (viscosity) and to conduct heat (conductivity). A traditional example is convection in a fluid layer of height h confined between two horizontal rigid plates and heated from below. The forcing is the temperature difference ΔT between the two plates, the lower one being the hotter of the two. At low temperature differences, the viscosity ν and heat diffusivity κ_T of the fluid prevent convective motions, the fluid remains at rest, and the heat is carried solely by molecular diffusion (conduction). As the temperature difference is increased, everything else remaining is unchanged, the hot fluid at the bottom will eventually float upward, and the cold fluid will sink from above.

If viscosity is the limiting factor, the amplitude of the convective velocities, w_*, can be estimated from a balance between the upward buoyancy force $-g\rho'/\rho_0 \sim \alpha g \Delta T$ (where α is the coefficient of thermal expansion) and the retarding frictional force $\nu \partial^2 w / \partial z^2 \sim \nu w_* / h^2$, yielding:

$$w_* \sim \frac{\alpha g \Delta T h^2}{\nu}. \tag{14.113}$$

Comparing the convective heat flux $\overline{w'T'} \sim w_* \Delta T \sim \alpha g \Delta T^2 h^2 / \nu$ to the conductive flux $\kappa_T \partial T / \partial z \sim \kappa_T \Delta T / h$, we form the ratio

$$Ra = \frac{\alpha g \Delta T^2 h^2 / \nu}{\kappa_T \Delta T / h} = \frac{\alpha g \Delta T h^3}{\nu \kappa_T}, \tag{14.114}$$

which is known as the *Rayleigh number*, in honor of British scientist Lord Rayleigh,[10] who first studied this problem quantitatively (1916).

---

[10]Rayleigh was a contemporary of Kelvin. See the joint photograph at end of Chapter 9.

Convection occurs, theories show (Chandrasekhar, 1961), when this number exceeds a critical value, which depends on the nature of the boundary conditions. For a fluid confined between two rigid plates, the critical Rayleigh number is $Ra = 1708$. At values slightly above the threshold, convection organizes itself in parallel two-dimensional rolls or in packed hexagonal cells. At higher values of the Rayleigh number, erratic time-dependent motions develop, and convection appears much less organized.

Geophysical fluids almost always fall in this last category because of the large heights involved and the small values of molecular viscosity and conductivity of air and water. In the atmospheric boundary layer, where the Rayleigh number typically exceeds 10^{15}, convection is manifested by the intermittent formation near the ground of warm pockets of air, called *thermals*, which then rise through the convective layer; the circuit is completed by a weak subsidence of colder air between the rising thermals. In such a situation, viscosity and heat diffusivity play secondary roles, and the main characteristics of the flow do not depend on them.

As an application, consider the development of an atmospheric boundary layer from an initial, stable stratification under the action of a constant heat flux supplied by the ground (Fig. 14.13). At time $t = 0$, the air is assumed linearly stratified with potential-temperature profile given by

$$\bar{T}(z) = T_0 + \Gamma z, \tag{14.115}$$

where T_0 is the initial potential temperature at the ground and Γ is the vertical potential-temperature gradient, corresponding to a stratification frequency $N = (\alpha g \Gamma)^{1/2}$. The upward heat flux at the ground, denoted by $\rho_0 C_p Q$, is assumed constant. After some time t, convection has eroded the stratification up to a height $h(t)$. The temperature $T(t)$ in the convective layer varies according

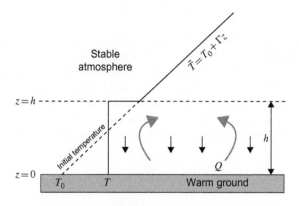

FIGURE 14.13 An unstable atmospheric boundary layer. The heat supplied at the ground surface generates convection, which progressively erodes the stratification above.

to the instantaneous distribution of thermals but, on the average, appears to be nearly constant with height. The heat budget for the intervening time requires that the change in heat content of the affected fluid be equal to the accumulated heat received from the ground,

$$\rho_0 C_p \int_0^h (T - \bar{T})\,dz = \int_0^t \rho_0 C_p Q\,dt, \tag{14.116}$$

and provides a first relation between the height of the convective layer and its temperature:

$$h(T - T_0) - \frac{1}{2}\Gamma h^2 = Q\,t. \tag{14.117}$$

Another relation between these two variables arises from the mechanical-energy budget. Because there is no source of mechanical energy, the sum of the kinetic and potential energies of the system decays with time under the action of frictional forces. In first approximation, to be verified a posteriori, the amount of kinetic energy and energy loss to friction are insignificant contributions compared with the potential-energy changes undergone by the system. So, it suffices to state in first approximation that potential energy, per unit area, at time t is equal to that at the initial time:

$$\int_0^{h(t)} \rho_0 \alpha Tgz\,dz = \int_0^{h(t)} \rho_0 \alpha \bar{T}gz\,dz, \tag{14.118}$$

which yields

$$T - T_0 = \frac{2}{3}\Gamma h. \tag{14.119}$$

Physically, this implies that the temperature rise at the ground is two-thirds of the temperature change over the height h according to the initial temperature gradient (Fig. 14.13). Oddly enough, the temperature in the upper third of the convective layer has decreased, whereas the fluid undergoes an overall heating. This is explained by the upward motion of colder air from below.

Together, Eqs. (14.117) and (14.119) provide the temporal evolution of the thickness and potential temperature of the atmospheric boundary layer:

$$h = \sqrt{\frac{6Qt}{\Gamma}} \tag{14.120}$$

$$T = T_0 + \sqrt{\frac{8\Gamma Qt}{3}}. \tag{14.121}$$

The atmospheric boundary layer thus grows according to the square root of time. This progressive erosion of the ambient stratification by convective motions is termed *penetrative convection*.

We are now in a position to estimate the contribution of kinetic energy. Because convection is accomplished by thermals that rise over the entire extent of the layer, the convective overturns are as deep as the layer itself, and the Monin–Obukhov length must be comparable to the layer thickness. Equating these two quantities, we write

$$\frac{w_*^3}{\mathcal{K}\alpha gQ} = h, \tag{14.122}$$

where the symbol w_* replaces u_* to indicate that the turbulent motions are not mechanically induced (such as by a shear stress) but are of thermal origin. This equality yields a measure of the turbulent velocity w_*:

$$w_* = (\mathcal{K}\alpha ghQ)^{1/3}, \tag{14.123}$$

which supersedes Eq. (14.113) when the Rayleigh number is so high that viscosity is no longer the dominant parameter. The kinetic energy is then estimated to be $\rho_0 w_*^2 h/2$, and its ratio to the instantaneous potential energy $\rho_0 \alpha g(T - T_0)h^2/2$ is

$$\frac{KE}{PE} \sim (Nt)^{-2/3}, \tag{14.124}$$

with the numerical coefficients discarded. In this last expression, $N = (\alpha g\Gamma)^{1/2}$ is the frequency of the undisturbed stratification. Because $1/N$ is typically on the order of a few minutes while the atmospheric boundary layer develops over hours, the product Nt is large, and we can justify the earlier neglect of the kinetic-energy contribution to the overall energy budget. A fortiori, the decay rate of kinetic energy by frictional forces is also unimportant in the overall energy budget. Finally, it is worth noting that if w_* is the velocity scale of the rising thermals, the heat flux $Q = \overline{w'T'}$ is carried by those thermals with their temperature differing from that of the descending fluid approximately by $T_* = Q/w_*$.

The preceding application is but a simple example of convection in the atmosphere. Generally, convective motions in the atmospheric boundary layer are affected by numerous factors, including winds, which they in turn affect. A sizable body of knowledge has been accumulated on the physics of the atmospheric boundary layer, and the interested reader is referred to Sorbjan (1989) or to Garratt (1992).

In numerical models, convection may or may not be resolved depending on its length scale compared with the size of the system. When convection occurs at scales too small to be resolved, numerical convective adjustment is used (see Section 11.4).

ANALYTICAL PROBLEMS

14.1. A stratified shear flow consists of two layers of depth H_1 and H_2 with respective densities and velocities ρ_1, U_1 and ρ_2, U_2 (left panel of Fig. 14.1). If the lower layer is three times as thick as the upper layer and the lower layer is stagnant, what is the minimum value of the upper layer velocity for which there is sufficient available kinetic energy for complete mixing (right panel of Fig. 14.1)?

14.2. In the ocean, a warm current ($T = 18°C$) flows with a velocity of 10 cm/s above a stagnant colder layer ($T = 10°C$). Both layers have identical salinities, and the thermal-expansion coefficient is taken as 2.54×10^{-4} K^{-1}. What is the wavelength of the longest unstable wave?

14.3. Formulate the Richardson number for a stratified shear flow with uniform stratification frequency N and linear velocity profile, varying from zero at the bottom to U at a height H. Then, relate the Richardson number to the Froude number and show that instabilities can occur only if the Froude number exceeds the value 2.

14.4. In an oceanic region far away from coasts and strong currents, the upper water column is stably stratified with $N = 0.015\,\text{s}^{-1}$. A storm passes by and during 10 hours exerts an average stress of 0.2 N/m^2. What is the depth of the mixed layer by the end of the storm? (For seawater, take $\rho_0 = 1028\,\text{kg/m}^3$.)

14.5. An air mass blows over a cold ocean at a speed of 10 m/s and develops a stable potential-temperature gradient of 8°C per kilometer in the vertical. It then encounters a warm continent and is heated from below at the rate of 200 W/m^2. Assuming that the air mass maintains its speed, what is the height of the convective layer 60 km inshore? What is then a typical vertical velocity of convection? (Take $\rho_0 = 1.20\,\text{kg/m}^3$, $\alpha = 3.5 \times 10^{-3}$ K^{-1}, and $C_p = 1005\,\text{J kg}^{-1}\,\text{K}^{-1}$.)

14.6. For the growing atmospheric boundary layer, show that thermals rise faster than the layer grows ($w_* > dh/dt$) and that thermals have a temperature contrast less than the temperature jump at the top of the layer $[T_* < (T - T_0)/2]$.

14.7. Why should eddy viscosity be considered positive? What happens to the energy budget if $\nu_E \leq 0$?

14.8. Consider the following governing equation for the turbulent kinetic energy k:

$$\frac{\partial k}{\partial t} = \nu_E M^2 - \kappa_E N^2 - \epsilon$$

with Eqs. (14.63), (14.64), and (14.42) for fixed M^2, N^2, l_m and $c_\mu / c'_\mu = 0.7$. Show that the solution is always nonnegative as long as the initial value of k is nonnegative.

NUMERICAL EXERCISES

14.1. Assuming the turbulent kinetic energy budget is dominated by local production and dissipation, how would you define a staggered grid for a one-dimensional model of a water column?

14.2. Implement a numerical method that keeps the turbulent kinetic energy k positive for decaying turbulence in a homogeneous k–ϵ model.

14.3. Show that for turbulence in statistical equilibrium, stability functions depend only on the Richardson number.

14.4. Revisit the estimate of the computing power needed to simulate geophysical fluid dynamics down to the dissipation range, with microscale in mind and for a typical value of $\epsilon = 10^{-3}$ W/kg.

14.5. What to you think that the requirement should be on the vertical grid spacing Δz compared with l_m?

14.6. Implement a 1D model including a k–ϵ closure scheme. If help is needed, look at kepsmodel.m, but do not cheat.

14.7. Use the program developed in Numerical Exercise 14.6 or kepsmodel.m to simulate the case of wind-induced mixing of Analytical Problem 14.4. In particular, consider the temporal evolution of the surface velocity in hodograph form (u, v axes), with or without Coriolis force. Then repeat the exercise but do not allow the wind to stop. Again, compare the situations with or without Coriolis force. To do so, trace in both cases, the mixed-layer depth evolution as shown in Fig. 14.14.

14.8. Prove that Eq. (14.100) is the sufficient condition to ensure that the numerical solution obtained with Eq. (14.98) converges toward the equilibrium value c^*, remains positive, and never crosses the value $c = c^*$.

14.9. Simulate a convection case in the ocean with a uniform initial stratification of $N = 0.015$ s^{-1}. Then apply a destabilizing heat loss of 200 W/m^2

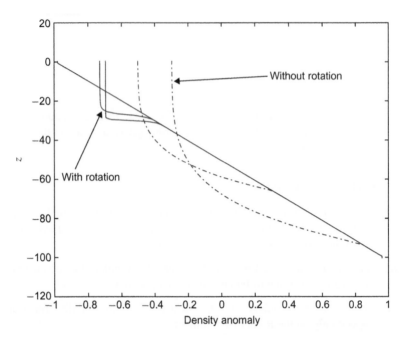

FIGURE 14.14   A uniform stratification being erased by a constant surface wind. In the absence of rotation, the mixed layer deepens (two profiles are shown in dashed lines). With rotation, the mixed layer stabilizes (two profiles in solid lines), and the pycnocline is sharper. See Numerical Exercise 14.7.

at the surface. Translate the heat flux into density anomaly flux and use the 1D model without rotation. Start from rest. Implement a method to detect the mixed-layer depth and trace its evolution over time. Do the same for the wind-mixing case of Numerical Exercise 14.7. Compare with the theoretical results of Section 14.7.

Lewis Fry Richardson
1881–1953

Unlike many scientists of his generation and the next, Lewis Fry Richardson did not become interested in meteorology because of a war. On the contrary, he left his secure appointment at the Meteorological Office in England during World War I to serve in a French ambulance convoy and tend the wounded. After the war, he returned to the Meteorological Office (see historical note at the end of Chapter 1), only to leave it again when it was transferred to the Air Ministry, deeply convinced that "science ought to be subordinate to morals."

Richardson's scientific contributions can be broadly classified in three categories: finite-difference solutions of differential equations, meteorology, and mathematical modeling of nations at war and in peace. The marriage of his first two interests led him to conceive of numerical weather forecasting well before computers were available for the task (see Section 1.9). His formulation of the dimensionless ratio that now bears his name is found in a series of landmark publications during 1919–1920 on atmospheric turbulence and diffusion. His mathematical theories of war and peace were developed in search of rational means by which nations could remain in peace.

According to his contemporaries, Richardson was a clear thinker and lecturer, with no enthusiasm for administrative work and a preference for solitude. He confessed to being "a bad listener because I am distracted by thoughts." (*Photo by Bassano and Vandyk, London*)

George Lincoln Mellor
1929–

George Mellor's career has been devoted to fluid turbulence in its many forms. His early interest in aerodynamics of jet engines and turbulent boundary layers soon yielded to a stronger interest in the turbulence of stratified geophysical flows. In the mid 1970s, he developed with Tetsuji Yamada a closure scheme to model turbulence in stratified flows, which is being used worldwide in atmospheric and oceanographic applications. Their joint 1982 publication in *Reviews of Geophysics and Space Physics* is one of the most widely cited papers in its field.

Mellor is also known as the architect of the so-called *Princeton Ocean Model*, nicknamed POM, which is used the world over to simulate ocean dynamics, particularly in coastal regions and wherever turbulent mixing is significant. He is the author of the textbook "Introduction to Physical Oceanography" (American Institute of Physics, 1996). (*Photo courtesy of Princeton University*)

Part IV

Combined Rotation and Stratification Effects

Chapter 15

Dynamics of Stratified Rotating Flows

ABSTRACT

Geostrophic motions can arise during the adjustment to density inhomogeneities and maintain a stratified fluid away from gravitational equilibrium. The key is a relationship between the horizontal density gradient and the vertical velocity shear, called the thermal-wind relation. Oceanic coastal upwelling is considered as it is a good example of rotating dynamics in a stratified fluid. Because large gradients and discontinuities (fronts) can form during geostrophic adjustment, the numerical section shows how to treat large gradients in computer models.

15.1 THERMAL WIND

Consider a situation where a cold air mass is wedged between the ground and a warm air mass (Fig. 15.1). The stratification has then both vertical and horizontal components. Mathematically, the density is a function of both height z and distance x (say, from cold to warm). Now, assume that the flow is steady, geostrophic, and hydrostatic:

$$-fv = -\frac{1}{\rho_0} \frac{\partial p}{\partial x} \tag{15.1}$$

$$\frac{\partial p}{\partial z} = -\rho g. \tag{15.2}$$

Here, v is the velocity component in the horizontal direction y, and p is the pressure field. Taking the z-derivative of Eq. (15.1) and eliminating $\partial p/\partial z$ with Eq. (15.2), we obtain

$$\frac{\partial v}{\partial z} = -\frac{g}{\rho_0 f} \frac{\partial \rho}{\partial x}. \tag{15.3}$$

Therefore, a horizontal density gradient can persist in steady state if it is accompanied by a vertical shear of horizontal velocity. Where density varies in both

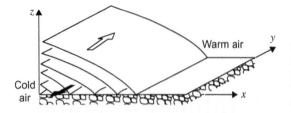

FIGURE 15.1 Vertical shear of a flow in the presence of a horizontal density gradient. The change of velocity with height is called thermal wind.

horizontal directions, the following also holds:

$$\frac{\partial u}{\partial z} = +\frac{g}{\rho_0 f}\frac{\partial \rho}{\partial y}. \tag{15.4}$$

These innocent-looking relations have profound meaning. They state that due to the Coriolis force, the system can be maintained in equilibrium, without tendency toward leveling of the density surfaces. In other words, the rotation of the earth can keep the system away from its state of rest without any continuous supply of energy.

Notice that the velocity field (u, v) is not specified, only its vertical shear, $\partial u/\partial z$ and $\partial v/\partial z$. This implies that the velocity must change with height. (In the case of Fig. 15.1, $\partial \rho/\partial x$ is negative and $\partial v/\partial z$ is positive.) For example, the wind speed and direction at some height above the ground may be totally different from those at ground level. The presence of a vertical gradient of velocity also implies that the velocity cannot vanish, except perhaps at some discrete levels. Meteorologists have named such a flow the *thermal wind*.[1]

In the case of pronounced density contrasts, such as across cold and warm fronts, a layered system may be applicable. In this case (Fig. 15.2), the system can be represented by two densities (ρ_1 and ρ_2, $\rho_1 < \rho_2$) and two velocities (v_1 and v_2). Relation (15.3) can be discretized into

$$\frac{\Delta v}{\Delta z} = -\frac{g}{\rho_0 f}\frac{\Delta \rho}{\Delta x},$$

where we take $\Delta v = v_1 - v_2$ and $\Delta \rho = \rho_2 - \rho_1$ to obtain

$$v_1 - v_2 = -\frac{g}{\rho_0 f}(\rho_2 - \rho_1)\frac{\Delta z}{\Delta x}. \tag{15.5}$$

The ratio $\Delta z/\Delta x$ is the slope of the interface. The equation is called the *Margules relation* (Margules, 1906), although a more general form of the relation for zonal flows was obtained earlier by Helmholtz (1888).

---

[1] Although thermal wind is a meteorological expression, oceanographers use it, too, to indicate a sheared current in geostrophic equilibrium with a horizontal density gradient.

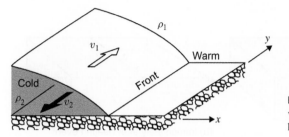

FIGURE 15.2 The layered version of Fig. 15.1, which leads to the Margules relation.

The thermal-wind concept has been enormously useful in analyzing both atmospheric and oceanic data because observations of the temperature and other variables that influence the density (such as pressure and specific humidity in the air, or salinity in seawater) are typically much more abundant than velocity data. For example, knowledge of temperature and moisture distributions with height and of the surface wind (to start the integration) permits the calculation of wind speed and direction above ground. In the ocean, especially in studies of large-scale oceanic circulation, for which sparse current-meter data may not be representative of the large flow due to local eddy effects, the basinwide distribution may be considered unknown. For this reason, oceanographers typically assume that the currents vanish at some great depth (e.g., 2000 m) and integrate the "thermal-wind" relations from there upward to estimate the surface currents. Although the method is convenient (the equations are linear and do not require integration in time), we should keep in mind that the thermal-wind relation of Eqs. (15.3) and (15.4) is rooted in an assumption of strict geostrophic balance. Obviously, this will not be true everywhere and at all times.

15.2 GEOSTROPHIC ADJUSTMENT

We may now ask how situations like the ones depicted in Fig. 15.1 and 15.2 can arise. In the atmosphere, the temperature gradient from the warm tropics to the cold polar regions creates a permanent feature of the global atmosphere, although storms do alter the magnitude of this gradient in time and space. Ocean currents can bring in near contact water masses of vastly different origins and thus densities. Finally, coastal processes such as freshwater runoff can create density differences between saltier waters offshore and fresher waters closer to shore. Thus, a variety of mechanisms exists by which different fluid masses can be brought in contact.

Oftentimes, the contact between different fluid masses is recent, and the flow has not yet had the time to achieve thermal-wind balance. An example is coastal upwelling: Alongshore winds create in the ocean an offshore Ekman drift, and the depletion of surface water near the coast brings denser water from below

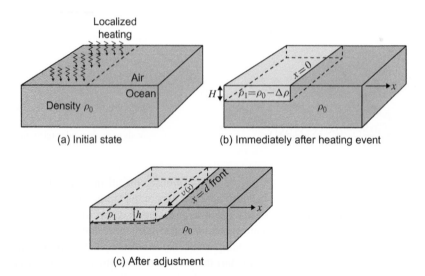

FIGURE 15.3   A simple case of geostrophic adjustment.

(see later section in this chapter). Such a situation is initially out of equilibrium and gradually seeks adjustment.

Let us explore in a very simple way the dynamical adjustment between two fluid masses recently brought into contact. Let us imagine an infinitely deep ocean that is suddenly heated over half of its extent (Fig. 15.3a). A warm upper layer develops on that side, whereas the rest of the ocean, on the other side and below, remains relatively cold (Fig. 15.3b). (We could also imagine a vertical gate preventing buoyant water from spilling from one side to the other.) After the upper layer has been created—or, equivalently, when the gate is removed— the ocean is not in a state of equilibrium, the lighter surface water spills over to the cold side, and an adjustment takes place. In the absence of rotation, spilling proceeds, as we can easily imagine, until the light water has spread evenly over the entire domain and the system has come to rest. But this scenario, as we are about to note, is not what happens when rotational effects are important.

Under the influence of the Coriolis force, the forward acceleration induced by the initial spilling creates a current that veers (to the right in the northern hemisphere) and can come into geostrophic equilibrium with the pressure difference associated with the density heterogeneity. The result is a limited spill accompanied by a lateral flow (Fig. 15.3c).

To model the process mathematically, we use the reduced-gravity model (12.19) on an f-plane and with reduced-gravity constant $g' = g(\rho_0 - \rho_1)/\rho_0$ according to the notation of Fig. 15.3b. We neglect all variations in the y-direction, although we allow for a velocity, v, in that direction, and write

$$\frac{\partial u}{\partial t} + u\frac{\partial u}{\partial x} - fv = -g'\frac{\partial h}{\partial x} \tag{15.6a}$$

$$\frac{\partial v}{\partial t} + u\frac{\partial v}{\partial x} + fu = 0 \tag{15.6b}$$

$$\frac{\partial h}{\partial t} + \frac{\partial}{\partial x}(hu) = 0. \tag{15.6c}$$

The initial conditions (i.e., immediately after the warming event) are $u = v = 0$, $h = H$ for $x < 0$, and $h = 0$ for $x > 0$. The boundary conditions are $u, v \to 0$ and $h \to H$ as $x \to -\infty$, whereas the velocity component u at the front is given by the material derivative $u = dx/dt$ where $h = 0$ at $x = d(t)$, the moving point where the interface outcrops. This nonlinear problem cannot be solved analytically, but one property can be stated. Fluid parcels governed by the preceding equations conserve the following form of the potential vorticity:

$$q = \frac{f + \partial v/\partial x}{h}. \tag{15.7}$$

Initially, all particles have $v = 0$, $h = H$ and share the same potential vorticity $q = f/H$. Therefore, throughout the layer of light fluid and at all times, the potential vorticity keeps the uniform value f/H:

$$\frac{f + \partial v/\partial x}{h} = \frac{f}{H}. \tag{15.8}$$

This property, it turns out, allows us to relate the initial state to the final state without having to solve for the complex, intermediate evolution.

Once the adjustment is completed, time derivatives vanish. Equation (15.6c) then requires that hu be a constant; since $h = 0$ at one point, this constant must be zero, implying that u must be zero everywhere. Equation (15.6b) reduces to zero equals zero and tells nothing. Finally, Eq. (15.6a) implies a geostrophic balance,

$$-fv = -g'\frac{dh}{dx}, \tag{15.9}$$

between the velocity and the pressure gradient set by the sloping interface. Alone, Eq. (15.9) presents one relation between two unknowns, the velocity and the depth profile. The potential-vorticity conservation principle (15.8), which still holds at the final state, provides the second equation, thereby conveying the information about the initial disturbance into the final state.

Despite the nonlinearities of the original governing Eqs. (15.6a)–(15.6c), the problem at hand, Eqs. (15.8) and (15.9), is perfectly linear, and the solution is relatively easy to obtain. Elimination of either $v(x)$ or $h(x)$ between the two equations yields a second-order differential equation for the remaining variable, which admits two exponential solutions. Discarding the exponential that grows for $x \to -\infty$ and imposing the boundary condition $h = 0$ at $x = d$ lead to

$$h = H\left[1 - \exp\left(\frac{x-d}{R}\right)\right] \tag{15.10}$$

$$v = -\sqrt{g'H}\,\exp\left(\frac{x-d}{R}\right), \tag{15.11}$$

where R is the deformation radius, defined by

$$R = \frac{\sqrt{g'H}}{f}, \tag{15.12}$$

and d is the unknown position of the outcrop (where h vanishes). To determine this distance, we must again tie the initial and final states, this time by imposing volume conservation.[2] Ruling out a finite displacement at infinity where there is no activity, we require that the depletion of light water on the left of $x = 0$ be exactly compensated by the presence of light water on the right, that is,

$$\int_{-\infty}^{0} (H - h)\,dx = \int_{0}^{d} h\,dx, \tag{15.13}$$

which yields a transcendental equation for d, the solution of which is surprisingly simple:

$$d = R = \frac{\sqrt{g'H}}{f}. \tag{15.14}$$

Thus, the maximum distance over which the light water has spilled in the adjusted state is none other than the radius of deformation, hence the name of the latter.

Notice that R has the Coriolis parameter f in its denominator. Therefore, the spreading distance, R, is less than infinity because f differs from zero. In other words, the spreading is confined because of the earth's rotation via the Coriolis effect. In a nonrotating framework, the spreading would, of course, be unlimited.

Lateral heterogeneities are constantly imposed onto the atmosphere and oceans, which then adjust and establish patterns whereby these lateral heterogeneities are somewhat distorted but maintained. Such patterns are at or near geostrophic equilibrium and can thus persist for quite a long time. This explains why discontinuities such as fronts are common occurrences in both the

---

[2] The reason why we use a volume conservation to determine the frontal position but let some energy be lost from the system, and not the reverse, is rooted in the very different nature of mass and energy propagation. The latter can be transported far away (to infinity) by waves without net displacement of fluid, while mass propagation demands advection by the flow.

atmosphere and the oceans. As the preceding example suggests, fronts and the accompanying winds or currents take place over distances on the order of the deformation radius. To qualify the activity observed at that length scale, meteorologists refer to the *synoptic scale*, whereas oceanographers prefer to use the adjective *mesoscale*.

We can vary the initial, hypothetical disturbance and generate a variety of geostrophic fronts, all being steady states. A series of examples, taken from published studies, is provided in Fig. 15.4. They are, in order, as follows: surface-to-bottom front on a flat bottom, which can result from sudden and localized heating (or cooling); surface-to-bottom front at the shelf break resulting from the existence of distinct shelf and deep water masses; double, surface-to-surface front; and three-layer front as a result of localized mixing of an otherwise two-layer stratified fluid. The interested reader is referred to

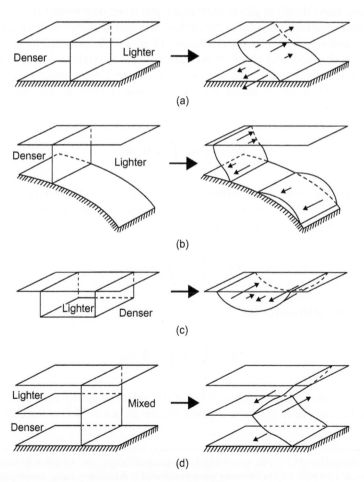

FIGURE 15.4   Various examples of geostrophic adjustment.

the original articles by Rossby (1937, 1938), the article by Veronis (1956), the review by Blumen (1972), and other articles on specific situations by Stommel and Veronis (1980), Hsueh and Cushman-Roisin (1983), and van Heijst (1985). Ou (1984) considered the geostrophic adjustment of a continuously stratified fluid and showed that if the initial condition is sufficiently away from equilibrium, density discontinuities can arise during the adjustment process. In other words, *fronts* can spontaneously emerge from earlier continuous conditions.

The preceding applications dealt with situations in which there is no variation in one of the two horizontal directions. The general case (see Hermann, Rhines & Johnson, 1989) may yield a time-dependent flow that is nearly geostrophic.

15.3 ENERGETICS OF GEOSTROPHIC ADJUSTMENT

The preceding theory of geostrophic adjustment relied on potential vorticity and volume conservation principles, but nothing was said of energy, which must also be conserved in a nondissipative system. All we can do, now that the solution has been obtained, is to check on the budget, where a surprise is awaiting us!

Initially, the system is at rest, and there is no kinetic energy ($KE_i = 0$), whereas the initial potential energy (per unit length in the transverse direction) is[3]

$$PE_i = \frac{1}{2} \rho_0 \int_{-\infty}^{0} g' H^2 \, dx. \qquad (15.15)$$

Although this expression is infinite, only the difference with the final potential energy will interest us. So, there is no problem. At the final state, the velocity u is zero, leaving the kinetic energy to be

$$KE_f = \frac{1}{2} \rho_0 \int_{-\infty}^{d} h v^2 \, dx, \qquad (15.16)$$

and the potential energy is

$$PE_f = \frac{1}{2} \rho_0 \int_{-\infty}^{d} g' h^2 \, dx. \qquad (15.17)$$

During the spreading phase, some of the lighter water has been raised and some heavier water has been lowered to take its place. Hence, the center of

---

[3] To verify that Eq. (15.15) is the correct form for the potential energy, solve Analytical Problem 12.2 and use as a template for demonstration the approach used in establishing energy conservation for the two-layer system in Section 12.6.

gravity of the system has been lowered, and we expect a drop in potential energy. Calculations yield

$$\Delta PE = PE_i - PE_f = \frac{1}{4}\rho_0 g' H^2 R. \tag{15.18}$$

Some kinetic energy has been created by setting a transverse current. The amount is

$$\Delta KE = KE_f - KE_i = \frac{1}{12}\rho_0 g' H^2 R. \tag{15.19}$$

Therefore, as we can see, only one-third of the potential-energy drop has been consumed by the production of kinetic energy, and we should ask: What has happened to the other two-thirds of the released potential energy? The answer lies in the presence of transients, which occur during the adjustment: some of the time-dependent motions are gravity waves (here, internal waves on the interface), which travel to infinity, radiating energy away from the region of adjustment. In reality, such waves dissipate along the way, and there is a net decrease of energy in the system. The ratio of kinetic-energy production to potential-energy release varies from case to case (Ou, 1986) but tends to remain between 1/4 and 1/2.

An interesting property of the geostrophically adjusted state is that it corresponds to the greatest energy loss and thus to a level of minimum energy. Let us demonstrate this proposition in the particular case at hand. The energy of the system is at all times

$$E = PE + KE = \frac{\rho_0}{2}\int_{-\infty}^{d}\left[g'h^2 + h(u^2 + v^2) \right]dx, \tag{15.20}$$

and we know that the evolution is constrained by conservation of potential vorticity:

$$f + \frac{\partial v}{\partial x} = \frac{f}{H}h. \tag{15.21}$$

Let us now search for the state that corresponds to the lowest possible level of energy, (15.20), under constraint (15.21) by forming the variational principle:

$$\mathcal{F}(h, u, v, \lambda) = \frac{\rho_0}{2}\int_{-\infty}^{+\infty}\left[g'h^2 + h\left(u^2 + v^2\right) - 2\lambda\left(f + \frac{\partial v}{\partial x} - \frac{fh}{H}\right) \right]dx \tag{15.22}$$

$$\delta\mathcal{F} = 0 \quad \text{for any } \delta h, \delta u, \delta v \text{ and } \delta\lambda. \tag{15.23}$$

Because expression (15.20) is positive definite, the extremum will be a minimum. The variations with respect to the three state variables h, u, and v and the

Lagrange multiplier λ yield, respectively,

$$\delta h \; : \; g'h + \frac{1}{2}\left(u^2 + v^2\right) + \frac{f}{H}\,\lambda = 0 \qquad (15.24a)$$

$$\delta u \; : \; hu = 0 \qquad (15.24b)$$

$$\delta v \; : \; hv + \frac{\partial \lambda}{\partial x} = 0 \qquad (15.24c)$$

$$\delta \lambda \; : \; f + \frac{\partial v}{\partial x} - \frac{f}{H}\,h = 0. \qquad (15.24d)$$

Equation (15.24b) provides $u = 0$, whereas the elimination of λ between Eqs. (15.24a) and (15.24c) leads to

$$\frac{\partial}{\partial x}\left(g'h + \frac{1}{2}\,v^2\right) + \frac{f}{H}\,(-hv) = 0,$$

or

$$g'\,\frac{\partial h}{\partial x} + v\left(\frac{\partial v}{\partial x} - \frac{f}{H}\,h\right) = 0.$$

Finally, use of Eq. (15.24d) reduces this last equation to

$$g'\,\frac{\partial h}{\partial x} - fv = 0.$$

In conclusion, the state of minimum energy is the state in which u vanishes, and the cross-isobaric velocity is geostrophic—that is, the steady, geostrophic state.

It can be shown that the preceding conclusion remains valid in the general case of arbitrary, multilayer potential-vorticity distributions, as long as the system is uniform in one horizontal direction. Therefore, it is a general rule that geostrophically adjusted states correspond to levels of minimum energy. This may explain why geophysical flows commonly adopt a nearly geostrophic balance.

15.4 COASTAL UPWELLING

15.4.1 The Upwelling Process

Winds blowing over the ocean generate Ekman layers and currents. The depth-averaged currents, called the Ekman drift, forms an angle with the wind, which was found to be 90° (to the right in the northern hemisphere) according to a simple theory (Section 8.6). So, when a wind blows along a coast, it generates an Ekman drift directed either onshore or offshore, to which the coast stands as an obstacle. The drift is offshore if the wind blows with the coast on its left (right) in the northern (southern) hemisphere (Fig. 15.5). If this is the case, water depletion occurs in the upper layers, and a low pressure sets in, forcing waters from below to move upward and replenish at least partly the space vacated by

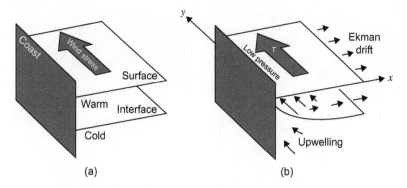

FIGURE 15.5   Schematic development of coastal upwelling.

the offshore drift. This phenomenon is called *coastal upwelling*. The upward movement calls for a replenishment at the lower levels, which is accomplished by an onshore flow at depth. To recapitulate, a wind blowing along the coast (with the coast on the left or the right in, respectively, the northern or the southern hemisphere) sets an offshore current in the upper levels, an upwelling at the coast, and an onshore current at lower levels.

This circulation in the cross-shore vertical plane is not the whole story, however. The low pressure created along the coast also sustains, via geostrophy, an alongshore current, while vertical stretching in the lower layer generates relative vorticity and a shear flow. Or, from a different perspective, the vertical displacement creates lateral density gradients, which in turn call for a thermal wind, the shear flow. The flow pattern is thus rather complex.

At the root of coastal upwelling is a divergent Ekman drift. And, we can easily conceive of other causes besides a coastal boundary for such divergence. Two other upwelling situations are noteworthy: one along the equator and the other at high latitudes. Along the equator, the trade winds blow quite steadily from east to west. On the northern side of the equator, the Ekman drift is to the right, or away from the equator, and on the southern side, it is to the left, again away from the equator (Fig. 15.6). Consequently, horizontal divergence occurs along the equator, and mass conservation requires upwelling (Gill, 1982, Chapter 11; Yoshida, 1959).

At high latitudes, upwelling frequently occurs along the ice edge, in the so-called marginal ice zone. A uniform wind exerts different stresses on ice and open water; in its turn, the moving ice exerts a stress on the ocean beneath. The net effect is a complex distribution of stresses and velocities at various angles, with the likely result that the ocean currents at the ice edge do not match (Fig. 15.6). For certain angles between wind and ice edge, these currents diverge, and upwelling again takes place to compensate for the divergence of the horizontal flow (Häkkinen, 1990).

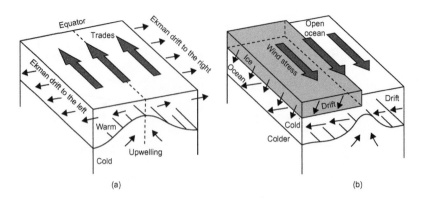

FIGURE 15.6 Other types of upwelling: (a) equatorial upwelling, (b) upwelling along the ice edge.

The upwelling phenomenon, especially the coastal type, has been the subject of considerable attention, chiefly because of its relation to biological oceanography and, from there, to fisheries. In brief, small organisms in the ocean (phytoplankton) proliferate when two conditions are met: sunlight and a supply of nutrients. In general, nutrients lie in the deeper waters, below the reach of sunlight, and so the waters tend to lack either nutrients or sunlight. The major exceptions are the upwelling regions, where deep, nutrient-rich waters rise to the surface, receive sunlight, and stimulate biological activity. Upwelling-favorable winds most generally occur along the west coasts of continents where the prevailing winds blow toward the equator. For a review of observations and a discussion of the biological implications of coastal upwelling, the interested reader is referred to the volume edited by Richards (1981).

15.4.2 A Simple Model of Coastal Upwelling

Consider a reduced-gravity ocean on an f-plane ($f > 0$), bounded by a vertical wall and subjected to a surface stress acting with the wall on its left (Fig. 15.5a). The upper moving layer, defined to include the entire vertical extent of the Ekman layer, supports an offshore drift current. The lower layer is, by virtue of the choice of a reduced-gravity model, infinitely deep and motionless. In the absence of alongshore variations, the equations of motion are

$$\frac{\partial u}{\partial t} + u\frac{\partial u}{\partial x} - fv = -g'\frac{\partial h}{\partial x} \tag{15.25a}$$

$$\frac{\partial v}{\partial t} + u\frac{\partial v}{\partial x} + fu = \frac{\tau}{\rho_0 h} \tag{15.25b}$$

$$\frac{\partial h}{\partial t} + \frac{\partial}{\partial x}(hu) = 0, \tag{15.25c}$$

where x is the offshore coordinate, τ is the alongshore wind stress, and all other symbols are conventional (Fig. 15.5b).

Despite its apparent simplicity, the preceding set of equations is nonlinear, and no analytical solution is known. We therefore linearize these equations by assuming that the wind stress τ and, in turn, the ocean's reaction are weak. Noting $h = H - a$, where H is the depth of the undisturbed upper layer and a the small upward displacement of the interface, we write

$$\frac{\partial u}{\partial t} - fv = g' \frac{\partial a}{\partial x} \tag{15.26}$$

$$\frac{\partial v}{\partial t} + fu = \frac{\tau}{\rho_0 H} \tag{15.27}$$

$$-\frac{\partial a}{\partial t} + H \frac{\partial u}{\partial x} = 0. \tag{15.28}$$

This set of equations contains two independent x-derivatives and thus calls for two boundary conditions. Naturally, u vanishes at the coast ($x = 0$) and a vanishes far offshore ($x \to +\infty$).

The solution to the problem depends on the initial conditions, which may be taken to correspond to the state of the rest ($u = v = a = 0$). Yoshida (1955) is credited with the first derivation of the problem's solution (extended to two moving layers). However, because of the fluctuating nature of winds, upwelling is rarely an isolated event in time, and we prefer to investigate the periodic solutions to the preceding linear set of equations. Taking $\tau = \tau_0 \sin \omega t$, where τ_0 is a constant in both space and time, we note that the solution must be of the type $u = u_0(x) \sin \omega t$, $v = v_0(x) \cos \omega t$ and $a = a_0(x) \cos \omega t$. Substitution and solution of the remaining ordinary differential equations in x yield

$$u = \frac{f\tau_0}{\rho_0 H (f^2 - \omega^2)} \left[1 - \exp\left(-\frac{x}{R_\omega}\right) \right] \sin \omega t \tag{15.29a}$$

$$v = \frac{\omega \tau_0}{\rho_0 H (f^2 - \omega^2)} \left[1 - \frac{f^2}{\omega^2} \exp\left(-\frac{x}{R_\omega}\right) \right] \cos \omega t \tag{15.29b}$$

$$a = \frac{-f R_\omega \tau_0}{\rho_0 g' H \omega} \exp\left(-\frac{x}{R_\omega}\right) \cos \omega t, \tag{15.29c}$$

where R_ω is a modified deformation radius defined as

$$R_\omega = \sqrt{\frac{g'H}{f^2 - \omega^2}}. \tag{15.30}$$

From the preceding solution, we conclude that the upwelling or downwelling signal is *trapped* along the coast within a distance on the order of R_ω. Far offshore ($x \to \infty$), the interfacial displacement vanishes, and the flow field includes the Ekman drift

$$u_{\text{Ek}} = \frac{\tau_0}{\rho_0 fH} \sin \omega t, \quad v_{\text{Ek}} = 0. \tag{15.31}$$

At long periods such as weeks and months ($\omega \ll f$), the distance R_ω becomes the radius of deformation, the vertical interfacial displacements become very large (indeed, the wind blows more steadily in one direction before it reverses), and the far-field oscillations become much smaller than the Ekman drift. Obviously, for very large vertical displacements and low frequency, we must ensure that the linearization hypothesis remains valid, that is, $|a| \ll H$. In terms of the forcing and with $R_\omega \simeq \sqrt{g'H}/f$, this condition translates into

$$\frac{\tau_0}{\rho_0 \omega H} \ll \sqrt{g'H}, \tag{15.32}$$

a condition for which an interpretation will soon be found.

At superinertial frequencies ($\omega > f$), the quantity R_ω becomes imaginary, indicating that the solution does not decay away from the coast but instead oscillates. Physically, the ocean's response is not trapped near the coast and inertia-gravity waves (Section 9.3) are excited. These radiate outward, filling the entire basin. Thus, depending on its frequency, the energy imparted by the wind to the ocean may either remain localized or be radiated away.

15.4.3   Finite-Amplitude Upwelling

If the wind is sufficiently strong or is blowing for a sufficiently long time, the density interface can rise to the surface, forming a front. Continued wind action displaces this front offshore and exposes the colder waters to the surface. This mature state is called *full upwelling* (Csanady, 1977). Obviously, the previous linear theory is no longer applicable.

Because of the added complications arising from the nonlinearities, let us now restrict our investigation to the final state of the ocean after a wind event of finite duration. Equation (15.25b), expressed as

$$\frac{\text{d}}{\text{d}t}(v + fx) = \frac{\tau}{\rho_0 h}, \tag{15.33}$$

where $\text{d}/\text{d}t = \partial/\partial t + u\partial/\partial x$ is the time derivative following a fluid particle in the offshore direction, can be integrated over time to yield:

$$(v + fx)_{\text{at end of event}} - (v + fx)_{\text{initially}} = I. \tag{15.34}$$

The *wind impulse I* is the integration of the wind-stress term, $\tau/\rho_0 h$, over time and following a fluid particle. Although the wind impulse received by every parcel cannot be precisely determined, it can be estimated by assuming that the wind event is relatively brief. The time integral can then be approximated by using the local stress value and replacing h by H:

$$I \simeq \frac{1}{\rho_0 H} \int_{\text{event}} \tau \, dt. \tag{15.35}$$

If the initial state is one of rest, relation (15.34) implies that a particle initially at distance X from the coast is at distance x immediately after the wind ceases and has an alongshore velocity v such that

$$v + fx - fX = I. \tag{15.36}$$

During the subsequent adjustment and until equilibrium is reached, Eq. (15.33) (with $\tau = 0$) implies that the quantity $v + fx$ remains unchanged, and relation (15.36) continues to hold after the wind has ceased.

If a spatially uniform wind blows over an ocean layer of uniform depth, the drift velocity, too, is uniform, and no vorticity is imparted to fluid parcels. Hence, potential vorticity is conserved during a uniform wind event over a uniform layer (see also Analytical Problem 15.9). After the event, in the absence of further forcing, potential vorticity remains conserved throughout the adjustment phase:

$$\frac{1}{h}\left(f + \frac{\partial v}{\partial x}\right) = \frac{f}{H}. \tag{15.37}$$

Once a steady state has been achieved, there is no longer any offshore velocity ($u = 0$), according to Eq. (15.25c). The remaining equation (15.25a), reduces to a simple geostrophic balance, which together with Eq. (15.37) provides the solution:

$$h = H - A \exp\left(-\frac{x}{R}\right) \tag{15.38}$$

$$v = A \sqrt{\frac{g'}{H}} \exp\left(-\frac{x}{R}\right), \tag{15.39}$$

where R is now the conventional radius of deformation ($\sqrt{g'H}/f$). The constant of integration A represents the amplitude of the upwelled state and is related to the wind impulse via Eq. (15.36). Two possible outcomes must be investigated: either the interface has not risen to the surface (Fig. 15.7, case I) or it has outcropped, forming a front and leaving cold waters exposed to the surface near the coast (Fig. 15.7, case II).

In case I, the particle initially against the coast ($X = 0$) is still there ($x = 0$), and relation (15.36) yields $v(x = 0) = I$. Solution (15.39) meets this condition if $A = I(H/g')^{1/2}$. The depth along the coast, $h(x = 0) = H - A$, must be positive requiring $A \leq H$; that is, $I \leq (g'H)^{1/2}$. In other words, the no-front situation or partial upwelling of case I occurs if the wind is sufficiently weak or sufficiently brief that its resulting impulse is less than the critical value $(g'H)^{1/2}$.

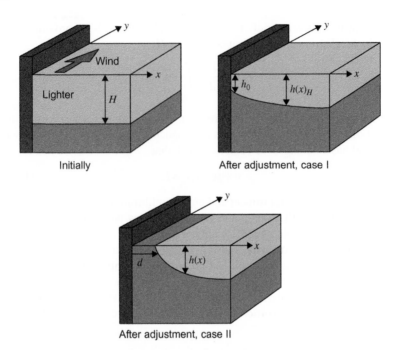

FIGURE 15.7   The two possible outcomes of coastal upwelling following an alongshore wind of finite duration. After a weak or brief wind (case I), the interface has upwelled but not to the point of reaching the surface. A strong or prolonged wind event (case II) causes the interface to reach the surface, where it forms a front; this front is displaced offshore, leaving cold waters from below exposed to the surface. This latter case corresponds to a mature upwelling that favors biological activity.

In the more interesting case II, the front has been formed, and the particle initially against the coast ($X = 0$) is now at some offshore distance ($x = d \geq 0$), marking the position of the front. There the layer depth vanishes, $h(x = d) = 0$, and solution (15.38) yields $A = H \exp(d/R)$. The alongshore velocity at the front is, according to (15.39), $v(x = d) = (g'H)^{1/2}$. Finally, relation (15.36) leads to the determination of the offshore displacement d in terms of the wind impulse:

$$d = \frac{I}{f} - R. \tag{15.40}$$

Since this displacement must be a positive quantity, it is required that $I \geq (g'H)^{1/2}$. Physically, if the wind is sufficiently strong or sufficiently prolonged, so that the net impulse is greater than the critical value $(g'H)^{1/2}$, the density interface rises to the surface and forms a front that migrates away from shore, leaving cold waters from below exposed to the surface. Note how the conditions for the realizations of cases I and II complement each other.

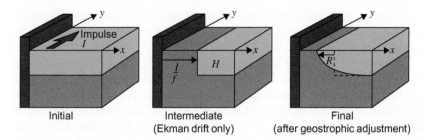

| Initial | Intermediate (Ekman drift only) | Final (after geostrophic adjustment) |

FIGURE 15.8 Decomposition of the formation of a coastal-upwelling front as a two-stage process: first, an offshore Ekman drift in response to the wind, followed by a backward geostrophic adjustment.

Note in passing how condition (15.32) can now be interpreted. Its left-hand side is the wind impulse over a time period $1/\omega$, which must be small compared to the critical value $(g'H)^{1/2}$, in order to remain far away from the outcropping situation, which would invalidate the linearization assumption.

Formula (15.40) has a simple physical interpretation as sketched in Fig. 15.8. The offshore Ekman velocity u_{Ek} is the velocity necessary for the Coriolis force to balance the alongshore wind stress:

$$u_{Ek} = \frac{\tau}{\rho_0 fh}, \tag{15.41}$$

according to Eq. (8.34a). Integrated over time, this yields a net offshore displacement proportional to the wind impulse

$$x_{Ek} = \frac{I}{f}. \tag{15.42}$$

If we were now to assume that the wind is responsible for an offshore shift of this magnitude, whereas the surface waters are moving as a solid slab, we would get the intermediate structure of Fig. 15.8. But such a situation cannot persist, and an adjustment must follow, causing an onshore spread similar to that considered in Section 15.2—that is, over a distance equal to the deformation radius. Hence, we have the final structure of Fig. 15.8 and formula (15.40).

15.4.4   Variability of the Upwelling Front

Up to this point, we have considered only processes operating in the offshore direction or, equivalently, an upwelling that occurs uniformly along a straight coast. In reality, the wind is often localized, the coastline not straight, and upwelling not at all uniform. A local upwelling sends a wave signal along the coast, taking the form of an internal Kelvin wave, which in the northern hemisphere propagates with the coast on its right. This redistribution of information not only decreases the rate of upwelling in the forced region but also generates upwelling in other, unforced areas. As a result, models of upwelling must retain a sizable portion of the coast and both spatial and temporal variations of the wind field (Crépon and Richez, 1982; Brink, 1983).

Because the upwelling front is a region of highly sheared currents, it is a likely region of instabilities. In the two-layer formulation presented in the previous section, this shear is manifested by a discontinuity of the current at the front. The warm layer develops anticyclonic vorticity (i.e., counter to the rotation of the earth) under the influence of vertical squeezing and flows alongshore in the direction of the wind. On the other side of the front, the exposed lower layer is vertically stretched, develops cyclonic vorticity (i.e., in the same direction as the rotation of the earth), and flows upwind. The currents on each side of the front thus flow in opposite directions, causing a large shear, which, as we have seen (Chapter 10), is vulnerable to instabilities. In addition to the kinetic-energy supply in the horizontal shear (barotropic instability), potential energy can also be released from the stratification by a spreading of the warm layer (baroclinic instability; see Chapter 17). Offshore jets of cold, upwelled waters have been observed to form near capes; these jets cut through the front, forge their way through the warm layer, and eventually split to form pairs of counter-rotating vortices (Flament, Armi & Washburn, 1985). This explains why mesoscale turbulence is associated with upwelling fronts (see, e.g., Fig. 15.9 and the article by Strub, Kosro & Huyer, 1991).

The situation is complex and demands careful modeling. Irregularities in the topography and coastline may play influential roles and require adequate spatial resolution, whereas accurate simulation of the instabilities is only possible if numerical dissipation is not excessive in the model.

15.5 ATMOSPHERIC FRONTOGENESIS

Atmospheric fronts are sharp boundaries between cold and warm air masses and have become familiar features of daily weather forecasts. A cold front, depicted in weather charts as a line with spikes (Fig. 15.10), occurs when a colder air mass overtakes a warmer air mass, thus lowering the temperature where it passes. In contrast, a warm front, depicted in weather charts as a line with semicircles, occurs when a warm air mass overtakes a cold air mass, thereby raising the local temperature. The process by which sharp temperature gradients naturally form in the atmosphere is called *frontogenesis* and is readily identified on temperature maps (Fig. 15.11). The word *front* was first coined by Vilhelm Bjerknes[4] who initiated the study of cyclone and front formation during World War I and suggested an analogy between the meeting of two atmospheric air masses and a military line, called a front. The study of frontogenesis has a long history, and the reader may wish to consult the seminal papers of Sawyer (1956),

---

[4]See biography at the end of Chapter 3.

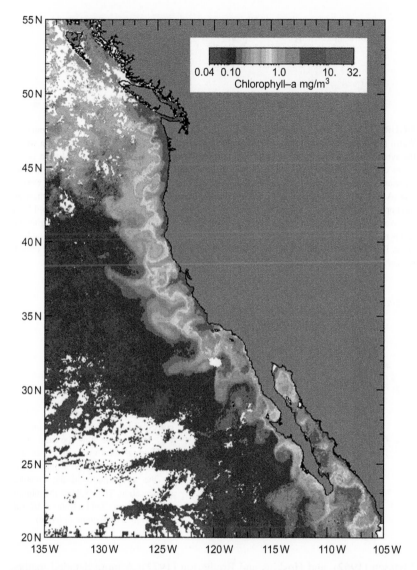

FIGURE 15.9   SeaWiFS satellite image of the North American Pacific coast showing the occurrence of coastal upwelling from Baja California (Mexico) to Vancouver Island (Canada). Shades indicate the amount of chlorophyll concentration in the water, with high values (lighter shades) in regions of high biological activity and low values (darker shades) in biologically inactive waters. Note how instabilities greatly distort the upwelling front. (*Composite image provided courtesy of Dr. Andrew Thomas, School of Marine Sciences, University of Maine, USA*). A color version can be found online at http://booksite.academicpress.com/9780120887590/

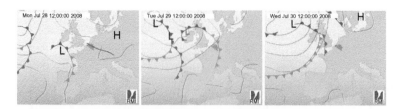

FIGURE 15.10 Typical evolution of fronts approaching Belgium (the dark-shaded country). Warm fronts are identified by semicircles, cold fronts by triangles. The side of the front on which the symbols are plotted indicates the direction of the frontal movement. For the cold front in the lower center of the first panel, there is cold air on the western side of the front, which is moving eastward. Indeed, by the next day (middle panel), the front has passed over Belgium from west to east. A day later (right panel), this front has disappeared from the map. Meanwhile, a warm front has appeared from the west followed by a cold front rapidly catching up with it. Once the cold front has overtaken the warm front, the warm air that was in the wedge between fronts has been lifted up and is no longer present at the surface. The new front, called an occluded front, has cold air on both sides. It is depicted by alternating semicircles and triangles (upper-central part of the right panel). (*Royal Meteorological Institute Belgium*)

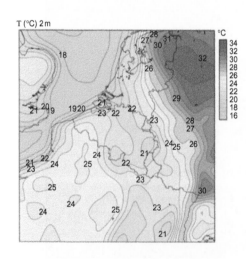

FIGURE 15.11 Temperature field corresponding to the middle panel of Fig. 15.10. Note how the cold front is chasing the warm air eastward. Some of the warm air is also lifted by the advancing cold air, creating condensation. This explains the rainfall accompanying cold fronts. (*Royal Meteorological Institute Belgium*)

Eliassen (1962), and Hoskins and Bretherton (1972). A more detailed mathematical presentation than the one given here can be found in Pedlosky (1987, Section 8.4).

The physical processes involved in frontogenesis are complex, and we will start the analysis with a kinematic study to understand how a given velocity field can lead to a deformation of a thermal distribution that intensifies temperature gradients. Because observations reveal that the generation of a front is a relatively fast process, typically taking no more than a day, we may neglect local heating effects. Also, creating a front by local differential heating would require heat fluxes that exhibit sharp gradients, an unlikely situation. Hence, we focus on temperature changes induced by advection only.

The simplest example (Fig. 15.12) assumes a horizontal velocity field given by

$$u = \omega x, \quad v = -\omega y \tag{15.43}$$

in which ω is a deformation rate. We note that this velocity fields satisfies volume conservation

$$\frac{\partial u}{\partial x} + \frac{\partial v}{\partial y} = 0, \tag{15.44}$$

implying that the vertical velocity is zero over a flat surface, which we assume.

Suppose now that this flow field advects a temperature field with initial gradient in the y-direction. Neglecting turbulent mixing, compressibility, and heating, temperature is conserved by individual fluid parcels, and the temperature field is is governed by the advection equation:

$$\frac{\mathrm{d}T}{\mathrm{d}t} = \frac{\partial T}{\partial t} + u\frac{\partial T}{\partial x} + v\frac{\partial T}{\partial y} = 0. \tag{15.45}$$

A differentiation of this equation with respect to x gives

$$\frac{\mathrm{d}}{\mathrm{d}t}\left(\frac{\partial T}{\partial x}\right) = -\omega\frac{\partial T}{\partial x}. \tag{15.46}$$

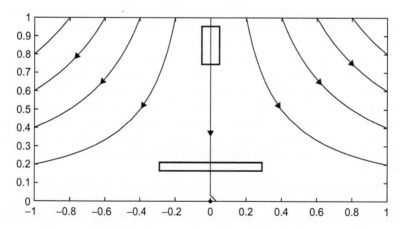

FIGURE 15.12 Frontogenesis induced by the flow field $u = \omega x$, $v = -\omega y$. A collection of fluid parcels forming a rectangle around $x = 0$, $y = 0.9$ stretches in the x-direction as it progresses downstream. By the time parcels have reached $y = 0.2$ (lower rectangle in the figure), convergence in the y-direction has squeezed the fluid parcels, as required by volume conservation to compensate for the divergence in the x-direction. The rear parcels have partly caught up with the front parcels. Without heating or cooling, temperature is conserved by individual fluid parcels, and any pre-existing temperature gradient in the y-direction is intensified. A front occurs when an infinite temperature gradient can be formed in a finite time. (`Frontogenesis.m` may be used to track other groups of fluid parcels.)

Since the initial temperature gradient was exclusively in the y-direction, it follows that $\partial T/\partial x$ was initially zero and remains zero at all subsequent times. Hence, the temperature gradient may change in intensity but not in direction.

More interestingly, differentiation of Eq. (15.45) with respect to y yields

$$\frac{d}{dt}\left(\frac{\partial T}{\partial y}\right) = \omega \frac{\partial T}{\partial y}, \tag{15.47}$$

which shows that the magnitude of the temperature gradient increases exponentially with time:

$$\frac{\partial T}{\partial y} = \left.\frac{\partial T}{\partial y}\right|_{t=0} e^{\omega t}, \tag{15.48}$$

following a fluid parcel. The evolution of the y position of a given air parcel is governed by

$$\frac{dy}{dt} = v = -\omega y \quad \Rightarrow \quad y = y_0 e^{-\omega t}. \tag{15.49}$$

Hence, all fluid parcels are converging toward $y = 0$. In other words, two parcels with identical x coordinates initially separated by a distance δy_0 see their distance shrink over time. Because each conserves its initial temperature, the temperature gradient increases accordingly.

We now understand how advection can intensify temperature gradients, but by keeping the flow field unchanged, we omitted to consider the fact that the increasing thermal gradient can in turn affect the dynamics. Indeed, a stronger thermal gradient is bound to produce a larger thermal wind, and this is expected to modify the wind velocity that advects the temperature. In other words, there is two-way coupling between velocity and temperature fields. This, it turns out, accelerates the process, and an infinite temperature gradient can be reached in a finite time.

As dynamics accelerate and shorter length scales arise, geostrophy is in jeopardy, and our model needs to retain the effects of nonlinear acceleration (inertia). However, frontal regions are characterized by strong spatial anisotropy, with steep variations across the front and weak variations along the front. Thus, our model may retain geostrophy in one direction. This leads to a semigeostrophic approach (Hoskins & Bretherton, 1972).

With the x-axis aligned with the front, the strong gradients are in the y-direction, and the geostrophic velocity component is u. The weaker v velocity is the one for which geostrophic breaks down. Density (function of temperature) is retained as a dynamically important variable, and the semigeostrophic equations on the f-plane are

$$\frac{du}{dt} - fv = -\frac{1}{\rho_0}\frac{\partial p}{\partial x} \tag{15.50a}$$

$$+fu = -\frac{1}{\rho_0}\frac{\partial p}{\partial y} \tag{15.50b}$$

$$-\alpha g T = -\frac{1}{\rho_0}\frac{\partial p}{\partial z} \tag{15.50c}$$

$$\frac{\partial u}{\partial x} + \frac{\partial v}{\partial y} + \frac{\partial w}{\partial z} = 0 \tag{15.50d}$$

$$\frac{dT}{dt} = 0, \tag{15.50e}$$

which forms a set of five equations for five variables, namely the three velocity components (u, v, w), pressure p and temperature T. Note that the density was eliminated by using a linear equation of state and that T is measured from the temperature at which density is ρ_0.

The acceleration term du/dt is kept next to the Coriolis term fv in the first equation because u is large and v small, breaking geostrophic balance in the x-momentum budget. Note also that the full material derivative is retained in the first and last equations:

$$\frac{d}{dt} = \frac{\partial}{\partial t} + u\frac{\partial}{\partial x} + v\frac{\partial}{\partial y} + w\frac{\partial}{\partial z}. \tag{15.51}$$

The thermal wind balance is obtained by combining the z-derivative of Eq. (15.50b) with the y-derivative of (15.50c):

$$\frac{\partial u}{\partial z} = -\frac{\alpha g}{f}\frac{\partial T}{\partial y}. \tag{15.52}$$

Next, we define the following quantity:

$$q = \left(f - \frac{\partial u}{\partial y}\right)\frac{\partial T}{\partial z} + \frac{\partial u}{\partial z}\frac{\partial T}{\partial y}, \tag{15.53}$$

which is a form of potential vorticity. This q variable is useful because it is conserved by moving fluid parcels. Indeed, some tedious algebra shows that the preceding equations yield the simple conservation equation:

$$\frac{dq}{dt} = 0. \tag{15.54}$$

To work with a model that is as simple as possible, we restrict our attention to a flow in which q is initially zero everywhere. With q conserved by fluid parcels over time, q remains zero everywhere at all subsequent times:

$$\left(f-\frac{\partial u}{\partial y}\right)\frac{\partial T}{\partial z}+\frac{\partial u}{\partial z}\frac{\partial T}{\partial y}=0. \tag{15.55}$$

As we will see shortly, this type of flow has simple attributes that facilitate mathematical developments, but it is not degenerate.

We now become more specific about the flow field, choosing the deformation field used earlier in this section with the addition of terms to reflect the fact that sharpening thermal gradients will affect the thermal wind balance and thus the flow field itself. We assume a solution of the type

$$u = +\omega x + u'(y,z,t) \tag{15.56a}$$

$$v = -\omega y + v'(y,z,t) \tag{15.56b}$$

$$p = -\rho_0 f \omega x y - \frac{1}{2}\rho_0 \omega^2 x^2 + p'(y,z,t) \tag{15.56c}$$

$$w = w(y,z,t) \tag{15.56d}$$

$$T = T(y,z,t). \tag{15.56e}$$

In writing these expressions, care was taken to include in the pressure field terms that are in geostrophic balance with the basic deformation field $(\omega x, -\omega y)$. Further, because of the anisotropy of the front, we anticipate that all components aside from the basic deformation field are independent of the coordinate x. Insertion into Eqs. (15.50) yields

$$\frac{du'}{dt}+\omega u'-fv'=0 \tag{15.57a}$$

$$fu'=-\frac{1}{\rho_0}\frac{\partial p'}{\partial y} \tag{15.57b}$$

$$\alpha\rho_0 gT=\frac{\partial p'}{\partial z} \tag{15.57c}$$

$$\frac{\partial v'}{\partial y}+\frac{\partial w}{\partial z}=0 \tag{15.57d}$$

$$\frac{dT}{dt}=0. \tag{15.57e}$$

No linearization was applied, and the material derivative in Eqs. (15.57a) and (15.57e) is the original one except for the x-derivative, which is now nil. We carefully note that the total velocity v appears in this material derivative. The thermal-wind relation (15.52) and $q=0$ equation (15.55) become

$$\frac{\partial u'}{\partial z} = -\frac{\alpha g}{f}\frac{\partial T}{\partial y} \qquad (15.58)$$

$$\left(f-\frac{\partial u'}{\partial y}\right)\frac{\partial T}{\partial z} + \frac{\partial u'}{\partial z}\frac{\partial T}{\partial y} = 0. \qquad (15.59)$$

Next, we define the so-called geostrophic coordinate

$$Y = y - \frac{u'}{f}, \qquad (15.60)$$

which combines the y coordinate along which gradients occur and the flow in the transverse direction. This quantity has a simple material derivative:

$$\frac{\mathrm{d}Y}{\mathrm{d}t} = -\omega Y \qquad (15.61)$$

because $\mathrm{d}y/\mathrm{d}t = v$. With this new variable substituting for u', Eq. (15.59) (expressing $q=0$) becomes

$$\frac{\partial Y}{\partial y}\frac{\partial T}{\partial z} - \frac{\partial Y}{\partial z}\frac{\partial T}{\partial y} = 0, \qquad (15.62)$$

which can be recast as

$$-\frac{\partial Y/\partial y}{\partial Y/\partial z} = -\frac{\partial T/\partial y}{\partial T/\partial z} = S. \qquad (15.63)$$

This last equation states that the slope S of the Y lines in the vertical plane (y, z) is everywhere equal to the slope of the T lines. This means that the lines of constant Y coincide with the lines of constant T (isotherms), and we can write

$$Y = Y(T,t), \qquad (15.64)$$

expressing the fact that in the (y, z) plane, the function Y is constant where T is constant. Here, time t plays the role of a parameter.

Exploiting the thermal-wind relation (15.58), the slope S of isotherms can be expressed in terms of Y and T as

$$S = -\frac{\partial T/\partial y}{\partial T/\partial z} = -\frac{f^2}{\alpha g}\frac{\partial Y/\partial z}{\partial T/\partial z} = -\frac{f^2}{\alpha g}\frac{\partial Y}{\partial T}, \qquad (15.65)$$

which takes advantage of the fact that Y is a function of T. This makes that S, too, is a function of only temperature T (and, parametrically, time t). Logic then imposes that if S is unchanging along an isotherm, this isotherm has uniform slope and is thus a straight line. It follows that all isotherms are straight lines.[5] Note that the slope may vary from isotherm to isotherm, with some more inclined than others, and that the slope of an individual isotherm may change over time.

Next, we return to Eq. (15.61) governing the temporal evolution of Y. Using the fact that Y is not a function of x but a function of only T and time, and the fact that $dT/dt = 0$, we obtain

$$\frac{dY}{dt} = \left.\frac{\partial Y}{\partial t}\right|_{T=\text{const}} + \frac{\partial Y}{\partial T}\frac{dT}{dt} \tag{15.66}$$

$$\left.\frac{\partial Y}{\partial t}\right|_{T=\text{const}} = -\omega Y. \tag{15.67}$$

Its solution is

$$Y = Y_0(T)e^{-\omega t}, \tag{15.68}$$

in which Y_0 is the initial distribution of Y, a function of T only, which we do not need to specify.

Passing from Y to S with Eq. (15.65), we have

$$S = -\frac{f^2}{\alpha g}\frac{dY_0}{dT}e^{-\omega t}. \tag{15.69}$$

The slope of each isotherm is thus reduced over time[6] but more so along certain isotherms than others. It now remains to determine how different isotherms can be compared to one another.

To obtain displacements, we first integrate the v velocity component vertically between horizontal and impermeable ($w = 0$) boundaries, for example, a flat land or sea surface below and the tropopause above (Fig. 15.13) Volume conservation (15.57d) dictates

$$\frac{\partial \bar{v}}{\partial y} = -\omega \tag{15.70}$$

where $\bar{v}$ is the vertical average of the full velocity component v (not just v'!). Assuming that the v velocity is the deformation field $-\omega y$ at large distances from the region of interest, the frontal region, Eq. (15.70) tells that the average velocity $\bar{v}$ is everywhere $-\omega y$. This means that a fluid column, on average, moves toward $y = 0$ and that the y distance between neighboring columns decreases exponentially over time.

---

[5] That isotherms are straight lines can be traced to the choice $q = 0$.
[6] This can also be proven by direct manipulation of the governing equations, see Analytical Problem 15.13.

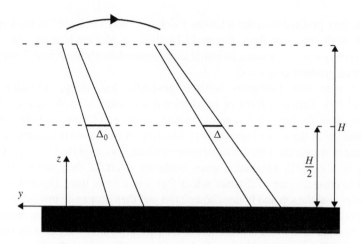

FIGURE 15.13 The distance Δ between two isopycnals at mid-height decreases over time as $\Delta = \Delta_0 e^{-\omega t}$.

Because tilting around $z = H/2$ does not change the volume between two isotherms, which are material surfaces since T is a conserved quantity, the distance Δ between two lines at midlevel $z = H/2$ must decrease exponentially according to $\Delta = \Delta_0 e^{-\omega t}$. This means that the y position of a given isotherm at midlevel $z = H/2$ is none other than the geostrophic coordinate Y. Further, by definition of the slope S, we are entitled to write explicitly

$$Y = y - \frac{z - H/2}{S}. \tag{15.71}$$

While this last equation appears to give Y in terms of y and z, it is best to see it as giving the (y, z) structure of the isotherms in terms of the variables Y and S, which depend only on T and time. Since we know how Y and S vary in time, we can determine the evolution of each isotherm from its initial state.

Given an initial, monotonic temperature distribution at $z = H/2$, say

$$T(y, z = H/2, t = 0) = F(y), \tag{15.72}$$

the inverse function $G = F^{-1}$, which exists because F is monotonic, provides the initial distribution of the geostrophic coordinate in terms of temperature

$$Y_0 = G(T), \tag{15.73}$$

since $Y = y$ at $z = H/2$. Note that, like F, the function G is monotonic, too. The initial slope of an isotherm is known to be

$$S_0(T) = -\frac{f^2}{\alpha g} \frac{dY_0}{dT}. \tag{15.74}$$

We then proceed with the relations $Y = Y_0 e^{-\omega t}$ and $S = S_0 e^{-\omega t}$ to track individual isotherms over time. Figure 15.14 shows a plot made using the code sgfrontogenesis.m using an initial temperature distribution F with a slightly enhanced gradient near $y = 0$.

We note that isotherms become gradually less steep, according to Eq. (15.69). This is a form of gravitational relaxation with the denser fluid (colder air in the lower left region) intruding under and lifting the lighter fluid (warmer air in the upper right region). Slacking of isotherms is accentuated in the center where the initial temperature gradient was slightly larger (smaller value of $|dY_0/dT|$). Gradually, some isotherms overtake their neighbors and begin to cross. A pair of discontinuities forms in a finite time. Discontinuities first appear at the top and bottom boundaries and then propagate inward, toward midlevel, where they eventually meet. Physically, a temperature discontinuity is interpreted as a front, a place where temperature varies very rapidly over a very short distance. Note that on the lower (upper) boundary, the discontinuity appears for a positive (negative) value of y, which is on the warm (cold) side of

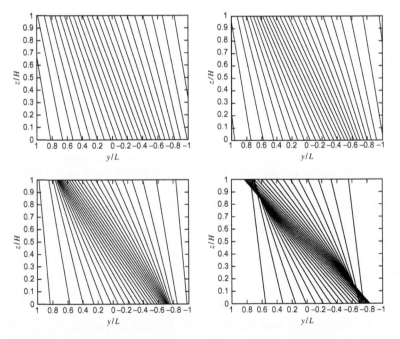

FIGURE 15.14 Evolution of isotherms in a vertical plane during frontogenesis. Note how isotherms become gradually less steep in the center. Eventually (last panel), some isotherms overtake their neighbors and overlap occurs. Physically, a discontinuity, which we call a front, has been formed in a finite time. Note that discontinuities first appear at the top and bottom boundaries. Later (not shown), they propagate inward, toward the midlevel.

the convergence region defined from the basic deformation flow $v = -\omega y$. The shift is attributed to the v' component of the flow field.

According to Eq. (15.71), the intersection of an isotherm with the bottom surface ($z = 0$) occurs at position

$$y_b = Y - \frac{H}{2S} = Y_0 e^{-\omega t} - \frac{H}{2S_0} e^{+\omega t}, \tag{15.75}$$

or, using Eq. (15.74),

$$
\begin{aligned}
y_b &= Y_0 e^{-\omega t} + \frac{\alpha g H}{2f^2} \frac{1}{dY_0/dT} e^{+\omega t} \\
&= Y_0 e^{-\omega t} + \frac{\alpha g H}{2f^2} \frac{dT}{dY_0} e^{+\omega t}.
\end{aligned} \tag{15.76}
$$

Two neighboring isotherms begin to intersect when their ground position coincides, that is, when they share the same ground position y_b while retaining their distinct temperatures. Mathematically, this is expressed by a vanishing variation of y_b for a nonzero variation of T, that is

$$\frac{\partial y_b}{\partial T} = 0. \tag{15.77}$$

Switching from the variable T to the variable Y_0, which is in monotonic relation with it, (15.73), we may transform the preceding condition into

$$\frac{\partial y_b}{\partial Y_0} = 0, \tag{15.78}$$

which yields

$$e^{-\omega t} + \frac{\alpha g H}{2f^2} \frac{d^2 T}{dY_0^2} e^{\omega t} = 0. \tag{15.79}$$

In this last equation, the second derivative $d^2 T/dY_0^2$ is known from the initial temperature distribution at midlevel, Eqs. (15.72) and (15.73). Therefore, for every isotherm for which this second derivative is negative, there exits a finite time t given by

$$t = \frac{1}{2\omega} \ln \left[\frac{2f^2}{\alpha g H (-d^2 T/dY_0^2)} \right] \tag{15.80}$$

for which condition (15.77) is met. At that time, the isotherm begins to intersect its neighbor, and a temperature discontinuity appears. A front occurs at time t_f that is the shortest of all possible times t given above:

$$t_f = \frac{1}{2\omega} \ln \left[\frac{2f^2}{\alpha g H |d^2 T/dY_0^2|_{\max}} \right]. \tag{15.81}$$

To make this result somewhat more concrete, imagine the coordinate y running northward and toward colder air (case of Fig. 15.14). In that case, dY_0/dT is negative and so is dT/dY_0. Finding the maximum of negative d^2T/dY_0^2 is then equivalent to selecting the isotherm marking the place where the initial midlevel temperature decreases fastest with latitude.

Once isotherms begin to intersect, the temperature field becomes multivalued, and the mathematical solution loses physical significance. In reality, dissipative process (dynamic instabilities, friction, and diffusion) become significant, keeping temperature as a unique function of space and thermal gradients large but finite.

15.6   NUMERICAL HANDLING OF LARGE GRADIENTS

A common characteristic of the preceding sections is the appearance and motion of strong gradients, which we call fronts. If we were to apply numerical techniques on a fixed grid to describe such fronts, we would immediately face the problem of needing very high spatial resolution. Indeed, to represent adequately a front in the horizontal, we would need to ensure $\Delta x \ll L$, where Δx is the horizontal grid spacing and L the frontal length scale to be resolved, and this length scale L can become very small across a front. High resolution spanning the model domain is most often very expensive computationally, and it would be preferable to restrict high resolution to the frontal region. This can be achieved by *nesting* approaches, that is, embedding higher resolution models into coarser resolution models where needed (e.g., Barth, Alvera-Azcárate, Rixen & Beckers, 2005; Spall & Holland, 1991). In this case, the abrupt change in grid size at the junction between models may lead to numerical problems and require particular care (Numerical Exercise 15.9). To improve the method, we can allow the grid spacing to vary gradually, rather than suddenly, across the domain.

Such a method based on variable resolution was already suggested when we discussed time discretization (see Fig. 4.10). At that point, we mentioned the problem of a sudden reduction in timescale, which we overcame with shorter time steps during the duration of the event. Our concern now is with space discretization, and we seek a method that uses nonuniform resolution in some optimal way.

We first distribute a series of points x_i according to a known function, say $f(x)$, which may or may not be directly related to one of the variables of the problem, and think that an optimal placement of points is such that, on average, differences in f are similar between adjacent points. In this way, regions of large variations of f will be more densely covered than regions of mild variations. To keep variations constant from point to point, that is

$$|f_{i+1} - f_i| = \text{const}, \tag{15.82}$$

where f_i stands for the known value of f at location x_i, we seek a monotonic coordinate transformation of the type

$$x = x(\xi, t) \tag{15.83}$$

with the variable ξ uniformly distributed while x is not. In terms of the new coordinate ξ, uniform variation in f means

$$\left| \frac{\partial f}{\partial \xi} \right| = \text{const}, \tag{15.84}$$

or, after taking the derivative with respect to ξ,

$$\frac{\partial}{\partial \xi} \left| \frac{\partial f}{\partial \xi} \right| = 0. \tag{15.85}$$

If we express the variations of f in terms of its original and physical variable x, the problem reduces to finding the function $x(\xi)$ that satisfies

$$\frac{\partial}{\partial \xi} \left(\left| \frac{\partial f}{\partial x} \right| \frac{\partial x}{\partial \xi} \right) = 0. \tag{15.86}$$

Since we are interested in discretized problems, we search for the discrete positions x_i that obey

$$\left| \frac{\partial f}{\partial x} \right|_{i+1/2} (x_{i+1} - x_i) - \left| \frac{\partial f}{\partial x} \right|_{i-1/2} (x_i - x_{i-1}) = 0, \tag{15.87}$$

in which we have taken $\Delta \xi = 1$ without loss of generality.

This equation is nonlinear because derivatives of f must be calculated at yet unknown locations. To overcome this quandary, an iterative method is used (see iterative solvers of Section 5.6):

$$x_i^{(k+1)} = x_i^{(k)} + \alpha \Delta t \left[\left| \frac{\partial f}{\partial x} \right|_{i+1/2} \left(x_{i+1}^{(k)} - x_i^{(k)} \right) - \left| \frac{\partial f}{\partial x} \right|_{i-1/2} \left(x_i^{(k)} - x_{i-1}^{(k)} \right) \right], \tag{15.88}$$

in which the superscript (k) is merely an index counting the iterations on the way to the solution, that is, a pseudo-time. If the method converges, $x_i^{(k+1)} = x_i^{(k)}$ eventually, and the vanishing of the bracketed part in Eq. (15.88) tells that the solution has been found.

The preceding iterative method can be interpreted as the numerical solution of a pseudo-evolution equation for the grid nodes:

$$\frac{\partial x}{\partial t} = \alpha \frac{\partial}{\partial \xi} \left(\left| \frac{\partial f}{\partial x} \right| \frac{\partial x}{\partial \xi} \right), \tag{15.89}$$

where the coefficient α is an adjustable numerical parameter that determines how quickly the solution is obtained. If the numerical calculation of the gradients of f proceeds with straightforward centered differences, then the rule of iteration (15.88) reduces to

$$x_i^{(k+1)} = x_i^{(k)} + \alpha \Delta t \left[\left| f_{i+1}^{(k)} - f_i^{(k)} \right| - \left| f_i^{(k)} - f_{i-1}^{(k)} \right| \right], \qquad (15.90)$$

in which $f_i^{(k)}$ stands for $f(x_i^{(k)})$. The algorithm is complemented with prescribed values of x on the known boundary positions.

A problem with this formulation appears when the function f is constant over large parts of the domain. By construction, such regions will be void of grid nodes because there are no variations of f there. The remedy is to induce a tendency toward a uniform point distribution where the gradient of f is weak, such as with

$$x_i^{(k+1)} = x_i^{(k)} + \alpha \Delta t \left[w_{i+1/2} \left(x_{i+1}^{(k)} - x_i^{(k)} \right) - w_{i-1/2} \left(x_i^{(k)} - x_{i-1}^{(k)} \right) \right] = 0 \quad (15.91)$$

with the function w replacing $|\partial f / \partial x|$ chosen as

$$w = \left| \frac{\partial f}{\partial x} \right| + w_0. \qquad (15.92)$$

In this approach, the parameter w_0 controls the tendency toward a uniform grid distribution. Ideally, its value should fall somewhere between the low and high values of the gradient of f. In this way, wherever the gradient of f is weak, w approaches w_0, and the algorithm leads to solving the equation $\partial^2 x / \partial \xi^2 = 0$, which yields a uniform grid. On the other hand, in places where the gradient of f is steep, w_0 becomes negligible, and we recover Eq. (15.88) that seeds grid points in proportion to the gradient of f. An example is shown in (Fig. 15.15).

The grid positions can thus be obtained by repeated application of a diffusion-type equation (15.91), which is not to be confused with physical diffusion. Here, only positions of grid nodes are calculated. Later, dynamic

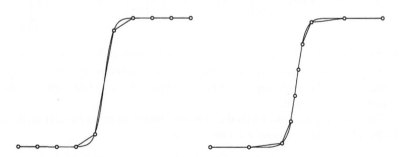

FIGURE 15.15 Grid nodes sampling a strongly varying function, shown by the smooth line. The left panel illustrates uniform sampling and the right panel an adapted grid with higher resolution in the steep region. A linear interpolation between nodes is also shown (line segments).

equations, possibly without any diffusion, may be discretized on this nonuniform grid.

With this adapting technique, we can follow strong gradients as long as the chosen function f effectively mimics the variations that are expected of the solution. Other techniques to distribute grid points nonuniformly exist (e.g., Liseikin, 1999; Thompson, Warsi & Mastin, 1985), always with the objective of reducing some measure of the discretization error.

Spatially nonuniform grids can be used in a frozen or adaptive way. In the frozen version, the grid is generated once at the beginning and then kept unchanged during the remainder of the calculations. This is done when the positions of steep gradients are known in advance, such as those attached to topographic features. Alternatively, it allows the modeler to zoom into a particular region of interest. In the adaptive version, the grid is allowed to move in time, following to the extent possible the dynamically relevant features (e.g., Burchard & Beckers, 2004). The challenge is then to find an effective rule of adaptation, which needs to be reflected in the discretization operators. The modification compared to standard methods on a fixed grid can be illustrated with the one-dimensional tracer equation

$$\frac{\partial c}{\partial t} + \frac{\partial (uc)}{\partial x} = \frac{\partial}{\partial x}\left(\mathcal{A}\frac{\partial c}{\partial x}\right). \tag{15.93}$$

The adaptive grid can be constructed via a coordinate transformation similar to the density coordinate substituting for depth. For the one-dimensional problem, we calculate $c(\xi(x,t),t)$ with ξ as the new coordinate. Since our grid generation provides $x(\xi,t)$, the rules for the transformation follow as in Section 12.1:

$$\frac{\partial \xi}{\partial t} = -\frac{\partial x/\partial t}{\partial x/\partial \xi}, \qquad \frac{\partial \xi}{\partial x} = \frac{1}{\partial x/\partial \xi}. \tag{15.94}$$

The equation for c in the new coordinate system (ξ, t) is then

$$\frac{\partial x}{\partial \xi}\frac{\partial c}{\partial t} - \frac{\partial x}{\partial t}\frac{\partial c}{\partial \xi} + \frac{\partial (uc)}{\partial \xi} = \frac{\partial}{\partial \xi}\left[\mathcal{A}\left(\frac{\partial x}{\partial \xi}\right)^{-1}\frac{\partial c}{\partial \xi}\right]. \tag{15.95}$$

All spatial derivatives with respect to ξ are performed in the new coordinate system, which is uniformly gridded, and standard discretization techniques can be applied. Furthermore, it is advantageous to use a flux form of the equation:

$$\frac{\partial}{\partial t}\left(\frac{\partial x}{\partial \xi}c\right) + \frac{\partial}{\partial \xi}\left[\left(u - \frac{\partial x}{\partial t}\right)c\right] = \frac{\partial}{\partial \xi}\left[\mathcal{A}\left(\frac{\partial x}{\partial \xi}\right)^{-1}\frac{\partial c}{\partial \xi}\right]. \tag{15.96}$$

The factor $\partial x/\partial \xi$ is readily interpreted as the grid spacing in physical space (Δx under the choice of $\Delta \xi = 1$), whereas the term $\partial x/\partial t$ is the velocity at which the grid nodes move. (The partial time derivative in the new coordinate system measures the x displacement per time unit, for fixed ξ.)

In the numerical space of the ξ grid, the advection term involves the velocity difference $u - \partial x/\partial t$, which is the velocity of the flow relative to the moving grid. This is indeed the velocity needed to advect the information relative to the nodes. Should we move the grid with the flow velocity, the relative velocity would be zero, and we would be using a Lagrangian method. The reader may have recognized that a particular case of an adaptive grid is the layered model of Section 12, where the vertical positions of discrete levels are moved in a Lagrangian fashion to follow density interfaces and the vertical velocity disappears from the formalism.

With an adaptive grid, however, the movement of the grid does not necessarily correspond to the flow velocity but must be chosen to follow large gradients. Care must be taken to ensure numerical stability of the scheme because the Courant number now includes the effective velocity, which differs from the actual velocity and may be larger if there are places where the drift speed of the grid is counter to the fluid velocity.

An alternative to performing a change of coordinate is to discretize the equation on a moving grid by directly applying the space integration in physical space between moving grid points. On integrating Eq. (15.93) between the two consecutive moving grid points $x_i(t)$ and $x_{i+1}(t)$, in a way similar to the finite-volume approach (Section 3.9), we can write

$$\int_{x_i(t)}^{x_{i+1}(t)} \frac{\partial c}{\partial t} dx + q_{i+1} - q_i = 0, \qquad q = uc - A\frac{\partial c}{\partial x}. \qquad (15.97)$$

Our goal is to make explicit the unknown, that is, the grid-averaged concentration, and this requires that we move the time derivative from inside to the outside of the integral. For this, we must be mindful that the integration boundaries vary in time and use Leibniz rule:

$$\frac{\partial}{\partial t} \int_{x_i(t)}^{x_{i+1}(t)} c\, dx + c(x_i, t) \frac{\partial x_i}{\partial t} - c(x_{i+1}, t) \frac{\partial x_{i+1}}{\partial t} + q_{i+1} - q_i = 0. \qquad (15.98)$$

Defining the modified flux

$$\hat{q} = \left(u - \frac{\partial x}{\partial t}\right) c - A\frac{\partial c}{\partial x}, \qquad (15.99)$$

the finite-volume equation on the moving grid reads

$$\frac{\partial}{\partial t} \int_{x_i(t)}^{x_{i+1}(t)} c\, dx + \hat{q}_{i+1} - \hat{q}_i = 0, \qquad (15.100)$$

which is similar to a straightforward finite-volume budget, except for the subtraction of the grid drift velocity $\partial x/\partial t$ from the flow velocity u. Defining $\tilde{c}_i$ as the cell-averaged concentration and relabeling the grid positions (by using half indices for clarity), the preceding equation can be recast as

$$\frac{\partial}{\partial t}\left[(x_{i+1/2}-x_{i-1/2})\tilde{c}_i\right]+\hat{q}_{i+1/2}-\hat{q}_{i-1/2}=0. \qquad (15.101)$$

In this form, the equation is now discrete in space but still continuous in time.

Generalization to three-dimensions proceeds along similar lines, and the outcome is again the subtraction in the advection terms of the grid drift speed from the physical velocity field. Issues in the implementation then concern the placement of nodes with respect to the grid cells (cell boundaries or interval centers in 1D, corners or centers of finite volumes in 2D and 3D; see Numerical Exercises 15.4 and 15.5).

Finally, it is also important to handle correctly the way the changing grid size is discretized in time. As usual, mathematical properties of the original budget equations are not necessarily shared by the numerical operators. In particular, we must ensure that the time discretization of Eq. (15.101) conserves the "volume" $\partial x/\partial \xi$ of the numerical grid in the sense that for a constant c, the time-discretized equation (15.101) is identically satisfied. If it is not, an artificial source of c will appear. This is similar to the advection problem in which the divergence operator of the fluxes has to be consistent with the one used in the physical volume conservation (Section 6.6).

15.7 NONLINEAR ADVECTION SCHEMES

Rather than chasing steep gradients (fronts) with a moving grid, we can also try to capture them with a fixed grid at the cost of appropriate numerical discretization. In this class of methods are the TVD (Total Variation Diminishing) advection schemes mentioned in Section 6.4. It is clear that advection schemes play a crucial role in the context of frontal displacements, and it is no surprise that intense research has been directed toward designing accurate advection schemes.

As we saw in Section 6.4, the basic upwind scheme is monotonic and does not create artificial extrema, but it rapidly smears out strong variations. In contrast, higher-order advection schemes better keep the gradients but at the cost of wiggles (unphysical extrema) in the numerical solution. Several attempts at designing schemes that are monotonic and more accurate than the upwind scheme can be mentioned. Flux-corrected transport (FCT) methods (Boris & Book, 1973; Zalesak, 1979) make two passes on the numerical grid, the first one with an upwind scheme and the second one adding as much anti-diffusion as possible (to restore the steep gradients) without generating wiggles. Flux-limiter methods (Hirsch, 1990; Sweby, 1984), presented hereafter in more detail, degrade the higher-order flux calculations toward upwind fluxes near problem zones. Finally, essentially nonoscillatory (ENO) methods (Harten, Engquist, Osher & Chakravarthy, 1987) adapt different interpolation functions near discontinuities.

The common characteristic of these methods (e.g., Thuburn, 1996) is that they allow the scheme to change its operation depending on the local solution itself. This feedback avoids the annoying consequence of the Godunov theorem (see Section 6.4), which states that the only linear scheme that is monotonic is the first-order upwind scheme, by violating its premise of linearity. Thus, we hope to find a monotonic scheme by introducing some clever nonlinearity in the formulation, even if the underlying physical problem is linear! The general strategy is the following: The nonlinearity is activated whenever over- or under-shooting is likely to occur, in which case the scheme increases numerical diffusion. When the solution is smooth, the scheme is allowed to remain of higher-order to be an improvement over the upwind scheme, of first order.

To design such an adaptive scheme in one dimension, we begin by defining a measure of the variation of the solution called *TV* for *Total Variation*:

$$TV^n = \sum_i |\tilde{c}_{i+1}^n - \tilde{c}_i^n|, \tag{15.102}$$

in which the sum is taken over all grid points of interest. A scheme is said TVD (Total Variation Diminishing) if

$$TV^{n+1} \le TV^n. \tag{15.103}$$

The *TV* value is meant to be a quantification of the wiggles that appear, such as those arising when the leapfrog or Lax–Wendrofff advection schemes are used.

Suppose now that the numerical scheme can be cast into the following form:

$$\tilde{c}_i^{n+1} = \tilde{c}_i^n - a_{i-1/2} \left(\tilde{c}_i^n - \tilde{c}_{i-1}^n \right) + b_{i+1/2} \left(\tilde{c}_{i+1}^n - \tilde{c}_i^n \right), \tag{15.104}$$

where the coefficients a and b may depend on $\tilde{c}$. We now prove that the so-defined scheme is TVD when

$$0 \le a_{i+1/2} \quad \text{and} \quad 0 \le b_{i+1/2} \quad \text{and} \quad a_{i+1/2} + b_{i+1/2} \le 1. \tag{15.105}$$

Note that $b_{i+1/2}$ appears in combination with $a_{i+1/2}$ in the TVD condition but with $a_{i-1/2}$ in the numerical scheme. Scheme (15.104) can also be written for point $i+1$:

$$\tilde{c}_{i+1}^{n+1} = \tilde{c}_{i+1}^n - a_{i+1/2} \left(\tilde{c}_{i+1}^n - \tilde{c}_i^n \right) + b_{i+3/2} \left(\tilde{c}_{i+2}^n - \tilde{c}_{i+1}^n \right),$$

from which we can subtract Eq. (15.104) to form a marching equation for the variations

$$\tilde{c}_{i+1}^{n+1} - \tilde{c}_i^{n+1} = \left(1 - a_{i+1/2} - b_{i+1/2} \right) \left(\tilde{c}_{i+1}^n - \tilde{c}_i^n \right)$$
$$+ b_{i+3/2} \left(\tilde{c}_{i+2}^n - \tilde{c}_{i+1}^n \right) + a_{i-1/2} \left(\tilde{c}_i^n - \tilde{c}_{i-1}^n \right).$$

We then take the absolute value of each side (remembering that the absolute value of a sum is smaller than the sum of the absolute values of its individual terms), assume that conditions (15.105) are satisfied, and sum over all grid points to obtain

$$\sum_i \left| \tilde{c}_{i+1}^{n+1} - \tilde{c}_i^{n+1} \right| \leq \sum_i \left(1 - a_{i+1/2} - b_{i+1/2} \right) \left| \tilde{c}_{i+1}^n - \tilde{c}_i^n \right|$$
$$+ \sum_i b_{i+3/2} \left| \tilde{c}_{i+2}^n - \tilde{c}_{i+1}^n \right| + \sum_i a_{i-1/2} \left| \tilde{c}_i^n - \tilde{c}_{i-1}^n \right|.$$

Ignoring boundary effects or assuming cyclic conditions, we may shift the index i in the last two sums in order to gather all sums with $|\tilde{c}_{i+1}^n - \tilde{c}_i^n|$. Then, utilizing the TVD conditions (15.105), we have

$$\sum_i \left| \tilde{c}_{i+1}^{n+1} - \tilde{c}_i^{n+1} \right| \leq \sum_i \left(1 - a_{i+1/2} - b_{i+1/2} \right) \left| \tilde{c}_{i+1}^n - \tilde{c}_i^n \right|$$
$$+ \sum_i b_{i+1/2} \left| \tilde{c}_{i+1}^n - \tilde{c}_i^n \right| + \sum_i a_{i+1/2} \left| \tilde{c}_{i+1}^n - \tilde{c}_i^n \right|$$
$$\leq \sum_i \left| \tilde{c}_{i+1}^n - \tilde{c}_i^n \right|.$$

The last inequality is none other than Eq. (15.103), proving that discretization (15.104) is TVD if conditions (15.105) are satisfied.

Advection does not increase variance (see Section 6.1), and discretizations with the TVD property appear interesting in this context. Let us therefore design nonlinear TVD schemes for the one-dimensional advection problem with positive velocity u. We combine explicit Euler schemes

$$\tilde{c}_i^{n+1} = \tilde{c}_i^n - \frac{\Delta t}{\Delta x} \left(\tilde{q}_{i+1/2} - \tilde{q}_{i-1/2} \right), \tag{15.106}$$

$$\tilde{q}_{i-1/2} = \tilde{q}_{i-1/2}^L + \Phi_{i-1/2} \left(\tilde{q}_{i-1/2}^H - \tilde{q}_{i-1/2}^L \right), \tag{15.107}$$

where the lower-order flux

$$\tilde{q}_{i-1/2}^L = u \tilde{c}_{i-1}^n$$

is the upwind flux, which is too diffusive, and the higher-order flux

$$\tilde{q}_{i-1/2}^H = u \tilde{c}_{i-1}^n + u \frac{1-C}{2} \left(\tilde{c}_i^n - \tilde{c}_{i-1}^n \right)$$

leads to a second-order nonmonotonic scheme (Section 6.4). The weighting factor Φ is allowed to vary with the solution by tending toward zero when excessive variations are present (exploiting the damping properties of the upwind scheme) but otherwise remaining close to 1 for maintaining accuracy where the solution is smooth. In other words, Φ controls the amount of antidiffusion applied to the scheme and bears the name of *flux limiter*. For the following, we assume that Φ

is positive, and the Courant number C satisfies the CFL condition ($C \leq 1$). Then this scheme with weighted flux can be expanded as

$$\tilde{c}_i^{n+1} = \tilde{c}_i^n - C\left[\tilde{c}_i^n + \Phi_{i+1/2}\frac{(1-C)}{2}\left(\tilde{c}_{i+1}^n - \tilde{c}_i^n\right)\right]$$
$$+ C\left[\tilde{c}_{i-1}^n + \Phi_{i-1/2}\frac{(1-C)}{2}\left(\tilde{c}_i^n - \tilde{c}_{i-1}^n\right)\right]. \tag{15.108}$$

It is readily seen that this is not yet a form that ensures TVD because the coefficient $b_{i+1/2}$ multiplying $\tilde{c}_{i+1}^n - \tilde{c}_i^n$ is always negative. However, we can group this term with the upwind part as follows:

$$\tilde{c}_i^{n+1} = \tilde{c}_i^n - C\left[1 - \Phi_{i-1/2}\frac{(1-C)}{2} + \Phi_{i+1/2}\frac{(1-C)}{2}\frac{\left(\tilde{c}_{i+1}^n - \tilde{c}_i^n\right)}{\left(\tilde{c}_i^n - \tilde{c}_{i-1}^n\right)}\right]\left(\tilde{c}_i^n - \tilde{c}_{i-1}^n\right).$$

This last form is of the type

$$\tilde{c}_i^{n+1} = \tilde{c}_i^n - a_{i-1/2}\left(\tilde{c}_i^n - \tilde{c}_{i-1}^n\right) \tag{15.109}$$

even though the coefficient $a_{i-1/2}$ depends on the solution. After all, we are designing a nonlinear method, and this dependence should be no surprise. The scheme can then be made TVD by imposing conditions (15.105), which reduce to $0 \leq a_{i-1/2} \leq 1$, with

$$a_{i-1/2} = C\left[1 - \Phi_{i-1/2}\frac{(1-C)}{2} + \Phi_{i+1/2}\frac{(1-C)}{2}\frac{\left(\tilde{c}_{i+1}^n - \tilde{c}_i^n\right)}{\left(\tilde{c}_i^n - \tilde{c}_{i-1}^n\right)}\right]. \tag{15.110}$$

We now try to find a simple strategy for specifying $\Phi_{i-1/2}$ and $\Phi_{i+1/2}$ such that the scheme remains TVD in all cases.

The function $\tilde{c}$ appears in parameter $a_{i-1/2}$ in the form of a ratio of differences:

$$r_{i+1/2} = \frac{\tilde{c}_i^n - \tilde{c}_{i-1}^n}{\tilde{c}_{i+1}^n - \tilde{c}_i^n}, \tag{15.111}$$

which is a measure of the variability of $\tilde{c}$: for $r_{i+1/2} = 1$, $\tilde{c}$ varies linearly over the three points involved, whereas for $r_{i+1/2} \leq 0$ there is a local extremum. The parameter $r_{i+1/2}$ will thus be involved in deciding the value of the local weighting Φ to be applied. If $r_{i+1/2}$ is negative (local extremum exists), we require $\Phi_{i+1/2} = 0$ because the local variation on the grid scale would create new extrema if the higher-order scheme were activated.

The TVD condition requires

$$0 \leq C + \frac{C(1-C)}{2}\left[\frac{\Phi_{i+1/2}}{r_{i+1/2}} - \Phi_{i-1/2}\right] \leq 1. \tag{15.112}$$

Ideally, we would like to choose a value of Φ at $i-1/2$ independently of its value at $i+1/2$, otherwise a simultaneous system of equations would have to be solved. To do this, we plan for the worst case, which happens when $\Phi_{i+1/2}=0$. We then need to ensure that $\Phi_{i-1/2}$ is not too large so that $a_{i-1/2}>0$, which implies

$$\Phi_{i-1/2} \leq \frac{2}{(1-C)}. \tag{15.113}$$

Reversing the roles of $\Phi_{i-1/2}$ and $\Phi_{i+1/2}$, the worst case happens when $\Phi_{i-1/2}=0$, demanding that

$$\frac{\Phi_{i+1/2}}{r_{i+1/2}} \leq \frac{2}{C} \tag{15.114}$$

to ensure that $a_{i-1/2} \leq 1$. Since both conditions must be satisfied for all i values, the following conditions on Φ are required to make a TVD scheme:

$$\Phi \leq \frac{2}{(1-C)} \quad \text{and} \quad \frac{\Phi}{r} \leq \frac{2}{C}, \tag{15.115}$$

in which the index has become unimportant. In practice, the parameter C varies and inequalities become cumbersome. To circumvent this, we resort again to sufficient conditions built on the worst cases. Since $0 \leq C \leq 1$, the sufficient conditions to ensure the TVD property are

$$\Phi \leq 2 \quad \text{and} \quad \frac{\Phi}{r} \leq 2. \tag{15.116}$$

Finally, we look for a function $\Phi(r)$ that meets this pair of conditions, and here various choices are open to us. We use this freedom at our advantage by trying to keep the scheme at the highest possible order. If r falls close to 1, the solution is smooth, behaving nearly as a straight line, and a second-order method should do well. So, we impose $\Phi(1)=1$. Incidentally, this is the case for both Lax–Wendroff and Beam–Warming schemes, but these schemes otherwise fail to meet the TVD conditions (Fig. 15.16).

Examples of acceptable limiters are (Fig. 15.16):

- van Leer: $\Phi = \frac{r+|r|}{1+|r|}$
- minmod: $\Phi = \max(0, \min(1, r))$
- Superbee: $\Phi = \max(0, \min(1, 2r), \min(2, r))$
- MC: $\Phi = \max(0, \min(2r, (1+r)/2, 2))$

Note that flux-limiter calculations depend on the direction of the flow, and the ratio r involved in the calculation of Φ must be adapted if the velocity changes sign (Numerical Exercise 15.8).

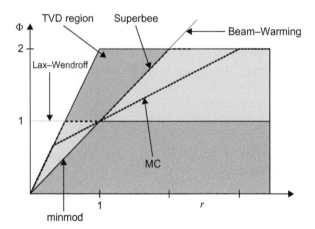

FIGURE 15.16 TVD domain and some standard limiters. The Lax–Wendroff scheme, which corresponds to $\Phi = 1$, has a portion that does not lie within the TVD region (for $r < 0.5$), and likewise the Beam–Warming scheme, with $\Phi = r$, does not lie entirely within the TVD region. The other limiters (minmod, MC, and Superbee) are TVD schemes.

After this lengthy exposition of the design of a TVD scheme, the attentive reader may ask what is the relation between this and a monotonic scheme, which was our original goal. From (15.109), we see that $\tilde{c}_i^{n+1}$ is obtained by linear interpolation of $\tilde{c}_i^n$ and $\tilde{c}_{i-1}^n$ because $0 \le a_{i-1/2} \le 1$. Therefore, $\min(\tilde{c}_i^n, \tilde{c}_{i-1}^n) \le \tilde{c}_i^{n+1} \le \max(\tilde{c}_i^n, \tilde{c}_{i-1}^n)$, and no new local extremum can be created (no over- or under-shooting in the jargon of numerical analysis), and if all values are initially positive, they are guaranteed to remain positive at all times.

This can be verified on the standard test case with the Superbee limiter (Fig. 15.17). The scheme, indeed, keeps the solution within the initial bounds, and results are quite improved compared with previous methods. The method is particularly well suited to GFD applications with large gradients and strong fronts. However, the occasional use of the upwind scheme during parts of the calculations tends to degrade the formal truncation error below second order. Also, absent large gradients when the solution is smooth, fourth-order methods generally outperform second-order TVD schemes. For this reason, fourth-order TVD schemes have been formulated (e.g., Thuburn, 1996). Clearly, the choice of one scheme over another is a question of modeling priorities (conservation, monotonicity, accuracy, ease of implementation, robustness, stability) and expected behavior on the part of the solution (strongly varying, gentle, steady state, etc.).

ANALYTICAL PROBLEMS

15.1. In a certain region, at a certain time, the atmospheric temperature along the ground decreases northward at the rate of 1°C every 35 km, and there are good reasons to assume that this gradient does not change much with

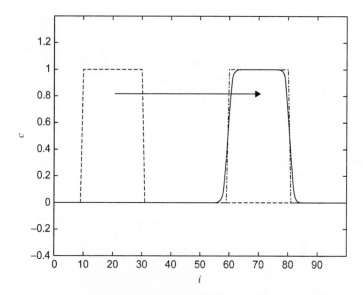

FIGURE 15.17   Advection scheme with TVD Superbee limiter applied to the transport of a "hat" signal with $C=0.5$ using 100 time steps. Note that no new extremum is created by the numerical advection and that the diffusion is drastically reduced compared to the upwind scheme.

height. If there is no wind at ground level, what are the wind speed and direction at an altitude of 2 km? To answer, take latitude $=40°$N, mean temperature $=290$ K, and uniform pressure on the ground.

15.2. A cruise to the Gulf Stream at $38°$N provided a cross-section of the current, which was then approximated to a two-layer model (Fig. 15.18) with a warm layer of density $\rho_1 = 1025$ kg/m^3 and depth $h(y) = H - \Delta H \tanh(y/L)$, overlying a colder layer of density $\rho_2 = 1029$ kg/m^3. Taking $H = 500$ m, $\Delta H = 300$ m, and $L = 60$ km and assuming that there is no flow in the lower layer and that the upper layer is in geostrophic balance, determine the flow pattern at the surface. What is the maximum velocity of the Gulf Stream? Where does it occur? Also, compare the jet width (L) to the radius of deformation.

15.3. Derive the discrete Margules relation (15.5) from the governing equations written in the density-coordinate system (Chapter 12).

15.4. Through the Strait of Gibraltar, connecting the Mediterranean Sea to the North Atlantic Ocean, there is an inflow of Atlantic waters near the top and an equal outflow of much more saline Mediterranean waters below. At its narrowest point (Tarifa Narrows), the strait is 11 km wide and 650 m deep. The stratification closely resembles a two-layer configuration with a relative density difference of 0.2% and an interface sloping from 175 m along the Spanish coast (north) to 225 m along the

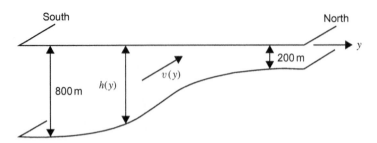

FIGURE 15.18 Schematic cross section of the Gulf Stream, represented as a two-layer geostrophic current (Analytical Problem 15.2).

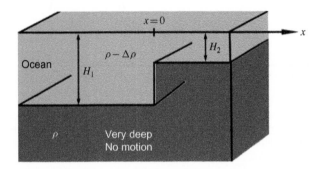

FIGURE 15.19 State prior to geostrophic adjustment (Analytical Problem 15.6).

African coast (south). Taking $f = 8.5 \times 10^{-5}$ s^{-1}, approximating the cross section to a rectangle, and assuming that the volumetric transport in one layer is equal and opposite to that in the other layer, estimate this volumetric transport.

15.5. Determine the geostrophically adjusted state of a band of warm water as depicted in Fig. 15.4c. The variables are ρ_0 = density of water below, $\rho_0 - \Delta\rho$ = density of warm water, H = initial depth of warm water, $2a$ = initial width of warm-water band, and $2b$ = width of warm-water band after adjustment. In particular, determine the value b, and investigate the limits when the initial half-width a is much less and much greater than the deformation radius R.

15.6. Find the solution for the geostrophically adjusted state of the initial configuration shown on Fig. 15.19, and calculate the fraction of potential-energy release that has been converted into kinetic energy of the final steady state.

15.7. In a valley of the French Alps ($\simeq 45°$N), one village (**A**) is situated on a flank 500 m above the valley floor and another (**B**) lies on the opposite side 200 m above the valley floor (Fig. 15.20). The horizontal distance

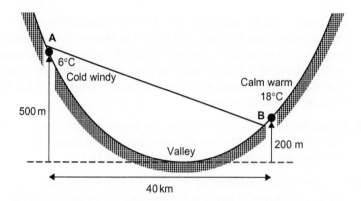

FIGURE 15.20 Air masses in a valley. What is the wind? (Analytical Problem 15.7)

between the two is 40 km. One day, a shepherd in the upper village, who is also a fine meteorologist, notes a cold wind with temperature 6°C. Upon calling her cousin, a blacksmith in the lower village across the valley, she learns that he is enjoying a calm afternoon with a comfortable 18°C! Assuming that the explanation of this perplexing situation resides in a cold wind blowing along one side of the valley (Fig. 15.20), she is able to determine a lower bound for its speed. Can you? Also, in which direction is the wind blowing? (*Hint*: Do not ignore compressibility of air.)

15.8. Using the linearized equations for a two-layer ocean (undisturbed depths H_1 and H_2) over a flat bottom and subject to a spatially uniform wind stress directed along the coast,

$$\frac{\partial u_1}{\partial t} - f v_1 = g' \frac{\partial a}{\partial x} - \frac{1}{\rho_0} \frac{\partial p_2}{\partial x}, \quad \frac{\partial u_2}{\partial t} - f v_2 = -\frac{1}{\rho_0} \frac{\partial p_2}{\partial x}$$

$$\frac{\partial v_1}{\partial t} + f u_1 = \frac{\tau}{\rho_0 H_1}, \quad \frac{\partial v_2}{\partial t} + f u_2 = 0$$

$$-\frac{\partial a}{\partial t} + H_1 \frac{\partial u_1}{\partial x} = 0, \quad \frac{\partial a}{\partial t} + H_2 \frac{\partial u_2}{\partial x} = 0,$$

study the upwelling response to a wind stress oscillating in time. The boundary conditions are no flow at the coast ($u_1 = u_2 = 0$ at $x = 0$), and no vertical displacements at large distances ($a \to 0$ as $x \to +\infty$). Discuss how the dynamics of the upper layer are affected by the presence of an active lower layer and what happens in the lower layer.

15.9. Investigate under which conditions it is valid to make the assertion made in the text above Eq. (15.37) that the potential vorticity is conserved if the wind stress is spatially uniform.

15.10. A coastal ocean at midlatitude ($f = 10^{-4}$ s^{-1}) has a 50-m-thick warm layer capping a much deeper cold layer. The relative density difference between the two layers is $\Delta\rho/\rho_0 = 0.002$. A uniform wind exerting a surface stress of 0.4 N/m^2 lasts for 3 days. Show that the resulting upwelling includes outcropping of the density interface. What is the offshore distance of the front?

15.11. Generalize to the two-layer ocean, the theory for the steady adjusted state following a wind event of given impulse. For simplicity, consider only the case of equal initial layer depths ($H_1 = H_2$).

15.12. Because of the roughness of the ice, the stress communicated to the water is substantially larger in the presence of sea ice than in the open sea. Assuming that the ice drift is at 20° to the wind, that the water drift is at 90° to the wind (in the open) and to the ice drift (under ice), and that the stress on the water surface is twice as large under ice than in the open sea, determine which wind directions with respect to the ice-edge orientation are favorable to upwelling.

15.13. Show from (15.67) that during frontogenesis with zero potential vorticity, the isopycnal slope

$$S = -\frac{\partial T/\partial y}{\partial T/\partial z} = -\frac{\partial Y/\partial y}{\partial Y/\partial z} = -\frac{f^2}{\alpha g}\frac{\partial Y/\partial y}{\partial T/\partial y} \tag{15.118}$$

evolves according to

$$\frac{dS}{dt} = -\omega S. \tag{15.119}$$

(*Hint*: Use the fact that $q = 0$.)

NUMERICAL EXERCISES

15.1. By looking into `upwelling.m`, guess which governing equations are being discretized. Then use the program to simulate coastal upwelling and see if the outcropping condition (15.40) is realistic. Examine the algorithm `flooddry.m` used to deal with the outcropping and explore what happens when you deactivate it.

15.2. Add a discretization of momentum advection to `upwelling.m` and redo Numerical Exercise 15.1. Then diagnose

$$I \simeq \frac{1}{\rho_0} \int_{\text{event}} \frac{\tau}{h}\,dt \tag{15.120}$$

during the simulation with `upwelling.m` and compare to the estimate (15.35). (*Hint*: Remember that I is calculated for an individual water parcel.)

15.3. Use `adaptive.m` and see how the linearly interpolated function using a uniform or adapted grid approximates the original function `functiontofollow.m`. Quantify the error by sampling the linear interpolations on a very high–resolution grid and calculate the rms error between this interpolation and the original function. Show how this error behaves during the grid-adaptation process.

15.4. Analyze file `adaptiveupwind.m` used to simulate an advection problem with upwind discretization and optional adaptive grid (Fig. 15.21). Explain how the grid size changes and how grid velocities must be discretized in a consistent way. Verify your analysis by using a constant value for c. Modify the parameters involved in the grid adaptation. Try to implement a Lagrangian approach by moving the grid nodes with the physical velocity. What problem appears in a fixed domain?

15.5. Redo Numerical Exercise 15.4 but define and move the grid nodes at the interface and calculate concentration-point positions at the center.

15.6. Prove that the Beam–Warming scheme of Section 6.4 can be recovered using a flux-limiter function $\Phi = r$. Implement the scheme and apply it to the standard problem of the top-hat signal advection. How does the solution compare to the Lax–Wendroff solution of Fig. 6.9?

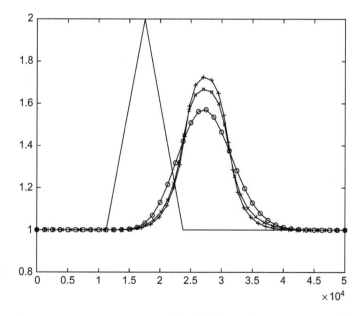

FIGURE 15.21 Upwind advection of a triangular distribution with fixed grid (most diffusive solution) and an adaptive grid with or without added Lagrangian-type advection. The best solution is obtained with the Lagrangian-type advection. See Numerical Exercise 15.4.

15.7. Apply the Superbee TVD scheme to the advection of the top-hat signal and then to the advection of a sinusoidal signal. What do you observe? Can you explain the behavior and verify that your explanation is correct by choosing another limiter? Experiment with `tvdadv1D.m`.

15.8. Write out the flux-limiter scheme in the 1D case for $u \leq 0$ by exploiting symmetries of the problem.

15.9. Implement a leapfrog advection scheme on a nonuniform grid with scalar c defined at the center of the cells of variable width. The domain of interest spans $x = -10L$ to $x = 10L$. For initialization and downstream boundaries, use an Euler upwind scheme. Use a spatial grid spacing of Δx for $x < 0$ and $r\Delta x$ for $x > 0$. Use a constant velocity $u = 1$ m/s. Simulate a time window of $15L/u$ and show the solution at every time step. Use a maximum value of $C = 0.5$. Advect a Gaussian distribution of width L initially located at $x = -5L$:

$$c(x, t = 0) = \exp\left[-(x + 5L)^2/L^2\right]. \qquad (15.121)$$

Use all combinations of $\Delta x = L/4, L/8, L/16$ with $r = 1, 1/2, 2, 1/10, 10$. In particular, observe what happens when the patch crosses $x = 0$.

George Veronis
1926–

An applied mathematician turned oceanographer, George Veronis has been a driving force in geophysical fluid dynamics since its early days. With Willem V. R. Malkus, he cofounded the GFD Summer Program at the Woods Hole Oceanographic Institution, which continues after more than 50 years to bring together oceanographers, meteorologists, physicists, and mathematicians to debate problems related to geophysical flows.

Veronis is best known for his theoretical studies on oceanic circulation, rotating and stratified fluids, thermal convection with and without rotation, and double-diffusion processes. His model of the circulation of the world ocean was an analytical study based on planetary geostrophic dynamics and the nonlinearity of thermal processes, in which he showed how western boundary currents cross the boundaries of wind gyres and connect all of the world's oceans into a single circulating system.

Veronis has earned a reputation as a superb lecturer, who can explain difficult concepts with amazing ease and clarity. (*Photo credit: G. Veronis*)

Kozo Yoshida
1922–1978

In the early years of his professional career, Kozo Yoshida studied long (tsunami) and short (wind) waves. Later, during a stay at the Scripps Institution of Oceanography, he turned to the investigation of the upwelling phenomenon, which was to become his lifelong interest. His formulations of dynamic theories for both coastal and equatorial upwelling earned him respect and fame. A wind-driven surface eastward current along the equator is called a Yoshida jet. In his later years, he also became interested in the Kuroshio, a major ocean current off the coast of Japan, wrote several books, and promoted oceanography among young scientists.

Known to be very sincere and logical, Yoshida did not shun administrative responsibilities and emphasized the importance of international cooperation in postwar Japan. (*Photo courtesy of Toshio Yamagata*)

Chapter 16

Quasi-Geostrophic Dynamics

ABSTRACT

At timescales longer than about a day, geophysical flows are ordinarily in a nearly geostrophic state, and it is advantageous to capitalize on this property to obtain a simplified dynamical formalism. Here, we derive the traditional quasi-geostrophic dynamics and present some applications in both linear and nonlinear regimes. The central component of quasi-geostrophic models, namely advection of vorticity, requires particular attention in numerical models, for which the Arakawa Jacobian is presented.

16.1 SIMPLIFYING ASSUMPTION

Rotation effects become important when the Rossby number is on the order of unity or less (Sections 1.5 and 4.5). The smaller is the Rossby number, the stronger is the rotation effect, and the larger is the Coriolis force compared with the inertial force. In fact, the majority of atmospheric and oceanic motions are characterized by Rossby numbers sufficiently below unity ($Ro \sim 0.2$ down to 0.01), enabling us to state that, in first approximation, the Coriolis force is dominant. This leads to geostrophic equilibrium (Section 7.1) with a balance struck between the Coriolis force and the pressure-gradient force. In Chapter 7, a theory was developed for perfectly geostrophic flows, whereas in Chapter 9 some near-geostrophic, small-amplitude waves were investigated. In each case, the analysis was restricted to homogeneous flows. Here, we reconsider near-geostrophic motions but in the case of continuously stratified fluids and nonlinear dynamics. Much of the material presented can be traced to the seminal article by Charney[1] (1948), which laid the foundation of quasi-geostrophic dynamics.

Geostrophic balance is a linear and diagnostic relationship; there is no product of variables and no time derivative. The resulting mathematical advantages explain why near-geostrophic dynamics are used routinely: The underlying assumption of near-geostrophy may not always be strictly valid, but the formalism is much simpler than otherwise.

---

[1] See biographical sketch at the end of the chapter.

Introduction to Geophysical Fluid Dynamics
521

Mathematically, a state of near-geostrophic balance occurs when the terms representing relative acceleration, nonlinear advection, and friction are all negligible in the horizontal momentum equations. This requires (Section 4.5) that the temporal Rossby number,

$$Ro_T = \frac{1}{\Omega T},$$ (16.1)

the Rossby number,

$$Ro = \frac{U}{\Omega L},$$ (16.2)

and the Ekman number,

$$Ek = \frac{\nu_E}{\Omega H^2},$$ (16.3)

all be small simultaneously. In these expressions, Ω is the angular rotation rate of the earth (or planet or star under consideration), T is the timescale of the motion (i.e., the time span over which the flow field evolves substantially), U is a typical horizontal velocity in the flow, L is the horizontal length over which the flow extends or exhibits variations, ν_E is the eddy vertical viscosity, and H is the vertical extent of the flow.

The smallness of the Ekman number (Section 4.5) indicates that vertical friction is negligible, except perhaps in thin layers on the edges of the fluid domain (Chapter 8). If we exclude small-amplitude waves that can travel much faster than fluid particles in the flow, the temporal Rossby number (16.1) is not greater than the Rossby number (16.2). (For a discussion of this argument, the reader is referred to Section 9.1.) By elimination, it remains to require that the Rossby number (16.2) be small. This can be justified in one of the two ways: Either velocities are relatively weak (small U) or the flow pattern is laterally extensive (large L). The common approach, and the one that leads to the simplest formulation, is to consider the first possibility; the resulting formalism bears the name of *quasi-geostrophic dynamics*. We ought to keep in mind, however, that some atmospheric and oceanic motions could be nearly geostrophic for the other reason, that is, large velocities on a large scale (Cushman-Roisin, 1986). Such motions are called frontal geostrophic and would be improperly represented by quasi-geostrophic dynamics.

16.2 GOVERNING EQUATION

To set the stage for the development of quasi-geostrophic equations, it is most convenient to begin with the restriction that vertical displacements of density surfaces be small (Fig. 16.1). In the (x, y, z) coordinate system, we write

$$\rho = \bar{\rho}(z) + \rho'(x, y, z, t) \quad \text{with} \quad |\rho'| \ll |\bar{\rho}|.$$ (16.4)

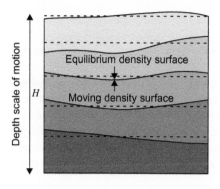

FIGURE 16.1   A rotating stratified fluid undergoing weak motions, which can be described by quasi-geostrophic dynamics.

Because the density surfaces are nearly horizontal, there is no real advantage to be gained by using the density-coordinate system, and we follow the tradition here by formulating the quasi-geostrophic dynamics in the (x, y, z) Cartesian coordinate system.

The density profile $\bar{\rho}(z)$, independent of time and horizontally uniform, forms the basic stratification. Alone, it creates a state of rest in hydrostatic equilibrium. We shall assume that such stratification has somehow been established, and that it is maintained in time against the homogenizing action of vertical diffusion. The quasi-geostrophic formalism does not consider the origin and maintenance of this stratification but only the behavior of motions that weakly perturb it.

The following mathematical developments are purposely heuristic, with emphasis on the exploitation of the main idea rather than on a systematic approach. The reader interested in a rigorous derivation of quasi-geostrophic dynamics based on a regular perturbation analysis is referred to Chapter 6 of the book by Pedlosky (1987).

The governing equations of Section 4.4 with $\rho = \bar{\rho}(z) + \rho'(x, y, z, t)$ and, similarly, $p = \bar{p}(z) + p'(x, y, z, t)$ are on the beta plane and, for simplicity, in the absence of friction and diffusion:

$$\frac{du}{dt} - f_0 v - \beta_0 y v = -\frac{1}{\rho_0}\frac{\partial p'}{\partial x} \tag{16.5a}$$

$$\frac{dv}{dt} + f_0 u + \beta_0 y u = -\frac{1}{\rho_0}\frac{\partial p'}{\partial y} \tag{16.5b}$$

$$0 = -\frac{\partial p'}{\partial z} - \rho' g \tag{16.5c}$$

$$\frac{\partial u}{\partial x} + \frac{\partial v}{\partial y} + \frac{\partial w}{\partial z} = 0 \tag{16.5d}$$

$$\frac{\partial \rho'}{\partial t} + u\frac{\partial \rho'}{\partial x} + v\frac{\partial \rho'}{\partial y} + w\frac{d\bar{\rho}}{dz} = 0, \tag{16.5e}$$

where the advective operator is

$$\frac{d}{dt} = \frac{\partial}{\partial t} + u\frac{\partial}{\partial x} + v\frac{\partial}{\partial y} + w\frac{\partial}{\partial z}. \tag{16.6}$$

The basic assumption that $|\rho'|$ is much less than $|\bar{\rho}|$ has been implemented in the density Eq. (16.5e) by dropping the term $w\partial\rho'/\partial z$. In writing that equation, we have also neglected density diffusion (the right-hand side of the equation) in agreement with our premise that the basic vertical stratification persists. Finally, because the basic stratification is in hydrostatic equilibrium with the pressure at rest, the corresponding terms cancel out, and only the perturbed pressure p', that due to motion, appears in the equations.

If the density perturbations ρ' are small, so are the pressure disturbances p', and by virtue of the horizontal momentum equations, the horizontal velocities are weak. While the Coriolis terms are small, the nonlinear advective terms, which involve products of velocities, are even smaller. For expediency, we shall use the phrase *very small* for these and all other terms smaller than the small terms. Thus, the ratio of advective to Coriolis terms, the Rossby number, is small. Let us assume now and verify *a posteriori* that the timescale is long compared with the inertial period $(2\pi/f_0)$, so the local-acceleration terms are, too, very small. Finally, to guarantee that the beta-plane approximation holds, we further require $|\beta_0 y| \ll f_0$. Having made all these assumptions, we take pleasure in noting that the dominant terms in the momentum equations are, as expected, those of the geostrophic equilibrium:

$$-f_0 v = -\frac{1}{\rho_0}\frac{\partial p'}{\partial x} \tag{16.7a}$$

$$+f_0 u = -\frac{1}{\rho_0}\frac{\partial p'}{\partial y}. \tag{16.7b}$$

As noted in Chapter 7, this geostrophic state is singular, for it leads to a zero horizontal divergence $(\partial u/\partial x + \partial v/\partial y = 0)$, which usually (e.g., over a flat bottom) implies the absence of any vertical velocity. In the case of a stratified fluid, this in turn implies no lifting and lowering of density surfaces, and thus no pressure disturbances and no variations in time.

To explore dynamics beyond such a simple state of affairs, we consider that the velocity includes a small ageostrophic (not geostrophic) correction and write

$$u = u_g + u_a, \quad v = v_g + v_a, \tag{16.8}$$

in which the first terms represent the geostrophic component, defined as

$$u_g = -\frac{1}{f_0\rho_0}\frac{\partial p'}{\partial y} \tag{16.9a}$$

$$v_g = +\frac{1}{f_0\rho_0}\frac{\partial p'}{\partial x}, \tag{16.9b}$$

and (u_a, v_a) are the ageostrophic corrections.

In the smaller time-derivatives, advection and beta terms of Eqs. (16.5a) and (16.5b), we replace the velocity by its geostrophic approximation (16.7), while we care to retain both geostrophic and ageostrophic components in the larger Coriolis terms. Because it is small compared with horizontal advection, itself already a small correction term compared with the Coriolis term, vertical advection is neglected, and we obtain the following:

$$-\frac{1}{\rho_0 f_0}\frac{\partial^2 p'}{\partial y \partial t}-\frac{1}{\rho_0^2 f_0^2}J\left(p',\frac{\partial p'}{\partial y}\right)-f_0 v-\frac{\beta_0}{\rho_0 f_0}y\frac{\partial p'}{\partial x}=-\frac{1}{\rho_0}\frac{\partial p'}{\partial x}\qquad(16.10a)$$

$$+\frac{1}{\rho_0 f_0}\frac{\partial^2 p'}{\partial x \partial t}+\frac{1}{\rho_0^2 f_0^2}J\left(p',\frac{\partial p'}{\partial x}\right)+f_0 u-\frac{\beta_0}{\rho_0 f_0}y\frac{\partial p'}{\partial y}=-\frac{1}{\rho_0}\frac{\partial p'}{\partial y}\qquad(16.10b)$$

The symbol $J(\cdot,\cdot)$ stands for the Jacobian operator, defined as $J(a,b)=(\partial a/\partial x)(\partial b/\partial y)-(\partial a/\partial y)(\partial b/\partial x)$.

From these equations, more accurate expressions for u and v can be readily extracted:

$$u=u_g+u_a=-\frac{1}{\rho_0 f_0}\frac{\partial p'}{\partial y}-\frac{1}{\rho_0 f_0^2}\frac{\partial^2 p'}{\partial t \partial x}$$

$$-\frac{1}{\rho_0^2 f_0^3}J\left(p',\frac{\partial p'}{\partial x}\right)+\frac{\beta_0}{\rho_0 f_0^2}y\frac{\partial p'}{\partial y}\qquad(16.11a)$$

$$v=v_g+v_a=+\frac{1}{\rho_0 f_0}\frac{\partial p'}{\partial x}-\frac{1}{\rho_0 f_0^2}\frac{\partial^2 p'}{\partial t \partial y}$$

$$-\frac{1}{\rho_0^2 f_0^3}J\left(p',\frac{\partial p'}{\partial y}\right)-\frac{\beta_0}{\rho_0 f_0^2}y\frac{\partial p'}{\partial x}\qquad(16.11b)$$

which, unlike Eq. (16.7), contain both the geostrophic flow and a first series of ageostrophic corrections. This improved estimate of the flow field has a nonzero divergence, which is small because it is caused solely by the weak velocity departures from the otherwise nondivergent geostrophic flow.

Upon substitution of these expressions in continuity equation (16.5d), we obtain

$$\frac{\partial w}{\partial z}=\frac{1}{\rho_0 f_0^2}\left[\frac{\partial}{\partial t}\nabla^2 p'+\frac{1}{\rho_0 f_0}J(p',\nabla^2 p')+\beta_0\frac{\partial p'}{\partial x}\right],\qquad(16.12)$$

where $\nabla^2=\partial^2/\partial x^2+\partial^2/\partial y^2$ is the two-dimensional Laplacian operator. We note that the right-hand arises only because of ageostrophic components. Thus, the vertical velocity is on the order of ageostrophic terms, and this justifies a posteriori our dropping of the w-terms from advection.

We now turn our attention to the density-conservation equation (16.5e). The first term is very small because ρ' is small, and the timescale is long. Likewise, the last term is very small because, as we concluded before, the vertical velocity arises from the ageostrophic corrections to the already weak horizontal velocity. The middle terms involve the density perturbation, which is small, and the horizontal velocities, which are also small. There is thus no need, in this equation, for the corrections brought by Eq. (16.11), and the geostrophic expressions (16.7) suffice, leaving

$$\frac{\partial \rho'}{\partial t} + \frac{1}{\rho_0 f_0} J(p', \rho') - \frac{\rho_0 N^2}{g} w = 0, \tag{16.13}$$

in which the stratification frequency, $N^2(z) = -(g/\rho_0)d\bar{\rho}/dz$, has been introduced. Dividing this last equation by N^2/g, taking its z-derivative, and using the hydrostatic balance (16.5c) to eliminate density, we obtain

$$\frac{\partial}{\partial t}\left[\frac{\partial}{\partial z}\left(\frac{1}{N^2}\frac{\partial p'}{\partial z}\right)\right] + \frac{1}{\rho_0 f_0} J\left[p', \frac{\partial}{\partial z}\left(\frac{1}{N^2}\frac{\partial p'}{\partial z}\right)\right] + \rho_0 \frac{\partial w}{\partial z} = 0. \tag{16.14}$$

Equations (16.12) and (16.14) form a two-by-two system for the perturbation pressure p' and vertical stretching $\partial w/\partial z$. Elimination of $\partial w/\partial z$ between the two yields a single equation for p':

$$\frac{\partial}{\partial t}\left[\nabla^2 p' + \frac{\partial}{\partial z}\left(\frac{f_0^2}{N^2}\frac{\partial p'}{\partial z}\right)\right] + \frac{1}{\rho_0 f_0} J\left[p', \nabla^2 p' + \frac{\partial}{\partial z}\left(\frac{f_0^2}{N^2}\frac{\partial p'}{\partial z}\right)\right]$$
$$+ \beta_0 \frac{\partial p'}{\partial x} = 0. \tag{16.15}$$

This is the quasi-geostrophic equation for nonlinear motions in a continuously stratified fluid on a beta plane. Usually, this equation is recast as an equation for the potential vorticity, and the pressure field is transformed into a streamfunction ψ through $p' = \rho_0 f_0 \psi$. The result is

$$\frac{\partial q}{\partial t} + J(\psi, q) = 0, \tag{16.16}$$

where q is the potential vorticity:

$$q = \nabla^2 \psi + \frac{\partial}{\partial z}\left(\frac{f_0^2}{N^2}\frac{\partial \psi}{\partial z}\right) + \beta_0 y. \tag{16.17}$$

Once the solution is obtained for q and ψ, the original variables can be recovered from Eqs. (16.7a), (16.7b), and (16.13):

$$u_g = -\frac{\partial \psi}{\partial y} \tag{16.18a}$$

$$v_g = +\frac{\partial \psi}{\partial x} \tag{16.18b}$$

$$u_a = -\frac{1}{f_0}\frac{\partial^2 \psi}{\partial t \partial x} - \frac{1}{f_0}J\left(\psi, \frac{\partial \psi}{\partial x}\right) + \frac{\beta_0}{f_0}y\frac{\partial \psi}{\partial y} \tag{16.18c}$$

$$v_a = -\frac{1}{f_0}\frac{\partial^2 \psi}{\partial t \partial y} - \frac{1}{f_0}J\left(\psi, \frac{\partial \psi}{\partial y}\right) - \frac{\beta_0}{f_0}y\frac{\partial \psi}{\partial x} \tag{16.18d}$$

$$w = -\frac{f_0}{N^2}\left[\frac{\partial^2 \psi}{\partial t \partial z} + J\left(\psi, \frac{\partial \psi}{\partial z}\right)\right] \tag{16.18e}$$

$$p' = \rho_0 f_0 \psi \tag{16.18f}$$

$$\rho' = -\frac{\rho_0 f_0}{g}\frac{\partial \psi}{\partial z}. \tag{16.18g}$$

If turbulent dissipation is retained in the formalism, the equation governing the evolution of potential vorticity becomes complicated, but an approximation suitable for most numerical applications is as follows:

$$\frac{\partial q}{\partial t} + J(\psi, q) = \frac{\partial}{\partial x}\left(\mathcal{A}\frac{\partial q}{\partial x}\right) + \frac{\partial}{\partial y}\left(\mathcal{A}\frac{\partial q}{\partial y}\right) + \frac{\partial}{\partial z}\left(\nu_E\frac{\partial q}{\partial z}\right), \tag{16.19}$$

where q remains defined by Eq. (16.17).

16.3   LENGTH AND TIMESCALE

Expression (16.17) indicates that q is a form of potential vorticity. Indeed, the last term represents the planetary contribution to the vorticity, whereas the first term, $\nabla^2\psi = \partial v/\partial x - \partial u/\partial y$, is the relative vorticity. The middle term can be traced to the layer-thickness variations in the denominator of the classical definition of potential vorticity [e.g., Eq. (12.21)]. Indeed, in view of Eq. (16.18g), this term measures vertical variations of ρ', directly related to thickness changes between density surfaces. It is thus a linear version of vertical stretching (see Analytical Problem 16.7). Although expression (16.17) for q does not have the same dimension as potential vorticity defined in Eq. (12.21), we shall follow common practice here and not coin another name but call it potential vorticity.

It is most interesting to compare the first two terms of the potential-vorticity expression, namely, relative vorticity and vertical stretching. With L and U as the horizontal length and velocity scales, respectively, the streamfunction ψ scales like LU by virtue of Eqs. (16.18a) and (16.18b). If H is the vertical length scale, not necessarily the fluid depth, the magnitudes of those contributions to potential vorticity are as follows:

$$\text{Relative vorticity} \sim \frac{U}{L}, \quad \text{Vertical stretching} \sim \frac{f_0^2 UL}{N^2 H^2}. \tag{16.20}$$

The ratio of the former to the latter is

$$\frac{\text{Relative vorticity}}{\text{Vertical stretching}} \sim \frac{N^2 H^2}{f_0^2 L^2} = Bu, \tag{16.21}$$

which is the Burger number defined in Section 11.6. For weak stratification or long length scale (i.e., small Burger number, $NH \ll f_0 L$), vertical stretching dominates, and the motion is akin to that of homogeneous rotating flows in nearly geostrophic balance (Chapter 7), where topographic variations are capable of exerting great influence. For large Burger numbers ($NH \gg f_0 L$), that is, strong stratification or short length scales, relative vorticity dominates, stratification reduces coupling in the vertical, and every level tends to behave in a two-dimensional fashion, stirred by its own vorticity pattern, independently of what occurs above and below.

The richest behavior occurs when the stratification and length scale match to make the Burger number of order unity, which occurs when

$$L = \frac{NH}{f_0}. \tag{16.22}$$

As noted in Section 12.2, this particular length scale is the internal radius of deformation. To show this, let us introduce a nominal density difference $\Delta\rho$, typical of the density vertical variations of the ambient stratification. Thus, $|d\bar{\rho}/dz| \sim \Delta\rho/H$ and $N^2 \sim g\Delta\rho/\rho_0 H$. Defining a reduced gravity as $g' = g\Delta\rho/\rho_0$, which is typically much less than the full gravity g, we obtain

$$N \sim \sqrt{\frac{g'}{H}}. \tag{16.23}$$

Definition (16.22) yields

$$L \sim \frac{\sqrt{g'H}}{f_0}. \tag{16.24}$$

Comparing this expression with definition (9.12) for the radius of deformation in homogeneous rotating fluids, we note the replacement of the full gravitational acceleration by a much smaller, reduced acceleration and conclude that motions in stratified fluids tend to take place on shorter scales than dynamically similar motions in homogeneous fluids.

Before concluding this section, it is noteworthy to return to the discussion of the timescale. Very early in the derivation, an assumption was made to restrict the attention to slowly evolving motions, namely, motions with timescale T much longer than the inertial timescale $1/f_0$ (i.e., $T \gg \Omega^{-1}$). This relegated the terms $\partial u/\partial t$ and $\partial v/\partial t$ to the rank of small perturbations to the dominant geostrophic balance. Now, having completed our analysis, we ought to check for consistency.

The timescale of quasi-geostrophic motions can be most easily determined by inspection of the governing equation in its potential-vorticity form. The balance of Eq. (16.16) requires that the two terms on its left-hand side be of the same order:

$$\frac{Q}{T} \sim \frac{UL}{L}\frac{Q}{L},$$

where Q is the scale of potential vorticity, regardless of whether it is dominated by relative vorticity ($Q \sim U/L$) or vertical stretching ($Q \sim f_0^2 UL/N^2H^2$), and LU is the streamfunction scale. The preceding statement yields

$$T \sim \frac{L}{U}, \tag{16.25}$$

in other words, the timescale is advective. The quasi-geostrophic structure evolves on a time T comparable to the time taken by a particle to cover the length scale L at the nominal speed U. For example, a vortex flow (such as an atmospheric cyclone) evolves significantly while particles complete one revolution.

Because the quasi-geostrophic formalism is rooted in the smallness of the Rossby number ($Ro = U/\Omega L \ll 1$), it follows directly that the timescale must be long compared with the rotation period:

$$T \gg \frac{1}{\Omega}, \tag{16.26}$$

in agreement with our premise. Note, however, that a lack of contradiction is proof only of consistency in the formalism. It implies that slowly evolving, quasi-geostrophic motions can exist, but the existence of other, non-quasi-geostrophic motions are certainly not precluded. Among the latter, we can distinguish nearly geostrophic motions of other types (Cushman-Roisin, 1986; Cushman-Roisin, Sutyrin & Tang, 1992; Phillips, 1963) and, of course, completely ageostrophic motions (see examples in Chapters 13 and 15). Whereas ageostrophic flows typically evolve on the inertial timescale ($T \sim \Omega^{-1}$), geostrophic motions of type other than quasi-geostrophic usually evolve on much longer timescales ($T \gg L/U \gg \Omega^{-1}$).

16.4   ENERGETICS

Because the quasi-geostrophic formalism is frequently used, it is worth investigating the approximate energy budget that is associated with it. Multiplying the governing equation (16.16) by the streamfunction ψ and integrating over the entire three-dimensional domain, we obtain, after several integrations by parts:

$$\frac{d}{dt} \iiint \frac{1}{2} \rho_0 |\nabla \psi|^2 dx\, dy\, dz$$
$$+ \frac{d}{dt} \iiint \frac{1}{2} \rho_0 \frac{f_0^2}{N^2} \left(\frac{\partial \psi}{\partial z} \right)^2 dx\, dy\, dz = 0. \tag{16.27}$$

The boundary terms have all been set to zero by assuming rigid bottom and top surfaces, and, in the horizontal, any combination of periodicity, vertical wall, or decay at large distances.

Equation (16.27) can be interpreted as a mechanical-energy budget: The sum of kinetic and potential energies is conserved over time. That the first integral corresponds to kinetic energy is evident once the velocity components have been expressed in terms of the streamfunction ($u^2 + v^2 = \psi_y^2 + \psi_x^2 = |\nabla \psi|^2$). By default, this leaves the second integral to play the role of potential energy, which is not as evident. Basic physical principles would indeed suggest the following definition for potential energy:

$$PE = \iiint \rho g z\, dx\, dy\, dz, \tag{16.28}$$

which by virtue of Eq. (16.18g) would yield a linear, rather than a quadratic, expression in ψ.

The discrepancy is resolved by defining the *available potential energy*, a concept first advanced by Margules (1903) and developed by Lorenz (1955). Because the fluid occupies a fixed volume, the rising of fluid in some locations must be accompanied by a descent of fluid elsewhere; therefore, any potential-energy gain somewhere is necessarily compensated, at least partially, by a potential-energy drop elsewhere. What matters then is not the total potential energy of the fluid but only how much could be converted from the instantaneous, perturbed density distribution. We define the available potential energy, *APE*, as the difference between the existing potential energy, as just defined, and the potential energy that the fluid would have if the basic stratification were unperturbed.

The situation is best illustrated in the case of a two-layer stratification (Fig. 16.2): A lighter fluid of density ρ_1 floats atop of a denser fluid of density ρ_2. In the presence of motion, the interface is at level a above the resting height H_2 of the lower layer. Because of volume conservation, the integral of a over the horizontal domain vanishes identically. The potential energy associated

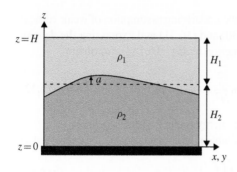

FIGURE 16.2 A two-layer stratification, for the illustration of the concept of available potential energy.

with the perturbed state is

$$PE(a)$$

$$= \iint \left[\int_0^{H_2+a} \rho_2 gz\,dz + \int_{H_2+a}^{H} \rho_1 gz\,dz \right] dx\,dy$$

$$= \iint \left[\frac{1}{2}\rho_1 gH^2 + \frac{1}{2}\Delta\rho gH_2^2 \right] dx\,dy$$

$$+ \iint \Delta\rho H_2 a\,dx\,dy + \iint \frac{1}{2}\Delta\rho ga^2\,dx\,dy,$$

where H is the total height, and $\Delta\rho = \rho_2 - \rho_1$ is the density difference. The first term represents the potential energy in the unperturbed state, whereas the second term vanishes because a has a zero mean. This leaves the third term as the available potential energy:

$$APE = PE(a) - PE(a=0)$$

$$= \iint \frac{1}{2}\Delta\rho ga^2\,dx\,dy. \tag{16.29}$$

Introducing the stratification frequency $N^2 = -(g/\rho_0)d\bar{\rho}/dz = g\Delta\rho/\rho_0 H$ and generalizing to three dimensions, we obtain

$$APE = \iiint \frac{1}{2}\rho_0 N^2 a^2\,dx\,dy\,dz. \tag{16.30}$$

In continuous stratification, the vertical displacement a of a fluid parcel is directly related to the density perturbation because the density anomaly at one point is created by moving to that point a particle that originates from a different vertical level:

$$\rho'(x,y,z,t) = \bar{\rho}[z - a(x,y,z,t)] - \bar{\rho}(z)$$

$$\simeq -a\frac{d\bar{\rho}}{dz} = \frac{\rho_0 N^2}{g}a. \tag{16.31}$$

This Taylor expansion is justified by the underlying assumption of weak vertical displacements. Combining Eqs. (16.30) and (16.31) and expressing the density perturbation in terms of the streamfunction by Eq. (16.18g), we obtain

$$APE = \iiint \frac{1}{2} \rho_0 \frac{f_0^2}{N^2} \left(\frac{\partial \psi}{\partial z} \right)^2 dx\,dy\,dz, \qquad (16.32)$$

which is the integral that arises in the energy budget, (16.27).

The time rate of change of the available potential energy can be expressed as

$$\frac{d}{dt} APE = g \iiint \rho' w \, dx\,dy\,dz, \qquad (16.33)$$

as can be verified by substitution of Eqs. (16.18e) and (16.18g) into Eq. (16.33) and an integration by parts of the Jacobian term. This shows that potential energy increases when heavy fluid parcels rise (ρ' and w both positive) and light parcels sink (ρ' and w both negative).

As a final note, we observe that a scale analysis provides the ratio of kinetic energy, KE, to available potential energy APE:

$$\frac{KE}{APE} \sim \frac{N^2 H^2}{L^2 f^2} \sim Bu, \qquad (16.34)$$

which gives another interpretation to the Burger number: It compares kinetic to potential energy.

16.5 PLANETARY WAVES IN A STRATIFIED FLUID

In Chapter 9, it was noted that inertia-gravity waves are superinertial ($\omega \geq f$) and that Kelvin waves require a fundamentally ageostrophic balance in one of the two horizontal directions [see Eq. (9.4b) with $u = 0$]. Therefore, the quasi-geostrophic formalism cannot describe these two types of waves. It can, however, describe the slow waves and, in particular, the planetary waves that exist on the beta plane.

It is instructive to explore the three-dimensional behavior of planetary (Rossby) waves in a continuously stratified fluid. The theory proceeds from the linearization of the quasi-geostrophic equation and, for mathematical simplicity only, the assumptions of a constant stratification frequency and no dissipation. Equations (16.16) and (16.17) then yield

$$\frac{\partial}{\partial t} \left(\nabla^2 \psi + \frac{f_0^2}{N^2} \frac{\partial^2 \psi}{\partial z^2} \right) + \beta_0 \frac{\partial \psi}{\partial x} = 0. \qquad (16.35)$$

We seek a wave solution of the form $\psi(x, y, z, t) = \phi(z) \cos(k_x x + k_y y - \omega t)$, with horizontal wavenumbers k_x and k_y, frequency ω, and amplitude $\phi(z)$. The vertical structure of the amplitude is governed by

$$\frac{d^2\phi}{dz^2} - \frac{N^2}{f_0^2}\left(k_x^2 + k_y^2 + \frac{\beta_0 k_x}{\omega}\right)\phi = 0, \tag{16.36}$$

which results from the substitution of the wave solution into Eq. (16.35). To solve this equation, boundary conditions are necessary in the vertical. For these, let us assume that our fluid is bounded below by a horizontal surface and above by a free surface. In the atmosphere, this situation would correspond to the troposphere above a flat terrain or sea and below the tropopause.

At the bottom (say, $z = 0$), the vertical velocity vanishes, and the linearized form of Eq. (16.18e) implies $\partial^2\psi/\partial z\partial t = 0$, or

$$\frac{d\phi}{dz} = 0 \quad \text{at} \quad z = 0. \tag{16.37}$$

At the free surface [say, $z = h(x, y, t)$], the pressure is uniform. Because the total pressure consists of the hydrostatic pressures due to the reference density ρ_0 (eliminated when the Boussinesq approximation was made; see Section 3.7) and to the basic stratification $\bar{\rho}(z)$, together with the perturbation pressure caused by the wave, we write:

$$P_0 - \rho_0 g z + g\int_z^h \bar{\rho}(z')\,dz' + p'(x, y, h, t) = \text{constant}, \tag{16.38}$$

at the free surface $z = h$. Because particles on the free surface remain on the free surface at all times (there is no inflow/outflow), we also state

$$w = \frac{\partial h}{\partial t} + u\frac{\partial h}{\partial x} + v\frac{\partial h}{\partial y} \quad \text{at} \quad z = h. \tag{16.39}$$

The preceding two statements are then linearized. Writing $h = H + \eta$, where the free-surface displacement $\eta(x, y, t)$ is small to justify linear wave motions, we expand the variables p' and w in Taylor fashion from the mean surface level $z = H$ and systematically drop all terms involving products of variables of the wave field. The two requirements then reduce to

$$-\rho_0 g\eta + p' = 0 \quad \text{and} \quad w = \frac{\partial\eta}{\partial t} \quad \text{at} \quad z = H. \tag{16.40}$$

Elimination of η yields $\partial p'/\partial t = \rho_0 g w$ and, in terms of the streamfunction,

$$\frac{\partial}{\partial t}\left(\frac{\partial\psi}{\partial z} + \frac{N^2}{g}\psi\right) = 0 \quad \text{at} \quad z = H \tag{16.41}$$

or, finally, in terms of the wave amplitude,

$$\frac{d\phi}{dz} + \frac{N^2}{g}\phi = 0 \quad \text{at} \quad z = H. \tag{16.42}$$

Together, Eq. (16.36) and its two boundary conditions, Eqs. (16.37) and (16.42), define an eigenvalue problem, which admits solutions of the form

$$\phi(z) = A \cos k_z z \tag{16.43}$$

already satisfying boundary condition (16.37). Substitution of this solution into Eq. (16.36) yields the dispersion relation linking the wave frequency ω to the wavenumber components, k_x, k_y, and k_z:

$$\omega = -\frac{\beta_0 k_x}{k_x^2 + k_y^2 + k_z^2 f_0^2/N^2}, \tag{16.44}$$

whereas substitution into boundary condition (16.42) imposes a condition on the wavenumber k_z:

$$\tan k_z H = \frac{N^2 H}{g}\frac{1}{k_z H}. \tag{16.45}$$

As Fig. 16.3 demonstrates graphically, there is an infinite number of discrete solutions. Because negative values of k_z lead to solutions identical to those with positive k_z values [see Eqs. (16.43) and (16.44)], it is necessary to consider only the latter set of values ($k_z > 0$).

A return to the definition $N^2 = -(g/\rho_0)d\bar{\rho}/dz$ reveals that the ratio $N^2 H/g$, appearing on the right-hand side of Eq. (16.45), is equal to $\Delta\rho/\rho_0$, where $\Delta\rho$ is the density difference between top and bottom of the basic stratification $\bar{\rho}(z)$. The factor $N^2 H/g$ is thus very small, implying that the first solution of Eq. (16.45) falls very near the origin (Fig. 16.3). There, $\tan k_z H$ can be approximated to $k_z H$, yielding

$$k_z H = \frac{NH}{\sqrt{gH}}. \tag{16.46}$$

The fraction on the right is the ratio of the internal gravity wave speed to the surface gravity wave speed, which is small. Note also that this mode disappears in the limit $g \to \infty$, which we would have obtained if we had imposed a rigid lid at the top of the domain.

Because $k_z H$ is small, the corresponding wave is nearly uniform in the vertical. Its dispersion relation, obtained from the substitution of the preceding value of k_z into Eq. (16.44),

$$\omega = -\frac{\beta_0 k_x}{k_x^2 + k_y^2 + f_0^2/gH}, \tag{16.47}$$

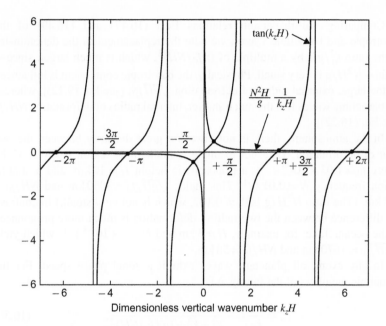

FIGURE 16.3   Graphical solution of Eq. (16.45). Every crossing of curves yields an acceptable value for the vertical wavenumber k_z. The pair of values nearest to the origin corresponds to a solution fundamentally different from all others.

is independent of the stratification frequency N and identical to the dispersion relation obtained for planetary waves in homogeneous fluids [see Eq. (9.27)]. Because it is almost uniform in the vertical, we conclude that this wave is the barotropic component of the set.

The remaining solutions for k_z can also be determined to the same degree of approximation. Because N^2H/g is small, the finite solutions of Eq. (16.45) fall very near the zeros of $\tan k_z H$ (Fig. 16.3) and are thus given approximately by

$$k_{zn} = n\frac{\pi}{H}, \quad n = 1, 2, 3, \ldots \qquad (16.48)$$

Unlike the barotropic wave, the waves with these wavenumbers exhibit substantial variations in the vertical and can be called *baroclinic*. Their dispersion relation,

$$\omega_n = -\frac{\beta_0 k_x}{k_x^2 + k_y^2 + (n\pi f_0/NH)^2}, \qquad (16.49)$$

is morphologically identical to Eq. (16.47), implying that they, too, are planetary waves. In summary, the presence of stratification permits the existence of an infinite, discrete set of planetary waves, one barotropic and all other baroclinic.

Comparing the dispersion relations Eqs. (16.47) and (16.49) of the barotropic and baroclinic waves, we note the replacement in the denominator of the ratio f_0^2/gH by a multiple of $(\pi f_0/NH)^2$, which is much larger, since—again—N^2H/g is very small. Physically, the barotropic component is influenced by the large, external radius of deformation $\sqrt{gH}/f_0$ [see Eq. (9.12)], whereas the baroclinic waves feel the much shorter, internal radius of deformation NH/f_0 [see Eq. (16.22)].

In the atmosphere, there is not always a great disparity between the two radii of deformation. Take, for example, a midlatitude region (such as 45°N, where $f_0 = 1.03 \times 10^{-4}$ s^{-1}), a tropospheric height $H = 10$ km, and a stratification frequency $N = 0.01$ s^{-1}. This yields $\sqrt{gH}/f_0 = 3050$ km and $NH/f_0 = 972$ km. (The ratio N^2H/g is then 0.102, which is not very small.) In contrast, the difference between the two radii of deformation is much more pronounced in the ocean. Take, for example, $H = 3$ km and $N = 2 \times 10^{-3}$ s^{-1}, which yield $\sqrt{gH}/f_0 = 1670$ km and $NH/f_0 = 58$ km.

In any event, all planetary waves exhibit a zonal phase speed. For the baroclinic members of the family, it is

$$c_n = \frac{\omega_n}{k_x} = -\frac{\beta_0}{k_x^2 + k_y^2 + (n\pi f_0/NH)^2}. \tag{16.50}$$

Because this quantity is always negative, the direction can only be westward.[2] Moreover, the westward speed is confined to the interval

$$-\beta_0 R_n^2 < c_n < 0, \tag{16.51}$$

with the lower bound approached by the longest wave $(k_x^2 + k_y^2 \to \infty)$. The lengths R_n, defined as

$$R_n = \frac{1}{n}\frac{NH}{\pi f_0}, \quad n = 1, 2, 3, \ldots \tag{16.52}$$

are identified as internal radii of deformation, one for each baroclinic mode. The greater the value of n, the greater the value of k_{zn}, the more reversals the wave exhibits in the vertical, and the more restricted is its zonal propagation. Therefore, the waves most active in transmitting information and carrying energy from east to west (or from west to east, if the group velocity is positive) are the barotropic and the first baroclinic component. Indeed, observations reveal that these two modes alone carry generally 80–90% of the energy in the ocean.

Let us now turn our attention to the spatial structure of a baroclinic planetary wave. For simplicity, we take the first mode ($n = 1$), which corresponds to a

---

[2] The meridional phase speed, ω_n/k_y, may be either positive or negative, depending on the sign of k_y.

wave with one reversal of the flow in the vertical, and we set k_y to zero to focus on the zonal profile of the wave. The streamfunction, pressure, and density distributions are as follows:

$$\psi = A \cos k_z z \cos (k_x x - \omega t) \tag{16.53a}$$

$$p' = \rho_0 f_0 \psi = \rho_0 f_0 A \cos k_z z \cos(k_x x - \omega t) \tag{16.53b}$$

$$\rho' = -\frac{\rho_0 f_0}{g} \frac{\partial \psi}{\partial z} = +\frac{\rho_0 f_0 k_z}{g} A \sin k_z z \cos(k_x x - \omega t). \tag{16.53c}$$

The geostrophic velocity component is

$$u_g = -\frac{\partial \psi}{\partial y} = 0 \tag{16.54a}$$

$$v_g = +\frac{\partial \psi}{\partial x} = -k_x A \cos k_z z \sin(k_x x - \omega t), \tag{16.54b}$$

and we immediately recognize that it cannot be responsible for the wave because it has no associated vertical velocity needed to displace density surfaces and allow energy conversion between kinetic and potential energy. Hence, the essence of the dynamics resides in the ageostrophic velocity component,

$$u_a = -\frac{1}{f_0} \frac{\partial^2 \psi}{\partial t \partial x}$$
$$= -\frac{k_x \omega}{f_0} A \cos k_z z \cos(k_x x - \omega t) \tag{16.55a}$$

$$v_a = -\frac{\beta_0}{f_0} y \frac{\partial \psi}{\partial x}$$
$$= +\frac{\beta_0 k_x}{f_0} y A \cos k_z z \sin(k_x x - \omega t) \tag{16.55b}$$

$$w = -\frac{f_0}{N^2} \frac{\partial^2 \psi}{\partial t \partial z}$$
$$= +\frac{f_0 \omega k_z}{N^2} A \sin k_z z \sin(k_x x - \omega t). \tag{16.55c}$$

We leave it as an exercise to the reader to verify that the three-dimensional divergence of the ageostrophic motion is zero as it should. The corresponding wave structure is displayed in Fig. 16.4 and can be interpreted as follows.

At the bottom, vertical displacements are prohibited, and there is no density anomaly. At the surface, vertical displacements would only be important if it were the barotropic (external) mode, but this is a baroclinic (internal) mode, and vertical displacements are negligible at the top. In the interior, however, vertical displacements are significant, with one maximum at midlevel for the lowest baroclinic mode (depicted here). Where the middle density surface rises,

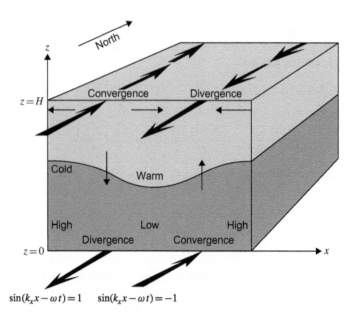

FIGURE 16.4 Structure of a baroclinic planetary wave. In constructing this diagram, we have taken $f_0 > 0$, $k_x > 0$, $k_y = 0$, and $k_z = \pi/H$, which yield $\omega < 0$ and a wave structure with a single reversal in the vertical.

heavier (colder) fluid from below is found, forming a cold anomaly. Similarly, a warm anomaly accompanies a subsidence, half a wavelength away. Because colder fluid is heavier and warmer fluid is lighter, the bottom pressure is higher under cold anomalies and lower under warm anomalies. At the lowest order of approximation, the resulting zonal pressure gradient drives an alternating geostrophic meridional flow v_g given by Eq. (16.54b). In the northern hemisphere (as depicted in Fig. 16.4), the bottom velocity has the higher pressure on its right and, therefore, assumes a southward direction east of the high pressures and a northward direction east of the low pressures. Because of the baroclinic nature of the wave, there is a reversal in the vertical, and the velocities near the top are counter to those below (Fig. 16.4).

On the beta plane, the variation in the Coriolis parameter causes this meridional flow to be convergent or divergent. In the northern hemisphere, the northward increase of f implies, under a uniform pressure gradient, a decreasing velocity and thus convergence of northward flow and divergence of southward flow. The resulting convergence–divergence pattern calls for transverse ageostrophic velocities, either zonal or vertical or both. According to Fig. 16.4, based on Eqs. (16.55c) and (16.55a), both transverse components come into play, each partially relieving the convergence–divergence of the meridional flow. The relative importance of vertical to horizontal convergence is

$$\frac{k_z W}{k_x U} = \frac{f_0^2 k_z^2}{N^2 k_x^2},$$
(16.56)

and we recover the inverse of the Burger number based on the length scales of the wave.

The ensuing vertical velocities at midlevel cause subsidence below a convergence and above a divergence, feeding the excess of the upper flow into the deficit of that underneath, and create uplifting half a wavelength away, where the situation is vertically reversed. Subsidence generates a warm anomaly, while uplifting generates a cold anomaly. As we can see in Fig. 16.4, this takes place a quarter of a wavelength to the west of the existing anomalies, thus inducing a westward shift of the wave pattern overtime. The result is a wave pattern steadily translating to the west.

16.6 SOME NONLINEAR EFFECTS

In its original form, the quasi-geostrophic equation (16.16) is quadratic in the streamfunction. An assumption of weak amplitudes was, therefore, necessary to explore the linear wave regime, and it is proper to ask now what role nonlinearities could play. For evident reasons, no general solution of the nonlinear equation is available. Nonlinearities cause interactions among the existing waves, generating harmonics and spreading the energy over a wide spectrum of scales. According to numerical simulations (McWilliams, 1989; Rhines, 1977), the result is a complicated unsteady state of motion, which has been termed *geostrophic turbulence*.

Although this topic will be more fully developed in a later chapter (Section 18.3), it is worth mentioning here the natural tendency of geostrophic turbulence to form coherent structures (McWilliams, 1984, 1989). These take the form of distinct and robust vortices that can be clearly identified and traced for periods of time long compared with their turn-around times. Figure 18.17 provides an example. The vortices contain a disproportionate amount of the energy available, being therefore highly nonlinear and leaving a relatively weak and linear wavefield in the intermediate space. In other words, a mature state of geostrophic turbulence displays a dichotomous pattern of nonlinear, localized vortices and linear, nonlocalized waves.

To explore nonlinear effects, let us seek a localized, vortex-type solution of finite amplitude. To simplify the analysis, we make the following assumptions of inviscid fluid and uniform stratification (N = constant). Furthermore, expecting a possible zonal drift reminiscent of planetary waves, we seek solutions that are steadily translating in the x-direction. Thus, we state

$$\frac{\partial q}{\partial t} + J(\psi, q) = 0 \tag{16.57}$$

$$q = \nabla^2 \psi + \frac{f_0^2}{N^2} \frac{\partial^2 \psi}{\partial z^2} + \beta_0 y, \tag{16.58}$$

in which $\psi = \psi(x - ct, y, z)$ is a function vanishing at large distances.

Because the variables x and t occur only in the combination $x - ct$, the time derivative can be assimilated to an x-derivative ($\partial/\partial t = -c\partial/\partial x$) and the equation becomes

$$J(\psi + cy, q) = 0,$$

which admits the general solution

$$q = \nabla^2 \psi + \frac{f_0^2}{N^2} \frac{\partial^2 \psi}{\partial z^2} + \beta_0 y = F(\psi + cy). \tag{16.59}$$

The function F is, at this stage, an arbitrary function of its variable, $\psi + cy$. Because the vortex is required to be localized, the streamfunction must vanish at large distances, including large zonal distances at finite values of the meridional coordinate y. From Eq. (16.59), this implies

$$\beta_0 y = F(cy),$$

and the function F is linear: $F(\alpha) = (\beta_0/c)\alpha$. Naturally, the function F may be multivalued, taking values along contours of $\psi + cy$ not connected to infinity (and, therefore, closed onto themselves within the confines of the vortex) that are different from the values along other, open contours of $\psi + cy$. In other words, the same $\psi + cy$ on two different contours could correspond to two distinct values of F.

Mindful of this possibility but restricting our attention for now to the region extending to infinity, where F is linear, we have from Eq. (16.59)

$$\nabla^2 \psi + \frac{f_0^2}{N^2} \frac{\partial^2 \psi}{\partial z^2} = \frac{\beta_0}{c} \psi. \tag{16.60}$$

Now, assuming the existence of rigid surfaces at the top and bottom, we impose $\partial\psi/\partial z = 0$ at $z = 0$ and H, restricting the number of vertical modes. The gravest baroclinic mode has the structure $\psi = a(r, \theta)\cos(\pi z/H)$, where (r, θ) are the polar coordinates associated with the Cartesian coordinates $(x - ct, y)$. The horizontal structure of the solution is prescribed by the amplitude $a(r, \theta)$, which must satisfy

$$\frac{\partial^2 a}{\partial r^2} + \frac{1}{r}\frac{\partial a}{\partial r} - \frac{1}{r^2}\frac{\partial^2 a}{\partial \theta^2} - \left(\frac{1}{R^2} + \frac{\beta_0}{c}\right)a = 0, \tag{16.61}$$

by virtue of Eq. (16.60). Here,

$$R = \frac{NH}{\pi f_0}, \qquad (16.62)$$

is the internal radius of deformation. (The factor π is introduced here for convenience.) Such an equation admits solutions consisting of sinusoidal functions in the azimuthal direction and Bessel functions in the radial direction.

Because the potential energy is proportional to the integrated square of the vertical displacements, a localized vortex structure of finite energy requires a streamfunction field (proportional to a) that decays at large distances faster than $1/r$. This requirement excludes the Bessel functions of the first kind, which decay only as $r^{-1/2}$ and leaves us with the modified Bessel functions, which decay exponentially:

$$a(r, \theta) = \sum_{m=0}^{\infty} (A_m \cos m\theta + B_m \sin m\theta) K_m(kr), \qquad (16.63)$$

where the factor k defined by

$$k^2 = \frac{1}{R^2} + \frac{\beta_0}{c} \qquad (16.64)$$

must be real. The condition that k be real implies, from Eq. (16.64) that the drift speed c of the vortex must be either less than $-\beta_0 R^2$ or greater than zero. In other words, c must lie outside of the range of linear planetary wave speeds [see Eq. (16.51)]. Because k enters the solution in multiplication with the radial distance r, its inverse, $1/k$, can be considered as the width of the vortex:

$$L = \frac{1}{k} = \frac{R}{\sqrt{1 + \beta_0 R^2/c}}. \qquad (16.65)$$

The faster the propagation (either eastward or westward), the closer L is to the deformation radius R. Eastward-propagating vortices ($c > 0$) are smaller than R, whereas westward-propagating ones ($c < 0$) are wider.

The Bessel functions K_m are singular at the origin, and solution (16.63) fails near the vortex center. The situation is remedied by requiring that in the vicinity of $r = 0$, the function F assumes another form than that used previously, changing the character of the solution there. Here, we shall not consider this solution, called the *modon*, and instead refer the interested reader to Flierl, Larichev, McWilliams and Reznik (1980).

Let us now consider disturbances on a zonal jet, such as the meanders of the atmospheric Jet Stream. Waves and finite-amplitude perturbations propagate zonally at a net speed that is their own drift speed plus the jet average velocity. They thus move away from their region of origin, such as a mountain range, by traveling either upstream or downstream, unless their net speed is about zero. In this last case, when the disturbance's own speed c is equal and opposite to the average jet velocity U, the disturbance is stationary and can persist for a much longer time. Typically, the zonal jet flows eastward (Jet Stream in the atmosphere, Gulf Stream in the ocean, and prevailing winds on Jupiter at the latitude of the Great Red Spot), and so we take U positive. The mathematical requirement is $c = -U$ (westward), and two cases arise: Either U is smaller than $\beta_0 R^2$ or it is not. For U less than $\beta_0 R^2$, c falls in the range of planetary waves, and the disturbance is a train of planetary waves, giving the jet a meandering character. By virtue of dispersion relation (16.50), applied to the gravest vertical mode ($n = 1$) and to the zero meridional wavenumber ($m = 0$), the zonal wavelength is

$$\lambda = \frac{2\pi}{k_x} = 2\pi R \sqrt{\frac{U}{\beta_0 R^2 - U}}. \tag{16.66}$$

If the jet velocity varies downstream, the wavelength adjusts locally, increasing with U. However, if U exceeds $\beta_0 R^2$, finite-amplitude, isolated disturbances are possible, and the jet may be strongly distorted. The preceding theory suggests the following length scale

$$L = \frac{1}{k} = R \sqrt{\frac{U}{U - \beta_0 R^2}}. \tag{16.67}$$

16.7 QUASI-GEOSTROPHIC OCEAN MODELING

Quasi-geostrophic models were at the core of the first weather-forecast systems (see biographies at the end of Chapter 5 and of the present chapter), and the reason for their success was their highly simplified mathematics and numerics while capturing the dynamics essential to weather forecasting. Because of limited computing power, the first few models were two-dimensional. Here, we illustrate the core numerical properties of these two-dimensional models because they are representative of the numerics used in the subsequent three-dimensional models.

In two dimensions, in the absence of friction and turbulence, the equation governing quasi-geostrophic dynamics reduces to

$$\frac{\partial q}{\partial t} + J(\psi, q) = 0 \tag{16.68}$$

with the potential vorticity q defined as

$$q = \nabla^2 \psi + \beta_0 y. \tag{16.69}$$

It is clear from Eq. (16.68) that the Jacobian J operator plays a central role in the mathematics. This Jacobian can be expressed mathematically in several different ways:

$$J(\psi, q) = \frac{\partial \psi}{\partial x}\frac{\partial q}{\partial y} - \frac{\partial \psi}{\partial y}\frac{\partial q}{\partial x} \tag{16.70a}$$

$$= \frac{\partial}{\partial x}\left(\psi\frac{\partial q}{\partial y}\right) - \frac{\partial}{\partial y}\left(\psi\frac{\partial q}{\partial x}\right) \tag{16.70b}$$

$$= \frac{\partial}{\partial y}\left(q\frac{\partial \psi}{\partial x}\right) - \frac{\partial}{\partial x}\left(q\frac{\partial \psi}{\partial y}\right) \tag{16.70c}$$

so that we can readily write the corresponding finite-difference forms, all of second order (see Fig. 16.5 for notation):

$$J^{++} = \frac{(\tilde{\psi}_4 - \tilde{\psi}_8)(\tilde{q}_6 - \tilde{q}_2) - (\tilde{\psi}_6 - \tilde{\psi}_2)(\tilde{q}_4 - \tilde{q}_8)}{4\Delta x\Delta y} \tag{16.71a}$$

$$J^{+\times} = \frac{\left[\tilde{\psi}_4(\tilde{q}_5 - \tilde{q}_3) - \tilde{\psi}_8(\tilde{q}_7 - \tilde{q}_1)\right] - \left[\tilde{\psi}_6(\tilde{q}_5 - \tilde{q}_7) - \tilde{\psi}_2(\tilde{q}_3 - \tilde{q}_1)\right]}{4\Delta x\Delta y} \tag{16.71b}$$

$$J^{\times+} = \frac{\left[\tilde{q}_6(\tilde{\psi}_5 - \tilde{\psi}_7) - \tilde{q}_2(\tilde{\psi}_3 - \tilde{\psi}_1)\right] - \left[\tilde{q}_4(\tilde{\psi}_5 - \tilde{\psi}_3) - \tilde{q}_8(\tilde{\psi}_7 - \tilde{\psi}_1)\right]}{4\Delta x\Delta y} \tag{16.71c}$$

With a multiplicity of discretizations at our disposal, we may wonder which leads to the best model. Because all are second order, we have to invoke other properties than truncation error to decide on the optimal discretization, such as conservation laws. We can, for example, identify the following integral constraints over a 2D domain of surface S within a close impermeable boundary

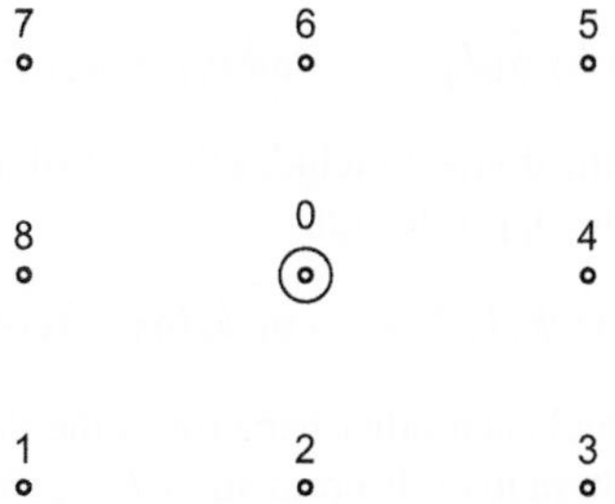

FIGURE 16.5 Grid notation for Jacobian $J(\psi, q)$ around the central point labeled 0. The discretization J^{++} (16.71a) uses ψ and q at side points 2, 4, 6, and 8, and $J^{+\times}$ (16.71b) takes ψ values at side points 2, 4, 6, 8 and q values at corner points 1, 3, 5, and 7, while $J^{\times+}$ switches the values of ψ and q.

(uniform ψ along the boundary) or with periodic boundaries:

$$\int_S J(\psi, q)\,d\mathcal{S} = 0, \tag{16.72}$$

$$\int_S qJ(\psi, q)\,d\mathcal{S} = 0, \tag{16.73}$$

$$\int_S \psi J(\psi, q)\,d\mathcal{S} = 0. \tag{16.74}$$

Expression (16.74) can be related to the evolution of kinetic energy. In addition, we have an antisymmetry property:

$$J(\psi, q) = -J(q, \psi). \tag{16.75}$$

Numerical discretization generally does not ensure conservation of the corresponding integral properties in the discrete solution. Akio Arakawa (see biography at the end of Chapter 9) had the brilliant idea of combining different versions of the discretized Jacobian in order to preserve those properties in the discrete formulation. The combination

$$J = (1 - \alpha - \beta)J^{++} + \alpha J^{+\times} + \beta J^{\times+} \tag{16.76}$$

for any value of α and β leads to a consistent discretization. We should, therefore, try to assign those values of α and β that ensure as many simultaneous conservation properties as possible.

Integral (16.74) in its discrete form sums up individual terms involving products $\psi_{i,j}J_{i,j}$ or, with the shorter notation of Fig. 16.5, terms such as

$$\Delta x \Delta y\, \psi_0 J_0. \tag{16.77}$$

For the Jacobian discretized according to Eq. (16.71a), this involves

$$4\Delta x \Delta y\, \tilde{\psi}_0 J_0^{++} = \tilde{\psi}_0 \tilde{\psi}_4(\tilde{q}_6 - \tilde{q}_2) + \cdots \tag{16.78}$$

The sum (integral) over the domain includes the contribution $\psi_4 J_4$, in which we find similar terms with the opposite sign:

$$4\Delta x \Delta y\, \tilde{\psi}_4 J_4^{++} = -\tilde{\psi}_0 \tilde{\psi}_4(\tilde{q}_5 - \tilde{q}_3) + \cdots \tag{16.79}$$

but the terms do not cancel each other because of the differing q values. However, if we look at the alternative discretization $J^{+\times}$, we do find the terms that do cause cancellation:

$$4\Delta x \Delta y\, \tilde{\psi}_0 J_0^{+\times} = \tilde{\psi}_0 \tilde{\psi}_4(\tilde{q}_5 - \tilde{q}_3) + \cdots \tag{16.80}$$

and

$$4\Delta x \Delta y\, \tilde{\psi}_4 J_4^{+\times} = -\tilde{\psi}_4 \tilde{\psi}_0 (\tilde{q}_6 - \tilde{q}_2) + \cdots \qquad (16.81)$$

So, if we add $(J^{++} + J^{+\times})$ and then integrate over the domain, the ψJ products cancel one another out in pairs. The same reasoning applies to other combinations of terms such as those between point 0 and 6. Thus, constraint (16.74) is respected if $(J^{++} + J^{+\times})/2$ is used for the discretization of the Jacobian.

There is better. Because $\psi J^{\times +}$ does not contain the terms $\tilde{\psi}_0 \tilde{\psi}_4$ and $\tilde{\psi}_0 \tilde{\psi}_6$ but contains terms with the product $\tilde{\psi}_0 \tilde{\psi}_5$

$$4\Delta x \Delta y\, \tilde{\psi}_0 J_0^{\times +} = \tilde{\psi}_0 \tilde{\psi}_5 (\tilde{q}_6 - \tilde{q}_4) + \cdots \qquad (16.82)$$

$$4\Delta x \Delta y\, \tilde{\psi}_5 J_5^{\times +} = \tilde{\psi}_0 \tilde{\psi}_5 (\tilde{q}_4 - \tilde{q}_6) + \cdots \qquad (16.83)$$

which cancel each other out, it turns out that we can dilute the sum $(J^{++} + J^{+\times})$ with any amount of $J^{\times +}$ (make $1 - \alpha - \beta$ equal to α while keeping β arbitrary).

Similarly, if we take the sum $(J^{++} + J^{\times +})$ or $J^{+\times}$ or combination, the sum of qJ over all grid points of the domain vanishes, respecting constraint (16.73). Thus, Eqs. (16.74) and (16.73) can be simultaneously respected if we take $1 - \alpha - \beta = \alpha = \beta$, which calls for $\alpha = \beta = 1/3$.

Let us now check on the antisymmetry condition (16.75). It is satisfied with J^{++} and with the sum $(J^{+\times} + J^{\times +})$. Thus, we happily note that our combination $(J^{++} + J^{+\times} + J^{\times +})/3$ already respects it. In conclusion, the values $\alpha = \beta = 1/3$ are ideal because they carry Eqs. (16.73)–(16.75) from the continuum representation over to the discretized formulation. This discretization, which is very popular, has become known as the *Arakawa Jacobian*.

The second essential ingredient in the quasi-geostrophic evolution is the relation between q and ψ. Its core term is

$$q = \frac{\partial^2 \psi}{\partial x^2} + \frac{\partial^2 \psi}{\partial y^2}, \qquad (16.84)$$

which is also the sole remaining term on the f-plane in two-dimensions. Because Eq. (16.68) provides the equation to advance q in time, Eq. (16.84) can be considered as the equation to be solved for ψ once an updated value has been calculated for q. Hence we need to invert a Poisson equation at each time step, a task already encountered in Section 7.8. In the first quasi-geostrophic models, inversion was done by successive over-relaxation, occasionally with a red–black approach on vector computers.

ANALYTICAL PROBLEMS

16.1. Derive the one-layer quasi-geostrophic equation

$$\frac{\partial}{\partial t}\left(\nabla^2\psi - \frac{1}{R^2}\psi\right) + J(\psi, \nabla^2\psi) + \beta_0\frac{\partial\psi}{\partial x} = 0, \qquad (16.85)$$

where $R = (gH)^{1/2}/f_0$, from the shallow-water model (7.17) assuming weak surface displacements. How do the waves permitted by these dynamics compare with the planetary waves exposed in Section 16.5?

16.2. Demonstrate the assertion made at the end of Section 16.4 that the time rate of change of available potential energy is proportional to the integral of the product of density perturbation with vertical velocity.

16.3. Elucidate in a rigorous manner the scaling assumptions justifying simultaneously the quasi-geostrophic approximation and the linearization of the equations for the wave analysis. What is the true restriction on vertical displacements?

16.4. Show that the assumption of a rigid upper surface (combined to the assumption of a flat bottom) effectively replaces the external radius of deformation by infinity. Also show that the approximate solutions for the vertical wavenumber k_z in Section 16.5 then become exact.

16.5. Explore topographic waves using the quasi-geostrophic formalism on an f-plane ($\beta_0 = 0$). Begin by formulating the appropriate bottom-boundary condition.

16.6. Establish the so-called *Omega Equation* on the f-plane for a quasi-geostrophic system without friction and with N^2 horizontally uniform. The Omega Equation provides the vertical velocity in a diagnostic form (i.e., without need for time integration). The formulation involves the geostrophic flow (u_g, v_g) associated with the (observed) density field:

$$N^2\frac{\partial^2 w}{\partial x^2} + N^2\frac{\partial^2 w}{\partial y^2} + f^2\frac{\partial^2 w}{\partial z^2} = \frac{\partial Q_x}{\partial x} + \frac{\partial Q_y}{\partial y} \qquad (16.86)$$

with

$$Q_x = +2f\left(\frac{\partial u_g}{\partial z}\frac{\partial v_g}{\partial x} + \frac{\partial v_g}{\partial z}\frac{\partial v_g}{\partial y}\right)$$

$$Q_y = -2f\left(\frac{\partial u_g}{\partial y}\frac{\partial v_g}{\partial z} + \frac{\partial u_g}{\partial z}\frac{\partial u_g}{\partial x}\right).$$

16.7. Show that potential vorticity q defined by Eq. (16.17) is a linearization of potential vorticity $\tilde{q}$ defined in Eq. (12.21) in the sense that for constant N^2

$$\tilde{q} = \frac{f_0}{h_0} + \frac{q}{h_0}, \tag{16.87}$$

in which h_0 is the unperturbed thickness of the layer, and the linearization assumes small vertical displacements and weak horizontal velocities. (*Hint*: The unperturbed height is directly related to N^2. For the linearization, express $1/h$ as a function of vertical density gradients.)

16.8. Take the reduced-gravity version of the quasi-geostrophic equation:

$$\frac{\partial q}{\partial t} = J(\psi, q) \quad \text{with} \quad q = \nabla^2 \psi - \frac{\psi}{R^2} + \beta_0 y \tag{16.88}$$

and show that the center of mass of a vortex patch propagates westward at a speed $\beta_0 R^2$, where the coordinates of the center of mass $(X(t), Y(t))$ are defined as

$$X = \frac{\iint x\psi \, dxdy}{\iint \psi \, dxdy}, \quad Y = \frac{\iint y\psi \, dxdy}{\iint \psi \, dxdy}, \tag{16.89}$$

(*Hint*: Calculate dX/dt and dY/dt.)

NUMERICAL EXERCISES

16.1. Verify numerical conservation (16.72) by adapting qgmodel.m in a closed two-dimensional domain of size L. Compare leapfrog and explicit Euler time discretizations. Initialize with a streamfunction given by

$$\psi = \omega_0 L^2 \sin\left(\frac{\pi x}{L}\right) \sin\left(\frac{\pi y}{L}\right). \tag{16.90}$$

On all four sides, $x = 0$, $x = L$, $y = 0$, and $y = L$, the boundaries are impermeable, and the streamfunction is kept zero. Use $\omega_0 = 10^{-5} \, \text{s}^{-1}$ and $L = 100 \, \text{km}$. For simplicity, also use zero vorticity along the perimeter.

16.2. Start with qgmodelrun.m and generalize the code to allow dynamics on the two-dimensional beta plane. Also add superviscosity (biharmonic diffusion) as shown in Section 10.6. Redo the simulation of Numerical Exercise 16.1 including the beta term. Take $\beta_0 = 2 \times 10^{-11} \, \text{m}^{-1} \text{s}^{-1}$ and $L = 3000 \, \text{km}$. Observe the evolution of the streamfunction. (*Hint*: Keep relative vorticity as the dynamic variable and express the beta effect as a forcing term in the governing equation for relative vorticity, e.g., within the Jacobian operator.)

16.3. Simulate the evolution of an eddy on the f-plane. Begin with an eddy centered at the origin with its streamfunction given by

$$\psi = -\omega_0 L^2 (r+1)e^{-r}, \qquad r = \frac{\sqrt{x^2+y^2}}{L}, \qquad (16.91)$$

with $L = 100$ km and $\omega_0 = 10^{-5}$ s^{-1}, and then perturb it by multiplying r used in the initial calculation of ψ by $1 + \epsilon \cos(2\theta)$, where θ is the azimuthal angle and ϵ a small parameter, for example, ~ 0.03. Perform the calculations in the square domain $[-10L, 10L] \times [-10L, 10L]$ with zero values for ψ along the perimeter.

16.4. Redo Numerical Exercise 16.3 with an initial eddy defined by

$$\omega = \begin{cases} -\omega_0 & \text{for } 0 < r/L < 1/\sqrt{2} \\ +\omega_0 & \text{for } 1/\sqrt{2} \le r/L < 1 \\ 0 & \text{for } 1 \le r/L \end{cases} \qquad (16.92)$$

and verify that you obtain the evolution shown in Fig. 16.6.

16.5. Adapt `qgmodel.m` to simulate the instability of the barotropic flow of Section 10.4 or analyze `shearedflow.m`. Initialize with the basic flow perturbed by an unstable wave. Instead of an infinite domain in the y–direction, prescribe zero values for the streamfunction at $y = \pm 10L$. Apply periodic boundary conditions in the x-direction. What boundary conditions do you use for the potential vorticity q? Which problem related to boundary conditions do you encounter if you want to use biharmonic diffusion (Section 10.6)? In any case, take a weak diffusion for

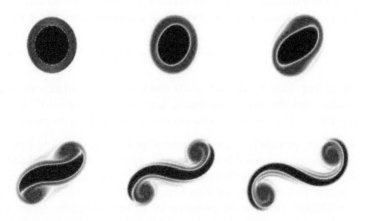

FIGURE 16.6 Evolution of a perturbed vorticity patch within a quasi-geostrophic framework. A color version can be found online at http://booksite.academicpress.com/9780120887590/

the simulations. (*Hint*: Expect to have time for a cup of coffee during the simulation.)

16.6. Implement a more efficient Poisson-equation solver using a conjugate-gradient approach (see Section 7.8) by using MATLAB™ routine pcg, and redo Numerical Exercise 16.5. Search the World Wide Web for a multi-grid version of the Poisson equation solver to further reduce calculation times if necessary.

16.7. Adapt the over-relaxation parameter to decrease the computation time of simulations in Numerical Exercise 16.5. Then simulate the schematized atmospheric jet stream of Analytical Problem 10.4, with a long wave-like perturbation. In a second experiment, reduce the intensity of the jet stream by a factor 4. In a third experiment, keep the lower velocity but disable the beta term. Discuss the stability in the context of the solution to Analytical Problem 10.4.

16.8. Simulate the evolution of the triangular jet of Analytical Problem 10.5, with $L = 50$ km and $U = 1$ m/s. (*Hint*: Perturb the zonal flow by a rather long wave in a sufficiently long domain.)

Jule Gregory Charney
1917–1981

A strong proponent of the idea that intelligent simplifications of a problem are not only necessary to obtain answers but also essential to understand the underlying physics, Jule Charney was a major contributor to dynamic meteorology. As a student, he studied the instabilities of large-scale atmospheric flows and elucidated the mechanism that is now called baroclinic instability (Chapter 17). His thesis appeared in 1947, and the following year, he published an article outlining quasi-geostrophic dynamics (the material of this chapter). He then turned his attention to numerical weather prediction, an activity envisioned by L. F. Richardson some 30 years earlier. The success of the initial weather simulations in the early 1950s is to be credited not only to J. von Neumann's first electronic computer, but also to Charney's judicious choice of simplified dynamics, the quasi-geostrophic equation. Later on, Charney was instrumental in convincing officials worldwide of the significance of numerical weather predictions, while he also gained much deserved recognition for his work on tropical meteorology, topographic instability, geostrophic turbulence, and the Gulf Stream. Charney applied his powerful intuition to systematic scale analysis. Scaling arguments are now a mainstay in geophysical fluid dynamics. (*Photo from archives of the Massachusetts Institute of Technology*)

Allan Richard Robinson
1932–2009

An avowed "phenomenologist," Allan Robinson is counted among the founding fathers of geophysical fluid dynamics because of his seminal contributions on the dynamics of rotating and stratified fluids, boundary-layer flows, continental shelf waves, and the maintenance of the oceanic thermocline. Underlying his accomplishments is the firm belief that "curiosity about nature is the primary driving force and rationalization for research." During the 1970s he chaired and cochaired a series of international programs that established the existence and importance of intermediate-scale eddies in the open ocean, the *internal weather of the sea*. His research led him to formulate numerical models for ocean forecasting and to emphasize the role of ocean physics in regulating biological activity. Robinson has also contributed significantly to the development of techniques for the assimilation of data in ocean-forecasting models. During the 1980s and 1990s, he led a group of international scientists, predominantly from bordering nations, to advance the science of the Mediterranean Sea. Later, he headed a program to synthesize knowledge of the interdisciplinary global coastal ocean. (*A. R. Robinson, Harvard University*)

Chapter 17

Instabilities of Rotating Stratified Flows

ABSTRACT

In a stratified rotating fluid, not all geostrophic flows are stable, for some are vulnerable to growing perturbations. This chapter presents two primary mechanisms by which instability may occur: motion of individual particles (called *inertial instability*) and organized motions across the flow (called *baroclinic instability*). In each case, kinetic energy is supplied to the disturbance by release of potential energy from the original flow. Baroclinic instability is at the origin of the midlatitude cyclones and anticyclones that make our weather so variable. Because the evolution of weather perturbations is essentially nonlinear, a two-layer quasi-geostrophic model is presented here to simulate the evolution of the baroclinic instability past the linear-growth phase.

17.1 TWO TYPES OF INSTABILITY

There are two broad types of flow instability. One is *local* or *punctual* in the sense that every particle in (at least a portion of) the flow is in an unstable situation. A prime example of this type is gravitational instability, which occurs in the presence of a reverse stratification (top-heavy fluid): if displaced, either upward or downward, a particle is subjected to a buoyancy force that pulls it further away from its original location and, since all other particles are individually subjected to a similar pull, the result is a catastrophic overturn of the fluid followed by mixing. In the absence of friction, there is no specific temporal and spatial scales for the event.

The second type of instability can exist only if the flow is stable with respect to the first kind. It is more gradual and relies on a collaborative action of many, if not all, particles and for this reason can be called *global* or *organized*. The instability is manifested by the temporal growth of a wave at a preferential wavelength that eventually overturns and forms vortices. An example is the barotropic instability encountered in Chapter 10 (see Section 10.4 in particular).

Rotating stratified flows can be subjected to either type of instability. If the instability is local, it is called *inertial instability*, and if it is global, *baroclinic*

TABLE 17.1 Contrasting Characteristics of the Two Types of Instability to which a Fluid Flow May Be Subjected

Local Instability	Global Instability
Particles act individually.	Particles act in concert.
Motion proceeds randomly.	Motion proceeds in a wave arrangement.
Instability criterion depends only on local properties of the flow.	Instability criterion depends on bulk properties of the flow and on wavelength of perturbation.
Instability is independent of boundary conditions.	Instability is sensitive to boundary conditions.
Instability is catastrophic (major overturn, mixing).	Instability is gradual (growing wave and vortex formation).
Example: overturning of a top-heavy fluid	Example: Kelvin–Helmholtz instability
In rotating stratified flow: inertial instability	In rotating stratified flow: mixed barotropic–baroclinic instability

instability. Table 17.1 summarizes the contrasting properties of the two types of instabilities.

Baroclinic instability is actually an end member of a more general instability, called *mixed barotropic–baroclinic instability*, which occurs when the flow is sheared in both horizontal and vertical directions. Baroclinic instability is the extreme when there is no shear in the horizontal, and barotropic instability (Chapter 10) is the other extreme, when the original flow has a no shear in the vertical.

17.2 INERTIAL INSTABILITY

In this section, we consider the possibility of catastrophic instability, namely one in which a fluid particle once displaced from its position of equilibrium keeps moving further away from that position. Such instability is catastrophic because, if one such particle migrates away from its initial position, all others can do so as well, and the ensuing situation is overturn, mixing and chaos.

This instability can be characterized also as *inertial* because acceleration is the crux of the growing displacement of the particles in the system. Finally, inertial instability is sometimes called *symmetric instability* (Holton, 1992) due to some symmetry in its formulation, as the following developments will shortly reveal.

Let us consider an inviscid steady flow in thermal-wind balance with variation across the vertical plane (x, z), with sheared velocity $v(x, z)$ in equilibrium with a slanted stratification $\rho(x, z)$. Such flow must be both geostrophic and hydrostatic:

$$-fv = -\frac{1}{\rho_0}\frac{\partial p}{\partial x} \tag{17.1a}$$

$$0 = -\frac{1}{\rho_0}\frac{\partial p}{\partial z} - \frac{g\rho}{\rho_0}. \tag{17.1b}$$

Elimination of pressure p between these two equations yields the thermal-wind balance

$$f\frac{\partial v}{\partial z} = -\frac{g}{\rho_0}\frac{\partial \rho}{\partial x}. \tag{17.2}$$

From these flow characteristics, let us define the stratification frequency N by

$$N^2 = -\frac{g}{\rho_0}\frac{\partial \rho}{\partial z} = \frac{1}{\rho_0}\frac{\partial^2 p}{\partial z^2}, \tag{17.3}$$

and, similarly, two quantities that will be become useful momentarily:

$$F^2 = f\left(f + \frac{\partial v}{\partial x}\right) = f^2 + \frac{1}{\rho_0}\frac{\partial^2 p}{\partial x^2} \tag{17.4}$$

$$fM = f\frac{\partial v}{\partial z} = -\frac{g}{\rho_0}\frac{\partial \rho}{\partial x} = \frac{1}{\rho_0}\frac{\partial^2 p}{\partial x \partial z}. \tag{17.5}$$

Note that the three quantities N^2, F^2, and fM all have the dimension of a frequency squared. But, although the first two are defined as squares, we ought to entertain the possibility that they may be negative.

Next, let us perturb such flow by adding time dependency and velocity components u and w within the x–z plane, while assuming still no variation in the perpendicular direction. For clarity of exposition, we further assume inviscid flow and restrict the attention to the f-plane, but we allow for possible nonhydrostaticity in the vertical, in anticipation of large vertical accelerations:

$$\frac{du}{dt} - fv = -\frac{1}{\rho_0}\frac{\partial p}{\partial x} \tag{17.6a}$$

$$\frac{dv}{dt} + fu = 0 \tag{17.6b}$$

$$\frac{dw}{dt} = -\frac{1}{\rho_0}\frac{\partial p}{\partial z} - \frac{g\rho}{\rho_0}, \tag{17.6c}$$

where d/dt stands for the material derivative (following particle movement).

In this flow, let us track an individual fluid particle with moving coordinates $[x(t), z(t)]$. Its velocity components in the vertical plane are

$$u = \frac{dx}{dt}, \qquad w = \frac{dz}{dt}, \tag{17.7}$$

which transform Eq. (17.6b) into

$$\frac{dv}{dt} + f\frac{dx}{dt} = 0. \tag{17.8}$$

Since f is constant in our model, the quantity $v + fx$ is an invariant of the motion,[1] and it follows that if the particle is displaced horizontally over a distance Δx it undergoes a change of transverse velocity Δv such that

$$\Delta v + f\Delta x = 0. \tag{17.9}$$

Turning our attention to Eqs. (17.6a) and (17.6c) and eliminating from them u and w by using (17.7), we obtain

$$\frac{d^2 x}{dt^2} - fv = -\frac{1}{\rho_0}\frac{\partial p}{\partial x} \tag{17.10a}$$

$$\frac{d^2 z}{dt^2} = -\frac{1}{\rho_0}\frac{\partial p}{\partial z} - \frac{g\rho}{\rho_0}. \tag{17.10b}$$

Note that in these equations the pressure terms on the right-hand side are complicated functions of the particle position (x, z).

Let us now imagine that the fluid particle under consideration is only moved from its original position by a small displacement Δx in the horizontal and Δz in the vertical: $x(t) = x_0 + \Delta x(t)$, $z(t) = z_0 + \Delta z(t)$, so that we may linearize the equations. Note that any displacement along y has no effect on the dynamic balance and can be ignored. Neglecting compressibility effects, we assume that the displacement causes no change in density for the particle. At its new position, the particle is out of equilibrium. In the vertical, it is subject to a buoyancy force, while in the horizontal, it is no longer in geostrophic equilibrium. These forces are reflected in the new, local values of the pressure gradient, which for a small displacement can be obtained from the original values by a Taylor expansion:

$$\frac{\partial p}{\partial x}\bigg|_{\text{at } x+\Delta x, z+\Delta z} = \frac{\partial p}{\partial x}\bigg|_{\text{at } x,z} + \Delta x \frac{\partial^2 p}{\partial x^2}\bigg|_{\text{at } x,z} + \Delta z \frac{\partial^2 p}{\partial x \partial z}\bigg|_{\text{at } x,z} \tag{17.11a}$$

$$\frac{\partial p}{\partial z}\bigg|_{\text{at } x+\Delta x, z+\Delta z} = \frac{\partial p}{\partial z}\bigg|_{\text{at } x,z} + \Delta x \frac{\partial^2 p}{\partial x \partial z}\bigg|_{\text{at } x,z} + \Delta z \frac{\partial^2 p}{\partial z^2}\bigg|_{\text{at } x,z}. \tag{17.11b}$$

---

[1] This is occasionally called the *geostrophic momentum*.

After subtraction of the unperturbed state, the equations governing the evolution of the displacement are

$$\frac{d^2\Delta x}{dt^2} - f\Delta v = -\frac{1}{\rho_0}\left(\frac{\partial^2 p}{\partial x^2}\right)\Delta x - \frac{1}{\rho_0}\left(\frac{\partial^2 p}{\partial x \partial z}\right)\Delta z \qquad (17.12a)$$

$$\frac{d^2\Delta z}{dt^2} = -\frac{1}{\rho_0}\left(\frac{\partial^2 p}{\partial x \partial z}\right)\Delta x - \frac{1}{\rho_0}\left(\frac{\partial^2 p}{\partial z^2}\right)\Delta z, \qquad (17.12b)$$

in which $\Delta v = -f\Delta x$ according to (17.9). The first equation tells that the force imbalance in the x-direction is due in part to the Coriolis force having changed by $f\Delta v$ and in part to immersion in a new pressure gradient. By Newton's second law, this causes a horizontal acceleration $d^2\Delta x/dt^2$. Likewise, the second equation states that the modified pressure environment causes an imbalance in the vertical. The new neighbors together exert a buoyancy force on our particle, and the latter acquires a vertical acceleration $d^2\Delta z/dt^2$.

Since the equations are now linear, we may seek solutions of the form

$$\Delta x = X \exp(i\omega t), \qquad \Delta z = Z \exp(i\omega t). \qquad (17.13)$$

If the frequency ω is real, the particle oscillates around its original position of equilibrium, and the flow can be characterized as stable. On the contrary, should ω be complex and have a negative imaginary part, the solution includes exponential growth, the particle drifts away from its original position, and the flow is deemed to be unstable.

Substitution on the solution type in the governing equations yields a two-by-two system for the amplitudes X and Z:

$$(F^2 - \omega^2)\Delta x + f M \Delta z = 0 \qquad (17.14a)$$

$$f M \Delta x + (N^2 - \omega^2)\Delta z = 0, \qquad (17.14b)$$

in which we introduced quantities defined in (17.3), (17.4), and (17.5). A nonzero solution exists only if ω obeys

$$(F^2 - \omega^2)(N^2 - \omega^2) = f^2 M^2, \qquad (17.15)$$

of which the ω^2 roots are

$$\omega^2 = \frac{F^2 + N^2 \pm \sqrt{(F^2 - N^2)^2 + 4f^2 M^2}}{2}. \qquad (17.16)$$

The question is whether one or both ω^2 values can be negative, in which case there is at least one ω root with a negative imaginary part.

Before proceeding with the general case, it is instructive to consider two extreme cases. First is the case of stratification without rotation (v is a constant and ρ is a function of z only; $F^2 = fM = 0$ and $N^2 \neq 0$), for which

$$\omega^2 = \frac{N^2 \pm \sqrt{N^4}}{2} = 0 \text{ or } N^2. \qquad (17.17)$$

All ω values are real if $N^2 \geq 0$, which corresponds to a density increasing downward ($d\rho/dz < 0$). Otherwise the fluid is top heavy and overturns. This is gravitational instability first encountered in Section 11.2.

The second extreme case is that of a pure shear (v is a function of x only and ρ is a constant: $F^2 \neq 0$ and $fM = N^2 = 0$), for which

$$\omega^2 = \frac{F^2 \pm \sqrt{F^4}}{2} = 0 \text{ or } F^2. \tag{17.18}$$

All ω values are real if $F^2 \geq 0$, which corresponds to $f(f + \partial v/\partial x) \geq 0$, that is, $(f + \partial v/\partial x)$ of the same sign as f. Should F^2 be negative, the flow mixes horizontally. This is inertial instability in a pure form.

This result is less intuitive than the first and begs for a physical explanation. So let us follow the evolution of a particle displaced laterally (Fig. 17.1). During the displacement Δx, it conserves its geostrophic momentum and sees its velocity v change according to (17.9). As it arrives at a new place, the Coriolis force exerted on the particle no longer matches the pressure gradient force, for the following two reasons: the particle's own velocity has changed and the pressure gradient is different at the new place. Hence the particle is

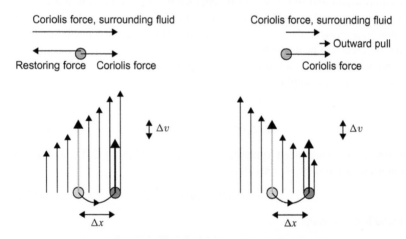

FIGURE 17.1 A fluid particle displaced to the right by a distance $\Delta x > 0$ conserves its geostrophic momentum and sees its velocity drop to $v(x) - f\Delta x$. In the case of the *left panel*, the new particle velocity is weaker than the ambient velocity $v(x + \Delta x) = v(x) + (dv/dx)\Delta x$ at its new place, and its own Coriolis force (from left to right) is insufficient to meet the local pressure-gradient force (from right to left). Consequently, the particle is subjected to a net residual force (from right to left) that pushes it back toward its original position, a restoring force. In the case of the *right panel*, the situation is reversed: the particle's new velocity exceeds the ambient velocity, and its Coriolis force (from left to right) is stronger than the local pressure-gradient force (from right to left), leaving a difference that pulls the particle from left to right and thus further away from its original place. The former case is stable, whereas the latter is unstable.

no longer in geostrophic equilibrium and undergoes a net acceleration in the x-direction. It is either pushed back toward its original location or further accelerated away from there, depending on how the particle's Coriolis force compares to the local pressure-gradient force. As shown in Fig. 17.1, in the northern hemisphere ($f > 0$), stability requires that a particle displaced to the right ($\Delta x > 0$) sees its new velocity $v - f\Delta x$ fall below the surrounding velocity $v + \Delta x(\partial v/\partial x)$ to be pushed back toward its original location, hence stability condition $f + \partial v/\partial x > 0$.

Returning to the general case, we realize that the switch between stability and instability occurs when $\omega^2 = 0$, which according to (17.16) occurs when

$$F^2N^2 = f^2M^2. \qquad (17.19)$$

Around this relation, the signs of the ω^2 roots are as depicted in Fig. 17.2. It is clear from this graph that stability demands three conditions[2]:

$$F^2 \geq 0, \quad N^2 \geq 0, \quad \text{and} \quad F^2N^2 \geq f^2M^2. \qquad (17.20)$$

The third condition is the most intriguing of the group and deserves some physical interpretation. For this, let us take F^2 and N^2 both positive and define the slope (positive downward) of the lines in the vertical (x, z) plane along which

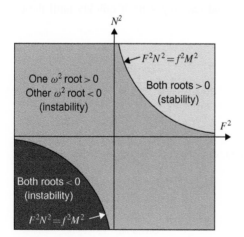

FIGURE 17.2 Stability diagram in the parameter space (F^2, N^2) for the inertial instability of a thermal-wind flow.

---

[2]Note that $F^2N^2 \geq f^2M^2$ alone is not sufficient because it could be obtained with F^2 and N^2 each negative.

the geostrophic momentum $v + fx$ and density ρ are constant:

$$S_{\text{momentum}} = \text{slope of line } v + fx = \text{constant}$$

$$= \frac{\partial(v + fx)/\partial x}{\partial(v + fx)/\partial z} = \frac{F^2}{fM} \tag{17.21}$$

$$S_{\text{density}} = \text{slope of line } \rho = \text{constant}$$

$$= \frac{\partial\rho/\partial x}{\partial\rho/\partial z} = \frac{fM}{N^2}. \tag{17.22}$$

The stability threshold $F^2 N^2 = f^2 M^2$ then corresponds to equal momentum and density slopes. Normally, the velocity varies strongly in x and weakly in z, whereas density behaves in the opposite way, varying more rapidly in z than in x. Typically, therefore, lines of equal geostrophic momentum are steeper than lines of equal density. It turns out that this is the stable case $F^2 N^2 > f^2 M^2$ (left panel of Fig. 17.3).

With increasing thermal wind, momentum lines become less inclined and density lines more steep, until they cross. Beyond this crossing, the steeper lines are the density lines, $F^2 N^2 < f^2 M^2$, and the system is unstable (right panel of Fig. 17.3). Particles quickly drift away from their initial position, and the fluid is vigorously rearranged until it becomes marginally stable, just as a top-heavy fluid ($N^2 < 0$) is gravitationally unstable and becomes mixed until its density is homogenized ($N^2 = 0$). In other words, a situation with density lines steeper than geostrophic lines cannot persist and rearranges itself quickly until these lines coincide.

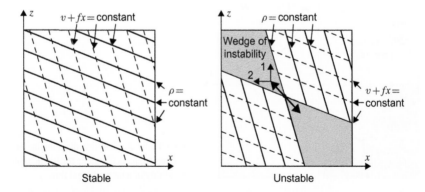

FIGURE 17.3 Left panel: Stability when lines of constant geostrophic momentum $v + fx$ are steeper than lines of constant density ρ. Right panel: Instability when lines of constant geostrophic momentum are less steep than lines of constant density; any particle (such as the one highlighted on the figure) displaced within the wedge is pulled further away by a combination of buoyancy force (1) and geostrophic imbalance (2).

In the unstable regime, it can be shown (see Analytical Problem 17.6) that growing particle displacements lie in the wedge between the momentum and density lines (right panel of Fig. 17.3). This justifies yet another name for the process: *wedge instability*.

17.3 BAROCLINIC INSTABILITY—THE MECHANISM

In thermal-wind balance, geostrophy and hydrostaticity combine to maintain a flow in equilibrium. Assuming that this flow is stable with respect to inertial instability (previous section), the equilibrium is not that of least energy, because a reduction in slope of density surfaces by spreading of the lighter fluid above the heavier fluid would lower the center of gravity and thus the potential energy. Simultaneously, it would also reduce the pressure gradient, its associated geostrophic flow and the kinetic energy of the system. Evidently, the state of rest is that of least energy (minimum potential energy and zero kinetic energy).

In a thermal wind, relaxation of the density distribution and tendency toward the state of rest cannot occur in any direct, spontaneous manner. Such an evolution would require vertical stretching and squeezing of fluid columns, neither of which can occur without alteration of potential vorticity.

Friction is capable of modifying potential vorticity, and under the slow action of friction a state of thermal wind decays, eventually bringing the system to rest. But there is a more rapid process that operates before the influence of friction becomes noticeable.

Vertical stretching and squeezing of fluid parcels is possible under conservation of potential vorticity if relative vorticity comes into play. As we have seen in Section 12.3, a column of stratified fluid that is stretched vertically develops cyclonic relative vorticity, and one that is squeezed acquires anticyclonic vorticity. In a slightly perturbed thermal-wind system, the vertical stretching and squeezing occurring simultaneously at different places generates a pattern of interacting vortices. Under certain conditions, these interactions can increase the initial perturbation, thus forcing the system to evolve away from its original state.

Physically, a partial relaxation of the density surfaces liberates some potential energy, while the concomitant stretching and squeezing creates new relative vorticity. The kinetic energy of the new motions can naturally be provided by the potential energy release. If conditions are favorable, these motions can then contribute to further relaxation of the density field and to stronger vortices. With time, large vortices can be formed at the expense of the original thermal wind. Vortices noticeably increase the amount of velocity shear in the system, greatly enhancing the action of friction. The evolution toward a lower energy level is therefore more effective via the transformation from potential energy into kinetic energy and generation of vortices than by friction acting on the thermal-wind flow.

Let us now investigate how a disturbance of a thermal-wind flow can generate a relative-vorticity distribution favorable to growth. For this purpose, a
two-fluid idealization, as depicted in Fig. 17.4, is sufficient. For the discussion,
let us also ignore the beta effect and align the x-direction with that of the thermal
wind $(U_1 - U_2)$. The interface then slopes upward in the y-direction (middle

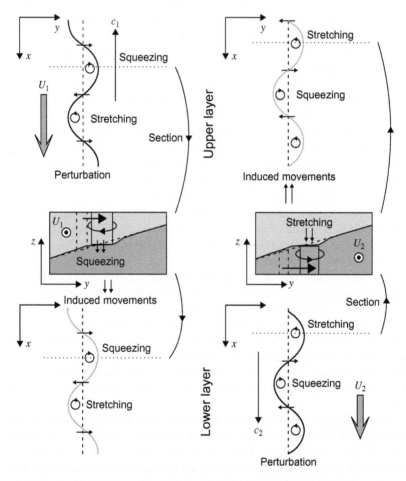

FIGURE 17.4   Patterns of squeezing and stretching caused by lateral displacements in a two-layer
flow in thermal-wind balance. Squeezing generates anticyclonic vorticity (clockwise motion in the
northern hemisphere), while stretching generates cyclonic vorticity (counterclockwise motion in the
northern hemisphere). The flexibility of the density interface distributes the squeezing and stretching
across both layers, and the result is that a cross-flow displacement in the upper layer (upper left of
the figure) causes an accompanying pattern of squeezing and stretching in the lower layer (lower
left of the figure). Vice versa, a cross-flow displacement in the lower layer (lower right of the figure)
causes a similar pattern of squeezing and stretching in the upper layer (upper right of the figure).
Growth occurs when the two sets of patterns mutually reinforce each other.

panels of Fig. 17.4). A perturbation of the upper flow causes some of its parcels to move in the $+y$-direction, into a shallower region (middle-left panel of the figure), and these undergo some vertical squeezing and thus acquire anticyclonic vorticity (clockwise in the figure). Because the density interface is not a rigid bottom but a flexible surface, it deflects slightly, relieving the upper parcels from some squeezing and creating a complementary squeeze in the lower layer. Thus, lower layer parcels, too, develop anticyclonic vorticity at the same location. Note that a lowering of the interface on the shallower side is also in the direction of a decrease of available potential energy.

Elsewhere, the disturbance causes upper layer parcels to move in the opposite direction—that is, toward a deeper region. There, vertical stretching takes place, and, again, because the interface is flexible, this stretching in the upper layer is only partial, the interface rises somewhat, and a complementary stretching occurs in the lower layer. Thus, parcels in both layers develop cyclonic relative vorticity (counterclockwise in the figure). Note that a lifting of the interface on the deeper side is again in the direction of a decrease of available potential energy. If the disturbance has some periodicity, as shown in the figure, alternating positive and negative displacements in the upper layer cause alternating columns of anticyclonic and cyclonic vorticities extending through both layers. Parcels lying between these columns of vortical motion are entrained in the directions marked by the arrows in the figure (upper left and lower left panels), creating subsequent displacements. Because these latter displacements occur not at but between the crests and troughs of the original displacements, they lead not to growth but to a translation of the disturbance.[3] Thus, a pattern of displacement in the upper layer generates a propagating wave. The direction of propagation (c_1 in upper left panel of Fig. 17.4) is opposite to that of the thermal wind ($U_1 - U_2$).

Similarly, cross-flow displacements in the lower layer (right panel of Fig. 17.4) generate patterns of stretching and squeezing in both layers. The difference is that, because of the sloping nature of the density interface, displacements in the $+y$-direction (middle-right panel in the figure) are accompanied by stretching instead of squeezing. Fluid parcels lying between vortical motions take their turn in being displaced, and the pattern again propagates as a wave (c_2 in lower right panel of Fig. 17.4), this time in the direction of the thermal wind.

By itself, each displacement pattern in a layer only generates a vorticity wave, but growth or decay of the whole can take place depending on whether the two separately induced patterns reinforce or negate each other. If the vorticity patterns induced by the upper layer and lower layer displacements are in quadrature with each other, the complementary vortical motions (upper right and upper left sides of Fig. 17.4, respectively) of one set fall at the crests and

---

[3] The mechanism here is identical to that of planetary and topographic waves, discussed in Section 9.6.

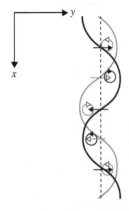

FIGURE 17.5 Interaction of displacement patterns and vortex tubes in the upper layer of a two-layer thermal-wind flow when displacements occur in both layers. The illustration depicts the case of a mutually reinforcing pair of patterns, when the vortical motions of one pattern act to increase the displacements of the other. A similar figure could be drawn for the lower layer, and it can be shown that, if the combination of patterns is self-reinforcing in one layer, it is self-reinforcing in the other layer, too. This is the essence of baroclinic instability.

troughs of the other set, and the ensuing interaction is either favorable or unfavorable to growth. If the spatial phase difference is such that the displacement pattern in one layer is shifted in the direction of the thermal-wind flow in that layer ($U_1 - U_2$ in the upper layer—the opposite in the lower layer), as depicted in Fig. 17.5, the vortical motions of one pattern act to increase the displacements of the other, and the disturbance in each layer amplifies that in the other. The system evolves away from its initial equilibrium.

The preceding description points to the need of a specific phase arrangement between the displacements in the two layers and emphasizes the role of vorticity generation. A further requirement is necessary for growth: the disturbance must have a wavelength that is neither too short nor too long and must be such that the vertical stretching and squeezing effectively generates relative vorticity. To show this in the simplest terms, let us consider the quasi-geostrophic form of the potential vorticity (16.17), on the f-plane:

$$q = \nabla^2 \psi + \frac{\partial}{\partial z} \left(\frac{f^2}{N^2} \frac{\partial \psi}{\partial z} \right), \qquad (17.23)$$

where ψ is the streamfunction, f the Coriolis parameter, N the stratification frequency, and ∇^2 the two-dimensional Laplacian. For a displacement pattern of wavelength L, the first term representing relative vorticity is on the the order of

$$\nabla^2 \psi \sim \frac{\Psi}{L^2}, \qquad (17.24)$$

where the streamfunction scale Ψ is proportional to the amplitude of the displacements. If the height of the system is H, the second term (representing vertical stretching) scales as

$$\frac{\partial}{\partial z} \left(\frac{f^2}{N^2} \frac{\partial \psi}{\partial z} \right) \sim \frac{f^2 \Psi}{N^2 H^2} = \frac{\Psi}{R^2}, \qquad (17.25)$$

where we have defined the deformation radius $R = NH/f$.

Now, if L is much larger than R, the relative vorticity cannot match the vertical stretching as scaled. This implies that vertical stretching will be inhibited, and the displacements in the layers will tend to be in phase in order to reduce squeezing and stretching of fluid parcels in each layer. On the other hand, if L is much shorter than R, relative vorticity dominates potential vorticity. The two layers become uncoupled, and there is insufficient potential energy to feed a growing disturbance. In sum, displacement wavelengths on the order of the deformation radius are the most favorable to growth.

Another requirement for the two layers to interact is related to their relative propagation speed. It is clear that the interaction described above must be persistent in order to allow the positive feedback mechanism to continue. With propagating patterns in each layer this is only possible if the two features are moving at the same speed with respect to a fixed observer. The upper layer perturbation moves with a retrograde speed c_1 with respect to the flow velocity U_1, thus at speed $U_1 - c_1$. The lower layer perturbation moves at a forward speed c_2 with respect to the flow velocity U_2, thus at speed $U_2 + c_2$. For the layers to interact in a persistent matter, we therefore expect

$$U_1 - c_1 = U_2 + c_2. \tag{17.26}$$

Neglecting momentarily the asymmetry in wave propagation due to the beta effect, c_1 and c_2 are wave speeds of topographic waves associated with the sloping interface between the layers. For identical layer thicknesses, symmetry dictates $c_1 = c_2$, and condition (17.26) gives $c_1 = c_2 = (U_1 - U_2)/2$. The absolute propagation speed of the instability is $U_1 - c_1 = U_2 + c_2 = (U_1 + U_2)/2$, the average flow speed.

Because fluctuations are so ubiquitous in nature, an existing flow in thermal-wind balance will continuously be subjected to perturbations. Most of these will have a benign effect because they do not have the proper phase arrangement or a suitable wavelength. But, sooner or later, a perturbation with both favorable phase and wavelength will occur, prompting the system to evolve irreversibly from its equilibrium state. We conclude that flows in thermal-wind balance are intrinsically unstable. Because their instability process depends crucially on a phase shift with height, the fatal wave must have a baroclinic structure. To reflect this fact, the process is termed *baroclinic instability*.

The cyclones and anticyclones of our midlatitude weather are manifestations of the baroclinic instability of the atmospheric jet stream. The person who first analyzed the stability of vertically sheared currents (thermal wind) and who demonstrated the relevance of the instability mechanism to our weather is J. G. Charney.[4] While Charney (1947) performed the stability analysis for a continuously stratified fluid on the beta plane, Eady (1949) did the analysis on the f-plane independently. The comparison between the two theories reveals that

---

[4] For a short biography, see the end of Chapter 16.

the beta effect is a stabilizing influence. Briefly, a change in planetary vorticity (by meridional displacements) is another way to allow vertical stretching
and squeezing while preserving potential vorticity. Relative vorticity is then
no longer as essential and, in some cases, sufficiently suppressed to render the
thermal wind stable to perturbations of all wavelengths.

17.4   LINEAR THEORY OF BAROCLINIC INSTABILITY

Numerous stability analyses have been published since those of Charney and
Eady, exemplifying one aspect or another. Phillips (1954) idealized the continuous vertical stratification to a two-layer system, a case which Pedlosky (1963,
1964) generalized by allowing arbitrary horizontal shear in the basic flow, and
Pedlosky and Thomson (2003) generalized to temporally oscillating basic flow.
Barcilon (1964) studied the influence of friction on baroclinic instability by
including the effect of Ekman layers, whereas Orlanski (1968, 1969) investigated the importance of non-quasi-geostrophic effects and of a bottom slope.
Later, Orlanski and Cox (1973), Gill, Green and Simmons (1974), and Robinson and McWilliams (1974) confirmed that baroclinic instability is the primary
cause of the observed oceanic variability at intermediate scales (tens to hundreds
of kilometers).

Here, we only present one of the simplest mathematical models, taken from
Phillips (1954), because it best exemplifies the mechanism described in the previous section. The fluid consists of two layers with equal thicknesses $H/2$ and
unequal densities ρ_1 on top and ρ_2 below, on the beta plane ($\beta_0 \neq 0$) over a flat
bottom (at $z = 0$) and under a rigid lid (at $z = H$, constant). The fluid is further
assumed to be inviscid ($\mathcal{A}$ and $\nu_E = 0$). The basic flow is taken uniform in the
horizontal and unidirectional but with distinct velocities in each layer:

$$\bar{u}_1 = U_1, \quad \bar{v}_1 = 0 \quad \text{for} \quad \frac{H}{2} \leq z \leq H \tag{17.27a}$$

$$\bar{u}_2 = U_2, \quad \bar{v}_2 = 0 \quad \text{for} \quad 0 \leq z \leq \frac{H}{2}. \tag{17.27b}$$

As we shall see, it is precisely the velocity difference $\Delta U = U_1 - U_2$ between
the two layers, the vertical shear, that causes the instability. For simplicity,
the dynamics are chosen to be quasi-geostrophic, prompting us to introduce
a streamfunction ψ and potential vorticity q that obey (16.16) and (16.17):

$$\frac{\partial q}{\partial t} + J(\psi, q) = 0, \tag{17.28a}$$

$$q = \nabla^2 \psi + \frac{f_0^2}{N^2} \frac{\partial^2 \psi}{\partial z^2} + \beta_0 y. \tag{17.28b}$$

Because of identical layer thicknesses, the stratification frequency may be considered uniform, in agreement with the layered model of Section 12.2, where

equal layer heights corresponded to a uniform stratification. The second equation contains derivatives in z, which must be "discretized" to conform with a two-layer representation. For this, we place values ψ_1 and ψ_2 at midlevel in each layer and two additional values ψ_0 and ψ_3 above and below at equal distances (Fig. 17.6). These latter values fall beyond the boundaries and are defined for the sole purpose of enforcing boundary conditions in the vertical. The flat bottom and rigid lid require zero vertical velocity at those levels, which by virtue of (16.18e) translate into $\partial\psi/\partial z = 0$. In discretized form, the boundary conditions are $\psi_0 = \psi_1$ and $\psi_3 = \psi_2$. The second derivatives may then be approximated as

$$\left.\frac{\partial^2\psi}{\partial z^2}\right|_1 \approx \frac{\psi_0 - 2\psi_1 + \psi_2}{\Delta z^2} = \frac{\psi_1 - 2\psi_1 + \psi_2}{(H/2)^2} = \frac{4(\psi_2 - \psi_1)}{H^2}$$

$$\left.\frac{\partial^2\psi}{\partial z^2}\right|_2 \approx \frac{\psi_1 - 2\psi_2 + \psi_3}{\Delta z^2} = \frac{\psi_1 - 2\psi_2 + \psi_2}{(H/2)^2} = \frac{4(\psi_1 - \psi_2)}{H^2}.$$

In a similar vein, we discretize the stratification frequency:

$$N^2 = -\frac{g}{\rho_0}\frac{d\rho}{dz} \approx -\frac{g}{\rho_0}\frac{\rho_1 - \rho_2}{\Delta z} = +\frac{2g(\rho_2 - \rho_1)}{\rho_0 H} = \frac{2g'}{H}, \tag{17.29}$$

for which we have defined the reduced gravity $g' = g(\rho_2 - \rho_1)/\rho_0$. It is also convenient to introduce the baroclinic radius of deformation as

$$R = \frac{1}{f_0}\sqrt{g'\frac{H_1 H_2}{H_1 + H_2}} = \frac{\sqrt{g'H}}{2f_0}. \tag{17.30}$$

The set of two governing equations can now be written as follows:

$$\frac{\partial q_1}{\partial t} + J(\psi_1, q_1) = 0 \tag{17.31a}$$

$$\frac{\partial q_2}{\partial t} + J(\psi_2, q_2) = 0, \tag{17.31b}$$

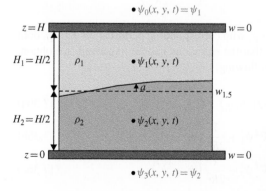

FIGURE 17.6 Representation of the vertical stratification by two layers of uniform density in a quasi-geostrophic model.

where the potential vorticities q_1 and q_2 are expressed in terms of the stream-functions ψ_1 and ψ_2 as

$$q_1 = \nabla^2 \psi_1 + \frac{1}{2R^2}(\psi_2 - \psi_1) + \beta_0 y \tag{17.32a}$$

$$q_2 = \nabla^2 \psi_2 - \frac{1}{2R^2}(\psi_2 - \psi_1) + \beta_0 y. \tag{17.32b}$$

From these quantities, the primary physical variables are derived as follows [see Eqs. (16.18)]

$$u_i = -\frac{\partial \psi_i}{\partial y}, \quad v_i = +\frac{\partial \psi_i}{\partial x} \tag{17.33a}$$

$$w_{1.5} = \frac{2f_0}{N^2 H}\left[\frac{\partial(\psi_2 - \psi_1)}{\partial t} + J(\psi_1, \psi_2)\right] \tag{17.33b}$$

$$p_i' = \rho_0 f_0 \psi_i, \tag{17.33c}$$

where $i = 1, 2$. The vertical displacement a of the density interface between the layers can be obtained from the hydrostatic balance $p_2' = p_1' + (\rho_2 - \rho_1)ga$, which in terms of the streamfunctions yields

$$a = \frac{f_0}{g'}(\psi_2 - \psi_1). \tag{17.34}$$

The same set of equations can be derived from the two-layer model of Section 12.4, in which the quasi-geostrophic approach is applied in each layer, following the perturbation technique of Chapter 16.

The basic-state values of ψ_i and q_i corresponding to (17.27) are

$$\bar{\psi}_1 = -U_1 y, \quad \bar{q}_1 = \left(\beta_0 + \frac{\Delta U}{2R^2}\right)y \tag{17.35a}$$

$$\bar{\psi}_2 = -U_2 y, \quad \bar{q}_2 = \left(\beta_0 - \frac{\Delta U}{2R^2}\right)y. \tag{17.35b}$$

Adding a perturbation ψ_i' to $\bar{\psi}_i$ with corresponding perturbation q_i' to $\bar{q}_i$, both of infinitesimal amplitudes so that the equations can be linearized, we obtain, from (17.31) and (17.32), the following:

$$\frac{\partial q_i'}{\partial t} + J(\bar{\psi}_i, q_i') + J(\psi_i', \bar{q}_i) = 0 \tag{17.36a}$$

$$q_1' = \nabla^2 \psi_1' + \frac{1}{2R^2}(\psi_2' - \psi_1') \tag{17.36b}$$

$$q_2' = \nabla^2 \psi_2' - \frac{1}{2R^2}(\psi_2' - \psi_1'). \tag{17.36c}$$

Elimination of q' and replacement of the basic-flow quantities with (17.35) yield a pair of coupled equations for ψ_1' and ψ_2':

$$\left(\frac{\partial}{\partial t} + U_1 \frac{\partial}{\partial x}\right)\left[\nabla^2 \psi_1' + \frac{1}{2R^2}(\psi_2' - \psi_1')\right] + \left(\beta_0 + \frac{\Delta U}{2R^2}\right)\frac{\partial \psi_1'}{\partial x} = 0, \quad (17.37a)$$

$$\left(\frac{\partial}{\partial t} + U_2 \frac{\partial}{\partial x}\right)\left[\nabla^2 \psi_2' - \frac{1}{2R^2}(\psi_2' - \psi_1')\right] + \left(\beta_0 - \frac{\Delta U}{2R^2}\right)\frac{\partial \psi_2'}{\partial x} = 0. \quad (17.37b)$$

Because both these equations have coefficients independent of x, y, and time, a sinusoidal function in those variables is a solution, and we write

$$\psi_i' = \Re\left[\phi_i e^{i\,(k_x x + k_y y - \omega t)}\right], \quad (17.38)$$

where ϕ_1 and ϕ_2 form a pair of unknowns giving the vertical structure of the wave perturbation, k_x and k_y are horizontal wavenumber components (both taken as real), and ω is the angular frequency. The symbol $\Re$ indicates that only the real part of what follows is retained. Should the frequency ω be complex with a positive imaginary part, exponential growth occurs in time, and the wave is unstable. Substitution in (17.37) leads to algebraic equations for ϕ_1 and ϕ_2:

$$(U_1 - c)\left[-k^2\phi_1 + \frac{1}{2R^2}(\phi_2 - \phi_1)\right] + \left(\beta_0 + \frac{\Delta U}{2R^2}\right)\phi_1 = 0 \quad (17.39a)$$

$$(U_2 - c)\left[-k^2\phi_2 - \frac{1}{2R^2}(\phi_2 - \phi_1)\right] + \left(\beta_0 - \frac{\Delta U}{2R^2}\right)\phi_2 = 0, \quad (17.39b)$$

in which we have defined $c = \omega/k_x$ and $k^2 = k_x^2 + k_y^2$. At this point, it is useful to decompose the ϕ values into barotropic and baroclinic components:

$$\text{Barotropic component:} \quad A = \frac{\phi_1 + \phi_2}{2} \quad (17.40a)$$

$$\text{Baroclinic component:} \quad B = \frac{\phi_1 - \phi_2}{2}. \quad (17.40b)$$

The sum and difference of the preceding equations then yield

$$\left[2\beta_0 - k^2(U_1 + U_2 - 2c)\right]A - k^2\Delta U B = 0 \quad (17.41a)$$

$$\left(\frac{1}{R^2} - k^2\right)\Delta U A + \left[2\beta_0 - \left(k^2 + \frac{1}{R^2}\right)(U_1 + U_2 - 2c)\right]B = 0. \quad (17.41b)$$

Note that a purely barotropic solution ($B = 0$, $A \neq 0$) is possible only in the absence of shear ($\Delta U = 0$), and for a wave speed $c = U - \beta_0/k^2$ easily interpreted as a barotropic planetary wave.

The preceding two equations form a homogeneous system of coupled linear equations for the constants A and B, the solution of which is trivially $A = B = 0$ unless the determinant of the system vanishes. This occurs when

$$R^2 k^2 (1 + R^2 k^2) \left(\frac{U_1 + U_2 - 2c}{\Delta U} \right)^2 - 2 \frac{\beta_0 R^2}{\Delta U} (1 + 2R^2 k^2) \left(\frac{U_1 + U_2 - 2c}{\Delta U} \right)$$
$$+ 4 \frac{\beta_0^2 R^4}{\Delta U^2} + R^2 k^2 (1 - R^2 k^2) = 0, \quad (17.42)$$

the c solution of which can be expressed as

$$\frac{U_1 + U_2 - 2c}{\Delta U} = \frac{\beta_0 R^2}{\Delta U} \frac{2R^2 k^2 + 1}{R^2 k^2 (R^2 k^2 + 1)}$$
$$\pm \frac{1}{R^2 k^2 (R^2 k^2 + 1)} \sqrt{\frac{\beta_0^2 R^4}{\Delta U^2} - R^4 k^4 (1 - R^4 k^4)}. \quad (17.43)$$

It is clear from this equation that the phase speed c of the wave is real as long as the quantity under the square root is positive, that is, as long as the wavenumber k satisfies the condition

$$R^4 k^4 (1 - R^4 k^4) \leq \left(\frac{\beta_0 R^2}{\Delta U} \right)^2. \quad (17.44)$$

The function $R^4 k^4 (1 - R^4 k^4)$ reaches a maximum of $1/4$ for $Rk = 1/2^{1/4} = 0.841$ (Fig. 17.7), and therefore, the condition is met for a perturbation of any wavenumber as long as

$$|\Delta U| \leq 2\beta_0 R^2 = \frac{\beta_0 g' H}{2 f_0^2}. \quad (17.45)$$

In other words, the system is stable to all small perturbations when the velocity shear ΔU is sufficiently weak not to exceed $2\beta_0 R^2$. Put another way, shear is destabilizing because the greater is ΔU, the higher is the likelihood that the threshold value will be exceeded. In contrast, the beta effect is stabilizing because the greater is β_0, the more generous is the threshold.

When the velocity shear exceeds the threshold value, condition (17.45) is not met, and not all wavenumbers satisfy condition (17.44). Perturbations of wavenumber $k = \sqrt{k_x^2 + k_y^2}$ within the interval $k_{min} < k < k_{max}$ are unstable,

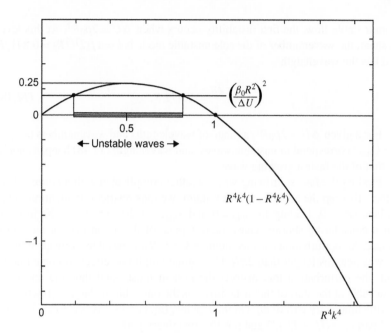

FIGURE 17.7   Instability interval for two-layer baroclinic instability. Small-amplitude waves with wavenumber k falling in the hatched interval are unstable and grow in time.

where

$$k_{min} = \left(\frac{1 - \sqrt{1 - 4\beta_0^2 R^4 / \Delta U^2}}{2R^4} \right)^{1/4} \qquad (17.46a)$$

$$k_{max} = \left(\frac{1 + \sqrt{1 - 4\beta_0^2 R^4 / \Delta U^2}}{2R^4} \right)^{1/4} . \qquad (17.46b)$$

Note that unstable waves not only grow but also propagate in time. According to (17.43), the real part of the wave speed is

$$\Re(c) = \frac{U_1 + U_2}{2} - \frac{\beta_0}{2k^2} \frac{1 + 2R^2 k^2}{1 + R^2 k^2} \qquad (17.47)$$

when c is complex, and thus, the zonal propagation speed is $(U_1 + U_2)/2$, or the average velocity of the basic flow, minus a (westward) planetary wave speed.

From Fig. 17.7 or from Eq. (17.45), we see that instability on the beta plane can only occur for a sufficiently large shear ΔU. With increasing ΔU starting from a stable flow, the first instability occurs when $\Delta U = 2\beta_0 R^2$. At this level of shear, the wavenumber of the sole unstable mode is $k = 1/(2^{1/4}R) = 0.841/R$ and has the wavelength

$$\lambda = \frac{2\pi}{k} = 7.472\, R. \tag{17.48}$$

For a given $\Delta U > 2\beta_0 R^2$, a range of wavelengths and wavenumbers ($k_{min} < k < k_{max}$) correspond to unstable waves, and wavelength (17.48) happens not to be that of the fastest growing wave.

Finding the fastest growing wave is rather complicated with a nonzero beta effect. To keep the analysis to a minimum, we thus restrict our attention to the f-plane ($\beta_0 = 0$, leading to $k_{min} = 0$ and $k_{max} = 1/R$), which can be justified when considering shorter scales, more typical of the ocean than of the atmosphere. All perturbations of wavenumber $k < 1/R$ are unstable, corresponding to all wavelengths longer than $2\pi R$. Until finite-amplitude effects become important, the perturbation that distorts the system most—and thus the one most noticeable at the start of the instability—is the one with the largest growth rate, ω_i (the imaginary part of ω). On the f-plane, Eq. (17.43) gives $c = c_r + i c_i$ with real part $c_r = (U_1 + U_2)/2$ and positive imaginary part

$$c_i = \frac{\Delta U}{2} \sqrt{\frac{1 - k^2 R^2}{1 + k^2 R^2}} \qquad (kR < 1) \tag{17.49}$$

The growth rate is $\omega_i = k_x c_i$ and reaches a maximum with respect to k_x and k_y for

$$k_x = \frac{\sqrt{\sqrt{2} - 1}}{R} = \frac{0.644}{R}, \qquad k_y = 0. \tag{17.50}$$

The wavelength of the fastest growing mode is $\lambda = 9.763\, R$.

It is interesting at this point to return to our initial considerations (Section 17.3) and to confirm them with the preceding solution. First and foremost, the fact that both the critical wavelength for instability ($2\pi R$) and the wavelength of the fastest growing perturbation (9.763 R) are proportional to R validates the argument that self-amplification requires a scale on the order of the deformation radius. Physically, it also verifies that the instability process involves a rearrangement of potential vorticity between relative vorticity and vertical stretching.

The necessary phase relationship between the transverse displacements of the upper and lower fluids can be checked as follows. We define the transverse displacement d, one for each layer, in terms of the meridional velocity by

$$v' = \frac{\partial d}{\partial t} + \bar{u}\,\frac{\partial d}{\partial x}, \tag{17.51}$$

after linearization. Expressing v' in terms of the streamfunction perturbation ($v' = \partial\psi'/\partial x$) and implementing the wave form $d_i = \Re[D_i \exp{i}(k_x x + k_y y - \omega t)]$, we then obtain

$$D_i = \frac{\phi_i}{U_i - c}, \tag{17.52}$$

from which we can deduce the ratio of transverse displacements in the upper and lower layers:

$$\frac{D_1}{D_2} = \frac{U_2 - c}{U_1 - c}\frac{A + B}{A - B}. \tag{17.53}$$

For the fastest growing wave on the f-plane in the case $U_1 > U_2$ (i.e., $\Delta U > 0$), the wavenumber is $k = 0.644/R$, and the ratio of displacements is found to be

$$\frac{D_1}{D_2} = 0.66 - i0.75$$

$$= \cos(49°) + i\sin(-49°). \tag{17.54}$$

Physically, the negative 49° angle corresponds to an advance (in the direction of the basic flow) of the top displacement over that in the bottom layer. The shift is not the 90° phase quadrature that was expected, but it is in the direction anticipated from the simple physical argument of the previous section.

From an observational point of view, however, the interest lies in the pressure field, which is proportional to the streamfunction [see (17.33c)]. Within an arbitrary multiplicative constant, which the linear theory is unable to determine, the pressure field associated with the fastest growing perturbation can be expressed in terms of the vertical structure of the streamfunction perturbation:

$$\frac{\phi_1}{\phi_2} = \frac{A + B}{A - B}$$

$$= \cos(66°) + i\sin(66°). \tag{17.55}$$

From this, we conclude that the crests and troughs of the pressure pattern at the top lag those of the bottom pattern by a fifth to sixth of a wavelength.

Finally, the maximum growth rate is

$$\omega_i = k_x c_i = \frac{\Delta U}{R}\frac{\sqrt{2} - 1}{2}. \tag{17.56}$$

If we assume the Rossby number Ro to be small, which must be by virtue of the quasi-geostrophic approximation, we find $\omega_i \lesssim 0.2 f Ro \ll f$. The timescale characteristic of the growth is therefore much longer than f^{-1} so that the solution is consistent within the quasi-geostrophic theory.

To conclude we briefly examine the confining effect of boundaries, taken as vertical walls at $y = 0$ and $y = L$. The basic flow, in the x-direction, satisfies the impermeability condition and needs no adjustment. As for the perturbation flow, the normal velocity must vanish at each boundary, which according to (17.33a) demands $\partial \psi / \partial x = 0$ at $y = 0$ and $y = L$. With ψ given by (17.38), both boundary conditions can be simultaneously met only when the sinusoidal structure in y of the streamfunction is such that $k_y L = n\pi$ with $n = 1, 2, \ldots$ In this case

$$k^2 R^2 = \left(k_x^2 + \frac{n^2 \pi^2}{L^2} \right) R^2. \tag{17.57}$$

Since according to (17.46), instability is only possible for $k^2 R^2 < 1$, the boundaries must be sufficiently distant from each other to satisfy

$$\frac{R^2}{L^2} \leq \frac{1}{\pi^2}. \tag{17.58}$$

This simply means that the domain must be wide enough to accommodate the instability, otherwise it cannot develop.

Interestingly, the preceding inequality, (17.58), shows that the Burger number associated with the basic flow of width L should not exceed 0.1, and this implies, according to (16.34), that the energy in the basic flow must be predominantly in the form of available potential energy in the basic flow. This reserve of potential energy feeds the growth of the instability. Note that, for the perturbation itself, the Burger number is always on the order of one, since the scale of the instability is the deformation radius.

17.5   HEAT TRANSPORT

The qualitative arguments developed in Section 17.3 revolved around the idea that if a flow in thermal-wind equilibrium is unstable, it will seek a lower level of energy by relaxation of density surfaces toward simple gravitational equilibrium. If we now think of the atmosphere, where the heavier fluid is colder air and the lighter fluid warmer air, relaxation implies a flow of warm air spilling over the colder air ($+y$-direction in Fig. 17.4) and of cold air intruding under the warmer air ($-y$-direction in Fig. 17.4). In other words, we expect a net heat flux and, because the atmospheric temperature typically increases toward the equator, a poleward heat flux. Let us examine what the preceding linear theory predicts.

The vertically integrated heat flux in the north-south direction (y-direction) per unit length of east-west direction (x-direction) is defined as

$$q = \rho_0 C_p \int_0^H \overline{vT} \, dz, \tag{17.59}$$

where C_p is the heat capacity of the fluid at constant pressure ($1005 \text{ J kg}^{-1} \text{ K}^{-1}$ for dry air, $4186 \text{ J kg}^{-1} \text{ K}^{-1}$ for seawater), T is temperature, and the overbar indicates an average over a wavelength in the x-direction. In the two-layer representation of Fig. 17.6, the vertical integration is straightforward:

$$q = \rho_0 C_p \left[\overline{v_2 T_2 (H_2 + a)} + \overline{v_1 T_1 (H_1 - a)} \right]$$
$$= \rho_0 C_p \left[\overline{v_2 a}\, T_2 - \overline{v_1 a}\, T_1 \right] \tag{17.60}$$

since the temperature is uniform within each layer and the integral over a wavelength yields $\overline{v_1} = 0$ and $\overline{v_2} = 0$. Using $v_i = v_i' = \partial \psi_i' / \partial x$ and $a = f_0 (\psi_2' - \psi_1') / g'$ and then exploiting $\overline{\partial \psi_i^2 / \partial x} = 0$ and $\overline{\partial (\psi_2 \psi_1) / \partial x} = 0$, we have successively

$$q = \frac{\rho_0 C_p f_0}{g'} \left[T_2 \overline{\frac{\partial \psi_2'}{\partial x} (\psi_2' - \psi_1')} - T_1 \overline{\frac{\partial \psi_1'}{\partial x} (\psi_2' - \psi_1')} \right]$$
$$= \frac{\rho_0 C_p f_0}{g'} \left[-T_2 \overline{\psi_1' \frac{\partial \psi_2'}{\partial x}} - T_1 \overline{\psi_2' \frac{\partial \psi_1'}{\partial x}} \right]$$
$$= \frac{\rho_0 C_p f_0}{g'} (T_1 - T_2) \overline{\psi_1' \frac{\partial \psi_2'}{\partial x}}. \tag{17.61}$$

Some rather lengthy algebra using the periodic structure (17.38) and the modal decomposition (17.40) successively provides

$$\overline{\psi_1' \frac{\partial \psi_2'}{\partial x}} = \frac{k_x}{2} [\Im(\phi_1) \Re(\phi_2) - \Re(\phi_1) \Im(\phi_2)] e^{2\Im(\omega)t}$$
$$= k_x [\Re(A) \Im(B) - \Im(A) \Re(B)] \, e^{2\Im(\omega)t}.$$

The real and imaginary parts of Eq. (17.41a) are

$$\left[2\beta_0 - k^2 (U_1 + U_2 - 2c_r) \right] \Re(A) - k^2 c_i \, \Im(A) = k^2 \Delta U \, \Re(B)$$
$$\left[2\beta_0 - k^2 (U_1 + U_2 - 2c_r) \right] \Im(A) + k^2 c_i \, \Re(A) = k^2 \Delta U \, \Im(B),$$

where c_r and c_i stand, respectively, for the real and imaginary parts of c. From these relations, it follows that

$$\Re(A)\Im(B) - \Im(A)\Re(B) = \frac{c_i}{\Delta U} |A|^2. \tag{17.62}$$

Putting it altogether, we finally obtain an expression for the heat flux

$$q = \frac{\rho_0 C_p f_0 \Im(\omega)}{g' \Delta U} (T_1 - T_2) |A|^2 e^{2\Im(\omega)t}. \tag{17.63}$$

It is clear from this expression that the heat flux is nonzero only when the wave is unstable (imaginary part of $\omega \neq 0$) and is positive, as anticipated by the earlier physical arguments. In the atmospheric case, this means that the heat flux is poleward.

Because the earth is heated in the tropics and cooled at high latitudes, the global heat budget requires a net poleward heat flux in each hemisphere. The flux is carried by both atmosphere and ocean. In the atmosphere, higher temperatures in the tropics and lower temperatures at high latitudes maintain an overall thermal wind system, which is baroclinically unstable. Vortices emerge on the scale of the baroclinic radius of deformation ($R \sim$ 1000 km), which carry the heat poleward and tend to relax the thermal-wind structure. The latter, however, is maintained by continuous heating in the tropics and cooling at high latitudes. As a consequence, the cyclones and anticyclones of our weather are the primary agents of meridional heat transfer in the atmosphere. Without baroclinic instability, they would not exist, and weather forecasting would be a much simpler task, but the tropical regions would be much hotter and the polar regions, much colder. Also, the dominance of zonal winds would preclude efficient mixing across latitudes, exacerbating certain problems by severely limiting, for example, the spread of volcanic ash. Moreover, less atmospheric variability would imply greatly reduced temperature and moisture contrasts and thus much less precipitation at midlatitudes. All in all, we must concede that baroclinic instability in our atmosphere is highly beneficial.

In the ocean, the situation is quite different. The presence of meridional boundaries prevents thermal-wind type currents from encircling the globe, and ocean circulation consists of large-scale gyres (Chapter 20). The meridional branches of these gyres, especially the western boundary currents (Gulf Stream in the North Atlantic, Kuroshio in North Pacific), are the main conveyers of heat toward high latitudes (e.g., Siedler, Church & Gould, 2001). This greatly reduces the need for poleward heat transfer by eddies. Baroclinic instability is active in regions of strong currents, such as the Antarctic Circumpolar Current, Gulf Stream, and Kuroshio extensions, but the eddies so created transport little net heat across latitudes. One should also remember here that the baroclinic radius of deformation is significantly shorter in the ocean than in the atmosphere, with the consequence that the aggregate effect of eddies in the ocean is more regional than planetary.

17.6 BULK CRITERIA

The theory exposed in Section 17.4 is admittedly a very simplified version of baroclinic-instability physics. Since it is not our purpose here to review the many advanced analyses that have been published over the years since the pioneering studies of Charney, Eady, and Phillips (the interested reader will find a survey in the book of Pedlosky, 1987), we will once again turn to integral relations, from which some necessary but not sufficient criteria for instability can be derived. We already used this approach in the study of horizontally sheared currents in homogeneous fluids (Section 10.2) and of vertically sheared currents in nonrotating stratified fluids (Section 14.2). Although a general presentation that would encompass the preceding two situations and baroclinic instability could be formulated, it is most instructive to emphasize the conditions necessary for baroclinic instability by basing the analysis on the quasi-geostrophic equation.[5] The following derivations are based on the work of Charney and Stern (1962).

We start again with Eqs. (17.28) but this time retain continuous variation in the vertical albeit with uniform stratification frequency. Adding a small perturbation to a basic zonal flow $\bar{u}(y, z)$ possessing both horizontal and vertical shear, we obtain

$$\frac{\partial q'}{\partial t} + J(\bar{\psi}, q') + J(\psi', \bar{q}) = 0 \tag{17.64a}$$

$$q' = \nabla^2 \psi' + \frac{f_0^2}{N^2} \frac{\partial^2 \psi'}{\partial z^2}, \tag{17.64b}$$

where $\bar{\psi}(y, z)$ is the streamfunction associated with the basic zonal flow ($\bar{u} = -\partial\bar{\psi}/\partial y$), and the basic potential vorticity is related to it by

$$\bar{q} = \frac{\partial^2 \bar{\psi}}{\partial y^2} + \frac{f_0^2}{N^2} \frac{\partial^2 \bar{\psi}}{\partial z^2} + \beta_0 y. \tag{17.65}$$

Substitution of (17.64b) and (17.65) into (17.64a) yields a single equation for the streamfunction perturbation ψ', which includes nonconstant coefficients depending on the basic flow structure via $\bar{\psi}$ and $\bar{q}$. Because those coefficients depend only on y and z, a waveform solution in x and time can be sought: $\psi'(x, y, z, t) = \Re[\phi(y, z)\exp(ik_x(x - ct))]$. The amplitude function $\phi(y, z)$ must obey

$$\frac{\partial^2 \phi}{\partial y^2} + \frac{f_0^2}{N^2} \frac{\partial^2 \phi}{\partial z^2} + \left(\frac{1}{\bar{u} - c}\frac{\partial \bar{q}}{\partial y} - k_x^2\right)\phi = 0, \tag{17.66}$$

with $\bar{q}$ defined in (17.65).

The upper and lower boundaries are once again taken as rigid horizontal surfaces, where the vertical velocity must vanish. According to (16.18e), this implies after splitting between basic flow and perturbation and linearizing:

$$(\bar{u} - c)\frac{\partial \phi}{\partial z} - \frac{\partial \bar{u}}{\partial z}\phi = 0 \quad \text{at} \quad z = 0, H. \tag{17.67}$$

In the meridional direction, we idealize the domain to a channel of width L between two vertical walls, where the meridional velocity $v' = \partial\psi'/\partial x$ vanishes. We thus impose

$$\phi = 0 \quad \text{at} \quad y = 0, L. \tag{17.68}$$

---

[5] Actually, this equation eliminates the Kelvin–Helmholtz instability but not barotropic instability.

Multiplying (17.66) by the complex conjugate ϕ^* of ϕ, integrating over the meridional and vertical extents of the domain, performing integrations by parts, and using the preceding boundary conditions, we obtain

$$
\int_0^H \int_0^L \left[\left| \frac{\partial \phi}{\partial y} \right|^2 + \frac{f_0^2}{N^2} \left| \frac{\partial \phi}{\partial z} \right|^2 + k_x^2 |\phi|^2 \right] dy \, dz
$$

$$
= \int_0^H \int_0^L \frac{1}{\bar{u} - c} \frac{\partial \bar{q}}{\partial y} |\phi|^2 \, dy \, dz
$$

$$
+ \int_0^L \left[\frac{f_0^2}{N^2} \frac{1}{\bar{u} - c} \frac{\partial \bar{u}}{\partial z} |\phi|^2 \right]_0^H dy. \tag{17.69}
$$

The imaginary part of this equation is

$$
c_i \left\{ \int_0^H \int_0^L \frac{|\phi|^2}{|\bar{u} - c|^2} \frac{\partial \bar{q}}{\partial y} \, dy \, dz + \int_0^L \left[\frac{f_0^2}{N^2} \frac{|\phi|^2}{|\bar{u} - c|^2} \frac{\partial \bar{u}}{\partial z} \right]_0^H dy \right\} = 0. \tag{17.70}
$$

A necessary condition for instability is that c_i not be zero (so that the disturbance grows in time). According to (17.70), this implies that the quantity within braces must vanish, and therefore conditions for instability are

1. $\partial \bar{q}/\partial y$ changes sign in the domain, or
2. the sign of $\partial \bar{q}/\partial y$ is opposite to that of $\partial \bar{u}/\partial z$ at the top, or
3. the sign of $\partial \bar{q}/\partial y$ is the same as that of $\partial \bar{u}/\partial z$ at the bottom.

A sufficient condition for stability is that none of the above three conditions is met.

Before proceeding, it is worth applying this result to the case of a uniform shear flow $\bar{u} = Uz/H$ in the absence of the beta effect ($\beta_0 = 0$). We then have $\bar{q} = 0$ and $\partial \bar{u}/\partial z = U/H$, reducing (17.70) to

$$
c_i \int_0^L \frac{f_0^2 U}{N^2 H} \left[\frac{|\phi(y, H)|^2}{|U - c|^2} - \frac{|\phi(y, 0)|^2}{|c|^2} \right] dy = 0, \tag{17.71}
$$

in which the integral is obviously not sign definite. Stability cannot be guaranteed, and this flow is unstable (Eady, 1949). Had we instead chosen a weak flow field with no vertical shear at the boundaries [e.g., $\bar{u}(z) = U(3z^2/H^2 - 2z^3/H^3)$] and on the beta plane ($\partial \bar{q}/\partial y \simeq \beta_0$), we would have concluded (after much lengthier mathematics) that this flow is stable to all perturbations. This points to the sensitivity of baroclinic instability to the structure of the basic flow field.

Another application of (17.70) is to laterally sheared but vertically uniform flow, $\bar{u}(y)$. Then, the potential-vorticity gradient is $d\bar{q}/dy = \beta_0 - d^2\bar{u}/dy^2$, and

(17.70) reduces to

$$
c_i \left[H \int_0^L \frac{|\phi|^2}{|\bar{u} - c|^2} \left(\beta_0 - \frac{d^2 \bar{u}}{dy^2} \right) dy \right] = 0. \tag{17.72}
$$

Here, we recover the result of barotropic instability obtained in Section 10.2 [see Eq. (10.13)] and conclude that the instability conditions stated above include both barotropic and baroclinic instability criteria. Put another way, barotropic and baroclinic instabilities are two end members of a more general barotropic–baroclinic mixed instability.

Charney and Stern (1962) explored the case when $\partial \bar{u}/\partial z$ vanishes at both upper and lower boundaries by assuming a vanishing thermal-wind there (e.g., uniform temperature) and/or taking the limits $H \to \infty$, $\bar{u}(H) \to 0$. Of (17.70), only the first integral remains, and the necessary condition for instability is that $\partial \bar{q}/\partial y$ vanishes somewhere in the domain, a statement identical in form to—but differing in content from—the barotropic-instability criterion of Section 10.2.

According to Gill et al. (1974), the presence of a bottom slope in the meridional direction modifies the last of the three conditions as follows:

3. The sign of $\partial \bar{q}/\partial y$ is the same as that of $\partial \bar{u}/\partial z - (N^2/f_0)db/dy$ at the bottom $z = b(y)$.

Therefore, a bottom slope may be either stabilizing or destabilizing. It is generally a stabilizing factor if it creates an ambient potential-vorticity gradient in the same direction as the beta effect (i.e., shallower fluid toward higher latitudes; see Fig. 9.6) and a destabilizing factor otherwise. However, the theory fails to take into account the zonal topographic gradients that are more common on Earth (e.g., the Rocky Mountains in North America for the atmosphere and the Mid-Atlantic Ridge along the North Atlantic for the ocean).

There exist a number of other studies in baroclinic instability. The interested reader is referred to Gill (1982, Chapter 13), Pedlosky (1987, Chapter 7), and Vallis (2006, Chapters 6 and 9).

17.7   FINITE-AMPLITUDE DEVELOPMENT

Once the instability is underway, exponential growth eventually leads to perturbations whose amplitudes are no longer small compared to the size of the basic flow. Linear theory then ceases to be valid, and we must deal with the nonlinear equations, resorting as usual to numerical methods. The task in front of us is solving Eqs. (17.32) and (17.31) using the quasi-geostrophic approximation. Since these equations are similar to those of the two-dimensional QG model of Section 16.7, we may start with the discretization of the latter and adapt it for our present purpose. In addition, for the study of baroclinic instability, it is useful to exploit the fact that, because the basic flow is stationary, only the perturbation variables need to be updated. So, we solve

$$\frac{\partial q_1'}{\partial t} + J(\psi_1, q_1) = 0, \tag{17.73a}$$

$$\frac{\partial q_2'}{\partial t} + J(\psi_2, q_2) = 0. \tag{17.73b}$$

Note that the Jacobian operator J involves the streamfunction and potential vorticity of the flow consisting of both basic flow and perturbation. Updating only perturbation components, therefore, does not involve any linearization. Equations (17.73) are readily discretized if we use the Arakawa Jacobian of Section 16.7. For time stepping, the simplest approach is an explicit scheme such as the predictor-corrector method, so that we can easily calculate the potential vorticities at time level $n+1$ knowing both streamfunction and vorticity at time level n.

Once perturbations q_1' and q_2' are obtained at the new time level, we have to invert a pair of Poisson equations (17.36b) and (17.36c) to calculate the respective streamfunction at the same moment, in preparation of the next time step. Solving these Poisson equations, however, is more complicated than in the two-dimensional case of Chapter 16 because we are now in the presence of two coupled equations. To overcome the added complexity, we generalize the iterative Gauss–Seidel approach, working jointly on ψ_1' and ψ_2'. Omitting the $'$ and referring to an iteration by superscript $^{(k+1)}$, we iterate in tandem as follows:

$$\psi_1^{(k+1)} = \psi_1^{(k)} + \alpha \left[\nabla^2 \psi_1 - q_1 + \frac{(\psi_2^{(k)} - \psi_1^{(k)})}{2R^2} \right]$$

$$\psi_2^{(k+1)} = \psi_2^{(k)} + \alpha \left[\nabla^2 \psi_2 - q_2 - \frac{(\psi_2^{(k)} - \psi_1^{(k+1)})}{2R^2} \right].$$

The spatial operator ∇^2 is calculated using the most recent values of ψ available on the discrete grid, and the iterations are performed at a frozen time level.[6] The parameter α contains discretization constants and over-relaxation parameters. This approach is easily implemented and can be generalized to more than two layers.

For the present two-layer model, another option is to decouple the equations by decomposing q_1 and q_2 into their barotropic and baroclinic parts. The sum and difference of (17.73a) and (17.73b) then yield two uncoupled equations that can be solved independently, possibly with different iteration schemes.

Once the iterations have converged, the two perturbation streamfunctions are known at the new time level, and the total (basic flow + perturbation)

---

[6] Do not confuse the time index ^{n} with the iteration index $^{(k)}$.

streamfunction and potential vorticity can be evaluated. The Arakawa Jacobian is then recalculated, and the time stepping continued.

To initialize the whole procedure, it is sufficient to provide initial conditions on either vorticity or streamfunction. If the streamfunction is provided as the initial condition, initial potential vorticity can be deduced from the streamfunction by definitions (17.36b)–(17.36c), and time stepping can start. If vorticity is provided as the initial condition, we have to begin with a solution (inversion) of the Poisson equations before time stepping can begin.

We also have to provide adequate boundary conditions. In the x-direction, the length of the domain is dictated by the wavelength of the perturbation whose stability is being investigated. Periodic conditions are then readily applied to both streamfunction and vorticity. In the y-direction, we assume a channel configuration with zonal boundaries at $y = 0$ and $y = L$. The condition of zero normal velocity forces the streamfunction to be uniform along each wall at a given moment. At $t = 0$, the values of the constants are dictated by the initial condition of the still unperturbed flow. In analytical studies, these initial constants are then kept fixed over time. This is justified by the fact that the analytical streamfunction perturbation possesses a wave structure in the x-direction, which demands that the amplitude of the wave be zero on the boundary.

With the nonlinear equations, the situation is different because all we need to ensure is that the instantaneous streamfunction, combining basic flow and perturbation, is constant along the wall. The value of the constant, however, is allowed to change with time. The physical reason lies in the possibility that the interface between the two layers flattens out, in which case the cumulated flow across the channel weakens in each layer. The problem we face is similar to the one encountered in Section 7.7, where values of the streamfunction on the periphery of islands had to be determined depending on the flow evolution itself. For the correct and subtle way of specifying such boundary conditions in a quasi-geostrophic model, we refer to McWilliams (1977). Here, we follow the simpler approach of Phillips (1954), which is appropriate to our configuration. The original equation for the velocity component u in an isopycnal model can be cast as

$$\frac{\partial u}{\partial t} + \frac{1}{2}\frac{\partial u^2}{\partial x} + v\frac{\partial u}{\partial y} = fv - \frac{1}{\rho_0}\frac{\partial P}{\partial x}. \tag{17.74}$$

For simplicity, there is no need to indicate to which layer we refer because the same type of equation holds for each layer. We can integrate the preceding equation over the wavelength λ in the x-direction. By virtue of periodicity, the second term of the left-hand side as well as the second term of the right-hand side will not contribute to the integral. As long as the wall coincides with the same x-direction, impermeability requires $v = 0$, and the third term of the left-hand side and the first term of the right-hand side vanish. All that remains is the integration of the first term, which must vanish on its own. In terms of the streamfunction ($u = -\partial\psi/\partial y$), we have

$$\int_0^\lambda \frac{\partial^2 \psi}{\partial y \partial t} \, dx = 0 \text{ on a wall parallel the } x\text{-axis.} \qquad (17.75)$$

In conclusion, the constant value to be ascribed to the streamfunction along an impermeable boundary is to be obtained from (17.75).

The implementation of this new boundary condition can be shown for the boundary at $y = 0$. To determine the value of $\psi_{i,1} = C_1$ along the boundary at the new time step, we discretize Eq. (17.75) as

$$C_1^{n+1} = C_1^0 + \frac{1}{m} \sum_{i=1}^m \left(\psi_{i,2}^{n+1} - \psi_{i,2}^0 \right) \qquad (17.76)$$

in which the sum covers the grid points along the wall. Hence, the value to be prescribed along the boundary depends on the yet unknown values $\psi_{i,2}^{n+1}$ in the interior of the domain, themselves depending on the boundary conditions. This circular dependence can be resolved by wrapping the evaluation of constant C_1 into the iterations of the Poisson solver, updating ψ not only in the interior but also on the boundary during the Gauss–Seidel iterations.

If the Poisson solver is applied to the perturbation ψ' only, then the boundary condition formulation is simpler: the initial value of $\int_0^\lambda \partial\psi/\partial y \, dx$ is fixed by the basic flow, and $\int_0^\lambda \partial\psi'/\partial y \, dx$ must be held at zero at all times.

The outlined numerical algorithm was implemented in `baroclinic.m` and is now used to simulate baroclinic instability as presented in Section 17.3 extended into the nonlinear regime. We show results of such a model simulation at different moments of the evolution (Fig. 17.8). For simplicity, we take $U_2 = -U_1$, so that the linear perturbation theory predicts a wave that amplifies in place.

Initially, the evolution follows the theoretical prediction: The perturbation grows without displacement, maintaining the expected phase shift between layers. After a while, however, the perturbation reaches a mature stage when its amplitude is comparable to the strength of the basic flow. We are then in the nonlinear regime. Now, the phase shift between layers diminishes, and the interface between layers relaxes. Together these changes indicate a barotropization of the flow, (Fig. 17.9) that is, the two layers begin to act more and more as if they were making a single layer. This is confirmed by the structure of the potential vorticity, which becomes almost vertically uniform by the end of the simulation. We conclude that baroclinic instability releases available potential energy and uses it to spin eddies and to strengthen the barotropic component of the flow.

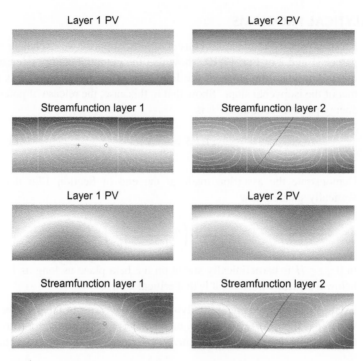

FIGURE 17.8 Evolution of a perturbed thermal wind. For each of the two snapshots during the evolution of the system, four panels show, in order of presentation: the upper layer potential vorticity, the lower layer potential vorticity, the upper layer streamfunction, and lower layer streamfunction. On the latter two, the contours depict the streamfunction perturbation. On the upper layer streamfunction plot, symbols have been added to represent the location of the maximum ψ' value of the upper (cross) and lower layer (circle). On the lower layer streamfunction plot, a line depicting the x-averaged position of the interface is shown. White separates the positive from negative values. For perturbations, the contoured values change over time. To view the evolution as computer animation, the reader should run `baroclinic.m` or look at the video provided with the files. A color version can be found online at http://booksite.academicpress.com/9780120887590/

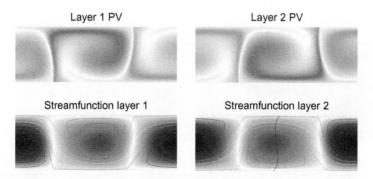

FIGURE 17.9 Further evolution of the instability showing barotropization of the flow. A color version can be found online at http://booksite.academicpress.com/9780120887590/

ANALYTICAL PROBLEMS

17.1. Irrespectively of momentum considerations, suppose we exchange fluid parcels 1 and 2 of Fig. 17.10. Show that the potential energy released is maximum when the slope of the line connecting points 1 and 2 is half that of the isopycnal slope. Show that in this case, the release of potential energy per unit volume ΔPE is

$$\Delta PE = \frac{\rho_0}{4} N^2 L^2 \qquad (17.77)$$

17.2. Demonstrate the assertion made at the end of Section 17.6 that the vertically sheared flow

$$\bar{u}(z) = U \left(3 \frac{z^2}{H^2} - 2 \frac{z^3}{H^3} \right)$$

in $0 \leq z \leq H$ is baroclinically stable on the beta plane as long as U falls below a critical value. What is that critical value?

17.3. Establish an energy budget involving quadratic forms of the perturbation variables. Then, derive an energy budget involving quadratic forms of the total velocity. Identify types of energy and the exchanges between them.

17.4. Compare the magnitudes of the potential and kinetic energies of the most unstable wave described in Section 17.4.

17.5. Assuming a general initial profile of the interface $a = \bar{a}(y)$, with zero total transport in each column, establish linearized equations a small perturbation of this situation must obey. Verify that for a linear interface, you retrieve formulation (17.37) with $U_1 = -U_2$.

17.6. Prove the assertion made at the end of Section 17.2 that the unstable regime of the wedge instability corresponds to particles moving along

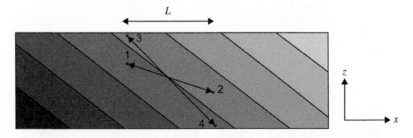

FIGURE 17.10 Exchange of fluid parcels across a system in thermal-wind balance. Exchange between parcels 1 and 2 leads to a release of potential energy, whereas exchange between parcels 3 and 4 would obviously lead to an increase in potential energy. Lighter shades represent lighter fluid.

surfaces wedged between geostrophic surfaces and isopycnals. (*Hint*: Solve for the equations of motion (17.14) for the vector displacement $(\Delta x, \Delta z)$ and study its direction.)

17.7. Apply the two-layer model of baroclinic instability to an atmospheric situation, where the domain height $H \sim 7$ km encompasses the troposphere with typical stratification $N = 10^{-2}\,\text{s}^{-1}$. Calculate the wavelength of the most unstable mode. Compare to the weather patterns of the synoptic maps you see any day in the newspaper or online. Then, calculate the growth rate and compare to the lead time of a typical weather forecast.

NUMERICAL EXERCISES

17.1. Program (17.43) and plot, as a function of ΔU, the wavenumber corresponding to the maximum growth rate and the growth rate itself for different values of β_0 and R.

17.2. Use `baroclinic.m` to see what happens if instead of boundary condition (17.75), the initial value of ψ is kept on the boundaries.

17.3. Explore with `baroclinic.m` the effect of changing the domain width.

17.4. Use `baroclinic.m` and add the beta term to the discretization. Apply the new program to a wide domain and analyze the impact of the beta effect by placing yourself in situations allowing to verify that (17.45) corresponds to the stability limit. If necessary, use the program you constructed for solving Numerical Exercise 17.1.

17.5. Include the possibility of more general interfacial profiles in `baroclinic.m` and analyze a localized baroclinic jet where the interface displacement around mid-depth is

$$a = \frac{f_0 U R}{g'} \tanh\left(\frac{y}{R}\right), \tag{17.78}$$

where R is the internal deformation radius, U the jet velocity, and y the cross-channel coordinate, centered in the middle of the channel.

17.6. Include diagnostics of energy evolution and transfers of energy into `baroclinic.m`. Check to which extent the numerical discretization conserves energy. Verify if another time discretization can improve the simulation.

Joseph Pedlosky
1938–

A student of J. G. Charney, Joseph Pedlosky first followed his mentor's footsteps and developed a fascination for baroclinic instability. He quickly became an authority on the subject, having derived new instability criteria and developed a nonlinear theory for growing baroclinic disturbances in nearly inviscid flow. He also made important contributions to the general theory of rotating stratified fluids, the oceanic thermocline, the Gulf Stream, and the general oceanic circulation. In 1979, Pedlosky published the first treatise on Geophysical Fluid Dynamics, which greatly helped codify the discipline.

Pedlosky's approach to research is first to find a problem that is simple enough to be solved completely, yet physically informative, and then to "worry a great deal about it until I could describe the results to an amateur." This incessant quest for clarity has won him great respect as a scientist and much admiration as a speaker. (*Photo credit: J. Pedlosky*)

Peter Broomell Rhines
1942–

Peter Rhines studied aerospace engineering, but his professors at MIT and Cambridge University exposed him to Rossby waves and potential vorticity, and he became hooked by geophysical fluid dynamics. Over his career, his interests have covered a wide span of geophysical phenomena, ranging from the general ocean circulation and oceanic eddies, to the dynamics of the atmosphere and climate. His approach to questions is equally diverse, replete with paradigm-shifting theories (on potential vorticity homogenization in the ocean), original laboratory experiments, incisive numerical simulations (in geostrophic turbulence), and challenging oceanographic cruises to "white and blue rim of the Arctic world."

   The numerous awards Rhines has received emphasize his "amazing physical insight and profound appreciation of observations" and honor his "elegant theoretical studies that have initiated new fields of inquiry." But Rhines remains modest, claiming that "in a sparsely populated discipline like geophysical fluid dynamics, a short life is long enough to work on many aspects of the field." (*Photo courtesy of Peter B. Rhines*)

Chapter 18

Fronts, Jets and Vortices

ABSTRACT

When the Rossby number is not small, the dynamics are nonlinear and nonquasi-geostrophic. Such regimes exhibit fronts and jets, the latter being related to the former through pressure gradients. Strong jets meander and shed vortices, which also populate this dynamical regime. The chapter ends with a brief discussion of geostrophic turbulence, the state of many interacting vortices under the influence of Coriolis effects. This problem is particularly well suited to introduce spectral methods for nonlinear problems.

18.1 FRONTS AND JETS

18.1.1 Origin and Scales

A common occurrence in the atmosphere and ocean is the encounter of two fluid masses that, due to separate origins, have distinct properties. The result is the existence of a local transitional region that is relatively narrow (compared with the dimensions of the main fluid masses) and where properties vary spatially more rapidly than on either side. Such a region of intensified gradients of fluid properties is called a *front*.

Typically, the adjacent fluid masses have different densities, and the front is accompanied by a relatively large pressure gradient. Under the action of Coriolis forces, the process of geostrophic adjustment is at work, leading to a relatively intense flow aligned with the front. The weaker density gradients in the main part of each fluid mass confine the motion to the frontal region, and the flow exhibits the form of a jet. The most notable jet in the atmosphere is the so-called polar-front jet stream found around a latitude of 45°N and a few kilometers above sea level (pressure around 300 millibars), at the boundary between subtropical and polar air masses (Fig. 18.1). From the thermal–wind relation

$$f\frac{\partial u}{\partial z} = \frac{g}{\rho_0}\frac{\partial \rho}{\partial y},\tag{18.1}$$

we can readily see that a weak velocity at sea level must intensify with height to become an intense eastward flow at high altitude. This is because of the general north–south gradient of temperature between the two air masses. In the ocean,

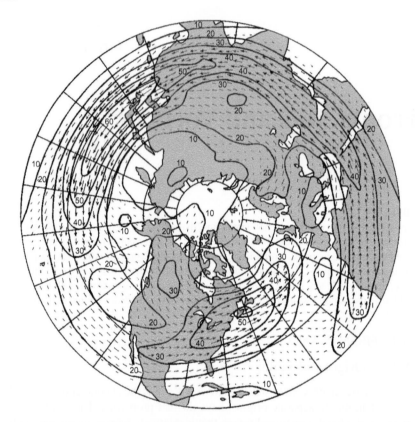

FIGURE 18.1   Monthly winds (in meters per second) over the northern hemisphere for January 1991 at the 300-mb pressure level. Note the jet stream around the 45°N parallel, except over the eastern North Pacific and eastern North Atlantic, where blockings are present. (*From National Weather Service, NOAA, Department of Commerce, Washington, D.C.*)

a surface-to-bottom front is often found in the vicinity of the shelf break owing to different water properties above the continental shelf and in the deep ocean; such a front is invariably accompanied by currents along the shelf (Fig. 18.2).

According to Section 15.1, the simultaneous presence of a horizontal gradient of density and a vertical gradient of horizontal velocity can yield a thermal–wind balance, which may persist for quite some time. Our earlier discussions of geostrophic adjustment (Section 15.2) demonstrated how such a balance can be achieved following the penetration of one fluid mass into another of different density and indicated that the width of the transitional region is measured by the internal radius of deformation, expressed as

$$R = \frac{NH}{f} \sim \frac{\sqrt{g'H}}{f} \tag{18.2}$$

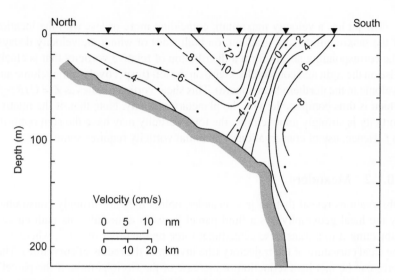

FIGURE 18.2   Monthly mean along-shelf currents for April 1979 across the shelf break on the southern flank of Georges Bank (41°N, 67°W). The units are centimeters per second, and positive values indicate flow pointing into the page. (*From Beardsley et al., 1983, as adapted by Gawarkiewicz & Chapman, 1992*)

in the respective cases of continuous stratification and layered configuration. Here f is the Coriolis parameter, H is an appropriate height scale (assuming large excursions of density surfaces in frontal systems), N is the stratification frequency, and g' is a suitable reduced gravity. If the density difference between the fluid masses is $\Delta\rho$, the accompanying pressure difference is $\Delta P \sim \Delta\rho \, gH = \rho_0 g' H$, and, through geostrophy, the velocity scale is

$$U = \frac{\Delta P}{\rho_0 fR} \sim \frac{g'H}{fR} = \sqrt{g'H}. \qquad (18.3)$$

From this follows that the internal radius of deformation may also be expressed as $R = U/f$, in which we recognize the inertial-oscillation radius (see Section 2.3). Here, the two coincide because we assume a frontal structure with $\Delta H = H$.

The Froude and Rossby numbers are, respectively

$$Fr = \frac{U}{NH} \sim \frac{\sqrt{g'H}}{fR} \sim 1, \qquad (18.4)$$

$$Ro = \frac{U}{fR} \sim \frac{\sqrt{g'H}}{fR} \sim 1, \qquad (18.5)$$

and thus both are on the order of unity, implying that the effects of stratification and rotation are equally important within the jet (see Section 11.6).

The jet has a velocity maximum, coinciding more or less with the location of the maximum density gradient, on both sides of which the velocity decays. The corresponding shears form a distribution of relative vorticity that is clockwise on the right and counterclockwise on the left (respectively, anticyclonic and cyclonic in the northern hemisphere). This shear vorticity scales as $Z = U/R \sim f$, which is thus comparable with the planetary vorticity. Note that, if the relative vorticity is strongly anticyclonic, the total vorticity may have the sign opposite to f. Hence, use of conservation of potential vorticity requires some care.

18.1.2   Meanders

Observations reveal that all jets meander, unless they are strongly constrained by the local geography. As a fluid parcel flows in a meander, its path curves, subjecting it to a transverse centrifugal force on the order of KU^2, where K is the local curvature of the trajectory (the inverse of the radius of curvature). This force can be met by a reduction or increase of the Coriolis force if the parcel's velocity adjusts by ΔU, such that $f\Delta U \sim KU^2$, or

$$\frac{\Delta U}{U} \sim \frac{KU}{f} \sim KR. \tag{18.6}$$

Note that the product KR is in essence the deformation radius divided by the curvature radius.

In the northern hemisphere $(f > 0)$, the Coriolis force acts to the right of the fluid parcel, and thus a rightward turn causing a centrifugal force to the left necessitates a greater Coriolis force and an acceleration $(\Delta U > 0)$ (see Fig. 18.3). Similarly, a leftward turn is accompanied by a jet deceleration $(\Delta U < 0)$. The reverse conclusions hold for the southern hemisphere, but in each case, the stronger the curvature, the larger the change in velocity, according to Eq. (18.6).

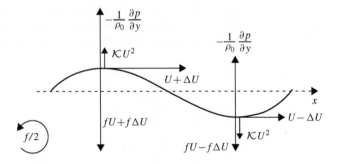

FIGURE 18.3   For the same pressure gradient, a rightward turn requires a larger velocity enabling Coriolis force to balance pressure and centrifugal force. In a left turn, the opposite takes place, and the velocity is reduced.

The same result can be obtained by considering changes in relative vorticity. Neglecting for the moment the beta effect as well as any vertical stretching or squeezing, the relative vorticity is conserved. It can be expressed locally as

$$\zeta = \frac{\partial v}{\partial x} - \frac{\partial u}{\partial y} = \frac{\partial V}{\partial n} - \mathcal{K}V, \tag{18.7}$$

where $V = (u^2 + v^2)^{1/2}$ is the flow speed (scaling as U), n is the cross-jet coordinate (measured positively to the right of the local velocity and scaling as R), and $\mathcal{K}$ is the jet's path curvature (positive clockwise). The first term, $\partial V/\partial n$, is the contribution of the shear and the second, $-\mathcal{K}V$, represents a vorticity due to the turning of the flow. We shall call these contributions shear vorticity and orbital vorticity (See Fig. 18.4 and Analytical Problem 11). In a rightward turn ($\mathcal{K} > 0$), the fluid parcel acquires clockwise orbital vorticity, on the order of $\mathcal{K}U$, which must be at the expense of shear vorticity, $\Delta U/R$. Equating $\mathcal{K}U$ to $\Delta U/R$ again leads to Eq. (18.6).

The change in shear vorticity implies a shift of the parcel with respect to the jet axis. To show this, let us take for example, the fluid parcel that possesses the maximum velocity (i.e., it is on the jet axis) upstream of the meander; there, it has no shear and no orbital vorticity. If this parcel turns to the right in the meander, it acquires clockwise orbital vorticity, which must be compensated by a counterclockwise shear vorticity of the same magnitude. Thus, the parcel must now be on the left flank of the jet. The parcel occupying the jet axis (having maximum velocity and thus no shear vorticity) is one that was on the right flank of the jet upstream and has exchanged its entire clockwise shear vorticity for an equal amount of clockwise orbital vorticity. From this, it is straightforward to conclude that all parcels are displaced leftward with respect to the jet axis in a rightward meander, and rightward with respect to the jet axis in a leftward turn. (This rule is easy to remember: Fluid parcels shift across the jet in the direction of the centrifugal force.)

A consequence of these vorticity adjustments created by meandering is that fluid parcels near the edges may separate from the jet or be captured by it.

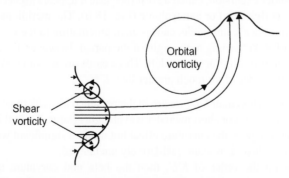

FIGURE 18.4   Difference between shear and orbital vorticity of a jet.

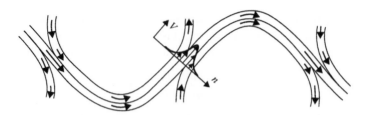

FIGURE 18.5 Separation and capture of fluid parcels along the edges of a meandering jet. This process occurs because the vorticity adjustment required by the meander causes marginal parcels to reverse their velocity. Parcels near the inside edge of the meander see their velocity reversed once curvature has ceased and effectively leave the jet. Similarly, parcels of fluid are joining the jet from the outside at the exit of a meander.

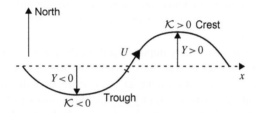

FIGURE 18.6 Meandering of an eastward jet on the beta plane (northern hemisphere). If the meridional displacement Y, curvature $\mathcal{K}$, and jet speed U are related by $\beta_0 Y \simeq \mathcal{K} U$, changes in planetary and orbital vorticity are comparable and opposite in sign, leaving the velocity profile of the jet (shear vorticity) relatively unperturbed.

Indeed, a parcel on the edge of the jet may have insufficient shear vorticity for trading with orbital vorticity (Fig. 18.5).

The preceding considerations ignored the beta effect, by which the Coriolis force is able to vary. Let us limit ourselves here to the case of an eastward *westerly* jet in the northern hemisphere, which is the case of the atmospheric jet stream and the Gulf Stream in the North Atlantic beyond Cape Hatteras. In a northern meander excursion, called a *crest* (because it appears higher on a map), the curvature is rightward or anticyclonic (Fig. 18.6). The meridional displacement Y, the meander's amplitude, causes an augmentation to the Coriolis force on the order of $\beta_0 YU$, acting to the right of the parcel. However, the centrifugal force on the order of $\mathcal{K} U^2$ acts to its left. Three cases are possible: $\beta_0 Y$ is much less than, on the order of, or much greater than $\mathcal{K} U$.

- If $\beta_0 Y$ is much less than $\mathcal{K} U$, we are in the presence of weak meander amplitudes (small Y) and/or short meander wavelengths (large $\mathcal{K}$). In this case, the beta effect mitigates the curvature effect but not in a significant way, and the previous conclusions remain qualitatively unchanged.
- If $\beta_0 Y$ is on the order of $\mathcal{K} U$, then the beta and curvature effects can balance each other, leaving the structure of the jet barely affected. For

a sinusoidal meander $Y(x) = A \sin k_x x$, where A is the meander amplitude, $\lambda = 2\pi/k_x$ is its wavelength, and x is the eastward coordinate, we deduce that at the meander's peak ($\sin k_x x = +1$), the meridional displacement Y is A, and the curvature $\mathcal{K} = -[d^2Y/dx^2]/[1 + (dY/dx)^2]^{3/2}$ is $k_x^2 A$. The balance $\beta_0 Y \sim \mathcal{K} U$ then yields $\beta_0 \sim k_x^2 U$ or

$$\lambda = \frac{2\pi}{k_x} = 2\pi \sqrt{\frac{U}{\beta_0}}. \tag{18.8}$$

From this emerges a particular length scale,

$$L_\beta = \sqrt{\frac{U}{\beta_0}}, \tag{18.9}$$

which we shall call the *critical meander scale*. Cressman (1948) noted its importance in relation to the development of long waves on the atmospheric jet stream, whereas Moore (1963) obtained a solution to an ocean-circulation model that exhibits meanders at that scale. Later, Rhines (1975) demonstrated how this same scale plays a pivotal role in the evolution of geostrophic turbulence on the beta plane.

- In very broad meanders, for which meridional displacements are large and curvatures are small ($\beta_0 Y \gg \mathcal{K}U$), the beta effect dwarfs the curvature effect, and the trade-off is almost exclusively between changes in planetary vorticity and shear vorticity. In a meander crest (greater f), the shear vorticity must become less cyclonic or more anticyclonic (see Fig. 18.7).

Meanders on a jet do not remain stationary but propagate, usually downstream and rarely upstream. The direction of propagation can be inferred from

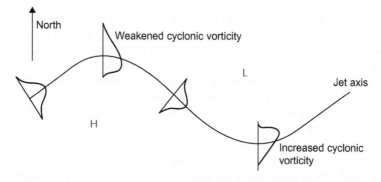

FIGURE 18.7   Changes in shear vorticity in very broad meanders. These are caused by the beta effect, which changes the Coriolis force with latitude. The figure is drawn for the case of south–north displacements in the northern hemisphere. The letters H and L indicate high- and low-pressure regions, respectively.

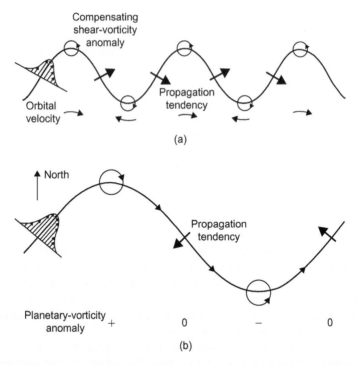

FIGURE 18.8   Schematic descriptions explaining why (a) curvature and (b) beta effects on an eastward jet induce meander-propagation tendencies that are, respectively, downstream and upstream.

vorticity considerations, as outlined previously. In the absence of the beta effect (or $\beta_0 Y \ll \mathcal{K}U$), leftward and rightward turns create, respectively, clockwise and counterclockwise shear vorticity. Picturing these vorticity anomalies as vortices at the meanders' tips (Fig. 18.8a), we infer that the entrainment velocities at the inflection points between meanders all have a downstream component and that the meander pattern translates downstream. On a westerly jet, this direction is eastward. At the opposite extreme of a large beta effect and negligible curvature ($\beta_0 Y \gg \mathcal{K}U$), the vorticity anomalies are cyclonic in troughs and anticyclonic in crests (Fig. 18.8b). The entrainment velocities at the inflection points all point westward. On a westerly jet, this is upstream. This mechanism is the same as that invoked in Section 9.4 to explain the westward phase propagation of planetary waves. (Compare Fig. 18.8b with Fig. 9.7)

    We note, therefore, that curvature and beta effects induce opposite meander-propagation tendencies on an eastward jet. Comparing $\beta_0 Y$ with $\mathcal{K}U$ — or, equivalently, the wavelength to the critical meander scale — we conclude that if the former is larger than the latter, the meander propagates upstream (westward) and in the opposite direction otherwise. The meander is stationary if the tendencies cancel each other, which occurs if its wavelength is near the critical meander scale. Since this scale is rather long (220 km in the ocean and 1600 km in the atmosphere, with $\beta_0 = 2 \times 10^{-11}$ m^{-1}s^{-1} and U ranging from 1 m/s to 50 m/s), observed meanders are usually of the curvature-type and propagate eastward.

18.1.3   Multiple Equilibria

Because the critical meander scale depends on the jet speed U, and also because the relation $\beta_0 Y \sim \mathcal{K}U$ depends on the shape of the meander (Y and $\mathcal{K}$ are not simply related if the meander is other than sinusoidal), the critical size for meander stationarity depends on the jet speed and the meander shape. This conclusion is the basis of one explanation for the bimodality of the Kuroshio (Fig. 18.9). The geography of coastal Japan and the regional bottom topography force this intense current of the western North Pacific to pass through two channels, south of Yakushima Island (30°N, 130°E) and between Miyake and Hachijo Islands near the Izu Ridge (34°N, 140°E).

Between these two points, the current is known to assume one of two preferential states: a relatively straight path or a curved path with a substantial southward excursion. Each pattern persists for several years, whereas the transition from one to the other is relatively brief. The theory (Masuda, 1982; Robinson & Taft, 1972) explains this bimodal character by arguing that a stationary meander with a half-wavelength meeting the geographical specification may or may not exist, depending on the jet velocity. Calculations show that the meander state occurs if the jet velocity does not exceed a certain threshold value. At any velocity below this value, there exists a stationary-meander shape that meets the geographical constraints. At larger velocities, no stationary meander is possible, and the jet must assume a straight path.

The atmospheric analog of this oceanic situation is known as *blocking*, a word now used in a sense different from that used in Chapter 11. Here, blocking is a midlatitude phenomenon characterized by the unusual persistence of a nearly stationary meander on an eastward jet over topographic irregularities (Fig. 18.1). The theory (Charney & DeVore, 1979; Charney & Flierl, 1981) again invokes multiplicity of equilibrium solutions, including the normal state (no meander) and the anomalous blocking configuration (with large meander).

18.1.4   Stretching and Topographic Effects

Up to here, our considerations of vorticity adjustments in a jet meander included exchanges among planetary, shear, and orbital vorticity for an unchanged total. This is correct only for barotropic jets over a flat bottom, whereas in a baroclinic jet, in which vertical stretching can occur, potential vorticity rather than vorticity is the conserved quantity.

A complete theory involving all relevant dynamics such as momentum and mass balances is beyond our scope, and we derive here only the vertical-stretching tendency experienced by a fluid parcel in a meander. Assuming that the trade-off is solely between orbital vorticity due to the meander's curvature

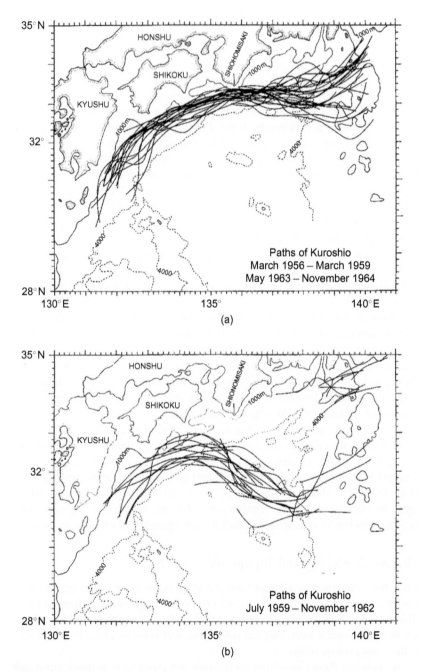

FIGURE 18.9 Observed Kuroshio paths: (a) straight jet and (b) stationary meander. (*From Robinson & Taft, 1972*)

and vertical stretching, we reason that a meander crest (with anticyclonic orbital vorticity) lowers the total vorticity and thus calls for a proportional decrease in the column's vertical thickness. In meander troughs, fluid columns are vertically stretched and shifted toward the anticyclonic side of the jet. In an oceanic surface jet such as the Gulf Stream, such a modification causes upwelling upon approaching crests and downwelling upon approaching troughs. Observations (Bower & Rossby, 1989) confirm such behavior, which can also be noted in numerical simulations (Fig. 18.10).

Just as meanders generate vertical stretching or squeezing, vertical stretching or squeezing induced by topography can cause meanders. To illustrate this, let us consider the case of a zonal jet (barotropic or baroclinic) on the beta plane that encounters a topographic step (Fig. 18.11). If the jet is flowing eastward (the usual situation) and enters a deeper region, the expansion in layer thickness translates first into a cyclonic deflection, away from the equator. As the Coriolis parameter increases away from the equator, this cyclonic vorticity is progressively exchanged downstream by a greater planetary vorticity, and the jet curvature weakens. Further poleward progression reverses the sense of orbital

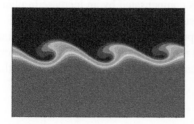 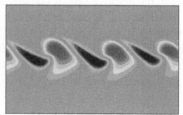

FIGURE 18.10   Frontal meander on a sea surface temperature field (left panel) and associated vertical-velocity distribution indicating upwelling and downwelling cells (right panel). Note that maxima of vertical velocity occur between the meander's crests and troughs (*From Rixen, Beckers & Allen, 2001*).

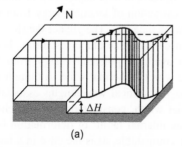 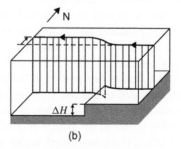

FIGURE 18.11   Eastward and westward jets passing over a topographic step: (a) the eastward jet develops an oscillatory behavior, whereas (b) the westward jet begins to feel the influence of the step upstream and executes a single meander. Both experience a net meridional shift $Y = f_0 \Delta H / \beta_0 h$, the sign of which depends on whether the step is up or down.

vorticity, and the jet oscillates back and forth about a new latitude (Fig. 18.11a). The average northward shift, Y, of the jet axis corresponds to an exchange between vertical stretching and increased planetary vorticity:

$$\frac{\beta_0 Y}{f_0} \sim \frac{\Delta H}{h}, \qquad (18.10)$$

where ΔH is the height of the topographic step, and h is the upstream thickness of the jet. Because the first meander is rooted at the location of the step, the meander must be stationary, and therefore the wavelength must be comparable to the critical meander scale.

The same argument can be invoked for an eastward jet entering a shallower region to conclude that the flow exhibits a stationary oscillation about a net equatorward shift, given by Eq. (18.10) where ΔH is now negative. However, the argument fails for westward jets. Upon entering a deeper region, a fluid parcel acquires cyclonic vorticity and turns equatorward, its planetary vorticity decreases, further increasing the orbital vorticity. Clearly, if this were the case, the jet would be looping onto itself. Instead, the jet begins to be distorted ahead of the topographic step (Fig. 18.11b), acquiring an anticyclonic curvature in which the negative orbital vorticity is compensated by an increase in planetary vorticity. The jet thus reaches the step at an oblique angle. The nature of the vorticity adjustments past the step progressively restores the jet's original zonal orientation. A net meridional shift remains, expressing a balance between changes in planetary vorticity and vertical thickness. The reader can verify that this shift is again given by Eq. (18.10).

18.1.5 Instabilities

In addition to their propagation, meanders on a jet also distort and frequently grow, close onto themselves, and form eddies that separate from the rest of the jet. Such a finite change to the jet structure results from an instability, the nature of which is barotropic (Chapter 10), baroclinic (Chapter 17), or mixed. Barotropic instability proceeds with the extraction of kinetic energy from the horizontally sheared flow to feed the growing meander. The greater the shear in the jet, the more likely is this type of instability. Baroclinic instability, however, is associated with a conversion of available potential energy from the horizontal density distribution in balance with the thermal wind. Although the example treated in Section 10.4 suggests that critical wavelengths associated with barotropic instability scale as the jet width, consideration of baroclinic instability points to the critical role of the internal radius of deformation [see Eq. (17.48)]. If the two length scales are comparable, as is the case in a baroclinic jet with finite Rossby number, both processes may be equally active, and the instability is most likely of the mixed type (Griffiths, Killworth & Stern, 1982; Killworth, Paldor & Stern, 1984; Orlanski, 1968). The beta effect further complicates the situation, occasionally facilitating the eddy detachment process:

The large meridional displacement of the growing meander induces a westward-propagation tendency, whereas the high-curvature regions where the meander attaches to the rest of the jet induce a downstream propagation tendency. The result is a complex situation in which the outcome sensitively depends on the relative magnitudes of the different effects (Flierl, Malanotte-Rizzoli & Zabusky, 1987; Robinson, Spall & Pinardi, 1988). The meandering and eddy shedding of the Gulf Stream manifest this complexity.

The development of synoptic-scale weather disturbances, a process now called *cyclogenesis*, is thought to be initiated mostly by baroclinic instability, whereas accompanying finer-scale processes, such as cold and warm fronts, are explained by ageostrophic dynamics. The interested reader is referred to the book by Holton (1992) and Section 15.5.

18.2   VORTICES

A vortex, or eddy, is defined as a closed circulation that is relatively persistent. By persistence, we mean that the turnaround time of a fluid parcel embedded in the structure is much shorter than the time during which the structure remains identifiable. A *cyclone* is a vortex in which the rotary motion is in the same sense as the earth's rotation, counterclockwise in the northern hemisphere and clockwise in the southern hemisphere. An *anticyclone* rotates the other way, clockwise in the northern hemisphere and counterclockwise in the southern hemisphere.

The prototypical vortex is a steady circular motion on the f-plane. Using cylindrical coordinates, we can express the balance of forces in the radial direction r (measured outward) as follows:

$$-\frac{v^2}{r} - fv = -\frac{1}{\rho_0}\frac{\partial p}{\partial r}, \qquad (18.11)$$

where v is the orbital velocity (positive counterclockwise), and p is the pressure (or Montgomery potential). Both v and p may be dependent upon the vertical coordinate, either height z or density ρ. This equation, called the *gradient-wind balance*, represents an equilibrium between three forces, the centrifugal force (first term), the Coriolis force (second term), and the pressure force (third term). Although the centrifugal force is always directed outward, the Coriolis and pressure forces can be directed either inward or outward, depending on the direction of the orbital flow and on the center pressure.

If we introduce the following scales, U for the orbital velocity, L for r (measuring the vortex radius), and ΔP for the pressure difference between the ambient value and that at the vortex center, we note that the terms composing (18.11) scale, respectively, as

$$\frac{U^2}{L}, \quad fU, \quad \frac{\Delta P}{\rho_0 L}. \qquad (18.12)$$

At low Rossby numbers ($Ro = U/fL \ll 1$), the first term is negligible relative to the second (i.e., the centrifugal force is small compared with the Coriolis force), the balance is nearly geostrophic, providing

$$fU = \frac{\Delta P}{\rho_0 L}, \tag{18.13}$$

and thus $U = \Delta P/(\rho_0 f L)$. Since the pressure difference is most likely the result of a density anomaly $\Delta \rho$, the hydrostatic balance provides $\Delta P = \Delta \rho g H = \rho_0 g' H$, where H is the appropriate height scale (thickness of vortex), and $g' = g \Delta \rho / \rho_0$ is the reduced gravity. This leads to $U = g' H / f L$ and

$$Ro = \frac{U}{fL} = \frac{g'H}{f^2 L^2} = \left(\frac{R}{L}\right)^2, \tag{18.14}$$

in which we recognize the internal deformation radius $R = (g'H)^{1/2}/f$. Thus, a small Rossby number occurs as a consequence of a horizontal scale large compared with the deformation radius. This is typically the case in the largest weather cyclones and anticyclones at midlatitudes and in large-scale oceanic gyres (Fig. 18.12-top). Note that in this analysis, the Rossby number coincides with the Burger number. Thus, its smallness implies that the vorticity in broad gyres is mostly constrained by vertical stretching rather than relative vorticity (see Section 16.3). Also, the energy of such gyres is dominated by available potential energy rather than kinetic energy as shown by Eq. (16.34).

At scales on the order of the deformation radius, L can be taken equal to R, the Rossby number is on the order of unity, the velocity scale is $U = (g'H)^{1/2}$, and the centrifugal force is comparable with the Coriolis force. Around a low pressure, the outward centrifugal force partially balances the inward pressure force, leaving the Coriolis force to meet only the difference. By contrast, the Coriolis force acting on the flow around a high pressure must balance both the outward pressure force and the outward centrifugal force (Fig. 18.12-middle). Consequently, the orbital velocity in an anticyclone is greater than that in a cyclone of identical size and equivalent pressure anomaly. Tropical hurricanes (Anthes, 1982; Emmanuel, 1991) and the so-called rings shed by the Gulf Stream (Flierl, 1987; Olson, 1991) fall in the category of vortices with length scale on the order of the deformation radius.

At progressively shorter radii, the centrifugal force becomes increasingly important, and for $L \ll R$, the Coriolis force becomes negligible. The cyclone-anticyclone nomenclature then loses its meaning, and the relevant characteristic is the sign of the pressure anomaly. The inward force around a low pressure is balanced by the outward centrifugal force regardless of the direction of rotation (Fig. 18.12-bottom). Such a state is said to be in *cyclostrophic balance*. Examples are tornadoes and bathtub vortices. A vortex with high-pressure center cannot exist because pressure and centrifugal forces would both be directed outward.

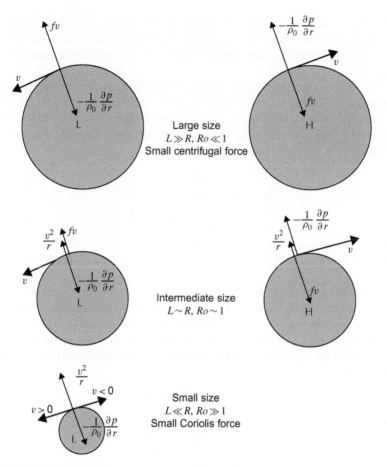

FIGURE 18.12   Balances between pressure gradient $-(1/\rho_0)\partial p/\partial r$, Coriolis fv, and centrifugal v^2/r forces in northern hemispheric circular vortices. The letters L and H indicate low and high pressures, respectively.

It is interesting to determine the minimum size for which an anticyclone of given pressure anomaly can exist. Returning to the gradient-wind balance where we introduce $v = -fr/2 + v'$, we write

$$\frac{f^2 r}{4} + \frac{1}{\rho_0}\frac{\partial p}{\partial r} = \frac{1}{r}v'^2 \geq 0. \tag{18.15}$$

Integrating over the radius a of the vortex and defining the pressure anomaly $\Delta p = p(r=0) - p(r=a)$, we obtain

$$a^2 \geq \frac{8\Delta p}{\rho_0 f^2}. \tag{18.16}$$

For a low-pressure center ($\Delta p < 0$), this inequality yields no constraint, whereas for a high-pressure center ($\Delta p > 0$), it specifies a minimum vortex radius. Below this minimum, high-pressure centers simply do not exist as isolated steady structures.

Let us now examine how an existing vortex can move within the fluid that surrounds it. To do this, we consider a vortex contained within a single layer of fluid, be it the lowest, the uppermost, or any intermediate layer in the fluid. If the local thickness of this layer is h, and the pressure (actually, Montgomery potential) is p, we write, in density coordinates,

$$\frac{\partial u}{\partial t} + u\frac{\partial u}{\partial x} + v\frac{\partial u}{\partial y} - fv = -\frac{1}{\rho_0}\frac{\partial p}{\partial x}, \tag{18.17a}$$

$$\frac{\partial v}{\partial t} + u\frac{\partial v}{\partial x} + v\frac{\partial v}{\partial y} + fu = -\frac{1}{\rho_0}\frac{\partial p}{\partial y}, \tag{18.17b}$$

$$\frac{\partial h}{\partial t} + \frac{\partial}{\partial x}(hu) + \frac{\partial}{\partial y}(hv) = 0. \tag{18.17c}$$

We further restrict ourselves to the f-plane. At large distances from the vortex center, in what can be considered the ambient fluid, we assume that there exists a steady uniform flow ($\bar{u}$, $\bar{v}$) and a uniform thickness gradient ($\partial\bar{h}/\partial x$, $\partial\bar{h}/\partial y$). According to Eqs. (18.17a) and (18.17b), this flow must be geostrophic, and according to Eq. (18.17c), it must be aligned with the direction of constant layer thickness:

$$-f\bar{v} = -\frac{1}{\rho_0}\frac{\partial\bar{p}}{\partial x}, \tag{18.18a}$$

$$+f\bar{u} = -\frac{1}{\rho_0}\frac{\partial\bar{p}}{\partial y}, \tag{18.18b}$$

$$\bar{u}\frac{\partial\bar{h}}{\partial x} + \bar{v}\frac{\partial\bar{h}}{\partial y} = 0. \tag{18.18c}$$

A thickness gradient is retained because, in some instances, a thermal wind in layers above or below may be accompanied by such a thickness variation. Also, if the vortex lies in the lowest layer, the thickness gradient may represent a bottom slope. The assumption of uniformity of $\bar{u}$, $\bar{v}$, and of the derivatives of $\bar{p}$ and $\bar{h}$ is justified if the ambient-flow properties vary over horizontal distances much longer than the vortex diameter. Defining the velocity components, pressure, and layer-thickness variations proper to the vortex as $u' = u - \bar{u}$, $v' = v - \bar{v}$, $p' = p - \bar{p}$, and $h' = h - \bar{h}$, we can transform Eqs. (18.17) as follows:

$$\frac{\partial u'}{\partial t} + (\bar{u}+u')\frac{\partial u'}{\partial x} + (\bar{v}+v')\frac{\partial u'}{\partial y} - fv' = -\frac{1}{\rho_0}\frac{\partial p'}{\partial x} \tag{18.19a}$$

$$\frac{\partial v'}{\partial t} + (\bar{u}+u')\frac{\partial v'}{\partial x} + (\bar{v}+v')\frac{\partial v'}{\partial y} + fu' = -\frac{1}{\rho_0}\frac{\partial p'}{\partial y} \tag{18.19b}$$

$$\frac{\partial h'}{\partial t}+\frac{\partial (h'\bar{u})}{\partial x}+\frac{\partial (h'\bar{v})}{\partial y}+\frac{\partial}{\partial x}\left[(\bar{h}+h')u'\right]+\frac{\partial}{\partial y}\left[(\bar{h}+h')v'\right]=0. \qquad (18.19c)$$

We then define the anomalous layer volume due to the vortex:

$$V=\iint h'\,dx\,dy, \qquad (18.20)$$

where the integration covers the entire horizontal extent of the layer. The perturbation h' induced by the vortex is assumed to be sufficiently localized to make the preceding integral finite. The use of continuity equation (18.19c) followed by integration by parts over several terms shows that the temporal derivative of this volume,

$$\frac{dV}{dt}=\iint \frac{\partial h'}{\partial t}\,dx\,dy \qquad (18.21)$$

vanishes, as we expect. Defining the coordinates of the vortex position by the volume-weighted averages

$$X=\frac{1}{V}\iint xh'\,dx\,dy, \quad Y=\frac{1}{V}\iint yh'\,dx\,dy, \qquad (18.22)$$

we can track the vortex displacements by calculating their temporal derivatives. For X, we obtain successively

$$\frac{dX}{dt}=\frac{1}{V}\iint x\frac{\partial h'}{\partial t}\,dx\,dy$$

$$=\frac{-1}{V}\iint \left\{x\bar{u}\frac{\partial h'}{\partial x}+x\bar{v}\frac{\partial h'}{\partial y}+x\frac{\partial}{\partial x}[(\bar{h}+h')u']+x\frac{\partial}{\partial y}[(\bar{h}+h')v']\right\}\,dx\,dy$$

$$=\frac{+1}{V}\iint [\bar{u}h'+(\bar{h}+h')u']\,dx\,dy$$

$$=\bar{u}+\frac{1}{V}\iint hu'\,dx\,dy. \qquad (18.23)$$

Similarly, we obtain for the other coordinate

$$\frac{dY}{dt}=\bar{v}+\frac{1}{V}\iint hv'\,dx\,dy. \qquad (18.24)$$

The preceding integrals cannot be evaluated without knowing the precise structure of the vortex. However, a second time derivative will bring forth the acceleration $(\partial u'/\partial t,\ \partial v'/\partial t)$, which is provided by the equations of motion,

(18.19a) and (18.19b). For the X-coordinate, we obtain

$$
\frac{d^2 X}{dt^2} = \frac{1}{V} \iint \left[\frac{\partial h'}{\partial t} u' + (\bar{h} + h') \frac{\partial u'}{\partial t} \right] , dx\, dy
$$

$$
= \frac{-1}{V} \iint \left[\frac{\partial}{\partial x}(huu') + \frac{\partial}{\partial y}(hvu') \right] dx\, dy
$$

$$
+ \frac{f}{V} \iint hv'\, dx\, dy - \frac{1}{\rho_0 V} \iint h \frac{\partial p'}{\partial x}\, dx\, dy. \tag{18.25}
$$

The pressure anomaly p' associated with the vortex motions can be related by hydrostatic balance to the layer-thickness anomaly. If other layers do not move and keep their pressure value over time, the pressure anomaly inside the vortex is given by integration of Eq. (12.14) where the pressure anomaly above the layer of interest is zero:

$$
p' = \rho_0 g' h', \tag{18.26}
$$

with a suitable definition of the reduced gravity g'. Note that if the vortex is contained in the lowest layer above an uneven bottom, the bottom elevation does not enter (18.26) but instead enters the corresponding hydrostatic balance for the mean-flow properties.

Noting that the first integral in Eq. (18.25) vanishes because u' and v' go to zero at large distances from the vortex, that the second integral can be eliminated by use of Eq. (18.24), and that the third integral, integrated by parts, can be simplified with use of Eq. (18.26), we obtain

$$
\frac{d^2 X}{dt^2} = f \frac{dY}{dt} - f\bar{v} + g' \frac{\partial \bar{h}}{\partial x}. \tag{18.27}
$$

A similar treatment of the second derivative of Y yields

$$
\frac{d^2 Y}{dt^2} = -f \frac{dX}{dt} + f\bar{u} + g' \frac{\partial \bar{h}}{\partial y}. \tag{18.28}
$$

Because the gradient of $\bar{h}$ is assumed uniform, and f, $\bar{u}$, and $\bar{v}$ are constants, the preceding two equations can be solved for the velocity of the vortex:

$$
\frac{dX}{dt} = \left(\bar{u} + \frac{g'}{f} \frac{\partial \bar{h}}{\partial y} \right) (1 - \cos ft) - \left(\bar{v} - \frac{g'}{f} \frac{\partial \bar{h}}{\partial x} \right) \sin ft \tag{18.29a}
$$

$$
\frac{dY}{dt} = \left(\bar{v} - \frac{g'}{f} \frac{\partial \bar{h}}{\partial x} \right) (1 - \cos ft) + \left(\bar{u} + \frac{g'}{f} \frac{\partial \bar{h}}{\partial y} \right) \sin ft, \tag{18.29b}
$$

where the constants of integration have been determined under the assumption that the vortex is not translating initially. In the preceding solution, we recognize inertial oscillations superimposed on a mean drift. This mean drift has two components:

$$c_x = \bar{u} + \frac{g'}{f}\frac{\partial \bar{h}}{\partial y}, \quad c_y = \bar{v} - \frac{g'}{f}\frac{\partial \bar{h}}{\partial x}. \tag{18.30}$$

The first contribution $(\bar{u}, \bar{v})$ indicates that the vortex is entrained by the ambient motion of its containing layer. Together, this entrainment and the inertial oscillations do not distinguish the vortex from a single fluid parcel. The cause of the second contribution, proportional to the gradient of $\bar{h}$, is less obvious and is what really distinguishes a vortex from a fluid parcel.

The existence of a thickness gradient in the vicinity of the vortex implies a nonuniform distribution of potential vorticity, which the swirling motion of the vortex redistributes; fluid parcels on the edge of the vortex are thus stretched and squeezed and develop vorticity anomalies that, in turn, act to displace the main part of the vortex. As the example in Fig. 18.13 illustrates, a northward decrease of layer thickness in the northern hemisphere causes squeezing on parcels moved northward and stretching on those moved southward. (The sense of rotation in the vortex is irrelevant here.) This causes the fluid on the northern flank of the vortex to acquire anticyclonic vorticity and that on the southern flank to acquire cyclonic vorticity. Both vorticity anomalies induce a westward displacement of the bulk of the vortex. Equations (18.30) confirm that those conditions $(\partial \bar{h}/\partial x = 0,\ \partial \bar{h}/\partial y < 0,\ f > 0)$ imply a negative c_x and a zero c_y. The general rule is that the vortex translates with the thin-layer side on its right in the northern hemisphere and on its left in the southern hemisphere.

Gradients in the vortex-containing layer can be caused by one of two reasons. If other layers, above or below, flow at speeds different from that of the vortex layer, there exists a thermal wind, which by virtue of the Margules

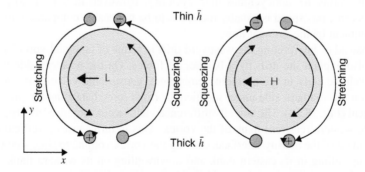

FIGURE 18.13 Lateral drift of a vortex embedded in layer of varying thickness. Advection by surrounding fluid induces cyclonic and anticyclonic vorticities, which combine to induce a drift of the vortex along lines of constant thickness. In the northern hemisphere (as drawn in the figure), the vortex moves with the thin-layer side on its right. Direction is opposite in the southern hemisphere.

relation [see (15.5)] requires sloping density surfaces and, therefore, varying layer thicknesses. It is left to the reader to show that in such a case the vorticity-induction mechanism described in the preceding paragraph amounts to a drift of the vortex in the same direction as the thermal wind. The other reason for layer-thickness variations is bottom topography. If the vortex is contained in the lower layer, bounded below by a sloping bottom, fluid parcels surrounding the vortex will be moved up or down this slope and undergo vorticity adjustments. The result (see Fig. 18.13 again) is a drift of the vortex with the shallower region to its right in the northern hemisphere and to its left in the southern hemisphere. Nof (1983) discusses this effect for cold eddy lenses on the ocean bottom.

Note that if the vortex starts from a resting position, its migration is not immediately transverse to the thickness gradient but is up-gradient, as solution (18.29) indicates for small values of time. In the case of a sloping bottom, this implies that the vortex first goes downhill, gradually acquiring a velocity in that direction, and under the action of the Coriolis force has its trajectory subsequently deflected in the direction transverse to the topographic gradient. (Compare this situation with that of Analytical Problem 2.9.)

Because of the analogy between a topographic slope and the beta effect (see Section 9.6), the preceding conclusions can be extrapolated to the motion of vortices on the beta plane. Regardless of their polarity (cyclonic or anticyclonic), vortices have a self-induced westward tendency. Repeating the argument made with Fig. 18.13, with the replacement of the thick-to-thin direction by the northward direction, we conclude that surrounding parcels entrained from the southern tip to the northern end acquire planetary vorticity and thus develop anticyclonic relative vorticity. Similarly, the surrounding parcels entrained from north to south develop cyclonic relative vorticity. The combined effect at the latitude of the vortex center is a westward drift. Theories (Cushman-Roisin, Chassignet & Tang, 1990 and references therein) show that the induced speed is on the order of $\beta_0 R^2$, where R is the internal radius of deformation, being slightly larger for anticyclones than cyclones. However, in both atmosphere and oceans, this speed is usually too weak to be noticeable compared with the entrainment by the ambient flow.

Instead of interpreting the westward drift in terms of potential vorticity, we can also explain the drift by a balance of forces. On the northern side of an anticyclonic eddy in the northern hemisphere, geostrophic velocity is smaller than on the southern side under identical Coriolis force balancing the pressure gradient (Fig. 18.14). The velocity difference yields a convergence (divergence) on the western (eastern) flank of the vortex. This in turn causes a vertical displacement of the density interface, causing the vortex volume to slide sideways with upwelling in its eastern flank and downwelling on its western flank. For the cyclone, similar reasoning yields again a westward displacement.

Implicit in the preceding derivations was the assumption that all variables related to the vortex decay sufficiently fast away from the vortex core to make all integrals finite. However, in the presence of a potential-vorticity gradient

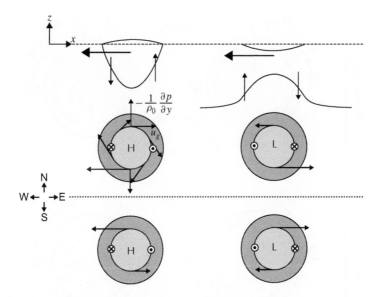

FIGURE 18.14   Alternative explication of the westward drift for an anticyclone (left side) and cyclone (right side). The vertical section at top shows an anticyclone's core of lighter fluid (top-left) and the reduced layer thickness associated with a cyclone (top-right). The plots below represent top views spanning the equator (dotted line) and show the velocity fields. The convergence and divergence pattern associated with north–south velocity differences cause the vortex to move westward regardless of its sense of rotation.

such as one created by a layer-thickness gradient (see the preceding text) or by the beta effect ($\beta_0 = df/dy$), waves are possible (Sections 9.4 and 9.5) and energy can be radiated away to large distances from the vortex, yielding non negligible eddy-related motions there. As it turns out, it is possible to predict, at least qualitatively, the effect of such waves by considering the early time evolution of the vortex. Figure 18.15 depicts the relative-vorticity adjustments brought to surrounding fluid parcels as they are moved by the vortex for the first quarter of their evolution. As for linear waves (Section 9.6), there is a direct analogy between the layer-thickness gradient and the beta effect: The thin-layer side and the poleward direction are dynamically similar, for they both point to an increase in potential vorticity. After a quarter turn, parcels surrounding the vortex acquire relative vorticity by stretching (or squeezing) or a decrease (or increase) in planetary vorticity. As Fig. 18.15 reveals, the cumulative effect in the northern hemisphere is a migration of cyclones toward decreasing layer thicknesses or northward; anticyclones migrate in the opposite direction. As vortices move in those directions, their own core fluid undergoes similar stretching or squeezing or planetary-vorticity changes. In all cases, the net result is a decrease in the absolute value of the relative vorticity and thus an overall spin-down of the vortex.

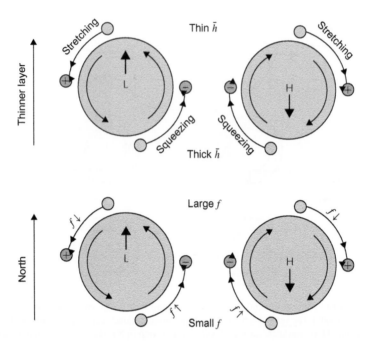

FIGURE 18.15 Secondary drift of vortices. The advection of surrounding fluid induces cyclonic and anticyclonic vortices on the flanks of the vortex, which combine to cause a drift as indicated. This drift component is perpendicular and in addition to that depicted in Fig. 18.13. Again, the figure is drawn for the northern hemisphere. In the southern hemisphere, cyclones still move in the direction of smaller layer thickness or poleward, and anticyclones move in the direction of greater layer thickness, or equatorward.

In the study of hurricane motion, Shapiro (1992) distinctly shows how the trajectory of the hurricane center (a low-pressure center and thus a cyclone) can be explained by the mechanisms just summarized. Here, the beta effect is relatively unimportant, but the presence of a westerly wind aloft and its accompanying layer-thickness gradient (thicker southward) combine to make the hurricane progress in the northwestward direction.

A discussion of geophysical vortices ought to address additional aspects such as axisymmetrization (assuming a nearly circular shape despite anisotropic birthing conditions), instabilities, secondary motions, frictional spin-down, wave radiation, and so on. Partly, because space does not permit a deeper discussion here but mostly because these aspects tend to be quite different in the atmosphere and ocean, the reader interested in atmospheric vortices is referred to the monograph by Anthes (1982), and the reader interested in oceanic vortices is referred to the book edited by Robinson (1983). Laboratory simulations of geophysical vortices have also been conducted; an interesting article on vortex instabilities is that by Griffiths and Linden (1981).

18.3 GEOSTROPHIC TURBULENCE

We alluded to geostrophic turbulence, which is the study of a large number of interacting vortices, in Section 16.6 when we introduced nonlinear effects in quasi-geostrophic dynamics. Here, we tackle the subject from the vortex point of view, without making the quasi-geostrophic assumption.

When several eddies are present and not too distant from one another, interactions are unavoidable. Vortices shear and peel off the sides of their neighbors and, at times, merge to create larger vortices. The sheared elements either curl onto themselves, forming new, smaller vortices, or dissipate under the action of friction. The net result is a combination of consolidation and destruction. When many vortices are simultaneously present, the situation is best described in a statistical sense.

A number of important properties can be derived rather simply by considering three integrals of motion, namely, the kinetic energy, the available potential energy, and the *enstrophy*, the latter being the integrated squared vorticity. We thus define the following:

$$\text{Kinetic energy:} \quad KE = \frac{1}{2}\rho_0 \iiint (u^2 + v^2)\,dx\,dy\,dz \quad (18.31a)$$

$$\text{Available potential energy:} \quad APE = \frac{1}{2}\rho_0 \iiint N^2 h^2\,dx\,dy\,dz \quad (18.31b)$$

$$\text{Enstrophy:} \quad S = \frac{1}{2}\iiint \left(\frac{\partial v}{\partial x} - \frac{\partial u}{\partial y}\right)^2 dx\,dy\,dz. \quad (18.31c)$$

In the formulation of the kinetic energy, the contribution of vertical velocity is usually insignificant. (It is insignificant whenever hydrostatic balance holds.) If the horizontal velocity scale is U, the domain depth is H, and the horizontal area is A, the size of KE is about $\rho_0 U^2 HA$. Available potential energy was defined in Eq. (16.29). If the vertical displacements of the density surfaces scale as ΔH ($\Delta H \leq H$, naturally) and if reduced gravity is introduced via $g' = N^2 H$ [see (18.2)], the available potential energy is on the order of $\rho_0 g' \Delta H^2 A$. For eddies of average size L, vorticity scales as U/L and enstrophy as $(U/L)^2 HA$. Finally, if we invoke geostrophy to set the velocity scale, we state $f_0 U \sim g' \Delta H / L$ (barring a substantial barotropic component) and write

$$KE \sim \rho_0 \left(\frac{g'\Delta H}{f_0 L}\right)^2 HA \quad (18.32a)$$

$$APE \sim \rho_0 g' \Delta H^2 A \quad (18.32b)$$

$$S \sim \left(\frac{g'\Delta H}{f_0 L^2}\right)^2 HA. \quad (18.32c)$$

The ratio of kinetic to potential energy is

$$\frac{KE}{APE} \sim \frac{g'H}{f_0^2 L^2} = \left(\frac{R}{L}\right)^2, \tag{18.33}$$

where R is the internal radius of deformation. We recognize here the Burger number of Eq. (16.34).

As the interactions among vortices proceed, the shearing and tearing of vortices introduce motions at ever shorter scales, until frictional dissipation becomes important. Because S increases much faster than KE with decreasing length scales, while APE is unaffected, friction removes disproportionally more enstrophy than kinetic energy and, surely, potential energy. In first approximation, we can assume that the total energy is conserved, while enstrophy decays with time. In a fixed domain ($HA = $ constant) and with constant f_0 and g' values (no heating or cooling), a decrease in enstrophy requires, by virtue of (18.32c), a decrease in the ratio $\Delta H/L^2$.

At short length scales ($L \ll R$), the energy consists primarily of kinetic energy, via (18.33), and its near conservation requires that $\Delta H/L$ remain approximately constant. The only possibility that satisfies both requirements is a steady increase of the length scale L, with a proportional increase in eddy amplitude ΔH. Thus, vortices become, on average, larger and stronger. Obviously, to conserve the total space allowed to them, they also become fewer. There is thus a natural tendency toward successive eddy mergers. With every merger, energy is consolidated into larger structures with concomitant enstrophy losses. Thus, the eddy field gradually assumes a dual pattern with ever fewer and larger vortices (coherent structures of increasing length scale) swimming in an increasingly sheared and disordered interstitial fluid (incoherent flow of decreasing length scale).

As the length scale of vortices increases toward the radius of deformation, the relative importance of potential energy increases. Because APE increases like ΔH^2, further increase in mean eddy amplitude ΔH requires a corresponding decrease in kinetic energy to preserve total energy, and $\Delta H^2/L^2$ must begin to decrease. The result is that ΔH and L continue to increase but no longer proportionally, L increasing faster than ΔH.

As the length scale continues to increase, indicating continued merging activity, it will eventually become much larger than R. Then, the energy is primarily in the form of potential energy, and its conservation requires a saturated value of ΔH. Further enstrophy decrease under frictional action is possible only with a further increase in length scale L (Rhines, 1975; Salmon, 1982). In sum, the interactions of a large number of vortices without addition of energy lead to an irreversible tendency toward fewer and larger vortices. This implies an emergence of coherent structures from a random initial vorticity field. As for the mean eddy amplitude, it increases only up to a certain point. The maximum possible eddy amplitude is achieved when almost all the energy available is in the form of available potential energy—that is,

$$\Delta H_{max} \sim \sqrt{\frac{E}{\rho_0 g' A}}, \qquad (18.34)$$

where E is the total energy present in the system ($E = KE + APE$), and A is the horizontal area of the system. Should this value exceeds the depth H of the domain, vortex amplitudes will be limited by the latter and not all the energy can be turned into potential energy; a certain portion of the energy must remain in the form of kinetic energy, implying a limit to the length scale L.

Pioneering results concerning the emergence of coherent vortices in quasi-geostrophic turbulence can be found in McWilliams (1984, 1989). Let us also note that the tendency toward successive merger is at the basis of the theories (Galperin, Nakano, Huang & Sukoriansky, 2004; Williams and Wilson, 1988, and references therein) that explain the persistence of the Great Red Spot in Jupiter's atmosphere (Fig. 1.5). This begs the question as to why no single dominating vortex occurs in our terrestrial atmosphere as on Jupiter. The answer lies in diabatic and orographic effects constantly acting to form and destroy existing atmospheric vortices. In other words, geostrophic turbulence in the earth's atmosphere is never freely evolving for very long. Similarly, wind forcing over the ocean and dissipation by internal waves and in coastal areas combine to prevent oceanic geostrophic turbulence from following its intrinsic evolution.

18.4   SIMULATIONS OF GEOSTROPHIC TURBULENCE

In the study of many interacting eddies and the statistical analysis of geostrophic turbulence, lateral boundaries are not the focus of the investigation, and may be ignored by invoking periodicity in space (as if the same domain were repeating itself in the several directions of space). When this is the case, a particular *meshless* spectral method can be used for numerical simulations. We already encountered a spectral technique for the solution of linear problems (see Section 8.8), and we now adapt it to a nonlinear problem.

For simplicity, we design the numerical scheme for a one-layer quasi-geostrophic system with rigid-lid and scale-selective biharmonic dissipation of vorticity (see Section 10.6 on filtering). The governing equations on the f-plane are thus

$$\frac{\partial q}{\partial t} + J(\psi, q) = -\mathcal{B}\left(\frac{\partial^4 q}{\partial x^4} + 2\frac{\partial^4 q}{\partial x^2 \partial y^2} + \frac{\partial^4 q}{\partial y^4}\right) \qquad (18.35a)$$

$$\frac{\partial^2 \psi}{\partial x^2} + \frac{\partial^2 \psi}{\partial y^2} = q \qquad (18.35b)$$

where the parameter $\mathcal{B}$ controls the strength of the damping. Both stream-function ψ and potential vorticity q are expanded as truncated series of sine and cosine functions spanning the periodic domain of interest $0 \leq x \leq L_x$ and

$0 \leq y \leq L_y$. For convenience, we use complex exponentials instead of sine and cosine functions and write

$$\tilde{\psi}(x, y, t) = \sum_k \sum_l \Psi_{kl}(t) e^{i \frac{2\pi kx}{L_x}} e^{i \frac{2\pi ly}{L_y}}, \qquad (18.36a)$$

$$\tilde{q}(x, y, t) = \sum_k \sum_l \mathcal{Q}_{kl}(t) e^{i \frac{2\pi kx}{L_x}} e^{i \frac{2\pi ly}{L_y}}. \qquad (18.36b)$$

The time-dependent coefficients Ψ_{kl} and $\mathcal{Q}_{kl}$, also called *spectral coefficients*, are the amplitudes of the spatial Fourier modes, and their value over time constitutes the solution to the problem. We obtain equations governing their evolution by multiplying (18.35) by $\exp(-2i\pi k'x/L_x)$ $\exp(-2i\pi l'y/L_y)$ and integrating over the extent of the domain. The orthogonality of the sine and cosine functions then isolates the time evolution of $\mathcal{Q}_{k'l'}$ and, after relabeling k' as k and l' as l, leads to

$$\frac{\partial \mathcal{Q}_{kl}}{\partial t} + \frac{1}{L_x L_y} \int_0^{L_x} \int_0^{L_y} J(\tilde{\psi}, \tilde{q}) e^{-i \frac{2\pi kx}{L_x}} e^{-i \frac{2\pi ly}{L_y}} \, \mathrm{d}y\mathrm{d}x = -\alpha_{kl} \mathcal{Q}_{kl} \qquad (18.37)$$

with

$$\alpha_{kl} = \mathcal{B} \left[\left(\frac{2\pi k}{L_x} \right)^2 + \left(\frac{2\pi l}{L_y} \right)^2 \right]^2. \qquad (18.38)$$

Note how the dissipation term involving derivatives was nicely transformed into an algebraic operation and how clearly the attenuation of amplitudes for different wavelengths can be seen in the value of α_{kl}.

Applying the same spectral projection to Eq. (18.35b) that serves as the definition of potential vorticity, we obtain the governing equation for streamfunction amplitudes:

$$-\left[\left(\frac{2\pi k}{L_x} \right)^2 + \left(\frac{2\pi l}{L_y} \right)^2 \right] \Psi_{kl} = \mathcal{Q}_{kl}. \qquad (18.39)$$

The solution of what was the Poisson equation is now trivial since the equation is now algebraic and retrieving the streamfunction from vorticity is reduced to a division. Note that for wavenumber $k = l = 0$, there is no need for a division by zero, because the streamfunction is defined up to an arbitrary constant Ψ_{00}, which plays no role in the dynamics. All linear operations are trivial to perform in the so-called *spectral space*, i.e., in the discrete (k, l) space associated with wavelengths L_x/k and L_y/l. Also, initialization of the fields from a given streamfunction in physical space can easily be translated into initial conditions for the spectral coefficients. Because periodic boundary conditions are already taken into account by the use of periodic functions in the truncated series, all we have to do is to calculate the time evolution of the amplitudes using Eq. (18.37). Once amplitudes are known, series (18.36) can be evaluated at any desired location (x, y) to obtain the solution in physical space.

There remains, however, to calculate the contribution from the nonlinear Jacobian term that appears inside the integral term of Eq. (18.37). Each derivative appearing in this term can be evaluated by means of the derivatives of the basis functions, for example

$$\frac{\partial \tilde{\psi}}{\partial x} = \sum_k \sum_l a_{kl} \, \mathrm{e}^{\mathrm{i} \frac{2\pi kx}{L_x}} \, \mathrm{e}^{\mathrm{i} \frac{2\pi ly}{L_y}}$$

$$a_{kl} = \mathrm{i} \frac{2\pi k}{L_x} \, \Psi_{kl} \tag{18.40}$$

and similarly for the other derivatives $\partial \tilde{\psi}/\partial y$, $\partial \tilde{q}/\partial x$, and $\partial \tilde{q}/\partial y$. The Jacobian term can then be expressed from products of these series-expansions and becomes

$$J(\tilde{\psi}, \tilde{q}) = \frac{4\pi^2}{L_x L_y} \sum_i \sum_j \sum_m \sum_n (jm - in) \, \Psi_{ij} \mathcal{Q}_{mn} \, \mathrm{e}^{\mathrm{i} \frac{2\pi (i+m)x}{L_x}} \, \mathrm{e}^{\mathrm{i} \frac{2\pi (j+n)y}{L_y}}. \tag{18.41}$$

Then, because of the orthogonality property of the basis functions, most of those terms after projection onto the (k, l) component in Eq. (18.37) vanish, except the terms for which $i + m = k$ and $j + n = l$. Using Eq. (18.39) to eliminate the streamfunction amplitudes and introducing the so-called interaction coefficients c_{mnkl}, we finally arrive at the equations governing the temporal evolution of the spectral coefficients of potential vorticity:

$$\frac{\partial \mathcal{Q}_{kl}}{\partial t} = -\alpha_{kl} \mathcal{Q}_{kl} - \sum_m \sum_n c_{mnkl} \, \mathcal{Q}_{mn} \mathcal{Q}_{k-m, l-n}. \tag{18.42}$$

For the time stepping of $\mathcal{Q}_{kl}$, any previous numerical method may be used since the interaction coefficients c_{mnkl} are known as well as parameters α_{kl}, which depend on the particular form of physical dissipation utilized in the model. The nonlinear term clearly reflects the physical interaction between wave components (eddies) at different scales.

The method has the advantage of automatically including periodic boundary conditions and avoiding the inversion of a Poisson equation at every time step. It further avoids another problem that can plague nonlinear numerical models, namely the aliasing of higher, unresolved wavenumbers created by interactions into lower, resolved wavenumbers (see Section 1.12 and 10.5). As we work in spectral space, we can easily disregard the higher wavenumbers that are created by nonlinear combinations by setting the corresponding interaction coefficients to zero.

The method is thus appealing but has a major drawback, its computational cost. If we retain N Fourier coefficients for each spatial direction, each sum in

Eq. (18.42) involves N terms, and a double sum requires N^2 operations. For each of the N^2 equations for Fourier amplitudes, we must therefore perform N^2 operations for the sums associated with the nonlinear term. The computational burden is then proportional to N^4 (M^2 in terms of the number M of unknowns). Because one aim of a geostrophic turbulence model is the study of turbulence per se, unrealistic subgrid-scale parameterizations must be avoided, and very high spatial resolution must be attempted. The computational cost of the present approach can rapidly become prohibitive.

A major breakthrough for the spectral approach came with the discovery of the so-called *Transform Method* (e.g., Orszag, 1970). The basic idea is to switch back and forth between physical space and spectral space, performing each task in the space in which it is least burdensome. Thus, one first calculates derivatives in spectral space by generating coefficients as in Eq. (18.40). The spatial derivative can then be calculated at any location from the Fourier series, on a regular grid spanning the physical domain of interest. Once all derivatives appearing in the Jacobian have been so calculated, the latter can be calculated in physical space from products at each grid node. With the numerical value of the Jacobian known on a regular grid, its spectral amplitudes are then obtained in the wavenumber space, which permit the calculation of time changes of the spectral coefficient of vorticity (Fig. 18.16). Because in the physical domain, the cost of operations associated with the Jacobian is only proportional to the number of grid points M, there is a net gain from the detour if the cost of the transformation falls below M^2.

For such method to be viable, the number of calculations involved in the switch back and forth between physical domain and spectral space must be lower than the number of calculations saved by performing the tasks in their respective space of greatest ease. This is the case because there exists a fast transformation method to pass back and forth between physical domain and spectral space.

For a one-dimensional case, the Fourier transformation can be achieved efficiently by the so-called Fast Fourier Transform (FFT, see Appendix C), which

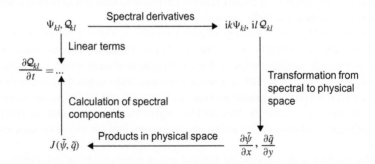

FIGURE 18.16   Schematic representation of the transform method applied to the evaluation of the Jacobian as the forcing term for spectral components $\mathcal{Q}_{kl}$.

demands only $N \log N$ operations for N Fourier modes and N grid points. In two dimensions, we first perform N FFTs along the x-direction, one for each of the y values. Each of these N transforms costs $N \log N$ operations, and we thus perform $N^2 \log N$ operations. We then perform N FFTs of the latter coefficients in the other direction, requiring another $N^2 \log N$ operations. In total, $N^2 \log N^2$, or in terms of the total number M of unknowns, $M \log M$ operations are needed for one passage from physical domain to spectral space or vice versa. With M operations needed for the calculation of the Jacobian by products in physical space, the chief cost is that of the transformation, but this is less than if the Jacobian had been computed in the spectral space, at the cost of M^2 calculations. The larger M (and M must be large indeed to achieve high resolution), the greater the reduction of operations.

The transform method can, of course, be generalized to any other term that is not easily calculated in spectral space, such as nonlinear source terms for tracers. All that needs to be done is to pass from spectral space to physical domain using an FFT scheme, calculate those complicated terms on the spatial grid and then return to spectral space by the inverse transformation.

Convergence of the truncated spectral series to exact solutions can be shown to be faster than any power of M as long as the solution is smooth for all derivatives. This makes the method extremely attractive. We note in passing that we found a way to solve a classical Poisson equation on a periodic domain with $M \log M$ operations through the transform approach.

Unfortunately, the advantage of the transform method is partially mitigated by a drawback, the aliasing due to products in physical space. As seen in Section 10.5, for a grid spacing of Δx, we can avoid any aliasing in the quadratic terms if modes with wavelengths between $2\Delta x$ and $3\Delta x$ are systematically removed from the solution. But removing modes downgrades resolution, and the requirement must be turned the other way round: For the shortest wavelength λ that needs to be resolved, we have to create a physical grid such that $\Delta x = \lambda/3$ instead of $\Delta x = \lambda/2$, the strict minimum needed to resolve λ. On this grid, the amplitudes of projected modes between $2\Delta x$ and $3\Delta x$ are absent by construction, and no aliasing can occur when computing quadratic terms. In other words, we simply have to use $3/2N$ grid points instead of N to be sure that the product in physical space does not lead to nonlinear numerical instabilities. In practice, such a generation of a higher-resolution sampling can be performed efficiently by first padding with $N/2$ zeros the arrays containing Fourier coefficients and then applying the transformation. This amounts to assigning zero amplitudes to higher-wavenumber signals and then calculating the resulting series expansions on a regular grid with $3N/2$ points using standard FFTs (see Appendix C). This method of proceeding permits to retain the advantage of a spectral method while avoiding aliasing.

An additional advantage of working in spectral space lies in the fact that spectral analysis of the results is trivial and that the power spectrum of initial conditions is easily controlled. This is particularly useful for statistical analysis of geostrophic turbulence which starts from a random field of known

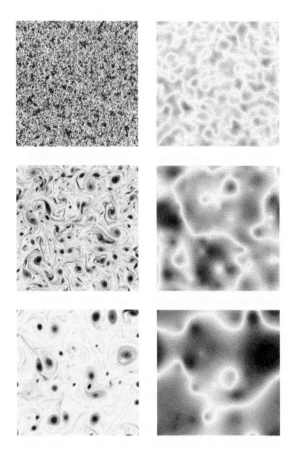

FIGURE 18.17 Emergence of isolated vortices from a random initial field. Vorticity (left) and streamfunction (right) in a periodic domain at three successive moments. The simulation was performed with a spectral method implemented in qgspectral.m. A color version can be found online at http://booksite.academicpress.com/9780120887590/

statistical spectral profile. Generally, the fields are generated as a realization of random streamfunctions with a Gaussian distribution of zero mean and variances depending on wavenumbers. Simulations then allow to see how vortices become organized under different dissipation conditions (Fig. 18.17) and how power spectra evolve.

ANALYTICAL PROBLEMS

18.1. Consider the center fluid parcel ($y = 0$) of the Gaussian jet $u(y) = U \exp(-y^2/2L^2)$ with $U = 10$ m/s and $L = 100$ km. On the f-plane, what is the shear vorticity acquired by that parcel in a rightward meander of curvature $\mathcal{K} = 1/800$ km? On the beta plane, what meridional displacement Y would permit the parcel to conserve its speed and maintain its center position?

18.2. For the one-layer reduced gravity model (12.19), express the gradient-wind balance for steady circular vortices on the f-plane. If the layer thickness is H at the center, and the density interface outcrops at radius a [i.e., $h(r=0)=H$, $h(r=a)=0$], show that H and a must satisfy the inequality

$$a \geq \frac{\sqrt{8g'H}}{f}. \tag{18.43}$$

18.3. Take a stretch of the jet profile $u(y)=U(1-|y|/a)$ in $|y| \leq a$, $u(y)=0$ elsewhere (see Fig. 10.13), and bend it to create a clockwise vortex. On the f-plane and in the absence of vertical variations, what is the orbital-velocity profile that preserves vorticity on every trajectory? How does the pressure anomaly in the vortex is compared with the pressure difference across the jet? Finally, show that the proportion of fluids with each vorticity is the same in the vortex as in the straight jet.

18.4. Determine the behavior of an eastward jet in the northern hemisphere flowing over a topographic step-up followed by a step-down of equal height. Is the flow oscillatory beyond the second step? Also discuss the cases where the distance between the two steps is short and long compared with the critical meander scale.

18.5. Redo Problem 18.4 for a westward jet in the southern hemisphere.

18.6. Hurricane Hugo (10–22 September 1989 in the western North Atlantic— see Fig. 18.18) had a maximum wind speed of 62 m/s and a low central pressure of 941.4 millibars during its passage over Guadeloupe on 17 September (Case & Mayfield, 1990). Assuming that the normal pressure outside the hurricane was 1010 millibars, estimate the storm's radius and importance of the centrifugal force relative to the Coriolis force (latitude $= 16°N$).

18.7. Using the gradient-wind balance (18.11) in a reduced-gravity model ($p = \rho_0 g'h$), explore lens-like vortex solutions where the interface exhibits a paraboloidal shape between a central maximum depth ($h=H$ at $r=0$) and a peripheral outcrop ($h=0$ at $r=R$). Show that the flow is in solid-body rotation. Relate vortex radius R to central depth H and discuss the limiting cases of wide/shallow and narrow/deep vortices. Do you recover an inequality of type (18.16)?

18.8. In first-approximation, the thick atmosphere of Jupiter may be modeled as a reduced-gravity system with $g' = 2.64 \, \text{m/s}^2$. Knowing that planet radius is 69,000 km, and that one Jovian day is only 10 Earth hours long,

FIGURE 18.18   Satellite visible image of Hurricane Hugo in the evening of 21 September 1989 as it approached the south-eastern coast of the United States. (*Courtesy of NOAA, Department of Commerce, Washington, D.C.*)

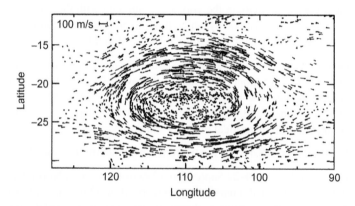

FIGURE 18.19   Velocity field on Jupiter in and around the Great Red Spot, obtained after tracking small cloud features in sequential images from *Voyager* spacecraft. The origin of each vector is indicated by a dot. (*From Dowling & Ingersoll, 1988*)

derive the thickness h of moving fluid for a few radial sections across the wind-velocity chart provided in Fig. 18.19.

18.9. A uniform eastward flow of velocity U over a flat bottom approaches a topographic step at right angle. Topography changes from H_0 to H_1 at $x = 0$. Determine the stationary streamfunction for $x > 0$ under the rigid-lid approximation and on the beta plane. Can you identify the critical meander scale in the solution? (*Hint*: On a streamline, potential vorticity is conserved. Express that at the step the relationship between the value of the streamfunction and vorticity is known and is thus valid beyond the step.)

18.10. Redo Analytical Problem 18.9 for a westward flow.

18.11. Prove Eq. (18.7). (*Hint*: Use Frenet coordinates.)

NUMERICAL EXERCISES

18.1. Use `qgspectral.m` to experiment with different eddy fields. Then include the beta effect and higher-order dissipation using sixth-order derivatives.

18.2. Include diagnostics on energy, enstrophy, and wavelength into `qgspectral.m` and simulate with different eddy viscosity values. Also include diagnostics on

$$k_e = \frac{\int k\,|k\Psi_k|^2\,\mathrm{d}k}{\int |k\Psi_k|^2\,\mathrm{d}k} \tag{18.44}$$

$$k_o = \frac{\int k\,|\mathcal{Q}_k|^2\,\mathrm{d}k}{\int |\mathcal{Q}_k|^2\,\mathrm{d}k}, \tag{18.45}$$

where integrals are performed over all wavenumbers $k = \sqrt{k_x^2 + k_y^2}$. Look at the time evolution of these quantities and provide an interpretation.

18.3. Generalize `qgspectral.m` to a two-layer system with equations given in (17.31). In particular, mind the vertical coupling in (17.32) and solve the coupled problem in spectral space exactly. Vary the parameter R to explore the effect of vertical stratification.

Melvin Ernest Stern
1929–2010

Melvin Stern was an important contributor to the GFD Summer Program at the Woods Hole Oceanographic Institution (see historical note at the end of Chapter 1) and had a major influence on the evolution of the field ever since the inception of that program. His early work in meteorology was followed by fundamental contributions to our understanding of baroclinic instability (work with Jule Charney) and of salt fingering (an oceanic small-scale diffusive process). After publishing a book titled *Ocean Circulation Physics* (Academic Press, 1975), Stern dedicated an increasing amount of time and effort to the investigation of vortices. He is credited with the *modon* solution (Section 16.6), and seminal studies of jets and jet-vortex interactions, complementing his theoretical results with original and illuminating laboratory experiments. (*Photo by the first author*)

Peter Douglas Killworth
1946–2008

A student of Adrian Gill (see Biography at the end of Chapter 13), Peter Killworth, made a career of touching on almost all aspects of ocean modeling. He is renowned for his contributions to both theoretical oceanography, including wave and stability analysis, and numerical model developments. His breadth of interests went as far as social networks, which he analyzed using mathematical and modeling skills that he cultivated while working on oceanographic problems.

In addition, Killworth gained a reputation as an extremely incisive and responsive editor from authors and reviewers of *Ocean Modelling*. He was awarded numerous honors "for his many far-reaching contributions to theoretical oceanography, which have significantly enlarged our understanding of the processes determining ocean circulation." (*Photo by Sarah Killworth*)

Part V

Special Topics

Chapter 19

Atmospheric General Circulation

ABSTRACT

This chapter briefly reviews the principal factors controlling the climate on our planet. We first summarize the global heat budget and then describe the large-scale convective cells and review the major wind systems. The chapter ends with weather forecasting and the particular challenge of simulating cloud dynamics. Ingredients of modern operational weather-forecast models are described.

19.1 CLIMATE VERSUS WEATHER

Climate is to be distinguished from *weather*. Whereas *weather* includes the detailed behavior of the atmosphere on a timescale of a day to a week, *climate* represents the prevailing or average weather conditions over a period of years. In other words, the climate of the earth can be regarded as the basic state of the atmosphere, subject to variations over years, centuries, millennia, and beyond, whereas the weather corresponds to its incessant and short-lived instabilities. The engine of climate is a global convection carrying heat from the warmer tropical belt to the much colder polar regions, and its primary manifestation is the distribution of prevailing winds over the globe.

Numerous books have been written on climate and weather dynamics. For texts presenting materials at a slightly deeper level than presented here, the reader is referred to the classic book by Gill (1982) and the highly readable textbook by Marshall and Plumb (2008), each written from the perspective of geophysical fluid dynamics.

19.2 PLANETARY HEAT BUDGET

Because the long-term gradual cooling of the earth's core contributes insignificantly to the heat input near the surface, the incoming solar radiation can be considered as the sole source of heat. From its hot surface ($T \simeq 5750$ K), the sun emits most of its energy in short wavelengths (200–4000 nm; 1 nm $= 10^{-9}$ m), of which about 40% is in the visible range (400–670 nm). According to the

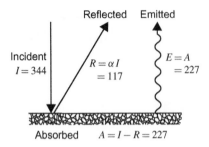

FIGURE 19.1 Simplest possible model of the earth's budget. Straight lines indicate short-wave radiation, whereas the wavy line represents long-wave radiation. (Fluxes are in watts per square meter.) Under this scenario, which does not account for the atmosphere, the earth's average temperature would be a freezing −21°C.

Stefan–Boltzmann law, a so-called *black body* (a perfect emitter and absorber of radiation) emits a radiative flux F depending on its temperature

$$F = \sigma T^4, \tag{19.1}$$

where σ is a constant equal to 5.67×10^{-8} W m^{-2} K^{-4}, and T is the absolute temperature. Idealizing the sun to a black body, we obtain $F_{sun} = 6.2 \times 10^7$ W/m^2 as the outgoing energy flux from the sun's surface. Given the size of the sun, the sun–earth distance, and the earth's area exposed to the sun, the earth receives only a minute fraction of the sun's output: 1376 W/m^2. Averaged over the entire earth's surface (equal to four times the projected area facing the sun), this incident flux amounts to $I = 344$ W/m^2.

Let us, at this point, first discard the thickness of the atmosphere and idealize the earth's land and sea surface plus atmosphere as a thin sheet insulated from below. Of the incident radiation, a fraction is reflected out to space by snow, ice, some types of clouds, and everything else that is bright. With α as the reflection coefficient, called the *albedo* ($\alpha \simeq 0.34$), the amount of radiation reflected is $R = \alpha I = 117$ W/m^2. The difference is the amount absorbed by the earth's surface: $A = I - R = (1 - \alpha)I = 227$ W/m^2 (Fig. 19.1). Because the earth is in overall thermal equilibrium[1] (its temperature is not constantly rising), its outgoing radiation matches absorption, and the earth emits a radiative flux E equal to A. This outgoing radiation is in the form of longer wavelengths than the incoming solar radiation and is termed long-wave radiation. Assuming as for the sun that the earth behaves as a black body and using the preceding values, we state

$$\sigma T^4 = E = 227 \text{ W/m}^2 \tag{19.2}$$

and deduce a mean temperature for the earth to be $T = 251$ K $= -21°$C. This value is obviously much less than the average temperature of the earth as we know it (about 15°C). The failure of this simple model resides in the neglect

---

[1] Some of the heat received by the sun is transformed into mechanical (wind) and chemical (photosynthesis) energies, but these eventually dissipate and turn back into heat.

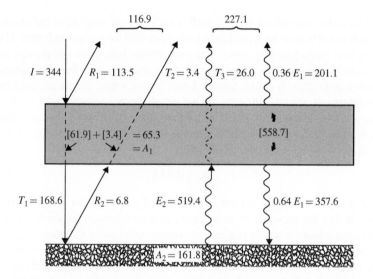

FIGURE 19.2 A second model of the earth's budget, which distinguishes the atmospheric layer from the earth's surface. All flux values are in watts per square meter. Under this scenario, the earth's average temperature would be a very warm $+36°C$. Here the greenhouse effect (flux loop between the earth's surface and the atmosphere) is present and exaggerated. Note how this effect causes the long-wave radiative fluxes from the earth and atmosphere to exceed the incident short-wave radiative flux from the sun.

of the atmospheric layer. The preceding value is more representative of the temperature at the top of the atmosphere than at ground level.

As a next step, we distinguish the atmosphere from the earth's surface (Fig. 19.2). The incident short-wave radiation from the sun is unchanged ($I = 344$ W/m^2); of it, the fraction α_1 ($= 0.33$) is reflected back to space, primarily by clouds and secondarily by particulate matter ($R_1 = \alpha_1 I = 113.5$ W/m^2), the fraction β_1 ($= 0.49$) is transmitted to the earth's surface ($T_1 = \beta_1 I = 168.6$ W/m^2), and the rest is absorbed by the atmosphere. The earth's surface (snow, ice, and so on) reflects a fraction α_2 ($= 0.04$) of what it receives ($R_2 = \alpha_2 T_1 = 6.8$ W/m^2) and absorbs the rest ($A_2 = T_1 - R_2 = 161.8$ W/m^2). Of the portion R_2 reflected from the earth's surface, the fraction β_1 is transmitted through the atmosphere and out to space ($T_2 = \beta_1 R_2 = 3.4$ W/m^2), whereas the rest is absorbed by the atmosphere. Thus the atmosphere absorbs short-wave radiation directly from the sun ($I - R_1 - T_1$) and indirectly from the earth below ($R_2 - T_2$), and the net is

$$
\begin{aligned}
A_1 &= (I - R_1 - T_1) + (R_2 - T_2) \\
&= [1 - \alpha_1 - \beta_1 + \beta_1 \alpha_2 (1 - \beta_1)]I \\
&= 65.3 \text{ W/m}^2.
\end{aligned}
\tag{19.3}
$$

Then both the atmosphere and the earth's surface emit long-wave radiation, in amounts equal to their total intakes of both short- and long-wave radiation. If the atmosphere emits a flux E_1, some of it goes upward into space and the rest goes downward to the earth. Because the top of the atmosphere, where the outgoing radiation originates, is colder than its lower layers, where the earthbound radiation originates, the two amounts are not equal; a representative split is 36% to space and 64% to the earth. Thus, the earth receives $0.64E_1$ of long-wave radiation from the atmosphere in addition to the amount A_2 received in short waves, and its emission E_2 must equal their sum:

$$E_2 = A_2 + 0.64E_1. \tag{19.4}$$

At this point, we still do not know either E_1 and E_2, but we can already conclude that the presence of atmospheric radiation toward the earth's surface establishes a loop, whereby the earth's surface emits some radiation, a portion of which returns to the earth. As a consequence, the earth's surface must emit more radiation in the presence of an atmosphere than in its absence and (according to the Stefan–Boltzmann law) must be correspondingly warmer. This is the *greenhouse effect*, so called because of its believed similarity with the trapping of long-wave infrared radiation by the glass panes of a greenhouse.[2]

Of the amount E_2 radiated by the earth's surface and entering the atmosphere, a fraction β_2 (= 0.05) is transmitted and lost to space ($T_3 = \beta_2 E_2$), with the remainder being absorbed by the atmosphere ($E_2 - T_3$). If the atmosphere absorbs the amounts A_1 and $E_2 - T_3$ in short- and long-wave radiations, respectively, its total emission must be equal to their sum; that is,

$$\begin{aligned} E_1 &= A_1 + E_2 - T_3 \\ &= A_1 + (1 - \beta_2)E_2. \end{aligned} \tag{19.5}$$

From Eqs. (19.4) and (19.5), we can obtain the emission fluxes E_1 and E_2 to find $E_1 = 558.7$ W/m^2 and $E_2 = 519.4$ W/m^2. Note that both are higher than the incident flux $I = 344$ W/m^2. Then, using the Stefan–Boltzmann law (19.1), we estimate the mean temperature of the earth's surface to be $T = (519.4/\sigma)^{1/4} =$ 309 K = 36°C. This temperature is higher than the first estimate, thanks to the capping effect of the atmosphere (greenhouse effect) but is unrealistically high.

In reality, the warming influence of the greenhouse effect is partially short-circuited by the hydrological cycle. As water evaporates over the ocean and land, latent heat is extracted from the earth's surface. (Latent heat is the heat required to change the phase of a substance, here to transform liquid water into water vapor. The latent heat of water is 2.5×10^6 J/kg.) This water vapor rises through the atmosphere, where it condenses in clouds before returning

---

[2] In fact this belief is incorrect because heat retention in a physical greenhouse is chiefly because of the elimination of convection by the glass barrier. Greenhouses covered with polyethylene plastic are about as effective as glass-covered greenhouses.

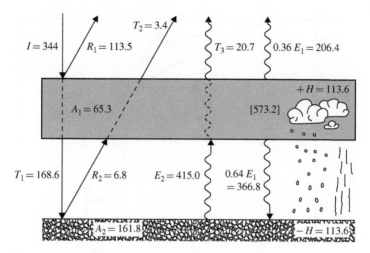

FIGURE 19.3   A third model of the earth's budget, with atmosphere and hydrological cycle. All flux values are in watts per square meter. This scenario includes the greenhouse effect tempered by the hydrological cycle, resulting in a realistic average temperature at the earth's surface of $+19°C$.

to the earth's surface as rain (liquid phase). Thus, the latent heat extracted from the earth's surface is released in the atmosphere, causing a net heat flux from the earth to the atmosphere that is not in the form of radiation. To this latent-heat flux is added a convective heat transfer. With an estimated total non-radiative heat flux $H = 113.6$ W/m^2, the earth and atmospheric balances, (19.4) and (19.5), must be amended as (Fig. 19.3):

$$E_2 = A_2 + 0.64\, E_1 - H, \tag{19.6a}$$

$$E_1 = A_1 + E_2 - T_3 + H, \tag{19.6b}$$

yielding $E_1 = 573.2$ W/m^2 and $E_2 = 415.0$ W/m^2. From the radiation law, we deduce a corrected estimate of the mean temperature at the earth's surface: $T = (415.0/\sigma)^{1/4} = 292$ K $= 19°C$. This third estimate is in good agreement with the seasonally and globally averaged temperature on the earth's surface. All in all, we conclude that the greenhouse effect resulting from the presence of the atmosphere (especially with regard to its near opacity to long-wave radiation) raises the temperature of the earth's surface and that the impact of this effect is partially canceled by the hydrological cycle.

19.3   DIRECT AND INDIRECT CONVECTIVE CELLS

The preceding considerations exposed the globally averaged heat budget, glossing over all spatial variations. However, that the tropical regions of the globe receive a disproportionate amount of solar radiation, because of their better exposure, is not to be overlooked. The earth receives considerably more heat

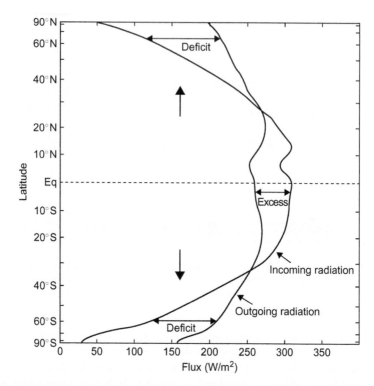

FIGURE 19.4 Averaged radiation flux by latitude, calculated from satellite data over the period 1974–1978. The latitude scale simulates the amount of surface area between latitude bands. Incoming radiation is the short-wave solar radiation absorbed by the earth and atmosphere. Outgoing radiation is the long-wave radiation leaving the atmosphere. (*From Winston et al., 1979*)

at low latitudes than near the poles, but its outgoing radiation is more uniformly spread, decreasing only slightly with latitude (Fig. 19.4). The resulting heat excess at low latitudes and deficit at high latitudes call for a poleward heat transfer. George Hadley[3] hypothesized that this transfer is accomplished by a giant thermally driven circulation: Warm tropical air rises and flows toward each pole, where it cools and sinks, returning to the tropics along the surface (Hadley, 1735). As it turns out, Hadley was partly correct, insofar as such convective circulations exist on both sides of the equator, and partly incorrect, insofar as these meridional circulations extend only to 30° of latitude.

Indeed, a single Hadley cell spanning equator to pole is unlikely to exist because of conservation of angular momentum. In the absence of friction, a torus of equatorial air mass m at rest with respect to the earth conserves its

---

[3] British physicist and meteorologist (1685–1768) who first explained the trade winds.

absolute angular momentum. Hence, when progressing poleward to latitude φ, it ought to conserve $ma^2\Omega = mr(\Omega r + u) = ma\cos\varphi\,(\Omega a\cos\varphi + u)$, where a is the earth's radius and $r = a\cos\varphi$ is the distance of the torus to the earth's axis of rotation. This would lead to unrealistically high wind velocities u at high latitudes.

North of 30°N and south of 30°S, different circulations are observed, up to 60°, beyond which circulations in the sense predicted by Hadley are again found. Because the convective circulations theorized by Hadley follow our intuition, they are generally called *direct cells*. Those direct cells bordering the equator are also called *Hadley cells*. In contrast, the reverse circulations found at midlatitudes bear the name of *indirect cells*. Our purpose here is to explain, in some qualitative manner, why such oppositely directed meridional circulations exist. The story is not simple, invoking the aggregate effect of the transient weather systems (storms) of the midlatitude regions.

To begin, we note that, although a single direct convective cell could theoretically span an entire hemisphere, such would be unstable. The strong zonal flow in thermal-wind balance with the large meridional temperature gradient would be baroclinically unstable. In fact, the more moderate zonal winds accompanying the alternating circulation structure that exists on the earth are themselves unstable, as the vagaries of the midlatitude weather show so well. According to our discussion of baroclinic instability (Section 17.4), such instabilities develop into coherent vortex systems, called cyclones and anticyclones, that are capable of transferring heat meridionally (Section 17.5). At midlatitudes, therefore, the transfer does not take place in a vertical loop, as in a Hadley cell, but through the horizontal circulation of each vortex moving warm air poleward on one side and cold air equatorward on the other. We will now show how the cumulative action of these weather systems at midlatitudes can perform the required poleward transfer of heat so effectively as to reverse the meridional circulation in the vertical plane.

The analysis starts with a few modifications of the governing equations. First, the density departure from the reference ρ_0 is expressed in terms of a temperature anomaly T measured from the temperature corresponding to the reference density: $\rho = -\rho_0 \alpha T$, where $\alpha = 1/T_0$ is the thermal-expansion coefficient. Then, viscosity and heat diffusivity are neglected, but a heat source or sink term is added in the temperature equation to represent the heat gain in the tropics and the heat loss at high latitudes. From Eqs. (4.21), we have

$$\frac{\partial u}{\partial t} + u\frac{\partial u}{\partial x} + v\frac{\partial u}{\partial y} + w\frac{\partial u}{\partial z} - fv = -\frac{1}{\rho_0}\frac{\partial p}{\partial x} \tag{19.7a}$$

$$\frac{\partial v}{\partial t} + u\frac{\partial v}{\partial x} + v\frac{\partial v}{\partial y} + w\frac{\partial v}{\partial z} + fu = -\frac{1}{\rho_0}\frac{\partial p}{\partial y} \tag{19.7b}$$

$$\frac{\partial p}{\partial z} = \rho_0 \alpha g T \tag{19.7c}$$

$$\frac{\partial u}{\partial x} + \frac{\partial v}{\partial y} + \frac{\partial w}{\partial z} = 0 \tag{19.7d}$$

$$\frac{\partial T}{\partial t} + u\frac{\partial T}{\partial x} + v\frac{\partial T}{\partial y} + w\frac{\partial T}{\partial z} = \frac{Q}{\rho_0 C_p}, \tag{19.7e}$$

where Q is the aforementioned thermal forcing (in watts per cubic meter). Focusing exclusively on the northern hemisphere, we take Q positive in the tropics (at lower values of y, the northward coordinate) and negative at high latitudes (higher values of y). Thus, the gradient $\partial Q/\partial y$ is negative. The choice of beta-plane equations based on a Cartesian coordinate system over more accurate equations in spherical coordinates is justified in the spirit of a highly simplified analysis aimed at highlighting physical processes in a qualitative way.

We next define the zonal average as the mean over the values of x at any given y and z levels and time t. The zonal averages of the linear equations (19.7c) and (19.7d) are immediate:

$$\frac{\partial \bar{p}}{\partial z} = \rho_0 \alpha g \bar{T} \tag{19.8}$$

$$\frac{\partial \bar{v}}{\partial y} + \frac{\partial \bar{w}}{\partial z} = 0, \tag{19.9}$$

where the overbar denotes this zonal average. With a prime denoting the departure from the average (e.g., $u = \bar{u} + u'$ etc.) and with some use of Eq. (19.7d), the zonal average of Eq. (19.7b) can be expressed as

$$\frac{\partial \bar{v}}{\partial t} + \bar{v}\frac{\partial \bar{v}}{\partial y} + \bar{w}\frac{\partial \bar{v}}{\partial z} + f\bar{u} = -\frac{1}{\rho_0}\frac{\partial \bar{p}}{\partial y} - \frac{\partial}{\partial y}\overline{v'^2} - \frac{\partial}{\partial z}\overline{v'w'}. \tag{19.10}$$

The large meridional pressure gradient $(\partial \bar{p}/\partial y)$ associated with the important northward decrease in temperature $(\partial \bar{T}/\partial y < 0)$ is balanced by a significant zonal flow $(\bar{u})$. In contrast, the meridional cell $(\bar{v}, \bar{w})$ is much weaker, as are the corresponding eddy fluxes $(\overline{v'^2}, \overline{v'w'})$. Thus, the preceding may be reduced to

$$f\bar{u} = -\frac{1}{\rho_0}\frac{\partial \bar{p}}{\partial y}. \tag{19.11}$$

Together, the hydrostatic balance, Eq. (19.8), and the geostrophic relation, Eq. (19.11), provide the thermal-wind relation

$$f\frac{\partial \bar{u}}{\partial z} = -\alpha g\frac{\partial \bar{T}}{\partial y}, \tag{19.12}$$

which relates the vertical shear of the average zonal wind to the average meridional temperature gradient. With the temperature decreasing northward in the northern hemisphere $(\partial \bar{T}/\partial y < 0, f > 0)$, the wind shear is positive $(\partial \bar{u}/\partial z > 0)$, indicating that the winds must become more westerly (eastward) with altitude.

Finally, we apply the zonal average to the remaining two equations, (19.7a) and (19.7e), to obtain:

$$\frac{\partial \bar{u}}{\partial t} + \bar{v}\frac{\partial \bar{u}}{\partial y} + \bar{w}\frac{\partial \bar{u}}{\partial z} - f\bar{v} = -\frac{\partial}{\partial y}\overline{u'v'} - \frac{\partial}{\partial z}\overline{u'w'} \tag{19.13a}$$

$$\frac{\partial \bar{T}}{\partial t} + \bar{v}\frac{\partial \bar{T}}{\partial y} + \bar{w}\frac{\partial \bar{T}}{\partial z} = \frac{\bar{Q}}{\rho_0 C_p} - \frac{\partial}{\partial y}\overline{v'T'} - \frac{\partial}{\partial z}\overline{w'T'}. \tag{19.13b}$$

According to our previous remarks, the eddy fluxes of momentum and heat associated with the horizontal circulations of the weather systems ($\overline{u'v'}$ and $\overline{v'T'}$) are anticipated to be important, and the corresponding terms are retained. However, the vertical eddy fluxes ($\overline{u'w'}$ and $\overline{w'T'}$) are neglected. Except for the mean vertical advection of temperature ($\bar{w}\partial\bar{T}/\partial z$) because there is a substantial vertical stratification, mean meridional and vertical advection is unimportant, compared with the meridional eddy transports. In the light of these considerations, the leading terms of the preceding two equations are

$$\frac{\partial \bar{u}}{\partial t} - f\bar{v} = -\frac{\partial}{\partial y}\overline{u'v'} \tag{19.14}$$

$$\frac{\partial \bar{T}}{\partial t} + \frac{N^2}{\alpha g}\bar{w} = \frac{\bar{Q}}{\rho_0 C_p} - \frac{\partial}{\partial y}\overline{v'T'}. \tag{19.15}$$

Here, we have introduced the stratification frequency N through $N^2 = \alpha g\partial\bar{T}/\partial z$. We shall assume that it does not vary significantly with y.

Forming f times the z-derivative of the first equation plus αg times the y-derivative of the second, to eliminate the time derivatives through Eq. (19.12), we obtain

$$\underbrace{\frac{\partial \bar{w}}{\partial y} - \frac{f^2}{N^2}\frac{\partial \bar{v}}{\partial z}}_{=\omega} = \frac{\alpha g}{\rho_0 C_p N^2}\frac{\partial \bar{Q}}{\partial y}$$

$$- \frac{\alpha g}{N^2}\frac{\partial^2}{\partial y^2}\overline{v'T'} - \frac{f}{N^2}\frac{\partial^2}{\partial y\partial z}\overline{u'v'}. \tag{19.16}$$

In this last equation, the sign of the left-hand side ω is directly related to the direction of the average circulation in the vertical plane. For simplicity, let us restrict our attention again to the northern hemisphere. In a direct cell (Fig. 19.5a), $\bar{w}$ decreases northward and $\bar{v}$ increases upward, together yielding a negative ω. On the other hand (Fig. 19.5b), an indirect cell corresponds to a positive left-hand side.

According to the right-hand side of Eq. (19.16), there are three competing mechanisms influencing the sense of the circulation. In the tropical regions, away from the midlatitude eddy activity, the dominant factor is heating ($\bar{Q}$ term). Because the rate of heating decreases northward ($\partial\bar{Q}/\partial y < 0$), this term is negative, and the circulation in the vertical plane is in the direct sense (as in

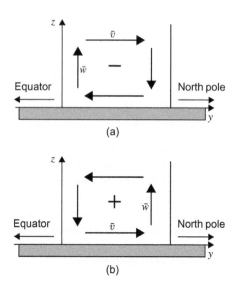

FIGURE 19.5 Atmospheric circulation in the meridional-vertical plane: (a) direct cell, also called Hadley cell, with $\partial \bar{w}/\partial y < 0$, $\partial \bar{v}/\partial z > 0$, and $\omega < 0$, and (b) indirect cell, also called Ferrel cell, with opposite circulation and positive ω.

Fig. 19.5a). This occurs up to about 30°N, and the circulation driven by thermal convection is the Hadley cell. The northerly (equatorward) winds along the surface ($\bar{v} < 0$) veer to the right under the action of the Coriolis force, resulting in easterly (westward) zonal winds ($\bar{u} < 0$). These form the *trade winds*.

North of approximately 30°N, where the eddy activity is most intense, the corresponding terms ($\overline{v'T'}$ and $\overline{u'v'}$) dominate the right-hand side of Eq. (19.16). Both induce an indirect circulation. This is easy to see with the $\overline{v'T'}$ term and a little harder with the $\overline{u'v'}$ term. The average product $\overline{v'T'}$ is proportional to the meridional heat flux of the eddies. Because this net heat flux must be northward, warm anomalies ($T' > 0$) are preferentially moved northward ($v' > 0$), while cold anomalies, are advected southward ($T' < 0$, $v' < 0$), both yielding a net positive $\overline{v'T'}$ correlation. Because the storm activity is most intense at midlatitudes, the term $\overline{v'T'}$ reaches a maximum there. Thus, the second derivative $\partial^2 \overline{v'T'}/\partial y^2$ must be negative. Preceded by a minus sign in Eq. (19.16), the corresponding term is positive.

The convergence of warm and cold air masses aloft creates a locally intensified gradient of temperature. In thermal-wind balance with this gradient is the polar-front jet stream (Fig. 18.1) that flows eastward. The maintenance of this jet in spite of eddy activity requires a continuous influx of eastward momentum (i.e., positive u' anomalies must be transported to that latitude). This is effected by the eddies, which import positive momentum anomalies from the south ($u' > 0$, $v' > 0$) and from the north ($u' > 0$, $v' < 0$). Thus, the average $\overline{u'v'}$ is

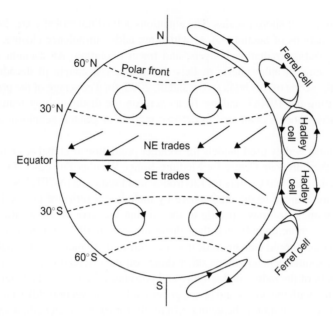

FIGURE 19.6 Sketch of the general atmospheric circulation, composed of direct (Hadley) and indirect (Ferrel) cells in the meridional direction and alternating winds in the zonal direction.

positive south of the jet and negative north of it, and the derivative $\partial \overline{u'v'}/\partial y$ must be negative. At the surface, where the jet stream is not found, the correlation $\overline{u'v'}$ is much less important, and we conclude that $\partial \overline{u'v'}/\partial y$ is increasingly negative with altitude, namely, $\partial^2 \overline{u'v'}/\partial y \partial z$ is negative. Preceded by a minus sign in Eq. (19.16), this term adds to the positive contribution of the other eddy-flux term, and together these terms overcome the $\bar{Q}$ term. The result is an indirect cell, called the *Ferrel cell*. A corresponding indirect cell is found in the southern hemisphere. These Ferrel cells extend to approximately 60°; beyond that latitude, eddy activity yields to a thermal circulation in the vertical and direct cells exist (Fig. 19.6).

The alternation of direct and indirect cells across latitudes leads to a similar alternation in surface zonal winds: from the easterly trades to the prevailing westerlies, to the polar easterlies (Fig. 19.6).

19.4   ATMOSPHERIC CIRCULATION MODELS

Atmospheric circulation models are generally at the forefront of developments in both parameterization of subgrid-scale processes and numerical aspects (see Section 1.9) and continue to change in terms of included physics and numerical discretizations. From the first operational models using a single-layer quasi-geostrophic approach, models have progressed to solving the primitive equations at ever shorter scales. The equations solved numerically are the governing equations of Section 4.4, to which are added turbulence closure, cloud parameterization, radiation budgets, and tracer evolution. An account of the improvements in atmospheric models during the last couple of decades lies beyond the scope of this book (see Randall, 2000, for a coverage of the progress made during the 1990s), and we focus here on the distinguishing features of atmospheric models compared to oceanic models or general geophysical-flow models.

The most widely used models for weather prediction at global scale include those of the European Centre for Medium-Range Weather Forecasts (ECMWF, EU) and the National Centers for Environmental Prediction (NCEP, USA). Both models are adapted to the atmosphere by including a series of physical models or parameterizations, specific to the air that we have on Earth. Radiation budgets are more complicated than those of Section 19.2, and in reality, the heat equation should contain a local source term due to radiation, the behavior of which should depend among other things on the orientation of the sun, the wavelength of the radiation, humidity and presence of aerosols, the latter being transported with the winds and hence governed by an advection-diffusion equation. Because radiation is behaving differently for each wavelength, a separate equation for radiative transfer should be used for each wavelength. In practice, wavelengths are lumped into spectral bands, and, at a minimum, two groups are distinguished: short- and long-wave radiation, as we did earlier in our simple globally averaged models. The models include absorption of radiation (and hence local heating) by water vapor, ozone, carbon dioxide, and clouds within the atmosphere itself and on the lower boundary, the earth's surface. Not only absorption must be calculated at each grid point, but also the scattering of radiation by aerosols and clouds, as well as the reflection by the earth's surface and clouds. One must further take care of the re-emission of long-wave radiation by ozone. These processes involve a series of parameterizations which make up the particular radiative transfer equations of the models.

The ECMWF model uses for example a radiation scheme based on Orcrette (1991): for clear-sky conditions, short-wave radiation is mainly constrained by aerosol scattering and the effects of the absorption by water vapor, ozone, oxygen, carbon monoxide, methane, and nitrous oxide. Clear-sky, long-wave radiation is modeled using absorptive properties of water vapor, carbon dioxide, and ozone, which are temperature and pressure dependent. Cloudy skies are handled separately, and their parameterization includes absorption and scattering properties of cloud droplets, with clouds being characterized by optical thickness and their scattering properties. Cloud physics are not only important for studies of radiative transfer but also in precipitation forecasts. Because of their short dimensions compared with typical model grid sizes, clouds need to be parameterized in some ways. This requires specific treatment (Section 19.6).

When weather is calculated over the entire planet, Atmospheric General Circulation Models (AGCMs) have to take into account the spherical nature of the domain and hence governing equations are most naturally written in spherical coordinates (Appendix A). This is both a source of complications and of simplifications. Complications arise because of the more complex nature of the equations (coefficients now depend on latitude φ) and of the mathematical singularities at each pole because of the presence of $1/\cos\varphi$ in certain coefficients. The latter causes numerical stability problems.

To understand why numerical stability problems may arise, let us imagine that discretization is performed by using a rectangular grid in the longitude–latitude (λ, φ) coordinates with $\Delta\lambda = 2\pi/M$ where M is the number of grid points used in the west–east direction. Then, the Euclidean distance between two neighboring grid points is $\Delta x = a\cos\varphi\,\Delta\lambda$, where a is the earth's radius. It is clear that this distance Δx vanishes at the poles. So, if there is any numerical stability condition of the type $U\Delta t \le \Delta x$, where U is a physical propagation velocity of similar magnitude across latitudes, the stability condition will be much more stringent near the pole than near the equator, even if the underlying physical process acts similarly in both locations. The overall numerical efficiency is then drastically reduced if a pole imposes its stability condition on the rest of the domain. This is called the problem of *convergence of meridians*, and it must be addressed. If finite-difference grids in longitude–latitude are used, an implicit scheme or filtering must be used in the vicinity of the poles.

A distinct simplification for any AGCM covering the planet is the absence of any lateral boundary, avoiding the need for open-boundary conditions. Regional models (so-called *Limited Area Models*, nicknamed LAMs) may also avoid the open-boundary problem by appropriate nesting within an AGCM running alongside. The ALADIN model (*Aire Limitée Adaptation dynamique Développement InterNational*[4]) is one such LAM used for downscaling processes from the global scale to the desired regional scale. ALADIN simulates smaller scale features such as the sea breeze and thunderstorms, using high-resolution orography and less parameterization.

Even if they do not have lateral boundaries, AGCMs need conditions at the vertical edges of the domain. The upper boundary of atmospheric models is generally taken at a given pressure level (e.g., 0.25 hPa) or at a given height (e.g., 70 km), well above the troposphere in which most weather phenomena are confined. It is also placed well above the tropopause to avoid unphysical reflections of waves at the boundary. This, however, is still an artificial boundary because air gradually rarefies with height, and there exists no distinct level where the atmosphere ends and space begins. Nevertheless, rigid-lid conditions are commonly assumed. At the lower boundary, the atmosphere interacts with oceans, land masses, and ice covers, which all demand specific definitions of fluxes of heat and momentum.

---

[4]Translation: Limited area, dynamical adaptation, international development.

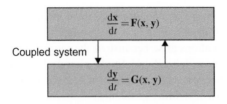

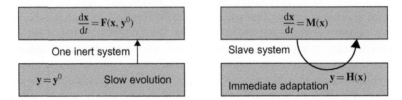

FIGURE 19.7 Depending on its intrinsic timescale compared with that of the atmosphere, the effect of an interacting system (such as vegetation or clouds) may be reduced to persistence or instantaneous adaptation. Only when the timescales of both systems are comparable is the fully coupled version to be retained.

Depending on the timescale of the processes at hand, those interactions between systems may be simplified (Fig. 19.7). If the system coupled to the atmosphere (e.g., glaciers or vegetation) is reacting comparatively slowly, there is no need to take into account temporal variations in the feedback mechanism, and persistence of the slower system can be assumed while the atmospheric model is moved forward for some time. For example, the ice cover of Antarctica does not change significantly within the few days covered by a weather forecast model, and the observation of the ice cover at the beginning of the forecast is sufficient to constrain the atmospheric evolution during the forecast period.

If on the contrary the coupled component reacts relatively fast compared with atmospheric conditions, it is often possible to establish quasi-equilibrium laws predicting those fast adaptations directly in terms of atmospheric parameters. For example, the albedo of the land surface changes in time following alterations to surface characteristics and should normally require a comprehensive land-use model, but, if the atmospheric model predicts snow fall, the albedo may be immediately updated for use in the atmospheric model so as to accommodate the change in reflection.

When the non-atmospheric system possesses a similar timescale to, and interacts with, the atmosphere, both models must be run in parallel, and outputs from each must inform the other. A good example of such a situation is the forecast of El Niño events (see Section 21.4). Such coupling, however, is not easy, and the first attempts at coupling AGCMs to ocean models for climate calculations demanded unphysical discontinuities in fluxes between the atmosphere and ocean. Ocean models are sensitive to errors in wind distribution over the sea surface, requiring that the wind field provided by the atmospheric model include the same timescales (spectral window) as those active in the ocean model. When fluxes were not subjected to one correction or another, the atmospheric and ocean models drifted away from each other, leading to unrealistic situations. To enable simulation of past climatic variations, it was then deemed necessary to inject information on the climatic average within the flux formulations, usually by way of relaxation toward a known state. This meant that a piece of the solution had to be incorporated into the model formulation. Such an unsatisfactory approach of feeding models with *a priori* knowledge was cause for objecting that climate forecasts based on such models could not be trusted. At this time, improvements have been made, and *flux correction* is no longer necessary. Coupled ocean-atmospheric models are now the core of so-called *global climate models* incorporating a vast number of physical, biological, and chemical components such as wind, ocean currents, ice cover, hydrological cycle, vegetation, land use, carbon and nutrients cycles, and so on.

The coupling of models allows feedback to be taken into account such as the melting of ice due to heating followed by a change in the albedo itself modifying heat budgets. Other feedback loops are possible through chemical reactions. The ozone layer, for example, changes under modified climate conditions, itself changing radiative budgets. Integrated models that include feedback loops are called *Earth simulators*. They incorporate as many processes as possible instead of considering them as forcing functions. Note that because of high computational demands, most of the submodels are optimized to exploit particular computer hardware (parallel, distributed, or shared), and making them work together is not trivial. Because of the physical coupling, this requires information exchange between models, called message passing on parallel computers. The implementation of exchanges is quite a challenging task both in terms of physical-interaction modeling as well as technical programming. Needless to say, such integrated models are also more and more demanding in terms of understanding the simulation results, particularly when grid resolution increases and more and more physical processes are resolved. After all, the model should behave almost as the real world does, which, as we know, is extremely difficult to comprehend. Hence, associated with most models, there is now a suite of statistical and graphical analysis tools to help the modeler grapple with the huge amount of information produced.

When speaking about atmospheric models, it should always be specified if the model is used for weather forecast, seasonal forecast, or climate-change scenarios. Indeed, a frequent argument invoked to disqualify climate-change studies is the inability to predict weather beyond a few days while germane models are used for investigating climate change. This argument simply disregards the difference between weather and climate (see Section 19.1). We may well be unable to predict next week's weather in New York but still be able to predict an increase of temperature over the United States over the next few years. The situation is similar to the unpredictability of individual eddies in a turbulent flow not preventing us from making meaningful statements about their aggregate effect on pollutant transport while tapping from the same family of governing principles and models. The problem is simply a question of scales of interest, and the reader may benefit from meditating on Fig. 1.7. As long as the model possesses prediction capability for the processes at temporal and spatial scales of interest, it does not matter whether its prediction capabilities are reduced for processes at other scales.

19.5 BRIEF REMARKS ON WEATHER FORECASTING

For weather forecasts, namely for forecasts not beyond of few days, most of the feedbacks with systems other than the atmosphere itself can be simplified and rendered relatively passive. For example, air–sea heat fluxes depend on sea surface temperature (SST), and the situation should ideally call for a fully coupled ocean-atmosphere model, but for weather forecast, the atmospheric model can rely on estimates of sea surface temperature, such as climatological sea surface temperature, observed SST over previous days, a simple mixed-layer model, or any combination of these.

The general approach of using observations to prescribe forcings explains why weather forecasting needs to rely on dense observational networks. To improve over time, however, prediction capabilities demand not only denser observational networks and better data-assimilation techniques but also more sophisticated physical parameterizations, better performing numerical methods, and ever more powerful computers to permit increased spatial resolution.

Nowadays, weather predictions are accompanied by calculated error bars around the predicted values. Thus, the weather reporter may speak about probability of occurrence of events, such rain probability. The error bars are not restricted to precipitation, however; they also apply to temperature, dew point, wind speed, pressure, cloud cover, snow fall, visibility, radiation, and so on. Although forecast capability has improved significantly over the last decades, weather remains somewhat elusive and difficult to predict, especially in the case of extreme events.

19.6 CLOUD PARAMETERIZATIONS

Probably the most difficult parameterizations encountered in atmospheric models are those related to clouds because several difficulties arise simultaneously: The physical processes at play are multiple and complex, and cover a broad range of scales. Clouds involve microphysics at crystal and water-drop level (millimeter scale), convection related processes at larger scales (hundreds of meters), and turbulence at all scales. Yet, none of these can possibly be represented in an explicit way in a global model. In one grid cell of a global model, even the largest cumulus clouds barely cover a fraction of the space.

At the shortest scale, physical processes leading to precipitation are complicated but can be explained by thermodynamics. The description involves water condensation on tiny solid particles (aerosols), collisions of incipient droplets, and interaction between water droplets and ice crystals (*Bergeron process*). The problem for models, of course, is that they are not able to represent these processes for each individual droplet, but models must nonetheless incorporate the aggregate effect at the level of the variables retained in the model. Mixing ratios[5] of cloud water content and ice content are typical variables to be computed. Because drop formation is controlled by the presence of condensation nuclei, aerosol concentration is also a pertinent variable. Then, in the calculation of water budget, ice and vapor content within a grid cell, exchanges between the three phases of water (ice, liquid water, and water vapor) must appear in the governing equations. The problem is to extrapolate these microphysical processes to the scale of the grid cell. For example, condensation and evaporation can coexist depending on the vapor pressure distribution within the cell (saturated vapor). Yet, the model can only calculate one value for the entire grid cell (Fig. 19.8).

Outlining a series of cloud parameterization techniques lies beyond the scope of the present introductory text, and the reader is referred to Randall, Khairoutdinov, Arakawa and Grabowski (2003) for a helpful review and further references. For examples of actual parameterizations, one can consult Sundqvist, Berge and Kristjansson (1989). The variables most commonly involved include humidity and temperature in cloud-free and cloud-covered regions, as well as fractional stratiform cloud coverage, mixing ratios of cloud water and ice, mixing ratios of rain and snow, and so on.

For some cloud types, parameterization is easier than for others. *Stratus*-type clouds are associated with large-scale global upward motion and stratiform condensation so that they can be captured by model grids. *Cumulus*-type clouds, on the contrary, are formed by shorter-scale convective motions that are nonhydrostatic. Towers of ascending buoyant air, involving *thermals*, cannot be captured by the grid and need a heavy dose of parameterization, which generally assumes that surrounding air is uniform with properties specified by the model solution at the resolved scale so that exchange laws between convective and ambient air can be formulated in terms of temperature, moisture, and so on of the ambient air and cloud.

The situation is even more complicated with heat budgets associated with phase changes. Warming by condensation and cooling through evaporation may take place side by side, over a distance shorter than the width of a grid cell. This creates thermal gradients and movement at the level of unresolved dynamics that are important because they modify cloud behavior, which in turn affects radiation, heat content, and so on. In other words, the system includes feedback

---

[5]Mixing ratios are expressed in gram or kilogram of the variable per kilogram of air.

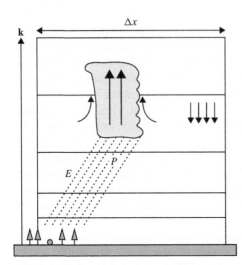

FIGURE 19.8 Numerical grids of global atmospheric models are too wide to represent individual clouds. The localized upward motion inside a cumulus cloud and the subsidence that surrounds it cannot be resolved by the grid, yet the cloud formed by condensation in upward motion of moist air and associated precipitation affects the water budget. The situation is further complicated by possible entrainment of surrounding unsaturated air, leading to evaporation, cooling, and weakening updraft. Precipitation originating in saturated air is normally modeled as a loss of water in the model. Yet, during its journey down to the earth surface, it can pass through unsaturated air and evaporate. This cools the air and, in the presence of downward entrainment creates a downdraft. These and other processes all take place on a spatial scale shorter than the size of a single grid cell.

mechanisms at unresolved scales. This is not only important for day-to-day prediction of rainfall, but clouds also play an essential role in climate dynamics. Because clouds provide shadow in the day but act as a thermal blankets during night, they are crucial in the global heat budget. Climate variations in turn modify the hydrological cycle and cloud coverage and yet another feedback exists. Because of the extreme complexity of cloud dynamics, the Intergovernmental Panel on Climate Change (IPCC, 2001) identified possible changes in cloud cover as one of the major uncertainties in predicting future states of the climate.

19.7   SPECTRAL METHODS

Numerical methods used in atmospheric models have evolved from quasi-geostrophic models using Arakawa grids for the discretization of the Jacobian operator and inversion of Poisson equations toward more sophisticated spectral models based on the primitive equations (i.e., no longer any QG approximation) with semi-Lagrangian tracer advection. Most modern global models are based on this approach, which we now outline.

Spectral models are based on the same technique as the spectral models presented in the quasi-geostrophic framework (see Section 16.7). They use a truncated series of orthogonal basis functions spanning the domain of interest. For global models, spherical coordinates do not lend themselves to a classical Fourier expansion of the solution in terms of trigonometric (sine and cosine) functions, and more complicated spherical functions must be used. Assuming that the vertical dependence is handled by standard finite-volume or finite-difference techniques, the dependence of a field u on longitude λ and latitude φ is expressed as a series of functions $Y_{m,n}$ called spherical harmonics:

$$u(\lambda, \varphi, t) = \sum_m \sum_n a_{m,n} Y_{m,n}(\lambda, \sin \varphi) \tag{19.17}$$

in which

$$Y_{m,n}(\lambda, \sin \varphi) = P_{m,n}(\sin \varphi) e^{i m \lambda}. \tag{19.18}$$

The expansion series is of Fourier type in longitude but involves Legendre functions $P_{m,n}$ of $\sin \varphi$:

$$P_{m,n}(x) = \sqrt{\frac{(2n+1)(n-m)!}{2(n+m)!}} (1-x^2)^m \frac{d^m}{dx^m} P_n(x). \tag{19.19}$$

These Legendre functions are in turn defined in terms of Legendre polynomials of degree n:

$$P_n(x) = \frac{1}{2^n n!} \frac{d^n}{dx^n} \left[(x^2 - 1)^n \right]. \tag{19.20}$$

Because P_n is a polynomial of degree n, Legendre functions differ from zero only when $m \le n$. Extension toward negative values of m is desirable to correspond to the full set of Fourier modes $\exp(i m \lambda)$, and so we extend the definition of Legendre functions with $P_{-m,n}(x) = (-1)^m P_{m,n}(x)$. With the well-chosen coefficient in front of Eq. (19.19), Legendre functions are orthonormal:

$$\int_{-1}^{1} P_{m,n}(x) P_{m,k}(x) dx = \delta_{n,k}, \tag{19.21}$$

which is $= 1$ if $n = k$ and $= 0$ otherwise.

On the surface S of the sphere of radius r, the elementary surface element $r^2 \cos \varphi \, d\varphi d\lambda$ may be written as $r^2 \, d\xi d\lambda$ with $\xi = \sin \varphi$ so that

$$\frac{1}{r^2} \int_S Y_{m,n} Y_{p,k}^* dS = \int_{-1}^{1} \int_0^{2\pi} Y_{m,n} Y_{p,k}^* d\lambda d\xi = 2\pi \delta_{m,p} \, \delta_{n,k}, \tag{19.22}$$

where the symbol $*$ stands for the complex conjugate. The horizontal Laplacian of the basis functions is in spherical coordinates:

$$\nabla^2 Y_{m,n} = -\frac{n(n+1)}{r^2} Y_{m,n} \tag{19.23}$$

so that inversion of the Poisson equation can be performed algebraically in the transformed space. Note that, surprisingly, the pseudo[6] wavenumber $\sqrt{n(n+1)}/r$ is independent of m.

Orthogonality of spherical harmonics can be used to obtain separate evolution equations for the amplitudes $a_{m,n}(t)$ by multiplying the governing equations by $Y_{m,n}^*$ and integrating over the global surface. To do so, we exploit the inverse transform (19.17) and the associated forward transformation that reads

$$a_{m,n} = \int_{-1}^{1} \left[\int_{0}^{2\pi} u(\lambda,\xi,t)e^{-im\lambda}\,d\lambda \right] P_{m,n}(\xi)\,d\xi \qquad (19.24)$$

as it follows from the orthogonality property.

Truncation of the sums in numerical schemes can be achieved in several ways (Fig. 19.10), the only constraint being that in all cases $|m| \leq n$ which can be achieved by

$$\tilde{u} = \sum_{m=-M}^{M} \sum_{n=|m|}^{N(m)} a_{m,n} Y_{m,n} \qquad (19.25)$$

The structure of spatial resolution depends on the value taken for $N(m)$. When $N(m) = M$, called a *triangular truncation*, uniform resolution is achieved on the sphere. Other truncations provide increased resolution in particular regions (Figs. 19.9 and 19.10).

The various possibilities are denoted according to the chosen truncation. For example, T256L60 signifies a triangular truncation with $M = 256$ spectral components.[7] The qualifier L60 stands for the vertical grid using 60 discrete levels.

In the ECMWF model, vertical levels are distributed according to a hybrid coordinate system where the vertical coordinate s depends on pressure p and surface pressure p_{surf} by $s = s(p, p_{surf})$ scaled so that $s(0, p_{surf}) = 0$ (top of the atmosphere) and $s(p_{surf}, p_{surf}) = 1$ (land or ocean surface). This is a generalization of a pressure coordinate, called σ coordinate, introduced by Phillips (1957) with $s = p/p_{surf}$ and which is used in NCEP. Because more general hybrid vertical coordinates are now used in ocean models, we will postpone the corresponding discussion until Section 20.6.1.

---

[6] In Cartesian coordinates with the Fourier decomposition $u = U\exp(ik_x x + ik_y y)$, we would have $\nabla^2 u = -(k_x^2 + k_y^2)\,u$, hence the analogy.

[7] Note that because m takes negative values, the actual wavenumbers resolved are indeed M, contrary to the standard FFT presentation, which uses only positive values and hence half of the wavenumbers (see Appendix C).

FIGURE 19.9   Real part of spherical harmonics $Y_{m,n}$ for (m, n) taking values $(0,1),(1,1)$ on the first row, $(0,2),(1,2),(2,2)$ on the second row, and $(0,3),(1,3),(2,3),(3,3)$ on the bottom row. The white color marks the separation between positive and negative values. Mode $(0,0)$ is not shown because its value is constant on the sphere.

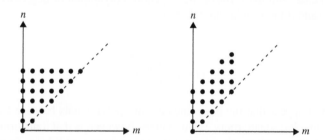

FIGURE 19.10   Triangular and rhomboidal truncations stipulating which values of m are retained for every value of n. Only positive m values are shown, the negative ones being symmetric. Triangular truncation is now more popular than the originally preferred rhomboidal version.

With discrete versions of integrals, orthogonality of basis functions is no longer ensured, and a direct transformation followed by an inverse transformation does not guarantee to a perfect return to the original function. For the discrete Fourier transform, orthogonality is maintained (see Appendix C and Section 18.4), so that we can evaluate the inner integral of the forward transform through an FFT and the corresponding inverse transform by an inverse

FFT. There only remains to ensure that the numerical treatment of the outer integral of Eq. (19.24) conserves orthogonality.

Unfortunately, there exists no numerical tool similar to the FFT that allows to perform transforms on the Legendre expansion as quickly, and we must resort to numerical quadrature of the integrals. First we can perform an FFT at a series of given latitudes $\varphi_j, j = 1, \ldots, J$ with $\xi_j = \sin \varphi_j$ to obtain the Fourier coefficients

$$b_m(\xi_j, t) = \int_0^{2\pi} u(\lambda, \xi_j, t) e^{-im\lambda} d\lambda \qquad (19.26)$$

defined at locations ξ_j. Then the time-dependent coefficients $a_{m,n}$ can be estimated by a numerical quadrature using the value of the integrand at those locations φ_j:

$$a_{m,n} = \sum_{j=1}^{J} w_j b_m(\xi_j, t) P_{m,n}(\xi_j). \qquad (19.27)$$

The weights w_j and locations ξ_j can be chosen so as to reduce integration errors. Gaussian quadrature can be shown to produce exact results when integrating polynomials of degree $(2J - 1)$ if the J points on which the integrand is evaluated coincide with the zeros of the Legendre polynomial of degree J, that is, $P_J(\xi_j) = 0$ and if the weights are taken as

$$w_j = \frac{2}{(1 - \xi_j^2)\left[\frac{dP_J}{d\xi}(\xi_j)\right]^2}. \qquad (19.28)$$

It would appear that the integrands are not polynomials because Legendre functions involve square roots. What matters, however, is that transforms of the nonlinear terms, such as $u\partial u/\partial \lambda$, are treated correctly, and these, fortunately, involve products of Legendre functions which turn into polynomials that can be integrated exactly. The number of points J must then be taken so as to integrate correctly polynomials of the highest degree as a consequence of nonlinear terms. The transform of such a term would require the evaluation of triplet of Legendre functions (one for each appearance of u and then the application of the transform itself involving the third Legendre function). For a highest degree, $m = M$, of Legendre functions, a polynomial of degree $3M$ appears, and we must therefore use $J > (3M + 1)/2$ points to integrate it exactly.

For the Fourier transform in longitude, we also face an aliasing problem, and the same analysis as in Section 18.4 applies. Because M modes use $(2M + 1)$

grid points,[8] avoidance of aliasing requires the use of $(3M + 1)$ evaluation points in longitude. Hence a model with 42 modes will use typically an underlying longitude–latitude grid of 128×64 points for the evaluation of the nonlinear terms (note the rounding toward powers of 2 to take advantage of efficient FFTs). This grid is called the *Gaussian grid* or *transform grid*. The calculation of grid spacing based on the number of Gaussian grid points overestimates actual resolution because it is designed to avoid aliasing on nonlinear interactions, and the actual, lower resolution is that which corresponds to the wavenumbers associated with the spectral decomposition.

The transform methods thus allows us to calculate some terms in spectral space (linear terms) and others (quadratic advection terms and nonlinear terms stemming from various parameterizations) in the transformed space so as to use the most appropriate technique for each term. In practice, it means that the model utilizes both spectral and grid representations of each variable.

The high convergence rate of spectral methods is inherited with the spherical harmonics, as long as the physical solution is sufficiently smooth. When fronts or jumps are present in the solution, however, spatial oscillations emerge near the place of rapid variation. This is known as Gibb's phenomena. The associated over- or undershooting on the physical grid can lead to spurious physical results. An overshooting of specific humidity, for example, may lead to the poetically named *spectral rain*.

Because of the calculation of some terms on the physical grid, the geometrical convergence of meridians toward the poles may also be a problem. For the advection part, this can be overcome by the semi-Lagrangian approach, which we describe next.

19.8   SEMI-LAGRANGIAN METHODS

To deal with advection, we again turn our attention to the passive-tracer concentration c, which is conserved along a trajectory of a parcel of fluid as long as diffusion remains negligible. The Lagrangian approach ensures exact conservation of the tracer value at the price of calculating its trajectory in time (see Section 12.8). As we have seen, however, the pure Lagrangian method leads sooner or later to an impractical distribution of particles, and it becomes impossible to determine concentration values in regions nearly void of particles. This is what happens when we follow the same set of particles over time: some of which flow out of the system or are caught in stagnation points. *Semi-Lagrangian* methods avoid this problem by using a different set of particles at each time step. The set is chosen at t^n so that at t^{n+1} the chosen particles arrive at the nodes of the numerical grid. This amounts to integrating trajectories

---

[8]Remember that the sum runs from $-M$ to M.

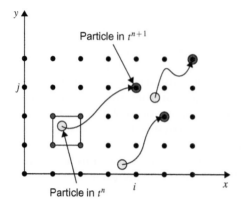

FIGURE 19.11 In a semi-Lagrangian method, trajectories are integrated backward in order to find an earlier location of the fluid particle that reaches grid node (i,j) by time t^{n+1}. Once this location is known, the value of the variable of interest, such as a temperature or concentration, at that location can be obtained by interpolation of nearby values. The interpolated value is then translated by advection to the new location (i,j) at time t^{n+1}.

backward for one time step in order to find where they originate. Once the past locations are determined, at t^n, the concentration in those locations is then determined by interpolation among known values on the grid (Fig. 19.11).

For simplification, let us consider first the one-dimensional case with positive velocity u and uniformly spaced grid (Fig. 19.12). The particle that lands at grid node x_i by time t^{n+1} was at the earlier time $t^n = t^{n+1} - \Delta t$ at position

$$x = x_i - u\Delta t. \tag{19.29}$$

On a uniform grid with spacing Δx, this position x most likely lies within a grid interval rather than, per chance, at an other grid point. This grid interval is given by

$$x_{i-1-p} \le x = x_i - u\Delta t \le x_{i-p}, \quad \text{with} \quad p = \text{integer part of } u\frac{\Delta t}{\Delta x}. \tag{19.30}$$

By virtue of advection without diffusion, the value of $\tilde{c}_i^{n+1}$ is none other than $\tilde{c}^n$ at x, a value which we can obtain by interpolation. Performing a linear interpolation, we obtain

$$\tilde{c}_i^{n+1} = \frac{(x_{i-p} - x)}{\Delta x}\tilde{c}_{i-1-p}^n + \frac{(x - x_{i-1-p})}{\Delta x}\tilde{c}_{i-p}^n = \tilde{C}\tilde{c}_{i-1-p}^n + (1 - \tilde{C})\tilde{c}_{i-p}^n, \tag{19.31}$$

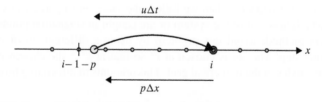

FIGURE 19.12 The semi-Lagrangian method in one dimension. The particle in light gray moves during time interval $[t^n, t^{n+1}]$ over a distance $u\Delta t$ to reach grid node labeled i at time t^{n+1}.

for which we define

$$\tilde{C} = \left(\frac{u\Delta t}{\Delta x} - p \right). \tag{19.32}$$

The scheme is monotonic and thus of first order. We can easily see that, as long as $u\Delta t \leq \Delta x$ (and thus $p=0$), the scheme is equivalent to the upwind scheme. However, contrary to the upwind scheme, no stability condition is necessary here because the method uses the correct grid interval from which to interpolate. Numerical diffusion persists, however, although it is reduced in the sense that large time steps can be used and the total number of time steps can be decreased. For a given simulation time, fewer time steps mean less numerical diffusion.

To decrease the amount of diffusion introduced by the interpolation, a better than linear interpolation can be used. A second-order, parabolic interpolation yields a scheme equivalent to the Lax–Wendroff method.[9]

In two dimensions, the approach is readily generalized with backward trajectories for a single time step followed by 2D spatial interpolation (either bilinear or biparabolic). The trajectory calculation needs to take into account that the flow field (u, v), and this may become quite complicated if $U\Delta t \geq \Delta x$. If the velocity field varies over the trajectory on a scale comparable to the grid scale Δx, intermediate time steps are necessary for the calculation of the backward trajectory in order to maintain accuracy, and the calculation cost increases rapidly.

However, if the velocity is relatively smooth on the scale of the numerical grid (which ought to be the case and will necessarily be the case near the pole), that is, $\Delta x \ll L$, simple trajectory integrations will suffice. In fact, if $\Delta x \ll U\Delta t \ll L$ the semi-Lagrangian approach is much more efficient than the Eulerian method because during each time step, a large number of grid points can be "jumped over" by advection. Hence, interpolation (and associated diffusion) is less frequent, and the spatial scale of the trajectories is correctly captured. This is the way to reap the maximum benefit for the Semi-Lagrangian approach.

If $\Delta x \sim L$, time steps are similar to the Eulerian approach. The major advantage in this case is the stability of the method in the face of the occasionally excessive time step. For higher accuracy, however, one should not use longer time steps than allowed by $U\Delta t \sim L$. If there are many different tracers to be advected simultaneously, as in air pollution studies or ecosystem modeling, the method presents considerable advantages because a single, common trajectory needs to be calculated for all tracers.

For the nonadvective terms, such as source/sink and diffusion terms, a fractional-step approach is possible, for example, by first using a semi-

---

[9]Note the difference: In Eulerian methods, we spoke about interpolating for flux calculations to be discretized subsequently; here we speak about interpolation of the solution itself.

Lagrangian advection scheme followed by a Eulerian diffusion scheme either on the physical grid or in spectral space. Alternatively, the evolution of source terms may be taken into account along the trajectory (e.g., McDonald, 1986). Contrary to the finite-volume approach of Eulerian methods, global conservation properties are more difficult to handle but can be respected (e.g., Yabe, Xiao & Utsumi, 2001).

ANALYTICAL PROBLEMS

19.1. Consider the regular gardening greenhouse and idealize the system as follows: The air plays no role, the ground absorbs all radiation and reradiates it as a black body, and the glass is perfectly transparent to short-wave (visible) radiation but totally opaque to long-wave (heat) radiation. Further, the glass emits its radiation inward and outward in equal parts. Compare the ground temperature inside the greenhouse with that outside. Then, redo the exercise for a greenhouse with two layers of glass separated by a layer of air.

19.2. Consider the long-wave radiation fluxes of Figs. 19.2 and 19.3. In each case, the upward flux from the ground (E_2) is greater than the downward flux from the atmosphere ($0.64\,E_1$). Can you explain why?

19.3. Consider the crudest heat budget for the earth (without atmosphere and hydrological cycle) and assume the following dependency of the albedo on temperature: At low temperatures, much ice and clouds cover the earth, yielding a high albedo, whereas at high temperatures, the absence of ice and clouds reduce the albedo to zero. Taking the functional dependence as

$$
\begin{aligned}
\alpha &= 0.5 && \text{for} \quad T \leq 250 \text{ K} \\
\alpha &= \frac{270 - T}{40} && \text{for} \quad 250 \text{ K} \leq T \leq 270 \text{ K} \\
\alpha &= 0 && \text{for} \quad 270 \text{ K} \leq T,
\end{aligned}
\tag{19.33}
$$

solve for the earth's average temperature T. Discuss the several solutions.

19.4. Using the global heat budget of the earth model, complete with an atmospheric layer and hydrological cycle, explore a worst-case scenario, whereby elevated concentrations of greenhouse gases completely block the transmission of long-wave radiation from the earth's surface, the intensity of the hydrological cycle is unchanged, and the anticipated global warming has caused the complete melting of all ice sheets, effectively eliminating all reflection by the earth's surface of short-wave solar radiation. What would then be the globally averaged temperature of the earth's surface? (Except for those transmission and reflection coefficients that need to be revised, use the parameter values quoted in the text.)

NUMERICAL EXERCISES

19.1. What is the spatial resolution (in kilometers) along the equator for a T256 spectral model? How many grid points must the underlying Gaussian grid have in order to avoid aliasing in the advection terms?

19.2. Use spherical.m to consider other basis functions $Y_{m,n}$ than those of Fig. 19.9.

19.3. Estimate the numerical cost of the forward and inverse transform associated with spectral harmonics.

19.4. In addition to the problem of decreasing grid spacing near the poles, which other problem can you identify at the poles for models that do not work with a spectral decomposition? (*Hint*: Think about boundary conditions for an AGCM, for longitude first and then for latitude.)

19.5. Exploiting properties of Legendre polynomials, given for example in Abramowitz and Stegun (1972), find the spectral coefficients of spatial derivatives, knowing the spectral coefficients of the function to be differentiated.

Edward Norton Lorenz
1917–2008

Edward Lorenz began his career as a weather forecaster for the U.S. army during World War II, before obtaining his doctorate in meteorology from MIT and later becoming professor of meteorology at the same institution. In the early 1960s, while using early numerical models of weather systems, Lorenz once introduced a small error in the value of a parameter that led to a completely different weather scenario. Puzzled by this, he proceeded to reduce the complex problem to an apparently simple set of equations (see Section 22.1) and sensitivity to very small perturbations persisted. The result was the discovery of chaos in atmospheric dynamics, published in a highly acclaimed paper (Lorenz, 1963). This contribution, among the most often cited scientific papers in the world, initiated a brand new field of research, that of deterministic chaos and strange attractors. Since then, the poetically named "butterfly effect" has become a standard metaphor for describing the sensitive dependence of a system's evolution on its initial conditions. (*Photo by Jane Loban*)

Joseph Smagorinsky
1924–2005

A native of New York City, Joseph Smagorinsky studied meteorology and began a career with the U.S. Weather Bureau. In 1955, he founded the *Geophysical Fluid Dynamics Laboratory*, which was first established in Washington and later relocated to Princeton University. The 1950s were exciting years when the prospect of computers gave hope that weather could be predicted by machines. Recognizing this opportunity, Smagorinsky developed numerical methods for predicting weather and climate, and by so doing profoundly influenced the practice of weather forecasting. In particular, he made the first attempt in 1955 to predict precipitation, and this led him to include compensating effects like radiation and to argue for the inclusion of a comprehensive "physics package," which became standard in all operational models.

Besides numerical methods and models, Smagorinsky also contributed to weather prediction by assuming a leading role in the establishment of a global observational network. While setting high goals for himself, Smagorinsky had an excellent sense of humor and a common touch. (*Photo by Michael Oort*)

Chapter 20

Oceanic General Circulation

ABSTRACT

The concepts of geostrophy, hydrostaticity, and potential vorticity are merged to study the large-scale baroclinic circulation in the midlatitude oceans. The results lead to the Sverdrup balance, the beta spiral, and a number of properties of large-scale oceanic motions. The numerical part of the chapter provides an overview of the issues raised in constructing a model of the 3D circulation at the scale of ocean basins or the planet.

20.1 WHAT DRIVES THE OCEANIC CIRCULATION

Ocean motions span a great variety of scales in both time and space. At one extreme, we find microturbulence, not unlike that in hydraulics, and at the other, the large-scale circulation, which spans ocean basins and evolves over climatic timescales. The latter extreme is the objective of this chapter.

There are multiple mechanisms that set oceanic water masses in motion: the gravitational pull exerted by the moon and sun, differences in atmospheric pressure at sea level, wind stress over the sea surface, and convection resulting from atmospheric cooling and evaporation. The moon and sun generate periodic tides with negligible permanent circulation, whereas differences in atmospheric pressure play no significant role. On the other hand, deep convection at high latitudes generates currents responsible for a very slow movement in the abyss called the *conveyor belt* (Fig. 20.1). This leaves the stress exerted by the winds along the sea surface as the main driving force of basin-wide circulations in the upper part of the water column.

Ocean waters respond to the wind stress because of their low resistance to shear (low viscosity, even after viscosity magnification by turbulence) and because of the relative consistency with which winds blow over the ocean. Good examples are the *trade winds* in the tropics; they are so steady that shortly after Christopher Columbus and until the advent of steam, ships chartered their courses across the Atlantic according to those winds; hence their name. Further away from the tropics are winds blowing in the opposite direction. While trade

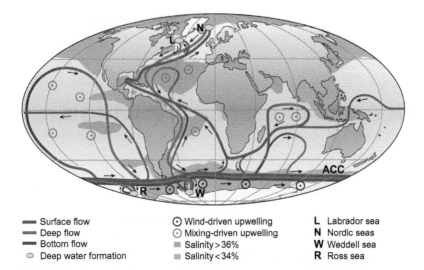

— Surface flow	⊙ Wind-driven upwelling	**L** Labrador sea
— Deep flow	⊙ Mixing-driven upwelling	**N** Nordic seas
— Bottom flow	▨ Salinity > 36%	**W** Weddell sea
◯ Deep water formation	▨ Salinity < 34%	**R** Ross sea

FIGURE 20.1   Cold and salty waters newly formed by deep convection at high latitudes are carried away by the conveyor belt across the ocean basins. These waters eventually resurface in warmer climes and return toward places of deep convection. The time to complete a loop is on the order of several hundred years to millennia. (*Kuhlbrodt et al., 2007*). A color version can be found online at http://booksite.academicpress.com/9780120887590/

winds blow from the east and slightly toward the equator (they are also more aptly called northeasterlies and southeasterlies, depending on the hemisphere), midlatitude winds blow from west to east and are called *westerlies* (Fig. 20.2). Generally, much more variable than the trades, these westerlies nonetheless possess a substantial average component, and the combination of the two wind systems drives significant circulations in all midlatitude basins: North and South Atlantic, North and South Pacific, and Indian Oceans.

In the ocean, the water column can be broadly divided into four segments (Fig. 20.3). At the top lies the *mixed layer* that is stirred by the surface wind stress. With a depth on the order of 10 m, this layer can be assimilated for the purpose of large-scale ocean circulation with the Ekman layer (see Chapter 8) and is characterized by $\partial \rho / \partial z \simeq 0$. Below lies a layer called the *seasonal thermocline*, a layer in which the vertical stratification is erased every winter by convective cooling. Its depth is on the order of 100 m. Below the maximum depth of winter convection is the *main thermocline*, which is fed by water left behind whenever the seasonal thermocline retreats; it is permanently stratified ($\partial \rho / \partial z \neq 0$), and its thickness is on the order of 500–1000 m. The rest of the water column, which comprises most of the ocean water, is the *abyssal layer*. It is very cold, and its movement is very slow.

When considered together, the main thermocline and the abyssal layer form the *oceanic interior*. While mesoscale motions exist in both these layers, under

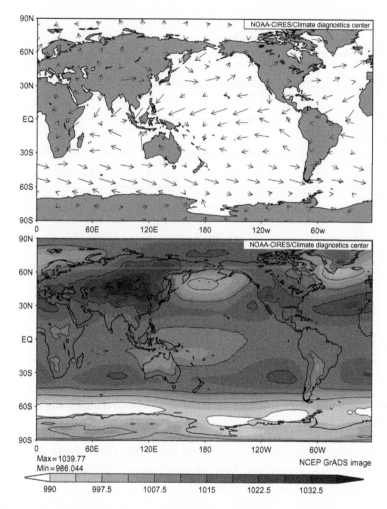

FIGURE 20.2 The major winds over the world ocean for the month of January averaged over the years 1968–1996 and associated sea surface pressure. (*From NCEP*)

the action of pressure fluctuations in the upper layers, it is believed that, in first approximation, the study of the slow background motion in the oceanic interior can proceed independently of the smaller scale, higher-frequency processes.

Although mariners have long been aware of the major ocean currents, such as the Gulf Stream,[1] ocean circulation theory was long in coming, chiefly for lack of systematic data below the surface. The discipline began in earnest

---

[1] Benjamin Franklin receives credit for publicizing and mapping the Gulf Stream in 1770.

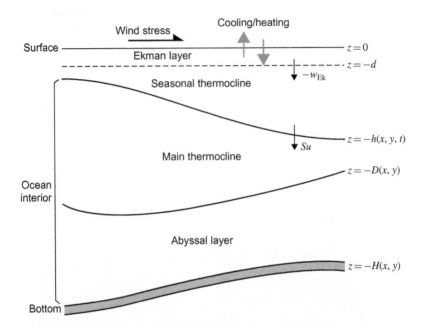

FIGURE 20.3 Vertical structure of the ocean from the point of view of large-scale circulation. Note that the relative thicknesses of the various layers are not to scale, for the abyss is far thicker than all other layers combined.

with the seminal works of Harald Sverdrup,[2] who formulated the equations of large-scale ocean dynamics (Sverdrup, 1947) and Henry Stommel[3], whose major contributions to ocean circulation are many and diverse, beginning with the first correct theory for the Gulf Strean (Stommel, 1948). Today, ocean circulation theory is a significant body of knowledge (Marshall & Plumb, 2008; Pedlosky, 1996; Warren & Wunsch, 1981).

20.2 LARGE-SCALE OCEAN DYNAMICS (SVERDRUP DYNAMICS)

Because oceanic basins have dimensions comparable to the size of the earth, model accuracy demands the use of spherical coordinates, but because the present book only intends to present an introduction to physical oceanography, clarity of exposition trumps accuracy, and we continue to use Cartesian coordinates, with inclusion nonetheless of the beta effect (see Section 9.4). Spherical

---

[2]Harald Ulrik Sverdrup (1888–1957), Norwegian oceanographer who made his greatest contributions while being director of the Scripps Institution of Oceanography in California. A unit of volumetric ocean transport bears his name: 1 sverdrup = 1 Sv = 10^6 m^3/s.

[3]See biography at end of this chapter.

coordinates (Appendix A) do not change the qualitative nature of the theoretical results exposed here (Pedlosky, 1996, Chapter 1).

Large-scale flows in the main thermocline and abyss are slow and nearly steady. Their long timescales allow us to ignore all time derivatives, whereas their low velocities over long distances make their Rossby number very small and allow us to neglect the nonlinear advection terms in the momentum equations. Furthermore, there is a strong indication that dissipation is not an important feature of large-scale dynamics, at least not at the leading order (Pedlosky, 1996, page 6). Without time derivatives, advection, and dissipation, the horizontal momentum equations reduce to the geostrophic balance:

$$-fv = -\frac{1}{\rho_0}\frac{\partial p}{\partial x} \qquad (20.1a)$$

$$+fu = -\frac{1}{\rho_0}\frac{\partial p}{\partial y}, \qquad (20.1b)$$

in which the Coriolis parameter f includes the so-called beta effect, which is important over the long length scales under consideration:

$$f = f_0 + \beta_0 y. \qquad (20.2)$$

The y-coordinate is thus directed northward, leaving the x-direction to point eastward. The coefficients (see Eq. 9.18) $f_0 = 2\Omega\sin\varphi$ and $\beta_0 = 2(\Omega/a)\cos\varphi$ both depend on the choice of a reference latitude φ, which may be taken as the middle latitude of the basin under consideration.

The geostrophic equations are complemented by the hydrostatic balance

$$\frac{\partial p}{\partial z} = -\rho g, \qquad (20.3)$$

the continuity equation (mass conservation for an incompressible fluid)

$$\frac{\partial u}{\partial x} + \frac{\partial v}{\partial y} + \frac{\partial w}{\partial z} = 0, \qquad (20.4)$$

and the energy equation, which states conservation of heat and salt but is expressed as conservation of density, again with neglect of the time derivative:

$$u\frac{\partial \rho}{\partial x} + v\frac{\partial \rho}{\partial y} + w\frac{\partial \rho}{\partial z} = 0. \qquad (20.5)$$

In the preceding equations, u, v, and w are the velocity components in the eastward, northward, and upward directions, respectively, ρ_0 is the reference density (a constant), ρ is the density anomaly, the difference between the actual density and ρ_0, p is the hydrostatic pressure induced by the density anomaly ρ, and g is the earth's gravitational acceleration (a constant). This set of five equations for five unknowns (u, v, w, p, and ρ) is sometimes referred to as *Sverdrup dynamics*.

Note that the problem is nonlinear owing to product of unknowns in the density equation (20.5). We now proceed with the study of some of its most immediate properties.

20.2.1 Sverdrup Relation

Elimination of pressure between the two momentum equations, by subtracting the y-derivative of Eq. (20.1a) from the x-derivative of Eq. (20.1b) yields

$$\frac{\partial}{\partial x}(fu) + \frac{\partial}{\partial y}(fv) = 0, \qquad (20.6)$$

or since f is a function of y but not of x,

$$f\left(\frac{\partial u}{\partial x} + \frac{\partial v}{\partial y}\right) + \beta_0 v = 0. \qquad (20.7)$$

With the use of continuity equation (20.4), it can be recast as

$$\beta_0 v = f\frac{\partial w}{\partial z}. \qquad (20.8)$$

This most simple equation, called the *Sverdrup relation*, has a clear physical meaning. The factor $\partial w/\partial z$ represents vertical stretching, and any stretching ($\partial w/\partial z > 0$) or squeezing ($\partial w/\partial z < 0$) demands a change in vorticity for the sake of potential-vorticity conservation, which holds in the absence of dissipation. There is no relative vorticity here,[4] and the only way for a parcel of fluid to change its vorticity is to adjust its planetary vorticity (Fig. 20.4). This requires a meridional displacement, and hence meridional velocity v, to reach the correct f value.

However, this is not the end of the story. We can go further with a vertical integration:

$$\beta_0 \int_{-H}^{-d} v\,dz = f[w(z=-d) - w(z=-H)], \qquad (20.9)$$

where $z = -H(x, y)$ represents the ocean bottom and $z = -d$ the base of the Ekman layer (Fig. 20.3). Note that in performing this integration, we have taken the liberty of including the seasonal thermocline under the assumption that although it is marked by seasonal variations, that time scale is long compared to the inertial period and its flow is nearly geostrophic.

Abyssal flow is extremely slow, and we may very well take $w(z=-H)=0$ at the bottom regardless of whether the bottom is sloping or is accompanied by

---

[4]Relative vorticity scales as U/L while planetary vorticity changes are on the order of βL. The former is much smaller than the latter because motion is slow and over a long scale, with $L^2 \gg U/\beta$.

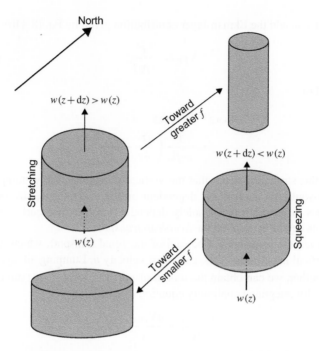

FIGURE 20.4 Meridional migration of fluid parcels induced by vertical stretching or squeezing in the large-scale oceanic circulation.

a bottom Ekman layer. At the base of the Ekman layer, the vertical velocity is the Ekman pumping

$$w_{Ek} = \frac{1}{\rho_0} \left[\frac{\partial}{\partial x} \left(\frac{\tau^y}{f} \right) - \frac{\partial}{\partial y} \left(\frac{\tau^x}{f} \right) \right], \tag{20.10}$$

as in Eq. (8.36).

20.2.2 Sverdrup Transport

If we define the meridional transport as the vertically integrated north-south velocity, $V = \int_{-H}^{-d} v \, dz$, from the ocean's bottom to the base of the Ekman layer, Eqs. (20.9) and (20.10) provide

$$V = \frac{f}{\beta_0} w(z = -d)$$

$$= \frac{f}{\beta_0} w_{Ek}$$

$$= \frac{f}{\rho_0 \beta_0} \left[\frac{\partial}{\partial x} \left(\frac{\tau^y}{f} \right) - \frac{\partial}{\partial y} \left(\frac{\tau^x}{f} \right) \right]. \tag{20.11a}$$

To this we can add the Ekman-layer contribution given by Eq. (8.34b):

$$V_{Ek} = -\frac{\tau^x}{\rho_0 f},$$ (20.11b)

for a total of

$$V_{total} = V + V_{Ek}$$
$$= \frac{1}{\rho_0 \beta_0}\left(\frac{\partial \tau^y}{\partial x} - \frac{\partial \tau^x}{\partial y}\right).$$ (20.11c)

We note this surprising result that the vertically cumulated flow component in the north-south direction is not dependent on the basin shape, size or overall wind-stress distribution but is solely dependent on the local curl of the wind stress. This equation is called the *Sverdrup transport*.

However, the same cannot be said of the zonal transport, which we define as the vertical integration of the east-west velocity u. Lumping all layers of the ocean together, we can obtain the total transport $U_{total} = \int_{-H}^{0} u\,dz$ directly from the vertically integrated continuity equation (20.4):

$$\frac{\partial U_{total}}{\partial x} + \frac{\partial V_{total}}{\partial y} = 0,$$ (20.12)

which yields

$$U_{total} = -\frac{1}{\rho_0 \beta_0}\int_{x_0}^{x}\frac{\partial}{\partial y}\left(\frac{\partial \tau^y}{\partial x} - \frac{\partial \tau^x}{\partial y}\right)dx,$$ (20.13)

where the starting point of integration ($x = x_0$) is to be selected wisely.

Ideally, we wish to impose a boundary condition on the flow at both eastern and western ends of the basin. For example, if we consider a basin limited on both eastern and western sides by north-south coastlines (a fair approximation of the major oceanic basins), the zonal flow and its vertical integral (U_{total}) ought to vanish at those ends. However, this is impossible to require simultaneously because there is only one constant (x_0) to adjust. If we set $x_0 = x_E$, the value of x at the eastern shore of the basin, then we enforce the impermeable-wall condition on the eastern side but make no provision for meeting any boundary condition on the western side, and vice versa if we take $x_0 = x_W$, the value of x at the western shore of the basin. The consequence is that the theory fails at one end of the domain, and as we will see in Section 20.3, a boundary layer must exist at one of the sides, which turns out to be western boundary.

20.2.3 Thermal Wind and Beta Spiral

In the ocean interior, the flow is approximately geostrophic, as we assumed when we wrote Eqs. (20.1). If we now take the vertical derivative of these equations and then eliminate $\partial p/\partial z$ by use of the hydrostatic balance (20.3), we obtain the thermal-wind relations:

$$\frac{\partial u}{\partial z} = +\frac{g}{\rho_0 f}\frac{\partial \rho}{\partial y} \tag{20.14a}$$

$$\frac{\partial v}{\partial z} = -\frac{g}{\rho_0 f}\frac{\partial \rho}{\partial x}. \tag{20.14b}$$

These are powerful relations in analyzing large-scale oceanic data. While oceanic velocities are difficult to measure,[5] density data are comparatively easy to obtain by dropping a Conductivity-Temperature-Depth (CTD) probe at repeated intervals from a ship cruising across the ocean. After some smoothing over mesoscale wiggles, the data provide the large-scale trends of density across the ocean basin, and it is relatively straightforward to determine the zonal and meridional gradients of density. Thus, we can consider $\partial \rho/\partial x$ and $\partial \rho/\partial y$ as known quantities and, from them, infer the velocity shear (20.14).

To obtain the actual velocity components u and v requires an additional assumption. The traditional approach is to assume a level of no motion, a deep horizon along which the pressure field is assumed to be uniform. Vertical integration of Eq. (20.14) from this level upward then provides the horizontal velocity up to the surface. This approach works well if the deep horizon is chosen within the relatively quiet abyssal layer, and the upward integration is performed to obtain the flow field in the main thermocline (Talley, Pickard, Emery & Swift, 2007).

Let us now return to the thermal-wind relations and extract from them an interesting property. For this, we decompose the horizontal velocity (u, v) in its magnitude U and azimuth θ:

$$u = U\cos\theta, \quad v = U\sin\theta. \tag{20.15}$$

The azimuth angle θ is measured counterclockwise from east and is related to the velocity components by $\theta = \arctan(v/u)$. Its vertical variation is given

$$\begin{aligned}\frac{\partial \theta}{\partial z} &= \frac{1}{u^2+v^2}\left(u\frac{\partial v}{\partial z} - v\frac{\partial u}{\partial z}\right)\\ &= \frac{-g}{\rho_0 f U^2}\left(u\frac{\partial \rho}{\partial x} + v\frac{\partial \rho}{\partial y}\right)\\ &= \frac{gw}{\rho_0 f U^2}\frac{\partial \rho}{\partial z},\end{aligned} \tag{20.16}$$

where the last step makes use of Eq. (20.5).

---

[5]There are several reasons why horizontal velocities are notoriously difficult to obtain directly from the deep ocean. First, they are almost always fluctuating at the mesoscale, making the time average easily fall within the noise level. Second, deep-water currentmeter moorings are expensive and wobble, thus adding an instrumental drift component to the velocity.

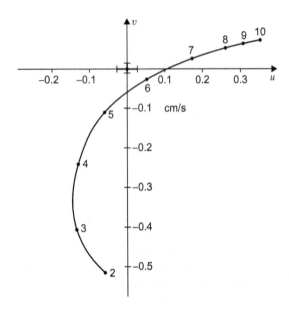

FIGURE 20.5   Beta spiral constructed from hydrographic data in the vicinity of 28°N, 36°W in the North Atlantic ocean. Numbers along the curve indicate depth in units of 100 m. Error bars are shown at the origin. (*Redrawn from Stommel & Schott, 1977*)

As we can see, there is a direct relation between the vertical velocity and the veering (twisting) of the horizontal velocity in the vertical. In the northern hemisphere $(f > 0)$ and in the presence of a gravitationally stable water column $(\partial \rho / \partial z < 0)$, $\partial \theta / \partial z$ has the opposite sign of w. Thus, in the midlatitudes, where the wind-stress curl is clockwise and Ekman pumping downward, the vertical velocity w is generally negative, and the vector of horizontal velocity turns clockwise with depth. This property has been dubbed the *beta spiral* (Schott & Stommel, 1978; Stommel & Schott, 1977). Figure 20.5 shows an example from the North Atlantic Ocean.

The veering implies that the waters at different levels in the vertical come from different directions and thus possess different origins. However, all levels of motion are under the tight constraint of the Sverdrup transport (20.11c). The local wind-stress curl appears therefore as a constraint on, rather than the forcing of, the flow.

20.2.4   A Bernoulli Function

In Sverdrup dynamics, the Montgomery potential defined as $P = p + \rho g z$ [see Eq. (12.4)] happens to play the role of a Bernoulli function. To show this, we need to prove that P is conserved along streamlines. Thus, we calculate its material derivative, which for steady flow contains only the spatial derivatives:

$$\frac{dP}{dt} = u \frac{\partial P}{\partial x} + v \frac{\partial P}{\partial y} + w \frac{\partial P}{\partial z}$$

$$= u\left(\frac{\partial p}{\partial x} + gz\frac{\partial \rho}{\partial x}\right) + v\left(\frac{\partial p}{\partial y} + gz\frac{\partial \rho}{\partial y}\right)$$
$$+ w\left(\frac{\partial p}{\partial z} + \rho g + gz\frac{\partial \rho}{\partial z}\right). \tag{20.17}$$

Use of the geostrophic relations (20.1) and hydrostatic balance (20.3) to eliminate all three derivatives of p leads to the cancellation of several terms, leaving

$$\frac{dP}{dt} = gz\left(u\frac{\partial \rho}{\partial x} + v\frac{\partial \rho}{\partial y} + w\frac{\partial \rho}{\partial z}\right), \tag{20.18}$$

which is identically zero by virtue of density conservation (20.5). Thus, the Montgomery potential P is conserved along the flow.[6] The same result could have been obtained more directly if we had taken the preliminary trouble of expressing the equations in density coordinate (Chapter 12).

While the preceding result holds some appeal, it is rarely useful because it is nearly impossible in the deep ocean to extract the dynamic signal of the pressure field from pressure measurements dominated by the hydrostatic component (that due to ρ_0!) and depth uncertainties caused by the variable sea surface. In contrast, potential vorticity, which depends on the density field, is of greater use with data.

20.2.5 Potential Vorticity

In the low Rossby number regime, momentum advection is negligible compared to the Coriolis acceleration, and consequently, the formulation of potential vorticity does not include relative vorticity next to planetary vorticity. We thus expect that in large-scale ocean dynamics, the expression of potential vorticity reduces to planetary vorticity over layer thickness. Specifically, the form is

$$q = \frac{f}{\Delta z/\Delta \rho}$$
$$= -\frac{f}{\partial z/\partial \rho} \tag{20.19}$$

in density coordinates, to become in depth coordinates

$$q = -f\frac{\partial \rho}{\partial z}. \tag{20.20}$$

---

[6]The more general expression of the Bernoulli function is $B = \rho_0(u^2 + v^2 + w^2)/2 + p + \rho gz$, but the kinetic-energy term is absent here as a consequence of the neglect of advection in the momentum equations, leaving only the last two terms, which together form the Montgomery potential.

To show that this expression is indeed conserved for Sverdrup dynamics, we begin by taking the vertical derivative of the density equation (20.5):

$$\frac{d}{dt}\left(\frac{\partial \rho}{\partial z}\right) + \frac{\partial u}{\partial z}\frac{\partial \rho}{\partial x} + \frac{\partial v}{\partial z}\frac{\partial \rho}{\partial y} + \frac{\partial w}{\partial z}\frac{\partial \rho}{\partial z} = 0. \qquad (20.21)$$

If we now eliminate the z-derivatives of u and v by using the thermal-wind relations (20.14) and the Sverdrup relation (20.8) to eliminate $\partial w/\partial z$, the middle terms cancel out, and we obtain

$$\frac{d}{dt}\left(\frac{\partial \rho}{\partial z}\right) + \frac{\beta_0 v}{f}\frac{\partial \rho}{\partial z} = 0. \qquad (20.22)$$

Then recognizing that $\beta_0 v = v\,(df/dy) = df/dt$, the equation can be recast as

$$f\frac{d}{dt}\left(\frac{\partial \rho}{\partial z}\right) + \frac{df}{dt}\frac{\partial \rho}{\partial z} = \frac{d}{dt}\left(f\frac{\partial \rho}{\partial z}\right) = 0, \qquad (20.23)$$

which means that the expression $f(\partial \rho/\partial z)$ is conserved along the flow. Thus, the potential vorticity defined above in Eq. (20.20) is conserved by individual water parcels. This conservation law, like conservation of the Bernoulli function shown in the previous subsection, holds wherever the flow is on a large scale and dissipation is weak, that is, in the main thermocline and abyssal layer.

It is interesting to return to the beta spiral and interpret it in the light of potential vorticity. The connection is illustrated in Fig. 20.6, which ties the following ingredients: conservation of potential vorticity, thermal wind, and the effect of

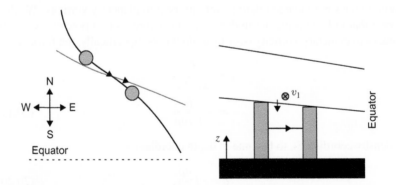

FIGURE 20.6 Along its journey toward the equator, a fluid column experiences a decrease of planetary vorticity f, which is compensated by a decrease in layer height. If we follow the flow in the vertical transect along the trajectory (right panel), there is, by definition of the trajectory, no transverse flow at the level of the water column being followed. Because the layer height decreases approaching the equator, a thermal wind must appear (v_1 on the right panel and gray trajectory in the left panel). With zero transverse flow below, the horizontal velocity vector turns clockwise with depth, in agreement with Eq. (20.16) and downward vertical movement.

vertical velocity. It shows how the necessary squeezing along a southward trajectory must be accompanied by a thermal wind that creates veering with depth
and hence the beta spiral.

We have seen that in the ocean's interior, three quantities are conserved
simultaneously by traveling water parcels: their density ρ, their Montgomery
potential P, and their potential vorticity q. Since it retains its ρ value along the
way, a moving parcel is confined to stay on an isopycnal surface. Similarly with
P, the same parcel is also confined to stay on a surface of constant P value.
Putting both constraints together, we conclude that a trajectory is an intersection line between a ρ surface and a P surface. Since all trajectories are also
lines of constant q, it follows that the potential vorticity q must be everywhere
a function of density ρ and Montgomery potential P:

$$q = Q(\rho, P). \tag{20.24}$$

In considering data from the North Atlantic Ocean, Williams (1991) found
that the density and depth of the seasonal thermocline at its late-winter coldest
and deepest stage are such that the waters deposited in the main thermocline are characterized by a potential vorticity that is nearly homogeneous
along isopycnal surfaces. This is particularly true for densities in the range
1026.4–1026.75 kg/m^3, corresponding to the main thermocline between $30°$ and
$40°$N. Where this occurs, we may reduce the preceding relation to a function of
a single variable:

$$q = Q(\rho). \tag{20.25}$$

Since both density and potential vorticity are easily observed variables, mapping these is a standard procedure in the construction of ocean circulation from
observations (see also Optimal Interpolation in Section 22.3). Mapping can
show isolines of potential vorticity on isopycnal surfaces. Because both density and potential vorticity are nearly conserved, potential vorticity then serves
as a tag to trace water movement.

20.3 WESTERN BOUNDARY CURRENTS

In commenting on Eq. (20.13), we concluded that the simplified Sverdrup
dynamics do not allow the simultaneous enforcement of impermeability boundary conditions at both eastern and western sides of an ocean basin. Thus,
Sverdrup dynamics must break down at one of the sides. The answer is that a
boundary layer with scale shorter than the basin scale must exist on the western
side of the basin. Let us now verify this.

Over the midlatitude ocean basins, the wind pattern is clockwise (counterclockwise) in the northern (southern) hemisphere because trade winds blow
from east to west in the tropics and westerly winds blow from west to east
further away from the equator (see Fig. 20.2). The resulting Ekman pumping is

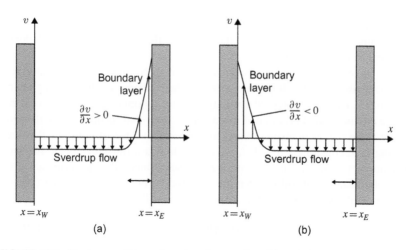

FIGURE 20.7   The two possible configurations for a northward boundary current to compensate the southward Sverdrup flow that exists across most of a midlatitude ocean basin of the northern hemisphere: (a) boundary current on the eastern side, (b) boundary current on the western side. The former is to be rejected on dynamic grounds, leaving the latter as the correct configuration.

downward, and the Sverdrup transport (20.11c) is equatorward (in both hemispheres). Conservation of mass requires that water flow toward the equator must be compensated elsewhere by poleward flow, but poleward flow violates Sverdrup dynamics. Thus, this return flow must exist on a scale shorter than the long length scale invoked in Sverdrup dynamics. In other words, it exists in the form of a narrow boundary current. With two meridional boundaries, one on each side of the basin, there are only two possibilities: either the poleward current follows the eastern boundary (Fig. 20.7a) or it follows the western boundary (Fig. 20.7b). In each case, connection with the equatorward Sverdrup flow in the basin's interior creates a velocity gradient (shear) and thus relative vorticity.

To make the argument easier, let us restrict our attention to the northern hemisphere. (The conclusion continues to hold for the southern hemisphere.) If the boundary layer lies along the eastern wall as shown in Fig. 20.7a, the northward flow has positive $\partial v/\partial x$. This derivative is large because the boundary layer is narrow and also the velocity must be large to accommodate the entire Sverdrup transport. By comparison, $\partial u/\partial y$ is very small. Thus, if flow is returning on the eastern side, it has positive relative vorticity ($\zeta > 0$). On the contrary, if the boundary layer lies along the western wall (Fig. 20.7b), the velocity shear $\partial v/\partial x$ is negative, and because $\partial u/\partial y$ is still negligible, the return flow has negative vorticity ($\zeta < 0$). But which type of vorticity is the return flow allowed to have?

When it makes its return journey back toward higher latitudes, water sees its planetary vorticity f increases, and conservation of potential vorticity

$(f+\zeta)/h$ demands that ζ or h or both change accordingly. Since relative vorticity ζ becomes important in the boundary current, the increase in f must be accompanied by a decrease in ζ, thus the value of ζ must drop from about zero to a negative value as fluid parcels exit the Sverdrup interior and enter the boundary current. The western boundary current does accomplish that, whereas the eastern boundary current does not. So, we reject the existence of a boundary layer along the eastern side of the basin and conclude that the necessary boundary layer must lie along the western side of the basin (Fig. 20.7b).

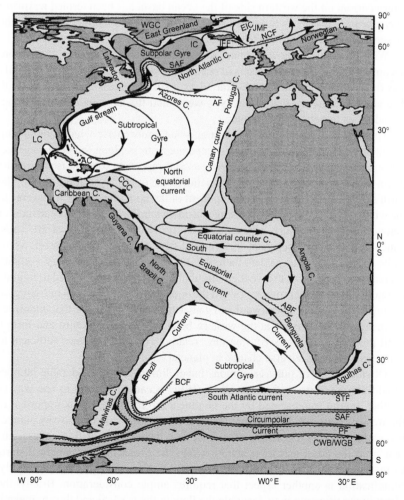

FIGURE 20.8   The average circulation in the Atlantic Ocean, showing equatorward Sverdrup flow in the midlatitudes of each hemisphere and western boundary currents. (*Tomczak & Godfrey, 2003*)

It is evident that the circulation takes the form of an asymmetric gyre, with a slow equatorward flow (the Sverdrup transport) occupying most of the domain and a swift narrow current on the western side that returns the water poleward (Fig. 20.8). The latter current is naturally identified with the Gulf Stream in the North Atlantic Ocean and the respective western-boundary currents in the other ocean basins: Brazil Current in the South Atlantic, Kuroshio in the North Pacific, East Australia Current in the South Pacific, and Agulhas Current in the south Indian Ocean. The circulation closes with zonal currents that connect the entrance and exit of the western boundary current with the interior Sverdrup flow. The funneling of the equatorward Sverdrup transport into a strong return current on the western side of the ocean basin has been termed *westward intensification* by Henry Stommel, who provided the first correct theory for the existence of the Gulf Stream (Stommel, 1948).

With the solution now at hand, let us recapitulate the results. As we are becoming aware, the mechanisms of ocean circulation are intricate and certainly less direct than a simple torque exerted by a surface stress on a viscous fluid. The chief reasons are that viscosity is weak, and the Coriolis effect is strong, including its variation with latitude.

The scenario is as follows. The large-scale atmospheric winds, comprising essentially the trades and westerlies, generate a stress along the ocean surface. Because seawater is only slightly viscous and planetary rotation is strong, the direct effect of the stress is limited to a thin (10 m or so) layer of the ocean. The earth's rotation also generates a component of the upper layer flow transverse to the winds, which converges, resulting in a downward flow into the ocean interior below (Ekman pumping). Although relatively weak on the order of 10^{-6} m/s, which is about 30 m per year, this vertical flow squeezes water parcels vertically. In reaction, fluid parcels flatten and widen, and their planetary vorticity decreases in order to conserve their circulation. They are thus forced to migrate equatorward. As they progress toward the equator, these waters run into a region of slower flow, veering westward and gathering into a zonal flow that intensifies downstream. On arriving at the western boundary, the waters turn and form a swift poleward flow, so swift that their relative vorticity becomes sufficient to compensate for the adverse change in planetary vorticity.

Obviously, all of our assumptions have eliminated a considerable number of additional processes that can all affect the ocean circulation in one way or another. Inertia (represented by the nonlinear advection terms) is important in the western boundary layer where the flow is swift and narrow (Rossby number becoming of order 1). The consequence is a detachment of this intense current from the coast and its penetration into the ocean interior, where it starts to meander rather freely. Barotropic instabilities (see Chapter 10) are likely. Stratification is another aspect that requires ample consideration. Briefly, the effect of stratification is to decouple the flow in the vertical and thus to make it respond less to bottom friction. On the other hand, a reserve of potential energy due to the presence of stratification causes baroclinic instabilities (Chapter 17). Barotropic and baroclinic instabilities generate eddies, and these in turn create a net horizontal mixing of momentum. Finally, because poleward western-boundary currents, such as the Gulf Stream, bring warm water masses to higher latitudes, an air–sea heat flux is created, resulting in the cooling of the ocean and a distortion of the circulation pattern. The interested reader will find additional information on ocean-circulation dynamics in the review article by Veronis (1981), the book by Abarbanel and Young (1987), and the article by Cushman-Roisin (1987a).

20.4   THERMOHALINE CIRCULATION

As stated in Section 20.1, the region below the seasonal thermocline is comprised of two subregions, the main thermocline and the abyssal layer, together called the ocean interior. The dynamics expounded in the previous section are applicable to both these regions. We now turn our attention more specifically to the upper of these two layers, the main thermocline.

In contrast to the abyss, which is fed by deep-water convection at high latitudes, the main thermocline is the region of the ocean in which the circulation is primarily caused by the wind-driven Ekman pumping received from the surface layer and is most pronounced at midlatitudes. Figure 20.9, compiled by Talley et al. (2007), summarizes the meridional distribution of density in the North and South Atlantic, North and South Pacific, and Indian Oceans. They reveal similar patterns in all five oceans. The pycnocline is very strong and shallow (100–200 m) at the equator; from there, it spreads vertically downward toward the poles, with a tendency to split into two branches: one surfacing around 25° latitude and the other plunging down to 1000 m around 35° before heading upward again and surfacing around 45° latitude.

20.4.1   Subduction

At the top of the unchanging ocean interior, water is exchanged with the seasonal thermocline. The process called *subduction*, if the water passes from the seasonal thermocline to the steady interior, or *entrainment*, if water is engulfed by the seasonal thermocline, deserves special attention.

There are several processes by which water enters the main thermocline: downward Ekman pumping injection, retreat of the convective mixed layer at the end of winter, and convergence of the flow inside the mixed layer (Cushman-Roisin, 1987b). The sum of these processes creates *subduction*, which can be defined as the deposition of fluid formerly belonging to the seasonal thermocline into the ocean interior. It is expressed mathematically as a volumetric flow rate per unit horizontal area (i.e., with dimensions of velocity, although it is not a vertical velocity).

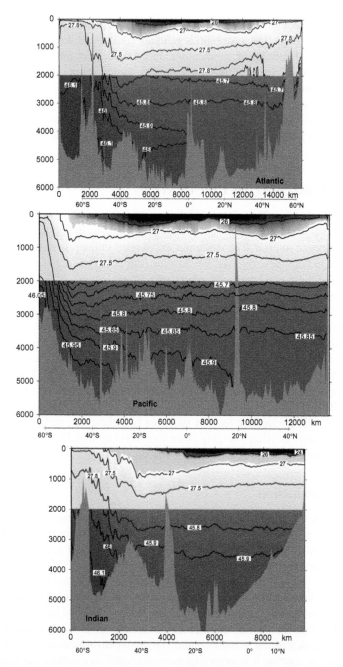

FIGURE 20.9 Meridional structure of potential density (σ_θ top and σ_4 bottom, in kg/m^{-3}) in the five major ocean basins. Top panel: North and South Atlantic. Middle panel: North and South Pacific. Bottom panel: Indian Ocean. Data were gathered during the World Ocean Circulation Experiment (1990–2002). The variables σ_θ and σ_4 are density anomalies corrected for the effect of compressibility under high pressure. *(Talley et al., 2007)*. A color version can be found online at http://booksite.academicpress.com/9780120887590/

Consider a fluid column of infinitesimal width and extending from the base of the seasonal thermocline upward to the base fluid of the Ekman layer. The density is this fluid column is vertically uniform because the seasonal thermocline is in a state of mixing (by definition). Denoting its height by $h(x, y, t)$, its volume budget is

$$\frac{\partial h}{\partial t} + \frac{\partial (h u_{ST})}{\partial x} + \frac{\partial (h v_{ST})}{\partial y} = -w_{Ek} - Su, \qquad (20.26)$$

where u_{ST} and v_{ST} are the eastward and northward velocity components inside the seasonal thermocline, w_{Ek} the Ekman pumping at its top (base of Ekman layer—see Fig. 20.3), and Su the subduction rate at its bottom. Positive Su represents subduction proper (flow into the main thermocline below), whereas negative Su corresponds to capture of ocean interior water by the seasonal thermocline, that is entrainment. A negative w_{Ek} brings water from the surface mixed layer down and into the seasonal thermocline.

If we now assume that the horizontal velocity (u_{ST}, v_{ST}) is characterized on a seasonal scale by low frequency and long length scale, it must be nearly geostrophic, and we may write

$$-\rho_0 f v_{ST} = -\frac{\partial p_{ST}}{\partial x}, \qquad +\rho_0 f u_{ST} = -\frac{\partial p_{ST}}{\partial y}, \qquad (20.27)$$

with $f = f_0 + \beta y$ and p_{ST} the pressure inside the seasonal thermocline. The divergence term of Eq. (20.26) becomes

$$\frac{\partial (h u_{ST})}{\partial x} + \frac{\partial (h v_{ST})}{\partial y} = J\left(\frac{p_{ST}}{\rho_0}, \frac{h}{f}\right), \qquad (20.28)$$

and Eq. (20.26) can then be recast as

$$Su = -\frac{\partial h}{\partial t} - w_{Ek} - J\left(\frac{p_{ST}}{\rho_0}, \frac{h}{f}\right), \qquad (20.29)$$

which shows that subduction is a combination of seasonal thermocline retreat, downward Ekman pumping, and convergence of geostrophic flow in the seasonal thermocline.

Before continuing with motion inside the main thermocline, it is worth making a few remarks concerning the temporal variability of subduction. As Stommel (1979) remarked, much of the water left behind during spring and summer when the thermocline retreats is recaptured the following fall and winter, because it has not had the time to sink deeply enough into the interior before the seasonal thermocline penetrates once again. Much of subduction is in vain, and effective feeding of the ocean interior by subduction occurs only during a relatively brief time interval in late winter. This explains why water properties in the ocean interior systematically reflect surface water properties of late winter and never those of summer (Stommel, 1979). Cushman-Roisin (1987b) explored in somewhat greater detail the kinematics of seasonal subduction and concluded that the time of effective subduction is not quite as brief as Stommel suggested but that about 30% of the volume subducted escapes recapture and feeds the ocean interior.

20.4.2   Ventilated Thermocline Theory

Early theories of the main thermocline attempted to explain the observed vertical temperature structure as a local advection–diffusion equilibrium between upwelling of cold abyssal water and downward diffusion of heat from the surface or sought analytical solutions based on assumptions dictated by mathematical convenience. Then, the paradigm changed in the 1980s with the publication of a paper by Luyten, Pedlosky and Stommel (1983) titled "The ventilated thermocline." The theory combines subduction from the mixed layer with advective descent into the stratified thermocline.

The scenario is as follows. In the midlatitude ocean, where Ekman pumping is downward, mixed-layer waters are subducted into the main thermocline where they slide along density surfaces, carrying with them their surface properties such as density and potential vorticity, which are set at subduction time. Layers of thermocline water that can be traced back to the base of the mixed layer where Ekman pumping is downward are said to be *ventilated*, and their intersection with the mixed layer where they are supplied is called the *outcrop line*. The vertical structure of the main thermocline in the ventilated area thus reflects the density distribution at the surface in winter (time of effective subduction), a fact long noted by Iselin (1938).

Under these premises, the thermocline problem reduces to solving for the vertical layering of density by tracking individual density layers upstream to their respective outcrop lines, where the (winter) surface density distribution is known. As the idea of subduction and subsequent sliding along density surface suggests a thermocline controlled by advection, a theory using an inviscid and nondiffusive fluid seems appropriate (Luyten et al., 1983). The theory was later extended to continuous stratification by Huang (1989). The situation is complicated by the fact that only a sector of the main thermocline can be traced back to outcrops where surface conditions are known; the remaining portions, on the eastern and western sides, form so-called *shadow zones*. In these zones, the flow is circulating without surface contact, and it has been speculated that it is characterized by a slow but effective homogenization of potential vorticity (Huang, 1989).

Instead of launching in an exposition of the ventilated-thermocline theory and the attending complication of shadow zones (for this, the reader is referred to chapter 4 of Pedlosky, 1996), we shall limit ourselves here to determining a scale for the vertical thickness of the main thermocline.

20.4.3 Scaling of the Main Thermocline

The dynamics of the circulation in the main ocean thermocline are governed by a small number of parameters, namely the constants f_0, β_0, ρ_0, and g that enter the governing equations, and a few external scales that enter through boundary conditions: L_x the width of the basin, W_{Ek} a typical magnitude of the Ekman pumping, and $\Delta\rho$ a typical density variation across the thermocline.

Following Welander (1975), the thermocline depth h_{scale} can be derived by balancing the various terms in the equations of Sverdrup dynamics. First, the scale for pressure is $\Delta P = g h_{scale} \Delta\rho$ from the hydrostatic balance (20.3), from which follow the north-south velocity scale through (20.1a): $v \sim \Delta P/(\rho_0 f_0 L_x) = g h_{scale}\Delta\rho/(\rho_0 f_0 L_x)$. The vertical velocity must necessarily scale like the Ekman pumping velocity because it is equal to it at the base of the Ekman layer. With scales for both meridional and vertical velocities known, the Sverdrup relation (20.8) implies

$$\beta_0 \frac{g h_{scale}\Delta\rho}{\rho_0 f_0 L_x} \sim f_0 \frac{W_{Ek}}{h_{scale}} \tag{20.30}$$

from which follows the depth scale h_{scale} of the main thermocline:

$$h_{scale} = \sqrt{\frac{f_0^2 L_x W_{Ek}}{\beta_0 g(\Delta\rho/\rho_0)}}. \tag{20.31}$$

To see whether this scale is reasonable, let us use numbers corresponding to the North Atlantic Ocean circa 35°N. At that latitude, the Coriolis parameters are $f_0 = 8.4 \times 10^{-5}$ s^{-1} and $\beta_0 = 1.9 \times 10^{-11}$ m^{-1}s^{-1}, while the width of the basin, stretching from 10° to 80°W, gives $L_x = 6400$ km. With a wind stress curl leading to an Ekman pumping on the order of $W_{Ek} = 2 \times 10^{-6}$ m/s and a relative density difference $\Delta\rho/\rho_0$ of about 0.002, we obtain $h_{scale} = 490$ m, about 500 m as observations indicate.

Because ocean basins are much deeper[7] than this scale, there is much water lying below the main thermocline that is not subject to surface Ekman pumping. This is the abyssal layer, which is the subject of the next section. As we shall see, it is driven by deep convection under atmospheric cooling at high latitudes.

20.5 ABYSSAL CIRCULATION

Since the wind stress only affects a relatively small portion of the water column, the bulk of ocean waters forming the abyssal layer must be driven by another

---

[7] The average depth in the ocean is 3720 m.

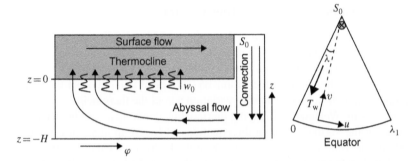

FIGURE 20.10 A highly simplified model of the abyssal circulation, with "longitude" arbitrarily taken as zero at the western boundary (Left panel: side view - Right panel: top view). The volumetric source S_0 is meant to represent dense-water formation at high latitudes.

mechanism. As it turns out, this greatest body of water is set in slow motion by vertical convection taking place over an exceedingly small portion of the ocean surface, narrow zones of extreme cooling at high latitudes.

The scenario is as follows. Exposure to a very cold and dry atmosphere at high latitudes causes both thermal contraction and evaporation. Evaporation takes distilled water away, leaving the salt in the ocean. Brine rejection from water freezing into ice further increases the salinity of the remaining liquid water, and the result is a water that is both very cold and very salty, thus significantly denser than typical seawater. This dense water sinks under the effect of gravity, slowly but effectively filling the abyss of the world ocean. Known areas of dense-water formation by deep convection are the northern reaches of the Atlantic Ocean at the entrance of the Arctic Ocean and several marginal seas along the periphery of Antarctica (Weddell Sea and Ross Sea).

To close the circulation, slow upwelling in lower latitudes must take place to return the waters back to the surface, where they are heated and can once again migrate to high latitudes and complete the circuit (Fig. 20.10). In three dimensions, the totality of this circulation system forms the so-called *conveyor belt* (Fig. 20.1), and a round trip along this path is believed to take thousands of years.

By way of exception, we work here in spherical coordinates because the present analysis is not much more complicated than with Cartesian coordinates, and its results are easier to identify with actual oceanic features. The analysis follows closely that of Stommel and Arons (1960a,b).

The dynamical balance reduces as usual to geostrophy, now in spherical coordinates [see (A.18)] and is complemented by volume conservation:

$$-fv = -\frac{1}{a\cos\varphi}\frac{\partial p}{\partial \lambda} \qquad (20.32a)$$

$$+fu=-\frac{1}{a}\frac{\partial p}{\partial \varphi} \tag{20.32b}$$

$$\frac{1}{a}\frac{\partial u}{\partial \lambda}+\frac{1}{a}\frac{\partial}{\partial \varphi}(v\cos\varphi)+\frac{\partial}{\partial z}(w\cos\varphi)=0, \tag{20.32c}$$

where a is the earth's radius, λ longitude, φ latitude, z the local vertical coordinate, and $f=2\Omega\sin\varphi$ the latitude-dependent Coriolis parameter. Elimination of pressure between the first two equations yields

$$\frac{\partial}{\partial \lambda}(fu)+\frac{\partial}{\partial \varphi}(fv\cos\varphi)=0, \tag{20.33}$$

from where, by using volume conservation (20.32c), we recover the Sverdrup relation, now in spherical coordinates

$$\beta v=f\frac{\partial w}{\partial z}, \tag{20.34}$$

with $\beta=2(\Omega/a)\cos\varphi$.

Vertical integration across the abyssal layer (from $-H$ to 0) from a flat bottom yields the Sverdrup transport

$$\int_{-H}^{0} v\,dz = V = aw_0\tan\varphi. \tag{20.35}$$

With positive velocity (upwelling) w_0 almost everywhere (since the sinking of dense water is confined to small corner regions), the abyssal flow must be northward in the northern hemisphere ($\varphi>0$) and southward in the southern hemisphere ($\varphi<0$), that is, everywhere poleward. We also note that there is no Sverdrup transport across the equator ($\varphi=0$). This is evidently problematic since intuition would lead us to believe that convection at polar latitudes should create flow away from, not toward, the poles. The solution to this paradox is that the flow coming from the high-latitude regions is confined to narrow strips along the western margins of the oceanic basins, and the broader abyssal flow consists of the return flow toward higher latitudes.

For the following discussion, we now restrict our attention to the northern hemisphere. The water budget from any northern latitude φ up to the pole, including the deep-water source, demands that the flow carried by the interior (boundary integral of V) plus the deep-water inflow (S_0) be compensated by the western boundary layer flow (T_w), which goes further south, and the cumulated effect of upwelling at the top (surface integral of w_0):

$$S_0+\int_{0}^{\lambda_1} V a\cos\varphi\,d\lambda = T_w(\varphi)+\int_{\varphi}^{\pi/2}\int_{0}^{\lambda_1} w_0 a^2\cos\varphi\,d\lambda\,d\varphi. \tag{20.36}$$

In the case of a uniform upwelling speed w_0 and with Eq. (20.35) to eliminate V, we have

$$S_0 + \sin\varphi\,\lambda_1 a^2 w_0 = T_w(\varphi) + (1 - \sin\varphi)\lambda_1 a^2 w_0. \qquad (20.37)$$

The transport in the western boundary layer must therefore be

$$T_w(\varphi) = S_0 + (2\sin\varphi - 1)\lambda_1 a^2 w_0, \qquad (20.38)$$

which is maximum at the pole ($\varphi = \pi/2$) and decreases toward the equator ($\varphi = 0$). Three cases can arise.

Case 1: $S_0 = \lambda_1 a^2 w_0$. In this case, upwelling from the equator to the pole exactly matches the source at the pole. The transport in the western boundary current is $T_w = 2S_0 \sin\varphi$, which vanishes at the equator ($\varphi = 0$). Since the Sverdrup transport is also zero at the equator, the two hemispheres are decoupled. We further note that near the pole, the boundary-layer flow is *twice* as strong as the source, while the northward Sverdrup flow is equal to the source implying that the half the flow is pure recirculation. This suprising feature seemed so counterintuitive to Stommel and Arons that they performed laboratory experiments (Stommel & Arons, 1960a,b) to verify their findings.

Case 2: $S_0 > \lambda_1 a^2 w_0$. The dense-water source is stronger than the distributed upwelling, and an excess transport spills across the equator into the other hemisphere. This is the situation encountered in the North Atlantic.

Case 3: $S_0 < \lambda_1 a^2 w_0$. The source is insufficient to sustain the required upwelling, and a northward boundary-layer flow across the equator is necessary to supply the difference. Such a situation is prevailing in the North Pacific.

The preceding solution can be extended to any w_0 distribution in the calculation of the zonal velocity u, requiring zero normal flow at the eastern boundary, consistent with the position of the unresolved boundary layer along the western boundary. From Eq. (20.32a), knowing the meridional velocity in a barotropic abyssal layer (20.35), we can calculate the pressure by imposing a constant value, taken as zero without loss of generality, along the eastern boundary:

$$p = \frac{2\Omega a^2}{H} \sin^2\varphi \int_{\lambda_1}^{\lambda} w_0 \, d\lambda, \qquad (20.39)$$

from which we deduce the zonal velocity according to Eq. (20.32b)

$$u = -\frac{a}{H\sin\varphi} \frac{\partial}{\partial\varphi} \left(\sin^2\varphi \int_{\lambda_1}^{\lambda} w_0 \, d\lambda \right). \qquad (20.40)$$

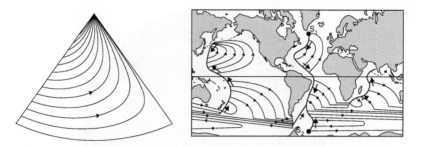

FIGURE 20.11   A few trajectories of the abyssal flow in a basin sector (left panel) and the abyssal circulation inferred by Stommel (1958) (right panel).

For uniform w_0, this reduces to

$$u = 2\frac{a}{H}w_0(\lambda_1 - \lambda)\cos\varphi, \qquad (20.41)$$

which is always directed eastward, implying that the boundary layer feeding this flow must be on the western side of the basin.

The velocity field (20.41) and (20.35) gives rise to the trajectories depicted in the left panel of Fig. 20.11, in the light of which we can understand the abyssal circulation proposed by Stommel (1958), right panel of Fig. 20.11.

The preceding theory serves as an interesting application of geophysical fluid dynamics but barely illustrates the complex dynamics of the abyssal circulation. Chief among our assumptions was that of a flat bottom. The oceanic bathymetry, as we all know, is rather fractured, with a multiplicity of ridges standing as obstacles and passages guiding the flow. Among the special features that ridges and passages inflict on the flow are concentrated zonal currents and recirculation patterns. The interested reader is referred to chapter 7 of Pedlosky (1996).

20.6   OCEANIC CIRCULATION MODELS

A milestone in numerical ocean modeling was the first *Ocean General Circulation Model*, or OGCM in short, developed by a team of scientists at Princeton University's Geophysical Fluid Dynamics Laboratory (Bryan, 1969; Bryan & Cox, 1972). The release to the scientific community of the source code of this model and its successive variants, of the type now called Modular Ocean Model (MOM), contributed to the model's widespread use, especially because the simple numerics lent themselves to various adaptations by many users. The code was based on straightforward second-order centered finite differencing of the governing equations in longitude-latitude coordinates with stepwise representation of coastlines and bottom topography. Time marching was leapfrog.

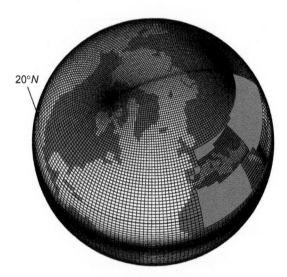

20°N

FIGURE 20.12 Ocean model grid with numerical grid poles over continents. (*LODYC, ORCA configuration; Madec, Delecluse, Imbard & Levy, 1998*)

This model enabled the first general circulation studies with the primitive equations (that is, without reliance on the quasi-geostrophic approximation, as it had been the case until then), but it eventually became apparent that a series of improvements were necessary. In particular, the polar convergence of meridians in spherical coordinates led to annoying singularities, creating the so-called *pole problem*. This was later avoided by shifting the spherical coordinate system to relocate its "poles" on continents or by using curvilinear orthogonal grids that maintain a topologically rectangular grid (Fig. 20.12). The stepwise topography (Fig. 20.13) was also ill fitted to simulate weak bottom slopes and the associated potential-vorticity constraint. In response, partially masked cells were introduced (e.g., Adcroft, Hill & Marshall, 1997). A further issue caused by stepwise topography was its poor representation of overflows typical of deep-water formation (Fig. 20.14). To tackle such situations, special algorithms have been developed (e.g., Beckmann & Döscher, 1997). Worthy of special note is DieCAST (Dietrich, 1998), which uses a modified Arakawa "A" grid with fourth-order accuracy in the horizontal directions. The unusually low dissipation of this model provides more accurate simulation of narrower features, such as boundary currents and mesoscale eddies.

In addition to these improvements, new generations of models are constantly being developed. Perhaps the most significant change in terms of numerical implementation is the move from structured to unstructured grids. The presence of continental boundaries remains an obstacle to the use of spectral models (preferred in modeling of the global atmosphere), while unstructured grids offer, by

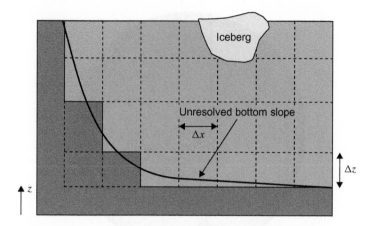

FIGURE 20.13 Masking of a regular grid allows to discretize topography. For rectangular grid boxes, weak slopes can only be resolved if $\Delta z < \Delta x |\partial b/\partial x|$ with bottom given by $z = b(x, y)$. If not, the weak slope is awkwardly approximated by a flat bottom stretching over several grid steps followed by an abrupt step. Masking can also be applied to the top boundary in the presence of icebergs or thick ice (thick compared to the vertical spacing near the surface).

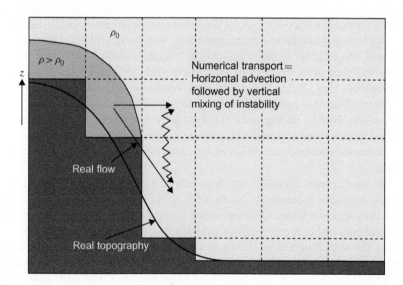

FIGURE 20.14 Vertical section of a schematic overflow where dense bottom water cascades down a slope. With stepwise topography representation, advection carries the dense water horizontally to a model grid point with lower density. It is tempting to remove the resulting unphysical gravitational instability by a mixing algorithm, but this has the unfortunate consequence of diluting the dense-water vein.

FIGURE 20.15   Example of a finite-element model grid for the world ocean. (*Legrand, Legat & Deleersnijder, 2000*)

design, the possibility of following complicated contour lines. Structured grids, natural for models discretized along Cartesian coordinates or longitude-latitude, are topologically similar to a rectangular grid, with every grid point having one and only one neighbor to the "east," "west," "north," "south," above and below. In unstructured grids, on the contrary, every grid cell has a variable number of neighbors (Fig. 20.15), providing great flexibility in terms of geographical coverage. Coastlines can be followed by adding small elements along the side, canyons resolved by adding small elements at the bottom, and open boundaries pushed further away by increasing the size of distant elements. Intense model developments continue unabated (see, e.g., Pietrzak, Deleersnijder & Schroeter, 2005).

   An unstructured finite-volume approach is a generalization of the finite-volume approach presented in Section 5.5, in which integration is performed over each finite volume, with the entire group covering the model domain. Physical coupling between the finite volumes arises naturally through the fluxes across the shared interfaces. Finite elements, in contrast, start with a totally different approach based on the Galerkin method (Section 8.8). The solution is expanded as a sum of non-orthogonal basis functions, and each governing equation is multiplied by each basis function before being integrated over the model domain. The basis functions (sometimes also called trial functions) used in finite-element methods are of a special nature, being nonzero only over a given element. Connection between elements then arises because functions are forced to obey some continuity requirement between elements. This necessitates solving a system of linear equations (i.e., inverting a matrix) at every time step, but the nature of the basis functions leads to relatively sparse matrices.

Beside the widespread finite-volume and finite-element methods, additional schemes have been implemented in ocean models such as the spherical cube grid (a three-dimensional generalization of the squaring of a circle, with special connectivity at a few nodes—Adcroft, Campin, Hill & Marshall, 2004) and spectral elements (Haidvogel, Wilkin & Young, 1991) a method in which a domain is covered by large elements, inside of which spectral series are used to approximate the solution.

The advantage of variable resolution exposes the fundamental problem of subgrid-scale parameterization. Scales resolved by the variable grid do change regionally, and therefore the nature of the parameterizations involved should change from place to place within the same model. Among processes requiring parameterization, deep-water formation is perhaps the most crucial example (see Section 11.4). The sinking of dense water is dominated by nonhydrostatic convection, forcing all hydrostatic models to have a parameterization of one form or another. Because the validity of the hydrostatic approximation is related to the aspect ratio of the flow (ratio of vertical scale to horizontal scale, see Section 4.3), nonhydrostatic effects are only relevant at extremely high resolution, and ocean general circulation models are rarely nonhydrostatic. Yet, highly localized dense-water formation does influence broader scale flow and needs to be included in these models. The brutal, and rather common, way to deal with the issue is *convective adjustment* (see Section 11.4). In this scheme, whenever the water column is found to have a density reversal between two vertically aligned grid points, temperature and salinity are merely replaced by their average values, and the process continues downward until no inversion remains. Despite many years of use, this scheme was eventually found to be flawed (Cessi, 1996), for it generates an unstable mode at the smallest resolved horizontal scale. A better way to handle convective overturn is by local mesh refinement with flexible horizontal grids, and today's hydrostatic ocean models often include a nonhydrostatic option, such as used in *MITgcm* (Marshall, Jones & Hill, 1998).

The upper boundary of an ocean circulation model requires special care because this is where atmospheric forcing is applied to the ocean. Rigid-lid models eliminate the fast surface gravity waves in order to permit longer time steps. However, there is a trend toward the use of free-surface models because of their added flexibility and wider range of applicability, including tidal representation. The trick to continue enjoying a reasonably large time step is to treat time stepping of the surface elevation either implicitly, semi-implicitly, or with a mode-splitting scheme (Section 12.7), while the rest of the scheme remains explicit.

The coupling with the atmosphere (see Section 19.4) remains a difficult problem because the ocean model must simulate both surface temperature and mixed-layer depth correctly. The former is essential for air–sea exchanges while the product of temperature and mixed-layer depth is directly related to the heat content of the water column and therefore also the heat budget. Because of the low heat capacity of air, the atmospheric component of the model can be very sensitive to a small error in the heat content of the upper ocean. Furthermore, because mixed-layer evolution critically depends on turbulence, special care is needed in the specification of the eddy viscosity and diffusivities, particularly if vertical grid spacing is coarse. Turbulence-closure schemes were described in Sections 14.3–14.5.

20.6.1   Coordinate Systems

Vertical resolution near the surface is essential to capture air–sea interactions, and most oceanic models use shorter vertical grid spacing near the surface. For the ocean interior, vertical gridding is again considered crucial, and a significant part of the problem is the separation between vertical and horizontal coordinates, which is necessary because of the small geometric aspect ratio and the vast difference in dynamics between the vertical and horizontal directions. We therefore assume that the horizontal discretization is performed using one among the methods outlined earlier in this text and focus now on vertical gridding.

Vertical gridding can be done quite freely, with variable resolution permitting higher resolution in sensitive segments (near the surface, across a pycnocline, and near the bottom) and lower resolution at less important levels for the sake of computational economy. About the only requirement is that a grid point topologically above (below) another grid point corresponds to a physical point also physically above (below) the other point. The nature of the vertical coordinate can very well change from top to bottom, switching from depth to density to a bottom-following coordinate. The generation of such hybrid-grid models can be achieved in one of two ways: we can subject the governing equations to the finite-volume integration technique over the desired vertical spacings and write a set of governing equations for each cell, or we can first perform a change of coordinates after which we discretize the equations along a uniform grid in that new coordinate system (Fig. 20.16; see also Section 15.6).

Direct integration in physical space exposes the need for parameterization because integrals of nonlinear terms cannot be expressed in terms of cell averages unless some assumptions are made. In contrast, a coordinate transformation followed by finite differencing runs the risk of masking the need for parameterization of subgrid-scale processes. Ultimately, however, the discrete equations that are obtained will be similar, and the choice of approach depends on the path preferred by the modeler.

Here we present the coordinate transformation approach, in close analogy with the isopycnal coordinate change of Chapter 12. In the original Cartesian system of coordinates (x, y, z, t), z is an independent variable, while in the transformed coordinate system (x, y, s, t) the new coordinate s replaces z as an independent variable and $z(x, y, s, t)$ becomes a dependent variable giving the depth at which the value s is found at location (x, y) and at time t. A surface along which s is constant is called a *coordinate surface*. From a differentiation

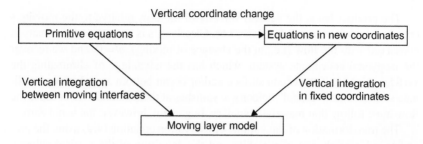

FIGURE 20.16 Equations for discrete layers can be obtained either by integrating the governing equations across each chosen layer or by a transformation to moving coordinates followed by discretization on a fixed grid. The discrete equations to be solved by the computer are the same.

of the expression $a = a(x, y, s(x, y, z, t), t)$, where a stands for any variable, the rules for the change of variables follow:

$$\frac{\partial}{\partial x} \longrightarrow \frac{\partial a}{\partial x}\Big|_z = \frac{\partial a}{\partial x}\Big|_s + \frac{\partial a}{\partial s}\frac{\partial s}{\partial x}\Big|_z$$

$$\frac{\partial}{\partial y} \longrightarrow \frac{\partial a}{\partial y}\Big|_z = \frac{\partial a}{\partial y}\Big|_s + \frac{\partial a}{\partial s}\frac{\partial s}{\partial y}\Big|_z$$

$$\frac{\partial}{\partial z} \longrightarrow \frac{\partial a}{\partial z} = \frac{\partial a}{\partial s}\frac{\partial s}{\partial z}$$

$$\frac{\partial}{\partial t} \longrightarrow \frac{\partial a}{\partial t}\Big|_z = \frac{\partial a}{\partial t}\Big|_s + \frac{\partial a}{\partial s}\frac{\partial s}{\partial t}\Big|_z$$

in analogy with the isopycnal coordinate system presented in Chapter 12. A noteworthy difference with isopycnal coordinates is that s is not necessarily a physical property conserved along the flow. An important expression in the coordinate transformation is the quantity

$$\hbar = \frac{\partial z}{\partial s}, \tag{20.42}$$

which denotes the change in z for a unit change in s. Hence, it is a measure of the *coordinate layer thickness*, analogous to the thickness of a density layer. It can be positive or negative depending on whether s increases upward or downward. (It is negative if $s = \rho$, the density coordinate of Chapter 12.)

The material derivative in the new coordinate system takes the form:

$$\frac{da}{dt} = \frac{\partial a}{\partial t} + u\frac{\partial a}{\partial x} + v\frac{\partial a}{\partial y} + \omega\frac{\partial a}{\partial s}, \tag{20.43}$$

where all derivatives are taken in the transformed space ($\partial a/\partial x$ is performed at constant s, *etc.*) and where ω substitutes for the vertical velocity. It is defined as

$$\omega = \frac{\partial s}{\partial t}\Big|_z + u\frac{\partial s}{\partial x}\Big|_z + v\frac{\partial s}{\partial y}\Big|_z + w\frac{\partial s}{\partial z}. \tag{20.44}$$

The product $\hbar\omega$ is the vertical velocity of the flow relative to the moving s coordinate surface (recall Section 15.6). Clearly, if s is density and if density is conserved with the flow (i.e., in the absence of mixing), $\omega = 0$, and we recover the isopycnal coordinate system, which has the advantage of eliminating the vertical velocity. If the ocean surface and/or ocean bottom are taken as coordinate surfaces, the "vertical" velocity ω vanishes at those boundaries because the flow must follow that material boundary. In general, however, $\hbar\omega$ is not zero.

The transformation of the volume-conservation equation (4.9) using the preceding rules of change of variables and the definition of the vertical velocity leads to

$$\frac{\partial \hbar}{\partial t} + \frac{\partial}{\partial x}(\hbar u) + \frac{\partial}{\partial y}(\hbar v) + \frac{\partial}{\partial s}(\hbar \omega) = 0 \tag{20.45}$$

with derivatives taken in s space. Interestingly, if we integrate from bottom to top in the case of s constant along both ocean surface and bottom, we obtain

$$\frac{\partial \eta}{\partial t} + \frac{\partial U}{\partial x} + \frac{\partial V}{\partial y} = 0 \tag{20.46}$$

because

$$\int_{\text{bottom}}^{\text{surface}} \hbar\, ds = z_{\text{surface}} - z_{\text{bottom}} \tag{20.47}$$

which, for a time-independent bottom, leads after temporal derivation to the time derivative of the surface elevation η. The other two terms involve U and V, which are the vertically integrated transports

$$U = \int u\hbar ds = \int u\, dz, \qquad V = \int v\hbar ds = \int v\, dz, \tag{20.48}$$

where the integration is performed from bottom to top. We recover a vertically integrated volume-conservation equation.

Irrespective of whether or not the ocean surface and bottom are taken as coordinate surfaces, volume conservation leads to a conservative form of the material derivative

$$\begin{aligned} \hbar\left(\frac{\partial a}{\partial t} + u\frac{\partial a}{\partial x} + v\frac{\partial a}{\partial y} + \omega\frac{\partial a}{\partial s}\right) \\ = \frac{\partial}{\partial t}(\hbar a) + \frac{\partial}{\partial x}(\hbar a u) + \frac{\partial}{\partial y}(\hbar a v) + \frac{\partial}{\partial s}(\hbar a \omega), \end{aligned} \tag{20.49}$$

which we can interpret as the evolution of the content of a within a layer of s. This form is particularly well suited for integration over a finite volume in the transformed space.

The vertical diffusion term is readily transformed in the new coordinate system:

$$\frac{\partial}{\partial z}\left(\nu_E \frac{\partial a}{\partial z}\right) = \frac{1}{\hbar}\frac{\partial}{\partial s}\left(\frac{\nu_E}{\hbar}\frac{\partial a}{\partial s}\right), \tag{20.50}$$

so that the governing equation for a vertically diffusing tracer c becomes

$$\frac{\partial}{\partial t}(\hbar c) + \frac{\partial}{\partial x}(\hbar c\,u) + \frac{\partial}{\partial y}(\hbar c\,v) + \frac{\partial}{\partial s}(\hbar c\,\omega) = \frac{\partial}{\partial s}\left(\frac{\kappa_E}{\hbar}\frac{\partial c}{\partial s}\right). \tag{20.51}$$

We could also transform the horizontal diffusion term, but generally this operation is combined with the parameterization of subgrid-scale processes (see next section).

In view of the isomorphy of (20.51) with a Cartesian-coordinate version, an s-model can thus be implemented in a general way without much additional work once the functional relationship $s(x, y, z, t)$ is specified. The choice is at the modeler's discretion.

Besides the isopycnal transform with $s = \rho$ or $s = -\rho/\Delta\rho$, another coordinate change is the so-called sigma coordinate (σ) system, a particular form of terrain-following coordinates. It is very popular in coastal modeling. This coordinate is defined as

$$s = \sigma = \frac{z-b}{h}, \quad \hbar = h, \tag{20.52}$$

so that it varies between 0 at the bottom $[z = b(x, y)]$ and 1 at the surface $[z = b(x, y) + h(x, y, t)]$, which are therefore coordinate surfaces (Fig. 20.17). All topographic slopes and free surface movements are naturally followed, avoiding the problem of discretizing equations in a changing domain. Also, calculation points are efficiently used because they all fall into the water column ($0 \le \sigma \le 1$), and vertical boundary conditions are straightforward. Moreover, and this is the main advantage in coastal modeling, grid points are more closely spaced in shallow water than in deep water, providing the highest vertical resolution where it is the most needed (Fig. 20.17).

However, the use of this coordinate transform in global-ocean models has raised some concern about the so-called pressure-gradient problem (e.g., Deleersnijder & Beckers, 1992; Haney, 1991). Although the problem was initially identified for the σ-coordinate, it is more general, and we describe it here in a general coordinate framework.

The horizontal pressure gradient, for example, along x, can be evaluated in the new coordinate system by using the following transformation rules:

$$\left.\frac{\partial p}{\partial x}\right|_z = \left.\frac{\partial p}{\partial x}\right|_s + \frac{\partial p}{\partial s}\left.\frac{\partial s}{\partial x}\right|_z, \tag{20.53}$$

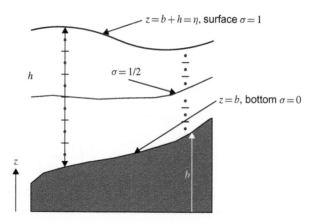

FIGURE 20.17   A sigma coordinate system divides the entire water column into an equal number of vertical grid cells regardless of local depth and surface elevation.

while the hydrostatic balance used to calculate pressure becomes

$$\frac{\partial p}{\partial z} = \frac{1}{h}\frac{\partial p}{\partial s} = -\rho g. \tag{20.54}$$

This allows us to write several, mathematically equivalent expressions for the horizontal pressure force

$$\frac{1}{\rho_0}\frac{\partial p}{\partial x}\Big|_z = \frac{1}{\rho_0}\frac{\partial p}{\partial x}\Big|_s + \frac{\rho g}{\rho_0}\frac{\partial z}{\partial x}\Big|_s \tag{20.55a}$$

$$= \frac{1}{\rho_0 h}J(p,z) \tag{20.55b}$$

$$= \frac{1}{\rho_0}\frac{\partial p}{\partial x}\Big|_{\text{surface}} + \frac{\rho g}{\rho_0}\frac{\partial \eta}{\partial x} - g\int_{\text{surface}}^{s} J(\rho,z)\,\mathrm{d}s \tag{20.55c}$$

$$= \frac{1}{\rho_0}\frac{\partial P}{\partial x}\Big|_s - \frac{gz}{\rho_0}\frac{\partial \rho}{\partial x}\Big|_s \tag{20.55d}$$

where the last expression uses the Montgomery potential $P = p + \rho g z$. For the second and third expressions, the Jacobian operator J is defined as $J(a,b) = (\partial a/\partial x)(\partial b/\partial s) - (\partial a/\partial s)(\partial b/\partial x)$ in the transformed space.

A standard test for terrain-following models is to prescribe density and pressure fields that depend solely on z (e.g., Beckmann & Haidvogel, 1993; Numerical Exercise 20.4). In this way, the horizontal pressure gradient should be identically zero, and no motion should be generated in the absence of external driving forces. The two terms on the right-hand side of Eqs. (20.55a) and (20.55d) cancel each other exactly in the continuous representation and should continue to do so after discretization. Due to the different nature of the two

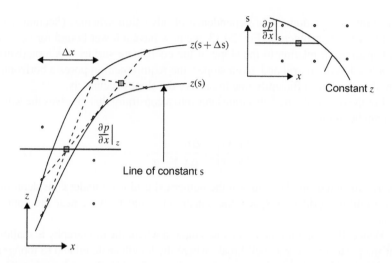

FIGURE 20.18 In the numerical grid (x,s), a standard finite differencing of the pressure gradient at the location of the squares involves the logical neighbors in the discrete mesh, connected with dashed lines in the physical space (left panel). For upper right point, the calculation points are physical neighbors but for the lower left point, the calculation uses far distant points and performs extrapolations.

terms, however, even very careful numerical discretization will most likely leave a residual that acts as a nonzero pressure force and hence generates unphysical horizontal motion. The main problem is that this error is most often not small because the vertical variation of pressure $p(z)$ is very large and taking the pressure gradient along a sloping s direction entails a large vertical component.

Should we argue that the problem would be eliminated with increased resolution, another problem would appear, the problem of so-called hydrostatic consistency: By increasing resolution more rapidly in the vertical than in the horizontal direction, (Fig. 20.18) the numerical stencil used in the calculation of a horizontal pressure gradient may involve grid points that are vertically too far away, and horizontal derivatives may be evaluated by extrapolation instead of interpolation. Such extrapolation leads to large relative errors (see Numerical Exercise 3.5), as well as to an inconsistency. The vertical gradients in Eq. (20.55a)–(20.55d) are then not calculated at the same depth as the horizontal gradient. For simple finite-difference schemes, extrapolations are avoided if the following criterion between slopes is met:

$$\left| \frac{\partial z/\partial x|_s}{\partial z/\partial s} \right| \leq \frac{\Delta s}{\Delta x}, \tag{20.56}$$

The slope of z lines in the transformed space (left-hand side) cannot be larger than the aspect ratio of the grid in the same space (right-hand side) so that lines of constant z remain within the local stencil. This constraint is not unlike the constraint on the domain of dependence of advection schemes (Section 6.4). Contrary to the problem in Fig. 20.13, we now have a lower bound for vertical grid spacing, in relation to the slope of the coordinate surfaces. Alternatively, for a fixed vertical grid and given slopes, the requirement imposes a horizontal grid that must be sufficiently fine to resolve the slopes accurately.

For the σ-coordinate,[8] this translates into a constraint that involves the water column height h:

$$\left| \frac{1}{h} \frac{\partial h}{\partial x} \right| \leq \frac{\Delta \sigma}{\Delta x} \frac{1}{1 - \sigma}, \tag{20.57}$$

where $\Delta \sigma$ and σ are the values at the numerical grid level under consideration. For a uniform grid in σ space, the constraint is most severe near the bottom layer.

Hence, the worst problems are encountered where the topography is shallow but steep, such as near a shelf break, where the length scale related to topography $|(1/h)(\partial h/\partial x)|^{-1}$ is shortest and must be resolved by the grid spacing. This length scale thus appears as an additional scale to be considered in the design of horizontal grids. Since stratification on the shelf break is typically intersecting topography, problems with the pressure gradient will be exacerbated there: regions of large variations in ρ coincide with regions of large $\partial z/\partial x|_s$, and the two contributions to the horizontal pressure force are large, leading to a significant numerical error. Solutions to this problem include higher-order finite differencing (using more grid points and being less prone to extrapolations— e.g., McCalpin, 1997), subtraction of average density profiles $\rho = \rho(z)$ before any pressure calculation (e.g., Mellor, Oey & Ezer, 1998), specialized finite differencing (e.g., Song, 1998), partial masking of topography (replacing slopes by vertical sections—e.g., Beckers, 1991; Gerdes, 1993), or, simply, smoothing the bottom topography.

With the aforementioned provision for the proper treatment of the pressure gradient, a generalized vertical coordinate can be very attractive, and several models have implemented the approach (e.g., Pietrzak, Jakobson, Burchard, Vested & Peterson, 2002), without actually using them at their full potential but prescribing a priori the position of coordinate surfaces. The general rule for the placement of coordinate surfaces is to match as closely as possible the surface on which physical properties remain smooth (Fig. 20.19; see also adaptive grids in Section 15.6). In ocean circulation models, grids should therefore follow closely density surfaces in the ocean interior, where mixing is weak.

Because a density coordinate is ill-suited to represent mixed layers, the use of the z-coordinate should be preferred near the surface. An implementation that nearly achieves this requirement is *HYCOM* (HYbrid Coordinate Ocean Model—e.g., Bleck, 2002), which is an extension of an earlier density model

---

[8] We neglect here any surface gradient for the sake of simplicity.

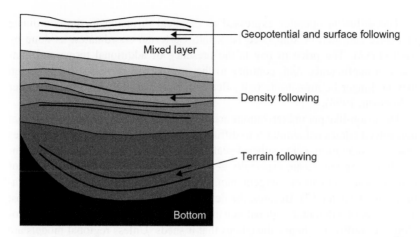

FIGURE 20.19 Most physically meaningful types of coordinate surfaces as a function of water depth. Near the surface and in the mixed layer, lateral coordinates run horizontally, or nearly horizontally, in order to follow the surface. In the ocean interior, where motion proceeds with little or no mixing, density becomes the best vertical coordinate, and coordinate surfaces follow density surfaces. Near the bottom, which the flow is forced to follow, a terrain-following coordinate is best.

allowing for an upper mixed layer spread over several z levels. For further improvement, the vertical coordinate near the bottom may be made to follow the topography and thus behave as a σ coordinate.

20.6.2   Subgrid-Scale Processes

Once the grid is defined, and the shortest resolved scale known, subgrid-scale processes must be considered. Up to this point, subgrid-scale processes other than turbulence were modeled by horizontal diffusion such as

$$\mathcal{D}(c) = \frac{\partial}{\partial x}\left(A\frac{\partial c}{\partial x}\right) + \frac{\partial}{\partial y}\left(A\frac{\partial c}{\partial y}\right) \qquad (20.58)$$

for any quantity c, be it a tracer concentration, temperature, salinity, or even a velocity component. One question that arises is whether the derivatives in this expression ought to be those in Cartesian coordinates or in any other set of coordinates. The choice of Cartesian coordinates implies that there exists a tendency to mix properties along the horizontal plane, but this may not always be representative of what is actually happening. The case in point is preferential mixing along (possibly sloping) density surfaces because this is the direction in which mixing motions are not inhibited by buoyancy forces. So, let us suppose that, to be faithful to the physics, mixing occurs along surfaces of constant density ρ, and the x- and y-derivatives in the preceding expression are to be taken at constant ρ.

The diffusion operator expressed as classical diffusion in the coordinate of choice can then be translated back into Cartesian coordinates, as done by Redi (1982). The price to pay is the presence of additional terms and non-constant coefficients, and, contrary to the original operator, its discretization may no longer be monotonic (e.g., Beckers, Burchard, Campin, Deleersnijder & Mathieu, 1998).

Diffusion-like parameterizations are, of course, based on the expectation that unresolved eddies act similarly to diffusion. However, depending on the scales under consideration, some subgrid-scale processes may not be considered randomly mixing the ocean, especially at the larger scales. The internal radius of deformation is a locus of energetic motions, in large part due to baroclinic instability (see Chapter 17). Because the deformation radius in the ocean is at most a few tens of kilometers, global ocean models typically do not resolve baroclinic instability and hence the eddies that it sheds. Unless regional models are used, coarse-resolution ocean models must parameterize the effect of mesoscale motions. Baroclinic instability releases potential energy by flattening density surfaces, a process quite different from pure mixing. Isopycnal diffusion cannot account for such a flattening since by construction it diffuses only along isopycnals, hence not forcing them to flatten out.

Instead of a diffusion-type parameterization, the so-called Gent–McWilliams parameterization (Gent & McWilliams, 1990; Gent, Willebrand, McDoughall & McWilliams, 1995) is recommended. This scheme proceeds by adding a so-called bolus velocity to the large-scale currents, the components of which, marked with a star, are

$$u^\star = -\frac{\partial Q_x}{\partial z}, \quad v^\star = -\frac{\partial Q_y}{\partial z}, \quad w^\star = \frac{\partial Q_x}{\partial x} + \frac{\partial Q_y}{\partial y}, \qquad (20.59)$$

with the pair (Q_x, Q_y) taken as

$$Q_x = -\frac{\kappa}{\rho_z}\frac{\partial \rho}{\partial x} = \kappa \left.\frac{\partial z}{\partial x}\right|_\rho \qquad (20.60a)$$

$$Q_y = -\frac{\kappa}{\rho_z}\frac{\partial \rho}{\partial y} = \kappa \left.\frac{\partial z}{\partial y}\right|_\rho \qquad (20.60b)$$

where $\rho_z = \partial\rho/\partial z$.

As it immediately appears, these quantities are proportional to the x- and y-slopes of the isopycnals, with the coefficient κ of proportionality being a tunable model parameter with same dimension as a diffusion coefficient (length squared per time). Since the derivatives are expressed in Cartesian coordinates, it is easily verified that the bolus velocity is divergence-free, a property that should be preserved by the numerical discretization. The bolus velocity effectively advects the density field and, with the chosen signs, leads to a reduction of the frontal slope (Fig. 20.20), substituting therefore for baroclinic instability. It is important to note that this advection is performed without any underlying dynamical

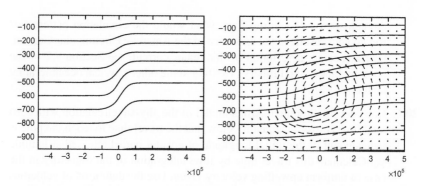

FIGURE 20.20 Vertical section across a density field with frontal structure as shown by sloping isopycnals (left panel) and the corresponding bolus velocity (right panel, with velocities originating from the dotted ends). Note how the bolus velocity acts to relax the density field in time so that some time later the front has indeed been weakened (right panel, solid lines). Since the bolus velocity depends on the existing slope, the flattening of isopycnals slows down over time.

equation, reflecting the fact that this is a parameterization of unresolved dynamics. The strength of the effect is controlled by the parameter κ, and Griffies (1998) shows how one may combine bolus advection with the isopycnal diffusion into a single operator as long as both "diffusion" coefficients are equal. For additional information and a recent review of ocean model developments, the reader is referred to Griffies et al. (2000) and references therein.

ANALYTICAL PROBLEMS

20.1. Derive the expression for the Sverdrup transport on an irregular bottom.

20.2. Derive the veering of the horizontal velocity with respect to depth, working with density ρ as the vertical coordinate. Show that one recovers Eq. (20.16) by changing to z-coordinates.

20.3. Given that the North Pacific Ocean is about twice as wide as the North Atlantic Ocean and that both basins are subjected to equally strong winds, compare their Sverdrup transport cumulated over the width of the basin. Express your answers in Sverdrup units (1 Sv $= 10^6$ m^3/s).

20.4. Demonstrate that western intensification would still occur if the global wind patterns were reversed, that is if Ekman pumping were upward and Sverdrup transport poleward at midlatitudes. In other words, show that the return flow from the Sverdrup transport must be squeezed against the western boundary regardless of the sign of the wind-stress curl over the ocean.

20.5. Imagine that a single ocean were covering the entire globe as the atmosphere does. With no western wall to support a boundary current returning the equatorward Sverdrup flow, what would the circulation pattern be? Relate your results to the existence of the Antarctic Circumpolar Current. [For a succinct description of this major current, see Section 7.2 of the book by Pickard and Emery (1990) or some other physical oceanography textbook.]

20.6. Consider the Stommel–Arons model of the abyssal circulation with zero total flow across the equator. Show that the travel time to reach the pole is independent of the initial longitude of a water parcel released at latitude φ_0. Calculate this travel time by time integration of the trajectory in the case of uniform upwelling velocity. (*Hint*: Use the definition of velocities in spherical coordinates from the material derivative (A.17) and integrate trajectories in the (λ, φ) domain.)

20.7. Consider the bolus velocity of equation (20.59), with constant diffusion coefficient κ. Investigate the possibility of stationary solutions for the density field solely advected by this bolus velocity in the (x, z) vertical plane for convenience. After finding a general condition on Q_x and ρ, assume a linear relationship between those variables and a uniform vertical stratification to determine a particular, stationary solution for ρ.

NUMERICAL EXERCISES

20.1. Take the density data used in `iwavemed` and calculate the geostrophic velocities. Work in z-coordinates with z levels corresponding to the data. First assume a level of no motion at 500 m and calculate currents at the surface and 2000 m down. Then repeat with level of no motion at 1500 m.

20.2. Experiment with `bolus` to see the flattening of isopycnals in a vertical section with flat bottom. Then implement a sigma-coordinate change in a vertical section with flat surface and topography given by

$$h(x) = h_0 + \alpha x, \tag{20.61}$$

where the domain extends from $x = -L/2$ to $x = L/2$ and where $\alpha L = h_0$. Start from the same physical density distribution as in the flat-bottom case. (*Hint*: Express the bolus advection term as a Jacobian in the vertical plane. Then use the rules of change of variables to express this Jacobian in the new coordinate system. Do not forget to calculate the slopes using the rules of changes of variables.)

20.3. Obtain a density section from somewhere and calculate bolus velocity using `bolus`. Which problems do you expect near boundaries? (*Hint*: You might consider κ as a calibration parameter that varies in space.)

20.4. Use `pgerror` to explore the pressure-gradient error for a fixed density anomaly profile depending only on z according to

$$\rho = \Delta\rho \tanh\left(\frac{z+D}{W}\right),\qquad(20.62)$$

in which D and W control the position and thickness of the pycnocline. Bottom topography is given by

$$h(x) = H_0 + \Delta H \tanh\left(\frac{x}{L}\right),\qquad(20.63)$$

where L and ΔH control the steepness of the slope. Calculate the error and associated geostrophic velocity for $f = 10^{-4}$ s^{-1}. Vary the number of vertical grid points, horizontal grid points, the position of the pycnocline, its depth, and strength. What happens if you increase only the number of vertical grid points? Implement the discretization of another pressure gradient expression of (20.55) in `bcpgr` and compare.

20.5. Bottom topography is generally smoothed before it is used in a model, by the repeated application of a Laplacian-type diffusion. In view of the hydrostatic consistency constraint, which adapted filter technique would you advocate? (*Hint*: Remember that a Laplacian filter applied to a function F decreases the norm of the gradient $\int\int \left[(\partial F/\partial x)^2 + (\partial F/\partial y)^2\right] dx\,dy$ over the domain.)

Henry Melson Stommel
1920–1992

At an early age, Henry Stommel considered a career in astronomy but turned to oceanography as a way to make a peaceful living during World War II. Having been denied admission to graduate school at the Scripps Institution of Oceanography by H. U. Sverdrup, then its director, Stommel never obtained a doctorate. This did not deter him; having soon realized that, in those years, oceanography was largely a descriptive science almost devoid of physical principles, he set out to develop dynamic hypotheses and to test them against observations. To him, we owe the first correct theory of the Gulf Stream (1948), theories of the abyssal circulation (early 1960s), and a great number of significant contributions on virtually all aspects of physical oceanography.

Unassuming and avoiding the limelight, Stommel relied on a keen physical insight and plain common sense to develop simple models that clarify the roles and implications of physical processes. He was generally wary of numerical models. Particularly inspiring to young scientists, Stommel continuously radiated enthusiasm for his chosen field, which, as he was the first to acknowledge, is still in its infancy. (*Photo by George Knapp*)

Kirk Bryan
1929–

As soon as computer mainframes became available for scientific research, in the 1960s, Kirk Bryan in collaboration with colleague Michael Cox (1941–1989) and student Bert Semtner began to develop codes for the simulation of oceanic circulation. This was truly pioneering work not only in the face of stringent hardware limitations but also because still little was known at the time about numerical stability, accuracy, spurious modes, etc. The so-called Bryan-Cox code of Princeton University's Geophysical Fluid Dynamics Laboratory quickly became a staple in oceanic modeling, often at the root of others' codes.

Concerns over climate change prompted Bryan later in his career to construct fully coupled atmosphere–ocean models, which are extremely challenging in view of their complexity and vastly different temporal scales. Rather than being daunted by this complexity, Bryan stresses the complementarity between atmospheric and oceanic processes and scales of motion.

Fame and numerous awards have come his way, but Kirk Bryan has retained a gentlemanly demeanor, with a kind word for all with whom he comes in contact. (*Photo courtesy of Princeton University*)

Chapter 21

Equatorial Dynamics

ABSTRACT

Because the Coriolis force vanishes along the equator, tropical regions exhibit particular dynamics. After an overview of linear waves that exist only along the equator, the chapter concludes with a brief presentation of the episodic transfer of warm waters from the western to the eastern tropical Pacific Ocean, a phenomenon called El Niño. The problem of its seasonal forecast then allows us to introduce another type of predictive tool, one based on empirical relationships.

21.1 EQUATORIAL BETA PLANE

Along the equator (latitude $\varphi = 0°$), the Coriolis parameter $f = 2\Omega \sin\varphi$ vanishes. Without a horizontal Coriolis force, currents cannot be maintained in geostrophic balance, and we expect dramatic dynamical differences between tropical and extratropical regions. The first question is the determination of the meridional extent of the tropical region where these special effects can be expected.

It is most natural here to choose the equator as the origin of the meridional axis. The beta-plane approximation to the Coriolis parameter (see Section 9.4) then yields

$$f = \beta_0 y, \tag{21.1}$$

where y measures the meridional distance from the equator (positive northward) and $\beta_0 = 2\Omega/a = 2.28 \times 10^{-11}\,\mathrm{m^{-1}\,s^{-1}}$ with Ω and a being, respectively, the earth's angular rotation rate and radius ($\Omega = 7.29 \times 10^{-5}\,\mathrm{s^{-1}}$, $a = 6371\,\mathrm{km}$). This representation of the Coriolis parameter bears the name of *equatorial beta-plane* approximation.

Our previous considerations of midlatitude processes (see Chapter 16, for example) point to the important role played by the internal deformation radius,

$$R = \frac{\sqrt{g'H}}{f} = \frac{c}{f}, \tag{21.2}$$

in governing the extent of dynamical structures. Here, g' is a suitable reduced gravity characterizing the stratification and H is a layer thickness. As f varies with y, so does R. If this distance from a given meridional position y includes the equator, equatorial dynamics must supersede midlatitude dynamics. Thus, a criterion to determine the width R_{eq} of the tropical region is (Fig. 21.1)

$$R_{eq} = R \quad \text{at} \quad y = R_{eq}. \tag{21.3}$$

Substituting Eq. (21.1) in Eq. (21.2), the criterion yields

$$R_{eq} = \sqrt{\frac{c}{\beta_0}}, \tag{21.4}$$

which is called the *equatorial radius of deformation*. For the previously quoted value of β_0 and for $c = (g'H)^{1/2} = 1.4\,\text{m/s}$, typical of the tropical ocean (Philander, 1990, Chapter 3), we estimate $R_{eq} = 248\,\text{km}$, or $2.23°$ of latitude. Because the stratification of the atmosphere is much stronger than that of the ocean, the equatorial radius of deformation is several times larger in the atmosphere. This implies that connections between tropical and temperate latitudes are different in the atmosphere and oceans.

Since c is a velocity (to be related shortly to a wave speed), we can define an *equatorial inertial time* T_{eq} as the travel time to cover the distance R_{eq} at speed c. Simple algebra yields

$$T_{eq} = \frac{1}{\sqrt{\beta_0 c}}, \tag{21.5}$$

which, for the previous values, is about 2 days.

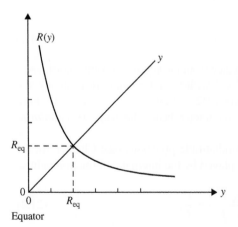

FIGURE 21.1   Definition of the equatorial radius of deformation.

21.2 LINEAR WAVE THEORY

Because of the important role they play in the so-called El Niño phenomenon, the focus of this section is on oceanic waves. The stratification of the equatorial ocean generally consists of a distinct warm layer separated from the deeper waters by a shallow thermocline (Fig. 21.2). Typical values are $\Delta\rho/\rho_0 = 0.002$ and thermocline depth $H = 100$ m, leading to the previously quoted value of $c = (g'H)^{1/2} = 1.4$ m/s. This suggests the use of a one-layer reduced-gravity model, which for the purpose of a wave theory is immediately linearized:

$$\frac{\partial u}{\partial t} - \beta_0 y v = -g' \frac{\partial h}{\partial x} \tag{21.6a}$$

$$\frac{\partial v}{\partial t} + \beta_0 y u = -g' \frac{\partial h}{\partial y} \tag{21.6b}$$

$$\frac{\partial h}{\partial t} + H\left(\frac{\partial u}{\partial x} + \frac{\partial v}{\partial y}\right) = 0. \tag{21.6c}$$

Here u and v are, respectively, the zonal and meridional velocity components, g' the reduced gravity $g\Delta\rho/\rho_0$ $(= 0.02 \text{ m/s}^2)$, and h the layer thickness variation (thickening counted positively and thinning counted negatively).

The preceding set of equations admits a solution with zero meridional flow. When $v = 0$, Eqs. (21.6a) and (21.6c) reduce to

$$\frac{\partial u}{\partial t} = -g'\frac{\partial h}{\partial x}, \quad \frac{\partial h}{\partial t} + H\frac{\partial u}{\partial x} = 0,$$

having any function of $x \pm ct$ and y as its solution. The remaining equation, (21.6b), sets the meridional structure, which for a signal decaying away from

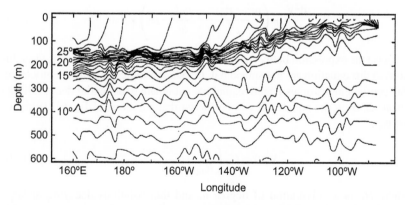

FIGURE 21.2 Temperature (in $^\circ$C) as a function of depth and longitude along the equator, as measured in 1963 by Colin, Henin, Hisard and Oudot (1971). Note the strong thermocline between 100 and 200 m.

the equator is given by

$$u = cF(x - ct)e^{-y^2/2R_{eq}^2} \qquad (21.7a)$$

$$v = 0 \qquad (21.7b)$$

$$h = HF(x - ct)e^{-y^2/2R_{eq}^2}, \qquad (21.7c)$$

where $F(\cdot)$ is an arbitrary function of its argument and $R_{eq} = (c/\beta_0)^{1/2}$ is the equatorial radius of deformation introduced in the preceding section. This solution describes a wave traveling eastward at speed $c = \sqrt{g'H}$, with maximum amplitude along the equator and decaying symmetrically with latitude over a distance on the order of the equatorial radius of deformation. The analogy with the coastal Kelvin wave exposed in Section 9.2 is immediate: wave speed equal to gravitational wave speed, absence of transverse flow, nondispersive behavior, and decay over a deformation radius. For this reason, it is called the *equatorial Kelvin wave*. Credit for the discovery of this wave, however, does not go to Lord Kelvin but to Wallace and Kousky (1968).

The set of equations (21.6) admits additional wave solutions, more akin to inertia-gravity (Poincaré) and planetary (Rossby) waves. To find these, we seek periodic solutions in time and zonal direction:

$$u = U(y)\cos(kx - \omega t) \qquad (21.8a)$$

$$v = V(y)\sin(kx - \omega t) \qquad (21.8b)$$

$$h = A(y)\cos(kx - \omega t). \qquad (21.8c)$$

Elimination of the $U(y)$ and $A(y)$ amplitude functions yields a single equation governing the meridional structure $V(y)$ of the meridional velocity:

$$\frac{d^2 V}{dy^2} + \left(\frac{\omega^2 - \beta_0^2 y^2}{c^2} - \frac{\beta_0 k}{\omega} - k^2 \right) V = 0. \qquad (21.9)$$

Because the expression between parentheses depends on the variable y, the solutions to this equation are not sinusoidal. In fact, for values of y sufficiently large, this coefficient becomes negative, and we anticipate exponential decay at large distances from the equator. It can be shown that solutions of Eq. (21.9) are of the type

$$V(y) = H_n \left(\frac{y}{R_{eq}} \right) e^{-y^2/2R_{eq}^2}, \qquad (21.10)$$

where H_n is a polynomial of degree n, and that solutions decaying at large distances from the equator exist only if

$$\frac{\omega^2}{c^2} - k^2 - \frac{\beta_0 k}{\omega} = \frac{2n+1}{R_{eq}^2}. \qquad (21.11)$$

Thus, the waves form a discrete set of modes ($n = 0, 1, 2, \ldots$). Equation (21.11) is the dispersion relation providing frequencies ω as a function of wavenumber k for each mode. As Fig. 21.3 shows, three ω roots exist for each n as k varies. (Important note: In this context, the phase speed ω/k of the wave is not necessarily equal to c, the speed of the Kelvin wave encountered previously.)

The largest positive and negative roots for $n \geq 1$ correspond to frequencies greater than the inverse of the equatorial inertial time. The slight asymmetry in these curves is caused by the beta term in Eq. (21.11). Without this term, the frequencies can be approximated by

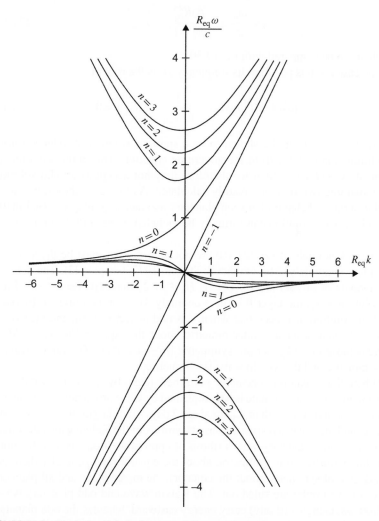

FIGURE 21.3   Dispersion diagram for equatorially trapped waves.

$$\omega \simeq \pm \sqrt{\frac{2n+1}{T_{eq}^2} + g'Hk^2}, \quad n \geq 1, \tag{21.12}$$

which is analogous to Eq. (9.17), the dispersion relation of inertia-gravity waves. These waves are thus the low-latitude extensions of the extratropical inertia-gravity waves (Section 9.3).

The third and much smaller roots for $n \geq 1$ correspond to subinertial frequencies and thus to tropical extensions of the midlatitude planetary waves (Section 9.4). At long wavelengths (small k values), these waves are nearly nondispersive and propagate westward at speeds

$$c_n = \frac{\omega_n}{k} \simeq -\frac{\beta_0 R_{eq}^2}{2n+1}, \quad n \geq 1, \tag{21.13}$$

which are to be compared with Eq. (9.30).

The case $n = 0$ is peculiar. Its frequency ω_0 is the root of

$$(\omega_0 + ck)\left(\omega_0 T_{eq} - \frac{1}{\omega_0 T_{eq}} - kR_{eq}\right) = 0. \tag{21.14}$$

The root $\omega_0 = -ck$ can be shown to be a spurious solution introduced during the elimination of $U(y)$ from the governing equation. This elimination indeed assumed $\omega_0 + ck \neq 0$, which we therefore may not accept as a valid solution. The remaining two roots are readily calculated. As Fig. 21.3 shows, this wave exhibits a mixed behavior between planetary and inertia-gravity waves. Finally, the Kelvin-wave solution can formally be included in the set by taking $n = -1$ (Fig. 21.3).

The polynomials of Eq. (21.10) are not arbitrary but must be the so-called Hermite polynomials (Abramowitz & Stegun, 1972, Chapter 22). The first few polynomials of this set are $H_0(\xi) = 1$, $H_1(\xi) = 2\xi$ and $H_2(\xi) = 4\xi^2 - 2$. From the solution $V(y)$, the layer thickness anomaly $A(y)$ can be retrieved by backward substitution. It is seen that when V is odd in y, A is even, and vice versa. Waves of even order are antisymmetric about the equator $[h(-y) = -h(y)]$, whereas those of odd order are symmetric $[h(-y) = h(y)]$. The mixed wave is antisymmetric and the Kelvin wave is symmetric.

When the equatorial ocean is perturbed (e.g., by changing winds), its adjustment toward a new state is accomplished by wave propagation. At low frequencies (periods longer than T_{eq}, or about 2 days), inertia-gravity waves are not excited, and the ocean's response consists entirely of the Kelvin wave, the mixed wave, and some planetary waves (those of appropriate frequencies). If, moreover, the perturbation is symmetric about the equator (and generally there is a high degree of symmetry about the equator), the mixed wave and all planetary waves of even order are ruled out. The Kelvin wave and odd planetary waves of short wavelengths (if any) carry energy eastward, whereas the odd planetary waves of long wavelengths carry energy westward. Figure 21.4 displays the

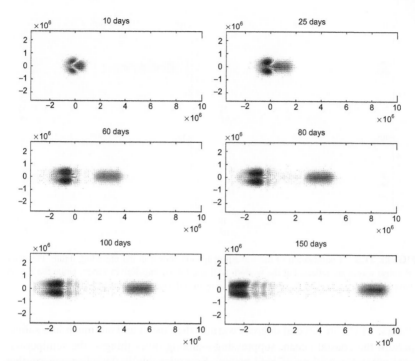

FIGURE 21.4 The dispersion of a perturbation generated by a 10-day wind anomaly imposed on a spot of the equatorial ocean. Clearly visible are the one-bulge Kelvin wave moving eastward and the double-bulge planetary (Rossby) wave propagating westward at a slower pace.

temporal dispersion of a thermocline displacement generated by a wind-stress anomaly imposed on a stretch of equatorial ocean. Clearly visible are the one-bulge Kelvin wave progressing eastward and the double-bulge lowest planetary wave ($n=1$) propagating westward. Although this case is obviously academic, it is believed that Kelvin waves and low-order planetary waves, together with wind-driven currents, are frequent in the equatorial ocean.

At this point, a number of interesting topics can be presented, such as the reflection of a Kelvin wave upon encountering an eastern boundary (Fig. 21.5), waves around islands, and the generation of equatorial currents by time-dependent winds. We shall, however, leave these matters for the more specialized literature (Gill, 1982; Philander, 1990; McPhaden & Ripa, 1990; and references therein) and limit ourselves to the presentation of the El Niño phenomenon.

21.3   EL NIÑO – SOUTHERN OSCILLATION (ENSO)

Every year, around the Christmas season, warm waters flow along the western coast of South America from the equator to Peru and beyond. These

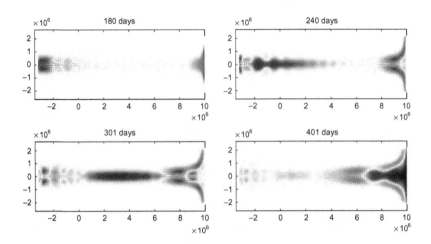

FIGURE 21.5 Continuation of the wave propagation shown in the preceding figure. Planetary (Rossby) waves are reflected at the western wall and turned into Kelvin waves, while the Kelvin wave generates a planetary wave after reflection against the eastern boundary.

waters, which are several degrees warmer than usual and are much less saline, perturb the coastal ocean, suppressing—among other things—the semipermanent coastal upwelling of cold waters. So noticeable is this phenomenon that early fishermen called it El Niño, which in Spanish means "the child" or more specifically the Christ Child, in relation to the Christmas season.

Regularly but not periodically (every 3 to 7 years), the amount of passing warm waters is substantially greater than in normal years, and life in those regions is greatly perturbed, for better and for worse. Anomalously abundant precipitations, caused by the warm ocean, can in a few weeks turn the otherwise arid coastal region of Peru into a land of plenty. But, suppression of coastal upwelling causes widespread destruction of plankton and fish. The ecological and economic consequences are noticeable. In Peru, the fish harvest is much reduced, sea birds (which prey on fish) die in large numbers, and, to compound the problem, dead fish and birds rotting on the beach create unsanitary atmospheric conditions.

In the scientific community, the name El Niño is being restricted to such anomalous occurrences and, by extension, the name La Niña has been used to signify the opposite situation, when waters are abnormally cold in the eastern tropical Pacific. Major El Niño events of the twentieth century occurred in 1904–05, 1913–15, 1925–26, 1940–41, 1957–58, 1972–73, 1982–83, 1986–88, 1991–95, 1997–98 (the strongest of all), and sofar in twenty-first century in 2002–2003, 2004–2005, and 2006–2007 (WMO, 1999; NOAA-WWW, 2006). Their cause remained obscure until Wyrtki (1973) discovered a strong correlation with changes in the central and western tropical Pacific Ocean, thousands of

kilometers away. It is now well established (Philander, 1990) that El Niño events
are caused by changes in the surface winds over the tropical Pacific, which
episodically release and drive warm waters, previously piled up by trade winds
in the western half of the basin, eastward to the American continent and southward along the coast. The situation is quite complex, and it took oceanographers
and meteorologists more than a decade to understand the various oceanic and
atmospheric factors.

Under normal conditions, winds over the tropical Pacific Ocean consist
of the northeast trade winds (northeasterlies) and the southeast trade winds
(southeasterlies) that converge over the *intertropical convergence zone* (ITCZ)
and blow westward (Section 19.3). Although it migrates meridionally in the
course of the year, the ITCZ sits predominantly in the northern hemisphere
(around 5° to 10°N). In addition to pushing and accumulating warm water in
the western tropical Pacific, the trade winds also generate equatorial upwelling
(Section 15.4) over the eastern part of the basin. Thus, in a normal situation, the
tropical Pacific Ocean is characterized by a warm water pool in the west and
cold surface waters in the east. This structure is manifested by the westward
deepening of the thermocline, as shown in Fig. 21.2.

The origin of an anomalous, El Niño event is associated with a weakening
of the trade winds in the western Pacific or with the appearance of a warm sea
surface temperature (SST) anomaly in the central tropical Pacific. Although one
may precede the other, they soon go hand in hand. A slackening of the western trades relaxes the thermocline slope and releases some of the warm waters;
this relaxation takes the form of a downwelling Kelvin wave, whose wake is
thus a warm SST anomaly. On the other hand, a warm SST anomaly locally
heats the atmosphere, creating ascending motions that need to be compensated
by horizontal convergence. This horizontal convergence naturally calls for eastward winds on its western side, thus weakening or reversing the trade winds
there (Gill, 1980). In sum, a relaxation of the trade winds in the western Pacific
creates a warm sea surface anomaly, and vice versa. Feedback occurs and the
perturbation amplifies. On the eastern side of the anomaly, convergence calls
for a strengthening of the trades that, in turn, enhances equatorial upwelling.
This cooling interferes with the eastward progression of the downwelling Kelvin
wave, and it is not clear which should dominate. During an El Niño event, the
anomaly does travel eastward while amplifying. Once the warm water arrives
at the American continent, it separates into a weaker northward branch and
a stronger southward branch, each becoming a coastal Kelvin wave (downwelling). The subsequent events are as described at the beginning of this section.

When an El Niño event occurs, its temporal development is strictly controlled by the annual cycle. The warm waters arrive in Peru around December,
and the seasonal variation of the general atmospheric circulation calls for a
northward return of the ITCZ and a reestablishment of the southeast trade winds
along the equator. The situation returns to normal.

This sequence of events is relatively well understood (Philander, 1990) and has been successfully modeled (Cane, Zebiak & Dolan, 1986). Today, models are routinely used to forecast the next occurrence of an El Niño event and its intensity with a lead time of 9 to 12 months. What remains less clear is the variability of the atmosphere-ocean system on the scales of several years. A strong connection with the *Southern Oscillation* has been made clear, and the broader phenomenon is called ENSO, for El Niño–Southern Oscillation (Rasmusson & Carpenter, 1982). The Southern Oscillation is a quasi-periodic variation of the surface atmospheric pressure and precipitation distributions over large portions of the globe (Bromwich et al., 2000; Troup, 1965).

Much hinges on variations of the so-called *Walker circulation*. This atmospheric circulation (Walker, 1924) consists of easterly trade winds over the tropical Pacific Ocean, low pressure and rising air above the western basin and Indonesia (with associated heavy precipitation) and, at the eastern end of the basin, high pressure, sinking air, and relatively dry climate. The strength of this circulation is effectively measured by the sea-level pressure difference Δp_{TD} between Tahiti (18°S, 149°W) and Darwin (in northern Australia, at 12°S, 131°E). In practice, the *Southern Oscillation Index* (SOI) is defined as (Troup, 1965):

$$SOI = 10 \; \frac{\text{monthly value of } \Delta p_{TD} - \text{long-term average of } \Delta p_{TD}}{\text{standard deviation of } \Delta p_{TD}}, \quad (21.15)$$

The nearly perfect negative correlation between these two pressures indicates that both are parts of a larger coherent system. The presence of a higher than normal pressure in Darwin with simultaneous lower pressure in Tahiti (negative *SOI* value) is intimately connected with an El Niño occurrence (Fig. 21.6). In its broad lines, the scenario unfolds as follows. A negative SOI value leads to a weakening of the Walker circulation, reduced strength of the easterly trade winds, especially in the western Pacific. The western warm water pool relaxes and begins to spill eastward as an equatorial Kelvin wave toward the central basin, accompanied by a similarly eastward displacement of the low atmospheric pressure above it. Feeding the low pressure from the west, anomalous westerly winds accelerate the eastward movement of the warm water pool. And so, the situation progresses eastward in an amplifying manner until the warm water pool reaches the coast of Peru and an El Niño event occurs. Because atmospheric pressure is then higher than normal on the western side, drought conditions occur over Indonesia and Australia, while South America experiences stronger precipitation than normal. For a more complete description of the many facets and ramifications of the event, the interested reader is referred to specialized books (D'Aleo, 2002; Diaz & Markgraf, 2000; Philander, 1990).

A conceptual model of the ENSO event that captures its major characteristics with a minimum of variables is the so-called recharge-discharge oscillator proposed by Jin (1997a,b). The situation being considered is defined by anomalies with respect to a mean climatological state, the latter being characterized by the Walker circulation, upwelling in the eastern Pacific, and warm water piled

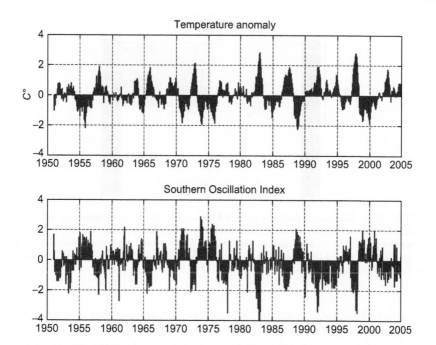

FIGURE 21.6   Time series of temperature anomalies over the central tropical Pacific Ocean and of the Southern Oscillation Index (3-month running means). There is a very strong correlation between higher than normal temperatures (El Niño events) and negative index values, indicating that El Niño is part of a global climatic variation. A spectral analysis of the temporal evolution of SOI reveals a peak in the interval of 3.5–4.5 years. (*NOAA, U.S. Department of Commerce, Washington, D.C.*)

up in the western Pacific. The variables are therefore the anomalies of the wind stress (positive for a westerly anomaly, corresponding to a weakening of the Walker circulation), of the oceanic surface layer thickness (positive for a thicker layer), and of the sea surface temperature (positive at the coast of Peru during an El-Niño event). The oceanic component is modeled as a single, reduced-gravity layer of average depth H. The model distinguishes two depth anomalies, one in the western basin and the other in the eastern basin of the Pacific, one value for each basin without further specification of the zonal and meridional structure. Since El Niño and La Niña are characterized, respectively, by positive and negative temperature anomalies in the eastern Pacific, we limit the temperature description to this anomaly (Fig. 21.7). With these few variables, we now proceed to model the key dynamics.

21.3.1   The Ocean

- *Dynamics*: Under changing wind stress, the ocean adjusts by means of wave propagation, but we may consider the time needed for such an adjustment to be short compared with the seasonal and longer time scales of an ENSO

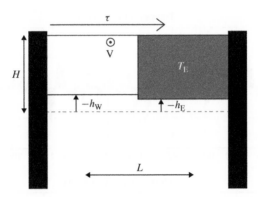

FIGURE 21.7   Schematic situation for a simple ENSO model. Two depth anomalies h_W and h_E are assigned respectively to the western and eastern Pacific basins and are counted positive downward. The water pool in the eastern Pacific, at temperature T_E, can become warmer or colder as a result of an anomaly. A wind blows above the surface, and τ denotes its windstress anomaly, which is responsible for a lateral transport V.

event. Therefore, following Jin (1997a), we suppose that a wind-stress anomaly τ in the zonal direction is instantaneously accompanied by an anomalous pressure difference between eastern and western basins. In the framework of a reduced-gravity model, the pressure-gradient anomaly is expressed as the difference in thermocline depth between East (h_E) and West (h_W), and equilibrium requires

$$g' \frac{h_E - h_W}{L} = \frac{\tau}{\rho_0 H}. \qquad (21.16)$$

For a given wind-stress anomaly, we can therefore calculate one of the two depth anomalies from the other. Volume conservation will be the means by which we can write a second relation between the two depth anomalies.

We further assume that the zonal velocity is proportional to the wind-stress anomaly:

$$U = \gamma \tau, \qquad (21.17)$$

in which U is the scale that provides the size of the zonal velocity anomaly and γ is an empirical coefficient of proportionality. This is justified for a wind stress that acts over an elongated nonrotating basin (e.g., Mathieu, Deleersnijder, Cushman-Roisin, Beckers & Bolding, 2002), like an equatorial strip.

- *Volume conservation*: During an ENSO event, the temperature anomaly is concentrated in the eastern Pacific, with the wind-stress anomaly occurring on its western flank. The volume budget is first established by integrating volume conservation Eq. (12.9) over the western pool, starting from no flow at the western boundary. Because of the balance of forces (21.16) in the zonal direction, there is no reason for persisting zonal transport between the western and eastern basins, and this leaves the meridional transport as the only way to export or import waters. Since the depth anomaly is characteristic of the equatorial strip of ocean water, we need to calculate the transport at the northern and southern limits of the equatorial band of interest, which we take as one equatorial radius of deformation from either side of the equator: $y = \pm R_{eq}$, with R_{eq} defined in Eq. (21.4). According to Eq. (20.11c), a meridional Sverdrup and Ekman transport must exist. With the wind anomaly aligned with the equator, the meridional transport is

$$V = -\frac{1}{\rho_0 \beta_0} \frac{\partial \tau^x}{\partial y}. \tag{21.18}$$

With atmospheric and oceanic anomalies tightly coupled, we may assume they are both concentrated around the equator in a strip of width $2R_{eq}$ and that the wind-stress anomaly decreases from τ at the equator to 0 at $y \sim \pm R_{eq}$, and we estimate

$$V \approx \pm \frac{1}{\rho_0 \beta_0} \frac{\tau}{R_{eq}}, \tag{21.19}$$

with the plus (minus) sign at the northern (southern) boundary. It follows that the flow divergence D between the two boundaries is

$$D \approx \frac{\tau}{\rho_0 \beta_0 R_{eq}^2}. \tag{21.20}$$

This divergence of flow in turn modifies the thermocline depth in the western basin, and the equation governing this thermocline depth is therefore

$$\frac{dh_W}{dt} = -D - r h_W, \tag{21.21}$$

where the last term on the right-hand side is intended to represent some damping of the ocean adjustment by lateral mixing and boundary-layer exchange.

- *Heat budget*: When the surface layer is modeled as a reduced-gravity model, a temperature anomaly is susceptible to modify the value of g' but not significantly. In the absence of anomalies, the heat budget of either basin is considered closed, and thus any departure from this unperturbed state is a function of the anomalies only. There are two ways by which the temperature in the eastern basin can be affected, either by zonal advection or by vertical advection.

The zonal velocity anomaly u caused by the wind-stress anomaly τ moves the background climatological temperature field in or out of the eastern basin,[1] and we may write

$$u\frac{\partial \bar{T}}{\partial x} \sim U\frac{T_E - T_W}{L}, \tag{21.22}$$

in which U is the previously defined scale for the zonal velocity anomaly and related to the wind-stress anomaly τ by Eq. (21.17), and T_E and T_W are the climatological temperatures in the eastern and western basins, respectively. For positive U (eastward flow under positive wind-stress anomaly τ), the temperature T_E in the east increases by import of warmer water from the western pool (because $T_W > T_E$).

For vertical advection, the situation is more subtle. Without anomaly, vertical advection of temperature in the eastern pool is

$$\bar{w}\frac{\partial \bar{T}}{\partial z} \sim \bar{w}\frac{\bar{T}_{surf} - \bar{T}_{-H}}{H}, \tag{21.23}$$

where the overbar refers to the climatological basic state. The strength of the upwelling $\bar{w}$ is proportional to the wind stress, $\bar{w} = -\alpha\bar{\tau}$, with the minus sign corresponding to upwelling under normal easterly trade winds.

A positive wind-stress anomaly τ reduces the upwelling intensity and thus creates a negative upwelling anomaly $\tilde{w}$. Therefore, less deep (and less cold) waters than usual are brought to the surface, and this corresponds to a positive heat flux anomaly. This flux anomaly can be estimated as the difference of Eq. (21.23) in which $\bar{w}$ is replaced by $\bar{w} + \tilde{w}$, and the climatological basic-state equation (21.23):

$$\tilde{w}\frac{T_{surf} - T_{-H}}{H} \sim -\alpha\tau\frac{\Delta_v T}{H}, \tag{21.24}$$

in which $\Delta_v T$ is the vertical temperature difference between climatological surface water and deep water.

If the eastern Pacific has a positive depth anomaly, the temperature at $y = -H$ is not the climatological value, but the climatological value found at $-H + h_E$, because a positive depth anomaly shifts the temperature profile downward. The surface temperature is the surface temperature augmented by the temperature anomaly, and taking the difference of Eq. (21.23) with these two modified temperature values and the reference (21.23) provides

$$-\bar{w}\left(\frac{T_E}{H} - \frac{\partial \bar{T}}{\partial z}\frac{h_E}{H}\right) \sim -\frac{\bar{w}}{H}T_E + \frac{\bar{w}\Delta_v T}{H^2}h_E, \tag{21.25}$$

in which we linearized $\bar{T}_{-H+h_E}$ around $\bar{T}_{-H}$. The interpretation of the sign in the term proportional to h_E is that a deeper thermocline results in an

---

[1]Even if the vertically averaged velocity is zero between the western and eastern pools, a wind-induced surface current can displace temperature anomalies.

upwelling of warmer waters than usual, creating a positive temperature anomaly.

Grouping the three contributions into the budget for temperature anomaly, we obtain

$$\frac{dT_E}{dt} = -\left(r' + \frac{\bar{w}}{H}\right) T_E + \frac{\bar{w}\Delta_v T}{H^2} h_E + \left(\gamma \frac{\Delta_h T}{L} + \alpha \frac{\Delta_v T}{H}\right)\tau, \qquad (21.26)$$

in which $\Delta_h T = T_W - T_E$ is the positive climatological temperature difference between western and eastern Pacific. In the first term on the right-hand side, we added a damping effect as for h_W and also included damping by exchanges with the atmosphere.

21.3.2   The Atmosphere

Because of the much lower inertia of the atmosphere compared with the ocean, we may assume that the sea surface anomaly in the east immediately creates a Walker circulation and write

$$\tau = \mu\, T_E. \qquad (21.27)$$

The coupling parameter can be calculated by a simplified atmospheric model (Gill, 1980).

21.3.3   The Coupled Model

Gathering the different pieces, we finally obtain the governing equations of the simplified model of an ENSO event:

$$\frac{dh_W}{dt} = -\frac{\mu}{\rho\beta_0 R_{eq}^2} T_E - r h_W \qquad (21.28)$$

$$\frac{dT_E}{dt} = \left[\mu\left(\gamma \frac{\Delta_h T}{L} + \alpha \frac{\Delta_v T}{H} + \frac{\bar{w} L \Delta_v T}{H^3 g' \rho_0}\right) - r' - \frac{\bar{w}}{H}\right] T_E + \frac{\bar{w}\Delta_v T}{H^2} h_W. \quad (21.29)$$

We note the positive temperature feedback in the temperature equation when the coupling parameter μ is large enough. As this term represents the coupling with the atmosphere and the advection feedback, it is clear that it models the amplification process described at the beginning of this section.

With realistic values for the model parameters (coded in jin.m), the solution exhibits a damped oscillation (Fig. 21.8; for an animation execute jinmodel.m) that models the mechanism depicted in Fig. 21.9. Although this model nicely captures ENSO oscillations with the phase shift between temperature and thermocline-depth anomalies, as well as a period of about 4 years, scientists debate whether ENSO has a natural oscillating cycle as described by the present model or it is triggered by some external effect (e.g., Kessler, 2002).

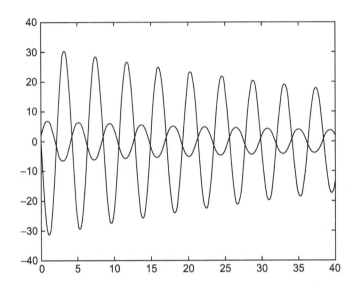

FIGURE 21.8   Temperature anomaly (low amplitude curve) and depth anomaly (high amplitude curve) as a function of time (in years). The period of the slightly damped oscillation is close to 4 years. Scales for temperature and depth are arbitrary.

21.4   ENSO FORECASTING

Forecasting the El Niño–Southern Oscillation (ENSO) event is of great interest to society because an impressive number of its consequences can affect daily life, ranging from changes in weather, appearance of droughts or floodings, to changes in crops, fish catch, and human health. These effects are not only felt in the equatorial region but also in remote locations, such as Australia. Hence, it is no surprise that reliable prediction of an El Niño or La Niña event can be of great help in preparing for the upcoming perturbations.

In forecasting ENSO, monthly averaged weather patterns are of interest because a given month in an El Niño year is quite different from the same month in a La Niña year. But since the forecast has to span several months, if not seasons, sea surface temperature (SST) cannot be considered known, and coupled atmosphere-ocean models are needed. This was recognized by Zebiak and Cane (1987), who succeeded in building the first coupled model to forecast ENSO. The importance of coupling between air and sea can be nicely shown by the following hindcast experiments: The use of observed variations in SST during an ENSO event generally helps to predict the atmospheric part correctly; similarly, using observed variations in atmospheric fluxes along the sea surface reveals the oceanic component of ENSO. Hence, both components are needed for the forecast.

ENSO forecasts differ from weather forecasts in that only average situations are predicted. While weather forecasts are mostly constrained by initial

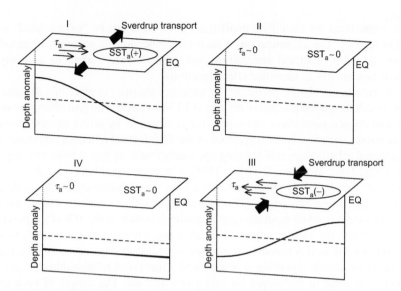

FIGURE 21.9 Discharge/recharge mechanism proposed by Jin (1997a, 1997b) as adapted by Meinen and McPhaden (2000). During a warm, El Niño phase (Phase I), the westerly wind-stress anomaly creates a diverging poleward Sverdrup flow. This removes water from the equatorial Pacific and, via the zonal dynamical balance, the anomalous warm water pool of the eastern Pacific. When the eastern warm water pool no longer exists, no east-west temperature anomaly remains, and the associated anomalous Walker circulation disapears (Phase II). During this stage, winds are those of the climatological Walker circulation, which are responsible for normal upwelling. Since the thermocline is shallower than usual, upwelling in the eastern Pacific brings up colder waters than usual, and the cold, La Niña phase begins. The reversal of the wind-stress anomaly creates an equatorward Sverdrup flow, which recharges the eastern pool (Phase III) until the mean thermocline is deeper than average (Phase IV). At this stage, climatological upwelling again brings up warmer water than usual, and the cycle repeats with a warm phase (Phase I).

conditions in the atmosphere and can be of only short duration, seasonal forecast benefits from the ocean's inertia, and its predictability extends over several months. Hence, seasonal forecast is constrained mostly by initial conditions in the ocean. Observing the tropical ocean is a crucial component of any ENSO forecast system, with the most important data being provided by the Tropical Atmosphere Ocean (TAO) array of moorings and by satellites measuring sea surface height and temperature.

Seasonal forecasting of ENSO is relatively successful because ENSO is known to be the largest single source of predictable internannual variability. Yet, even with a strong signal, models must be able to extract the information amidst the dominant high-frequency signal of atmospheric variability. With unavoidable model uncertainties, this is a challenging task, and one way to reduce uncertainties is to perform model intercomparisons (e.g., Mechoso et al., 1995; Neelin et al., 1992). Models are also used to identify teleconnections that is, correlations between dynamics in distant regions and ENSO events. If such teleconnections are identified, predictions of El Niño can be "extrapolated" to other regions. The identification of such teleconnections is generally obtained from statistics of model simulations and observations, leading to as many prediction models as identified teleconnections.

Forecasts can also be performed by means of statistics in place of a dynamical model. Empirical predictive models of El Niño begin with past observations of well-chosen parameters and search for correlations by fitting curves across data points, such as a linear regression of the *SOI*. Instead of a priori choosing a functional relationship, self-learning approaches such as neural networks (e.g., Tangang, Tang, Monahan & Hsieh, 1998) or genetic algorithms (e.g., Alvarez, Orfila & Tintore, 2001) select on their own the "best" functions. To do so, data are separated into two sets, a learning and a validation set. From the learning set, the model is given input data called predictors (such as the *SOI* of the previous year) and the known output value called predictand (such as the prediction of *SOI* for the next 6 months). If enough input-output pairs are available, the network or genetic algorithm is able to find a functional relationship that minimizes the error in the output for this given data set. The danger of such an approach is that overfitting may occur: If the functional relationship contains more adjustable parameters than independent data to be fitted, one can always find a "perfect" fit. The latter, however, will work only on this particular data set. An independent validation data set is required on which the model must be tested after the learning phase. If the performance in forecasting degrades significantly after switching from the learning set to the validation set, the model is unreliable. However, if validation is successful, such models offer predictions at extremely small computational costs compared with primitive equation models.

The simplest models, if they have any predictive skill, are therefore valuable in defining base forecasts against which forecats from more complex models may be compared, and to justify their use in operational forecasts, the highly complex dynamical models must demonstrate their superiority in prediction ability compared with statistical models (see Fig. 21.10). At the time of this writing, it appears that dynamical models are better predictors of early stages of an El Niño, but once the event is under way, statistical models are quite satisfactory. The lesson is that the trigger is relatively difficult to identify but that, once an ENSO event begins, it unfolds according to a repeatable pattern.

The search for empirical relationships can of course be guided by physical considerations. For El Niño, wave propagation along the equator and reflection at continental boundaries provide a delayed feedback mechanism on the system. This can be formally translated into a delayed oscillator model (Suarez & Schopf, 1988), the governing equation of which can be expressed in the form:

$$\frac{dT}{dt} = aT(t) - aT^3(t) - bT(t-\delta), \tag{21.30}$$

where T stands for a normalized temperature anomaly associated with El Niño, the term aT models the positive feedback of the initial Kelvin wave with the

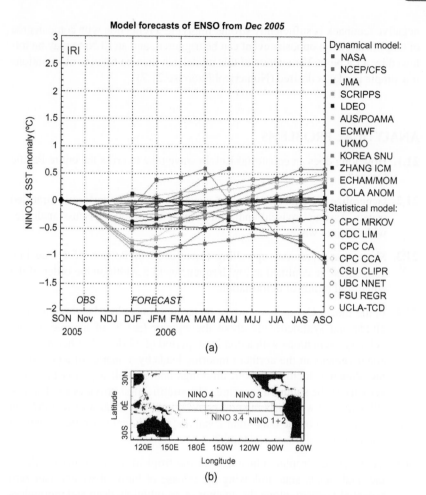

FIGURE 21.10 Prediction of a future El Niño event by means of sea surface temperature (SST) anomaly starting from December 2005 by several models, including coupled ocean-atmosphere primitive equation models and statistical models. Predictions were made for different subregions in a central strip of the basin (see insert), the probablity of El Niño in 2006 was deemed low. (*IRI, International Research Institute for Climate and Society*)

moving atmospheric perturbation, and the cubic term $-aT^3$ represents damping to keep the solution bounded in time.[2] Finally, the last term $-bT(t-\delta)$ is introduced to include the negative feedback of the initial westward Rossby wave that is then reflected as a Kelvin wave of opposite amplitude. The delay δ is readily interpreted in terms of wave travel time along the equator. If the

---

[2]The temperature T can always be scaled so that the cubic term appears with the same coefficient as the linear feedback.

negative feedback is sufficiently strong, implying a reflection with some degree of amplification, an opposite event can be triggered, and an El Niño may be followed by a La Niña. Parameters of this model can then be fitted to observations if a simple model is desired (Numerical Exercise 21.7).

ANALYTICAL PROBLEMS

21.1. How long does an equatorial Kelvin wave take to cross the entire Pacific Ocean?

21.2. Generalize the equatorial Kelvin-wave theory to the uniformly stratified ocean. Assume inviscid and nonhydrostatic motions. Discuss analogies with internal waves.

21.3. Show that equatorial upwelling (mentioned in Section 15.4; see Fig. 15.6) must be confined at low frequencies to a width on the order of the equatorial radius of deformation.

21.4. In the Indian Ocean, two current-meter moorings placed at the same longitude and symmetrically about the equator ($\pm 1.5°$ of latitude) record velocity oscillations with a dominant period of 12 days. Furthermore, the zonal velocity at the northern mooring leads by a quarter of a period the meridional velocities of both moorings and by half a period the zonal velocity at the southern mooring. The stratification provides $c = 1.2$ m/s. What kind of wave is being observed? What is its zonal wavelength? Can a comparison of the maximum zonal and meridional velocities provide a confirmation of this wavelength?

21.5. Consider geostrophic adjustment in the tropical ocean. What would be the final steady state following the release of buoyant waters with zero potential vorticity along the equator of an infinitely deep and motionless ocean? For simplicity, assume zonal invariance and equatorial symmetry.

21.6. What kind of initial conditions are needed for the delayed oscillator model (21.30)?

21.7. Search for information to check whether the forecast provided in December 2005 of an unlikey El Niño in 2006 came true.

21.8. Show that the linearization of the governing equations for a Kelvin wave is valid as long as the function F is small enough, $|F| \ll 1$.

21.9. Study equatorial upwelling on a beta plane with an arbitrary wind-stress field. Use the approach of Section 8.6 but include a linear friction term in addition to vertical diffusion. What happens to the solution if this linear friction term is dropped? (*Hint*: Do not calculate explicitly the vertical structure of the Ekman layer but integrate vertically.)

NUMERICAL EXERCISES

21.1. Design a numerical solver for the delayed oscillator equation (21.30). Find the solution with $a^{-1} = 50$ days, $\delta = 400$ days, and $b^{-1} = 90$ days for different initial conditions. Then, change to $a^{-1} = 100$ days and $b^{-1} = 180$ days.

21.2. Design a numerical version of the linear reduced-gravity model (21.6), to which a spatially varying zonal wind stress is added, of the form

$$\tau = \tau_0 e^{-(x^2 + y^2)/L^2}. \tag{21.31}$$

Use a finite-difference approach on the Arakawa C-grid and time stepping of your choice. Start with a situation at rest and then apply the wind stress for 30 days. For the wind-stress amplitude and length scale, take $\tau_0 = 0.1$ N/m^2 (directed eastward) and $L = 300$ km. The reduced-gravity model's parameters are $\Delta\rho/\rho_0 = 0.002$ with a thermocline depth $H = 100$ m. For a first simulation, use a closed domain with boundaries at $x = -3000$ km, $x = 10000$ km and $y = \pm 2000$ km. Simulate the evolution for 600 days.

21.3. For the conditions set in Numerical Exercise 21.2 with closed southern and northern boundaries, the perturbation eventually propagates along these boundaries. Which physical process is responsible for this? Modify the southern and northern boundary conditions by opening the domain and apply $v = \pm\sqrt{g'H}\, h$ there. Choose a physically reasonable sign for each boundary by considering a physical interpretation of such boundary conditions.

21.4. Change the topology of the domain in Numerical Exercise 21.3 by adding land points in the southwestern (lower left) and northeastern (upper right) corners to represent continents on each side of the Pacific Ocean and redo the simulations. Can you identify the modes that are now present compared with the symmetric case?

21.5. Search for a spatial discretization of the Coriolis term on the Arakawa C-grid that does not create mechanical work in the sense that, after multiplying the evolution equation for $u_{i-1/2,j}$ by itself and adding a similar product of the $v_{i,j-1/2}$ equation by itself, the Coriolis terms cancel out. (*Hint*: Analyze which products of u and v appear, similar to the analysis of the Arakawa Jacobian performed in Section 16.7 and find how to average by taking into account the variation of y.)

21.6. Using the sea-surface temperature (SST) anomalies and Southern Oscillation Index (SOI) values from 1991 to 2005 included in soi.m, perform a linear regression over data windows and find out whether the extrapolation of these regressions is able or unable to predict the SST or SOI at later times. First, use a data window of 4 months and try to extrapolate for the next month. Plot the prediction error over time when applying the method over all possible data windows. To decide whether the prediction is useful, compare with the prediction error corresponding to simple persistence (constant anomaly). Then, try to change the data window and lead time to improve the prediction. Instead of a linear regression, higher-order polynomial fits may also be tried.

21.7. Do the same as in Numerical Exercise 21.6, but try to calibrate the delayed oscillator model (21.30) for the temperature anomaly. Use the calibrated model for the extrapolation. If necessary, use data windows of several years.

S. George H. Philander
1942–

Born in South Africa and son of a poet, George Philander studied applied mathematics and physics before going to Harvard University to obtain his doctorate and embarking on a career in oceanography. His seminal studies of El Niño and, from there, also the Southern Oscillation earned him a position of prominence in Geophysical Fluid Dynamics. From unraveling the global connection between ocean and atmosphere, the study of global warming and climate change necessarily became the next scientific pursuit. Philander is known as a "teacher to his bones," passionate about sharing knowledge with the next generation, and he is praised for his clarity of thought and elegance of expression.

Philander is also a prolific writer, having written multiple books on the subject of El Niño and climate, for both experts and nonexperts alike, including *Our Affair with El Niño – How We Transformed an Enchanting Peruvian Current into a Global Climate Hazard* (2006) an acclaimed book aimed at a broad readership. (*Photo credit: Princeton University*)

Paola Malanotte Rizzoli

Paola Malanotte Rizzoli had obtained a doctorate in quantum mechanics and was well on her way to a distinguished career in physics when a massive flood of Venice, where she worked at the time, made her change her mind. She switched to physical oceanography and obtained a second doctorate. Her contributions to this field have been significant and varied, spanning the theory of long-lived geophysical structures, such as eddies and hurricanes, numerical modeling of the Atlantic Ocean and Gulf Stream system, the Black Sea ecosystem, data assimilation, and tropical-subtropical interactions.

Professor Rizzoli teaches at the Massachusetts Institute of Technology and lectures across the world. She is known as a dynamic speaker and an inspiring scientist. In addition to her teaching and research, she has served the oceanographic community in a number of capacities, at both national and international levels.

Never abandoning her love for Venice, Paola Rizzoli was instrumental in developing a system of sea gates to protect the city from future floods and sea level rise. This protection system is currently under construction. (*Photo MIT archives*)

Chapter 22

Data Assimilation

ABSTRACT

This chapter outlines the problem of predictability and the methods used to blend, in an optimal way, observations with model computations in order to guide the latter and to produce improved simulations of geophysical fluid phenomena. The methods invoke physical as well as statistical reasoning and rely on certain approximations that facilitate their implementation in operational forecast models.

22.1 NEED FOR DATA ASSIMILATION

Personal experience teaches us that a weather forecast is reliable for only a few days from the moment when it is issued. The time up to which the prediction is made is called *lead time*, and a reliable forecast can only be had if the lead time is not too long, typically no more than a week for midlatitude weather. Predictions further into the future are so imprecise that very simple prediction methods, such as the use of climatological values for the day concerned or persistence of today's weather, may perform as well as sophisticated weather forecast systems. Before discussing the reasons why forecasting errors increase with lead time, we anticipate that any forecast system will need periodic reinitialization if predictions are needed on a regular basis. Such a reinitialization must certainly take into account the most recent observations in order to infer a correct present state of the system, an operation known as *field estimation* in forecast jargon. From this better estimate, a forecast can be restarted on a better basis.

For weather forecasts, field estimates are *sequential* in the sense that they only use existing data, that is, from the past up to the day on which the forecast begins. For other applications, the best field estimate of a past situation is constructed, in which case data from moments after the forecast has begun can also be incorporated, and a nonsequential method is used. A typical example in which all available data are used is a so-called *re-analysis*, which incorporates the best fields over a given time period together with the physics; these provide the best picture of reality at any moment.

The melding of physical laws and observations, be it in a sequential or nonsequential way, is carried out through so-called *data assimilation*. Data assimilation may be performed intermittently, for example every day using data from the previous day, or continuously, as data become available.

Since data assimilation exploits observational data, it is also possible to quantify the forecast errors up to a point once new data have arrived. Forecast errors can then be used to assess the skill of the forecast system, a measure of its predictive capability. Here, it is customary to compare the error of the forecast with the error of an elementary forecast. Rudimentary forecasts are one of the following: Persistence (tomorrow the weather will be the same as today), climatology (next week's weather will be the average weather of the last twenty years at the same time of year), or random forecast (e.g., one of the two preceding methods with an added random noise of zero average and prescribed variance). One measure of forecasting skill is the Brier skill score, S, which is calculated from an error measure ϵ^f of the actual forecast system and the corresponding error measure ϵ^r of a rudimentary (or reference) forecast system (Brier, 1950; see also Wilks, 2005):

$$S = 1 - \frac{\epsilon^f}{\epsilon^r}, \tag{22.1}$$

Clearly, if the forecasting system has an error (ϵ^f) equal to the rudimentary approach error (ϵ^r), its skill is nil. On this scale, the unattainable perfect forecasting skill corresponds to a score of unity (100%). Should the skill of the forecast system falls below zero, the forecast system should be considered as no better than the most basic forecast system, although there might still be some useful information in the forecast. The skill score is often used to quantify the improvement of a new forecast system over an older version, in which case ϵ^r is taken as the error of that older system.

Clearly the skill score depends on the nature of the chosen error norm ϵ, which is, for example, the root-mean-square (rms) error between two fields, the error on the maximum temperature, the error on the hours of sunshine etc., but more importantly the skill varies with lead time. The further the forecast extents into the future, the lower the skill score tends to be, and we naturally come back to the question of why it is so difficult to make accurate long-range forecasts. The previous chapters might have biased our perception of geophysical fluid dynamics toward a system governed by equations, the solutions of which uniquely follow from an adequate set of initial and boundary conditions. This is the case in theory, but we ought to accept the idea that with imperfect models and inaccurate boundary and, especially, initial conditions, errors can have a tendency to accumulate over the course of a long-range forecast and thus to reduce skill with lead time. As a matter of fact, the situation is more dramatic than that. Even if we could control the initial errors below any arbitrary value, which obviously we would never be able to do, some dynamical equations, by their very nature, lead to solutions that diverge rapidly for even extremely small changes in the initial conditions. The famous *Lorenz equations*

(Lorenz, 1963, page 135) provide an archetype of a system of equations that exhibit such a behavior:

$$\frac{dx}{dt} = \sigma (y - x), \tag{22.2a}$$

$$\frac{dy}{dt} = rx - y - xz, \tag{22.2b}$$

$$\frac{dz}{dt} = xy - bz, \tag{22.2c}$$

in which σ, r, and b are fixed parameters, and x, y, and z are temporal variables. The equations form a low-order truncation of a spectral model of atmospheric motions and, despite their innocent look, are known to generate chaotic trajectories such that two very close initial conditions will lead to completely different solutions after some time (Fig. 22.1). Clearly, there is a *predictability limit*.

More generally, the accumulation of errors, even when starting with arbitrarily small errors in the initial conditions, can result in the existence of a predictability limit in strongly nonlinear systems. This limit is estimated to be one to two weeks for the global atmosphere and on the order of one month for midlatitude ocean eddies. It is therefore not surprising that forecast skill decreases with lead time approaching the predictability limit of the system (Fig. 22.2). An idea of the predictability limit can be obtained by considering the autocorrelation of the solution for time delay $\Delta t > 0$:

$$\rho(\Delta t) = \frac{\frac{1}{T}\int_{\Delta t}^{T} u(t)u(t - \Delta t)\, dt}{\sqrt{\frac{1}{T}\int_{\Delta t}^{T} u(t)^2\, dt}\sqrt{\frac{1}{T}\int_{\Delta t}^{T} u(t - \Delta t)^2\, dt}} \tag{22.3}$$

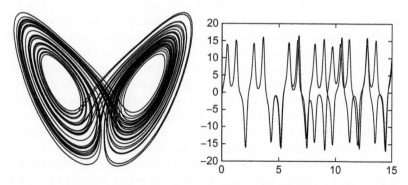

FIGURE 22.1   Trajectory in (x, z) space (left panel) showing the vacillation of the solution between two cycles. For two slightly different initial conditions, the corresponding two solutions $x(t)$ track each other for some time and then diverge (right panel). Graph obtained by using chaos.m that solves the Lorenz equations (22.2) with $\sigma = 10$, $r = 28$ and $b = 8/3$.

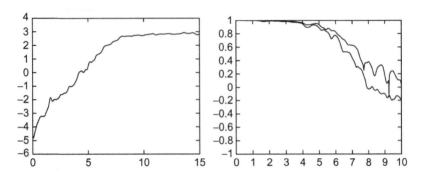

FIGURE 22.2 For incorrect initial conditions, the logarithm of the forecast error increases as a function of lead time (left panel). Skill score decreases as a function of lead time for two different base forecasts (right panel). For each curve an ensemble of 200 simulations with the Lorenz equations was averaged.

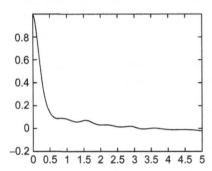

FIGURE 22.3 Autocorrelation as a function of time delay Δt for a solution $x(t)$ of the Lorenz equations (22.2). The curve, obtained by averaging 200 different trajectories, shows that, on average, after a few time units, the solution is not well correlated with its own previous values.

with time $T \to \infty$. This function measures the extent to which the solution at a given moment is on average close to the solution at a moment Δt earlier. In this sense, the delay Δt for which the autocorrelation value ρ approaches zero defines the time interval over which the solution no longer resembles its own past, i.e., the time over which it "decorrelates" from itself. Put another way, the value of Δt for which ρ falls close to zero is the memory time of the dynamical process. For a purely random function, one without any past memory, the autocorrelation is zero for any delay Δt. For the solution to the Lorenz equations, the predictability time can inferred from Fig. 22.3. It is important to note, however, that the system may still be considered as deterministic as every initial condition determines a unique evolution. The gradual loss of predictability means that, as time goes on, we become less able to identify this unique trajectory even with the finest numerical surgery and measurement tools.

In geophysical fluid dynamics, motions are not only controlled by initial conditions but also by boundary conditions, and the predictability limit depends on the relative importance of boundary conditions to initial conditions. If the system mostly responds to forcing applied at a boundary, such as an semi-enclosed shallow sea with a strong ocean tide at its opening, the future behavior can be predicted very far ahead, and the model's skill is essentially measured by the accuracy of the forcing. Since forcings and boundary conditions are typically well known, skill remains high for long periods. The predictability issue manifests itself most severely with systems essentially controlled by initial conditions. The global atmosphere, which has no lateral boundary, is the quintessential example, and its initial conditions must be set with utmost care, and, nonetheless, the prediction skill may start very high but degrade quite quickly. Between those two extremes, the predictability of a system depends on the relative importance of initial over boundary conditions (Fig. 22.4).

Despite the inherent problem of predictability, we can increase forecast skills by reducing to a maximum the uncertainties in the model and its initial and boundary conditions, so as to push further away the predictability limit of the forecasting system. We can try to keep some of the errors under control, especially those already classified as modeling errors (see Section 4.8). Within the pages of this book, we encountered various levels of model simplifications, such as the hydrostatic approximations, the quasi-geostrophic approximation, and the shallow-water model. For all of these, numerical discretization in space and time added further sources of errors. For initial and boundary conditions based on observed fields, we can also distinguish several other kinds of error, with the most obvious one being instrumental error (of generally known and relatively low standard deviation). In Section 1.8, we also encountered the representativity error due to the fact that a point measurement (e.g., a temperature measurement in a city) is not the variable we actually would like to observe (the temperature at the 50 km scale). Synopticity errors (i.e., lack of simultaneity among several measurements) can be a concern when observations are binned into time slots for analysis and assimilation (e.g., assembling into a single "snapshot" view of the ocean the data gathered during the span of a cruise) because waves propagate, and temporal shifts can lead to severe Doppler effects (Rixen, Beckers & Allen, 2001). In general, any data treatment before assimilation into the model, such as interpolation, must be taken into account when assessing the errors associated with the "observation."

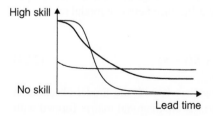

FIGURE 22.4   Predictability behavior for boundary-condition dependent systems (flat line), initial-condition dependent systems (steep line) and mixed situations (intermediate line).

The distinction between modeling and observational errors is not always clear, and the discrete sampling necessitated by the numerical grid can be considered either a modeling error (truncation of continuous operators) or an observational error (inaccurate because incomplete data). In any case, we face the problem that some of the information, both from model and observations, is incomplete and corrupted by errors. The main objective of data assimilation is the reduction of the influence of those errors on the simulation by utilizing data to guide the model in the best possible way. This improved analysis will generally extend the lead time for which predictions are reliable. This, however, is not the sole advantage of data assimilation; others will be discussed at the end of the chapter.

The methods described below were for the most part developed within the context of atmospheric modeling (e.g., Bengtsson, Ghil & Kallen, 1981; Ghil & Childress, 1987; see also Navon, 2009 for a review) and later adapted by the oceanographic community (e.g., Evensen, 1994; Ghil, 1989; Ghil & Malanotte-Rizzoli, 1991). The presentation uses the unified notation proposed by Ide, Courtier, Ghil and Lorenc (1997) also adopted in the reference book of Kalnay (2003), with a few exceptions to stay consistent with the notation used elsewhere in this book. Further seminal textbooks are those by Bennett (1992), Malanotte-Rizzoli (1996), Robinson, Lermusiaux and Sloan (1998), and Wunsch (1996).

22.2   NUDGING

Among the first methods used to guide numerical simulations by data injection is the nudging method. It starts from the governing equations of the state vector **x**, which is the collection of variables in the forecasting system,

$$\frac{d\mathbf{x}}{dt} = \mathcal{Q}(\mathbf{x}, t),$$

where the operator $\mathcal{Q}$ bearing on the set of variables stands for the model equations. Assuming the observations are distributed in exactly the same way as the state-vector components (i.e., observations are collected on the same grid as the numerical model), we group them into a vector **y** (**y** being what **x** should have been if the model had been correct). The nudging method then simply adds a correction term proportional to the difference between model results and observations:

$$\frac{d\mathbf{x}}{dt} = \mathcal{Q}(\mathbf{x}, t) + \mathbf{K}(\mathbf{y} - \mathbf{x}). \tag{22.4}$$

The additional term is the product of a matrix **K** with the model-observation misfit. For the nudging method, this matrix is the diagonal matrix formed with a set of so-called time scales of *relaxation* noted τ_i, which may differ from variable to variable: $K_{ii} = 1/\tau_i$ and $K_{ij} = 0$ for $i \neq j$. Since the difference between the model simulation $\mathbf{x}$ and observations $\mathbf{y}$ is zero where there is no error, the additional term only acts as a correction wherever necessary, "nudging" (i.e., pulling) the solution towards the observations.

The "strength" of the nudging depends intimately on the values ascribed to the relaxation time scales τ_i. If the time scale is long compared with the time scale of evolution of the corresponding variable, the correction is kept small and qualified as background relaxation. Such background relaxation very often uses climatological values in place of "observations." When no observation is present for a variable, the corresponding relaxation time is simply set to infinity. When observations are only available at certain moments, the relaxation time scale is made to decrease from a large to a small value when approaching the moment t^o at which data are available to ensure a smooth temporal incorporation of data. An example of time-dependent function is

$$\frac{1}{\tau} = K \exp\left[-(t - t^o)^2 / T^2\right] \qquad (22.5)$$

with T being an appropriate time scale over which the observation can nudge the simulation. The time relaxation can also be made spatially dependent when different dynamical regimes can be identified for physical reasons. One particular form of nudging is surface relaxation in ocean models, where the simulated sea surface is nudged towards the observed surface fields. Such nudging is often maintained with a low intensity even when full atmospheric fluxes are applied onto the ocean in order to avoid any drift (e.g., Pinardi et al., 2003). At the other extreme, when the relaxation time is taken very short and comparable with the time step, the nudging method basically replaces the simulated value with the corresponding observation, a process called *direct insertion* when carried out in a single time step.

Nudging was widely used in the early days of data assimilation but has now been superseded by more sophisticated techniques (see below). Nonetheless nudging remains popular near boundaries, where it can be interpreted as a boundary condition corrected by observations. In this case, the time-continuous nature of the correction is beneficial by avoiding sudden shocks in the model.

22.3   OPTIMAL INTERPOLATION

The previous method, though robust and found useful in the past, is rather an ad-hoc approach, and we will now present a method that is based on sound statistical optimization. In particular, relationships between different variables will be exploited in order to enhance the assimilation. For this, we will use a notation that is slightly different from the discretization notation used until now. Because sequential assimilation cycles perform an analysis at not every time step of the

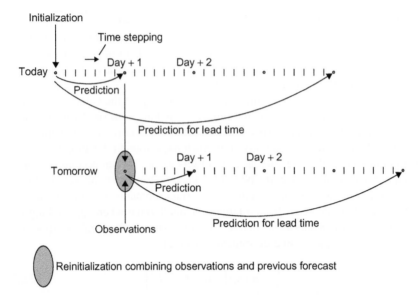

FIGURE 22.5 Schematic representation of reinitialization through sequential data assimilation on a daily basis. On a given day, the model provides a forecast for several days ahead. The next day, two pieces of information become available, the forecast for the day and the daily observation. Combining the two generates a new initial condition for the model, and an improved forecast can be provided for the following days.

model (Fig. 22.5), we shall denote by $\mathbf{x}_n$ a particular cycle of the assimilation.[1] We also should keep in mind that in practical applications, it is advantageous to use a state vector defined by anomalies (i.e., departure from a reference state), normalized so that each element of the state vector should be comparable with the others. The state vector then does not gather velocity and temperature but normalized versions of them. Because it contains variables of different types, we are heading for a so-called *multivariate* approach. Forecast values will be referenced by superscript f and the "analyzed" variables, obtained after combining forecast with observations by the superscript a.

For the purpose of illustration, we start with the elementary problem of having at our disposal at a given moment two pieces of information about a temperature of unknown true state T^t. The information can originate from either measurement or model and can include an error ϵ. For the two given values T_1 and T_2, we therefore write

$$T_1 = T^t + \epsilon_1, \quad \langle \epsilon_1 \rangle = 0, \quad T_2 = T^t + \epsilon_2, \quad \langle \epsilon_2 \rangle = 0, \tag{22.6a}$$

---

[1] Note the difference with the notation $\mathbf{x}^n$ previously used to refer to the variable at time step n.

where we assume that, on average, denoted by brackets $\langle\ \rangle$, errors vanish. In other words, we suppose the values to be unbiased. We can estimate the unknown, true temperature by a linear combination of the two available values:

$$T = w_1 T_1 + w_2 T_2 = (w_1 + w_2)T^t + (w_1 \epsilon_1 + w_2 \epsilon_2) \tag{22.7}$$

and on average, this estimate will take the value

$$\langle T \rangle = (w_1 + w_2)T^t, \tag{22.8}$$

so that we obtain an unbiased estimate of the true state if we take $w_1 + w_2 = 1$. In this case, we perform a de-facto weighted average among the two pieces of available information, an intuitive approach. An unbiased estimate, or *analysis*, T^a of the true state is therefore

$$T^a = (1 - w_2) T_1 + w_2 T_2 = T_1 + w_2 (T_2 - T_1) \tag{22.9}$$

while in reality there is an error

$$T^a - T^t = (1 - w_2)\epsilon_1 + w_2 \epsilon_2. \tag{22.10}$$

This error is zero on average, but its variance is not zero:

$$\langle (T^a - T^t)^2 \rangle = (1 - w_2)^2 \langle \epsilon_1^2 \rangle + w_2^2 \langle \epsilon_2^2 \rangle + 2(1 - w_2)w_2 \langle \epsilon_1 \epsilon_2 \rangle. \tag{22.11}$$

The actual errors ϵ_1 and ϵ_2 are not known, otherwise we would have had ready access to T^t. However, given some basic information on the source of the errors, we can assess the so-called *error variance* $\langle \epsilon_1^2 \rangle$ or, equivalently, the standard deviation $\sqrt{\langle \epsilon_1^2 \rangle}$. If the two pieces of information (observed and/or modeled) leading to T_1 and T_2 are independent of each other, we may reasonably suppose that the errors ϵ_1 and ϵ_2 are uncorrelated, which in statistical terms means $\langle \epsilon_1 \epsilon_2 \rangle = 0$. Hence, the error variance $\langle \epsilon^2 \rangle$ of the analysis is

$$\langle \epsilon^2 \rangle = (1 - w_2)^2 \langle \epsilon_1^2 \rangle + w_2^2 \langle \epsilon_2^2 \rangle. \tag{22.12}$$

Naturally, the best estimate of T^t is the one with the lowest error variance. Finding the value of w_2 that minimizes the right-hand side of the preceding equation, we obtain

$$w_2 = \frac{\langle \epsilon_1^2 \rangle}{\langle \epsilon_1^2 \rangle + \langle \epsilon_2^2 \rangle} \tag{22.13}$$

and obtain the minimal error variance:

$$\langle \epsilon^2 \rangle = \frac{\langle \epsilon_1^2 \rangle \langle \epsilon_2^2 \rangle}{\langle \epsilon_1^2 \rangle + \langle \epsilon_2^2 \rangle} = \left(1 - \frac{\langle \epsilon_1^2 \rangle}{\langle \epsilon_1^2 \rangle + \langle \epsilon_2^2 \rangle} \right) \langle \epsilon_1^2 \rangle, \tag{22.14}$$

while the temperature estimate is

$$T^a = T_1 + \left(\frac{\langle \epsilon_1^2 \rangle}{\langle \epsilon_1^2 \rangle + \langle \epsilon_2^2 \rangle} \right) (T_2 - T_1). \tag{22.15}$$

We observe that the error variance of the combination of T_1 and T_2 is smaller than each of $\langle \epsilon_1^2 \rangle$ and $\langle \epsilon_2^2 \rangle$. Using information from two sources, even if one of the two has a relatively large error, reduces on average the uncertainty. This is the idea underlying data assimilation using error analysis to reduce overall uncertainty. If optimization such as the minimization of Eq. (22.12) is used, the process can be quite effective at decreasing the forecast error.

We can reach the same optimal estimate by finding the T value that minimizes a weighted measure of the differences between the analysis and the available information, with weights inversely proportional to the error variance of the information:

$$\min_T J = \frac{(T - T_1)^2}{2 \langle \epsilon_1^2 \rangle} + \frac{(T - T_2)^2}{2 \langle \epsilon_2^2 \rangle}. \tag{22.16}$$

In other words, we do not mind that the analysis departs from the relatively uncertain observation, but we do require that the analysis falls closer to the observation if the latter happens to be more accurate. The minimum of Eq. (22.16) is reached when T takes the value T^a of Eq. (22.15).

The optimal reduction in error can be used to blend observations with a model forecast through the technique called *Optimal Interpolation*.[2] Typically, there are many more model data than available observations[3] so that the size M of $\mathbf{x}$ is much larger than the size P of the vector $\mathbf{y}$ containing observations.

---

[2] Sometimes the term *objective analysis* is used instead. This is a poor choice of words, however, as the latter is generally no more than a mathematical interpolation that stands opposed to the historical *subjective analysis* of weather patterns by pencil and paper.

[3] In 2006, The European Centre for Medium-Range Weather Forecasts, ECMWF, used $M = 3 \times 10^7$ state variables for its operational ensemble T256 weather forecast model and assimilated $P = 3 \times 10^6$ observations every cycle of 12 hours. The Mercator ocean model PSY3v1 operates with $M = 10^8$ but "only" $P = 0.25 \times 10^6$ data were assimilated once a week.

To blend observations into a model simulation at a certain point in time, we construct the $\mathbf{x}^a$, called the *analysis*, as a linear combination of the forecast $\mathbf{x}^f$ up to that point in time with the observations $\mathbf{y}$ that have become available:

$$\mathbf{x}^a = \mathbf{x}^f + \mathbf{K}\left(\mathbf{y} - \mathbf{H}\mathbf{x}^f\right). \tag{22.17}$$

This procedure uses a linear *observation operator* $\mathbf{H}$ that selects or interpolates the simulated state variables to the points of observation in order to quantify the model-observation misfit $(\mathbf{y} - \mathbf{H}\mathbf{x}^f)$ (e.g., the temperature forecast is interpolated to the position in which a meteorological station measures temperature). In some cases, the matrix $\mathbf{H}$ may contain mathematical operations that relate multiple forecast fields to observed parameters not directly predicted. For example, the so-called Gelbstoff (literally, yellow matter) measured in the ocean by satellites is an aggregate value of organic compounds in the water column, but the corresponding dispersion model typically calculates the three-dimensional structure of the organic matter, and the matrix $\mathbf{H}$ has to add the values at the various grid points spanning the height of the water column at the point of observation. Occasionally the relationship between model variables and observations is nonlinear, and the correction term, called *innovation* vector and noted $\mathbf{d}$, should be replaced by

$$\mathbf{d} = \mathbf{y} - \mathcal{H}\left(\mathbf{x}^f\right) \tag{22.18}$$

where $\mathcal{H}$ stands for a nonlinear function. Here, we only consider a matrix $\mathbf{H}$ assuming a linear relationship between observed and modeled values.

We can also mention that the interpretation of observation errors depends on how the data are prepared, and how the observation operator is constructed. Altimetric data can, for example, be assimilated along tracks, and the observational error estimated as a combination of instrumental, representativity (mix of spatial scales), synopticity (time-slot binning), and interpolation errors when sampling the model along the tracks using $\mathbf{H}$. If, for convenience, the tracks are gridded beforehand on the same mesh as the model grid, this interpolation, itself performed, for example, with a spatial Optimal Interpolation version (Numerical Exercise 22.2), has an associated error covariance, which must be taken into account in prescribing the observational error. Generally, interpolation from the model to data location (through $\mathbf{H}$) introduces fewer errors because there are typically many more model grid points than observation points.

The matrix $\mathbf{K}$ of size $M \times P$, where M is the number of model variables and P the number of observations, is determined in order to obtain the "best" analysis, that is, an optimal blending of data into model. This analysis depends on the error fields of both forecast and observations. The forecast error is

$$\epsilon = \mathbf{x} - \mathbf{x}^t, \tag{22.19}$$

that is, the difference between the calculated field $\mathbf{x}$ and the (unknown) true state $\mathbf{x}^t$. Likewise, the observational error is

$$\epsilon^o = \mathbf{y} - \mathbf{y}^t. \tag{22.20}$$

From these errors, even though we do not know their actual values because the true states are not known, we can define statistical averages and variances. Obviously, for unbiased models and observations—which we both assume—the averages (first-order moments) $\langle \epsilon \rangle$ and $\langle \epsilon^o \rangle$ are zero. The error-covariance matrix for observations

$$\mathbf{R} = \langle \epsilon^o \epsilon^{oT} \rangle \tag{22.21}$$

is square matrix with nonzero elements on the diagonal if each observation is subject to its own error. On the diagonal, we therefore find the error variance of each observation. Off-diagonal terms are nonzero whenever two separate observations are correlated, as it occasionally happens with satellite observations. Note that off-diagonal terms are symmetric and that for any vector $\mathbf{z}$, the quadratic form $\mathbf{z}^T \mathbf{R} \mathbf{z} = \langle (\mathbf{z}^T \epsilon^o)^2 \rangle$ is never negative so that a covariance matrix is always semi-positively defined. If errors are random and span the whole state-vector space, the covariance matrix is strictly positively defined.

The analysis step (22.17) can be expressed as

$$\mathbf{x}^t + \epsilon^a = \mathbf{x}^t + \epsilon^f + \mathbf{K}(\epsilon^o - \mathbf{H}\epsilon^f) + \mathbf{K}\underbrace{(\mathbf{y}^t - \mathbf{H}\mathbf{x}^t)}_{=0}. \tag{22.22}$$

The last term is zero because, by definition, a perfect model would perfectly match the observations of the true state. It follows that the error on the analysis field is

$$\epsilon^a = \epsilon^f + \mathbf{K}(\epsilon^o - \mathbf{H}\epsilon^f), \tag{22.23}$$

from which we can proceed to construct the error covariance $\langle \epsilon^a \epsilon^{aT} \rangle$ of the analysis by multiplying (22.23) by its own transposed and take the average:

$$\langle \epsilon^a \epsilon^{aT} \rangle = \langle \epsilon^f \epsilon^{fT} \rangle + \mathbf{K} \langle (\epsilon^o - \mathbf{H}\epsilon^f) \epsilon^{fT} \rangle$$
$$+ \langle \epsilon^f (\epsilon^{oT} - \epsilon^{fT}\mathbf{H}^T) \rangle \mathbf{K}^T$$
$$+ \mathbf{K} \langle (\epsilon^o - \mathbf{H}\epsilon^f)(\epsilon^{oT} - \epsilon^{fT}\mathbf{H}^T) \rangle \mathbf{K}^T. \tag{22.24}$$

Defining covariance matrices

$$\mathbf{P} = \langle \epsilon \epsilon^T \rangle \tag{22.25}$$

for both forecast and analysis error fields (with superscripts ^{f} and ^{a}, respectively) and making the reasonable assumption that there is no correlation[4] between observational error and forecast error, i.e., $\langle \epsilon^o \epsilon^{fT} \rangle = 0$, we can rewrite the error-covariance matrix after analysis as

$$\mathbf{P}^a = \mathbf{P}^f - \mathbf{KHP}^f - \mathbf{P}^f \mathbf{H}^T \mathbf{K}^T + \mathbf{K} \left(\mathbf{R} + \mathbf{HP}^f \mathbf{H}^T \right) \mathbf{K}^T$$

$$= \mathbf{P}^f - \mathbf{P}^f \mathbf{H}^T \mathbf{A}^{-1} \mathbf{HP}^f + \left(\mathbf{P}^f \mathbf{H}^T - \mathbf{KA} \right) \mathbf{A}^{-1} \left(\mathbf{HP}^f - \mathbf{AK}^T \right) \quad (22.26)$$

in which matrix $\mathbf{A}$, defined for convenience as

$$\mathbf{A} = \mathbf{HP}^f \mathbf{H}^T + \mathbf{R} \quad (22.27)$$

is symmetric and can most likely be inverted.[5]

If state variables are suitably scaled to enable comparison of, say, temperature errors with velocity errors, the overall error ϵ^a of the analyzed field may be taken as the expected norm of the error vector:

$$\epsilon^a = \langle \epsilon^{aT} \epsilon^a \rangle. \quad (22.28)$$

This, however, is nothing other than the trace of the covariance matrix $\langle \epsilon^a \epsilon^{aT} \rangle$, and an overall measure of the analysis error is thus

$$\epsilon^a = \text{trace} \left(\mathbf{P}^a \right). \quad (22.29)$$

Since the $M \times P$ matrix $\mathbf{K}$ is still unspecified, a reasonable choice is to choose it such that it minimizes the global error. One way of proceeding is to take the trace of Eq. (22.26) and differentiate the trace with respect to all components of $\mathbf{K}$ in order to find the extremal value of the global error; because (22.26) is a quadratic form in terms of $\mathbf{K}$, and because $\mathbf{A}^{-1}$ is positive defined if it exists, the extremum is assured to be a minimum. Alternatively, we can think of the error as being a function $\epsilon^a(\mathbf{K})$ of the matrix $\mathbf{K}$ and search for an optimal $\mathbf{K}$ such that

$$\epsilon^a(\mathbf{K+L}) - \epsilon^a(\mathbf{K}) = 0 \quad (22.30)$$

for an arbitrary, small departure matrix $\mathbf{L}$. After linearization with respect to $\mathbf{L}$, we thus require

$$\text{trace} \left(-\mathbf{L} \left(\mathbf{HP}^f - \mathbf{AK}^T \right) - \left(\mathbf{P}^f \mathbf{H}^T - \mathbf{KA} \right) \mathbf{L}^T \right) = 0. \quad (22.31)$$

---

[4]Decorrelation between observational error and modeling error is justified by the very different origin of the information.

[5]Because $\mathbf{P}$ and $\mathbf{R}$ are semipositive defined matrices, the chances are not low. In addition, when observations and state variables are covering a wide domain, the covariances between distant points are generally small so that the matrices have a tendency to be diagonally dominant.

Since the two terms are the transpose of each other and thus share the same trace, it is sufficient that this common trace vanishes

$$\text{trace}\left(\left(\mathbf{P}^f\mathbf{H}^\mathsf{T}-\mathbf{KA}\right)\mathbf{L}^\mathsf{T}\right)=0.$$

Since $\mathbf{L}$ is arbitrary, its matrix coefficient must be zero, and the $\mathbf{K}$ matrix that minimizes the analysis error is

$$\mathbf{K}=\mathbf{P}^f\mathbf{H}^\mathsf{T}\mathbf{A}^{-1}$$
$$=\mathbf{P}^f\mathbf{H}^\mathsf{T}\left(\mathbf{HP}^f\mathbf{H}^\mathsf{T}+\mathbf{R}\right)^{-1}. \tag{22.32}$$

We note that matrix $\mathbf{A}$ must be invertible in order to reach the error minimum. The $\mathbf{K}$ matrix, which combines model forecast with data, is the analog of Eq. (22.13) and is called the *Kalman gain matrix*.

The minimum error covariance of the analysis is obtained by inserting (22.32) into (22.26):

$$\mathbf{P}^a=(\mathbf{I}-\mathbf{KH})\,\mathbf{P}^f=\left(\mathbf{I}-\mathbf{P}^f\mathbf{H}^\mathsf{T}\left(\mathbf{HP}^f\mathbf{H}^\mathsf{T}+\mathbf{R}\right)^{-1}\mathbf{H}\right)\mathbf{P}^f \tag{22.33}$$

which is the analog of Eq. (22.14). We note that neither optimal Kalman gain matrix nor minimum error covariance depends on the *value* of the observations or the forecast state vector but only on their statistical error covariance. The only field that depends on the actual field values is, of course, the state vector itself, which after optimal analysis becomes

$$\mathbf{x}^a=\mathbf{x}^f+\mathbf{P}^f\mathbf{H}^\mathsf{T}\left(\mathbf{HP}^f\mathbf{H}^\mathsf{T}+\mathbf{R}\right)^{-1}\left(\mathbf{y}-\mathbf{Hx}^f\right). \tag{22.34}$$

The use of Eq. (22.32) in Eq. (22.17) to combine forecast and observations with respective error covariances $\mathbf{P}^f$ and $\mathbf{R}$ is known as *Optimal Interpolation* (OI).

With forecast and observation covariances given, an alternative derivation of Optimal Interpolation is a variational approach called *3D-Var*, in which the objective is to find the state vector that minimizes the error measure J given by

$$J(\mathbf{x})=\frac{1}{2}\left(\mathbf{x}-\mathbf{x}^f\right)^\mathsf{T}\mathbf{P}^{f-1}\left(\mathbf{x}-\mathbf{x}^f\right)+\frac{1}{2}(\mathbf{Hx}-\mathbf{y})^\mathsf{T}\mathbf{R}^{-1}\left(\mathbf{Hx}-\mathbf{y}\right) \tag{22.35}$$

In other words, the procedure is to search for the state vector closest to both model forecast and observations and which penalizes less the more accurate information, in close analogy with Eq. (22.16). Finding the optimum state vector leads to the same analyzed field as in Eq. (22.34). The demonstration is left as an exercise (Analytical Problem 22.4).

Optimal Interpolation can also be cast in terms of the maximum likelihood estimator of the true field, i.e., the field that has the highest probability of matching reality, which is also given by Eq. (22.34) if the *probability density function* (pdf) of each errors follows the normal (Gaussian) distribution (e.g., Lorenc, 1986). In any case, it is necessary to quantify all error variances, which is not an easy task (see for example Lermusiaux et al., 2006).

22.4   KALMAN FILTERING

In the formulation of Optimal Interpolation, it turns out that the dynamical model used to provide the forecast does not appear explicitly. Only its forecast error covariance matrix $\mathbf{P}^f$ is needed besides the forecast itself. Thus, very little of what is known about the model is actually used.

Knowing, for example, that a fluid-flow model has a tendency to propagate errors on state variables along preferential directions, such as the flow itself or a wave guide, or may be amplified by unstable modes, we ought to ask how data assimilation could take into account certain model properties. For this, we start from the fact that between times n and $n+1$ when assimilation is performed the model advances the state vector according to

$$\mathbf{x}_{n+1} = \mathcal{M}(\mathbf{x}_n) + \mathbf{f}_n + \boldsymbol{\eta}_n, \tag{22.36}$$

where $\mathbf{f}_n$ represents the external forcing between times n and $n+1$, and $\boldsymbol{\eta}_n$ the error introduced by the model as it marches through a multiplicity of time steps from time level n of the last data assimilation to time level $n+1$ when the next data assimilation is scheduled to take place. The operator $\mathcal{M}$ stands for the inner machinery of the model, which calculates the state vector at time level $n+1$, over multiple time steps, from its previous value at time level n. Assuming the model forecast starts at time level n with the analyzed field produced from data assimilation at that time and also assuming a linearized model (for pure convenience because it is not true!), we write with a certain level of approximation:

$$\mathbf{x}_{n+1}^f = \mathbf{M}\mathbf{x}_n^a + \mathbf{f}_n + \boldsymbol{\eta}_n, \tag{22.37}$$

where the matrix $\mathbf{M}$ replaces the nonlinear operator $\mathcal{M}$ of Eq. (22.36). Such a matrix is actually never constructed in operational models but is only introduced here to provide an elegant presentation of the method. The unknown, true state evolves similarly but without modeling error and thus obeys

$$\mathbf{x}_{n+1}^t = \mathbf{M}\mathbf{x}_n^t + \mathbf{f}_n \tag{22.38}$$

so that the forecast error $\boldsymbol{\epsilon}^f = \mathbf{x}^f - \mathbf{x}^t$ is

$$\boldsymbol{\epsilon}_{n+1}^f = \mathbf{M}\,\boldsymbol{\epsilon}_n^a + \boldsymbol{\eta}_n. \tag{22.39}$$

Multiplying this last equation to the right by its transpose and performing the statistical average to obtain error covariance, we obtain the so-called *Lyapunov equation*, which allows for the advancement in time of the error covariance:

$$\mathbf{P}_{n+1}^f = \mathbf{M}\,\mathbf{P}_n^a\mathbf{M}^\mathsf{T} + \mathbf{Q}_n = \mathbf{M}\left(\mathbf{M}\mathbf{P}_n^a\right)^\mathsf{T} + \mathbf{Q}_n \tag{22.40}$$

with the following definition of the model-error covariance matrix:

$$\mathbf{Q}_n = \left\langle \boldsymbol{\eta}_n\boldsymbol{\eta}_n^\mathsf{T} \right\rangle. \tag{22.41}$$

As earlier, errors of different origins are assumed to be uncorrelated. Since the forcing $\mathbf{f}$ disappears from the error evolution equation, we will not keep it during later developments. Note that, even if we do not write down the matrix $\mathbf{M}$ explicitly, the evolution of the error-covariance matrix $\mathbf{P}^a$ can still be calculated by using instead the actual model on each of its columns $\mathbf{c}$, as shown by operations such as $\mathbf{M}\mathbf{c}$ involved in the error-covariance matrix updates. In order to start the calculation of the error evolution, we need to know the initial value of $\mathbf{P}$, which is related to the error on initial conditions:

$$\mathbf{P}_0 = \left\langle \left(\mathbf{x}_0 - \mathbf{x}_0^t\right)\left(\mathbf{x}_0 - \mathbf{x}_0^t\right)^\mathsf{T} \right\rangle. \tag{22.42}$$

Now, we have a method by which we can calculate the evolution of the error-covariance, and the Kalman filter assimilation is summarized in Fig. 22.6, which includes an extension towards a nonlinear model with linearized error propagation (Extended Kalman Filter, EKF). The analysis step itself is unchanged from Optimal Interpolation, but the error covariance is updated at each assimilation step. In sum, what *Kalman Filtering* adds to Optimal Interpolation is that not only state variables but also their errors are marched in time by the forecast model.

Two extremes are noteworthy. At one extreme, the time between two consecutive assimilations is very short compared with the time scale of evolution of the dynamical process being modeled, and the state variables and their error remain virtually unchanged. The model can then be considered as persistent, with $\mathbf{M}\sim\mathbf{I}$. In this case

$$\mathbf{P}_{n+1}^f \sim \mathbf{P}_n^a + \mathbf{Q}_n. \tag{22.43}$$

In other words, the forecast error is the error of the previous analysis, without advection or any other modification, augmented by the model error introduced during the simulation between assimilations. The last error is relatively small for a well-constructed model because the time integration is short compared with the time scale of interest.

At the other extreme, when assimilation takes place only very infrequently, the model is likely to reach its limit of predictability, and its forecast becomes little more than random. Put another way, the forecast itself is the modeling error. Mathematically this amounts to write $\mathbf{M}\sim\mathbf{0}$ and hence

$$\mathbf{P}_{n+1}^f \sim \mathbf{Q}_n, \tag{22.44}$$

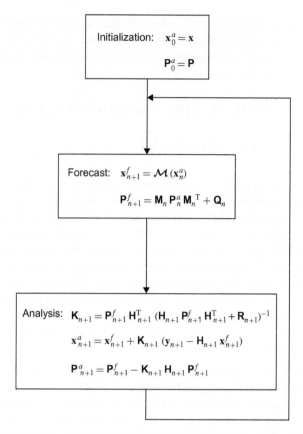

FIGURE 22.6 Sequence of steps followed in the Extended Kalman Filter assimilation scheme with changing observation network (changing matrix **H**), nonlinear model forecast ($\mathcal{M}$), and model linearization for the error forecast between times of assimilation. Note that strongly nonlinear models, such as those found in ecosystem studies, lead to special problems (e.g., Robinson & Lermusiaux, 2002).

implying that the error field no longer has any memory of the previous error, a fact consistent with the trespassing of the predictability limit, but is entirely due to the simulation. Obviously, $\mathbf{Q}_n$ in this case is much larger than in the previous case because of the long-term integration and error amplifications inherent to the predictability limit.

In order to illustrate the structure of Kalman filtering in a general case, we first note that for the analysis step the error covariance matrix only appears in the combination $\mathbf{P}^f \mathbf{H}^\mathsf{T}$,

$$\mathbf{P}^f \mathbf{H}^\mathsf{T} = \left\langle \boldsymbol{\epsilon}^f \boldsymbol{\epsilon}^{f\mathsf{T}} \right\rangle \mathbf{H}^\mathsf{T} = \left\langle \left(\mathbf{x}^f - \mathbf{x}^t \right) \left(\mathbf{H}\mathbf{x}^f - \mathbf{H}\mathbf{x}^t \right)^\mathsf{T} \right\rangle \tag{22.45}$$

which is the covariance between the observed quantities and all others. Because it is the matrix that ultimately multiplies a vector of the size of the observational data set (P), it effectively propagates information from the data locations into the model grid. We also have

$$
\mathbf{A} = \mathbf{H}\mathbf{P}^f\mathbf{H}^{\mathsf{T}} + \mathbf{R} = \left\langle \left(\mathbf{H}\mathbf{x}^f - \mathbf{H}\mathbf{x}^t \right) \left(\mathbf{H}\mathbf{x}^f - \mathbf{H}\mathbf{x}^t \right)^{\mathsf{T}} \right\rangle + \mathbf{R}
$$

$$
= \left\langle \left(\mathbf{H}\mathbf{x}^f - \mathbf{y} \right) \left(\mathbf{H}\mathbf{x}^f - \mathbf{y} \right)^{\mathsf{T}} \right\rangle
\tag{22.46}
$$

which can be interpreted in terms of error variance of the forecast in the observed part, combined with the corresponding observational error, reminiscent of $\langle \epsilon_1^2 \rangle + \langle \epsilon_2^2 \rangle$ in Eq. (22.13). This also shows that $\mathbf{A}^{-1}$ exists according to the statistics regarding the innovation vector $\mathbf{H}\mathbf{x}^f - \mathbf{y}$. Should a component of this innovation vector always be zero making $\mathbf{A}$ singular, this would mean that the corresponding model part would never need a correction and that it should therefore be excluded from the data analysis procedure. Also, note that the Kalman gain matrix $\mathbf{K}$ gives greater weight to observations with higher accuracy and that it transmits the corresponding information to other locations.

It is instructive to look at the assimilation of a single observation on the kth component of the state vector $\mathbf{x}$. In this case

- $\mathbf{P}^f\mathbf{H}^{\mathsf{T}}$ is an $M \times 1$ matrix with components P_{ik}^f, $i = 1, \ldots, M$, and it is thus responsible for transferring the innovation learned from observation in location k to the other components of the state vector. The covariance matrix thus appears as the matrix allowing the correction of the fields using remote information. The structure of the covariance depends on the problem at hand (Fig. 22.7). In the example of Fig. 22.7, the error covariance after the analysis is reduced at the data location and then advected by the flow. Hence, downstream of the data point, error covariances are lower and reflect the remote influence of the data.

- $\left(\mathbf{H}\mathbf{P}^f\mathbf{H}^{\mathsf{T}} + \mathbf{R} \right)^{-1}$ reduces to a scalar value $\left(P_{kk}^f + \epsilon_0^2 \right)^{-1}$ in which ϵ_0^2 is the variance of the observational error and P_{kk}^f the forecast error covariance at the same location.

The error-covariance matrix thus allows the radiation of information from the data location into other regions of the domain and onto other state variables taking into account the relative error of observations and model. Thus, assimilating data from atmospheric temperature profiles might well serve to change winds elsewhere, and sea surface height measurement by an altimeter satellite may well be used to adjust density fields in deeper ocean layers. In contrast to the atmosphere for which a large network of ground-based observation systems and vertical profile soundings by means of balloons have been deployed at reasonable cost, it is far more expensive at sea to maintain a fixed network of moorings or to perform periodic ship cruises in order to sample the ocean

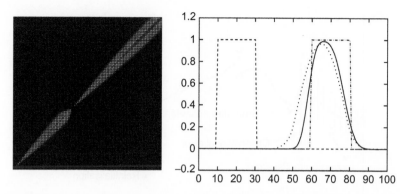

FIGURE 22.7 Assimilation experiment in an advection problem in which data are provided at grid point $i = 40$ at every time step. *Left panel*: The error-covariance matrix (left panel with lower values in black and higher values in lighter shades). The covariance matrix has the largest values along its diagonal, symptomatic of locality. Note, however, that these diagonal values are lower downstream of the data point because the accurate data reduce errors there. *Right panel*: The advected field, initial pattern (square signal on the left), what it should have been at a later time (square signal on the right), what it actually turned out to be with inaccurate velocity and numerical diffusion (dotted curve), and what the prediction became with the help data assimilation (solid curve). See `kalmanupwadv.m` for details of the implementation.

interior. In consequence, satellite data are immensely valuable in ocean forecasts, and it is essential to be able to utilize these surface data to inform density fields and currents in the interior of the ocean.

Because of our minimization of errors using linear combinations and the hypotheses of zero bias, the full Kalman filter is called a *Best Linear Unbiased Estimation* (BLUE) of the true state. Other linear methods presented previously, such as nudging, are suboptimal. It is interesting to note that the Kalman filter approach encompasses in some way the other assimilation methods. For example, if we were to prescribe a priori the forecast error covariance in the Kalman filter instead of letting it be calculated by the model, we would downgrade the Kalman filter to an Optimal Interpolation. If, furthermore, the error and observation covariance matrices were not just prescribed but also taken diagonal, then we recover the nudging scheme. Finally, should the nudging time scale be decreased towards zero, the procedure would be reduced to direct insertion (mere replacement of state variables by corresponding observations). In all cases, the term filter method is appropriate because only past data are utilized to estimate local fields.

22.5   INVERSE METHODS

The Kalman filter operation relies only on past information, the only type of information that is available in an operational forecast mode. Separately, it acknowledges that the dynamic model is not perfect. The procedure of data assimilation optimized with the Kalman filter, however, presents a major

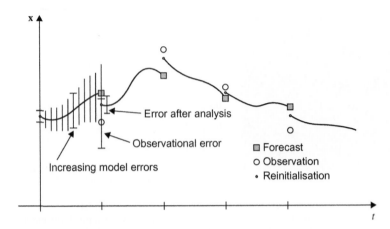

FIGURE 22.8 The Kalman Filter approach leads to a model trajectory that is interrupted each time data assimilation is performed. Vertical bars depicting the error level show that errors creep up during forecast intervals but are suddenly reduced every time assimilation takes place.

drawback in terms of model simulation analyzed over long time periods: The simulated model states (trajectories of state variables) are no longer continuous but exhibit a jump each time data assimilation is performed (Fig. 22.8), and these sudden variations in the values of dynamical variables may cause unphysical shocks. For some applications, it is desirable to avoid any jump to obtain a simulation that preserves the continuity of the physical system (Fig. 22.9).

When we are concerned with obtaining a continuous model trajectory, it is reasonable to assume the dynamical model is perfect except that errors may be introduced by inaccurate initial conditions, incorrect forcing, and possibly also incorrect model parameters such as an eddy-viscosity value. Those errors are then considered responsible for model predictions not conforming with reality. The idea we will now follow is that of optimizing those model components so that the model evolution remains as close as possible to the observations over an extended period of time. The state of the system is at any moment influenced not only by prior data but also by future data. Needless to say, this is not possible in a forecasting mode, but is nonetheless a useful method for re-analysis of past observations with an eye on model improvement. Such a method using data over an interval of time is termed a *smoother*.

With an *inverse model*, the goal is the minimization of model-data misfit over a finite time interval with N data sets $\mathbf{y}_n$ available along the way by searching for an optimal set of initial conditions $\mathbf{x}_0$. The method proceeds by minimizing a so-called *cost function* defined as

$$J = \sum_{n=0}^{N-1} \frac{1}{2} \left(\mathbf{H}_n \mathbf{x}_n - \mathbf{y}_n \right)^\top \mathbf{R}_n^{-1} \left(\mathbf{H}_n \mathbf{x}_n - \mathbf{y}_n \right) + J_b. \qquad (22.47)$$

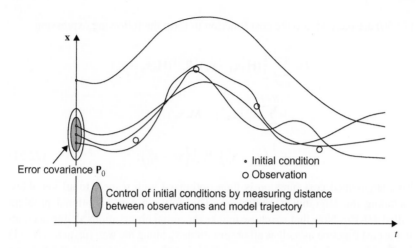

FIGURE 22.9   The adjoint method selects the model trajectory that best fits the observations over time. The initial conditions used to create different trajectories are drawn with a probability density function centered on a background state $\mathbf{x}_0^b$.

The quadratic first term is understood as a weighted sum of squares of misfits between observation ($\mathbf{y}_n$) and corresponding model realizations ($\mathbf{H}_n\mathbf{x}_n$). To this has been added a term J_b, which may become useful should we want to avoid a set of initial conditions $\mathbf{x}_0$ departing too much from a reference (background) field $\mathbf{x}_0^b$. In such case, this extra term takes the form

$$J_b = \frac{1}{2}\left(\mathbf{x}_0 - \mathbf{x}_0^b\right)^{\mathsf{T}}\mathbf{P}_0^{-1}\left(\mathbf{x}_0 - \mathbf{x}_0^b\right),\tag{22.48}$$

the minimization of which "pulls" the initial conditions toward their background values. Such a background state $\mathbf{x}_0^b$ of initial conditions can be supplied for example by previous forecasts or climatology.

In addition to the minimization of the cost function J, we need to enforce the constraint that the solution $\mathbf{x}$ is a model result satisfying

$$\mathbf{x}_{n+1} = \mathcal{M}(\mathbf{x}_n).\tag{22.49}$$

For simplicity, however, we again resort to a linearized version,

$$\mathbf{x}_{n+1} = \mathbf{M}_n\mathbf{x}_n.\tag{22.50}$$

An elegant way to satisfy the combination of constraints is to use the method of Lagrange multipliers. By means of these multipliers $\boldsymbol{\lambda}_n$, the set constraints

(22.50) are merged into the cost function to form the following expression:

$$
\begin{aligned}
J = &\sum_{n=0}^{N-1} \frac{1}{2} (\mathbf{H}_n \mathbf{x}_n - \mathbf{y}_n)^\mathsf{T} \mathbf{R}_n^{-1} (\mathbf{H}_n \mathbf{x}_n - \mathbf{y}_n) \\
&+ \sum_{n=0}^{N-1} \boldsymbol{\lambda}_n^\mathsf{T} (\mathbf{x}_{n+1} - \mathbf{M}_n \mathbf{x}_n) \\
&+ \frac{1}{2} \left(\mathbf{x}_0 - \mathbf{x}_0^b \right)^\mathsf{T} \mathbf{P}_0^{-1} \left(\mathbf{x}_0 - \mathbf{x}_0^b \right).
\end{aligned}
\tag{22.51}
$$

This augmented cost function is then optimized with respect to all variables, including the Lagrangian multipliers. This forms a new variational problem called *4D-Var*. Note that we chose to take the initial observations $\mathbf{x}_0^b$ into account in the cost function as well as all observations $\mathbf{y}_n$ along the way (up to $n = N - 1$) but not the simulation outcome $\mathbf{x}_N$, which can thus be regarded as the forecast based on an evolution that is constrained to pass as close as possible to the N previous observations.

The optimization of (22.51) with respect to the Lagrange multipliers (i.e., derivatives with respect to $\boldsymbol{\lambda}_n$ set to zero) returns the model constraints of Eq. (22.50) for $n = 0$ to $N - 1$. Variation with respect to the initial state $\mathbf{x}_0$ leads to

$$
\nabla_{x_0} J = \mathbf{P}_0^{-1} \left(\mathbf{x}_0 - \mathbf{x}_0^b \right) + \mathbf{H}_0^\mathsf{T} \mathbf{R}_0^{-1} \left(\mathbf{H}_0 \mathbf{x}_0 - \mathbf{y}_0 \right) - \mathbf{M}_0^\mathsf{T} \boldsymbol{\lambda}_0,
\tag{22.52}
$$

which must vanish for the optimal solution. Variation of Eq. (22.51) with respect to each intermediate state $\mathbf{x}_m$ must also be zero. Realizing that $\mathbf{x}_m$ appears in each sum for both $n = m$ and $n = m - 1$, we obtain the following condition in which we switched from m back to n:

$$
\mathbf{H}_n^\mathsf{T} \mathbf{R}_n^{-1} (\mathbf{H}_n \mathbf{x}_n - \mathbf{y}_n) - \mathbf{M}_n^\mathsf{T} \boldsymbol{\lambda}_n + \boldsymbol{\lambda}_{n-1} = 0 \qquad n = 1, \dots, N - 1.
\tag{22.53}
$$

Finally, variation with respect to the final state $\mathbf{x}_N$ simply provides $\boldsymbol{\lambda}_{N-1} = 0$.

The different conditions can be recast into the following algorithm. We start with an estimate $\mathbf{x}_0$ of the initial condition and then perform the following operations

$$
\mathbf{x}_{n+1} = \mathbf{M}_n \, \mathbf{x}_n, \qquad\qquad\qquad\qquad n = 0, \dots, N - 1 \tag{22.54a}
$$

$$
\boldsymbol{\lambda}_{N-1} = 0 \tag{22.54b}
$$

$$
\boldsymbol{\lambda}_{n-1} = \mathbf{M}_n^\mathsf{T} \boldsymbol{\lambda}_n - \mathbf{H}_n^\mathsf{T} \mathbf{R}_n^{-1} (\mathbf{H}_n \mathbf{x}_n - \mathbf{y}_n), \qquad n = N - 1, \dots, 1. \tag{22.54c}
$$

In this process, we note that model-data misfits $\mathbf{H}_n \mathbf{x}_n - \mathbf{y}_n$ are driving the values of the Lagrange multipliers. All stationary conditions imposed on Eq. (22.51) are now satisfied except that $\nabla_{x_0} J$ given by Eq. (22.52) is not yet zero. The recurrence (22.54c) on the Lagrange multiplier can formally be extended to $n = 0$ so that $\boldsymbol{\lambda}_{-1}$ takes on the value of $-\nabla_{x_0} J$ if no background field is used ($\mathbf{P}_0^{-1} = 0$).

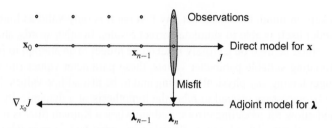

FIGURE 22.10 The forward integration starts from an initial guess of the control (adjustable) parameters and then provides the state variable over the simulation window. Misfits between observations and model results are stored, and error norms of the cost function J defined by Eq. (22.35) are accumulated. Then, the adjoint model is integrated backward in time, with the misfits used as forcing. Upon returning to the start $n = 0$, the gradient (22.52) can be calculated. If it is not zero, the optimum is not reached and an improved guess of the control parameters can be calculated by minimization tools using the cost function value J and its gradient with respect to the control parameters.

But because we now have access to a value of J and its gradient with respect to the variables on which we optimize our solution, the initial condition, we may use any mathematical minimization tool that searches for such minimum using gradients. The steepest descent or the most efficient *conjugate gradient* methods (see Section 7.8) are iterative methods that generate a succession of states (here for x_0) that decrease the value of the cost function J depending on the gradient of the function. We therefore now have a relatively simple way of calculating those gradients by performing a forward integration of the model, called the direct model, and a backward integration to evaluate the Lagrange multipliers, called the inverse model, and finally the gradients (Fig. 22.10).

This apparently simple procedure masks several practical problems, however. The equations for the Lagrange multipliers are very similar to those of the direct model with the seemingly innocent difference that, instead of matrix **M**, its transpose appears, and instead of applying **M** we apply its adjoint to λ_n to create the time series. Therefore, we speak of the *adjoint model* when referring to the backward integration for the Lagrange multipliers. In practice, since a numerical model never explicitly creates the matrix **M**, this means that actual programming of an adjoint, nonlinear model is necessary, the action of which on λ is equivalent to applying the transpose model matrix. Another practical problem for time-varying models is the need to store intermediate model results, or to regenerate them on a need basis, over the entire simulation interval because of the backward integration. For more details on implementation aspects of the adjoint method, the reader is referred to Courtier, Thepaut and Hollingsworth, 1994 and for early theoretical developments to Talagrand and Courtier (1987).

The method is readily extended to optimizations of parameter values such as eddy viscosity and boundary conditions. Parameters to be optimized, called *control parameters*, can, for example, be introduced as additional state variables into the state vector, with an evolution equation of persistence. In all cases, it must be kept in mind, however, that any inversion is only valid as long as the direct model itself is able to simulate correct results. In other words, should an inverse method be used on a grossly inadequate model in an attempt to improve it by estimating suitable parameter values, these parameter values run the risk of no longer having any physical meaning and to be just ad-hoc values. It might therefore be necessary instead to relax the hypothesis of a dynamically correct model and allow for modeling errors, as done when a Kalman filter is used.

Allowing the model solution to deviate somewhat from a pure model solution is achieved by replacing the strong constraint (22.50) with a so-called *weak constraint*, penalizing only strong departures from Eq. (22.50), with the stiffest penalties reserved for models that are most trustworthy. Doing so, we can adapt the cost function of the inverse model to form a *generalized inverse model* by minimizing

$$
J = \sum_{n=0}^{N-1} \frac{1}{2} (\mathbf{H}_n \mathbf{x}_n - \mathbf{y}_n)^{\mathsf{T}} \mathbf{R}_n^{-1} (\mathbf{H}_n \mathbf{x}_n - \mathbf{y}_n)
$$
$$
+ \sum_{n=0}^{N-1} \frac{1}{2} (\mathbf{x}_{n+1} - \mathbf{M}_n \, \mathbf{x}_n)^{\mathsf{T}} \mathbf{Q}_n^{-1} (\mathbf{x}_{n+1} - \mathbf{M}_n \, \mathbf{x}_n)
$$
$$
+ \frac{1}{2} \left(\mathbf{x}_0 - \mathbf{x}_0^b \right)^{\mathsf{T}} \mathbf{P}_0^{-1} \left(\mathbf{x}_0 - \mathbf{x}_0^b \right), \tag{22.55}
$$

where $\mathbf{Q}$ is the error covariance of the model.

Differentiation with respect to the initial condition provides

$$
\nabla_{x_0} J = \mathbf{P}_0^{-1} \left(\mathbf{x}_0 - \mathbf{x}_0^b \right) + \mathbf{H}_0^{\mathsf{T}} \mathbf{R}_0^{-1} \left(\mathbf{H}_0 \mathbf{x}_0 - \mathbf{y}_0 \right) - \mathbf{M}_0^{\mathsf{T}} \mathbf{Q}_0^{-1} (\mathbf{x}_1 - \mathbf{M}_0 \, \mathbf{x}_0), \quad (22.56)
$$

while differentiation with respect to intermediate states ($n = 1, \ldots, N-1$) leads to the additional conditions

$$
\mathbf{H}_n^{\mathsf{T}} \mathbf{R}_n^{-1} (\mathbf{H}_n \mathbf{x}_n - \mathbf{y}_n) -
$$
$$
\mathbf{M}_n^{\mathsf{T}} \mathbf{Q}_n^{-1} (\mathbf{x}_{n+1} - \mathbf{M}_n \, \mathbf{x}_n) + \mathbf{Q}_{n-1}^{-1} (\mathbf{x}_n - \mathbf{M}_{n-1} \, \mathbf{x}_{n-1}) = 0, \tag{22.57}
$$

and differentiation with respect to the final state provides $\mathbf{x}_N = \mathbf{M}_{N-1} \, \mathbf{x}_{N-1}$.

This system of equations can be written in a more familiar way as

$$
\mathbf{x}_n = \mathbf{M}_{n-1} \mathbf{x}_{n-1} + \mathbf{Q}_{n-1} \lambda_{n-1}, \qquad n = 1, \ldots, N \tag{22.58a}
$$
$$
\lambda_{N-1} = 0 \tag{22.58b}
$$
$$
\lambda_{n-1} = \mathbf{M}_n^{\mathsf{T}} \lambda_n - \mathbf{H}_n^{\mathsf{T}} \mathbf{R}_n^{-1} (\mathbf{H}_n \mathbf{x}_n - \mathbf{y}_n) \qquad n = N-1, \ldots, 1 \tag{22.58c}
$$

with the need for gradient (22.56) to be zero. These equations are very similar to those of the adjoint method (22.54) with the additional term involving $\mathbf{Q}$ in the direct model allowing the propagation of model errors. Though similar, the practical solution is now more complicated than for the adjoint method and can be

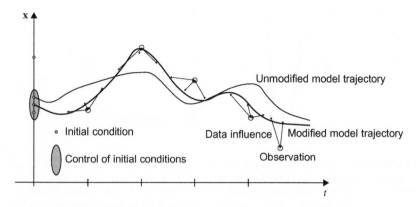

FIGURE 22.11   Generalized inverse methods allow the simultaneous optimization of initial conditions and the minimization of the departure of the model from observations.

obtained with the so-called *representer method* (Bennett, 1992). The generalized inverse method thus exhibits a solution that is not a true solution of the direct model but one that effectively reaches a compromise between observations and a true model solution (Fig. 22.11).

Variational approaches are attractive because they permit efficient model calibration, but error covariances of observations and model need to be prescribed a priori, which is a handicap. More importantly, except for the value of the cost function, which is an overall measure, the analysis is not accompanied by an error estimate on the final analysis contrary to the Kalman filter (22.33). In principle, it is possible to obtain such error estimates with variational methods by having recourse to the second derivatives of the cost function with respect to the control parameters, the Hessian matrix. Intuitively, if the cost function is decreasing sharply (i.e., if it can be said that it exhibits a "deep well"), the optimum is much better constrained than for a relatively flat cost function (Fig. 22.12). The Hessian (curvature) matrix allows therefore the calculation of the error covariance, a property demonstrated in Rabier and Courtier (1992).

Another way to obtain smooth trajectories with error estimates is to generalize the Kalman filter to include not only past values but also future ones. The so-called *Kalman smoother* does this and provides, for linear systems, identical results to those of the generalized inverse approach over the whole simulation interval.

Similarly it can be shown that the standard 4D-Var method and Kalman filter are equivalent to each other for a perfect model with identical data over a given time interval and initial error covariance, in the sense that they lead to the same final analysis for linear models and linear observation operators.

Although equivalence has been proven in some cases, the methods tend to lose their overlaps when the underlying hypotheses, such as linearity of the

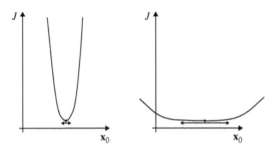

FIGURE 22.12 When the Hessian matrix of J—the set of its second derivatives with respect to adjustable parameters—is large, the minimum tends to be well constrained ("deep well"—left panel), and errors in the final state are small. On the contrary, errors in the optimal solution will be larger if the cost function exhibits a wide range over which the adjustable parameters may vary without penalizing significantly the value of J ("shallow well"—right panel).

model and unbiased information, are violated. The different methods then lead to different outcomes. Also, the practical implementation can differ significantly from method to method and can widely affect their algorithmic performance.

22.6   OPERATIONAL MODELS

Practical implementation of the aforementioned data assimilation methods is a concern when considering the typical number of operations that are to be performed routinely with an operational model. The nudging method barely adds a linear term to existing equations, and the overhead cost associated with this procedure is negligible. For Optimal Interpolation in its original form, there arises at least the need to invert a matrix of size proportional to the number of observations.[6] Since the matrix $\mathbf{H}$ is generally mostly composed of zeros, we need not pay attention to costs associated with multiplications by $\mathbf{H}$. Nevertheless, for a full matrix $\mathbf{HPH}^\mathsf{T} + \mathbf{R}$, the cost of its inversion is roughly P^3 operations. In addition, we need to store $\mathbf{HP}$ of size $M \times P$. This amounts to a major burden on computer resources if P is large. If we want to calculate the error covariance after the analysis, the required matrix multiplications demand PM^2 operations. If only the diagonal of the covariance matrix, i.e., if only the local error variances are sought, the number of operations drops to PM, still a large number.

The computational demand, of course, rises if we update the error covariance over time as needed in Kalman filtering. Granted, we never construct the matrix $\mathbf{M}$ to march the model forward, but multiplication by this matrix with a vector actually represents a model integration. Hence, if the matrix $\mathbf{M}$ multiplies another matrix of size $M \times M$, the cost estimate is that of M model integrations,

---

[6]This is assuming the covariance matrix $\mathbf{P}$ or, rather, $\mathbf{HP}$ can be easily calculated.

which is therefore the cost of updating the forecast error covariance, which must be compared with the cost of a single model integration as when no assimilation is performed.

For the 3D-Var method, the situation is not any better because it necessitates M^3 operations to invert each covariance matrix, unless their inverses are provided. In that case, we need M^2 operations to calculate each cost function, M operations to calculate its gradient, and K iterations to find the minimum. In order to reach the minimum, $K = M$ iterations are needed in theory, but good approximations can be found for K much less than M.

For 4D-Var, similar estimates apply with the need to perform a direct model integration and its adjoint, reverse integration for each evaluation of the gradient needed during the minimization. For K iterations, KM model integrations are required, while each evaluation of the cost function along the way requires M^2 operations, unless the error covariances have special forms (e.g., a diagonal matrix).

In view of the numbers M (on the order of 10^7 or 10^8) of model variables and P (on the order of 10^6) of observations in use for operational models, we clearly see a need for simplification. Otherwise, for the sizes of state vectors and observation arrays currently in use, Kalman filters and 3D-Var are out of the question unless covariance matrices are of a very special form. Reducing the complexity of an assimilation method to a computationally affordable version while retaining its advantages is where the "art" of modeling comes into play, by sifting the unnecessary elements from the essential features.

The type and quantity of observation greatly affects the possibility of simplifications. In the ocean, for operational purposes, observations are mostly surface data from satellites (sea-level anomaly with respect to an average dynamical position, sea surface temperature, color, sea ice and, hopefully in the future, salinity as well) and coastal data (sea level at tide gages), with Argo floats (e.g., Taillandier, Griffa, Poulain & Béranger, 2006) complementing the information with a few deep profiles. In the atmosphere, observational networks are much denser than at sea, and, in particular, daily radiosoundings from all over the globe contribute to data assimilation. Simplified assimilation methods generally take into account the type of data incorporated. For the ocean, one approach is to perform assimilation in two steps, one incorporating only surface information and the other profile data, with moreover specific simplifications brought to the covariances.

Reduction of complexity can be justified in most cases because the system's evolution includes a series of physically damped modes, which do not require correction since they fade away. Damped modes occur through the existence of attractors such as geostrophic equilibrium. However, unstable modes, so characteristic of our weather patterns, must absolutely be followed by the assimilation process.

Another avenue for simplifications is the size of the state vector. The very large number of numerical state variables results mostly from our numerical need to have the grid size short compared with the scales of interest, i.e., $\Delta x \ll L$ and $\Delta z \ll H$. But this immediately indicates that we are using many more calculation points than the actual number of degrees of freedom that are needed. We can therefore try to downsize the burden of the data assimilation procedure from the number of computational points to the significantly lower number of significant degrees of freedom in the dynamics or number of observations.

One of the most popular approaches to simplify the assimilation procedure is the use of a *reduced rank* covariance matrix, by writing

$$\mathbf{P} \sim \mathbf{S}\mathbf{S}^{\mathsf{T}} \qquad (22.59)$$

in which the reduced matrix $\mathbf{S}$ is of size $M \times K$, which is significantly less than $M \times M$ because K is much smaller than M. It is easily demonstrated that the rank[7] of $\mathbf{S}\mathbf{S}^{\mathsf{T}}$ is at most K, hence the name reduced rank. When adopting this simplification, we no longer need to store $\mathbf{P}$ because we store the much smaller matrix $\mathbf{S}$ and any matrix multiplication involving $\mathbf{P}$ reduces to successive matrix multiplications by $\mathbf{S}$ and its transpose. A multiplication of $\mathbf{P}$ with a square matrix of identical size $M \times M$ no longer requires M^3 operations but only $2KM^2$ operations.

We can also achieve reduction by saving computations at the level of matrix inversion. The effect of a reduced rank can be exemplified most easily with a diagonal matrix $\mathbf{R}$ of uncorrelated observational errors with the same error variance μ^2 at all locations. Defining the matrix $\mathbf{U} = \mathbf{HS}$ of dimension $P \times K$ with $K \ll P$, we have

$$\mathbf{P}\mathbf{H}^{\mathsf{T}}\left(\mathbf{HS}\mathbf{S}^{\mathsf{T}}\mathbf{H}^{\mathsf{T}} + \mathbf{R}\right)^{-1} = \mathbf{S}\mathbf{U}^{\mathsf{T}}\left(\mathbf{U}\mathbf{U}^{\mathsf{T}} + \mu^2\mathbf{I}\right)^{-1} = \mathbf{S}\left(\mathbf{U}^{\mathsf{T}}\mathbf{U} + \mu^2\mathbf{I}\right)^{-1}\mathbf{U}^{\mathsf{T}}.$$
$$(22.60)$$

The last equality can be demonstrated directly by matrix operations or by using a special case of arguably the most useful matrix identity in data assimilation, the Sherman-Morisson formula (Analytical Problem 22.8). The last operation in Eq. (22.60) transforms the matrix to be inverted from a $P \times P$ matrix into a much smaller $K \times K$ one. This is where a major gain in computing is obtained. The same gain is attained when $\mathbf{R}$ is not diagonal but has an inverse that can be calculated easily.

In an ensemble forecast approach (e.g., Houtekamer & Mitchell, 1998), the direct model is used to create a series of simulations, each one being a slightly perturbed version of the others, with perturbations introduced through initial conditions, forcings, parameter values, or even topographic modifications. The ensemble of model results thus obtained permits statistical parameters to be estimated from the ensemble members. In practice, the convergence of variance estimations from K samples converges only as $1/\sqrt{K}$, and we therefore

---

[7]The rank of a matrix is the number of linearly independent columns.

anticipate needing a large ensemble or somehow creating an ensemble with optimal distributions of its members (e.g., Evensen, 2004).

Combining ensemble approaches and reduced rank approximations leads to a series of data assimilation variants with different implementations (e.g., Barth, Alvera-Azcárate, Beckers, Rixen & Vandenbulcke, 2007; Brasseur, 2006; Lermusiaux & Robinson, 1999; Pham, Verron & Roubaud, 1998; Robinson et al., 1998). The ensemble approach can even be extended to include not only simulation results of a single model with perturbed setups but also different models, aimed at modeling the same properties (super-ensemble approach) or even using different physical parameterizations and governing equations (hyper-ensemble approach). Such combinations can dramatically reduce errors, in particular biases (e.g., Rixen & Coelho, 2007).

Other simplifications are based on the reduction of the size of the state vector on which assimilation works. One possibility is to propagate error covariances with coarser grids or simplified models (e.g., Fukumori & Malanotte-Rizzoli, 1995). Particular dynamical balances can also be taken into account through covariances. If only one component of such a balance is observed, for example, sea surface height in a geostrophically balanced system, velocity does not need to be included in the assimilation step. Once a correction of sea surface height and density is made, corrections to velocity can be calculated by using geostrophy.

Operational models have been in use for some applications for quite some time, with weather forecasts based on numerical models having been initiated in the postwar period. These are now widespread with two major centers providing global forecasts (ECMWF and NCEP). Operational tidal models, hurricane predictions, and tsunami warning systems in the Pacific are also well established, incorporating data from observational networks at dedicated institutes. Ocean circulation forecasts began to appear (Mersea, Hycom, Hops etc.) well before public demand for tsunami predictions in December 2004.

A common aspect of operational models is that data assimilation was initially developed to reduce errors. It is fair, however, to request that operational models not only include forecasts but also provide the associated uncertainty by means of confidence intervals, which assimilation now permits to do. From a scientific point of view, we might well argue that the forecast corrections do not provide new insight into physical processes. In reality, the analysis of assimilation cycles can help in understanding error sources and verify that model results are statistically consistent with observations. Also verifying that the innovations (model-data departures) and error estimates are compatible with the statistical models in use allows us to identify inconsistencies. For example, innovations should on average be zero, lest a bias arises in the system that needs correction. Forecast verifications (Jolliffe & Stephenson, 2003) can teach us not only about dynamics but also about model and observation errors and can help to identify the model or observing system components that are most in need of improvement. Finally, strategies for adaptive sampling can emerge from such studies (e.g., Lermusiaux, 2007).

ANALYTICAL PROBLEMS

22.1. Analyze the exact solution of the following equations

$$\frac{du}{dt} = +\tilde{f}v - \frac{u - u_o(t)}{\tau} \tag{22.61}$$

$$\frac{dv}{dt} = -\tilde{f}u - \frac{v - v_o(t)}{\tau} \tag{22.62}$$

with $u_o(t) = \cos(ft)$, $v_0 = -\sin(ft)$ and a relaxation time $\tau = 1/(\alpha\tilde{f})$. These equations represent nudging to follow an inertial oscillation of frequency f while the "model" has a tendency to generate inertial oscillations with incorrect frequency $\tilde{f}$. Identify how nudging corrects an error on the initial condition, distinguishing the effect of an error on amplitude and phase. Then investigate how a difference between $\tilde{f}$ and f affects the solution.

22.2. Establish and describe an Optimal Interpolation method to interpolate in space knowing the spatial covariance of the true field and data errors at given locations.

22.3. An analysis method provides both optimal fields and their corresponding error estimates, as shown in Fig. 22.13 under the assumption that the background (or forecast) field has a uniform error variance. In view of the data distribution, the error field, and spatial structure of the analysis, how would you rank the following measures

- observational error
- forecast error

and include an estimate of their values? In your opinion, what is a rough estimate of a correlation length used for the analysis of the field? (*Hint*: Think about the way observations are propagated into the domain and the significance of covariance functions. In particular, investigate how the analysis behaves far away from any observation and then near a single observation far away from any other data location. For better reading of the fields displayed in Fig. 22.13, refer to `divashow.m`.)

22.4. Prove that the minimization of Eq. (22.35) leads to the same analysis as the Optimal Interpolation. What is the value of J at the optimum?

22.5. A weather forecast is typically limited to one or two weeks, after which the forecast skill is nil. Can you justify why climate models, which use the same governing equations of geophysical fluid dynamics, may be used to predict decades ahead? (*Hint*: Think about the significance of state variables and parameterizations.)

22.6. Bottom friction is generally parameterized by a quadratic law $\tau_b = \rho_0 C_d u^2$. Show that, even for an unbiased estimate of u, the bottom-stress estimate itself is biased. What is the error variance?

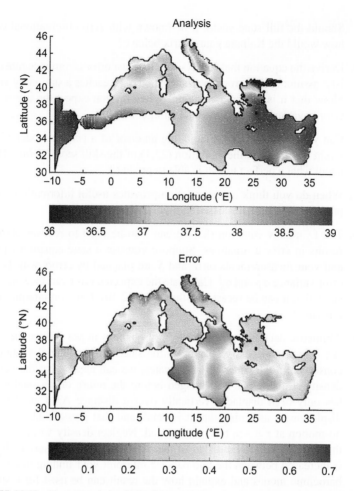

FIGURE 22.13 *Upper panel*: *An Optimal Interpolation of surface-salinity data in the Mediterranean Sea with data locations indicated by white dots. Lower panel*: The corresponding analysis error (in the same units as salinity), assuming uniform observational errors and a uniform forecast error. Note the impact of data distribution on the error estimates. (*Data Interpolating Variational Analysis (Diva), Brasseur, Blayo & Verron, 1996*). A color version can be found online at http://booksite.academicpress.com/9780120887590/

22.7. Prove the following identity:

$$L^T\left(LL^T+\mu^2 I\right)^{-1} = \left(L^T L+\mu^2 I\right)^{-1} L^T. \qquad (22.63)$$

Are there conditions to be satisfied for this to hold true?

22.8. Prove the Sherman-Morrison formula:

$$\left(A+UV^T\right)^{-1} = A^{-1} - A^{-1}U\left(I+V^T A^{-1}U\right)^{-1} V^T A^{-1}. \qquad (22.64)$$

22.9. Should the full state vector **x** be known with zero observational error, how would the Kalman gain matrix behave?

22.10. Derive the equation that steady-state forecast error covariance must satisfy, assuming the error-covariance matrix **P** reaches a stationary value. Show that it is a Ricatti equation. When do you expect the use of this error covariance to become interesting?

22.11. Can you see a reason for including an error of a perfect model ϵ^P different from zero in the definition (22.1) of the skill score? (*Hint*: Think about what a perfect model is able to predict.)

22.12. When do you think autocorrelation provides useful information on the predictability limit discussed in Fig. 22.4?

22.13. The Lyapunov equation (22.40) can also be used to recover classical results in error estimations. Suppose you use a state equation $\rho(T, S)$ and your measurements on T and S are plagued by errors with known error variance σ_T^2 and σ_S^2. Calculate the expected error variance on ρ and show that it can be recast into the form (22.40). Can you interpret **Q** in this case?

22.14. Altimetric data allow the detection of changes in sea surface elevation $\Delta\eta$ between successive passes of the satellite. If potential vorticity is conserved in the ocean at large scales, we can expect that the vertical density structure has not changed before the return of the satellite but has only been displaced vertically over a distance Δz. Calculate this displacement assuming that it is small and that pressure at a level of no-motion at $z = -z_0$ has not changed. Neglect density variations near the surface. What is the sign of the displacement if sea surface height is increased between the two passes? Interpret your finding in terms of baroclinic modes and explain how the result can be used for a simple data assimilation scheme (see Cooper & Haines, 1996).

NUMERICAL EXERCISES

22.1. Implement a nudging method and solve numerically Analytical Problem 22.1. When you decrease τ, what specific action do you need to take during time stepping? Investigate what happens when you add noise to the "observations" u_0 and v_0 and decrease the sampling rate of the "pseudo data."

22.2. Work with an Optimal Interpolation scheme on a one-dimensional gravity-wave problem already discretized in `oigrav.m` with 61 interior calculation points in a 61 km long and 100 m deep domain. Perform a twin experiment with a reference simulation starting from rest with a sinusoidal elevation of 40 cm amplitude and wavelength equal to the domain size. From this simulation, you can extract pseudo data in order to compare at least two assimilation strategies:

- A sampling point for surface elevation η located near the left boundary, at one third of the total domain length, sampled, and assimilated every 10 seconds.
- An observation of the surface elevation at all grid points sampled and assimilated every 10 minutes.

In both cases, add random noise to your "data" with a standard deviation representative of altimetric precision (2 cm). Assume the noise is uncorrelated in time and space. For the assimilation experiment, start from initial conditions at rest and zero sea-level height.

Prescribe the forecast covariance as a function of distance between points x_i and x_j, proportional to the following correlation function

$$c(x_i, x_j) = \exp\left[-(x_i - x_j)^2 / R^2\right] \qquad (22.65)$$

where R is a correlation length. The covariance is obtained by multiplying this correlation function by the estimated variance of the forecast.

The observational error covariance can be specified according to the perturbation added. Look at the evolution of the simulations and quantify the error individually on η and velocity.

Which variance of the background field would you advocate in view of the initial conditions? Change the value and assess the effect on the assimilation behavior. Then, change the value of correlation length and finally try other combinations of space-time coverage. (*Hint*: For ease of programming, the two simulations (reference run and assimilation run) may be advanced simultaneously, and you can diagnose the error as you go, before assimilating, however.)

22.3. Experiment with `kalmanupwadv.m` by changing the error specifications, sampling rate, and sampling locations.

22.4. Develop an adjoint method for Numerical Exercise 22.3 and optimize the initial condition using the same "observations" as for the Kalman filter of Numerical Exercise 22.3. Use the exact advection velocity with $C = U \Delta t / \Delta x = 0.2$ and then $C = 1$ in order to witness the effect of model errors, remembering that $C = 1$ corresponds to perfect dynamics because the upwind scheme is exact in this case. What is the effect of reducing the number of observations and disregarding the background initial conditions? (*Hint*: Think about underdetermined and overdetermined problems.)

22.5. Estimate the computer memory needed to store **P** for today's weather forecast systems.

22.6. Implement an ensemble Kalman filter for the Lorenz equations (22.2) and the assimilation of observations of $x(t)$ only. Investigate the effect of the observation frequency by performing assimilation steps at 0.001, 0.01, 0.1, and 1 time intervals.

22.7. Work to improve the Optimal Interpolation of Numerical Exercise 22.2 by calculating and using covariance functions estimated from an ensemble run in which you perturb the initial condition.

22.8. Improve the assimilation scheme of Numerical Exercise 22.2 by implementing the full Kalman filter with an updated covariance matrix. For ease of programming, you might consider constructing explicitly the transition matrix $\mathbf{M}$.

Michael Ghil
1944–

Born in Budapest, Hungary, Michael Ghil spent his high-school years in Bucharest, Romania and then acquired an engineering education in Israel, where he served as a naval officer. After moving to the United States, he obtained his doctorate at the Courant Institute of Mathematical Sciences of New York University under Professor Peter Lax (see biography at end of Chapter 6).

His scientific work includes seminal contributions to climate system modeling, chaos theory, numerical and statistical methods, data assimilation, and mathematical economics. He provided self-consistent theories of Quaternary glaciation cycles, of the low-frequency variability of extratropical atmospheric flows, and of the midlatitude oceanic interannual variability. He is a prolific writer, with his name attached to a dozen books and well over two hundred research and review articles.

Professor Ghil takes turns teaching at the École Normale Supérieure in Paris and at the University of California in Los Angeles. He has enjoyed learning from and working with a large number of students and younger colleagues, with whom he often stays in contact. Many of these have attained considerable professional achievements in their own right, on three continents. Ghil speaks six languages fluently. (*Photo by Philippe Bruère, Compagnie des Guides, Chamonix—Mont Blanc*)

Eugenia Kalnay
1942–

Eugenia Kalnay was awarded a Ph.D in Meteorology from the Massachusetts Institute of Technology under the direction of Jule Charney (see biography end of Chapter 16). Following a position as Associate Professor in the same department, she became Chief of the Global Modeling and Simulation Branch at the NASA Goddard Space Flight Center, where she developed the accurate and efficient "NASA Fourth Order Global Model." Later as Director of the Environmental Modeling Center of the US National Weather Service (NWS), she spearheaded major improvements in the forecast skills of the NWS models. Many successful projects such as ensemble forecasting, 3D and 4D variational data assimilation, advanced quality control, seasonal and interannual dynamical predictions, were started under her leadership. She also directed the NCEP/NCAR 50-year Reanalysis Project, and the resulting re-analysis paper of 1996 is one of the most cited paper in all of geosciences. Moving to academia, Professor Kalnay cofounded at the University of Maryland the Weather-Chaos Group, a leader in Ensemble Kalman Filter methods.

Over the years, Kalnay has received numerous awards, including the prestigious IMO prize of the World Meteorological Organization, for her contributions to numerical weather prediction, data assimilation, and predictability. Kalnay, a key figure in this field, has pioneered many of the essential techniques, which she describes in her book *Atmospheric Modeling, Data Assimilation, and Predictability* (2003). (*Photo by Zhao-Xia Pu, used with permission*)

Part VI

Web site Information

The content is organized by folders numbered according to the chapters of the book. In each folder, the MATLAB™ programs mentioned in each chapter are found and can be edited. The files contain comments beginning with % like % this is a comment. This helps identifying the parameters to be changed by the students and the meaning of variables and loops.

To execute the programs, the Current Folder of MATLAB™ must be set to the chapter under investigation.

The distribution contains MATLAB™ scripts rather than programs with a graphical user interface. This is a choice made in order to get students used to programming and automatic chaining of operations encountered in leading modeling centers. The programs are not designed to exploit and optimize MATLAB™ features but rather serve as illustrations of the numerical schemes and not programming languages. Hence, sometimes loops are spelled out in the programs and sometimes more efficient direct matrix operations are used. The MATLAB™ expert users certainly would like to modify some of the codes to replace loops of finite differencing over the domain by a sparse matrix multiplication, where the sparse matrix contains the discretisation constants. Also sometimes short programs are written for which MATLAB™ functions exist.

When the execution of programs to prepare animations is time consuming, the distribution contains some precalculated movies in Quicktime .mov format in folder animations.

For students not having access to MATLAB™, a freely available clone exists and is called OCTAVE (http://octave.sourceforge.net). Most operations used in the programs are portable between MATLAB™ and OCTAVE, only plotting parts might require adaptations.

Input data are stored predominantly in NETCDF format in folder nc. Files in NETCDF format are very common in the ocean and atmosphere modeling community because they are platform independent and self-explaining. For NETCDF support installations and testing, please consult the README.txt file.

Because the programs write out some results and images on disk as in real modeling centers, the user cannot run MATLAB™ from a CD but needs to execute programs in a folder with writing access.

Finally, the authors may be contacted for updated versions of the distribution content and additional explanations on the programs.

Elements of Fluid Mechanics

ABSTRACT

Basic principles of fluid mechanics are recalled and summarized. It is shown how budgets can be established on infinitesimal volumes. The distinction is made between Eulerian and Lagrangian approaches to fluid dynamics. For reference, a few equations and operators are expressed in cylindrical and spherical coordinates. Finally, the link between vorticity and rotation is outlined.

A.1 BUDGETS

Most physical principles of fluid mechanics can be cast as budgets of one quantity or another, with the simplest budget being the one for mass conservation. We begin here with the one-dimensional (1D) version, from which the 3D generalization is immediate.

For a 1D budget, we consider a very short (infinitesimal) segment of fluid, of length dx, along the x-axis of the system (Fig. A.1), for which we state that the mass within this segment at one moment, say time $t + dt$, is the mass that was there at a previous moment, say time t, augmented by the amount of inflow on the left, say at $x - dx$, minus the amount of outflow from the right, say x, during the elapsed time dt:

$$\underbrace{\rho(x, t+dt)dx}_{\text{mass at time } t+dt} = \underbrace{\rho(x,t)dx}_{\text{mass at time } t} + \underbrace{u(x-dx,t)dt\,\rho(x-dx,t)}_{\text{mass entering}} - \underbrace{u(x,t)dt\,\rho(x,t)}_{\text{mass exiting}}.$$

$$(A.1)$$

After division by the time interval dt and space interval dx, this budget can be recast as

$$\frac{\rho(x,t+dt) - \rho(x,t)}{dt} + \frac{\rho(x,t)u(x,t) - \rho(x-dx,t)u(x-dx,t)}{dx} = 0. \quad (A.2)$$

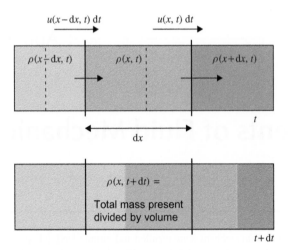

FIGURE A.1   One-dimensional mass conservation.

In the limit of vanishing d*t* and d*x*, the differences become derivatives, and the one-dimensional mass conservation equation is obtained:

$$\frac{\partial \rho}{\partial t} + \frac{\partial}{\partial x}(\rho u) = 0. \tag{A.3}$$

Note that we assumed such an infinitesimal limit exists, meaning that "infinitesimal" is extremely small compared with the scales of macroscopic properties, yet large compared with the size of the molecules constituting the fluid for which the budget is established. This is the essence of continuum mechanics.

Similarly, for a three-dimensional domain (Fig. A.2), the budget calculation yields:

$$\rho(x, y, z, t+dt)\,dx\,dy\,dz = \rho(x, y, z, t)\,dx\,dy\,dz$$
$$+ u(x-dx, y, z, t)\,dt\,dy\,dz\,\rho(x-dx, y, z, t) - u(x, y, z, t)\,dt\,dy\,dz\,\rho(x, y, z, t)$$
$$+ v(x, y-dy, z, t)\,dt\,dx\,dz\,\rho(x, y-dy, z, t) - v(x, y, z, t)\,dt\,dx\,dz\,\rho(x, y, z, t)$$
$$+ w(x, y, z-dz, t)\,dt\,dx\,dy\,\rho(x, y, z-dz, t) - w(x, y, z, t)\,dt\,dx\,dy\,\rho(x, y, z, t).$$

Division by the infinitesimal volume d*x*d*y*d*z* and time interval d*t* provides in the continuous limit:

$$\frac{\partial \rho}{\partial t} + \frac{\partial}{\partial x}(\rho u) + \frac{\partial}{\partial y}(\rho v) + \frac{\partial}{\partial z}(\rho w) = 0. \tag{A.4}$$

This is the mass conservation equation, also called the *continuity equation*.

Newton's second law of physics stating that mass times acceleration is equal to the sum of forces can likewise be cast as a budget, this time with

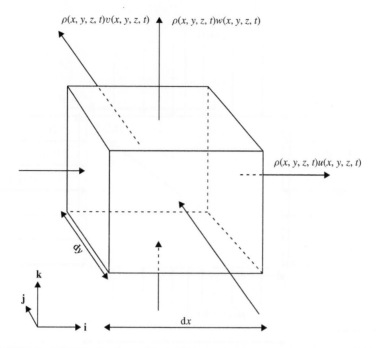

FIGURE A.2 Infinitesimal volume with mass inflow and outflow across boundaries for the three-dimensional mass budget.

momentum (mass times velocity, here per unit volume) being the quantity for which the budget is written and with forces (per unit volume) acting as sources. For clarity, the budget is established here in the two-dimensional case (Fig. A.3).

A suitable departure point is the mass budget equation (A.1) in which we replace density by the product of density with velocity. The sources are forces per volume. We also progress from one to two dimensions. Thus, we write

$$\rho u|_{\text{at } x,y,t+dt}\, dx\, dy = \rho u|_{\text{at } x,y,t}\, dx\, dy$$

$$+ \rho u u|_{\text{at } x-dx,y}\, dy\, dt - \rho u u|_{\text{at } x,y}\, dy\, dt$$

$$+ \rho u v|_{\text{at } x,y-dy}\, dx\, dt - \rho u v|_{\text{at } x,y}\, dx\, dt$$

$$+ \text{Sum of forces in the } x\text{-direction} \qquad (A.5a)$$

$$\rho v|_{\text{at } x,y,t+dt}\, dx\, dy = \rho v|_{\text{at } x,y,t}\, dx\, dy$$

$$+ \rho v u|_{\text{at } x-dx,y}\, dy\, dt - \rho v u|_{\text{at } x,y}\, dy\, dt$$

$$+ \rho v v|_{\text{at } x,y-dy}\, dx\, dt - \rho v v|_{\text{at } x,y}\, dx\, dt$$

$$+ \text{Sum of forces in the } y\text{-direction.} \qquad (A.5b)$$

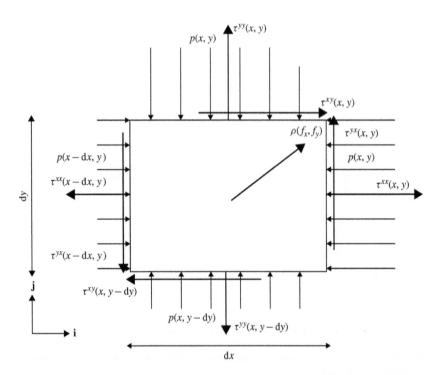

FIGURE A.3   Two-dimensional element subject to the typical forces encountered in a fluid flow: pressure p (a normal force per unit area), shear stress τ (a tangential force per unit area), and an internal (body) force $\rho(f_x, f_y)$, which is usually the gravitational force.

The forces applied to the fluid element are as follows:

$$
\begin{aligned}
\text{Sum of forces in the } x\text{-direction} = {}& p|_{\text{at } x-dx,y}\, dy - p|_{\text{at } x,y}\, dy \\
& - \tau^{xx}|_{\text{at } x-dx,y}\, dy + \tau^{xx}|_{\text{at } x,y}\, dy \\
& - \tau^{xy}|_{\text{at } x,y-dy}\, dx + \tau^{xy}|_{\text{at } x,y}\, dx \\
& + \rho f_x\, dx\, dy \qquad\qquad (A.6a)
\end{aligned}
$$

$$
\begin{aligned}
\text{Sum of forces in the } y\text{-direction} = {}& p|_{\text{at } x,y-dy}\, dx - p|_{\text{at } x,y}\, dx \\
& - \tau^{yx}|_{\text{at } x-dx,y}\, dy + \tau^{yx}|_{\text{at } x,y}\, dy \\
& - \tau^{yy}|_{\text{at } x,y-dy}\, dx + \tau^{yy}|_{\text{at } x,y}\, dx \\
& + \rho f_y\, dx\, dy. \qquad\qquad (A.6b)
\end{aligned}
$$

Here, the force (f_x, f_y) is the body force per unit mass, so that the product $\rho(f_x, f_y)$ is the body force per unit volume. Note that the stresses τ depend on the nature of the fluid and its flow, and that the tangential stresses τ^{xy} and τ^{yx} act in different directions (Fig. A.3) but must have the same strength. This equality

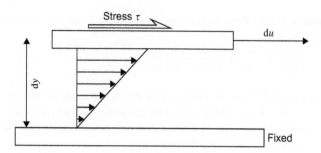

FIGURE A.4   Creeping flow with stress proportional to shear: $\tau \propto du/dy$.

of stresses, $\tau^{xy}(x, y) = \tau^{yx}(x, y)$, proceeds from the fact that, if this were not the case, the infinitesimal element would be subjected to an uncompensated torque.

A so-called constitutive equation must relate the stress components to the fluid flow, usually its velocity shear (see Fig. A.4).

With these forces and dividing by $dx\,dy$, we obtain the momentum budget in the x-direction.

$$\frac{\partial}{\partial t}(\rho u) + \frac{\partial}{\partial x}(\rho uu) + \frac{\partial}{\partial y}(\rho uv) +$$
$$= \rho f_x - \frac{\partial p}{\partial x} + \frac{\partial \tau^{xx}}{\partial x} + \frac{\partial \tau^{xy}}{\partial y}, \tag{A.7}$$

and similarly in the y-direction. Generalization to three dimensions is straightforward and leads to Eq. (3.2) with Eq. (3.3).

Note that in solid mechanics, Newton's second law is presented by following a given mass along its path (called the Lagrangian approach) rather than by performing a budget over a fixed part of space (called the Eulerian approach). Because the physical law is the same, we should be able to reach the same governing equations by either approach. To show that this is possible, we express the Eulerian derivative of a field $F(x, y, z, t)$, which may be any property of the fluid or flow field, as

$$\frac{\partial F}{\partial t} = \text{derivative of } F \text{ with respect to } t, \text{ at fixed } x, y, z. \tag{A.8}$$

In other words, this is the time change of F as perceived by an observer at a fixed location. In contrast, the Lagrangian approach considers the change moving with a fluid parcel, the position of which changes over time, $(x, y, z) = [x(t), y(t), z(t)]$. This time dependence of the coordinates describes the trajectory of the fluid parcel. The time change of F, taking the displacement over time into account, is the *total* time derivative of F:

$$\frac{dF}{dt} = \text{derivative of } F(x(t), y(t), z(t), t) \text{ with respect to } t. \tag{A.9}$$

This change of F for a fluid parcel is obtained by the chain rule of derivatives:

$$\frac{dF}{dt} = \frac{\partial F}{\partial x}\frac{dx}{dt} + \frac{\partial F}{\partial y}\frac{dy}{dt} + \frac{\partial F}{\partial z}\frac{dz}{dt} + \frac{\partial F}{\partial t}. \tag{A.10}$$

Because $[x(t), y(t), z(t)]$ is the trajectory of the fluid parcel, the change in position over time dx/dt is nothing else than the parcel velocity u, and similarly $dy/dt = v$ and $dz/dt = w$, so that we can express the Lagrangian derivative dF/dt, also called the *material derivative*, as

$$\frac{dF}{dt} = \frac{\partial F}{\partial t} + u\frac{\partial F}{\partial x} + v\frac{\partial F}{\partial y} + w\frac{\partial F}{\partial z}. \tag{A.11}$$

This relates the Lagrangian derivative to the Eulerian derivative, permitting a switch from one approach to the other. The difference between the two expressions, that is, the sum of terms with velocity components, is the *advection* contribution.

The passage from Eulerian to Lagrangian formulation also permits a manipulation of the mass-conservation equation (A.4) by using the material derivative (A.11):

$$\frac{1}{\mathrm{v}}\frac{d\mathrm{v}}{dt} = -\frac{1}{\rho}\frac{d\rho}{dt} = \frac{\partial u}{\partial x} + \frac{\partial v}{\partial y} + \frac{\partial w}{\partial z}, \tag{A.12}$$

with $\mathrm{v} = 1/\rho$ being the volume per unit mass. The expression $\partial u/\partial x + \partial v/\partial y + \partial w/\partial z$ is the divergence of the flow field. It is positive when the flow diverges and negative when it converges. It follows that a fluid volume is dilated (shrinking) and density drops (increases) when the flow diverges (converges).

In two dimensions, we can proceed one step further by relating the divergence to the area S containing the fluid element:

$$\frac{1}{S}\frac{dS}{dt} = \frac{\partial u}{\partial x} + \frac{\partial v}{\partial y}. \tag{A.13}$$

This last expression becomes useful in the study of vorticity. It is left as an exercise to the reader to formulate the momentum equation in a Lagrangian way and to interpret the resulting equation.

A.2 EQUATIONS IN CYLINDRICAL COORDINATES

The preceding equations assumed a rectangular (Cartesian) system of coordinates, but in geophysical fluid dynamics, we occasionally encounter circular structures, such as vortices, for which the use of cylindrical coordinates is more convenient. The three coordinates of space are then the radial distance r, the azimuthal angle θ (in radians), and the vertical coordinate z.

In cylindrical coordinates, the material derivative becomes

$$\frac{d}{dt} = \frac{\partial}{\partial t} + u\frac{\partial}{\partial r} + \frac{v}{r}\frac{\partial}{\partial\theta} + w\frac{\partial}{\partial z}. \tag{A.14}$$

In this notation, u is the radial velocity, v the azimuthal velocity (positive for a parcel turning in the trigonometric sense, increasing θ), and w the vertical velocity.

Mass conservation and horizontal components of the momentum equations are as follows:

$$\frac{\partial\rho}{\partial t} + \frac{1}{r}\frac{\partial}{\partial r}(\rho r u) + \frac{1}{r}\frac{\partial}{\partial\theta}(\rho v) + \frac{\partial}{\partial z}(\rho w) = 0 \tag{A.15a}$$

$$\rho\left(\frac{du}{dt} - \frac{v^2}{r} - fv + f_*w\right) = -\frac{\partial p}{\partial r} + F_r \tag{A.15b}$$

$$\rho\left(\frac{dv}{dt} + \frac{uv}{r} + fu\right) = -\frac{1}{r}\frac{\partial p}{\partial\theta} + F_\theta \tag{A.15c}$$

$$\rho\left(\frac{dw}{dt} - f_*u\right) = -\frac{\partial p}{\partial z} - \rho g + F_z \tag{A.15d}$$

where F_r, F_θ, and F_z are the stress terms. The Laplacian of a scalar field ψ reads

$$\nabla^2\psi = \frac{1}{r}\frac{\partial}{\partial r}\left(r\frac{\partial\psi}{\partial r}\right) + \frac{1}{r^2}\frac{\partial^2\psi}{\partial\theta^2} + \frac{\partial^2\psi}{\partial z^2}. \tag{A.16}$$

Polar coordinates are cylindrical coordinates in two dimensions, with the z dependence dropped.

A.3 EQUATIONS IN SPHERICAL COORDINATES

When the dimension of the domain is comparable to the earth's radius, and especially when the entire globe is the domain, spherical coordinates are preferred. The three coordinates of space are then the radial distance r from the center of the earth (which is often cropped to z along the local vertical and measured form the mean sea level), longitude λ, and latitude[1] φ (both expressed in radians rather than degrees). The material derivative becomes

$$\frac{d}{dt} = \frac{\partial}{\partial t} + \frac{u}{r\cos\varphi}\frac{\partial}{\partial\lambda} + \frac{v}{r}\frac{\partial}{\partial\varphi} + w\frac{\partial}{\partial r}. \tag{A.17}$$

---

[1] Contrary to classical spherical coordinates, we do not use the polar angle but latitude.

Equations (3.1) through (3.3) become:

$$\frac{\partial}{\partial t}(\rho\cos\varphi)+\frac{\partial}{\partial\lambda}\left(\frac{\rho u}{r}\right)+\frac{\partial}{\partial\varphi}\left(\frac{\rho v\cos\varphi}{r}\right)+\frac{1}{r^2}\frac{\partial}{\partial r}\left(r^2\rho w\cos\varphi\right)=0 \quad \text{(A.18a)}$$

$$\rho\left(\frac{du}{dt}-\frac{uv\tan\varphi}{r}+\frac{uw}{r}-fv+f_*w\right)=-\frac{1}{r\cos\varphi}\frac{\partial p}{\partial\lambda}+F_\lambda \qquad \text{(A.18b)}$$

$$\rho\left(\frac{dv}{dt}+\frac{u^2\tan\varphi}{r}+\frac{vw}{r}+fu\right)=-\frac{1}{r}\frac{\partial p}{\partial\varphi}+F_\varphi \qquad \text{(A.18c)}$$

$$\rho\left(\frac{dw}{dt}-\frac{u^2+v^2}{r}-f_*u\right)=-\frac{\partial p}{\partial r}-\rho g+F_r, \qquad \text{(A.18d)}$$

in which $f=2\Omega\sin\varphi$ and $f_*=2\Omega\cos\varphi$. The components F_λ, F_φ, and F_r of the frictional force have complicated expressions and need not be reproduced here. For a detailed development of these equations, the reader is referred to Chapter 4 of the book by Gill (1982). The Laplacian of a scalar field ψ reads

$$\nabla^2\psi=\frac{1}{r^2\cos\varphi^2}\frac{\partial^2\psi}{\partial\lambda^2}+\frac{1}{r^2\cos\varphi}\frac{\partial}{\partial\varphi}\left(\cos\varphi\frac{\partial\psi}{\partial\varphi}\right)+\frac{1}{r^2}\frac{\partial}{\partial r}\left(r^2\frac{\partial\psi}{\partial r}\right). \quad \text{(A.19)}$$

It is worth noting that since the radius of the earth is much longer than the thickness of either atmosphere or ocean, some vertical derivatives may be approximated as

$$\frac{1}{r^2}\frac{\partial(r^2a)}{\partial r}\simeq\frac{\partial a}{\partial z} \qquad \text{(A.20a)}$$

$$\frac{1}{r^2}\frac{\partial}{\partial r}\left(r^2\frac{\partial a}{\partial r}\right)\simeq\frac{\partial^2 a}{\partial z^2}. \qquad \text{(A.20b)}$$

A.4 VORTICITY AND ROTATION

Vorticity, as it name indicates, quantifies the rotation rate of a fluid parcel. Because rotation is also defined by an axis around which the spin occurs, vorticity ought to be a vector. For simplicity, however, we start by considering the case of a flow in the horizontal plane, so that rotation takes place around the vertical axis, and the vorticity vector is directed along this axis. Only its intensity matters, which is defined as

$$\zeta=\frac{\partial v}{\partial x}-\frac{\partial u}{\partial y}. \qquad \text{(A.21)}$$

First let us consider a flow in solid-body rotation around the origin of the axes (left part of Fig. A.5). The flow field is then $(u=-\Omega y, v=+\Omega x)$, and the vorticity defined by Eq. (A.21) is $\zeta=2\Omega$, twice the rotation rate of the flow, and except for the factor 2, this seems intuitive.

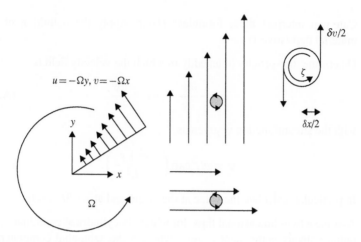

FIGURE A.5   Vorticity and rotation.

Not only flows with curved trajectories have vorticity; rectilinear shear flows, too, possess vorticity, as depicted in the middle of Fig. A.5. Take for example the flow $v(x)$ in which fluid parcels located at different x positions travel at different velocities in the y-direction, some overtaking others in a slipping movement. A stick placed across the flow would see one tip proceeding faster than the other and would effectively be rotated by the flow. This rotation is expressed mathematically by the vorticity:

$$\zeta = \frac{dv}{dx}.$$ (A.22)

The sign of vorticity is such that it is positive for rotation in the trigonometric (counterclockwise) sense seen downward along the vertical axis.

In three-dimensions, vorticity is the curl of the vector velocity, and its three components are as follows:

$$\zeta_x = \frac{\partial w}{\partial y} - \frac{\partial v}{\partial z}$$ (A.23)

$$\zeta_y = \frac{\partial u}{\partial z} - \frac{\partial w}{\partial x}$$ (A.24)

$$\zeta_z = \frac{\partial v}{\partial x} - \frac{\partial u}{\partial y}.$$ (A.25)

ANALYTICAL PROBLEMS

A.1. Verify that the velocity components in cylindrical coordinates are

$$u = \frac{dr}{dt}, \quad v = r\frac{d\theta}{dt}, \quad w = \frac{dz}{dt}.$$ (A.26)

Can you interpret these formulas? (*Hint*: Apply the definition of the material derivative.)

A.2. Determine the vorticity of an eddy in which the velocity field is

$$u = -\frac{\partial \psi}{\partial y}, \quad v = +\frac{\partial \psi}{\partial x} \tag{A.27}$$

with the streamfunction ψ given as

$$\psi = \omega L^2 \exp\left(-\frac{x^2 + y^2}{L^2}\right). \tag{A.28}$$

In particular, calculate the value at the origin and at $x = 3L, y = 0$.

A.3. Assume a two-dimensional flow, for which, in cylindrical coordinates, the radial velocity component is zero, whereas the azimuthal component is only depending on r:

$$v = v(r). \tag{A.29}$$

Calculate the circulation[2] around a circle of radius R, centered at the origin. Relate the result to the vorticity distribution within the surface delimited by the circle. (*Hint*: Show that vorticity is $\zeta_z = (1/r)(d/dr)(rv)$.)

A.4. Knowing that the divergence of a flow is the relative change of density over time, can you derive the expression of the divergence operator in cylindrical and spherical coordinates? (*Hint*: Look at the mass-conservation equation.)

NUMERICAL EXERCISE

A.1. Plot the velocity and vorticity fields of Analytical Problem A.2.

---

[2] Circulation is tangential velocity integrated along the chosen path, here simply the product of the azimuthal velocity by the circumference of the circle.

Wave Kinematics

ABSTRACT

Because numerous geophysical flow phenomena can be interpreted as waves, some understanding of basic wave properties is required in the study of geophysical fluid dynamics. The concepts of wavenumber, frequency, dispersion relation, phase speed and group velocity are introduced and given physical interpretations.

B.1 WAVENUMBER AND WAVELENGTH

For simplicity of presentation and easier graphical representation, we will consider here a two-dimensional plane wave, namely, a physical signal occupying the (x, y) plane, evolving with time t and with straight crest lines. The prototypical wave form is the sinusoidal function, and so we assume that a physical variable of the system, denoted by a and being pressure, a velocity component or whatever, evolves according to

$$a = A\cos(k_x x + k_y y - \omega t + \phi). \tag{B.1}$$

The coefficient A is the wave *amplitude* $(-A \leq a \leq +A)$, whereas the argument

$$\alpha = k_x x + k_y y - \omega t + \phi \tag{B.2}$$

is called the *phase*. The latter consists of terms that vary with each independent variable and a constant ϕ, called the reference phase. The coefficients k_x, k_y, and ω of x, y, and t, respectively, bear the names of *wavenumber in x*, *wavenumber in y*, and *angular frequency*, most often abbreviated to *frequency*. They indicate how rapidly the wave undulates in space and how fast it oscillates in time.

Equivalent expressions for the wave signal are

$$a = A_1 \cos(k_x x + k_y y - \omega t) + A_2 \sin(k_x x + k_y y - \omega t), \tag{B.3}$$

where $A_1 = A \cos\phi$ and $A_2 = -A \sin\phi$, and

$$a = \Re\left[A_c e^{i\,(k_x x + k_y y - \omega t)}\right], \tag{B.4}$$

where the notation $\Re[\ ]$ means "the real part of" and $A_c = A_1 - iA_2 = A\,e^{i\phi}$ is a complex amplitude coefficient. The choice of mathematical representation is generally dictated by the problem at hand. Formulation (B.3) is helpful in the discussion of problems exhibiting coexisting signals that are either in perfect phase or in quadrature, whereas formulation (B.4) is preferred when a given dynamical system is subjected to a wave analysis. Here, we will use formulation (B.1).

A wave *crest* is defined as the line in the (x, y) plane and at time t along which the signal is maximum $(a = +A)$; similarly, a *trough* is a line along which the signal is minimum $(a = -A)$. These lines and, in general, all lines along which the signal has a constant value at an instant in time are called *phase lines*. In a plane wave, as the one considered here, all crests, troughs, and other phase lines are straight lines. Figure B.1 depicts a few phase lines in the case of positive wavenumbers k_x and k_y.

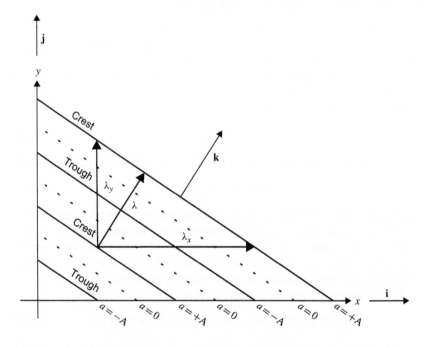

FIGURE B.1 Instantaneous phase lines of a plane two-dimensional wave signal. The lines are straight and parallel. The distances from crest to nearest crest along the x- and y-axes are λ_x and λ_y, respectively, whereas the wavelength λ is the shortest diagonal distance from crest line to nearest crest line.

Because of the oscillatory behavior of the sinusoidal function, crest lines recur at constant intervals, thus giving the wavy aspect to the signal. The distance over which the signal repeats itself in the x-direction is the distance over which the phase portion $k_x x$ increases by 2π, that is,

$$\lambda_x = \frac{2\pi}{k_x}. \tag{B.5}$$

Similarly, the distance over which the signal repeats itself in the y-direction is

$$\lambda_y = \frac{2\pi}{k_y}. \tag{B.6}$$

The quantities λ_x and λ_y are called the wavelengths in the x- and y-directions. They are the wavelengths seen by an observer who would detect the signal only through slits aligned with the x- and y-axes. The actual *wavelength*, λ, of the wave is the shortest distance from the crest to nearest crest (Fig. B.1) and is, therefore, smaller than both λ_x and λ_y. Elementary geometric considerations provide

$$\frac{1}{\lambda^2} = \frac{1}{\lambda_x^2} + \frac{1}{\lambda_y^2} = \frac{k_x^2 + k_y^2}{4\pi^2}$$

or

$$\lambda = \frac{2\pi}{k}, \tag{B.7}$$

where k, called the *wavenumber*, is defined as

$$k = \sqrt{k_x^2 + k_y^2}. \tag{B.8}$$

Note that since λ^2 is not the sum of λ_x^2 and λ_y^2, the pair (λ_x, λ_y) does not make a vector. However, the pair (k_x, k_y) can be used to define the *wavenumber vector*

$$\mathbf{k} = k_x \mathbf{i} + k_y \mathbf{j}, \tag{B.9}$$

where $\mathbf{i}$ and $\mathbf{j}$ are the unit vectors aligned with the axes (Fig. B.1). In this fashion, the wavenumber k is the magnitude of the wavenumber vector $\mathbf{k}$.

By definition, phase lines at any given time correspond to lines of constant values of the expression $k_x x + k_y y = \mathbf{k} \cdot \mathbf{r}$, where $\mathbf{r} = x\mathbf{i} + y\mathbf{j}$ is the vector position. Geometrically, this means that a phase line is the locus of points whose vectors from the origin share the same projection onto the wavenumber vector. These points form a straight line perpendicular to $\mathbf{k}$, and therefore the wavenumber vector points perpendicularly to all phase lines (Fig. B.1), that is, in the direction of the undulation.

B.2 FREQUENCY, PHASE SPEED, AND DISPERSION

Let us now turn our attention to the temporal evolution of the wave signal. At a fixed position (x and y given), an observer sees an oscillatory signal. The interval of time between two consecutive instants at which the signal is maximum is the time taken for the portion ωt of the phase to increase by 2π. It is called the *period*, which is

$$T = \frac{2\pi}{\omega}. \tag{B.10}$$

Let us now follow a particular crest line ($a = A$) from a certain time t_1 to a later time t_2 and note the time interval $\Delta t = t_2 - t_1$. During this time interval, the wave crest has progressed from one position to another (Fig. B.2). The intersection with the x-axis has translated over the distance $\Delta x = \omega t_2 / k_x - \omega t_1 / k_x = \omega \Delta t / k_x$ in time Δt. This defines the propagation speed of the wave along the x-direction:

$$c_x = \frac{\Delta x}{\Delta t} = \frac{\omega}{k_x}. \tag{B.11}$$

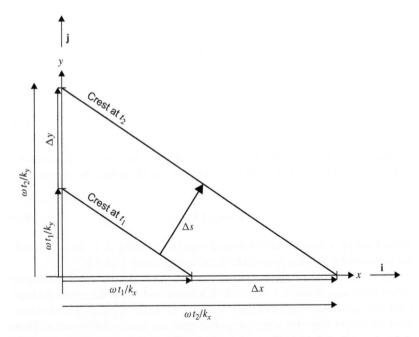

FIGURE B.2 Progress of a wave crest from time t_1 to time t_2. The ratio of the distance traveled, Δs, to the time interval $\Delta t = t_2 - t_1$ is the phase speed.

Similarly, the propagation speed along the y-direction is the distance $\Delta y = \omega t_2/k_y - \omega t_1/k_y$ divided by the time interval Δt, or

$$c_y = \frac{\Delta y}{\Delta t} = \frac{\omega}{k_y}. \tag{B.12}$$

But, these speeds are only speeds in particular directions. The true propagation speed of the wave is the distance Δs, measured perpendicularly to the crest line (Fig. B.2), covered by this crest line during the time interval Δt. Again, elementary geometric considerations provide

$$\frac{1}{\Delta s^2} = \frac{1}{\Delta x^2} + \frac{1}{\Delta y^2},$$

from which we deduce

$$\Delta s = \frac{\omega \Delta t}{k},$$

where k is the wavenumber defined in Eq. (B.8). The propagation speed of the crest line is thus

$$c = \frac{\Delta s}{\Delta t} = \frac{\omega}{k}. \tag{B.13}$$

Because all phase lines propagate at the same speed (so that the wave preserves its sinusoidal form over time), the quantity c is called the *phase speed*. Note that because c^2 is not equal to $c_x^2 + c_y^2$ (in fact, c is less than both c_x and c_y!), the pair (c_x, c_y) does not constitute a physical vector. The direction of phase propagation, as discussed before, is parallel to the wavenumber vector $\mathbf{k}$.

In general, the expression (B.1) of the wave signal appears as the solution to a particular dynamical system. Therefore, it must somehow be constrained by the physics of the problem, and not all its parameters can be varied independently. Let us suppose that the system under consideration is initially unchanging in time (state of rest or steady flow) and that at time $t = 0$, it is perturbed spatially according to a sinusoidal distribution of wavenumbers k_x and k_y in the x- and y-directions, respectively, and of amplitude A for the variable a. Intuition leads us to anticipate that subsequent to this perturbation, the system will react in a time-dependent fashion. If this reaction takes the form of a wave, it will have a frequency ω determined by the system. Therefore, the frequency can be viewed as dependent upon the wavenumber components k_x and k_y and the amplitude A. In most instances, the system's response is a wave because the set of equations representing the physics is linear, and when this is the case, the mathematical analysis yields a frequency that is independent of the amplitude of the perturbation. Therefore, ω is typically a function of k_x and k_y only.

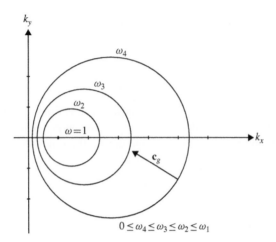

FIGURE B.3 Graphic representation of the dispersion relation $\omega = 2k_x/(k_x^2 + k_y^2 + 1)$ by curves of constant ω values in the (k_x, k_y) wavenumber plane.

If the frequency is a function of the wavenumber components, so is the phase speed:

$$c = \frac{\omega(k_x, k_y)}{\sqrt{k_x^2 + k_y^2}} = c(k_x, k_y).$$

Physically, this implies that the various waves of a composite signal will all travel at different speeds, causing a distortion of the signal over time. In particular, a localized burst of activity, which by virtue of the Fourier-decomposition theorem contains waves of many different wavelengths, will be progressively less localized as time goes on. This phenomenon is called *dispersion*, and the mathematical function that relates the frequency ω to the wavenumber components k_x and k_y bears the name of *dispersion relation*.

The dispersion relation can be represented, in two dimensions, as a set of curves in the (k_x, k_y) plane along which ω is a constant. Figure B.3 provides an example. At one dimension ($k_x = k$, $k_y = 0$) or at two dimensions when the physical system is isotropic (ω function of k only), a single ω-versus-k curve suffices to represent the dispersion relation.

In some special physical systems, the dispersion relation reduces to a single proportionality between frequency ω and wavenumber k. The phase speed is then the same for all wavenumbers, all waves travel in perfect accord, and the total signal retains its shape as time evolves. Such a wave is called a *nondispersive* wave.

B.3 GROUP VELOCITY AND ENERGY PROPAGATION

In general, a wave pattern consists of more than a single wave. A series of waves are superimposed, leading to constructive and destructive interference.

In areas where the waves are interfering constructively, the wave amplitude is greater, and the energy level is higher than in areas where the waves interfere destructively. Therefore, energy distribution is a property of a set of waves rather than of a single wave. (It can be said that a single wave has a uniform energy distribution.) Energy propagation by a set of waves depends on how interference patterns move about and is generally not the average speed of the waves present. To illustrate the principles and determine the speed at which energy propagates, let us restrict our attention to two one-dimensional waves and, more precisely, to two waves of equal amplitude and nearly equal wavenumbers:

$$a = A\cos(k_1 x - \omega_1 t) + A\cos(k_2 x - \omega_2 t), \tag{B.14}$$

where the wavenumbers k_1 and k_2 are close to their average $k = (k_1 + k_2)/2$, and the difference $\Delta k = k_1 - k_2$ is much smaller ($|\Delta k| \ll |k|$). Because both waves obey the single dispersion relation of the dynamical system, $\omega = \omega(k)$, the two frequencies $\omega_1 = \omega(k_1)$ and $\omega_2 = \omega(k_2)$ are close to their average $\omega = (\omega_1 + \omega_2)/2$, which is much larger than their difference $\Delta\omega = \omega_1 - \omega_2$ ($|\Delta\omega| \ll |\omega|$). In expression (B.14), the two reference phases were set to zero, which can always be done under suitable choices of space and time origins.

A trigonometric manipulation transforms expression (B.14) into

$$a = 2A\cos\left(\frac{\Delta k}{2}x - \frac{\Delta\omega}{2}t\right)\cos(kx - \omega t), \tag{B.15}$$

which now appears as the product of two waves. The second cosine function represents an average wave, of wavenumber and frequency between those of the two individual waves comprising the signal. The first cosine function, however, involves a much smaller wavenumber (i.e., much longer wavelength) and a much lower frequency. Over the cycle of the shorter (k, ω) wave, the longer wave appears almost unchanging. In other words, the (k, ω) wave appears modulated; its amplitude, $2A\cos[(\Delta k x - \Delta\omega t)/2]$, is slowly varying in space and time, as seen in Fig. B.4.

Although the wave signal exhibits a wavelength from crest to trough to the next crest equal to $\lambda = 2\pi/k$, the envelope has a much longer wavelength, $\lambda' = \frac{1}{2}[2\pi/(\Delta k/2)] = 2\pi/\Delta k$. The wave pattern is a succession of wave bursts, each of length λ'. Within each burst, the wave propagates at the phase speed $c = \omega/k$, while the burst travels at the speed $c' = \Delta\omega/\Delta k$.

Considering an infinitesimal wavenumber difference, we are led to define

$$c_g = \frac{d\omega}{dk}. \tag{B.16}$$

Because this is the propagation speed of a burst, or group of similar waves, it is called the *group velocity*. Energy is associated with each group, and so the group velocity is also the velocity at which energy is carried by the waves.

The preceding wave description relies on the existence of two waves of identical amplitudes. When two waves do not have equal amplitude, say A_1

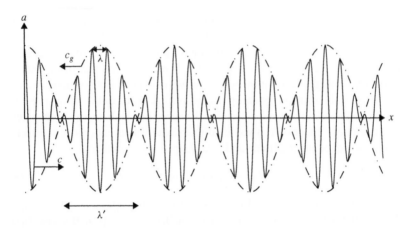

FIGURE B.4 The interference pattern of two one-dimensional waves with close wavenumbers. While the wave crests and troughs propagate at the speed $c = \omega/k$, the envelope (dashed line) propagates at the group velocity $c_g = d\omega/dk$.

and A_2, destructive interference is nowhere complete (the weak wave cannot completely cancel the strong wave), and there is no clear pinch-off between wave bursts. Rather, the modulating envelope undulates between the values $A_1 + A_2$ and $|A_1 - A_2|$ on the positive side and $-(A_1 + A_2)$ and $-|A_1 - A_2|$ on the negative side. It remains, however, that regions of constructive interference, and thus of higher energy level, propagate at the group velocity.

The theory can easily be extended to multidimensional waves. At two dimensions, for example, we define the group velocities in the x- and y-directions, respectively, as

$$c_{gx} = \frac{\partial \omega}{\partial k_x}, \quad c_{gy} = \frac{\partial \omega}{\partial k_y}, \qquad (B.17)$$

given the dispersion relation $\omega(k_x, k_y)$. Because these expressions are the components of the gradient of the function ω in the (k_x, k_y) wavenumber space, they can be interpreted as the components of a physical vector depicting the group velocity

$$\mathbf{c}_g = \nabla_k \omega, \qquad (B.18)$$

where ∇_k stands for the gradient operator with respect to the variables k_x and k_y. On the two-dimensional diagram (Fig. B.3), this vector group velocity points perpendicularly to the ω curves, toward the higher values of ω. Aligning the k_x- and k_y-axes with the x- and y-axes of the plane provides the direction of energy propagation in space.

Generalization to the three-dimensional space is immediate. An example is the internal wave discussed extensively in Chapter 13.

ANALYTICAL PROBLEMS

B.1. In waters deeper than half the wavelength, surface gravity waves obey the dispersion relation $\omega = \sqrt{gk}$, where g is the gravitational acceleration ($g = 9.81$ m/s^2). For these waves, show that the wavelength is proportional to the square of the period. At which speed does a 10-m-long wave travel?

B.2. Show that the group velocity of deep-water waves (see Problem B.1) is always less than the phase speed.

B.3. A former sea captain recounts a stormy night in the middle of the North Atlantic when he observed waves with wavelengths of a few meters passing his 51-m-long ship in less than 3 s. Should you believe him?

B.4. Suppose that you are in the middle of the ocean and off in the distance you see a storm. A little while later, you observe the passage of surface gravity waves of wavelength 5 m. Two hours later, you still observe gravity waves, but now their wavelength is 2 m. How far away was the storm?

B.5. Find the frequency ω of a Kelvin wave of wavenumber k (Section 9.2). Is the Kelvin wave dispersive?

B.6. Show that for inertia-gravity waves [$\omega^2 = f^2 + gH(k_x^2 + k_y^2)$; Section 9.3], the group velocity is always less than the phase speed. In which limit does the group velocity approach the phase speed?

B.7. Demonstrate that when the frequency ω is a function of the ratio k_x/k_y, the energy propagates perpendicularly to the phase.

B.8. Given the dispersion relation of internal waves in a vertical plane (see Section 13.2),

$$\omega = N \frac{k_x}{\sqrt{k_x^2 + k_z^2}},$$

where N is a constant, k_x is the horizontal wavenumber and k_z is the vertical wavenumber, show that phase and energy always propagate in the same horizontal direction but in opposite vertical directions.

NUMERICAL EXERCISES

B.1. Using animated graphics, display a time sequence ($t = 0$ to 10π by steps of $\pi/4$) of the double wave

$$a(x, t) = A_1 \cos(k_1 x - \omega_1 t) + A_2 \cos(k_2 x - \omega_2 t)$$

with $A_1 = A_2 = 1$, $k_1 = 1.9$, $k_2 = 2.1$, $\omega_1 = 2.1$, $\omega_2 = 1.9$, and for x ranging from 0 to 100. A suggested step in x is 0.25. Notice how the short waves [of wavelength $= 4\pi/(k_1 + k_2) = \pi$] travel toward increasing x

at the speed $c = (\omega_1 + \omega_2)/(k_1 + k_2) = +1$, while the wave envelope [of wavelength $= 2\pi/(k_2 - k_1) = 10\pi$] travels in the opposite direction at speed $c_g = (\omega_1 - \omega_2)/(k_1 - k_2) = -1$. This unequivocally demonstrates the nonintuitive fact that the energy propagation may well propagate in the direction opposite to the advancing crests and troughs. In other words, it is not impossible for energy to be transported up-wave.

Variations of this exercise can include uneven amplitudes (e.g., $A_1 = 1$ and $A_2 = 0.5$) and modified values for the wavenumbers and frequencies.

B.2. Using animated graphics, use the same function as in exercise B.1 with $k_1 = 0.35, k_2 = 0.5, \omega_1 = 0.5$, and $\omega_2 = 0.35$ the other values unchanged. Show the evolution of a and then of $a^2/2$. Can you explain the apparently shorter waves?

B.3. Given a dispersion relation

$$\omega = \frac{k}{(k^2 + 1)},$$

analyze the signal composed of two waves

$$a(x, t) = A_1 \cos(k_1 x - \omega(k_1)t) + A_2 \cos(k_2 x - \omega(k_2)t),$$

where ω is calculated using the dispersion relation. As before, show the evolution for $A_1 = A_2 = 1$ in the following situations:

- $k_1 = k_2 = 0.5$,
- $k_1 = k_2 = 2$
- $k_1 = 1.95, k_2 = 2.05$
- $k_1 = 0.45, k_2 = 0.55$

Can you explain the behavior? (*Hint*: Plot the dispersion relation.)

B.4. Redo exercise B.1 with $k_1 = 1$, $A_1 = 1$, $A_2 = 0$, and $\omega_1 = 1$. Then change the step in x to $\pi/4$ and $\pi/2$. Finally when using a step of $4\pi/3$, what do you observe?

Recapitulation of Numerical Schemes

ABSTRACT

This appendix gathers the most common numerical schemes for comparison and in order to facilitate their implementation.

C.1 THE TRIDIAGONAL SYSTEM SOLVER

As a special case of the general **LU** decomposition (e.g., Riley, Hobson & Bence, 1977), an efficient tridiagonal system solver, based on the so-called Thomas algorithm, can be constructed. We begin by assuming that there exists a decomposition for which the lower (**L**) and upper (**U**) matrices possess the same bandwidth of 2, that is, nonzero elements exist only along two diagonals:

$$
\begin{pmatrix}
a_1 & c_1 & 0 & 0 & \cdots & 0 \\
b_2 & a_2 & c_2 & 0 & \cdots & 0 \\
0 & b_3 & a_3 & c_3 & & 0 \\
\vdots & \vdots & \ddots & \ddots & \ddots & \vdots \\
0 & 0 & \cdots & b_{m-1} & a_{m-1} & c_{m-1} \\
0 & 0 & \cdots & 0 & b_m & a_m
\end{pmatrix}
=
\begin{pmatrix}
1 & 0 & 0 & 0 & \cdots & 0 \\
\beta_2 & 1 & 0 & 0 & \cdots & 0 \\
0 & \beta_3 & 1 & 0 & \cdots & 0 \\
\vdots & \vdots & \ddots & \ddots & \ddots & \vdots \\
0 & 0 & \cdots & \beta_{m-1} & 1 & 0 \\
0 & 0 & \cdots & 0 & \beta_m & 1
\end{pmatrix}
$$

$$
\times
\begin{pmatrix}
\alpha_1 & \gamma_1 & 0 & 0 & \cdots & 0 \\
0 & \alpha_2 & \gamma_2 & 0 & \cdots & 0 \\
0 & 0 & \alpha_3 & \gamma_3 & \cdots & 0 \\
\vdots & \vdots & \ddots & \ddots & \ddots & \vdots \\
0 & 0 & \cdots & 0 & \alpha_{m-1} & \gamma_{m-1} \\
0 & 0 & \cdots & 0 & 0 & \alpha_m
\end{pmatrix}
\tag{C.1}
$$

where the first matrix is the original tridiagonal matrix to be decomposed (i.e., elements a_1 etc. are known), the second matrix is **L** with one line of nonzero elements below the diagonal, and the last matrix is **U** with one line of nonzero elements above the diagonal.

Introduction to Geophysical Fluid Dynamics

Performing the product of matrices, we identify elements $(k, k-1)$, (k, k), and $(k, k+1)$ of the product as

$$b_k = \beta_k \alpha_{k-1} \tag{C.2a}$$

$$a_k = \beta_k \gamma_{k-1} + \alpha_k \tag{C.2b}$$

$$c_k = \gamma_k. \tag{C.2c}$$

These relations can be solved for the components of **L** and **U** by observing that $\gamma_k = c_k$ for all k. Since the first row demands $a_1 = \alpha_1$, subsequent rows provide α_k and β_k recursively from

$$\beta_k = \frac{b_k}{\alpha_{k-1}}, \quad \alpha_k = a_k - \beta_k c_{k-1}, \quad k = 2, \ldots, m \tag{C.3}$$

provided that no α_k is zero, otherwise the matrix is singular and cannot be decomposed. Note that there is no β_1.

The tridiagonal matrix **A** has now been decomposed into the product of lower and upper triangular matrices. The solution of $\mathbf{Ax} = \mathbf{LUx} = \mathbf{f}$ is then obtained by first solving $\mathbf{Ly} = \mathbf{f}$

$$
\begin{pmatrix}
1 & 0 & 0 & 0 & \cdots & 0 \\
\beta_2 & 1 & 0 & 0 & \cdots & 0 \\
0 & \beta_3 & 1 & 0 & \cdots & 0 \\
\vdots & \vdots & \ddots & \ddots & \ddots & \vdots \\
0 & 0 & \cdots & \beta_{m-1} & 1 & 0 \\
0 & 0 & \cdots & 0 & \beta_m & 1
\end{pmatrix}
\begin{pmatrix}
y_1 \\ y_2 \\ y_3 \\ \vdots \\ y_{m-1} \\ y_m
\end{pmatrix}
=
\begin{pmatrix}
f_1 \\ f_2 \\ f_3 \\ \vdots \\ f_{m-1} \\ f_m
\end{pmatrix}
\tag{C.4}
$$

by proceeding from the first row downward and then solving $\mathbf{Ux} = \mathbf{y}$,

$$
\begin{pmatrix}
\alpha_1 & \gamma_1 & 0 & 0 & \cdots & 0 \\
0 & \alpha_2 & \gamma_2 & 0 & \cdots & 0 \\
0 & 0 & \alpha_3 & \gamma_3 & \cdots & 0 \\
\vdots & \vdots & \ddots & \ddots & \ddots & \vdots \\
0 & 0 & \cdots & 0 & \alpha_{m-1} & \gamma_{m-1} \\
0 & 0 & \cdots & 0 & 0 & \alpha_m
\end{pmatrix}
\begin{pmatrix}
x_1 \\ x_2 \\ x_3 \\ \vdots \\ x_{m-1} \\ x_m
\end{pmatrix}
=
\begin{pmatrix}
y_1 \\ y_2 \\ y_3 \\ \vdots \\ y_{m-1} \\ y_m
\end{pmatrix},
\tag{C.5}
$$

by proceeding from the bottom row upward. The solution is

$$y_1 = f_1, \quad y_k = f_k - \beta_k y_{k-1}, \quad k = 2, \ldots, m \tag{C.6}$$

$$x_m = \frac{y_m}{\alpha_m}, \quad x_k = \frac{y_k - \gamma_k x_{k+1}}{\alpha_k}, \quad k = m-1, \ldots, 1. \tag{C.7}$$

In practice, the α values can be stored in vector **a** initially holding the values a_k, since once α_k is known, a_k is no longer needed. Similarly, β values can be stored in vector **b** initially holding b_k and γ values in a vector **c**. Also the y and f values can share the same vector $\mathbf{f}$ as the f_k value is no longer needed once the y_k value has been computed. With the additional vector $\mathbf{x}$ for the solution, only five vectors are required, and the solution is obtained with only three loops over m points. This demands approximately $5m$ floating-point operations instead of the m^3 operations that a brutal matrix inversion would have required. The algorithm is implemented in MATLAB™ file thomas.m.

C.2 1D FINITE-DIFFERENCE SCHEMES OF VARIOUS ORDERS

TABLE C.1 Standard Finite-difference Operators for Uniform Grids

Forward Difference $\mathcal{O}(\Delta t)$

	u^n	u^{n+1}	u^{n+2}	u^{n+3}	u^{n+4}
$\Delta t \frac{\partial u}{\partial t}$	-1	1			
$\Delta t^2 \frac{\partial^2 u}{\partial t^2}$	1	-2	1		
$\Delta t^3 \frac{\partial^3 u}{\partial t^3}$	-1	3	-3	1	
$\Delta t^4 \frac{\partial^4 u}{\partial t^4}$	1	-4	6	-4	1

Forward Difference $\mathcal{O}(\Delta t^2)$

	u^n	u^{n+1}	u^{n+2}	u^{n+3}	u^{n+4}	u^{n+5}
$2\Delta t \frac{\partial u}{\partial t}$	-3	4	-1			
$\Delta t^2 \frac{\partial^2 u}{\partial t^2}$	2	-5	4	-1		
$2\Delta t^3 \frac{\partial^3 u}{\partial t^3}$	-5	18	-24	14	-3	
$\Delta t^4 \frac{\partial^4 u}{\partial t^4}$	3	-14	26	-24	11	-2

Backward Difference $\mathcal{O}(\Delta t)$

	u^{n-4}	u^{n-3}	u^{n-2}	u^{n-1}	u^n
$\Delta t \frac{\partial u}{\partial t}$				-1	1
$\Delta t^2 \frac{\partial^2 u}{\partial t^2}$			1	-2	1
$\Delta t^3 \frac{\partial^3 u}{\partial t^3}$		-1	3	-3	1
$\Delta t^4 \frac{\partial^4 u}{\partial t^4}$	1	-4	6	-4	1

Backward Difference $\mathcal{O}(\Delta t^2)$

	u^{n-5}	u^{n-4}	u^{n-3}	u^{n-2}	u^{n-1}	u^n
$2\Delta t \frac{\partial u}{\partial t}$				1	-4	3
$\Delta t^2 \frac{\partial^2 u}{\partial t^2}$			-1	4	-5	2
$2\Delta t^3 \frac{\partial^3 u}{\partial t^3}$		3	-14	24	-18	5
$\Delta t^4 \frac{\partial^4 u}{\partial t^4}$	-2	11	-24	26	-14	3

Central Difference $\mathcal{O}(\Delta t^2)$

	u^{n-2}	u^{n-1}	u^n	u^{n+1}	u^{n+2}
$2\Delta t \frac{\partial u}{\partial t}$		-1	0	1	
$\Delta t^2 \frac{\partial^2 u}{\partial t^2}$		1	-2	1	
$2\Delta t^3 \frac{\partial^3 u}{\partial t^3}$	-1	2	0	2	1
$\Delta t^4 \frac{\partial^4 u}{\partial t^4}$	1	-4	6	-4	1

Central Difference $\mathcal{O}(\Delta t^4)$

	u^{n-3}	u^{n-2}	u^{n-1}	u^n	u^{n+1}	u^{n+2}	u^{n+3}
$12\Delta t \frac{\partial u}{\partial t}$		1	-8	0	8	-1	
$12\Delta t^2 \frac{\partial^2 u}{\partial t^2}$		-1	16	-30	16	-1	
$8\Delta t^3 \frac{\partial^3 u}{\partial t^3}$	1	-8	13	0	-13	8	-1
$6\Delta t^4 \frac{\partial^4 u}{\partial t^4}$	-1	12	-39	56	-39	12	-1

Adapted from Chung (2002)

C.3 TIME-STEPPING ALGORITHMS

TABLE C.2 Standard Time-stepping Methods for $du/dt = Q(t, u)$

Euler Methods

	Scheme	Order
Explicit	$\tilde{u}^{n+1} = \tilde{u}^n + \Delta t Q^n$	Δt
Implicit	$\tilde{u}^{n+1} = \tilde{u}^n + \Delta t Q^{n+1}$	Δt
Trapezoidal	$\tilde{u}^{n+1} = \tilde{u}^n + \frac{\Delta t}{2}(Q^n + Q^{n+1})$	Δt^2
General	$\tilde{u}^{n+1} = \tilde{u}^n + \Delta t((1-\alpha)Q^n + \alpha Q^{n+1})$	Δt

Multistage Methods

	Scheme	Order
Runge-Kutta	$\tilde{u}^{n+1/2} = \tilde{u}^n + \frac{\Delta t}{2} Q(t^n, \tilde{u}^n)$ $\tilde{u}^{n+1} = \tilde{u}^n + \Delta t Q(t^{n+1/2}, \tilde{u}^{n+1/2})$	Δt^2
Runge-Kutta	$\tilde{u}_a^{n+1/2} = \tilde{u}^n + \frac{\Delta t}{2} Q(t^n, \tilde{u}^n)$ $\tilde{u}_b^{n+1/2} = \tilde{u}^n + \frac{\Delta t}{2} Q(t^{n+1/2}, \tilde{u}_a^{n+1/2})$ $\tilde{u}^\star = \tilde{u}^n + \Delta t\, Q(t^{n+1/2}, \tilde{u}_b^{n+1/2})$ $\tilde{u}^{n+1} = \tilde{u}^n + \Delta t\left(\frac{1}{6}Q(t^n, \tilde{u}^n) + \frac{2}{6}Q(t^{n+1/2}, \tilde{u}_a^{n+1/2})\right.$ $\left. + \frac{2}{6}Q(t^{n+1/2}, \tilde{u}_b^{n+1/2}) + \frac{1}{6}Q(t^{n+1}, \tilde{u}^\star)\right)$	Δt^4

Multistep Methods

	Scheme	Truncation Order
Leapfrog	$\tilde{u}^{n+1} = \tilde{u}^{n-1} + 2\Delta t Q^n$	Δt^2
Adams-Bashforth	$\tilde{u}^{n+1} = \tilde{u}^n + \frac{\Delta t}{2}(-Q^{n-1} + 3Q^n)$	Δt^2
Adams-Moulton	$\tilde{u}^{n+1} = \tilde{u}^n + \frac{\Delta t}{12}(-Q^{n-1} + 8Q^n + 5Q^{n+1})$	Δt^3
Adams-Bashforth	$\tilde{u}^{n+1} = \tilde{u}^n + \frac{\Delta t}{12}(5Q^{n-2} - 16Q^{n-1} + 23Q^{n+1})$	Δt^3

Predictor-Corrector Methods

	Scheme	Order
Heun	$\tilde{u}^\star = \tilde{u}^n + \Delta t Q(t^n, \tilde{u}^n)$ $\tilde{u}^{n+1} = \tilde{u}^n + \frac{\Delta t}{2}(Q(t^n, \tilde{u}^n) + Q(t^{n+1}, \tilde{u}^\star))$	Δt^2
Leapfrog-Trapezoidal	$\tilde{u}^\star = \tilde{u}^{n-1} + 2\Delta t Q^n$ $\tilde{u}^{n+1} = \tilde{u}^n + \frac{\Delta t}{2}(Q^n + 5Q(t^{n+1}, \tilde{u}^\star))$	Δt^2
ABM	$\tilde{u}^\star = \tilde{u}^n + \frac{\Delta t}{2}(-Q^{n-1} + 3Q^n)$ $\tilde{u}^{n+1} = \tilde{u}^n + \frac{\Delta t}{12}(-Q^{n-1} + 8Q^n + 5Q(t^{n+1}, \tilde{u}^\star))$	Δt^3

C.4 PARTIAL-DERIVATIVES FINITE DIFFERENCES

On a regular grid $x = x_0 + i\Delta x$, $y = y_0 + j\Delta y$, the following expressions are of second order

- Jacobian $J(a, b) = \dfrac{\partial a}{\partial x}\dfrac{\partial b}{\partial y} - \dfrac{\partial b}{\partial x}\dfrac{\partial a}{\partial y}$

$$J_{i,j}^{++} = \frac{(a_{i+1,j} - a_{i-1,j})(b_{i,j+1} - b_{i,j-1}) - (b_{i+1,j} - b_{i-1,j})(a_{i,j+1} - a_{i,j-1})}{4\Delta x\Delta y}$$

$$J_{i,j}^{+\times} = \frac{\left[a_{i+1,j}(b_{i+1,j+1} - b_{i+1,j-1}) - a_{i-1,j}(b_{i-1,j+1} - b_{i-1,j-1})\right]}{4\Delta x\Delta y}$$
$$- \frac{\left[a_{i,j+1}(b_{i+1,j+1} - b_{i-1,j+1}) - a_{i,j-1}(b_{i+1,j-1} - b_{i-1,j-1})\right]}{4\Delta x\Delta y}$$

$$J_{i,j}^{\times+} = \frac{\left[b_{i,j+1}(a_{i+1,j+1} - a_{i-1,j+1}) - b_{i,j-1}(a_{i+1,j-1} - a_{i-1,j-1})\right]}{4\Delta x\Delta y}$$
$$- \frac{\left[b_{i+1,j}(a_{i+1,j+1} - a_{i+1,j-1}) - b_{i-1,j}(a_{i-1,j+1} - a_{i-1,j-1})\right]}{4\Delta x\Delta y}$$

- Cross derivatives

$$\left.\frac{\partial^2 u}{\partial x\partial y}\right|_{i+1/2,j+1/2} \simeq \frac{u_{i+1,j+1} - u_{i+1,j} + u_{i,j} - u_{i,j+1}}{\Delta x\Delta y}$$

$$\left.\frac{\partial^2 u}{\partial x\partial y}\right|_{i,j} \simeq \frac{u_{i+1,j+1} - u_{i+1,j-1} + u_{i-1,j-1} - u_{i-1,j+1}}{4\Delta x\Delta y}$$

- Laplacian

$$\left.\frac{\partial^2 u}{\partial x^2} + \frac{\partial^2 u}{\partial y^2}\right|_{i,j} \simeq \frac{u_{i+1,j} + u_{i-1,j} - 2u_{i,j}}{\Delta x^2} + \frac{u_{i,j+1} + u_{i,j-1} - 2u_{i,j}}{\Delta y^2}$$

C.5 DISCRETE FOURIER TRANSFORM AND FAST FOURIER TRANSFORM

In a periodic domain in which x varies between 0 and L, a complex function $u(x)$ may be expanded in Fourier modes according to

$$u(x) = \sum_{n=-\infty}^{+\infty} a_n e^{i n \frac{2\pi x}{L}}. \tag{C.8}$$

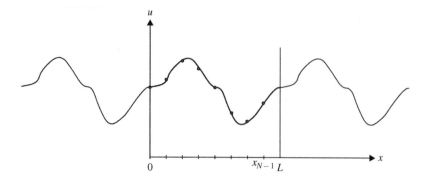

FIGURE C.1   Periodic signal sampled with N evenly spaced points x_0 to x_{N-1}. Note that no x_N is needed since by periodicity the value of the function at x_N is the same as at x_0.

From the orthogonality among Fourier modes,[1]

$$\frac{1}{L} \int_0^L e^{i\,(n-m)\frac{2\pi x}{L}}\, dx = \delta_{nm}, \tag{C.9}$$

the complex coefficients a_n are readily obtained by multiplying (C.8) by $\exp(-i\,m2\pi x/L)$ and integrating over the interval:

$$a_m = \frac{1}{L} \int_0^L u(x)\, e^{-i\,m\frac{2\pi x}{L}}\, dx. \tag{C.10}$$

Note that we could have defined $a_n = b_n/L$ such that b_n were the integral without the factor $1/L$. Different authors use different notations.

To be an exact representation of the periodic function, the sum must cover an infinity of Fourier modes, but this is not possible on finite computers. The discrete Fourier transform (DFT) simply truncates the infinite series by limiting it to its first N terms. The procedure begins with the sampling of the function $u(x)$ at N equidistant points: $u_j = u(x_j)$, with $x_j = j\Delta x$, $j = 0, \ldots, N-1$ ($\Delta x = L/N$). The Fourier coefficients are then calculated from

$$a_n = \frac{1}{N} \sum_{j=0}^{N-1} u_j\, e^{-i\,nj\frac{2\pi}{N}}, \tag{C.11}$$

---

[1] The symbol δ_{ij} is called the Kronecker delta. Its value is 1 if $i = j$, and 0 if i differs from j.

and the sampled function can be reconstructed by summing over the first N modes[2]:

$$\tilde{u}(x) = \sum_{n=0}^{N-1} a_n e^{in\frac{2\pi x}{L}}. \tag{C.12}$$

Obviously, this amounts to a discrete and finite version of the set (C.8) and (C.10). The interesting point is that the coefficients a_n are exact in the sense that if we evaluate $\tilde{u}(x_j)$ with those coefficients, we recover the sampled values at grid points $u(x_j)$. The proof that $\tilde{u}(x_j) = u(x_j)$, which is not trivial, begins by evaluating $\tilde{u}(x_j)$ of (C.12) with the a_n coefficients given by (C.11). We obtain successively

$$\tilde{u}(x_j) = \sum_{n=0}^{N-1} a_n e^{in\frac{2\pi x_j}{L}} \tag{C.13}$$

$$= \sum_{n=0}^{N-1} \frac{1}{N} \sum_{m=0}^{N-1} u_m e^{-inm\frac{2\pi}{N}} e^{inj\frac{2\pi}{N}}$$

$$= \frac{1}{N} \sum_{m=0}^{N-1} u_m \left[\sum_{n=0}^{N-1} e^{i(j-m)n\frac{2\pi}{N}} \right]$$

$$= \frac{1}{N} \sum_{m=0}^{N-1} u_m \left[\sum_{n=0}^{N-1} \rho^n \right] \tag{C.14}$$

where

$$\rho = e^{i(j-m)\frac{2\pi}{N}}. \tag{C.15}$$

For $j \neq m$, $\rho \neq 1$, and the geometric sum takes the value

$$\sum_{n=0}^{N-1} \rho^n = \frac{1-\rho^N}{1-\rho}, \tag{C.16}$$

which turns out to vanish because $\rho^N = 1$ with $j - m$ being an integer. When $j = m$, the sum between brackets is simply the sum of ones and is equal to N, which leads to $\tilde{u}(x_j) = u(x_j)$, the desired result.

Transformation (C.11) is called the direct discrete Fourier transform (DFT), whereas (C.13) is called the inverse discrete Fourier transform (IDFT). Note that the direct and inverse transforms are very similar in form, with two essential

---

[2]The truncation from 0 to $N-1$ is quite arbitrary. Sometimes a series with mode number running from $-N/2$ to $N/2$ is preferred.

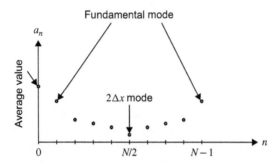

FIGURE C.2   Coefficients a_n contain the amplitudes of the different modes. With the summation running from 0 to $N-1$, $a_{N/2}$ contains the amplitude of the shortest wave.

differences, the sign in the exponential and the factor $1/N$. As for the infinite Fourier series, the scaling factor (then $1/L$, now $1/N$) is occasionally placed in front of the inverse transform rather than in front of the direct transform. It is a matter of choice.

Once the Fourier coefficients are known, Eq. (C.12) effectively provides a continuous and differentiable interpolation of the function $u(x)$ known only from its sampled values u_j. This interpolation may be used to calculate the value of the function at any intermediate location and also to evaluate derivatives to any order. The first derivative is

$$\frac{d\tilde{u}}{dx} = \sum_{n=0}^{N-1} i k_n a_n e^{i k_n x}, \quad k_n = \frac{2\pi n}{L}. \tag{C.17}$$

Thus, all we have to do to calculate the first derivative is to create a set of Fourier coefficients b_n that are simply the function's Fourier coefficients a_n multiplied by a factor: $b_n = i k_n a_n$. This simplicity is the key to the usefulness of the Fourier transform. Schematically, we have

$$u(x) \xrightarrow{\text{sampling}} u(x_j) \xrightarrow{\text{DFT}(u_j)} a_n \xrightarrow{\text{derivative}} b_n = i k_n a_n \xrightarrow{\text{IDFT}(b_n)} \left.\frac{d\tilde{u}}{dx}\right|_{x_j}.$$

It is worth noting that the wavenumber k_{N-n} associated with mode $N-n$ in Eq. (C.12) corresponds to the same wavelength as wavenumber k_n of mode n. This is because the two exponentials differ by the factor $\exp(i 2\pi N x/L) = \exp(i 2\pi x/\Delta x)$, which is equal to 1 at every sampling point. From this follows that the shortest wavelength resolved by the truncated series corresponds to $n=N/2$ and has wavenumber $k=N\pi/L$ or wavelength $2L/N=2\Delta x$. Indeed, the shortest wave in numerical terms is the simple oscillation $(+1, -1, +1, -1,$ etc.) that repeats every $2\Delta x$ interval. The coefficient a_0 corresponds to the zero-wavenumber component and thus holds the average value; the pair (a_1, a_{N-1}) contains information on wavelength L (amplitude and phase), (a_2, a_{N-2}) on

wavelength $L/2$, and so on up to $a_{N/2}$ for which there is no phase information.[3] We can see why some presentations of the DFT define the series with an
index n running between $-N/2$ and $N/2$ instead of running from 0 to N. The
information content, or $N+1$ numbers, is the same.

Up to now, the functions we considered were complex functions. For a real-
valued function u, there is a redundancy among the coefficients a_n: a_{N-n} is
equal to the complex conjugate of a_n. This is easily seen by equating the Fourier
transforms of the function and of its complex conjugate, which must be the same
because a real number is always its own complex conjugate.

Some algorithms exploit this redundancy. Cosine Fourier transforms (CFT)
and sine Fourier transforms (SFT) perform in a way similar to DFT by using
only cosine or sine functions, and they are particularly useful if the solution of a
problem is known to satisfy particular boundary conditions. For homogeneous
Neumann conditions at $x=0$ and $x=L$ (i.e., derivative of function vanishes at
both ends), CFT is the method of choice, whereas for homogeneous Dirichlet
conditions (i.e., function itself vanishes at both ends), expansion in terms of sine
functions guarantees automatic satisfaction of the boundary conditions.

The annoying aspect about the discrete Fourier transform is that a straightforward implementation as a sum of exponentials demands a data array of length
N, and for each coefficient a_n the calculation of N exponentials, necessitating a
total of N^2 calculations of exponential functions (or sine or cosine functions).
Recognizing that this is prohibitively expensive even for a modicum of spatial
resolution, Cooley and Tukey (1965) introduced a clever method that probably
ranks among the most celebrated numerical algorithms of all times. It reduces
the computational cost of the DFT from N^2 to $N \log_2 N$ operations. Interestingly
enough, the original idea of the method, now called Fast Fourier Transform,
goes back to Gauss in 1805 (see collected reprints Gauss, 1866), long before
computational methods could exploit it.

The Fast Fourier Transform (FFT) is a practical calculation of the discrete
Fourier transform that starts with the observation that if N is even, the series can
be split between its even and odd terms in j:

$$
N a_n = \sum_{\substack{j=0 \\ j\,\text{even}}}^{N-1} u_j e^{-i n j \frac{2\pi}{N}} + \sum_{\substack{j=0 \\ j\,\text{odd}}}^{N-1} u_j e^{-i n j \frac{2\pi}{N}}
$$

$$
= \sum_{m=0}^{N/2-1} u_{2m} e^{-i n m \frac{2\pi}{(N/2)}} + e^{-i n \frac{2\pi}{N}} \sum_{m=0}^{N/2-1} u_{2m+1} e^{-i n m \frac{2\pi}{(N/2)}} \tag{C.18}
$$

in which we set $j=2m$ in the first sum where j is even and $j=2m+1$ in the
second sum where j is odd. Now, if N is divisible by 2, the new exponentials are

---

[3]For the $2\Delta x$ wave, the only possible phase shift is the shift by Δx, which is accommodated by a
sign change in amplitude.

those of a discrete Fourier transform with only $N/2$ points, that is, half as many points. In turn, if $N/2$ is itself divisible by 2, the coefficients can be obtained from a transform with yet half as many points, $N/4$. Quite obviously, the greatest advantage is obtained when N is repeatedly divisible by 2, that is, when it is a power of 2, but computational efficiency can also be achieved, to a lesser extent, with N values that have factors other than 2.

To estimate the computational cost $C(N)$ for a transformation with N data, we note that it consists of the cost of two transformations of size $N/2$ and a multiplication by $\exp(-i\, n(2\pi/N))$ for each of the N coefficients a_n. Hence,

$$C(N) = 2C(N/2) + N. \tag{C.19}$$

For a single point ($N = 1$), only one operation is needed ($C(1) = 1$) so that we have $C(2) = 4$, $C(4) = 12$, $C(8) = 32$, $C(16) = 80$, etc. leading to an asymptotic behavior growing with N as

$$C(N) \sim N\log_2 N. \tag{C.20}$$

There is thus a substantial gain compared to the brute-force approach, and this is particularly interesting in the context of spectral methods (Section 18.4) when direct and inverse transforms are performed repeatedly.

Interpolation of a function knowing its Fourier series coefficients could be obtained by brute-force summation of many exponential functions (or sines and cosines) as prescribed by (C.12) at the various sampling locations. But we can do better. Suppose that we have at our disposal values of a function on a certain regular grid $x_j = (j/N)L$, $j = 0, ..., N-1$ and that we need to interpolate onto a finer regular grid $x_k = (k/M)L$, $k = 0, ..., M-1$ with $M > N$. One method of interpolation is to perform the discrete Fourier transform of the function on the original grid, add a string of zeros for the amplitudes of the higher modes permitted by the finer grid but absent from the original grid, and then take the inverse Fourier transform. This back and forth transformation in and out of spectral space may first appear as a wasteful detour, but given the computational efficiency of the Fast Fourier Transform, the procedure is actually advantageous. The procedure is called *padding*.

ANALYTICAL PROBLEMS

C.1. Find the truncation errors of the two Adams–Bashforth schemes by means of Taylor expansions.

C.2. Which of the two second-order approximations of the cross derivative has the lowest truncation error?

C.3. Generate a finite-difference approximation of a Jacobian at a corner point $i+1/2, j+1/2$, using only values from the nearest grid points.

C.4. Prove that the discrete Fourier transform is exact in the sense that the inverse transform of the transform returns exactly the initial set of values.

C.5. Prove that $C(N) \sim N \log_2 N$ is the asymptotic behavior of recursive relation $C(N) = 2\, C(N/2) + N$ for large N.

NUMERICAL EXERCISES

C.1. Compare the behavior of the second-order Adams–Bashforths method and leapfrog-trapezoidal method in the numerical simulation of an inertial oscillation.

C.2. Discretize the function $u(x, y) = \sin(2\pi x/L)\cos(2\pi y/L)$ on a regular grid in the plane (x, y). Calculate the numerical Jacobian between

- $\tilde{u}$ and $\tilde{u}$
- $\tilde{u}$ and $\tilde{u}^2$
- $\tilde{u}$ and $\tilde{u}^3$

and interpret your results.

C.3. Perform an FFT on $f(x) = \sin(2\pi x/L)$ between $x = 0$ and $x = L$ by sampling with 10, 20, or 40 points. Using the spectral coefficients obtained from the FFT, plot the Fourier series using very fine resolution in x (say, 200 points) and verify that you recover the initial function. Then repeat with the function $f(x) = x$. What do you observe?

C.4. Redo Numerical Exercise C.3 using the padding technique instead of the brute-force evaluation to plot the Fourier series expansion.

References

Abarbanel, H. D. I., & Young, W. R. (Eds.) (1987). *General Circulation of the Ocean*. New York: Springer-Verlag, 291 pages.

Abbott, P. L., (2004). *Natural Disasters* (4th ed.). Boston: McGraw-Hill, 460 pages.

Abramowitz, M., & Stegun, I. A. (Eds.) (1972). *Handbook of Mathematical Functions*. New York: Dover, 1046 pages.

Adcroft, A., Campin, J.-M., Hill, C., & Marshall, J. (2004). Implementation of an atmosphere-ocean general circulation model on the expanded spherical cube. *Monthly Weather Review*, *132*, 2845–2863.

Adcroft, A., Hill, C., & Marshall, J. (1997). Representation of topography by shaved cells in a height coordinate ocean model. *Monthly Weather Review*, *125*, 2293–2315.

Akerblom, F. (1908). Recherches sur les courants les plus bas de l'atmosphère au-dessus de Paris. *Nova Acta Regiae Societatis Scientiarum*, Uppsala, Ser. 4, *2*, 1–45.

Alvarez, A., Orfila, A., & Tintore, J. (2001). DARWIN: An evolutionary program for nonlinear modeling of chaotic time series. *Computer Physics Communications*, *136*, 334–349.

Anthes, R. A. (1982). *Tropical Cyclones: Their Evolution, Structure and Effects*, Meteorological Monographs, 19(41), Boston: American Meteorological Society, 208 pages.

Arakawa, A., & Lamb, V. R. (1977). Computational design of the basic dynamical processes of the UCLA general circulation model. *Methods in Computational Physics*, *17*, 173–265.

Arneborg, L., & Liljebladh, B. (2001). The internal seiches in Gullmar Fjord. Part I: Dynamics. *Journal of Physical Oceanography*, *31*, 2549–2566.

Asselin, R. (1972). Frequency filter for time integrations. *Monthly Weather Review*, *100*, 487–490.

Backhaus, J. O. (1983). A semi-implicit scheme for the shallow water equations for application to shelf sea modelling. *Continental Shelf Research*, *2*, 243–254.

Barcilon, V. (1964). Role of Ekman layers in the stability of the symmetric regime in a rotating annulus. *Journal of the Atmospheric Sciences*, *21*, 291–299.

Barth, A., Alvera-Azcárate, A., Beckers, J.-M., Rixen, M., & Vandenbulcke, L. (2007). Multigrid state vector for data assimilation in a two-way nested model of the Ligurian Sea. *Journal of Marine Systems*, *65*, 41–59.

Barth, A., Alvera-Azcárate, A., Rixen, M., & Beckers, J.-M. (2005). Two-way nested model of mesoscale circulation features in the Ligurian Sea. *Progress in Oceanography*, *66*, 171–189.

Batchelor, G. K. (1967). *An Introduction to Fluid Dynamics*. London and New York: Cambridge University Press, 615 pages.

Beardsley, R. C., Mills, C. A., Vermersch, J. A., Jr., Brown, W. S., Pettigrew, N., Irish, J., Ramp, S., Schlitz, R., & Butman, B. (1983). *Nantucket Shoals Flux Experiment (NSFE'79). Part 2: Moored array data report*. Woods Hole Oceanographic Institution Tech. Rep. No. WHOI-83-13, 140 pages.

Beckers, J.-M. (1991). Application of a 3D model to the Western Mediterranean. *Journal of Marine Systems*, *1*, 315–332.

Beckers, J.-M. (1999). On some stability properties of the discretization of the damped propagation of shallow-water inertia-gravity waves on the Arakawa B-grid. *Ocean Modelling*, *1*, 53–69.

Beckers, J.-M. (1999b). Application of Miller's theorem to the stability analysis of numerical schemes; some useful tools for rapid inspection of discretisations in ocean modelling. *Ocean Modelling*, *1*, 29–37.

Beckers, J.-M. (2002). Selection of a staggered grid for inertia-gravity waves in shallow water. *International Journal of Numerical Methods in Fluids*, *38*, 729–746.

Beckers, J.-M., Burchard, H., Campin, J.-M., Deleersnijder, E., & Mathieu, P.-P. (1998). Another reason why simple discretisations of rotated diffusion operators cause problems in ocean models. Comments on "Isoneutral diffusion in a z-coordinate ocean model". *Journal of Physical Oceanography*, *28*, 1552–1559.

Beckers, J.-M., Burchard, H., Deleersnijder, E., & Mathieu, P.-P. (2000). Numerical discretisation of rotated diffusion operators in ocean models. *Monthly Weather Review*, *128*, 2711–2733.

Beckers, J.-M., & Deleersnijder, E. (1993). Stability of a FBTCS scheme applied to the propagation of shallow-water inertia-gravity waves on various space grids. *Journal of Computational Physics*, *108*, 95–104.

Beckmann, A., & Döscher, R. (1997). A method for improved representation of dense water spreading over topography in geopotential-coordinate models. *Journal of Physical Oceanography*, *27*, 581–591.

Beckmann, A., Haidvogel, D. B. (1993). Numerical simulation of flow around a tall isolated seamount. Part I: Problem formulation and model accuracy. *Journal of Physical Oceanography*, *23*, 1736–1753.

Bender, C. M., & Orszag, S. A. (1978). *Advanced Mathematical Methods for Scientists and Engineers*, International Series in Pure and Applied Mathematics, McGraw-Hill, 593 pages.

Bengtsson L., Ghil, N., & Kallen, E. (1981). *Dynamic Meteorology: Data Assimilation Methods*, Applied Mathematical Sciences, New York: Springer Verlag, 36, 330 pages.

Bennett, A. (1992). *Inverse Methods in Physical Oceanography*, Cambridge Monographs on Mechanics and Applied Mathematics, New York: Cambridge University Press, 346 pages.

Betts, A. K. (1986). A new convective adjustment scheme. Part I: Observational and theoretical basis. *Quarterly Journal of the Royal Meteorological Society*, *112*, 677–691.

Bjerknes, V. (1904). Das Problem von der Wettervorhersage, betrachtet vom Standpunkte der Mechanik und der Physik (The problem of weather prediction considered from the point of view of Mechanics and Physics). *Meteorologische Zeitschrift*, *21*, 1–7. (English translation by Yale Mintz, 1954, reprinted in *The Life Cycles of Extratropical Cyclones*, M. A. Shapiro and S. Grønås, eds., American Meteorological Society, 1999, pages 1–4.)

Blayo, E., & Debreu, L. (2005). Revisiting open boundary conditions from the point of view of characteristic variables. *Ocean Modelling*, *9*, 231–252.

Bleck, R. (2002). An oceanic general circulation model framed in hybrid isopycnic-cartesian coordinates. *Ocean Modelling*, *4*, 55–88.

Bleck, R., Rooth, C., Hu, D., & Smith, L. T. (1992). Ventilation patterns and mode water formation in a wind- and thermodynamically driven isopycnic coordinate model of the North Atlantic. *Journal of Physical Oceanography*, *22*, 1486–1505.

Blumen, W. (1972). Geostrophic adjustment. *Reviews of Geophysics and Space Physics*, *10*, 485–528.

Booker, J. R., & Bretherton, F. P. (1967). The critical layer for internal gravity waves in a shear flow. *Journal of Fluid Mechanics*, *27*, 513–539.

Boris, J., & Book, D. (1973). Flux-corrected transport. I. SHASTA, a fluid transport algorithm that works. *Journal of Computational Physics*, *11*, 38–69.

Bower, A. S., & Rossby, T. (1989). Evidence of cross-frontal exchange processes in the Gulf Stream based on isopycnal RAFOS float data. *Journal of Physical Oceanography*, *19*, 1177–1190.

Brasseur, P. (2006). Ocean data assimilation using sequential methods based on the Kalman Filter. In J. Verron, & E. Chassignet (Eds.), *Ocean Weather Forecasting: An Integrated View of Oceanography* (pp. 271–316). Dordrecht: Springer, (Part 4).

Brasseur, P., Blayo, E., & Verron, J. (1996). Predictability experiments in the North Atlantic Ocean: Outcome of a quasigeostrophic model with assimilation of TOPEX/POSEIDON altimeter data. *Journal of Geophysical Research, 101*, 14161–14174.

Brier, G. W. (1950). Verification of forecasts expressed in terms of probability. *Monthly Weather Review, 78*, 1–3.

Brink, K. H. (1983). The near-surface dynamics of coastal upwelling. *Progress in Oceanography, 12*, 223–257.

Bromwich, D. H., Rogers, A. N., Kållberg, P., Cullather, R. I., White, J. W. C., & Kreutz, K. J. (2000). ECMWF analyses and reanalyses depiction of ENSO signal in Antarctic precipitation. *Journal of Climate, 13*, 1406–1420.

Brunt, D. (1934). *Physical and Dynamical Meteorology*. Upper Saddle River, NJ: Prentice Hall, 411 pages. (Reedited in 1952)

Bryan, K. (1969). A numerical method for the study of the circulation of the World Ocean. *Journal of Computational Physics, 4*, 347–376.

Bryan, K., & Cox, M. D. (1972). The circulation of the world ocean: A numerical study. Part I, A homogeneous model. *Journal of Physical Oceanography, 2*, 319–335.

Buchanan, G. (1995). *Schaum's Outline of Finite Element Analysis*. New York: Mc Graw-Hill, 264 pages.

Burchard, H. (2002). *Applied Turbulence Modelling in Marine Waters*. Berlin: Springer, 229 pages.

Burchard, H., & Beckers, J.-M. (2004). Non-uniform adaptive vertical grids in one-dimensional numerical ocean models. *Ocean Modelling, 6*, 51–81.

Burchard, H., & Bolding, K. (2001). Comparative analysis of four second-moment turbulence closure models for the oceanic mixed layer. *Journal of Physical Oceanography, 31*, 1943–1968.

Burchard, H., Deleersnijder, E., & Meister, A. (2003). A high-order conservative Patankar-type discretisation for stiff systems of production-destruction equations. *Applied Numerical Mathematics, 47*, 1–30.

Burchard, H., Deleersnijder, E., & Meister, A. (2005). Application of modified Patankar schemes to stiff biogeochemical models for the water column. *Ocean Dynamics, 10*, 115–136.

Burger, A. P. (1958). Scale consideration of planetary motions of the atmosphere. *Tellus, 10*, 195–205.

Cane, M. A., Zebiak, S. E., & Dolan, S. C. (1986). Experimental forecasts of El Niño. *Nature, 321*, 827–832.

Canuto, V. M., Howard, A., Cheng, Y., & Dubovikov, M. S. (2001). Ocean turbulence. Part I: One-point closure model. Momentum and heat vertical diffusivities. *Journal of Physical Oceanography, 31*, 1413–1426.

Canuto, C., Hussaini, M. Y., Quarteroni, A., & Zang, T. A. (1988). *Spectral Methods in Fluid Dynamics*. Springer-Verlag, 558 pages.

Case, B., & Mayfield, M. (1990). Atlantic Hurricane Season of 1989. *Monthly Weather Review, 118*, 1165–1177.

Cessi, P. (1996). Grid-scale instability of convective adjustment schemes. *Journal of Marine Research, 54*, 407–420.

Chandrasekhar, S. (1961). *Hydrodynamic and Hydromagnetic Stability*. London and New York: Oxford University Press, 652 pages.

Charney, J. G. (1947). The dynamics of long waves in a baroclinic westerly current. *Journal of Meteorology*, *4*, 135–163.

Charney, J. G. (1948). On the scale of atmospheric motions. *Geofysiske Publikasjoner Oslo*, *17*(2), 1–17.

Charney, J. G., & DeVore, J. G. (1979). Multiple-flow equilibria in the atmosphere and blocking. *Journal of the Atmospheric Sciences*, *36*, 1205–1216.

Charney, J. G., Fjörtoft, R., & von Neumann, J. (1950). Numerical integration of the barotropic vorticity equation. *Tellus*, *2*, 237–254.

Charney, J. G., & Flierl, G. R. (1981). Oceanic analogues of large-scale atmospheric motions. In B. A. Warren, & C. Wunsch (Eds.), *Evolution of Physical Oceanography* (pp. 504–548). Cambridge, Massachusetts: The MIT Press.

Charney, J. G., & Stern, M. E. (1962). On the stability of internal baroclinic jets in a rotating atmosphere. *Journal of the Atmospheric Sciences*, *19*, 159–172.

Chen, D., Cane, M. A., Kaplan, A., Zebiak, S. E., & Huang, D. (2004). Predictability of El Niño over the past 148 years. *Nature*, *428*, 733–736.

Chung, T. J. (2002). *Computational Fluid Dynamics*. New York: Cambridge University Press, 1012 pages.

Colella, P. (1990). Multidimensional upwind methods for hyperbolic conservation laws. *Journal of Computational Physics*, *87*, 171–200.

Colin, C., Henin, C., Hisard, P., & Oudot, C. (1971). Le Courant de Cromwell dans le Pacifique central en février. *Cahiers ORSTOM, Serie Oceanographie*, *9*, 167–186.

Conway, E. D., & the Maryland Space Grant Consortium (1997). *An Introduction to Satellite Image Interpretation*. Baltimore: The Johns Hopkins University Press, 264 pages.

Cooley, J., & Tukey, J. (1965). An algorithm for the machine calculation of complex Fourier series. *Mathematics of Computation*, *9*, 297–301.

Cooper, M., & Haines, K. (1996). Altimetric assimilation with water property conservation. *Journal of Geophysical Research*, *101*, 1059–1078.

Courant, R., Friedrichs, K. P., & Lewy, H. (1928). Über die partiellen Differenzengleichungen der mathematischen Physik. *Mathematische Annalen*, *100*, 32–74.

Courant, R., & Hilbert, D. (1924). *Methoden der mathematischen Physik I*. Berlin: Springer Verlag, 470 pages.

Courtier, P., Thepaut, J. N., & Hollingsworth, A. (1994). A strategy for operational implementation of 4D-Var, using an incremental approach. *Quarterly Journal of the Royal Meteorological Society*, *120*, 1367–1387.

Cox, M. (1984). A primitive three-dimensional model of the ocean. Rep. 1, Ocean Group, GFDL, Princeton University.

Crank, J. (1987). *Free and Moving Boundary Problems*. New York: The Clarendon Press, Oxford University Press, 424 pages.

Crépon, M., & Richez, C. (1982). Transient upwelling generated by two-dimensional atmospheric forcing and variability in the coastline. *Journal of Physical Oceanography*, *12*, 1437–1457.

Cressman, G. P. (1948). On the forecasting of long waves in the upper westerlies. *Journal of Meteorology*, *5*, 44–57.

Csanady, G. T. (1977). Intermittent 'full' upwelling in Lake Ontario. *Journal of Geophysical Research*, *82*, 397–419.

Curry, J. A., & Webster, P. J. (1999). *Thermodynamics of Atmospheres and Oceans*. London: Academic Press, 467 pages.

Cushman-Roisin, B. (1986). Frontal geostrophic dynamics. *Journal of Physical Oceanography*, *16*, 132–143.

Cushman-Roisin, B. (1987a). On the role of heat flux in the Gulf Stream–Sargasso Sea–Subtropical Gyre system. *Journal of Physical Oceanography, 17,* 2189–2202.

Cushman-Roisin, B. (1987b). Subduction. In P. Müller, & D. Henderson (Eds.), *Dynamics of the Oceanic Surface Mixed Layer,* Proc. Hawaiian Winter Workshop *'Aha Huliko'a* (pp. 181–196). Hawaii Institute of Geophysics Special Publication.

Cushman-Roisin, B., Chassignet, E. P., & Tang, B. (1990). Westward motion of mesoscale eddies. *Journal of Physical Oceanography, 20,* 758–768.

Cushman-Roisin, B., Esenkov, O. E., & Mathias, B. J. (2000). A particle-in-cell method for the solution of two-layer shallow-water equations. *International Journal for Numerical Methods in Fluids, 32,* 515–543.

Cushman-Roisin, B., Gačić, M., Poulain, P.-M., & Artegiani, A. (2001). *Physical Oceanography of the Adriatic Sea: Past, Present, and Future.* Dordrecht: Kluwer Academic Publishers, 230 pages.

Cushman-Roisin, B., & Malačič, V. (1997). Bottom Ekman pumping with stress-dependent eddy viscosity. *Journal of Physical Oceanography, 27,* 1967–1975.

Cushman-Roisin, B., Sutyrin, G. G., & Tang, B. (1992). Two-layer geostrophic dynamics. Part I: Governing equations. *Journal of Physical Oceanography, 22,* 117–127.

Dahlquist, G., & Björck, A. (1974). *Numerical Methods.* Englewood Cliffs: Prentice-Hall, 573 pages.

D'Aleo, J. S. (2002). *The Oryx Guide to El Niño and La Niña.* Westport, CT: Oryx Press, 230 pages.

D'Asaro, E. A., & Lien, R.-C. (2000). Lagrangian measurements of waves and turbulence in stratified flows. *Journal of Physical Oceanography, 30,* 641–655.

Davies, A. (1987). Spectral models in continental shelf sea oceanography. In N. Heaps (Ed.), *Three-Dimensional Coastal Ocean Models* (pp. 71–106). Washington, DC: American Geophysical Union.

Deleersnijder, E., & Beckers, J.-M. (1992). On the use of the σ-coordinate system in regions of large bathymetric variations. *Journal of Marine Systems, 3,* 381–390.

Deleersnijder, E., Hanert, E., Burchard, H., & Dijkstra, H. A. (2008). On the mathematical stability of stratified flow models with local turbulence closure schemes. *Ocean Dynamics, 58,* 237–246.

Dewar, W. K. (2001). Density coordinate mixed layer models. *Monthly Weather Review, 129,* 237–253.

Diaz, H. F., & Markgraf, V. (Eds.) (2000). *El Niño and the Southern Oscillation. Multiscale Variability and Global and Regional Impacts* (p. 512). Cambridge, UK: Cambridge University Press.

Dietrich, D. E. (1998). Application of a modified Arakawa 'A' grid ocean model having reduced numerical dispersion to the Gulf of Mexico circulation. *Dynamics of Atmospheres and Oceans, 27,* 201–217.

Dongarra, J. J., Duffy, I. A., Sorensen, D. C., & van der Vorst, H. A. (1998). *Numerical linear algebra on high-performance computers.* Society for Industrial and Applied Mathematics, 342 pages.

Doodson, A. T. (1921). The harmonic development of the tide-generating potential. *Proceedings of the Royal Society A, 100,* 304–329.

Dowling, T. E., & Ingersoll, A. P. (1988). Potential vorticity and layer thickness variations in the flow around Jupiter's Great Red Spot and White Oval BC. *Journal of the Atmospheric Sciences, 45,* 1380–1396.

Dritschel, D. G. (1988). Contour Surgery: A topological reconnection scheme for extended integrations using contour dynamics. *Journal of Computational Physics, 77,* 240–266.

Dritschel, D. G. (1989). On the stabilization of a two-dimensional vortex strip by adverse shear. *Journal of Fluid Mechanics, 206,* 193–221.

Ducet, N., Le Traon, P.-Y., & Reverdin, G. (2000). Global high resolution mapping of ocean circulation from Topex/Poseidon and ERS-1 and -2. *Journal of Geophysical Research, 105,* 19477–19498.

Dukowicz, J. (1995). Mesh effects for Rossby waves. *Journal of Computational Physics, 119,* 188–194.

Durran, D. (1999). *Numerical Methods for Wave Equations in Geophysical Fluid Dynamics.* New York: Springer, 465 pages.

Eady, E. T. (1949). Long waves and cyclone waves. *Tellus, 1*(3), 33–52.

Ekman, V. W. (1904). On dead water. *Scientific Results Norwegian North Polar Expedition 1893–96, 5*(15), 152 pages.

Ekman, V. W. (1905). On the influence of the earth's rotation on ocean currents. *Archives of Mathematics, Astronomy and Physics, 2*(11), 1–53.

Ekman, V. W. (1906). Beiträge zur Theorie der Meeresströmungen. *Annalen der Hydrographie und maritimen Meteorologie,* 9: 423–430; 10: 472–484; 11: 527–540; 12: 566–583.

Eliassen, A. (1962). On the vertical circulation in frontal zones. *Geofysicke Publicasjoner, 24,* 147–160.

Emmanuel, K. (1991). The theory of hurricanes. *Annual Review of Fluid Mechanics, 23,* 179–196.

Esenkov, O. E., & Cushman-Roisin, B. (1999). Modeling of two-layer eddies and coastal flows with a particle method. *Journal of Geophysical Research, 104,* 10959–10980.

Evensen, G. (1994). Sequential data assimilation with a nonlinear quasi-geostrophic model using Monte Carlo methods to forecast error statistics. *Journal of Geophysical Research, 99,* 10143–10162.

Evensen, G. (2004). Sampling strategies and square root analysis schemes for the EnKF. *Ocean Dynamics, 54,* 539–560.

Fernando, H. J. S. (1991). Turbulent mixing in stratified fluids. *Annual Review of Fluid Mechanics, 23,* 455–493.

Ferziger, J. M., & Perić, M. (1999). *Computational Methods for Fluid Dynamics.* Berlin: Springer Verlag, 390 pages.

Flament, P., Armi, L., & Washburn, L. (1985). The evolving structure of an upwelling filament. *Journal of Geophysical Research, 90,* 11765–11778.

Flierl, G. R. (1987). Isolated eddy models in geophysics. *Annual Review of Fluid Mechanics, 19,* 493–530.

Flierl, G. R., Larichev, V. D., McWilliams, J. C., & Reznik, G. M. (1980). The dynamics of baroclinic and barotropic solitary eddies. *Dynamics of Atmospheres and Oceans, 5,* 1–41.

Flierl, G. R., Malanotte-Rizzoli, P., & Zabusky, N. J. (1987). Nonlinear waves and coherent vortex structures in barotropic beta-plane jets. *Journal of Physical Oceanography, 17,* 1408–1438.

Fornberg, B. (1998). *A Practical Guide to Pseudospectral Methods.* Cambridge: Cambridge University Press, 242 pages.

Fox, R. W., & McDonald, A. T. (1992). *Introduction to Fluid Mechanics* (4th ed.). New York: John Wiley & Sons, 829 pages.

Fukumori, I., & Malanotte-Rizzoli, P. (1995). An approximate Kalman filter for ocean data assimilation; an example with an idealized Gulf Stream model. *Journal of Geophysical Research, 100,* 6777–6793.

Galperin, B., Kantha, L. H., Hassid, S., & Rosati, A. (1988). A quasi-equilibrium turbulent energy model for geophysical flows. *Journal of the Atmospheric Sciences, 45,* 55–62.

Galperin, B., Kantha, L. H., Mellor, G. L., & Rosati, A. (1989). Modeling rotating stratified turbulent flows with applications to oceanic mixed layers. *Journal of Physical Oceanography, 19,* 901–916.

Galperin, B., Nakano, H., Huang, H.-P., & Sukoriansky, S. (2004). The ubiquitous zonal jets in the atmospheres of giant planets and Earth's oceans. *Geophysical Research Letters, 31,* L13303.

Gardiner, C. W. (1997). *Handbook of Stochastic Methods.* Springer Series in Synergetics (Vol. 13, 2nd ed.), Berlin: Springer, 442 pages.

Garratt, J. R. (1992). *The Atmospheric Boundary Layer.* Cambridge: Cambridge University Press, 316 pages.

Garrett, C., & Munk, W. (1979). Internal waves in the ocean. *Annual Review of Fluid Mechanics, 11,* 339–369.

Gauss, C. F. (1866). Theoria interpolationis methodo nova tractata. *Werke Band, Nachlass 3,* 265–327 (Königliche Gesellschaft der Wissenschaften, Göttingen 1866).

Gawarkiewicz, G., & Chapman, D. C. (1992). The role of stratification in the formation and maintenance of shelf-break fronts. *Journal of Physical Oceanography, 22,* 753–772.

Gent, P. R., & McWilliams, J. C. (1990). Isopycnal mixing in ocean circulation models. *Journal of Physical Oceanography, 20,* 150–155.

Gent, P. R., Willebrand, J., McDougall, T. J., & McWilliams, J. C. (1995). Parameterizing eddy-induced tracer transports in ocean circulation models. *Journal of Physical Oceanography, 25,* 463–474.

Gerdes, R. (1993). A primitive equation model using a general vertical coordinate transformation. Part 1: Description and testing of the model. *Journal of Geophysical Research, 98,* 14683–14701.

Ghil, M. (1989). Meteorological data assimilation for oceanographers. I- Description and theoretical framework. *Dynamics of Atmospheres and Oceans, 13,* 171–218.

Ghil, M., & Childress, S. (1987). *Topics in Geophysical Fluid Dynamics: Atmospheric Dynamics, Dynamo Theory and Climate Dynamics.* New York: Springer-Verlag, 504 pages.

Ghil, M., & Malanotte-Rizzoli, P. (1991). Data assimilation in meteorology and oceanography. *Advances in Geophysics, Academic Press, 33,* 141–266.

Gibson, M. M., & Launder, B. E. (1978). Ground effects on pressure fluctuations in the atmospheric boundary layer. *Journal of Fluid Mechanics, 86,* 491–511.

Gill, A. E. (1980). Some simple solutions for heat-induced tropical circulation. *Quarterly Journal of Royal Metereological Society, 106,* 447–462.

Gill, A. E. (1982). *Atmosphere-Ocean Dynamics.* New York: Academic Press, 662 pages.

Gill, A. E., Green, J. S. A., & Simmons, A. J. (1974). Energy partition in the large-scale ocean circulation and the production of mid-ocean eddies. *Deep-Sea Research, 21,* 499–528.

Glantz, M. H. (2001). *Currents of Change: Impacts of El Niño and La Niña on Climate and Society* (2nd ed.). Cambridge University Press, 252 pages.

Godunov, S. K. (1959). A difference scheme for numerical solution of discontinuous solution of hydrodynamic equations. *Mathematics Sbornik, 47,* 271–306. (Translation: US Joint Publ. Res. Service, JPRS 7226, 1969).

Goldstein, S. (1931). On the stability of superposed streams of fluids of different densities. *Proceedings of the Royal Society London A, 132,* 524–548.

Golub, G. H., & Van Loan, C. F. (1990). *Matrix Computations.* Baltimore: The Johns Hopkins University Press, 728 pages.

Gottlieb, D., & Orszag, S. A. (1977). *Numerical Analysis of Spectral Methods: Theory and Applications*. Philadelphia, PA: Society for Industrial and Applied Mathematics, 170 pages.

Grant, H. L., Stewart, R. W., & Moilliet, A. (1962). Turbulence spectra from a tidal channel. *Journal of Fluid Mechanics*, *12*, 241–268.

Griffies, S. M. (1998). The Gent-McWilliams skew flux. *Journal of Physical Oceanography*, *28*, 831–841.

Griffies, S. M., Böning, C., Bryan, F. O., Chassignet, E. P., Gerdes, R., Hasumi, H., Hirst, A., Treguier, A.-M., & Webb, D. (2000). Developments in ocean climate modelling. *Ocean Modelling*, *2*, 123–192.

Griffiths, R. W., Killworth, P. D., & Stern, M. E. (1982). Ageostrophic instability of ocean currents. *Journal of Fluid Mechanics*, *117*, 343–377.

Griffiths, R. W., & Linden, P. F. (1981). The stability of vortices in a rotating, stratified fluid. *Journal of Fluid Mechanics*, *105*, 283–316.

Gustafson, T., & Kullenberg, B. (1936). Untersuchungen von Trägheitsströmungen in der Ostsee. *Svenska Hydrogr. Biol. Komm. Skrifter, Hydrogr.*, No. 13, 28 pages.

Hack, J. J. (1992). Climate system simulation: Basic numerical & computational concepts. In K. E. Trenberth (Ed.), *Climate System Modeling* (pp. 283–318). Cambridge: Cambridge University Press.

Hackbusch, W. (Ed.), (1985). *Multi-Grid Methods and Applications*. Springer Series in computational mathematics, 4, 377 pages.

Hadley, G. (1735). Concerning the cause of the general trade-winds. *Philosophical Transactions of the Royal Society London*, *39*, 58–62.

Hageman, L. A., & Young, D. M. (2004). *Applied Iterative Methods*. New York: Dover Publications, 386 pages.

Haidvogel, D. B., & Beckmann, A. (1999). *Numerical Ocean Circulation Modeling*. Series on Environmental Science and Management, World Scientific Publishing Co., 318 pages.

Haidvogel, D. B., Wilkin, J. L., & Young, R. (1991). A semi-spectral primitive equation ocean circulation model using vertical sigma and orthogonal curvilinear horizontal coordinates. *Journal of Computational Physics*, *94*, 151–185.

Häkkinen, S. (1990). Models and their applications to polar oceanography. In W. O. Smith, Jr. (Ed.), *Polar Oceanography, Part A: Physical Science* (pp. 335–384). Orlando: Academic Press, Chapter 7.

Hallberg, R. W. (1995). *Some Aspects of the Circulation in Ocean Basins with Isopycnals Intersecting the Sloping Boundaries*, Ph.D. Thesis, Seattle: University of Washington, 244 pages.

Hanert, E., Legat, V., & Deleersnijder, E. (2003). A comparison of three finite elements to solve the linear shallow water equations. *Ocean Modelling*, *5*, 17–35.

Haney, R. L. (1991). On the pressure gradient force over steep topography in sigma coordinate ocean models. *Journal of Physical Oceanography*, *21*, 610–619.

Harten, A., Engquist, B., Osher, S., & Chakravarthy, S. R. (1987). Uniformly high order accurate essentially non-oscillatory schemes. *Journal of Computational Physics*, *71*, 231–303.

Heaps, N. S. (Ed.) (1987). *Three-dimensional Coastal Ocean Models* . Coastal and Estuarine Series Volume 4, Washington, DC: American Geophys. U., 208 pages.

Helmholtz, H. von. (1888). Über atmosphärische Bewegungen I. *Sitzungsberichte Akademie Wissenschaften Berlin*, *3*, 647–663.

Hendershott, M. C. (1972). The effects of solid earth deformation on global ocean tides. *Geophysical Journal of the Royal Astronomical Society*, *29*, 389–402.

Hermann, A. J., Rhines, P. B., & Johnson, E. R. (1989). Nonlinear Rossby adjustment in a channel: beyond Kelvin waves. *Journal of Fluid Mechanics*, *205*, 469–502.

Hirsch, C. (1990). *Numerical Computation of Internal and External Flows. Vol. 2: Computational Methods for Inviscid and Viscous Flows.* Chichester: John Wiley, 714 pages.

Hodnett, P. F. (1978). On the advective model of the thermocline circulation. *Journal of Marine Research, 36,* 185–198.

Holton, J. R. (1992). *An Introduction to Dynamic Meteorology* (3rd ed.). San Diego: Academic Press, 511 pages.

Hoskins, B., & Bretherton, F. (1972). Atmospheric frontogenesis models: Mathematical formulation and solution. *Journal of the Atmospheric Sciences, 29,* 11–37.

Hoskins, B. J., McIntyre, M. E., & Robertson, A. W. (1985). On the use and significance of isentropic potential vorticity maps. *Quarterly Journal of Royal Meteorological Society, 111,* 877–946.

Houtekamer, P. L., & Mitchell, H. L. (1998). Data assimilation using an ensemble Kalman filter technique. *Monthly Weather Review, 126,* 796–811.

Howard, L. N. (1961). Note on a paper of John W. Miles. *Journal of Fluid Mechanics, 10,* 509–512.

Hsueh, Y., & Cushman-Roisin, B. (1983). On the formation of surface to bottom fronts over steep topography. *Journal of Geophysical Research, 88,* 743–750.

Huang, R. X. (1989). On the three-dimensional structure of the wind-driven circulation in the North Atlantic. *Dynamics of Atmospheres and Oceans, 15,* 117–159.

Hurlburt, H. E., & Thompson, J. D. (1980). A numerical study of loop current intrusions and eddy-shedding. *Journal of Physical Oceanography, 10,* 1611–1651.

Hunkins, K. (1966). Ekman drift currents in the Arctic Ocean. *Deep-Sea Research, 13,* 607–620.

Ide, K., Courtier, P., Ghil, M., & Lorenc, A. C. (1997). Unified notation for data assimilation: Operational, sequential and variational. *Practice, 75,* 181–189.

Ingersoll, A. P., Beebe, R. F., Collins, S. A., Hunt, G. E., Mitchell, J. L., Muller, P., Smith, B. A., & Terrile, R. J. (1979). Zonal velocity and texture in the Jovian atmosphere inferred from Voyager images. *Nature, 280,* 773–775.

Intergovernmental Panel on Climate Change (IPCC). (2001). *Climate Change 2001: The Scientific Basis,* Contribution of working group I to the third report assessment of the intergovernmental panel on climate change. J. T. Houghton, Y. Ding, D. J. Griggs, M. Noguer, P. J. van der Linden, X. Dai, K. Maskell and C. A. Johnson, eds. Cambridge University Press, 892 pages.

Iselin, C. O'D. (1938). The influence of vertical and lateral turbulence on the characteristics of the waters at mid-depths. *Transactions, American Geophysical Union, 20,* 414–417.

Ito, S. (1992). *Diffusion equations,* Translations of mathematical monographs, 114, American Mathematical Society, 225 pages.

Jin, F. F. (1997a). An equatorial recharge paradigm for ENSO. I. Conceptual model. *Journal of the Atmospheric Sciences, 54,* 811–829.

Jin, F. F. (1997b) An equatorial recharge paradigm for ENSO. II. A stripped-down coupled model. *Journal of the Atmospheric Sciences, 54,* 830–847.

Jolliffe, I. T., & Stephenson, D. B. (2003). *Forecast Verification: A Practitioner's Guide in Atmospheric Science.* Chichester: John Wiley and Sons, 240 pages.

Jones, W. L. (1967). Propagation of internal gravity waves in fluids with shear flow and rotation. *Journal of Fluid Mechanics, 30,* 439–448.

Kalnay, E. (2003). *Atmospheric Modeling, Data Assimilation and Predictability.* Cambridge: Cambridge University Press, 341 pages.

Kalnay, E., Lord, S. J., & McPherson, R. D. (1998). Maturity of operational numerical weather prediction: Medium range. *Bulletin of the American Meteorological Society, 79,* 2753–2769.

Kantha, L. H., & Clayson, C. A. (1994). An improved mixed layer model for geophysical applications. *Journal of Geophysical Research, 99,* 25235–25266.

Kessler, W. S. (2002). Is ENSO a cycle or a series of events? *Geophysical Research Letters*, *29*, 2125.

Killworth, P. D., Paldor, N., & Stern, M. E. (1984). Wave propagation and growth on a surface front in a two-layer geostrophic current. *Journal of Marine Research*, *42*, 761–785.

Killworth, P. D., Stainforth, D., Webb, D. J., & Paterson, S. M. (1991). The development of a free-surface Bryan-Cox-Semtner ocean model. *Journal of Physical Oceanography*, *21*, 1333–1348.

Kolmogorov, A. N. (1941). Dissipation of energy in locally isotropic turbulence. *Doklady Akademii Nauk SSSR*, *32*, 19–21 (in Russian).

Kraus, E. B. (Ed.), (1977). *Modelling and Prediction of the Upper Layers of the Ocean*. Oxford: Pergamon, 325 pages.

Kreiss, H.-O. (1962). Über die Stabilitätsdefinition für Differenzengleichungen die partielle Differentialgleichungen approximieren. *Nordisk Tidskrift Informationsbehandling (BIT)*, *2*, 153–181.

Kuhlbrodt, T., Griesel, A., Montoya, M., Levermann, A., Hofmann, M., & Rahmstorf, S. (2007). On the driving processes of the Atlantic meridional overturning circulation. *Reviews of Geophysics*, *45*, RG2001.

Kundu, P. K. (1990). *Fluid Mechanics*. New York: Academic Press, 638 pages.

Kuo, H. L. (1949). Dynamic instability of two-dimensional nondivergent flow in a barotropic atmosphere. *Journal of Meteorology*, *6*, 105–122.

Kuo, H. L. (1974). Further studies of the parameterization of the influence of cumulus convection on large-scale flow. *Journal of the Atmospheric Sciences*, *31*, 1232–1240.

Lawrence, G. A., Browand, F. K., & Redekopp, L. G. (1991). The stability of a sheared density interface. *Physics of Fluids A, Fluid Dynamics*, *3*, 2360–2370.

Lax, P. D., & Richtmyer, R. D. (1956). Survey of the stability of linear finite difference equations. *Communications on Pure Applied Mathematics*, *9*, 267–293.

LeBlond, P. H., & Mysak, L. A. (1978). *Waves in the Ocean*. Elsevier Oceanography Series, 20, Amsterdam: Elsevier, 602 pages.

Legrand, S., Legat, V., & Deleersnijder, E. (2000). Delaunay mesh generation for an unstructured-grid ocean general circulation model. *Ocean Modelling*, *2*, 17–28.

Lermusiaux, P. F. J. (2007). Adaptive modeling, adaptive data assimilation and adaptive sampling. *Physica D*, *230*, 172–196.

Lermusiaux, P. F. J., Chiu, C.-S., Gawarkiewicz, G. G., Abbot, P., Robinson, A. R., Miller, R. N., Haley, P. J., Leslie, W. G., Majumdar, S. J., Pang, A., & Lekien, F. (2006). Quantifying uncertainties in ocean predictions. *Oceanography*, *19*, 92–105.

Lermusiaux, P. F. J., & Robinson, A. R. (1999). Data assimilation via error subspace statistical estimation. Part I: Theory and schemes. *Monthly Weather Review*, *127*, 1385–1407.

Lindzen, R. S. (1988). Instability of plane parallel shear flow. (Toward a mechanistic picture of how it works.) *Pure and Applied Geophysics*, *126*, 103–121.

Liseikin, V. (1999). *Grid Generation Methods*. Berlin-Heidelberg: Springer, 362 pages.

Liu, C.-T., Pinkel, R., Hsu, M.-K., Klymak, J. M., Chen, H.-W., & Villanoy, C. (2006). Nonlinear internal waves from the Luzon Strait. *Eos, Transactions of the American Geophysical Union*, *87*, 449–451.

Long, R. R. (1997). Homogeneous isotropic turbulence and its collapse in stratified and rotating fluids. *Dynamics of Atmospheres and Oceans*, *27*, 471–483.

Long, R. R. (2003). Do tidal-channel turbulence measurements support $k^{-5/3}$? *Environmental Fluid Mechanics*, *3*, 109–127.

Lorenc, A. C. (1986). Analysis methods for numerical weather prediction. *Quarterly Journal of Royal Meteorological Society*, *112*, 1177–1194.

Lorenz, E. N. (1955). Available potential energy and the maintenance of the general circulation. *Tellus*, *7*, 157–167.

Lorenz, E. N. (1963). Deterministic nonperiodic flow. *Journal of the Atmospheric Sciences*, *20*, 130–141.

Love, A. E. H. (1893). On the stability of certain vortex motions. *Proceedings of the London Mathematical Society Series 1*, *25*, 18–42.

Lueck, R. G., Wolk, F., & Yamazaki, H. (2002). Oceanic velocity microstructure measurements in the 20th century. *Journal of Oceanography*, *58*, 153–174.

Lutgens, F. K., & Tarbuck, E. J. (1986). *The Atmosphere. An Introduction to Meteorology* (3rd ed.). Upper Saddle River, NJ: Prentice-Hall, 492 pages.

Luyten, J. R., Pedlosky, J., & Stommel, H. (1983). The ventilated thermocline. *Journal of Physical Oceanography*, *13*, 292–309.

Lvov, Y. V., & Tabak, E. G. (2001). Hamiltonian formalism and the Garrett-Munk spectrum of internal waves in the ocean. *Physical Review Letters*, *87*, 168501.

Madec G., Delecluse, P., Imbard, M., & Lévy, C. (1998). *OPA 8.1 Ocean general circulation model reference manual. Note du Pôle de Modélisation* (No. 11). Institut Pierre-Simon Laplace, 91 pages.

Madsen, O. S. (1977). A realistic model of the wind-induced Ekman boundary layer. *Journal of Physical Oceanography*, *7*, 248–255.

Malanotte-Rizzoli, P. (1996). *Modern Approaches to Data Assimilation in Ocean Modeling*. New York: Elsevier, 468 pages.

Margules, M. (1903). Über die Energie der Stürme. *Jahrb. Zentralanst. Meteorol. Wien*, *40*, 1–26. (English translation in Abbe, 1910, *The Mechanics of the Earth's Atmosphere. A Collection of Translations*. Misc. Collect. No. 51, Smithsonian Institution, Washington, D.C.)

Margules, M. (1906). Über Temperaturschichtung in stationär bewegter und ruhender Luft. *Meteorologische Zeitschrift*, *23*, 243–254.

Marotzke, J. (1991). Influence of convective adjustment on the stability of the thermohaline circulation. *Journal of Physical Oceanography*, *21*, 903–907.

Marshall, J., Jones, H., & Hill, C. (1998). Efficient ocean modeling using non-hydrostatic algorithms. *Journal of Marine Systems*, *18*, 115–134.

Marshall, J., & Plumb, R. A. (2008). *Atmosphere, Ocean, and Climate Dynamics: An Introductory Text*. New York: Academic Press, 319 pages.

Marshall, J., & Schott, F. (1999). Open-ocean convection: observations, theory and models. *Reviews of Geophysics*, *37*, 1–64.

Marzano, F. S., & Visconti, G. (Eds.), (2002). *Remote Sensing of Atmosphere and Ocean from Space: Models, Instruments and Techniques*. Dordrecht: Kluwer Academic Publishers, 246 pages.

Masuda, A. (1982). An interpretation of the bimodal character of the stable Kuroshio path. *Deep-Sea Research*, *29*, 471–484.

Mathieu, P.-P., Deleersnijder, E., Cushman-Roisin, B., Beckers, J.-M., & Bolding, K. (2002). The role of topography in small well-mixed bays, with application to the lagoon of Mururoa. *Continental Shelf Research*, *22*, 1379–1395.

McCalpin, J. D. (1994). A comparison of second-order and fourth-order pressure gradient algorithms in σ-coordinate ocean model. *International Journal for Numerical Methods in Fluids*, *128*, 361–383.

McDonald, A. (1986). A semi-Lagrangian and semi-implicit two time-level integration scheme. *Monthly Weather Review*, *114*, 824–830.

McPhaden, M. J. & Ripa, P. (1990). Wave-mean flow interactions in the equatorial ocean. *Annual Review of Fluid Mechanics*, *22*, 167–205.

McWilliams, J. C. (1977). A note on a consistent quasi-geostrophic model in a multiply connected domain. *Dynamics of Atmospheres and Oceans, 1*, 427–441.

McWilliams, J. C. (1984). The emergence of isolated coherent vortices in turbulent flow. *Journal of Fluid Mechanics, 146*, 21–43.

McWilliams, J. C. (1989). Statistical properties of decaying geostrophic turbulence. *Journal of Fluid Mechanics, 198*, 199–230.

Mechoso, C. R., Robertson, A. W., Barth, N., Davey, M. K., Delecluse, P., Gent, P. R., Ineson, S., Kirtman, B., Latif, M., Letreut, H., Nagai, T., Neelin, J. D., Philander, S. G. H., Polcher, J., Schopf, P. S., Stockdale, T., Suarez, M. J., Terray, L., Thual, O., & Tribbia, J. J. (1995). The seasonal cycle over the tropical pacific in coupled ocean-atmosphere general-circulation models. *Monthly Weather Review, 123*, 2825–2838.

Meinen, C. S., & McPhaden, M. J. (2000). Observations of warm water volume changes in the equatorial Pacific and their relationship to El Niño and La Niña. *Journal of Climate, 13*, 3551–3559.

Mellor, G., Oey, L.-Y., & Ezer, T. (1998). Sigma coordinate pressure gradient errors and the seamount problem. *Atmospheric and Oceanic Technology, 15*, 1112–1131.

Mellor, G. L., & Yamada, T. (1982). Development of a turbulence closure model for geophysical fluid problems. *Reviews of Geophysics and Space Physics, 20*, 851–875.

Mesinger, F., & Arakawa, A. (1976). Numerical methods used in the atmospheric models. *GARP Publications Series*, No. 17, International Council of Scientific Unions, World Meteorological Organization, 64 pages.

Miles, J. W. (1961). On the stability of heterogeneous shear flows. *Journal of Fluid Mechanics, 10*, 496–508.

Mofjeld, H. O., & Lavelle, J. W. (1984). Setting the length scale in a second-order closure model of the unstratified bottom boundary layer. *Journal of Physical Oceanography, 14*, 833–839.

Mohebalhojeh, A. R., & Dritschel, D. G. (2004). Contour-advective semi-Lagrangian algorithms for many layer primitive equation models. *Quarterly Journal of Royal Meteorological Society, 130*, 347–364.

Montgomery, R. B. (1937). A suggested method for representing gradient flow in isentropic surfaces. *Bulletin of the American Meteorological Society, 18*, 210–212.

Moore, D. W. (1963). Rossby waves in ocean circulation. *Deep-Sea Research, 10*, 735–748.

Munk, W. H. (1981). Internal waves and small-scale processes. In B. A. Warren, & C. Wunsch (Eds.), *Evolution of Physical Oceanography* (pp. 264–291). Cambridge, Massachusetts: The MIT Press, Chapter 9.

Navon, I. M. (2009). Data assimilation for numerical weather prediction: A review. In S. K. Park, & L. Xu (Eds.), *Data Assimilation for Atmospheric, Oceanic, and Hydrologic Applications* (pp. 21–65). Berlin, Heidelberg: Springer.

Nebeker, F. (1995). *Calculating the Weather: Meteorology in the 20th Century*. San Diego: Academic Press, 251 pages.

Neelin, J. D., Latif, M., Allaart, M. A. F., Cane, M. A., Cubasch, U., Gates, W. L., Gent, P. R., Ghil, M., Gordon, C., Lau, N. C., Mechoso, C. R., Meehl, G. A., Oberhuber, J.-M., Philander, S. G. H., Schopf, P. S., Sperber, K. R., Sterl, A., Tokioka, T., Tribbia, J. J., & Zebiak, S. E. (1992). Tropical air-sea interaction in general-circulation models. *Climate dynamics, 7*, 73–104.

Nezu, I., & Nakagawa, H. (1993). *Turbulence in Open-channel Flows*. International Association for Hydraulic Research Monograph Series, Rotterdam: Balkema, 281 pages.

Nihoul, J. C. J. (Ed.) (1975). *Modelling of Marine Systems*, Elsevier Oceanography Series (Vol. 10), Amsterdam: Elsevier, 272 pages.

Nof, D. (1983). The translation of isolated cold eddies on a sloping bottom. *Deep-Sea Research, 39*, 171–182.

O'Brien, J. J. (1978). El Niño – An example of ocean/atmosphere interactions. *Oceanus, 21*(4), 40–46.

Okubo, A. (1971). Oceanic diffusion diagrams. *Deep-Sea Research, 18*, 789–802.

Okubo, A., & Levin, S. A. (2002). *Diffusion and Ecological Problems* (2nd ed.). New York: Springer, 488 pages.

Olson, D. B. (1991). Rings in the ocean. *Annual Review of Earth and Planetary Sciences, 19*, 283–311.

Orcrette, J. J. (1991). Radiation and cloud radiative properties in the ECMWF operational weather forecast model. *Journal of Geophysical Research, 96*, 9121–9132.

Orlanski, I. (1968). Instability of frontal waves. *Journal of the Atmospheric Sciences, 25*, 178–200.

Orlanski, I. (1969). The influence of bottom topography on the stability of jets in a baroclinic fluid. *Journal of the Atmospheric Sciences, 26*, 1216–1232.

Orlanski, I., & Cox, M. D. (1973). Baroclinic instability in ocean currents. *Geophysical Fluid Dynamics, 4*, 297–332.

Orszag, S. A. (1970). Transform method for calculation of vector-coupled sums: Application to the spectral form of the vorticity equation. *Journal of the Atmospheric Sciences, 27*, 890–895.

Osborn, T. R. (1974). Vertical profiling of velocity microstructure, *Journal of Physical Oceanography, 4*, 109–115.

Ou, H. W. (1984). Geostrophic adjustment: A mechanism for frontogenesis. *Journal of Physical Oceanography, 14*, 994–1000.

Ou, H. W. (1986). On the energy conversion during geostrophic adjustment. *Journal of Physical Oceanography, 16*, 2203–2204.

Patankar, S. V. (1980). *Numerical Heat Transfer and Fluid Flow*. New York: McGraw-Hill, 198 pages.

Pavia, E. G., & Cushman-Roisin, B. (1988). Modeling of oceanic fronts using a particle method. *Journal of Geophysical Research, 93*, 3554–3562.

Pedlosky, J. (1963). Baroclinic instability in two-layer systems. *Tellus, 15*, 20–25.

Pedlosky, J. (1964). The stability of currents in the atmosphere and oceans. Part I. *Journal of the Atmospheric Sciences, 27*, 201–219.

Pedlosky, J. (1987). *Geophysical Fluid Dynamics* (2nd ed.). New York: Springer Verlag, 710 pages.

Pedlosky, J. (1996). *Ocean Circulation Theory*. Berlin: Springer, 453 pages.

Pedlosky, J. (2003). *Waves in the Ocean and Atmosphere: Introduction to Wave Dynamics*. Springer, 260 pages.

Pedlosky, J., & Thomson, J. (2003). Baroclinic instability of time-dependent currents. *Journal of Fluid Mechanics, 490*, 189–215.

Pham D. T., Verron, J., & Roubaud, M. C. (1998). Singular evolutive extended Kalman filter with EOF initialization for data assimilation in oceanography. *Journal of Marine Systems, 16*, 323–340.

Philander, S. G. (1990). *El Niño, La Niña, and the Southern Oscillation*. Orlando, Florida: Academic Press, 289 pages.

Phillips, N. A. (1954). Energy transformations and meridional circulations associated with simple baroclinic waves in a two-level, quasi-geostrophic model. *Tellus, 6*, 273–286.

Phillips, N. A. (1956). The general circulation of the atmosphere: A numerical experiment. *Quarterly Journal of Royal Meteorological Society, 82*, 123–164.

Phillips N. A. (1957). A coordinate system having some special advantages for numerical forecasting. *Journal of Meteorology, 14*, 184–1851.

Phillips, N. A. (1963). Geostrophic motion. *Reviews of Geophysics*, *1*, 123–176.

Pickard, G. L., & Emery, W. J. (1990). *Descriptive Physical Oceanography: An Introduction* (5th ed.). New York: Pergamon Press, 320 pages.

Pietrzak, J. (1998). The use of TVD limiters for forward-in-time upstream-biased advection schemes in ocean modeling. *Monthly Weather Review*, *126*, 812–830.

Pietrzak, J., Deleersnijder, E., & Schroeter, J. (Eds.). (2005). The second international workshop on unstructured mesh numerical modelling of coastal, shelf and ocean flows (Delft, The Netherlands, September 23–25, 2003). *Ocean Modelling*, *10*, 1–252.

Pietrzak, J., Jakobson, J., Burchard, H., Vested, H.-J., & Petersen, O. (2002). A three-dimensional hydrostatic model for coastal and ocean modelling using a generalised topography following co-ordinate system. *Ocean Modelling*, *4*, 173–205.

Pinardi, N., Allen, I., De Mey, P., Korres, G., Lascaratos, A., Le Traon, P.-Y., Maillard, C., Manzella, G., & Tziavos, C. (2003). The Mediterranean Ocean Forecasting System: first phase of implementation (1998–2001). *Annals of Geophysics*, *21*, 3–20.

Pollard, R. T., Rhines, P. B., & Thompson, R. O. R. Y. (1973). The deepening of the wind-mixed layer. *Geophysical Fluid Dynamics*, *4*, 381–404.

Pope, S. B. (2000). *Turbulent Flows*. Cambridge: Cambridge University Press, 771 pages.

Price, J. F., & Sundermeyer, M. A. (1999). Stratified Ekman layers. *Journal of Geophysical Research*, *104*, 20467–20494.

Proehl, J. A. (1996). Linear stability of equatorial zonal flows. *Journal of Physical Oceanography*, *26*, 601–621.

Proudman, J. (1953). *Dynamical Oceanography*. London: Methuen, and New York: John Wiley, 409 pages.

Rabier, F., & Courtier, P. (1992). Four-dimensional assimilation in the presence of baroclinic instability. *Quarterly Journal of Royal Meteorological Society*, *118*, 649–672.

Randall, D. (Ed.), (2000). *General Circulation Model Development. Past, Present, and Future*. International Geophysics Series (Vol. 70), Academic Press, 807 pages.

Randall, D., Khairoutdinov, M., Arakawa, A., & Grabowski, W. (2003). Breaking the cloud parameterization deadlock. *Bulletin of the American Meteorological Society*, *84*, 1547–1564.

Rao, P. K., Holmes, S. J., Anderson, R. K., Winston, J. S., & Lehr, P. E. (1990). *Weather Satellites: Systems, Data, and Environmental Applications*. Boston: American Meteorological Society, 503 pages.

Rasmusson, E. M., & Carpenter, T. H. (1982). Variations in tropical sea surface temperature and surface wind fields associated with the Southern Oscillation/El Niño. *Monthly Weather Review*, *110*, 354–384.

Rayleigh, Lord (John William Strutt) (1880). On the stability, or instability, of certain fluid motions. *Proceedings of the London Mathematical Society*, *9*, 57–70. (Reprinted in *Scientific Papers by Lord Rayleigh*, Vol. 3, 594–596).

Rayleigh, Lord (John William Strutt) (1916). On convection currents in a horizontal layer of fluid, when the higher temperature is on the under side. *Philosophical Magazine*, *32*, 529–546 (Reprinted in *Scientific Papers by Lord Rayleigh*, Vol. 6, 432–446).

Redi, M. H. (1982). Oceanic isopycnal mixing by coordinate rotation. *Journal of Physical Oceanography*, *12*, 1154–1158.

Reynolds, O. (1894). On the dynamical theory of incompressible viscous flows and the determination of the criterion. *Philosophical Transactions of the Royal Society London A*, *186*, 123–161.

Rhines, P. B. (1975). Waves and turbulence on the beta-plane. *Journal of Fluid Mechanics*, *69*, 417–443.

Rhines, P. B. (1977). The dynamics of unsteady currents. In E. D. Goldberg et al. (Eds.), *The Sea* (Vol. 6, pp. 189–318). New York: Wiley.

Rhines, P. B., & Young, W. R. (1982). A theory of the wind-driven circulation. I. Mid-ocean gyres. *Journal of Marine Research, 40*(suppl.), 559–596.

Richards, F. A. (Ed.), (1981). *Coastal Upwelling*. Coastal and Estuarine Sciences (Vol. 1). Washington, DC: American Geophysical Union, 529 pages.

Richardson, L. F. (1922). *Weather Prediction by Numerical Process*. Cambridge University Press. (Reprinted by Dover Publications, 1965, 236 pp.).

Richtmyer, R. D., & Morton, K. W. (1967). *Difference Methods for Initial-value Problems* (2nd ed.). New York: Interscience, John Wiley and Sons, 405 pages.

Riley, K. F., Hobson, M. P., & Bence, S. J. (1997). *Mathematical Methods for Physics and Engineering*. Cambridge University Press, 1008 pages.

Ripa, P. (1994). *La Increíble Historia de la Malentendida Fuerza de Coriolis*, La Ciencia/128 desde México, 101 pages.

Rixen, M., Beckers, J.-M., & Allen, J. T. (2001). Diagnosis of vertical velocities with the QG-omega equation: a relocation method to obtain pseudo-synoptic data sets. *Deep-Sea Research, 48*, 1347–1373.

Rixen, M., & Ferreira-Coelho, E. (2007). Operational surface drift prediction using linear and non-linear hyper-ensemble statistics on atmospheric and ocean models. *Journal of Marine Systems, 65*, 105–121.

Robinson, A. R., Tomasin, A., & Artegiani, A. (1973). Flooding of Venice: Phenomenology and prediction of the Adriatic storm surge. *Quarterly Journal of Royal Meteorological Society, 99*, 688–692.

Robinson, A. R. (Ed.), (1983). *Eddies in Marine Science*. Berlin: Springer-Verlag, 609 pages.

Robinson, A. R. (1965). A three-dimensional model of inertial currents in a variable-density ocean. *Journal of Fluid Mechanics, 21*, 211–223.

Robinson, A. R., & Lermusiaux, P. F. J. (2002). Data assimilation for modeling and predicting coupled physical-biological interactions in the sea. *The Sea, 12*, 475–536.

Robinson A. R., Lermusiaux, P. F. J., & Sloan, N. Q., III, (1998). Data assimilation. *The Sea, 10*, 541–594.

Robinson, A. R., & McWilliams, J. C. (1974). The baroclinic instability of the open ocean. *Journal of Physical Oceanography, 4*, 281–294.

Robinson, A. R., Spall, M. A., & Pinardi, N. (1988). Gulf Stream simulations and the dynamics of ring and meander processes. *Journal of Physical Oceanography, 18*, 1811–1853.

Robinson, A. R., & Taft, B. (1972). A numerical experiment for the path of the Kuroshio. *Journal of Marine Research, 30*, 65–101.

Robinson, I. (2004). *Measuring the Oceans from Space: The Principles and Methods of Satellite Oceanography*. Chichester and Heidelberg: Springer-Praxis, 670 pages.

Rodi, W. (1980). *Turbulence Models and their Application in Hydraulics*. Delft, The Netherlands: International Association for Hydraulic Research.

Roll, H. U. (1965). *Physics of the Marine Atmosphere*. New York: Academic Press, 426 pages.

Rossby, C. G. (1937). On the mutual adjustment of pressure and velocity distributions in certain simple current systems. I. *Journal of Marine Research, 1*, 15–28.

Rossby, C. G. (1938). On the mutual adjustment of pressure and velocity distributions in certain simple current systems. II. *Journal of Marine Research, 2*, 239–263.

Roussenov, V., Williams, R. G., & Roether, W. (2001). Comparing the overflow of dense water in isopycnic and cartesian models with tracer observations in the eastern Mediterranean. *Deep-Sea Research, 48*, 1255–1277.

810   References

Saddoughi, S. G., & Veeravalli, S. V. (1994). Local isotropy in turbulent boundary layers at high Reynolds number. *Journal of Fluid Mechanics*, *268*, 333–372.

Saffman, P. G. (1968). Lectures on homogeneous turbulence. In N. J. Zabusky (Ed.), *Topics in Nonlinear Physics* (pp. 485–614). Berlin: Springer Verlag.

Salmon, R. (1982). Geostrophic turbulence. In A. R. Osborne & P. Malanotte-Rizzoli (Eds.), *Topics in Ocean Physics*, Proc. Int. School of Phys. Enrico Fermi LXXX (pp. 30–78). North-Holland: Elsevier Sci. Publ.

Sawyer, J. (1956). The vertical circulation at meteorological fronts and its relation to frontogenesis. *Proceedings of the Royal Society London A*, *234*, 346–362.

Schmitz, W. J., Jr. (1980). Weakly depth-dependent segments of the North Atlantic circulation. *Journal of Marine Research*, *38*, 111–133.

Schott, F., & Stommel, H. (1978). Beta spirals and absolute velocities in different oceans. *Deep-Sea Research*, *25*, 961–1010.

Shapiro, L. J. (1992). Hurricane vortex motion and evolution in a three-layer model. *Journal of the Atmospheric Sciences*, *49*, 140–153.

Siedler, G., Church, J., & Gould, J. (Eds.), (2001). *Ocean Circulation and Climate: Modelling and Observing the Global Ocean* . San Diego: Academic Press, 715 pages.

Smagorinsky, J. (1963). General circulation experiments with the primitive equations. I. The basic experiment. *Monthly Weather Review*, *91*, 99–164.

Song, T. (1998). A general pressure gradient formulation for ocean models. Part I: Scheme design and diagnostic analysis. *Monthly Weather Review*, *126*, 3213–3230.

Sorbjan, Z. (1989). *Structure of the Atmospheric Boundary Layer*. Englewood Cliffs, New Jersey: Prentice Hall, 317 pages.

Spagnol S., Wolanski, E., Deleersnijder, E., Brinkman, R., McAllister, F., Cushman-Roisin, B., & Hanert, E. (2002). An error frequently made in the evaluation of advective transport in two-dimensional Lagrangian models of advection-diffusion in coral reef waters. *Marine Ecology Progress Series*, *235*, 299–302.

Spall, M. A., & Holland, W. R. (1991). A nested primitive equation model for oceanic applications. *Journal of Physical Oceanography*, *21*, 205–220.

Spiegel, E. A., & Veronis, G. (1960). On the Boussinesq approximation for a compressible fluid. *The Astrophysical Journal*, *131*, 442–447.

Spivakovskaya, D., Heemink, A. W., & Deleersnijder, E. (2007). The backward Ito method for the Lagrangian simulation of transport processes with large space variations of the diffusivity. *Ocean Science*, *3*, 525–535.

Stacey, M. W., Pond, S., & LeBlond, P. H. (1986). A wind-forced Ekman spiral as a good statistical fit to low-frequency currents in coastal strait. *Science*, *233*, 470–472.

Stern, A. C., Boubel, R. W., Turner, D. B., & Fox, D. L. (1984). *Fundamentals of Air Pollution*. Academic Press, 530 pages.

Stigebrandt, A. (1985). A model for the seasonal pycnocline in rotating systems with application to the Baltic Proper. *Journal of Physical Oceanography*, *15*, 1392–1404.

Stoer, J., & Bulirsh, R. (2002). *Introduction to Numerical Analysis*. Texts in Applied Mathematics (Vol. 12, 3rd ed.), New York: Springer-Verlag, 744 pages.

Stommel, H. (1948). The westward intensification of wind-driven ocean currents. *Transactions American Geophysical Union*, *29*, 202–206.

Stommel, H. (1958). The abyssal circulation. *Deep-Sea Research* (Letters), *5*, 80–82.

Stommel, H. (1979). Determination of water mass properties of water pumped down from the Ekman layer to the geostrophic flow below. *Proceedings of the National Academy of Sciences USA*, *76*, 3051–3055.

Stommel, H., & Arons, A. B. (1960a). On the abyssal circulation of the world ocean – I. Stationary planetary flow patterns on a sphere. *Deep-Sea Research, 6,* 140–154.

Stommel, H., & Arons, A. B. (1960b). On the abyssal circulation of the world ocean – II. An idealized model of the circulation pattern and amplitude in oceanic basins. *Deep-Sea Research, 6,* 217–233.

Stommel, H., Arons, A. B., & Faller, A. J. (1958). Some examples of stationary planetary flow patterns in bounded basins. *Tellus, 10,* 179–187.

Stommel, H., & Moore, D. W. (1989). *An Introduction to the Coriolis Force.* Irvington, New York: Columbia University Press, 297 pages.

Stommel, H., & Schott, F. (1977). The beta spiral and the determination of the absolute velocity field from hydrographic station data. *Deep-Sea Research, 24,* 325–329.

Stommel, H., & Veronis, G. (1980). Barotropic response to cooling. *Journal of Geophysical Research, 85,* 6661–6666.

Strang, G. (1968). On the construction and comparison of difference schemes, *SIAM Journal on Numerical Analysis, 5,* 506–517.

Strub, P. T., Kosro, P. M., & Huyer, A. (1991). The nature of cold filaments in the California Current system. *Journal of Geophysical Research, 96,* 14743–14768.

Stull, R. B. (1988). *Boundary-Layer Meteorology.* Dordrecht, The Netherlands: Kluwer Academic Publishers, 666 pages.

Stull, R. B. (1991). Static stability – An update. *Bulletin of the American Meteorological Society, 72,* 1521–1529 (Corrigendum: *Bull. Am. Met. Soc., 72,* 1883).

Stull, R. B. (1993). Review of nonlocal mixing in turbulent atmospheres: Transilient turbulence theory. *Boundary-Layer Meteorology, 62,* 21–96.

Sturm, T. W. (2001). *Open Channel Hydraulics.* New York: McGraw-Hill, 493 pages.

Suarez, M., & Schopf, P. (1988). A delayed action oscillator for ENSO. *Journal of the Atmospheric Sciences, 45,* 3283–3287.

Sundqvist, H., Berge, E., & Kristjansson, J. E. (1989). Condensation and cloud parameterization studies with a mesoscale numerical weather prediction model. *Monthly Weather Review, 117,* 1641–1657.

Sutyrin, G. G. (1989). The structure of a monopole baroclinic eddy. *Oceanology, 29,* 139–144 (English translation).

Sverdrup, H. U. (1947). Wind-driven currents in a baroclinic ocean, with application to the equatorial currents of the eastern Pacific. *Proceedings of the National Academy of Sciences USA, 33,* 318–326.

Sweby, P. K. (1984). High resolution schemes using flux-limiters for hyperbolic conservation laws. *SIAM Journal on Numerical Analysis, 21,* 995–1011.

Taillandier, V., Griffa, A., Poulain, P.-M., & Béranger, K. (2006). Assimilation of Argo float positions in the north western Mediterranean Sea and impact ocean circulation simulations. *Geophysical Research Letters, 33,* 11604.

Talagrand, O., & Courtier, P. (1987). Variational assimilation of meteorological observations with the adjoint vorticity equation. I: Theory. *Quarterly Journal of Royal Meteorological Society, 113,* 1311–1328.

Talley, J. D., Pickard, G. L., Emery, W. J., & Swift, J. (2007). *Descriptive Physical Oceanography* (6th ed.). Academic Press, 500 pages.

Tangang, F. T., Tang, B., Monahan, A. H., & Hsieh, W. W. (1998). Forecasting ENSO events: a neural network—extended EOF approach. *Journal of Climate, 11,* 29–41.

Taylor, G. I. (1921). Tidal oscillations in gulfs and rectangular basins. *Proceedings of Royal Society London A, 20,* 148–181.

Taylor, G. I. (1923). Experiments on the motion of solid bodies in rotating fluids. *Proceedings of Royal Society London A, 104,* 213–218.

Taylor, G. I. (1931). Effect of variation in density on the stability of superposed streams of fluid. *Proceedings of Royal Society London A, 132,* 499–523.

Tennekes, H., & Lumley, L. J. (1972). *A First Course in Turbulence.* Cambridge, Massachusetts: The MIT Press, 300 pages.

Thompson, J. F., Warsi, Z. U. A., & Mastin, C. W. (1985). *Numerical Grid Generation: Foundations and Applications.* North Holland, 483 pages.

Thomson, W. (Lord Kelvin) (1879). On gravitational oscillations of rotating water. *Proceedings of Royal Society Edinburgh, 10,* 92–100. (Reprinted in *Phil. Mag., 10,* 109–116, 1880; *Math. Phys. Pap.,* **4,** 141–148, 1910).

Thuburn, J. (1996). Multidimensional flux-limited advection schemes. *Journal of Computational Physics, 123,* 74–83.

Tomczak, M., & Godfrey, J. S. (2003). *Regional Oceanography: An Introduction* (2nd ed.). Delhi: Daya Publishing House, 390 pages.

Troup, A. J. (1965). The Southern Oscillation. *Quarterly Journal of Royal Meteorological Society, 91,* 490–506.

Turner, J. S. (1973). *Buoyancy Effects in Fluids.* Cambridge: Cambridge University Press, 367 pages.

Umlauf, L., & Burchard, H. (2003). A generic length-scale equation for geophysical turbulence models. *Journal of Marine Research, 61,* 235–265.

Umlauf, L., & Burchard, H. (2005). Second-order turbulence closure models for geophysical boundary layers. A review of recent work. *Continental Shelf Research, 25,* 795–827.

Vallis, G. K. (2006). *Atmospheric and Oceanic Fluid Dynamics: Fundamentals and Large-scale Circulation.* Cambridge: Cambridge University Press, 745 pages.

Van Dyke, M. (1975). *Perturbation Methods in Fluid Mechanics.* Stanford, CA: Parabolic Press, 271 pages.

van Heijst, G. J. F. (1985). A geostrophic adjustment model of a tidal mixing front. *Journal of Physical Oceanography, 15,* 1182–1190.

Verkley, W. T. M. (1990). On the beta-plane approximation. *Journal of the Atmospheric Sciences, 47,* 2453–2460.

Veronis, G. (1956). Partition of energy between geostrophic and non-geostrophic oceanic motions. *Deep-Sea Research, 3,* 157–177.

Veronis, G. (1963). On the approximations involved in transforming the equations of motion from a spherical surface to the β-plane. I. Barotropic systems. *Journal of Marine Research, 21,* 110–124.

Veronis, G. (1967). Analogous behavior of homogeneous, rotating fluids and stratified, non-rotating fluids. *Tellus, 19,* 326–336.

Veronis, G. (1981). Dynamics of large-scale ocean circulation. In B. A. Warren, & C. Wunsch (Eds.), *Evolution of Physical Oceanography* (pp. 140–183). Cambridge, Massachusetts: The MIT Press.

Vosbeek, P. W. C., Clercx, H. J. H., & Mattheij, R. M. M. (2000). Acceleration of contour dynamics simulations with a hierarchical-element method. *Journal of Computational Physics, 161,* 287–311.

Walker, G. T. (1924). Correlation in seasonal variations of weather, IX. A further study of world weather. *Memoirs of the India Meteorological Department, 24,* 275–333.

Wallace, J. M., & Kousky, V. E. (1968). Observational evidence of Kelvin waves in the tropical stratosphere. *Journal of the Atmospheric Sciences, 25,* 900–907.

Warren, B. A., & Wunsch, C. (1981). *Evolution of Physical Oceanography: Scientific Surveys in Honor of Henry Stommel*. Cambridge, Massachusetts: The MIT Press, 623 pages.

Weatherly, G. L., & Martin, P. J. (1978). On the structure and dynamics of the ocean bottom boundary layer. *Journal of Physical Oceanography, 8*, 557–570.

Wei, T., & Willmarth, W. W. (1989). Reynolds number effects on the structure of a turbulent channel flow. *Journal of Fluids Mechanics, 204*, 57–95.

Welander, P. (1975). Analytical modeling of the oceanic circulation. In *Numerical Models of Ocean Circulation: Proceedings of a Symposium Held at Durham, New Hampshire, October 17–20, 1972* (pp. 63–75). Washington: National Acad. Sci.

Wilks, D. S. (2005). *Statistical Methods in the Atmospheric Sciences* (2nd ed.). Academic Press, 468 pages.

Williams, G. P., & Wilson, R. J. (1988). The stability and genesis of Rossby vortices. *Journal of the Atmospheric Sciences, 45*, 207–241.

Williams, R. G. (1991). The role of the mixed layer in setting the potential vorticity of the main thermocline. *Journal of Physical Oceanography, 21*, 1803–1814.

Winston, J. S., Gruber, A., Gray, T. I., Jr., Varnadore, M. S., Earnest, C. L., & Mannello, L. P. (1979). *Earth-Atmosphere Radiation Budget Analyses from NOAA Satellite Data June 1974– February 1978* (Vol. 2). Washington, DC: National Environmental Satellite Service, NOAA, Dept. of Commerce.

WMO. (1999). *WMO Statement on the Status of the Global Climate in 1998*, WMO – No. 896. Geneva: World Meteorological Organization, 12 pages.

Woods, J. D. (1968). Wave-induced shear instability in the summer thermocline. *Journal of Fluid Mechanics, 32*, 791–800 + 5 plates.

Wunsch, C. (1996). *The Ocean Circulation Inverse Problem*. Cambridge: Cambridge University Press, 437 pages.

Wyrtki, K. (1973). Teleconnections in the equatorial Pacific Ocean. *Science, 180*, 66–68.

Yabe, T., Xiao, F., & Utsumi, T. (2001). The constrained interpolation profile method for multi-phase analysis. *Journal of Computational Physics, 169*, 556–593.

Yoshida, K. (1955). Coastal upwelling off the California coast. *Records of Oceanographic Works in Japan, 2*(2), 1–13.

Yoshida, K. (1959). A theory of the Cromwell Current and of the equatorial upwelling – An intepretation in a similarity to a coastal circulation. *Journal of the Oceanographical Society of Japan, 15*, 154–170.

Zabusky, N. J., Hughes, M. H., & Roberts, K. V. (1979). Contour dynamics for the Euler equations in two dimensions. *Journal of Computational Physics, 30*, 96–106.

Zalesak, S. T. (1979). Fully multidimensional flux-corrected transport algorithms for fluids. *Journal of Computational Physics, 31*, 335–362.

Zebiak, S. E., & Cane, C. A. (1987). A model El-Niño southern oscillation. *Monthly Weather Review, 115*, 2262–2278.

Zienkiewicz, O. C., & Taylor, R. L. (2000). *Finite Element Method: Volume 1. The Basis* (5th ed.). Butterworth-Heinemann, 712 pages.

Zienkiewicz, O. C., Taylor, R. L., & Nithiarasu, P. (2005). *The Finite Element Method for Fluid Dynamics* (6th ed.). Butterworth-Heinemann, 400 pages.

Zilitinkevich, S. S. (1991). *Turbulent Penetrative Convection*. Aldershot: Avebury Technical, 179 pages.

Index

Page numbers followed by "*f*" indicates figures, "*t*" indicates tables and "*n*" indicates footnotes.

A

A-grid model, 294, 296*f*
Absolute velocity, 42, 43
Abyssal circulation, 677–681
Abyssal layer, 658
Accuracy, 118, 119
  and errors, 120–125
Acoustic Doppler current profilers, 18
Adams–Bashforth method, 66, 67*f*
Adaptive time-stepping, 125
Adiabatic conservation law, 350
Adiabatic lapse rate, 351
Adjoint method, 745*f*
Adjoint model, 747
Adriatic
  density profile, 13
  seiches, 278, 381
Advection, 193*f*, 194*f*, 768
  1D, 191, 192*f*
  2D, 186–187, 190*f*
  and diffusion, 163–167
  multidimensional approach,
    186–196
  relative importance of, 167–168
  with TVD scheme, 195*f*
Advection schemes
  centered, 169–176
  numerical, 187
  standard test for, 175
  upwind, 176–183, 177*f*, 178*f*
Advection–diffusion
  equation, 163
    properties of, 165
  with sources and sinks, 183–186
Advective instability, 418
AGCMs, *see* Atmospheric General
    Circulation Models
Ageostrophic motion, 280, 524–525, 529,
    532, 537, 601
Aire Limitée Adaptation dynamique
    Développement InterNational
    (ALADIN) model, 639

Air–sea interactions, 5
ALADIN model, *see* Aire Limitée
    Adaptation dynamique
    Développement InterNational model
Albedo, 628, 640
Algebraic Reynolds-stress models, 453
Aliasing, 33–35, 329–330
Alternating direction implicit (ADI)
    methods, 156
Amphidromic points, 308, 381
Amplification factor, 145, 146, 150
Angular frequency, 773
Anticyclones, 20, 601, 609*f*
  of midlatitude weather, 565
Anticyclonic vorticity, 490
Arakawa, Akio, 295, 315
Arakawa Jacobian, 545
Arakawa's grids, 289*f*, 296
  A-grid model, 294, 296*f*
  C-grid, 295, 297*f*
  Coriolis force, 291, 292, 300
  numerical models, 289
  second-order method, 290
  staggered grid, 291, 293
Artificial diffusion, *see* Numerical diffusion
Asselin filter, 334
Atlantic Ocean, 27
Atmosphere, 715
  length, velocity and time scales, 15*t*
  motion in, 14–17
  processes and structures in, 16*f*
Atmospheric boundary layer, 257, 462
Atmospheric circulation models, 637–642
Atmospheric convection, 463–465
  parameterization, 355, 643
Atmospheric frontogenesis, 490–502,
    491*f*–493*f*, 499*f*, 500*f*
  physical processes in, 492
  temperature gradient, 494
Atmospheric General Circulations Models
    (AGCMs), 19, 638–639
Atmospheric stratification, 349–354

Printed and bound by CPI Group (UK) Ltd, Croydon, CR0 4YY

13/10/2024

01773603-0001